Contents

PART IV: STRATEGIC ENVIRONMENTAL ASSESSMENT / 145

PART V: STRATEGIC ANALYSIS / 265

PART VI: INFORMATION SYSTEMS FOR STRATEGIC MANAGEMENT / 385

PART VII: IMPLEMENTING STRATEGY / 437

Preface

The decades of the sixties and seventies saw tremendous growth in the development and application of the tools and techniques of strategic planning. Late in that period, the notion of "strategic management"—an activity that encompassed strategic planning but went well beyond it in scope—began to emerge and take form.

In the 1980s, there has been something of a reaction to the fast growth of the past two decades. Some companies have cut back on their strategic planning staffs and there is the appearance of a decline in the field.

Nothing could be further from the truth! Strategic planning has been decentralized resulting in dislocations of planners at the corporate levels of organizations. This is, however, a manifestation of the success of strategic planning, not of its declining importance.

Strategic management responsibility today lies more with line managers than with staff planners. The tools and techniques of strategic planning have become integrated into the everyday management of business units; they are no longer esoteric novelties that are the sole province of an elitist group of staff planners.

This decentralization and integration have been recognized as of fundamental importance to effective planning for many years. For example, in our 1978 book, *Strategic Planning and Policy,* one of our fundamental precepts is that "managers should perform the planning." So, planners have, to some degree, worked themselves out of a job by successfully integrating planning ideas into every facet of the organization.

However, the task is not done. Strategic management, as contrasted with strategic planning, is still an underdeveloped area. The basic premise of strategic management is that strategic planning is not enough. Formal strategic planning processes and techniques are useful, but not sufficiently robust and comprehensive to encompass all the creative, entrepreneurial, innovative, and leadership dimensions that are necessary if the organization is to create its own future rather than merely react to the changing times.

Thus, the domain of strategic management is broader than that of strategic planning, and its mission is larger and more important. Yet, effective strategic management requires that strategic planning be employed to good effect in combination with other elements of effective management.

This handbook is intended to provide both specific guidance and more general insight. The tools and techniques of strategic planning are explained and evaluated in various chapters. At the same time, many of the broader elements of strategic management, which cannot be dealt with at the level of tools and techniques, are described and discussed.

The contributors have attempted to make their presentations as practical and useful as possible. However, because of the dual nature of the area (strategic planning and management), it is not possible for each topic to be dealt with at the same level of specificity. However, the reader will find little that is too theoretical or overly academic here. Contributors were selected on the basis of their proven practical experience, and each has striven to address an area in a pragmatic way.

The coverage and organization are the responsibility of the editors. We have organized the handbook to make it useful to read from beginning to end as a basic primer for the subject. At the same time, we have provided descriptive chapter titles, an index, and ample references so that it can be used as a reference book by readers who desire information on a specific topic.

We gratefully acknowledge the substantial help of the contributors and of Olivia Harris and Claire Zubritzky whose long-standing support has a great impact on the quality of our undertakings.

WILLIAM R. KING
DAVID I. CLELAND

Reference

King, W. R., and D. I. Cleland, 1978, *Strategic Planning and Policy,* Van Nostrand Reinhold, New York, 374p.

Contributors

M. Robert Ackelsberg, Department of Management and Marketing, Shippensburg University of Pennsylvania, Shippensburg, PA 17257
Russell D. Archibald, 343 Medio Drive, Los Angeles, CA 90049
Albert V. Bruno, Leavey School of Business, University of Santa Clara, Santa Clara, CA 95053
John C. Camillus, Graduate School of Business, Mervis Hall, University of Pittsburgh, Pittsburgh, PA 15260
Archie B. Carroll, Department of Management, University of Georgia, Athens, GA 30602
Ravi Chinta, Department of Management, Louisiana State University, Baton Rouge, LA 70803
David I. Cleland, 1035 Benedum Hall, University of Pittsburgh, Pittsburgh, PA 15261
Malcolm B. Coate, Bureau of Economics, Federal Trade Commission, 400-M Gelman Building, 6th and Pennsylvania Ave., N.W., Washington, D.C. 20580
William H. Davidson, School of Business Administration, Department of Management and Organization, University of Southern California, Los Angeles, CA 90089-1421
George S. Day, University of Toronto, 246 Bloor St., West Toronto, Ontario M5S 1V4
Robert L. Eskridge, Growth Management Center, 796 Via Del Monte, Palos Verdes Estates, CA 90274
Liam Fahey, Boston University, 12 Bay Street Road, Boston, MA 02215
Harold W. Fox, 2001 Forrest Drive, Edinburg, TX 78539
John H. Grant, Graduate School of Business, Mervis Hall, University of Pittsburgh, Pittsburgh, PA 15260
John Hall, Department of Management, University of Georgia, Athens, GA 30602
Philippe Haspeslagh, INSEAD, Boulevard de Constance, 77305 Fontainebleau Cedex, France
Dominique Héau, INSEAD, Boulevard de Constance, 77305 Fontainebleau Cedex, France
LaRue T. Hosmer, School of Business Administration, University of Michigan, Ann Arbor, MI 48109
Marianne Moody Jennings, College of Business Administration, Arizona State University, Tempe, AZ 85287

Per Jenster, McIntyre School of Commerce, Munroe Hall, University of Virginia, Charlottesville, VA 22903
Patricia E. Jones, 5 Rydal Place, Montclair, NJ 07082
William R. King, Graduate School of Business, 222 Mervis Hall, University of Pittsburgh, Pittsburgh, PA 15260
Joel Leidecker, Leavey Graduate School of Business, University of Santa Clara, Santa Clara, CA 95053
Milton Leontiades, Faculty of Business Studies, Rutgers University, Camden, NJ 08102
Robert E. MacAvoy, Easton Consultants, Inc., 1200 Summer St.,Stamford, CT 06905
Michael A. McGinnis, Department of Management and Marketing, Shippensburg University of Pennsylvania, Shippensburg, PA 17257
Ian C. MacMillan, New York University, 100 Trinity Place, New York, NY 10006
Spyros G. Makridakis, INSEAD, Boulevard de Constance, 77305 Fontainebleau Cedex, France
Aubrey L. Mendelow, 137 Phillips Place, Pittsburgh, PA 15217
Malcolm C. Munro, Faculty of Management, University of Calgary, 2500 University Drive, N.W., Calgary, Alberta T2N 1N4, Canada
V. K. Narayanan, School of Business, University of Kansas, 307 Summerfield Hall, Lawrence, KS 66045
Jack Pearce II, College of Business Administration, University of South Carolina, The Francis M. Hipp Building, Columbia, SC 29208
John E. Prescott, Graduate School of Business, 252 Mervis Hall, University of Pittsburgh, Pittsburgh, PA 15260
Vasudevan Ramanujam, 311B Wickenden, Case Western Reserve University, Cleveland, OH 44106
Alfred Rappaport, Nathanial Leverone Hall, Northwestern University, Evanston, IL 60201
Lawrence C. Rhyne, J. L. Kellogg Graduate School of Management, Northwestern University, Leverone Hall, 2001 Sheridan Road, Evanston, IL 60201
David R. Rink, R #1, Box 502, Ellen Drive, Genoa, IL 60135
Richard B. Robinson, Jr., College of Business Administration, The Francis M. Hipp Building, University of South Carolina, Columbia, SC 29208
Kendall Roth, College of Business Administration, The Francis M. Hipp Building, University of South Carolina, Columbia, SC 29208
Narendra K. Sethi, 194 Greenway South, Forest Hills, NY 11375
Frank M. Shipper, College of Business Administration, Arizona State University, Tempe, AZ 85287
A. Graham Sterling, Analog Devices, Two Technology Way, Norwood, MA 02062-0280
Bernard Taylor, Henley, The Management College, Greenlands, Henley-on-Thames, Oxon RG9 3AU, England
N. Venkatraman, #52-553 Sloan School of Management, Massachusetts Institute of Technology, 50 Memorial Drive, Cambridge, MA 02139
Roy Wernham, British Telecom Center, Rm. C321, 81 Newgate Street, London, ED1A 7AJ, England

I

Introduction

This introductory section discusses strategic planning and management from a modern perspective, delineates various approaches to strategic planning, and deals with two general concepts that are central to strategic management: innovation and leadership.

Chapter 1, a somewhat philosophical introduction by Spyros Makridakis and Dominique Héau, discusses the evolution of strategic planning and management, the changing role of techniques in strategic management, and the evolving viewpoints that underlie these changes. The authors develop an analogy between various kinds of strategic management and the processes of biological and cultural evolution. In their view, strategic management can operate at any of three modes in dealing with the evolving and uncertain environments of business.

In Chapter 2, Bernard Taylor describes five pragmatic styles of strategic planning, one of which is the broader notion of strategic management. He distinguishes between these planning modes in terms of the focus, central ideas, important constituent parts, and techniques associated with each.

One of these five planning styles emphasizes innovation, which is the topic of Chapter 3 by Michael A. McGinnis and M. Robert Ackelsberg. In this chapter the emphasis is on innovation as a pervasive goal of strategic management and on the factors that contribute to innovation. McGinnis and Ackelsberg emphasize an integrated approach to how strategic planning and management may be performed to best achieve the goal of organizational innovation.

In Chapter 4, LaRue T. Hosmer provides a strategic view of leadership. After reviewing classic ideas of leadership, Hosmer develops a leadership view that is integral to strategic management. According to this view, effective strategy formulation and implementation—when viewed as organizational processes—produce the sort of leadership that is needed in modern complex organizations.

These four chapters contain the bases for developing the more detailed and specific ideas and tools of strategic planning and management in subsequent parts of the book. In their focus on alternative viewpoints, they serve to identify the domain of the field and to emphasize that there is no single best approach to it.

CHAPTER 1

The Evolution of Strategic Planning and Management

Spyros Makridakis and Dominique Héau

Spyros Makridakis is a professor of management science at INSEAD. He received his undergraduate degree from the School of Industrial Studies in Greece and the M.B.A. and Ph.D. from New York University. He has published extensively in the areas of general systems and forecasting and has coauthored *Computer-Aided Modeling for Managers* (Addison-Wesley, 1972), *Forecasting, Methods for Management*, 4th ed. (Wiley, 1985), *Interactive Forecasting*, 2nd ed. (Holden-Day, 1978), and *Forecasting: Methods and Applications*, 2nd ed. (Wiley-Hamilton, 1983). He also coedited Volume 12 of TIMS *Studies in Management Science*, and is the editor in chief of *The International Journal of Forecasting*.

A professor of business policy at INSEAD, Dominique Héau is a graduate of HEC and IEP in Paris. He holds an M.B.A. and a D.B.A. from Harvard University. He has been a consultant with a number of European and North American companies and with the Government of Quebec (for industrial policy). His current research activities involve planning practices in diversified multinational firms. Professor Héau has taught numerous management development programs and has done case research and written extensively, particularly in the field of international competition. He was chairman of the Third Annual Strategic Management Society Conference held in 1983.

"I am interested in the future because I will spend the rest of my life there." This is an intellectual view of the future; the majority of people view the future with a mixture of curiosity blended with awe and fear. The future is unknown, a source of anxiety. It can bring misery, disease, or death in a stroke of a second. We have little or no control over many future events. Fame, richness, and luxury are also part of the future and await the few who can discover the lucky key that will open the door to the world of the rich and famous.

People are different from animals in that they can take concrete actions in the present to deal with future events. These actions are forecasting, planning, and strategy. The purpose of forecasting is to predict future events. However, there usually is a lead time between when a need arises and its fulfillment. Planning therefore is concerned with doing something before a need arises so that it can be fulfilled with no delay. A car manufacturer cannot wait until someone wants to buy a car before starting to produce it. Since it takes almost four years from the time a car is designed until it comes off the production line, it is necessary to forecast the demand of specific car models several years in advance. Concrete actions must then be taken so that such car models are ready and in dealer showrooms when customers want to buy them. The four year forecasting-planning cycle in the automobile industry is a long enough period of time in which things can and do go wrong. The forecasts of the demand for cars can be erroneous. Car models can be produced that people do not like and, therefore, do not buy. A recession can hit the economy, resulting in reduced incomes and inability to buy cars. Moreover, a big increase in the price of gasoline can drastically change driving patterns and car-buying attitudes. In other words, many unexpected things can happen that can make plans obsolete. It obviously would be disastrous if every time a plan went wrong the organization involved went bankrupt. Organizations (and living systems in general) have the ability to weather difficulties. They can deal with changes in their plans caused by bad forecasts, new or unexpected events, competitive actions, and so forth. Thus strategy is a necessity that is the cornerstone of dealing with the future. This is true particularly for the long term, when forecasting can be highly inaccurate and planning far off target.

This chapter discusses strategy and relates it to forecasting and planning. It looks first at the historical development of corporate strategy and then proposes a somewhat different approach to strategy, which is based on analogies with biological and cultural evolution. As the strategy of living systems and human cultures is the product of an evolutionary process of many millions of years, it is believed that it can serve as a model for organizational strategy. Such strategy is based on two *simultaneous and seemingly contradictory types of behavior*: (1) dominance over existing competitors, which is achieved through specialization; and (2) ability to deal with environmental changes, requiring adjustments and adaptation over time, which is inhibited by specialization. Finally, the types of forecasting and planning required (and their limitations) are discussed and ways of integrating forecasting, planning, and strategy are proposed.

EVOLUTION OF STRATEGIC PLANNING AND MANAGEMENT

The first attempt to formalize the manner in which to deal with the future of organizations was advocated by Fayol[1] in 1916 and subsequently practiced at the Dupont Company[2] and General Motors.[3] Before that, corporate success was linked to the entrepreneurial abilities of the firm's owner.

After some thirty-odd years of little concern—both academically and in business organizations—and even fewer applications, strategy became a fashionable topic in the 1950s and 1960s when large firms and expanding business opportunities necessitated looking more systematically at the future. This took the form of long-range planning, the purpose of which was to first define the firm's objectives, then establish some plans in order to achieve those objectives, and finally to allocate resources through capital budgeting. A key aspect of the planning process was to reduce the gap that often occurred between the firm's aspirations and plans and the extrapolation of existing trends.

Such use of long-range planning, as a way of formulating strategy, lost its appeal when it became evident that forecasting existing trends into the future did not produce accurate results, that growth, as it was experienced in the 1950s and 1960s, could slow down or be interrupted, and that new opportunities as well as threats, which nobody had envisaged beforehand, were possible.[4] Furthermore, it became evident that closing the planning gap was not necessarily the most critical aspect of strategy formulation. Volatile markets, overcapacity, and resource constraints have become dominant management considerations since the early 1970s. Consequently, long-range planning was replaced by strategic planning, which incorporated accepting possible changes in trends and was not based on the assumption that adequate growth could be assured. In addition, strategic planning was focusing more closely at market competition since limited expansion of markets and products could not support the growth objectives of the various competitors. Whereas long-range planning was essentially concerned with detailed sequences of actions required to produce expected objectives, thereby resembling a pluriannual budget, the focus of strategic planning was on the environmental assumptions that underlie any plan; it came to be realized that number-crunching gave way to "what if?" inquiries.

Although the names changed, both long-range planning and strategic planning shared three key assumptions: (1) they assumed that environmental forecasting was sufficiently accurate to predict the future; (2) strategy formulation was viewed as a normative process[5] where objectives could be formulated in hierarchical order, information could be readily available, and alternatives could be neatly identified and optimized in the rational manner advocated by economic theory or management science; and (3) factors such as politics, self-serving interests, and psychological traits were ignored.

Strategic planning came under attack by both academicians and practitioners. Empirical research has shown that some of the assumptions underlying the long-range and strategic planning approaches do not hold. Forecasting and, in particular, long-term forecasting could be very inaccurate.[6] Product life cycles, a major ingredient of long-range and strategic planning, cannot be predicted.[7] Furthermore, factors such as organizational politics and personal ambitions play an important role in the formulation and implementation of strategy.[8] In addition, strategists are not immune from judgmental biases,[9] nor are they free from the considerable judgmental limitations that characterize humans.[10] Strategists are not *and* cannot be assumed to be either rational or superpeople simply because they are concerned with important decisions whose effect is critical to the survival of the organization.[11] On the contrary, research has shown that the quality of decision making can deteriorate under the stress caused by the impact of such decisions.[12]

Business users, after a period of enthusiasm, became disillusioned with strategic planning. They found that formal planning sometimes lacked relevance and became an empty ritual. Strategic plans often got buried in drawers, and key decisions were reached outside the formal planning process. By the mid-seventies a crisis had emerged as to the purpose, usefulness, and applicability of strategic planning. This gave rise to a new set of perspectives that explicitly recognized greater environmental uncertainty. In order to deal with such uncertainty the following types of approaches were emphasized: (1) scenario management, contingency planning, and weak-signal monitoring; and (2) discovery of competitive rules and principles through emphasis on industry analysis.

The new directions taken by strategic planning and management may be useful. However, they have not solved the basic criticisms levied upon strategy. Scenario management and contingency planning are interesting exercises aimed at helping management, or policy makers, better understand the extent of future uncertainty. Such contingency planning has been facilitated by the use of microcomputers and decision support systems. Yet, contingency plans are of little help in deciding which of the alternatives to choose. In real life, one cannot wait: decisions must be made, investments must be engaged that lock the organization into irreversible commitments and require single forecasts.

Monitoring weak signals and surprise management[13] still remain academic ideas of little practical value. It is obvious that an infinite number of weak signals constantly exist in the environment. To pick up those signals whose influence on the organization is critical requires considerable abilities far beyond present technological capabilities.

Finally, in the face of increasing uncertainty, strategic management has focused on discovering some general principles that govern competition. The widespread use of the concepts of *experience curves* and *generic strategies* are attempts to specify such principles. They aim at introducing greater discipline in strategic thinking and highlighting the actions to be taken to improve competitive posture. The analytical tool of *business portfolio*, first introduced by the Boston Consulting Group, aims at clarifying strategic choice. However, the conceptual simplicity of these analytical tools is also their main weakness. They make several assumptions of dubious validity,[14] assume a fairly predictable environment, ignore interdependence among activities, and do not take into consideration the organizational, political, and psychological aspects of strategy. Critics have questioned seriously the relevance and usefulness of such analytical tools.[15] As stated by a Matsushita executive:

> In America, we've become infatuated with this "portfolio concept" in which certain divisions are "harvested" and their resources shifted to more promising "stars." We think those ideas are simplistic. . . . The principal fallacy of the portfolio concept is that all that frequently stands between a division being viewed as a cash cow or a star is management creativity in seeing how to reposition their products in tune with the marketplace.[16]

In contrast to the concept of strategy described above, which is based on normative principles, a completely different school, looking at strategy from a descriptive point of view, has developed since the mid-1960s.[17] It acknowledges that the *raison d'être* of firms is to allocate resources. The pattern in such an allocation process,whether done explicitly or implicitly, becomes the *de facto* strategy of the firm. If we define the *de facto* strategy of a firm as the underlying pattern according to which it allocates and commits resources, all firms then have a strategy that can be inferred *ex post facto* by analyzing and understanding how resources have been allocated within the organization.[18]

Empirical studies of resource allocation decisions in large companies have shown that a full-fledged strategy rarely exists at the outset;[19] in other words, strategic decisions are seldom the result of a planned sequence of moves. Rather, the resource allocation process evolves slowly, step by step, on a trial-and-error basis.[20] This involves successful actions being further pursued and unsuccessful ones avoided, which is typified in the behavior of Japanese managers.[21] The end result is a stream of resource commitments that reflect not only the competitive imperative but also the needs and aspirations of those individuals who have the power to commit resources. Eventually, the pattern that emerges through the resource allocation process becomes the strategy of the firm. The usefulness of such a process-based approach to strategy lies in gaining a better understanding of *how* resources are being committed and, consequently, being able to *influence* this process so as to align the *de facto* strategy with the *intended* one.

An obvious question of interest, however, is to what extent understanding *how* strategy evolves *over time* can help *management* make better decisions and take more appropriate actions *at present* for *future events*. The question posed here is similar to that of the need to study and understand history. Can this help us create a better future? No doubt a keener sense of the past might help us better appreciate the present and foresee more correctly the unfolding events of the future. On the other hand, history never repeats itself in the same fashion. Furthermore, history—particularly recent history—is often used to rationalize decisions or to justify mistakes. Thus, unless *objective* ways are found to analyze the past,[22] little can be gained by analyzing strategy in historical terms, that is, as the pattern of committing resources over time.

If the criticisms of strategic planning are valid, they will have to be dealt with if it is to become more than a fad advocated by academicians and sold by consultants. Ignoring the criticisms is to condemn strategic planning to be irrelevant. Yet, what is the alternative? Can one do *without* a strategy? Can a firm consider its strategy only after the fact and use the word *strategy* in a historical sense to rationalize decisions or to justify mistakes? Can opportunism and muddling through be the only guide to strategic planning?

For strategy to be relevant and applicable it needs (1) to be used proactively; (2) to accept our limited ability to predict environmental changes; (3) to take into account the organizational, political, and psychological dimensions of corporate life; and (4) to be accepted by a majority of those concerned with strategy as a realistic relevant tool for more effectively coping with the future.

EVOLUTIONARY PROCESS OF LIVING SYSTEMS AND HUMANS: AN ANALOGY

In the Darwinian interpretation of evolution the key element is competition in a hostile environment. The survival of the fittest is paramount and is achieved through *specialization*, by finding some niche that brings competitive advantages. These advantages allow the living system to successfully compete in the environment and survive. There *are very few*, if any, advanced living systems that have managed to *survive* by any means other than specialization.

In a dynamic, constantly changing environment, specialization may become a double-edged knife. If a drastic change takes place in the environment, the living system runs the risk of not being able to adapt to this change. There are many species that have disappeared because they were unable to adapt to environmental changes. This was, for instance, the case with the dinosaur.

The evolution of living species thus indicates that *ultimate survival depends on two simultaneous and seemingly contradictory types of behavior:* (1) *dominance* over existing competitors achieved through specialization; and (2) *ability to deal* with environmental changes. However, the greater the specialization (in the form of irreversible characteristics and skills) to a certain environment, the higher the risk of being unable to adapt to permanent environmental changes or to deal with temporary ones.

To counteract the risks of specialization, living systems have acquired a host of evolutionary abilities enabling them to deal more effectively with environmental changes. These include the mechanism of homeostasis, developing better senses, reacting effectively to dangers, increasing the ability to move more freely, achieving a safer shelter, and diversifying the sources of food. These abilities are *not* part of specialization; on the contrary, they are purposeful efforts to master and decrease dependence on single aspects of the environment.

Basic principles governing evolution of living systems can provide useful insights to the strategist, yet modern biology does not blindly accept the Darwinian interpretation of evolution.[23] Even though competition is an important element of evolution, modern biology maintains that it is but the starting point in achieving as high an environmental mastery as possible. Competition, by itself, cannot explain the rate of evolutionary progress in the biological world. Current evolutionary theories[24] advocate that the challenge to master the environment is more important than competition. Living systems are driven much more by this *challenge* than by the fear of competition.

Furthermore, human cultures are driven much more by challenge than by fear of competition. The Athenians of Ancient Greece did not achieve the height of their civilization because of fear of their enemies, the Spartans. On the contrary, had competition been their main preoccupation, they would have had to put their energies and resources into preparing for forthcoming wars rather than advancing the arts and sciences. The same is true with individual artists, scientists, or inventors. An element of competition may exist, driving their behavior and actions, but this is by no means the dominant factor that motivates these people to excel. Conflicts and wars might be influential and necessary factors for survival; however, progress does not come from wars nor is it one of their consequences. Evolution is mostly achieved by the *challenge* to master the environment and the *need* to excel and much less by competition. This is an important observation with far-reaching implications as far as strategy is concerned.

Finally, several key differences between biological and cultural evolution need to be noted. First, humans can influence the future, that is, they are active actors in *shaping* the environment. The majority of animal species, on the other hand, can only react to events and get prepared in a general way to minimize the effects of unwanted change. In addition, animals have little or no anticipation of the future. Animal life is governed by several stability mechanisms (for example, the territorial imperative) that regulate their interaction with the environment and allow them to respond to environmental changes. Second, biological evolution takes place in slow steps, each adaptation lasting thousands of years. Cultural evolution usually is much faster, a fact influenced by the human ability to shape the environment in desired directions. Moreover, humans are often driven by ambitions, emotions, personality traits, and the like. This is not the case with animals, whose needs are mostly physical.

Any analogy of the theory of evolution with corporate behavior must therefore proceed with caution and take into account the uniqueness of human endeavor. The fact still remains that firms, like any other type of organism, will survive only if they are effective at adapting their resources to environmental changes.

Type and Duration of Environmental Changes

The environment can change in many different ways. Successful adaptation therefore depends on the type of change involved.

Environmental changes can be classified as either random or systematic. Random changes cannot, by definition, be predicted. Systematic changes, on the other hand, imply some kind of continuation and can further be classified as temporary or permanent, depending on the extent of their duration. No living system can survive unless it has the intrinsic capability to deal with the entire range of environmental changes.

Reacting, Adjusting, and Adapting to Changes

How do living systems deal with environmental change if they cannot always accurately forecast it? *Random* changes are dealt with by certain types of reactions. For instance, the pores of the body contract if a person is suddenly in a cold environment. *Systematic but temporary* changes require adjustments of the body. For instance, the color of the skin becomes lighter in the winter than in the summer in order to let less heat escape from the body. Finally, *systematic permanent* changes require adaptation. For example, the human body has lost the hair that used to cover it. Similarly, people living in northern countries have much lighter skin color than people from the south.

Three aspects are critical in the biological processes of reacting, adjusting, and adapting to changes in the environment. First, the mechanisms for reacting and adjusting are automatic and general. The following example from immunization can illustrate this point.[25] Until recently, biologists thought that living systems reacted to foreign bodies that entered the system by recognizing them and generating a specific antibody to counteract each invader. This kind of contingency theory approach, however, cannot explain how new synthetic substances, which did not exist before, could be recognized and neutralized by living systems. As experiments have shown, living systems are capable of counteracting a broad range of invading bodies, even new synthetic substances that never existed before. Similarly, there is a natural process that changes the color of the skin when exposed to sun or that raises the number of red blood cells when vertebrates are in high altitudes. These adjustment mechanisms work when systematic changes in the environment of the organism take place and are different from the reactions that result when random fluctuations occur.

Second, since the process of adaptation in biological systems to permanent environmental changes is very slow, the effects of permanent environmental changes need to be *weathered* by temporary adjustments until adaptation takes place. In the end, however, unless a living system can adapt to systematic changes in the environment, it cannot survive in the long run.

Third, living systems can react and adjust to a certain range of environmental changes only. For instance, a living system can tolerate high temperatures up to a certain point, but when some physiological limits are exceeded death is inevitable.[26] Environmental changes thus introduce uncertainty for living systems. In order to reduce such uncertainty and enhance their chances of survival, living systems have sought to minimize the adverse effects of unwanted change. Choosing an appropriate shelter reduces the effects of temperature fluctuations. In a similar manner, hibernation eliminates the problem of not finding food during the winter. Thus, in addition to responsive actions, several purposeful *preventive* ones are taken to facilitate responding to environmental changes, thereby minimizing their negative impact.

Dealing with Change by Planning

Living systems, with a few exceptions, can only respond to environmental changes or take preventive actions to facilitate dealing with such change. Humans and a few insects (for example, ants and bees) are capable of taking concrete actions in advance by anticipating the future or some aspects of it. This allows them to *plan* in expectation of forthcoming changes. Ants, for example, gather food in anticipation of the winter, and humans save and/or buy into pension funds, knowing that when they become old they cannot work. This certainly is planning in its plain dictionary definition ("a method of thinking out acts and purposes beforehand"),[27] which increases the ability to deal with environmental changes or to minimize their discomfort.

Dealing with Change by Controlling the Environment

Human cultures can go further in their ability to influence the future than any other species. Humans have been active actors in *shaping* their environment and controlling their destiny. The difference between ants collecting food for the winter and humans inventing and using agriculture is enormous. In the former case, the ants have *no* control over the environment when, for some reason, they cannot find enough food to store. In the latter case, humans *shape* the environment in a desired direction so that food availability is much more consistent. Humans exhibit a clear tendency toward decreasing, as much as possible, their dependence on environmental changes. Agriculture, domestication of animals, construction of appropriate houses, building dams, development of transportation, storage of foods, savings, building social institutions, and a plethora of other actions are purposeful efforts aimed at controlling the environment. The purpose of this proactive behavior is to prevent some unwanted environmental changes from occurring, to reduce their intensity, or to facilitate some desired change to take place.

ORGANIZATIONAL STRATEGIC MANAGEMENT

From a capitalistic economic point of view, business failures are inevitable and perfectly acceptable, because they ensure efficient allocation of resources. Financial resources (through capital markets) and labor (through shift in employment) will adapt, in the long run, to market or technological changes and result in more efficient and effective firms as exemplified by the recent crisis in industrialized countries. Firms such as Chrysler, Ford, and Caterpillar, that have survived this era are much leaner and more efficiently managed today than they were before. In a condition of perfect competition, excess profits cannot exist and long-term equilibrium is ensured by the creation of new enterprises and the disappearance of inefficient ones. However, what is good from a societal standpoint may not be good from the point of view of an individual firm. First, markets do not operate as efficiently as is advocated by theory. Second, managers have motives of their own and are not interested entirely in total social welfare but the survival of their own firms.[28] The major question, then, becomes how to enhance the long-term survival of their firms. This is where organizational strategy becomes very important, assuming that long-term survival is *the* ultimate organizational objective.

A great deal can be gained by relating organizational strategic management to that of living systems and human cultures, because they are all concerned with long-term survival. By analogy, this will require *two* types of behavior: (1) the organization must be capable of effectively competing in the existing environment, mainly through specialization; and (2) changes in the environment must be dealt with effectively. Because the topic of competition has received wide coverage in the strategy literature,[29] the remainder of this chapter will concentrate mostly on environmental changes. Note that the emphasis presently placed on competition (based on disputed Darwinian reasoning) does not focus on the most important aspect of strategy: too much is made of industry analysis (the boundaries of industries are not static but constantly changing) and of predicting competitive behavior (for example, through competitive signaling). There are obvious limitations in attempting to monitor competitors and predicting their actions and reactions. Competitors can spread inaccurate information, give false signals about what they intend to do, even initiate actions to mislead those attempting to predict them. Furthermore, overemphasis on gaining competitive advantages through specialization might lead to excessive rigidity and hence reduce the ability to deal with environmental changes.[30] One must therefore accept that, in addition to competition, long-term survival greatly depends on the ability to deal (effectively) with environmental changes. In our opinion, this aspect of strategy has been neglected in the literature and needs to be carefully considered.

By using an analogy to living systems and human cultures, organizational strategists have several options: (1) dealing with change can be based on *responsive or preventive* actions and/or reactions, as is the case with all living systems, humans included; (2) environmental *changes can be planned for*, assuming that accurate forecasting is possible, as is the case with some animals and all human cultures; (3) *proactive actions* can be taken toward possible environmental changes, which is the sole prerogative of humans; and (4) any combination of responsive and/or preventive, planning, or proactive strategies is possible. The various strategic options and the several available strategic alternatives are listed in Tables 1-1, 1-2, and 1-3 and will be discussed briefly below.

Responsive or Preventive Strategic Management

In a responsive or preventive mode of strategic management, the emphasis is on reacting to environmental changes, whether or not the changes have been anticipated. In other words, uncertainty is taken as a *given*, and this mode implies having sufficient flexible or idle resources to be able to cope with new environmental requirements. It is a modus operand essentially based on opportunism.

Table 1-1 presents the strategic alternatives for responsive or preventive actions to envi-

Table 1-1. Strategic Alternatives for Responding to Environmental Change: Responsive Mode

Type and Duration of Environmental Change		*Strategic Variable*	*Trade-Offs*
Random		Amount of Slack	Cost of slack vs. ability to react adequately to random changes
Systematic	Temporary	Degree of flexibility	Opportunity cost of unused resources vs. ability to withstand (adjust to) temporary systematic changes
	Permanent	Extent of specialization	Lower efficiency and higher costs vs. ability to adapt to permanent changes

ronmental changes. The controllable variables are the amount of slack, the degree of flexibility, and the extent of specialization. For instance, a random breakdown of machinery, forcing production to be halted, can be dealt with by having extra inventories (that is, slack). A cash shortfall during a recession can be met by selling marketable securities (that is, financial flexibility).[31] Furthermore, a permanent environmental change can be weathered if the organization is not overspecialized; it can change its structure and eventually adapt to such change. Such a robust strategy allows the organization to redeploy its resources to cope with a wide range of occurrences.[32] What is important to accept, however, is that having slack costs money, that flexibility requires unused resources, usually earning little return, and that lesser specialization might mean reduced efficiency and higher costs in the short term, which might in turn reduce the competitive position of the organization.

An illustration of a company following a reactive and/or preventive strategy is Philip Morris. Philip Morris carries sufficient inventories of green tobacco and maintains idle production capacity in order to cope with random changes in demand. Although this has forced the company to increase its debt and to incur higher financial costs, it has allowed it to respond quickly to market changes. In addition, Philip Morris has coped with the uncertainty related to new product introduction by letting competitors take the risk of launching new types of cigarettes (filter, menthol, low-tar). When a new product proves successful (that is, it becomes clear that the change is not temporary but permanent) Philip Morris aggressively promotes a similar brand of its own. As stated by its president, George Weissman: "We have not often been first with a new cigarette . . . . [An] essential element in our success is our ability to react rapidly to changes in market or competitive position." Finally, Philip Morris has reacted to the risk of a permanent decline in cigarette smoking by diversifying first into beer and then into soft drinks.

Planning-oriented Strategic Management

The planning-oriented strategic management mode is based on the belief that some environmental changes can be predicted. In this case, strategists may not have to wait and react, rather they might wish to make specific decisions and take actions in anticipation of the forthcoming change(s). The decision of whether or not to follow a planning strategy depends on several strategic variables related to the accuracy of forecasting, the ability and willingness to act, and the strategies of competitors (see Table 1-2).

The ability to forecast accurately is central

Table 1-2. Strategic Alternatives for Planning Forthcoming Environmental Change: Planning Mode

Type and Duration of Environmental Change	*Strategic Variables*	*Trade-offs*
Random	None (By definition random changes *cannot* be predicted. Planning is, thus, impossible.)	Appropriate reactive and/or preventive strategies must exist. Otherwise, organizations might not be capable of dealing with random changes.
Systematic	*Accuracy of forecasting* Ability to distinguish between random and systematic changes Ability to predict timing, intensity, and duration of change	Benefits gained by being prepared (because of planning and acting) to deal with a forthcoming change that does occur vs. costs (real and opportunity) of planning and acting for a change that does not occur or
	Ability and willingness to act Availability of resources Adequacy of lead times Overcoming resistance to change	Costs (real and opportunity) avoided for not planning for a change that does not occur vs. opportunity costs of waiting (e.g., being unprepared) until a change occurs *and* the real costs of having to deal with such change by reactive strategies.
	Competitive planning strategies Accuracy of competitor's forecasts Ability and willingness of competitors to act	Costs and benefits are mostly relative. They depend on the uniqueness of forecasts, plans, and actions vis-à-vis the competitor.

to profitable planning strategies. If the forecasts turn out to be wrong, the real costs and opportunity costs can be considerable. For instance, if an organization assumes a recession that does not occur (for example, Montgomery Ward after World War II), it can incur high opportunity costs. On the contrary, if no recession is predicted and one occurs, the losses can be substantial, as so many shipping companies found out as a result of the prolonged 1973–1975 recession. The accuracy and reliability of forecasting cannot be assumed to be perfect. Thus, some form of the responsive and/or preventive strategies examined above should also be considered in conjunction with and to supplement planning. The extent to which responsive and/or preventive strategies will be required will depend on the type of change. For instance, seasonal changes can be predicted more accurately than others.

Successful strategic management in this mode not only implies accurate forecasts but also the ability to act on them effectively. Does the organization have the necessary resources? Is the lead time sufficient to plan adequately? Can the organizational resistance to change be overcome? This last point is particularly crucial in case of expected permanent change as it forces the organization to significantly alter its modus operandi.

A last element to be considered is the competitor's likely strategies. Is their forecasting ability good? To what extent do they have the advantage of proprietary information? Given their own resources and culture, how likely is it that their plans will be more efficient than ours? In fact, there is no advantage in correctly predicting a change if the same prediction can be made by better-equipped competitors (see also Table 1-2). Many companies, in their planning effort, try to identify alternative *scenarios*. A scenario is a consistent view of the industry future based on explicitly stated assumptions. One way to deal with uncertainty is to assign different probabilities to scenarios. A firm may then decide to plan according to the most likely scenario if its adaptative capability compares favorably with that of competitors. Conversely, a firm may prefer to bet on a less likely scenario if its internal resources give it a competitive edge to take advantage of such occurrence. Plans that can accommodate several scenarios entail lesser risk, but their likely return will be limited.[33]

Sears, for example, anticipated and planned for a long period of economic growth after World War II, whereas its competitor Montgomery Ward assumed a prolonged recession similar to the prewar pattern of the 1930s. The result was a relative increase in the size of Sears of about three times that of Montgomery Ward within a single decade.

Proactive Strategic Management

The third major strategic management option is proactive. It accepts that a wide range of environmental changes are *unpredictable*, and it attempts to *shape* the environment into some desired direction so that unwanted changes would be less likely to occur or their undesirable consequences on the organization be reduced. Alternatively, purposeful actions can be taken by the organization to bring about some desired change that would not have occurred (or would have happened later) otherwise. The controllable variables and the major trade-offs involved with proactive strategies are summarized in Table 1-3.

Market Demand. An important proactive strategic variable concerns the firm's attitude toward the market. Traditionally, market demand has been viewed as being beyond the control of a firm. Strategists therefore have relied on a concept such as the product life cycle. Part of the success of Japanese firms, however, can be attributed to their proactive attitude vis-à-vis market developments. In their view, market conditions are partly a consequence of self-fulfilling and self-defeating prophecies. The large U.S. market for leisure motorcycles, for instance, would never have come into existence if Honda had not aggressively advertised a socially acceptable image of the product ("You meet the nicest people on a Honda"), invested in efficient distribution channels, and constantly introduced new models (47 new models in 1983 alone). In doing so Honda succeeded in creating a new market for motorcycles. Similarly, the decline of several products can be attributed to self-fulfilling prophecies that cause premature "milking" of what is perceived to be mature products or industries and that make such maturity happen by reducing advertising and R&D expenditures, raising prices, or cutting investment.

Reducing Dependence. Heavy dependence on single aspects of the environment involves danger. Something can go wrong, or those in control of such an aspect can decide to take advantage of their powerful position (for example, the 1973-1974 oil embargo caused by the West's overdependence on OPEC oil). To avoid such dependency and to discourage such an event from happening, a firm might diversify its sources of supply, customers, and so forth. Alternatively, it might decide to integrate vertically, if possible, to secure stable sources of supplies or captive customers. The disadvantage of reducing one's dependence on single aspects of the environment may mean inefficiencies and higher costs. Furthermore, vertical integration increases specialization and decreases the chances of adaptation to a permanent envi-

Table 1-3. Strategic Alternatives for Controlling Environmental Change: Proactive Mode

Type and Duration of Environmental Change	*Strategic Variables*	*Trade-offs*
Random	Reducing exposure to random changes (e.g., long-term contractual agreements to reduce price fluctuations)	Benefits derived from consistency and the ability to better plan vs. opportunity costs when price of resources and/or material goes below that of contractual agreement (e.g., oil bought on a long-term contract based on 1979 price while present oil price is much cheaper).
Systematic	*Market demand* Creating one's own demand	Benefits of higher (or more stable) demand vs. costs of attempting to create such demand and risks of not succeeding
Temporary	*Customers and suppliers* Degree of dependence	Benefits from lesser dependence vs. costs of more suppliers (e.g., lesser economies of scale) and risks of integration
	R&D expenditure for new products Amount of innovation Extent of commitment	Benefits from successful R&D projects vs. cost of R&D investments when failed
	Competition Barriers to entry	Benefits from less competition both real and opportunity vs. cost of erecting barriers (both real and opportunity)
Permanent	*Labor market* Labor availability Labor Costs	Benefits of political stability at home (with high labor costs) vs. risks of possible low politicald stability abroad (with low labor costs)
	Financial strength Retained earnings	Benefits from retained earnings (less dependence on shareholders) vs. less rate of return to shareholders

ronmental change, which could render obsolete the material or process on which such integration is based.

Research and Development. R&D expenditure aims at developing new technologies ahead of competitors or at least keeping up with them. R&D discoveries subsequently need to be implemented by investing in new projects and/or ventures aimed at providing new blood for the organization. The obvious trade-offs are related to the pay-offs of R&D expenditures and the success of the new projects and/or ventures. Pioneers can reap huge profits if nothing goes wrong, but they also run high risks that their investments might not pay off. Furthermore, the degree of commitment to new projects and/or ventures can often make a big difference between success and failure.[34] This is particularly true when long lead times are involved and large amounts of capital are needed. In such cases, the heavier the initial commitment, the bigger the risk, but also the greater the chance of success by shaping the future in the desired direction. IBM's pioneering introduction of third-generation computers has been referred to as the "5 billion dollar gamble." Proactive strategy proved successful in this specific case.

IBM is a clear example of a firm following a proactive strategy. Thanks to heavy commitment in R&D and marketing, it has developed a market for data processing, multiplying customers and applications to the point where a competitor referred to IBM "not as a competitor, but as the environment"; similarly, Apple's early success is due largely to its ability to develop a primary demand for microcomputers by stressing "user friendliness." On the other hand, the losses might be larger if the anticipated change does not occur. Presently, ITT bet over $1 billion on the development of a radically new digital switching system that would not be fully introduced for several years and where the technology was still partly untested. If such a system had proven successful, however, it would have ensured ITT a dominant position in the market. However, in 1986, ITT announced that it would abandon the U.S. market and write off $115 million, leaving doubt about the overall success of the huge investment. Often there is talk about incrementalism, but such a step-by-step approach is not always possible. Go/no-go decisions, involving huge investment, must often be made in order to achieve important technological breakthroughs or large penetration of markets.

Barriers to Entry. Strategists must carefully consider their ability to build effective barriers to entry. It is important to distinguish between short-term and long-term barriers since experience has shown that in the long run few barriers hold. No matter how clever those building barriers are, there is no guarantee that someone will not find a more ingenious way to destroy them. The trade-off involved here is between devoting organizational resources and energy to building barriers versus attempting to better the organization, making it competitive in the marketplace in a way that can directly meet competition rather than attempting to shut it off. An interesting case is AT&T before and after telecommunications were deregulated.

Labor Costs. Labor costs can be a large component of overall production cost, which makes them a critical factor for profitability and long-term survival. The problem is exaggerated by the fact that wages vary considerably among various countries. Thus, operating in a cheap-labor country can result in substantial decreases in the labor bill and can provide considerable competitive advantages to labor-intensive firms. On the other hand, third-world countries, in which cheap labor is available, could be politically unstable. This presents a source of uncertainty, making investments more risky than at home, hence the trade-off. Furthermore, many companies that invested abroad in order to capitalize on cheap labor have not sufficiently automated their manufacturing process and have been unable to improve product quality. The U.S. color television industry is a good illustration.

Financial Strength. Retained earnings can be used by corporations to reduce their dependence on shareholders and increase their financial flexibility. The disadvantage of high retained earnings is that shareholders might not be getting sufficient return on their investments and borrowed capital might be more desirable in increasing the rate of return.

UNCERTAINTY AND STRATEGIC MANAGEMENT

In a stable environment, the role of strategic management as a way of dealing with change will, by definition, be small; changes in the relative position of the various participants will be minimal. Planning for lead-time requirements will only be needed, and few surprises or changes in the existing competitive structure will be taking place. However, as mentioned earlier, such environmental stability is most unlikely in the long term. In a changing but predictable environment the role of strategy as planning will be the only one needed. If *perfect* forecasting were possible, plans could be made to consider the effects of present decisions and actions in order to obtain full benefits of the future. It is the changing environment *and* the inaccuracy and unreliability of forecasting that necessitate the widest utilization of strategy.

There are definite limits to a human's ability to predict the future. This means that non-predicted events will occur and that some of the predicted changes will *not* materialize. In both cases, the organization will find itself in a situation where inappropriate plans or none at all exist and where some actions (responses) will have to be made.[35] To increase its chances of long-term survival, an organization should *always* be ready to face any and all types of unanticipated change. Practically, however, this is impossible. The cost associated with preparedness is not insignificant. A major strategic issue is, therefore, the extent to which the organization should be prepared to face unanticipated events and deal with unwanted change. There are risks for not being prepared or trying to shape the environment, and costs for being prepared or attempting to control change.

The extent of uncertainty is related to five things: (1) the intensity of unanticipated change; (2) its type (normal, unusual, unexpected, inconceivable); (3) its character (random, systematic); (4) its duration (temporary, permanent); and (5) its predictability. The more intense and unexpected the unanticipated change, the larger the uncertainty and risk involved. Moreover, the more permanent the change and the more difficult it is to predict, the higher the uncertainty and risks involved and the more difficult it is to deal with such a change. Furthermore, it might be prohibitively expensive (for example, by buying insurance covering *all* possible risks) to be prepared to deal with all possible types of change, or it might even be useless (for example, the case of a full-scale nuclear war).

Assessing Uncertainty

Strategists need to consider how much uncertainty they are willing to live with and then assess its implications for strategic management. This will have to be done in conjunction with the trade-offs of the various responsive, planning, and proactive strategies considered above. For instance, how much financial flexibility should be available to deal with economic downturns? Should a company be prepared to deal with a depression? Should plants be built to accommodate drastic changes in technology? Should companies spend R&D money to discover extremely unlikely technological breakthroughs? Should firms attempt to discourage all competition? Obviously, there are limits to the amount of uncertainty that can be considered since, in the final analysis, everything is possible. A way of attempting to assess environmental change and the uncertainties and risks involved is by categorizing change in a manner similar to that shown in Table 1-4. Table1-4

Table 1-4. Examples of Environmental Changes

Characters of Change	*Type and Duration of Change*		
	Random	*Systematic*	
		Temporary	*Permanent*
Normal	A machine breakdown	Seasonal fluctuations in demand for product *A*	Shift in consumer preferences (e.g., small cars)
Unusual	A big snowstorm	A recession	A nuclear plant explosion
Unexpected	A big fire	A severe recession	The fall of the Shah in Iran
Inconceivable	A meteorite hits the Earth and destroys all physical facilities of company XYZ	A major conventional war in Europe	An all-out nuclear war that completely destroys civilization

classifies changes as normal, unusual, unexpected, and inconceivable and gives some examples of events in each category. Table 1-5 briefly lists the uncertainties involved in each subdivision of Table 1-4 and lists their implications for forecasting, planning, and strategy.[36] Obviously, the possibilities involved are immense.

Tables 1-4 and 1-5 imply that everything can happen. How much of this information can be usefully utilized by strategists? First, they must accept that everything is possible; second,they must decide how many of the possible changes they are willing to consider. Do they stop with unusual events, or are they willing to consider even unexpected changes in some key areas of critical importance? How much will it cost to be prepared, plan, or attempt to control change?

Table 1-6 provides a formalized framework for assessing change and conceptualizing the uncertainties involved. An environmental fluctuation can be random or a change can be systematic, either temporary or permanent. Strategists will have to make choices and plans or take actions assuming random, temporary, or permanent changes. The interaction of the three kinds of changes, together with the three possible planning or action alternatives, makes the matrix shown in Table 1-6. The choices involved are critical from a strategic point of view, but the outcome is uncertain. The organization can make the wrong decision in assuming that a change will occur that does not occur. This obviously can cost money (for example, it was expected that demand for Product *B* would increase substantially, thus a new plant was built, but actually it remains the same). Alternatively, if the organization does nothing, some of the competitors might (for example, no new plant is built and demand for Product *B* increases substantially), thereby gaining an advantage over the organization (for example, higher market share or lower production costs).

Finally, it is important to note that accurate forecasting is not necessarily beneficial and should not, therefore, be overemphasized as a strategic option. If the competitors have also predicted that the same change might occur, the end result might be high competition and eventual losses.[37] For instance, many people would predict today that computers will bring about a big change in our society and business practices, that industrial robots will change the way factories operate, that existing patterns of communication will be modified with the use of satellites, that lasers will substantially change the way materials are transformed, or that biochemistry and genetic engineering have high potential for the future of humans. Strategists have to figure out how such anticipations can be taken into account and acted upon in the form of planning. A major element of strategy will also have to judge how competitors will react to available forecasts and how the future environment will change because of planning and proactive strategies of existing competitors or new entrants. Accurate forecasting does not necessarily decrease—it might even increase—the uncertainty that surrounds strategists in their attempt to deal with future environmental change.

SUMMARY

In this chapter, effective strategic management is linked to biological and cultural evolution, that is to the evolution of living systems and human cultures. We advocate that strategic management must be concerned with two simultaneous and seemingly contradictory types of behavior: (1) dominance over existing competitors; and (2) ability to deal with environmental changes. Even though competitive strategy is important, we advocate that the most critical aspect of strategy is to deal successfully with environmental changes. This involves several strategic options aimed at allowing the organization to react to environmental changes and to facilitate its adjustment and adaptation.

Three types of strategic management are proposed: the first responsive or preventive, the second based on planning, and the third proactive. Responsive or preventive strategies are equated with the reactions, adjustments,

Table 1-5. Environmental Changes, Uncertainty, and Implications for Planning and Strategy

Character of Change	*Type and Duration of Change*		
	Random	*Systematic*	
		Temporary	*Permanent*
Normal	Extent of uncertainty can be assessed and incorporated into planning. Role of strategists is to establish some general policies of dealing with random fluctuations	Extent of uncertainty can be statistically assessed but timing of occurrence cannot be known. Actual and opportunity costs of being prepared for such uncertainties could be high. Consequences of costs and benefits need to be considered. Role of strategy is to decide extent of flexibility required to deal with such changes.	Timing cannot be known. Extent of uncertainty is difficult to estimate. Strategy must facilitate organizational adaptation. Ability to assess existing trends is critical. Role of anticipation is important. Open mindness is necessary. Ability to initiate actions and overcome resistance to change is critical. Strategy needs to be proactive.
Unusual	Uncertainty can be estimated, but cost of being prepared for the eventuality of unusual occurrences can be high. Role of strategy is to look at the trade-offs of simply responding to such changes versus the cost of being prepared to face them.	Uncertainty can be estimated but cost of being prepared for the eventuality of unusual occurrences can be extremely high. Consequences of costs, efficiency, returns, and competitiveness need to be considered. Role of strategy would be to establish the level of flexibility that should exist to deal with such occurrences.	Timing and extent of uncertainty is extremely difficult to estimate. Impossible to be prepared for all types of change. Strategy must be concerned with building an organization that can adjust to changes of this type and take actions to allow for eventual adaptation.
Unexpected	Although one can conceive of unexpected occurrences, it is impossible to know when they will occur. Costs and benefits should be considered. Role of strategy might be limited to deciding what types of insurance policies should be bought to protect against unexpected, random eventualities.	Difficult to predict impossible to plan for, unless great number of contingencies can exist. Role of strategy is to prepare the organization (with general competence and flexibility) to be capable of dealing with such changes.	Impossible to predict accurately, assess uncertainty, or be prepared. Strategy must be concerned with developing a general ability so that the organization can weather such (often abrupt) changes and be capable of eventually adapting to them.
Inconceivable	Forecasting, planning, and strategy as they are currently perceived are not relevant. Proactive strategies, even if they exist, would probably be of little value when inconceivable events occur.		

Table 1-6. Costs and Benefits ot Taking Actions for Different Types of Change

Action Alternatives	*Type of Environmental Change*		
	Random	*Systematic*	
		Temporary	*Permanent*
No action	Correct match	Opportunity or real cost of not planning or be prepared for a systematic, temporary change. Extent of cost is function of the intensity of systematic change.	Large opportunity or real costs can result from not being prepared or having planned systematic permanent change. Major disadvantages for the organization can result.
Temporary Change	A mistake has been made by assuming a systematic temporary change when this was not the case. Cost of planning, and being prepared on assumed change are involved, also possible opportunity costs.	Correct match	A permanent change has been thought to be only a temporary one. Important opportunity or real costs are involved for not having planned or been prepared for such change. Important negative consequence for the organization.
Permanent Change	An extremely serious mistake has been made for planning or being prepared for something that did not occur. Large costs and major negative consequences for the organization usually can result from such a mistake.	A critical mistake has been made. Costs (opportunity or real) can be substantial. The cost relates to planning or being prepared for something that did not turn out to be a permanent change.	Correct match

and adaptations of living systems; planning equates with the ability of some animals and all humans to anticipate certain aspects of the future, and proactive strategies are analogous to the human need to shape and control the environment. Furthermore, we indicate that no matter how accurate the forecasting or how good management's ability to anticipate, plan, or control future events, uncertainty will always be associated with the future. This uncertainty must be considered by examining the whole range of possible changes (see Tables 1-4 and 1-5). Finally, a framework is proposed for assessing uncertainty and considering the costs and benefits involved (see Table 1-6) when dealing with possible environmental change.

REFERENCES

1. H. Fayol, *General and Industrial Management* (London: Pitman 1949).
2. A. Chandler, *Strategy and Structure* (Cambridge, Mass.: MIT Press, 1962).
3. A. Sloan, *My Years at General Motors* (New York: Doubleday, 1963).
4. S. Makridakis and S. Wheelright, *The Handbook of Forecasting: A Manager's Guide* (New York: Wiley, 1982).
5. G. A. Steiner, (*Top Management Planning* New York: Macmillan, 1969); I. Ansoff, *Corporate Strategy* (New York: McGraw-Hill, 1965).
6. W. Ascher, *Forecasting: An Appraisal for Policy-Makers and Planners* (Baltimore, Md.: Johns Hopkins University Press, 1978).
7. N. K. Dhalla and S. Yuspeh, "Forget the

Product Life Cycle Concept," *Harvard Business Review* **54**(1976):102-112.
8. J. Bower, *Managing the Resource Allocation Process* (Cambridge, Mass.: Division of Research, Harvard Business School, 1970); M. F. R. Kets de Vries, *Organizational Paradoxes: Clinical Approaches to Management* (London: Tavistock, 1980).
9. A. Tversky and D. Kahneman, "Judgment under Uncertainty: Heurishes and Biasses," *Science* **185**(1974):1124-1131.
10. R. M. Hogarth and S. Makridakis, "Forecasting and Planning, an Evaluation," *Management Science* **27**(1981):115-138.
11. I. L. Janis and M. Mann, *Decision Making: A Psychological Analysis of Conflicts, Choice and Commitment* (New York: Free Press, 1977).
12. O. R. Holsti, "Crisis Stress and Decision-Making," *International Social Science Journal* **23**(1971):23-67.
13. I. Ansoff, "Managing Strategic Surprise by Response to Weak Signals," *California Management Review* **18**(1975):21-33.
14. T. Naylor, "Forecasting and the Strategic Process," *Journal of Forecasting,* **2**(1983) 109-118.
15. W. Kiechel, "The Decline of the Experience Curve," *Fortune* (October 1981):139-146; W. Kiechel, "Oh Where, Oh Where Has My Little Dog Gone? Or My Cash Gone? Or My Star?" *Fortune* (November 1982):148-153; R. Wensley, "PIMS and BCG: New Horizons or False Dawn," *Strategic Management Journal* **3**(1982):147-158; Naylor, "Forecasting and the Strategic Process."
16. R. Pascale and A. Athos, *The Art of Japanese Management* (Warner Books, 1982).
17. R. Cyert and G. March, *A Behavioral Theory of the Firm* (Englewood Cliffs, N.J.: Prentice-Hall, 1963).
18. Bower, *Managing the Resource Allocation Process;* H. Mintzberg, "Patterns in Strategy Formulation," *Management Science* **24**(1978):937-948.
19. E. Wrapp, "Good Managers Don't Make Policy Decisions," *Harvard Business Review* (Sept-Oct 1967):91-99; J. B. Quinn, *Strategies for Change* (Homewood, Ill.: Richard D. Irwin, 1980).
20. C. E. Lindblom, "The Science of 'Muddling Through'," *Public Administration Review* (Spring 1959):79-88; Quinn, *Strategies for Change;* R. Miles, *Coffin Nails and Corporate Strategy* (Englewood Cliffs, N.J.: Prentice-Hall, 1982).
21. Pascale and Athos, *Art of Japanese Management;* J. P. Lehman and Y. Doz, "The Strategic Management Process in Japanese Firms," INSEAD Working Paper, September 1982.
22. E. Carr, *What Is History?* (New York: Random House, 1961).
23. P. P. Grassé, *L'Evolution du Vivant* (Paris: Abbrie Michel, 1973); J. Ruffie, *De la Biologie à la Culture* (Paris: Flamarion, 1974).
24. Grassé, *L'Evolution du Vivant.*
25. C. Faucheux and S. Makridakis, "Automation or Autonomy in Organizational Design," *International Journal of General Systems* **5**(1979)213-220; F. Valera, "On Being Autonomous: The Lessons of Natural History for System Theory," in *Applied General Systems Research: Recent Developments and Trends,* ed. G. J. Klir (New York: Plenum Press, 1978), pp. 74-84.
26. W. R. Ashley, *Design for a Brain* (London: Science Paperbacks, 1960).
27. *The Random House Dictionary of the English Language,* college ed., s.v. "planning."
28. W. J. Baumol, *Economic Theory and Operations Analysis* (Englewood Cliffs, N.J.: Prentice-Hall, 1972); J. K. Galbraith, *The Affluent Society* (Boston: Houghton Mifflin, 1976).
29. M. E. Porter, *Competitive Strategy: Techniques for Analyzing Industries and Competitors* (New York: Free Press, 1980).
30. W. J. Abernathy and K. Wayne, "Limits of the Learning Curve," *Harvard Business Review* (Sept-Oct 1974):109-119.
31. G. Donaldson, *Strategy for Financial Mobility* (Cambridge, Mass.: Division of Research, Harvard Business School, 1969).
32. M. E. Porter, *Competitive Advantage* (New York: Free Press, 1985).
33. Ibid.
34. R. Biggadike, "The Risky Business of Diversification," *Harvard Business Review* (May-June 1979):103-111.
35. S. Makridakis, "If We Cannot Forecast How Can We Plan," *Long Range Planning Journal* **14**(1981):10-20.
36. R. M. Hogarth and S. Makridakis, "Limits to Predictability: From Superstition to Science," INSEAD Working Paper, September 1982.
37. Wensley, "PIMS and BCG."

CHAPTER 2

An Overview of Strategic Planning Styles

Bernard Taylor

Bernard Taylor is a professor of business policy at Henley—the Management College, which is one of Britain's leading business schools. He is editor of *Long Range Planning Journal* and a consultant on strategic planning to major companies and governments. He has also written twelve books and a number of articles on strategy and planning.

Over the past decade the practice of strategic planning has matured and developed in response to pressures from inside and outside business. What started as a unique system based on a simple model of problem solving and decision making has evolved into a broad range of philosophies and techniques designed to help the executive build an organization that is adaptable and responsive in a rapidly changing environment.

Each style of planning has a philosophy, a school of adherents, and a range of techniques that have been tested in practice, and each can provide management with a sensible approach to changing the orientation of a business.

A small- or medium-size firm may adopt only one of these styles—typically, a system controlling the allocation of resources or a framework for generating strategies for new ventures. However, in large corporations, such as General Electric, or Shell International, most or all of these approaches will be present. The philosophies and techniques are largely compatible and complementary.

The main styles or modes of strategic planning that have emerged in recent years are:

1. *Central Control System:* the view of planning as a system for acquiring and allocating resources.
2. *Framework for Innovation:* the idea that planning should provide a framework for the generation of new products and new processes and the entry into new markets and new businesses.
3. *Strategic Management:* the notion that planning should be concerned not just with formulating strategies but with developing the commitment, the skills, and the talents required to implement the strategies.
4. *Political Planning:* the perspective that sees planning as a process for resolving conflicts between interest groups and organizations inside and outside the business.
5. *Futures Research:* the concept of planning as exploring and creating the future. Futurists believe that the future cannot be forecast, therefore decision makers should consciously assess the uncertainties, then develop and work toward a vision of the future.

Adapted with permission from B. Taylor, "Strategic Planning: Which Style do you Need?" in John Grieve Smith (ed.), *Strategic Planning in Nationalized Industries,* Macmillan, London, 1984.

Strategic planning has a central role in the management of the modern corporation. It provides a practical approach to changing the way an enterprise is managed. For strategic planning to succeed, however, it needs to be seen not just as a set of techniques but as part of a coherent program of change.

This chapter describes these five basic approaches to strategic planning. Each represents an important school of thought in management thinking and practice. Each view has a large body of supporters, both academics and practitioners, and each offers a coherent philosophy and a range of practical systems and techniques for implementing them.

In determining their approach to planning, the chief executive and his or her planning staff need to examine the different available methodologies to discover which system best meets their needs. They should then adapt the approach to suit their own organization. For corporate planning systems do not come ready-made. They have to be tailor-made to fit each enterprise. The decision is important because typically it takes two or three years to introduce a particular strategic planning approach, and, if it is to be effective, it requires wholehearted commitment from both the board and from operating management.

In any large organization it is likely that several different strategic planning approaches will be present at any one time, and, in one part of the business, planning is likely to move through various phases—with management adopting different planning philosophies at different stages in the development of the firm. Strategic planning, like other managerial activities, is a process that grows and evolves—and sometimes has major setbacks and needs to be relaunched. It is my hope that this chapter will help the reader to assess the state of planning in his or her own organization and to suggest areas where the activity may be strengthened or improved. It is rare to find an enterprise where all the available planning approaches are being employed equally effectively.

Table 2-1 sets out the five main views of planning in broad outline under four headings:

1. *The Focus*: the main objective or purpose.
2. *Important Ideas*: the characteristic view or philosophy of planning.
3. *The Elements*: the key steps or stages in the process.
4. *The Techniques*: some widely used techniques.

As with any classification system, the categories are not watertight, but they do represent distinct traditions in current thinking and practice.[1]

STRATEGIC PLANNING AS A CENTRAL CONTROL SYSTEM

One of the main drives behind the development of strategic planning has been the desire of top management to have better control over allocation of capital and other key resources.

The Philosophy

The philosophy of planning and control is fundamental to management. Early thinkers on management, for example, Henri Fayol, described the management process in terms of "planning, command, co-ordination and control."[2] An analogy is often made with an army or other hierarchical organization, where decisions are taken at the top, instructions are passed down through the enterprise, and the leaders get back information that enables them to measure actual results against the plan.

The business enterprise is also frequently compared with a machine that can be regulated by an engineering control system. Automatic control systems, such as the domestic thermostat, contain certain basic elements: a sensor, a standard of performance, a collator, which compares actual performance with the standard, and an actor, which takes action to make up any deficiency in performance or to change the standard.

Table 2-1. Strategic Planning: Five Basic Approaches

	Central Control System	*Framework for Innovation*	*Strategic Management*	*Political Planning*	*Futures Research*
The focus	Allocation and control of resources	Developing new business	Managing organizational change	Mobilizing power and influence	Exploring the future
Important ideas	A rational decision-making and control process	A vehicle for commercializing innovation	A community with common values and culture	Interest groups and organizations competing for resources	A management with a real awareness of future uncertainty
The elements	Specific objectives A balanced portfolio of investments Action programs and budgets Monitoring and control	Commitment to innovation Funds for new development Strategies for corporate development Organizing project teams and action programs	Organization development Staff development Organization structure Management systems	Monitoring and forecasting social and political trends Assessing the impact on the firm Organizing and implementing action programs	Developing alternative futures Assessing social and economic impact Defining key decisions
The techniques	SWOT analysis Business portfolio analysis Gap analysis Extrapolative forecasting Extended budgeting	Programs for: Divestment Diversification Acquisition New product development Market penetration and development	Group work on: Stakeholder analysis SWOT analysis Portfolio analysis	Public affairs Civil affairs Employee communication Social issue analysis Country risk analysis Media relations	Scenarios Delphi studies Cross impact analysis Trend analysis Computer simulation Contingency planning

Writers and practitioners on planning have seen strategic planning as a comprehensive control system that could be used to regulate the operations of a whole firm—a logical extension of departmental control systems such as stock control, sales control, and production control. They were also attracted by the idea of the business as a total system with an integrated information and control system.

It is perhaps natural that accountants should have seen strategic planning as an adjunct to the budgetary control system. There is, however, an important distinction to be drawn between strategic planning and management control—though they are obviously related. Strategic planning includes, for example, choosing company objectives, planning the organizational structure, setting policies for personnel, finance, marketing and research, choosing new product lines, acquiring a new division, and deciding on nonroutine capital expenditures. Management control is concerned with formulating budgets, determining staff levels, formulating personnel, marketing, and research programs, deciding on routine capital expenditures, measuring, appraising and improving management performance, and so forth.[3]

The Processes

The rise of corporate planning in the 1960s coincided with a period of diversification and international expansion in large companies. In many cases these same firms were divided into product divisions, which were designated as profit centers or cost centers. Divisional general managers were appointed and each was instructed to manage his or her division as if it were an independent business. Unfortunately, some of these executives took the instruction too literally and top management saw their subordinates riding off in all directions. Strategic planning was seized upon as a technique that might enable the main board to reestablish some control over the situation. Traditional budgeting systems proved woefully inadequate to the task of controlling a multidivisional business—particularly when the divisional managers usually formed a majority on the main board and sat in judgement on their own capital projects.[4]

The solution commonly adopted was:

1. to restructure the board to reduce the power of the divisional managers by bringing in heads of functional departments, nonexecutive directors, and others who could form a board representing the whole corporation rather than specific local interest.
2. to require the divisions to put forward divisional strategies. This enabled top management and the central staff groups to debate various options for each product group before divisional plans became embedded in detailed project plans as the one true way to go.

The corporate staff groups (finance, personnel, manufacturing, and so forth) and the planners themselves, are involved in the corporate planning process through (1) preparing the planning guidelines for divisions; (2) reviewing divisional strategies and plans; (3) advising the board or an executive committee in approving the plans; and (4) monitoring divisional performance against the plans.

The Problems

The close association between planning and financial control has led to problems. In particular there has been a tendency (1) to confuse strategic planning with extended budgeting; (2) to produce three-year or five-year plans simply by extrapolating or pushing forward the present operations; (3) to prepare company plans by merely consolidating the operational plans of divisions and subsidiaries; and (4) to stress the numbers rather than the quality of the thinking.

This still goes on. It is common to find corporate plans that consist of comprehensive and detailed operational plans and budgets without any discussion of objectives, organization structure, or alternative strategies.

Another problem with the five-year plan and budget is that it can easily degenerate into a sterile but time-consuming routine. Corporate planning has provided many examples of this: highly structured planning systems that required many hours to build and maintain resulting in plans that were wrong to three points of decimals.

Nevertheless, resource allocation is at the core of most planning systems, and the operational plan and budget is the basic planning document. Other qualitative and informal approaches to planning have been developed to compensate for its inflexibility and its narrow scope.

STRATEGIC PLANNING AS A FRAMEWORK FOR INNOVATION

One powerful reason for the growth of strategic planning was the need to establish a central steering mechanism for the direction and coordination of large, diverse, multinational operations. An equally strong and opposite motivation was the desire to promote initiative at the local level—in particular, to prevent centralization and bureaucracy from stifling creativity and innovation.

Over time the need for continuous change and innovation has become accepted by many leading businesspersons and writers on business as an article of faith. To quote Peter Drucker, the businessperson's philosopher:

> In a world buffeted by change, faced daily with new threats to its safety, the only way to conserve is by innovating. The only stability possible is stability in motion.[5]

The implications of this philosophy were spelled out for businesspersons, politicians, and public officials by John Gardner, former U.S. Secretary of Health Education and Welfare, in his best-selling book *Self-Renewal.*

> A society whose maturing consists simply of acquiring more firmly established ways of doing things is headed for the graveyard—even if it learns to do these things with greater and greater skill. In the ever-renewing society what matures is a system or framework within which continuous innovation, renewal and rebirth can occur.[6]

For a competitive business this process of self-renewal is fundamental to survival. In the short term, management can make profits by mortgaging the future—and many managements are tempted to do this in the present crisis. However, in a rapidly changing situation, unless there is a continual reinvestment in staff training, market development, new products, and up-to-date equipment, companies are likely to find themselves overtaken by their competitors. As the Boston Consulting Group consultants concluded in their inquiry into the failure of the British motorcycle industry

> The result of the British industry's historic focus on short-term profitability has been low profits and now losses in the long term. The long term result of the Japanese industry's historic focus on market share and volume, often at the expense of short-term profitability, has been the precise opposite: high and secure profitability.[7]

This process of entrepreneurship has long been acknowledged as a central function of the businessperson. It involves identifying a market opportunity, developing a product to match it, raising the necessary finance and matching the risk to the opportunity, mobilizing the staff and the other resources necessary to provide the required service, and producing and distributing the product at a profit. To quote Donald Schon:

> The firm defines itself as a vehicle for carrying out a special kind of process. It defines itself through its engagement in entrepreneurship, the launching of new ventures, or in commercializing what comes out of development. The principal figure in the firm becomes the manager of the corporate entrepreneurial process; and the question is this: what are the potentials in development for new commercial ventures?[8]

In a one-person business the owner can be his or her own entrepreneur, but in a large corporation the process has to be formalized and systematized. To quote Peter Drucker again:

> Every one of the great business builders we know of—from the Medici to the founders of the Bank

of England down to Thomas Watson in our days—had a definite idea, a clear "theory of business" which informed his actions and decisions. Indeed a clear simple and penetrating "theory of the business" rather than "intuition" characterizes the truly successful entrepreneur, the man who not only amasses a large fortune but builds an organization that can endure and grow long after he is gone.

But the individual entrepreneur does not need to analyze his concepts and to explain his "theory of business" to others, let alone spell out the details. He is in one person thinker, analyst and executor. Business enterprise, however, requires that entrepreneurship be systematized, spelled out as a discipline and organized as work.[9]

In many corporations strategic planning is regarded as a form of organized entrepreneurship. Patrick Haggerty, the former chairman of Texas Instruments, described their planning system as a framework for innovation.

Self Renewal at Texas Instruments begins with deliberate, planned innovations in each of the basic areas of industrial life—creating, making and marketing. With our long range planning system we attempt to manage this innovation so as to provide a continuing stimulus to the company's growth.[10]

The Management of Corporate Development

Most management systems are concerned with operational problems. Operational plans start with the present situation and push it forward in terms of sales quotas, production targets, stock levels, budgets, and so forth. The horizon is typically one year, occasionally two years or perhaps three to five years for a specific product or facility. Other management systems—performance appraisal, salaries and incentives, promotions, and career development—also help to focus managers' ideas on the short term. A major problem for the large corporation is how to persuade staff to spend some of their time thinking and planning for new products, new markets, new production and administrative processes, and maybe entirely new kinds of businesses—joint ventures, mergers and acquisitions, new social and political initiatives.

How, for a start, can top management produce a strategy and a program for the development of the business? Typically, this involves the formation of *ad hoc* groups that report directly to the board: project teams, venture groups, a diversification task force, a group to deal with acquisitions or international expansion, and so forth. In the present recession we have also seen task forces formed to look at closures, divestments, rationalization, and organizational restructuring.

The challenge is to develop a *vision of success* for the total enterprise and its parts and then to produce action plans, budgets, and timetables to realize the vision.

The techniques in common use provide broad frameworks for discussion and analysis. For example,

1. *Gap Analysis* describes the planning task by identifying the gap between the company's objectives and its likely achievement in terms of profits, sales, cash flow, and so forth. Management is invited to (1) set an objective in quantitative terms—for example, rate of return on investment or market share; (2) forecast the momentum line for the present business assuming no major changes; and (3) plan to fill the gap with projects for increased efficiency, expansion, and diversification.
2. *SWOT analysis* provides a series of check lists for auditing the company's strengths and weaknesses and the opportunities and threats in the business environment. The business is assessed against leading competitors in world markets in terms of its technology, market position, financial base, production efficiency, management, and organization. The opportunities and threats are considered in light of trends in the environment—economic, sociopolitical, technological, and competitive. Then the two analyses are compared to see which market opportunities match the firm's resources, what new resources are required, and so forth.
3. *Business portfolio analysis* where the process of funds allocation (that is, allocation of both fixed and working capital) is frequently discussed on the basis of a matrix showing the pattern of business in the company's portfolio. Many large companies have their own matrix or screen, typically displaying on one scale the prospects

for the industry and on the other the strength of the company's market position. The criteria for the industry's attractiveness might include: growth potential, expected changes in markets and in technology, the strength of competition from existing competitors and possible newcomers, and government and environmental constraints. The analysis of one's own company strengths requires a comparison with leading competitors in terms of market share, production capability, relative costs, technical expertise, patent position, marketing, distribution and service, and government support.

4. *PIMS (Profit Impact of Market Strategy)* data base, which was set up originally by General Electric, is derived from approximately a thousand businesses in the United States and Western Europe over a period of up to ten years. The program collects 300 items of information about each business and attempts to discover which factors have most effect on profitability (return on investment) for example, market share, product quality, marketing expenditure, capital investment vs. sales, and so forth. The data base is used primarily by holding companies in assessing the performance of divisions and subsidiaries and in making decisions about investment and divestment.

In practice, the majority of managers find the task of strategic planning difficult, and they require a good deal of help. This is partly a matter of temperament. Operating managers tend to be chosen for their ability to get things done, and it has been well said that a man of action, forced into a state of thought, is unhappy until he can get out of it. It is largely the size and complexity of the task—to try to plan for the long-term development of the total enterprise in all its dimensions. It is also the problem of planning with little solid information in a situation of great uncertainty where all the elements interact. Inevitably, the manager has to rely on his or her judgment and imagination much more than in operational management.

Faced with these practical difficulties in generating new strategies, leading companies such as General Electric in the United States and Philips in Europe have started to think not just in terms of strategic planning but in terms of strategic management, that is, changing the whole management system.

STRATEGIC MANAGEMENT

This increasingly popular approach takes the view that policy making is a learning process, and strategic planning is the specific activity through which the members of an organization learn to adapt to radical changes in the external environment.

The Philosophy of Strategic Management

Consider the changes taking place and their impact on human institutions: fluctuations in supply and demand, the advent of new technologies, the appearance of social and political movements, the rise and fall of governments. All these trends are rendering established institutions and traditional ways of thinking and acting obsolete. In a rapidly changing world, organizations must adapt or go under. To quote Donald Schon:

> Our society and all of its institutions are in continuing processes of transformation . . . we must learn to understand guide, influence and manage these transformations.
>
> We must invent and develop institutions which are "learning systems"—systems capable of bringing about their own transformation.[11]

This is the theory of natural selection again: the view that organizations must adapt or be replaced by others better suited to their environment.

How then can we build institutions that learn to adapt to their environment? Is it possible to develop management's ability to cope with change? Can we help organizations or teams of people to set objectives, to be more aware of the changes taking place around them, and to develop their own plans for the future? Can management learn to do this on a continuing basis?

Those who support the view of planning as a part of a process of social change usually reject the theory that planning is a logical search for solutions, a cognitive decision-making process that establishes the area of search and certain performance criteria, collects and analyzes data, assesses alternatives, and makes an optimal choice. We are after all dealing not with inert objects but with people who have their own ideas, beliefs, and motivations.[12]

In place of this model of strategic planning as rational and sequential, behavioral scientists frequently present it as a trial-and-error process. Managers and administrators are encouraged to adopt an experimental approach, not looking for comprehensive solutions or great leaps forward but attempting to engineer incremental changes with the top managers and their advisers not moving too far ahead of the group.[13]

The Approach of Strategic Management

This *behavioral* view of strategic planning is more human, less comprehensive, more easily related to the organizations we all know and in which we work.

Planning is seen as a *process* through which individuals and teams can learn to cope with an unpredictable and rapidly changing environment. The fact that a forecast or a plan turns out to be wrong therefore is not an indication that management is incompetent or that planning is infeasible but rather a confirmation that we are living in an uncertain world and we need to reassess our situation continually. However, we should learn by experience, and our involvement in forecasting, strategy making, planning, programming, and budgeting should help us to get a better feel for the trends in the environment and should improve the organization's capacity to respond to them.

Planning is seen as one element in a wider program of organizational change. This may involve many other measures.

1. *Moves affecting individual managers:* retiring or retraining existing managers, recruiting new managers, promoting and developing staff for new roles.
2. *Changes in organization structure:* these might include, for example, (1) dividing the company into semiautonomous units such as product divisions; (2) establishing new groupings to coordinate policy by geographical regions, product groups, or strategic business units (that is, parts of the organization that have a common business strategy); or (3) reorganizing the board and revising the capital investment procedures to strengthen the role of the board as a policy-making body.
3. *Changes in management systems:* these might include (1) changing the procedures for staff appraisal, promotion, and payment to encourage management to give higher priority to new companywide programs, for example, social responsibility, new business development, or staff development; (2) introducing improved information systems for finance, manpower, and production; (3) developing planning and control systems to focus management attention on cash flow, productivity, planning for manpower, and so forth; or (4) providing new procedures for environmental assessment to give managers a better understanding of the external trends likely to affect their business.

In adopting the organization learning mode of planning, the planner takes on the role of a *change agent.* His or her task is not merely to produce a product—the plan—but rather to intervene in the process, that is, to work with management at various levels to help them define their problems and produce new programs directed at changing the orientation of the organization to fit new circumstances.

Sometimes the firm is engaged in a slow evolution. Occasionally there is a major crisis. Often the problem is to help a management team to adjust to some kind of radical change or discontinuity: (1) the integration and rationalization of several companies into a larger, divisionalized operation, following a program of diversification or a series of mergers; (2) the closure or sale of a number of businesses, and the slimming down of central staff functions following a reduction in demand or expropriation by government; (3) developing the capacity to design, sell, and manage total systems or turn-key projects as opposed to selling individual products to developing countries or to the communist world; or (4) the introduction of a sea-change in technology such as containerization in shipping or the use of microprocessors and fiber optics in telecommunications.

One of the problems with the *organizational learning* approach is that these radical changes occur infrequently in the life of an individual firm. It therefore is difficult for a manager to gain experience of closures and divestments, or mergers and acquisitions, within one company, except in a large multidivisional business. In cases of radical change, therefore, it often is necessary for top management to bring in consultants or to recruit managers from outside the company who have acquired the relevant experience in other businesses.

However, the strategic planner who adopts a social learning approach does not act merely as a change agent, intervening as and when required to carry out an attitude study or diagnose an organizational problem to improve working relationships between individuals and departments. He or she is concerned with development of the competence of management teams in various parts of the organization to take a *strategic* view of their businesses, to identify the key issues for decisions, and to take the action they regard as necessary for the survival and growth of the enterprise.

This usually involves (1) taking a comprehensive and realistic view of the organization from various perspectives—the world market, competitors, long-term trends in technology and in society; (2) assessing in comparative terms the business' overall performance—its levels of costs, productivity, product quality, price, customer service; and (3) considering feasible alternative futures for the organization—making established activities more efficient and more productive, developing new technologies, and building new businesses.

STRATEGIC PLANNING AS A POLITICAL PROCESS

A fourth approach to planning consists of a kind of *realpolitik*—a view that says strategic planning is essentially concerned not with logic, innovation, or learning but with *power*. Planning after all is a process that allocates resources. Planning decisions affect people's lives. Strategic planning determines where investments are made and where businesses stop investing; where jobs are created and where employees are made redundant; which new projects go forward and which existing projects are terminated. Dividends, wages and salaries, promotion and advancement, recognition and status—these are what planning is about.

The Underlying Political Philosophy

The supporters of this viewpoint share the notion that life is a struggle for survival, a continual battle between competing groups.

In the political analyst's eyes, society is made up of organizations and interest groups that are continually competing for support from the public, from politicians, and from other decision makers in public and private organizations. Various groups in society are engaged in a struggle for power. Sometimes the opposing lines are drawn according to social level in a type of class conflict. On other occasions or on other issues, they may be arranged by nationality, religious creed, race, or sex.

Each political party or pressure group also consists of warring factions, all clamoring for the attention of those in power. The business, too, is seen not as a homogeneous unit, a hierarchy led by the board, or a single culture with a common purpose. Instead the firm is conceived as a model or miniature of society itself in which department vies with department, region with region and product division with product division to gain a greater share of the firm's resources and the power that goes with them.

A major danger with this political game is that it can take everyone's eyes off the business of creating wealth. In their own interests and in the interests of society, managers and employees should be concerned primarily with building businesses: introducing new products, increasing productivity, and expanding markets at home and abroad. If the political battle inside and outside the firm becomes too intense, then the energies of business leaders and trade union officials can become absorbed in continual in-fighting and negotiation. Too much effort is spent in dividing the cake and too little time is left for the battle to keep ahead of international competition.

The Changing Political Environment

Nevertheless, the businessperson has much to learn from politicians, trade unionists, and the leaders of political pressure groups when it comes to influencing public opinion and using the media.

Management's authority is being challenged continually by trade unions and groups of workers, government agencies, and pressure groups acting on behalf of consumers, conservationists, women's liberation, and various ethnic minorities. Inside and outside the business the objectives, strategies, and plans of management are being questioned. Groups of employees, local politicians, and social action groups are rejecting or vetoing the plans of management, demanding the right to be informed or consulted—or to participate in the planning process.

These interest groups are in practice asking: "Whose objectives?" "Whose plans?" They oppose the idea of unilateral planning by management and claim the right to employee participation or public participation. Trade unions, committees of shop stewards, and action committees working on behalf of local communities are also putting forward their own alternative plans and requesting government assistance in putting forth their case.

The process of strategic planning in the political arena needs to be studied by management. A number of key elements are clear.

1. *Group Action:* planning in the public arena takes place largely *between* organizations and the most successful groups are those that are well organized. In many cases businesses must forget their traditional animosities in working for their common good—to influence government or to make a case when challenged by social action groups.
2. *Influence and Coercion:* In the public arena it is rarely possible for one party to control the activities of another. Each group has to operate by influence and persuasion and occasionally by threatening sanctions.
3. *Communications:* It becomes essential, therefore, for management to put their case in plain terms to company staff at all levels, to local communities, to particular interest groups, to national governments, and to the general public.
4. *Building Networks:* Another central activity of the top management team is to deal with the external relations or foreign policy of the firm. This means carrying on a diplomatic campaign. Maintaining liaison within professional and industrial associations, making contacts with political and social interest groups, and forming alliances within the industries and in the regions where the business operates.
5. *Liaison with the Media:* Continuing contact with the press, radio, and television is vital. Demonstrations, protests, marches, petitions—these are the stock-in-trade of the political activist. Industry has to be prepared to put forth its case like other interest groups through policy statements, manifestoes, national conferences, surveys, and reports.
6. *Contacts with Governments:* Links with government bodies need to be established on a continuing basis, directly in the case of a large firm, indirectly via a trade association for a smaller business. In either case it is essential to know how decisions will be made, who are the decision makers, and who will influence the decisions. Also it is necessary to identify key social and political issues important to the company, to put forward proposals that are constructive and politically feasible—if possible speaking not just for one firm but on behalf of a sector or subsector of industry or a region.
7. *Contact with Trade Unions:* Employee organizations need to be studied in the same way as government agencies to determine the political strength of various groups, the framework of regulations and practices, the arrangements for electing officials, the ambitions and policies of various individuals, and so forth.

Also it is necessary to establish communications with trade union officials outside the process of wage bargaining, if possible, in normal times so that an effective relationship can become established without the pressure of a crisis.

FUTURES RESEARCH

The futures movement grew up in the late 1960s but strategic planning in terms of *alternative futures* only became fashionable in large companies in the late 1970s.[14] Managers in business and administrators in government are now using scenarios and other futures research techniques to try to cope with what they perceive as discontinuities. The year of the oil crisis, 1974, is seen as a watershed marking the end of an era of relative stability and affluence and the beginning of a period of turbulence and economic stagnation. In this new environment a number of trends—political, social, economic, and technological—seem to be gathering momentum and interacting to create a highly volatile business environment.

Scenario Planning

As management witnessed successive plans being rendered obsolete by unforeseen changes, they began to doubt the value of traditional forecasting and planning techniques based on extrapolation budgeting and looked for other approaches better suited to a complex and turbulent environment. They were also convinced of the need to expand their planning and forecasting procedures to cover not only economic and market trends but also social and political changes that might be reflected in legislation and in the activities of trade unions and social pressure groups.

The result was a spate of experiments in the use of modern forecasting techniques: Delphi Studies, Cross Impact Analyses, Trend Impact Analyses, and so forth. Also there was an increase in the use of simple financial models aimed at examining the sensitivity of company plans to changes in assumptions about prices, levels of sales, costs of raw materials, wages and salaries, interest rates, and so forth. And companies began to make tentative contingency plans—confidentially and informally to provide for major risks such as a strike, action by a social pressure group, a change of government, a new piece of legislation, the shortage or nonavailability of a key raw material or component, or a substantial delay in the construction of a new facility.

However, the most impressive of these changes in planning techniques has been the increasing use of scenarios. In the late 1960s, Herman Kahn and Anthony Wiener defined scenarios as "hypothetical sequences of events constructed for the purpose of focusing attention on causal processes and decision points."[15] As used in business, scenarios usually take the form of "qualitative descriptions of the situation of a company, an industry, a nation or a region at some specified time in the future."[16] Recent scenarios developed in the United Kingdom cover, for example, the British economy, unemployment, supply and demand for energy, banking, the chemical industry, television, the world pharmaceutical industry, and the future for Japan.

Coping with an Uncertain Environment

Scenario planning has been criticized on the grounds that it is a practice without a discipline, that scenarios lack the exactness of traditional economic forecasting techniques, and that there is no proof of their effectiveness. On the other hand, it is the very precision and the bogus authority of conventional approaches to forecasting that have led operating managers and those involved in forecasting to search for other methods that reflect the real uncertainty in the environment. The supporters of scenarios assert that it is better to be approximately right than precisely wrong. To quote Alvin Toffler:

> Linear extrapolation, otherwise known as straight-line thinking is extremely useful and it can tell us many important things. But it works best between revolutionary periods, not in them.[17]

Scenarios are not intended to predict the future. They are designed to help executives deal with a highly uncertain environment: to assist the executive who is used to extrapolative forecasting and budgeting in coping with the unexpected.

Scenarios are not supposed to provide an accurate picture of the future; they are designed to challenge the imagination: to encourage managers to plan, not just for the most likely future but also for less likely alternative futures.

Scenario planning should help managers to be more flexible in various ways.

1. *Environmental Scanning:* It should stimulate managers to scan the business environment for *weak signals*, especially social and political changes that might foreshadow a crisis.
2. *Robust Plans:* It should encourage executives to produce *robust* plans, that is, that may not be optimal but would keep the business profitable under a wide range of conditions.
3. *Contingency Planning:* It should prompt managers to be prepared for contingencies, for example, strikes, revolutions, or a slump in demand.
4. *Awareness of Risk:* It should make decision makers more realistic about the social, political, technological, and competitive risks to their business and persuade them to minimize the risk to the business from overdependence on any one source—a customer group, a technology, a range of products, a national or regional market.
5. *Concern with Flexibility:* Scenario planning also invites managers to consider the advantages of building flexibility into their operations, that is, (1) designing facilities that can be used in different ways; (2) training staff for a broad range of tasks; (3) consciously carrying slack resources (skilled staff, extra stocks, back-up generators, and so forth) in case of a crisis or a new opportunity; and (4) diversifying their operations so as to have businesses, suppliers, production facilities, stock holding, or computers in more than one country or region.

A possible danger of scenario planning is that managers may become too preoccupied with uncertainty and risk—which is inseparable from business. As a result, they may play safe while their less sophisticated competitors are taking new initiatives, accepting or ignoring the risks, and capitalizing on opportunities for profit and growth.

Planning without Information

Futures research is a way of helping managers think creatively about the future. This is especially important in a business where a technology, a market, or the sociopolitical situation is changing quickly. In such an environment, managers have little useful information. They are planning in a vacuum. Often there are no historical data, the technology could develop in several directions, the market may not exist, the regulatory framework may not yet have been developed. Today, a surprisingly large number of businesses face this type of situation in relation to new technologies: biotechnology, cable television, telecommunications, the next generation of computers, and so forth, and also in international markets such as Brazil, Mexico, Nigeria, Hong Kong, and the Middle East. In these cases the only sensible way of planning seems to be in terms of alternative futures.

The Construction of Scenarios

The writing of scenarios typically involves using a number of futures research techniques. For example, the approach recommended by General Electric (shown in Table 2-2) includes the use of a Delphi Study, Trend Analysis, Trend Extrapolation, Trend Impact Analysis, and Cross Impact Analysis.

A number of the techniques most commonly used in the development of scenarios are:

1. *Trend Analysis:* This involves scanning and analyzing publications and other sources of information on a regular basis to plot long-term trends.

Table 2-2. Constructing Scenarios for an Industrial Sector (General Electric)

1. *Prepare background*
   Assess environmental factors—social, regulatory, technological, economic, and competitive
   Develop crude systems model of the industry
2. *Select critical indicators*
   Key indicators (trends)
   Future events affecting key indicators (literature search)
   Delphi panel to evaluate industry's future
3. *Establish past behavior for each indicator*
   Historical performance
   Reasons for past behavior of each trend
   Delphi questionnaire
4. *Verify potential future events*
   Delphi panel
   Past trends, future events, impact/probability, future trends
   Assumptions for forecasts, rationale for future trends
5. *Forecast each indicator*
   Range of future values for each indicator
   Results from literature search and Delphi study
   Trend Impact Analysis and Cross Impact Analysis
6. *Write scenarios*
   Guidelines for strategic business units
   Annual revision

Source: Based on R. O'Connor, *Planning Under Uncertainty* (New York: Conference Board, 1978), 8.

2. *Computer Simulations:* This entails building a computer model of an enterprise or an industry and making projections on different assumptions.
3. *Decision Analysis:* Using this technique, the analyst creates a road map of decisions relating to a particular issue or project. At each step the analyst plots the alternatives available to the decision maker, the estimated payoff or loss for each course of action, and the probability of success or failure.

   The technique is useful in determining the broad dimensions of a decision and as a means of keeping various options open. In analyzing real decisions, however, the range of alternatives available is often far too wide for a planner to carry out a comprehensive quantitative analysis.
4. *Sensitivity Analysis:* One of the commonest ways to explore alternative futures is to analyze the sensitivity of a plan to variations in the assumptions. For this purpose it is helpful to have access to a simple computer model. Thus the planner can produce an operating statement, a cash flow analysis, or a balance sheet based on different assumptions about investments, sales, costs, prices, interest rates, and so forth. Many companies require their subsidiaries or divisions to explore the effects of a 10 percent or 15 percent increase or decrease in the major assumptions underpinning any major new project.
5. *Delphi Study:* This is a systematic way of carrying out a poll among experts. The experts are asked a series of questions, usually concerning the likelihood of certain events taking place. Then the results are fed back to the panel and they are asked a further set of questions.

   Experience to date suggests that the technique is valuable in eliciting the opinion of specialists on a narrow subject such as the probability of a breakthrough in a particular technology they know well. It seems to be less useful in exploring much less structured social and political issues where there are fewer experts. However, General Electric has used this technique to explore likely trends in population, employment, education, and so forth.
6. *Impact Analysis:* This implies setting up a matrix of events likely to affect other events (Cross Impact Analysis) or exploring the various impacts that a particular trend may have (Trend Impact Analysis). These techniques involve weighing the likely effects and then assessing which are the most important and the most urgent.

SUMMARY

In the modern business corporation strategic planning is a widespread and highly diverse activity. Conventional budgeting is being supplemented by a *funds allocation* procedure based on a systematic evaluation of each business, its general environment, its competitive situation, and its strategy for the future. Also, instead of operating as a simple financial holding company, top management of multi-industry, multinational businesses are setting out to manage their investments as a *portfolio of businesses.*

It is being recognized that the process of corporate development—improving the competitive performance of existing businesses, generating new products, penetrating new markets, expanding internationally, and creating or acquiring new businesses—is a prime task of top management and needs to be fostered and managed with separate budgets, plans, project teams, task forces, and so forth.

Executives who have tried to implement these kinds of changes have found that it is often not enough to set demanding targets and to ask for new strategies. It is a major problem to recruit and train managers who have the capacity to *think strategically*. Usually a team of managers needs to work together over a period of time. They have to develop a new information system that relates to strategic issues rather than operational problems. Often the organization structure has to be changed to pull out the separate businesses or to coordinate strategies internationally in worldwide product divisions. The way managers are appraised and rewarded also needs to be adjusted to demonstrate that the development of new strategies and new businesses is just as important as the achievement of this year's targets.

The managements of most large business corporations now find themselves in a continual dialogue with governments, international agencies, trade unions, social pressure groups, and the media. To handle this sociopolitical area they now have public affairs departments and external consultants who monitor social trends, forecast emerging sociopolitical issues, and formulate action programs to safeguard the interests of the business and to help to contribute to the solution of social problems such as unemployment and the decline of city centers.

In businesses that have to cope with a great deal of uncertainty, for example, in new and growing technologies, in politically unstable countries, or in fluctuating international markets—managers have been forced to plan in terms of alternative futures using simulations, scenarios, and contingency plans rather than traditional forecasting based on extrapolation.

Dealing with Crises

Strategic planning processes all take time to put into operation. But what about the firm that is in crisis? Often the company in a turnaround situation is there because its top management has been unable to think strategically, to anticipate international competition or the appearance of a new technology, and to develop new products or enter new markets in good time.

The most urgent need of a company in crisis usually is to improve cash flow and buy time by closing loss-making businesses, cutting overheads, reducing staff, selling off assets, and so forth. The next step should then be to produce a strategy for the future and to buy or build new businesses. It is interesting to note that in these crisis situations, top managers in such firms as Fisons, the Burton Group, Scandinavian Airlines, and Electrolux have discovered that *strategic thinking* or *strategic management* is extremely effective without the usual apparatus of five-year planning, portfolio analysis, scenarios, and so forth.

REFERENCES

1. For a list of the publications in each of these schools of thought see B. Taylor, "New Dimensions in Corporate Planning," *Long Range Planning Journal* **9** (1976):80-106.
2. H. Fayol, *General and Industrial Management* (trans.) London: Pitman, (1961).
3. R. N. Anthony, *Planning and Control Systems—A Framework for Analysis* (Cambridge, Mass.: Harvard University, 1965), 67.
4. J. Bower, *Managing the Resource Allocation Process* (Cambridge, Mass.: Harvard University, 1970), 54.
5. P. F. Drucker, *Landmarks of Tomorrow*, (London: Heinemann, 1959).
6. J. W. Gardner, *Self-Renewal—The Individual and the Innovative Society* (New York: Harper & Row, 1963), 5.
7. Boston Consulting Group, *Strategy Alternatives for the British Motorcycle Industry*, vol.1 (London: Department of Industry, Her Majesty's Stationery Office, 1975), p. 4.
8. D. A. Schon, *Beyond the Stable State* (London: Temple Smith, 1971), 67.
9. P. F. Drucker, "Entrepreneurship in Business Enterprise," *JournalofBusinessPolicy* **1** (1970): 1-18.
10. Patrick E. Haggerty, *Management Philos-*

ophies & Practices of Texas Instruments, Inc. Texas Instruments, Inc., 1965, p. 16.
11. Schon, *Beyond the Stable State*, 30.
12. D. N. Michael, *On Learning to Plan and Planning to Learn* (London: Jossey-Bass, 1973), 19.
13. J. Friedman, "The Future of Comprehensive Urban Planning: A Critique," *Public Administration Review* **31**(1971):325.
14. See R. O'Connor, *Planning under Uncertainty: Multiple Scenarios and Contingency Planning* (New York: The Conference Board, 1978), 1-7.
15. H. Kahn and A. Wiener, *The Year 2000* (New York: Macmillan, 1967), 6.
16. Ibid.
17. U.S. Senate Committee Report, *Choosing Our Environment: Can We Anticipate the Future?* (Washington, D.C.: Government Printing Office, 1976), 7.

CHAPTER 3

Innovation and the Strategic Management Process

Michael A. McGinnis and M. Robert Ackelsberg

Michael A. McGinnis is a professor of marketing and transportation at Shippensburg University of Pennsylvania. Dr. McGinnis holds B.S. and M.S. degrees from Michigan State University and the D.B.A. degree from the University of Maryland. His professional interests include strategic management, innovation, and physical distribution management. His articles have appeared in such journals as *Sloan Management Review, Journal of Business Strategy, Business Horizons, Transportation Journal, The Journal of Business Logistics,* and *Journal of Marketing Research.*

M. Robert Ackelsberg is a professor of management at Shippensburg University of Pennsylvania. Dr. Ackelsberg holds the B.Ch.E. degree from City University of New York, the M.S. degree from Columbia University, the M.B.A. degree from New York University, and the Ph.D. degree from City University of New York. His professional interests include strategic management, international management, and entrepreneurship. His articles have appeared in *Decision Sciences* and *Journal of Business Strategy.*

The enhancement of innovation in organizations is a primary goal of strategic planning and management. Prior to the 1970s strategic planning emphasized the development of structural methodologies for dealing with the environment, the firm's strengths and weaknesses, and development of appropriate strategies. Beginning in the 1970s, these methodologies came into question as competitive pressures eroded traditional oligopolies in the auto, steel, chemical, forest products, and appliance industries, among others. In addition, changes in public policy led to deregulation that ended the sheltered existence of the banking, stock brokerage, telecommunications, airline, trucking, and railroad industries.

These changes in the environment, together with the inability of traditional strategic planning approaches to cope with these changes, has led some managers and management academicians to question the ability of strategic planning to produce innovative strategies. In particular, one author describes the problem of "paralysis of analysis," in which

Adapted from M. A. McGinnis and M. R. Ackelsberg, "Effective Innovation Management: Missing Link in Strategic Management?" *The Journal of Business Strategy* (Summer 1983):59-66.

too much emphasis is placed on quantitative analysis, market trends, and the company's own economics, and too little emphasis is placed on probable competitor strategies, unorthodox strategic alternatives, and challenging previous strategic planning assumptions.[1] The result of paralysis of analysis is that innovation is stifled, with the consequence that strategic plans merely reflect conventional management wisdom. Opportunities are missed and the firm becomes vulnerable to competitors that are willing to pursue innovative strategies. For example, the domestic automobile industry saw its market share erode as foreign competitors developed product lines that better met consumer needs.

A major contributor to paralysis of analysis in American companies has been the neglect of issues that affect innovation. Although a substantial body of knowledge was developed during the 1960s on such topics as *managing innovation, managing change,* and *technology transfer*, few of the results found their way into the conventional wisdom of practicing managers. Although the reasons for this lack of assimilation by managers can only be speculated on, two come to mind. First, a business-as-usual attitude existed among American executives because foreign competition had not yet emerged to the point of becoming a major threat. Growth potential seemed to be limitless and universal. This condition no longer exists. Over ten years of strong foreign competition and rapidly changing technologies, together with the emergence of energy and raw material shortages, high rates of inflation, and a sluggish economy, have altered these attitudes.

Second, the results of research in the area of innovation were never synthesized to the point where managerially useful insights were developed. The purpose of this chapter is to address this problem. Five key works from the field of innovation are used to develop a typology that places innovation into an overall strategic perspective. We then discuss the climate for innovation, a means for improving innovation management, and innovation's relation to strategic planning.

ISSUES IN INNOVATION

The first problem in a discussion of innovation is one of definition. Perhaps the most useful definition is that which treats innovation as having occurred when a firm either "learns to do something it did not know how to do before, and then proceeds to do it in a sustained way . . . or . . . learns to not do something that it formerly did, and proceeds not to do it in a sustained way."[2] Executives should think of innovation in the broadest context. Whether an idea is internal or external to the firm does not really matter. What matters is that the idea is new to the firm.

Although there is substantial literature on innovation, five key articles provide an excellent overall perspective of innovation research.[3] The resulting typology of these five authors' works is summarized in Figure 3-1. As shown in the figure, the variables affecting innovation may be classified into environmental, organizational, and individual.

ENVIRONMENTAL VARIABLES

Of the six environmental variables shown in Figure 3-1, three (*market uncertainty, supply uncertainty, cause and effect uncertainty*) have the effect of increasing the level of environmental uncertainty. The environment appears to be favorable to innovation during the 1980s because of uncertainty in sources, markets, and relationships between cause and effect. There seems to be an adequate supply of *adaptable innovations* at this time. In fact, most Americans are disturbed by the number of technological and managerial advances invented in this country that are being exploited by Japanese firms.

The fifth environmental variable is *social attitudes that favor innovation*. Attitudes in the United States seem to be more open to innovation now than at any other time since World War II. Quality circles, renegotiation of work rules in many industries, profit-sharing provisions in collective bargaining agree-

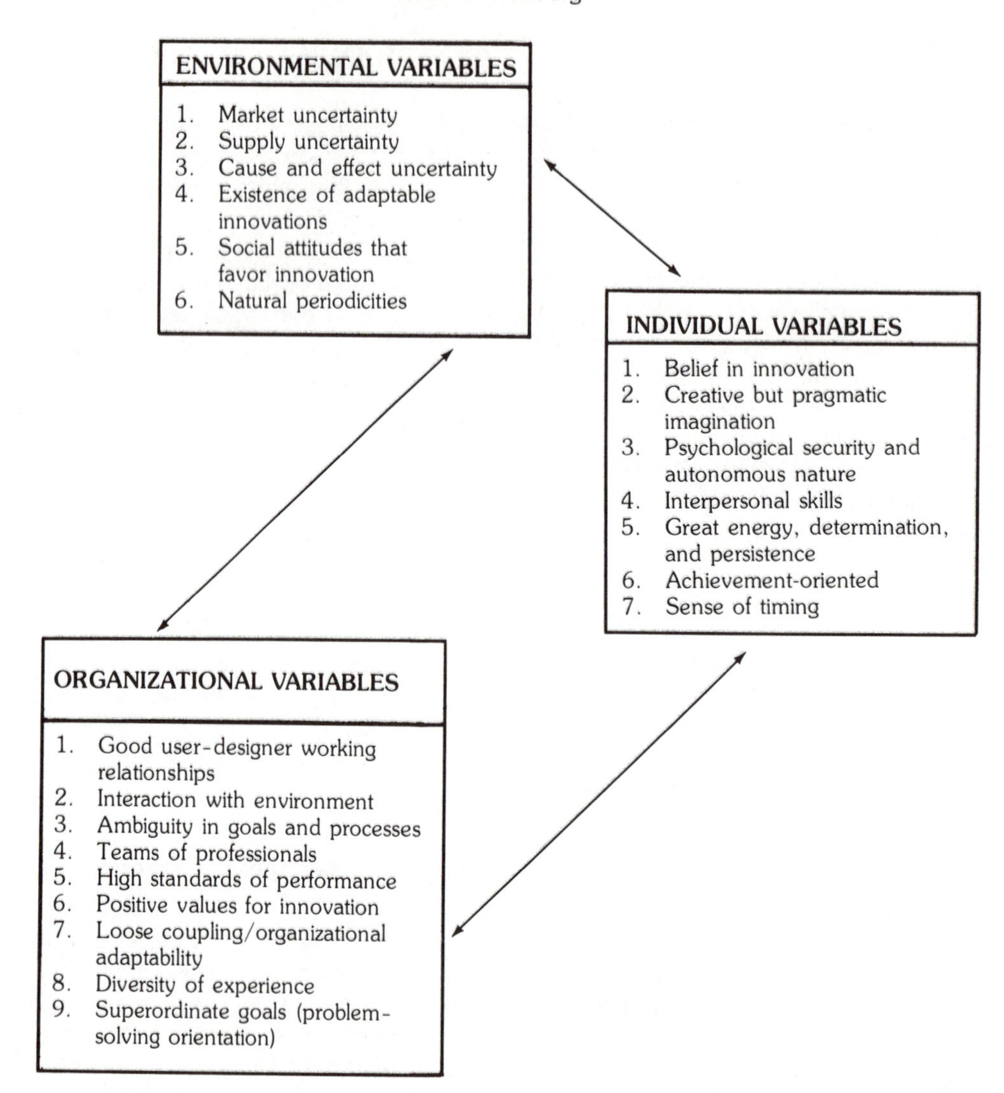

Figure 3-1. Innovation in perspective (from M. A. McGinnis and M. R. Ackelsberg, "Effective Innovation Management: A Missing Link in Strategic Management?" *Journal of Business Strategy*, Summer 1983:60. Copyright © 1983 Warren, Gorham and Lamont, Inc. Used with permission).

ments, and union representation on boards of directors represent an openness on the part of management, labor, and shareholders to explore new and different approaches to solving problems.

The final environmental variable, *natural periodicities,* refers to natural changes that occur in the environment, such as seasonal changes in demand patterns, periodic new product innovation, regular technological innovation, periodic changes in the relative costs of alternate input variables, and industrywide annual model changes. Many industries, including automotive, appliance, toy, garment, electronics, and aircraft, exhibit such cycles.

The current environment appears to be particularly conducive to innovation. Environmental uncertainties together with the existence of adaptable innovations and a more favorable social climate for innovation indicate that a lack of innovation is more likely the result of organizational and/or individual shortcomings.

Organizational Variables

As indicated in Figure 3-1, nine organizational variables affect innovation. All these variables are affected by the firm's policies and by the example set by top management. An absence of commitment to innovation by top management invariably will mean that the firm will not be particularly innovative. Conversely, a commitment by top management will result in innovation only if the firm exhibits most of the organizational characteristics shown in Figure 3-1.

The first organizational variable, *good user-designer working relationships*, refers to the ability of the firm to interact effectively with its constituencies. This interaction is crucial to identification of new opportunities and solving problems. Frequent monitoring of customer needs helps such companies as IBM and 3M to maintain strong positions in their respective markets. Similarly, the second organizational variable, *interaction of the firm with its environment*, increases the number of innovations to which the organization is exposed.

Ambiguity in goals and processes, the third organizational variable, creates a sense of uncertainty that frees the organization to search for and experiment with better techniques. Although some direction is desirable, a high level of rigidity in defining the company's goals and the processes for attaining those goals will reduce the number of options considered. For example, Polaroid's singular commitment to instant photography closed the door to strategic opportunities resulting from its technological developments in chemicals and electronics.

Teams of professionals are important because many innovations cut across departmental lines. Although we can attempt to define the word *professional* in terms of education, experience, or professional certification, we prefer to think of a professional as someone who is competent to the extent of being able to exercise excellent judgment when there is a lack of agreement in knowledge about cause and effect. Building heterogeneous teams of professionals can help to blur distinctions between functional (professional) areas. This blurring of distinctions between functions can help to reduce the empire-building symptoms of protecting one's turf, sharply defining boundaries of influence, scapegoating, blaming, and overemphasis on specialized (usually quantitative and financial) analysis techniques. The fifth organizational variable, *high standards of performance*, creates a performance gap between actual and expected performance. This gap helps to create motivation that leads to a search for innovative ideas. High standards of performance are particularly important if employee motivation is to be maintained in successful firms. For example, the need to simultaneously improve automobile fuel economy and reduce pollution resulted in research that has increased the knowledge of internal combustion in automotive engines.

Positive values for innovation are also important in creating organizational climate favorable to innovation. Although management may profess a commitment to innovation, employees are more likely to respond to their managers' actions. If the actions of management do not reflect a positive value for innovation, then the employees will not believe management supports innovation regardless of what is said. For instance, management might profess a desire to diversify or expand via innovative means but might always choose merger or acquisition as the means to accomplish this objective, even if internally developed innovative approaches are available. In such circumstances, employees might very well perceive that innovation is neither wanted nor valued. A reward system that actively recognizes the fruits of innovation will be perceived as management support of and commitment to innovation. For example, one company's professed desire for innovation was in contrast to management's practice of rewarding those who followed *safe* strategies, those who maximized profitability in the short run, and those who did not rock the boat of corporate convention. Management's desire for innovation must be communicated throughout the organization by word and deed. Unless there is a clear perception of management's desire for innovation, little will ensue.

Both environmental interaction and good working relationships can be implemented through the seventh organizational variable, *loose coupling and organizational adaptability*. Loose coupling is the free crossing of boundaries within the company as the situation dictates. Rigid lines of responsibility often stifle innovation when expertise from more than one department is needed to solve problems. Although the placing of credit or blame may become more difficult as a result of loose coupling, the more open, informal, and cooperative atmosphere that develops will mean an emphasis on results rather than on turf maintenance.

Loose coupling and organizational adaptability can be achieved by forming *ad hoc* teams to handle the early stages of implementation of an innovative strategy. Such teams are most likely to enhance innovation when there is a free flow of specialists in and out of these groups and when the team is unencumbered by rigid reporting, budgeting, and planning procedures. Additionally, management can encourage the formation of informal *ad hoc* groups to solve problems that cut across departmental lines.

Diversity of experience, the eighth organizational variable, refers to employees having a wide range of experience, both inside and outside the firm (in the form of education and work experiences). Programs that provide employees with a diversity of experience include informational assignments to other departments, regular interaction with other functional areas, participation in professional organizations, rotation of employees through job assignments in different functional areas, membership in trouble-shooting and problem-solving *ad hoc* groups, and international assignments. These programs should develop employees with broader experience, a greater range of perspectives, and greater receptiveness to ideas from other functional areas.

The final organizational variable, *superordinate goals* (problem-solving orientation), refers to the ability of the firm to work as a team to solve problems. This is in contrast to a win-lose or bargaining orientation. Corporatewide goals encourage functional departments to resolve their parochial interests and cooperate in the interest of common companywide objectives. For example, a major strength of Japanese firms has been their ability to combine marketing expertise and technology as a means of developing competitive strategies through the practice of setting companywide (superordinate) goals.

Although they are discussed individually, it is important to point out that the organizational variables must work together to create a climate favorable to innovation. For example, good user-designer relationships and interaction with its environment opens the firm to new ideas and concepts that improve the likelihood that innovation will occur. In a similar manner, ambiguity in goals and processes in combination with high standards of performance help to develop a gap between expectations and actual performance. This gap then can motivate those in the organization to seek innovations to narrow the expectation-performance gap. Teams of professionals together with a diversity of experience create a range of competencies that can bring an array of perspectives to bear on problems. Finally, loose coupling (with the resulting free crossing of boundaries), superordinate goals, and a climate of openness and problem solving help to reduce internal organizational conflict and focus efforts on external problems rather than on internal squabbling.

Individual Variables

As shown in Figure 3-1, the individual variables that affect innovation include a *belief in innovation*; a *creative but pragmatic imagination*; *psychological security and autonomous nature; good interpersonal skills; great energy, determination, and persistence; achievement orientation;* and *a sense of timing*.

These variables provide an interesting image of a secure, personable, energetic, imaginative, and pragmatic individual who has a proactive orientation. To a significant extent, the innovative individual is more likely to march to the beat of his or her own drum. If innovative individuals are to be recruited, developed, and retained, then a more open approach to leadership and motivation should be used. Fortunately, the organizational and individual variables that affect innovation are not in conflict.

INNOVATION IN PERSPECTIVE

The innovation process is best described as resulting from the interaction of all of the factors described in Figure 3-1. Certainly, if the environment is anti-innovation (such as in Medieval Europe), then innovative organizations will not evolve. Since, the environment generally is conducive to innovation at this time, it is important to focus on the organizational and individual issues affecting innovation.

The innovation process focuses on managing the organization and on creating a climate that enables innovative individuals to realize their full potential. The steps for improving innovation management are presented in the model for holistic innovation management, as shown in Table 3-1. As implied by the title of the model, management of innovation is an integrated process. This means that all components must be present and interacting if a firm is going to reach its innovative potential.

The three components of holistic innovative management are: (1) create an organizational climate favoring innovation; (2) create an innovative organization; and (3) harness individual innovation.

Of crucial importance are the roles played by the top management team in general and the chief executive officer in particular. Creation of the organizational climate and the innovative organization can only be achieved by the highest levels of management. This is crucial. Without a commitment to innovation as an organizational goal by top management, innovation will not be accomplished.

Table 3-1. Holistic Innovation Management

Create an Organizational Climate Favoring Innnovation	*Create an Innovative Organization*	*Harness Individual Innovation*
Open the firm to new ideas and concepts Good user-designer working relationships Interact with environment Create a Performance Gap Ambiguity in goals and processes High standards of performance Positive values for innovation	Develop competence Teams of professionals Diversity of experience Focus efforts on external challenges Loose Coupling/organizational adaptability Superordinate goals Problem-solving orientation Foster open communication	Recruit innovative people Provide opportunities for diversity Limit rules Reward innovators

Source: From M. A. McGinnis and M. R. Ackelsberg, "Effective Innovation Management: A Missing Link in Management?" *Journal of Business Strategy* (Summer, 1983): 60. Copyright ©1983 Warren, Gorham and Lamont, Inc. Used with permission.

Creating the Organizational Climate

Creating the organizational climate favoring innovation requires that two objectives be achieved. First, the firm must become open to new ideas and new concepts. This is achieved by fostering good relationships between designers and users and by developing a corporate ethic of interacting with the environment. Second, the firm must create a performance gap between current and expected achievements. Large, successful firms do not rest on their laurels. Rather, they constantly challenge themselves to improve their performance in all areas.

This performance gap is achieved through ambiguity in goals and processes, high standards of performance, and positive values for innovation. In particular, ambiguity plays an important role. When relationships between cause and effect are well understood (or thought to be well understood) there is very little need for experimentation and exploration of new ideas. On the other hand, when cause-and-effect relationships are ambiguous, searching for innovation is not inhibited by conventional wisdom or doctrine. For this reason, the maintenance of ambiguity plays an important but sophisticated role in creating performance gaps. High standards of performance and positive values for innovation also play key roles in creating the necessary performance gap.

Creating the Innovative Organization

Creating an innovative organization means that the firm must develop its competence and focus its efforts on external challenges. Developing corporate competence requires both diversity in skills and individual competence. A highly competent but homogeneous corporation will not be as creative as a firm that includes a competent array of individuals, each of whom possesses a wide range of experiences in his or her background. This diversity is achieved through the hiring and employee development processes.

Focusing efforts on external challenges and superordinate goals requires the utmost in top management leadership. The executive who looks internally to place blame, rather than looking outward for challenges, should not expect his or her firm to be innovative. In a similar manner, the executive who expects the organization to be a model of structure and formality should not expect much in the way of innovation. The next two issues, achieving loose coupling and/or organizational adaptability and focusing on superordinate goals, have been significant weaknesses of American firms and of American executives. Excessive emphasis on the organizational chart and following channels, inordinate efforts to pinpoint blame, and blaming failure on the outside environment are examples of behavior in noninnovative firms. Becoming more innovative is severely hampered by excessive bureaucracy and by too much focus on internal organizational mechanics.

The final two components relative to focusing on external challenges are open communication and a problem-solving orientation. Open communications relates to loose coupling, discussed earlier. In order for the firm to be able to proactively seek innovative solutions to problems, horizontal, open, spontaneous communications are imperative. A problem-solving orientation implies a willingness to take a proactive and pragmatic approach to decision making, a willingness to try new ideas, and a willingness to discard previous approaches that no longer work.

Harnessing Individual Creativity

Harnessing individual creativity sounds like a contradiction of terms. In order to harness an individuals creativity, the firm in some ways must *let go* and provide employees with the freedom needed to be innovative. As in the cases of creating an organizational climate favoring innovation and creating an innovative organization, harnessing individual creativity begins at the executive level. Regardless of what executives might say about corporate commitments to innovation, executive behavior provides the primary cues to employees. Innovators within the organization should be identified, nurtured, rewarded, and integrated into the planning process.

Four issues that are important to harnessing individual innovativeness are (1)recruiting innovative people; (2)providing opportunities for diversity; (3) limiting rules; and (4) rewarding innovators.

The hiring decision is one of the major influences of individual innovativeness within a firm. Selecting creative applicants improves the likelihood that the firm will have more innovative employees. An indication of innovative potential can be obtained from the résumé (hobbies, interests, patents), personal interviews, references, and past performance.

Employees should be provided with the opportunity to expand their horizons to allow them to develop broader ranges of experiences. Employees who have had a wide range of experiences (across departmental lines and products, line vs. staff assignments) are more likely to be innovative. Why? Because diverse experiences often help the person to look at a situation from various perspectives and see solutions in terms of a multitude of dimensions. Employee diversity can be enhanced through job assignment among functional areas, temporary assignments in other departments, assignment to diverse projects, the opportunity to communicate freely across departmental lines, and with professional contacts outside the organization.

Guaranteed inhibitors to individual creativity are rules that limit the employee's flexibility to communicate. Restrictions on employee freedom to gather information, discuss ideas, and seek assistance can inhibit innovation. Innovative contributions can be increased by liberating creative employees from some of their more routine assignments and providing them with blocks of free time to follow their interests. Innovators should be relieved of the need to justify each new thought, freeing them from bureaucratic systems that find it easy to say "no" but difficult to say "yes" to new ideas.

The final issue in harnessing individual creativity is the employee reward system. Positive reinforcement is essential if individuals are expected to be creative. Innovative individuals can be rewarded through pay, title, incentive awards, recognition, and increased autonomy.

IMPLICATIONS FOR STRATEGIC MANAGEMENT

Awareness of organizational issues that affect innovation provides a basis for understanding how these factors can be used to improve strategic management and the strategic planning process.

Regardless of which approach a company chooses, an integrated approach to the strategic planning process will lead to more innovative strategies. By this we mean there should be a free flow of ideas among functional departments. If top management recognizes that (1) goals and processes should be initially ambiguous; (2) a loose (open) organizational structure can enhance innovation; (3) good working relationships among departments favor innovation (as opposed to playing one department off against another); (4) meaningful consideration of alternative points of view enhances innovation; and (5) it must value innovation, then there is a better chance that the strategic planning process will lead to more integrated strategic plans. By *integrated* we mean strategic plans that use the coordinated abilities of all areas of the firm to effectively identify and attain company goals.

Setting companywide goals can improve the results of strategic planning. Whether a top-down or a bottom-up approach to strategic planning is used, a set of companywide goals must be identified before a company can begin to make strategic choices (let along begin to implement those choices). In addition, the process of setting companywide goals can be a means for minimizing and/or resolving conflicts within the firm. Third, management must recognize that ambiguity in goals and processes is healthy, particularly during the early stages of the strategic planning process.

Understanding innovation can help managers develop a more mature perspective of strategic planning. A danger of any planning process is that the process becomes an end in itself rather than a means of establishing and attaining company goals. The mature perspective of strategic planning recognizes that the process may have to adapt as conditions change, so that the resulting goals and strategies are those that are best for the company.

Finally, understanding innovation can help management recognize that a team approach is essential to developing creative, effective strategies. Inputs from all parts of the firm may slow the process and may be time consuming, but meaningful participation by all areas of the firm will bring a wide range of experiences to the strategic planning process. Such diversity can increase the innovative potential of company goals and strategies.

Few would argue with the statement that in many companies strategic planning has not lived up to its expectations. This criticism centers on the inability of the resulting strategies to be adequately creative. In some instances, management may have expected strategic planning to substitute for their own lack of creativity. In other instances, management may have looked at the strategic planning process as an end in itself rather than as a means for setting company goals and developing strategies.

In an increasingly dynamic business environment, company management can no longer maintain a mechanical orientation toward strategic planning. Increased awareness of the organizational and individual innovation issues can help enhance the innovativeness of many companies' strategic planning.

REFERENCES

1. J. Q. Hunsicker, "The Malaise of Strategic Planning," *Management Review* **69**(1980):8-14.
2. H. A. Shepard, "Innovation Resisting and Innovation Producing Organizations," *Journal of Business* **40**(1967):470-477.
3. T. Burns and G. M. Stalker, The Management of Innovations (London: Tavistock, 1961); R. L. Daft and S. W. Becker, *Innovation in Organizations* (New York: Elsevier North-Holland, 1978) L. B. Mohr, "Determinants of Innovation in Organization," *American Political Science Review* **63**(1969):111-126; H. A. Shepard, "Innovation Resisting and Innovation Producing Organizations," *Journal of Business* **40**(1967):470-477; J. M. Utterback, "The Process of Technological Innovation Within the Firm," *Academy of Management Journal* **14**(1971):75-88.

CHAPTER 4

A Strategic View of Leadership

LaRue T. Hosmer

LaRue T. Hosmer is a professor of policy and control at the Graduate School of Business Administration, University of Michigan. He holds A.B., M.B.A., and D.B.A. degrees from Harvard University and was the founder and president of a company that manufactured heavy equipment for sawmills and papermills. He has been teaching business policy, small business management, and entrepreneurship at the University of Michigan since 1972, with visiting appointments during that period at Stanford University and Yale University. His research interests are in managerial ethics, corporate responsibility, and strategic implementation. He is the co-author of *The Entrepreneurial Function* (Prentice-Hall, 1977) and the author of *Strategic Management: Text and Cases on Business Policy* (Prentice-Hall, 1982), *Formation Planning* (McGraw-Hill, 1984), *Managerial Ethics* (in press), and *Making Strategy Work* (in press).

Leadership is not an obsolete concept from some less scientific, more romantic age, nor is it limited to military heroes and presidential candidates. It is a vital component in the successful performance of business organizations. Someone has to direct the members of each firm as they attempt to *adapt* to rapid environmental change. Someone has to encourage those members as they try to *integrate* their diverse functional and technical activities during that period of adaptation and change. Achieving adaptation and integration are primary functions of business leadership; certainly, both are essential for the long-term profitability of business firms.

These leadership functions will become even more important in the future than they have been in the past. There will be a greater need for organizational adaptation as the pace of technological and market change quickens, as the degree of industrial competitiveness intensifies, and as the complexity of international transactions is heightened.

Events such as the advent of the *superchip* with more than one million transistors and the potential introduction of the *geometric* chip with over four million transistors will have a major impact not only on data processing, telecommunication, and consumer electronics companies but also on firms in the service industries such as banking, distribution, and transportation. The basic industries of steel, automobiles, and appliances will also be affected by such innovations and will have to change their production processes if not their product designs and market segments.

Product, market, and process changes will be driven by the availability of such new technologies as well as by the competitiveness of the industry participants. Foreign entrants in many U.S. industries have provided customers with options and prices they did not formerly have. American companies will no longer be able to delay the installation of product and process improvements that promise to reduce cost and/or increase quality, and those improvements will no longer be made on a country-by-country basis. Instead, they will be made on a global basis, with a need to consider the source and marketing restrictions of host governments and the capital and personnel savings of joint ventures. The world is becoming more dynamic, more competitive, and more complex. Organizational adaptation is therefore becoming more important.

Organizational integration is becoming equally important. Each response to external change has brought internal differentiation. I use the terms *differentiation* and *integration* here in the well-known sense originally proposed by Paul Lawrence and Jay Lorsch.[1] Differentiation refers to the differences in attitude and behavior that are formed between the functional and technical departments of a firm by the influence of environmental change. Integration refers to the coordination of the activities of the members of those departments, despite their differences in attitude and behavior. People think differently and act differently doing different functional and technical jobs, and those differences have been heightened recently by the magnitude of market, technological, competitive, and international changes. The marketing and technical people who have to respond to the changes in their areas now have very different educational backgrounds and managerial experiences due to the degree of sophistication and specialization needed to understand market and technological problems. Their attitudes and patterns of behavior, in turn, differ markedly from those of manufacturing personnel who are primarily concerned, given the recent competitive intensity, with matters of cost, productivity, and quality. Add to this the problem of diverse national cultures imposed by the international operations of many firms and the need for integration is obvious.

THE ESSENCE OF LEADERSHIP

We all have an intuitive definition of leadership. At the most elementary level, we recognize that a leader is a person with followers who ideally are enthusiastic and eager to contribute to the activities of the group being led. The enthusiasm is, perhaps, not essential, but the members of the group have to be willing to contribute something extra—some additional effort for the benefit of the organization. Unless this "something extra" is present, I think that most would agree that the individual directing the organization is an administrator rather than a leader.

Richard Pascale and Anthony Athos caught the essence of this willingness to contribute "something extra" very succinctly in their book *The Art of Japanese Management.*[2] Edward Carlson, ex-president of United Airlines, is quoted as saying, "The president of a company has a constituency much like a politician. The employees may not actually go to the polls, but each one does elect to do his or her job in a better or worse fashion every day."[3] Doing a job in a better fashion is what I mean by the term *something extra*. It is evidence of commitment to the success of the company, and of course it applies to all members of the organization: those in functional, technical, and managerial positions as well as to the hourly employees.

We also understand that, in business organizations, the leader has to be successful and that the extra efforts contributed by the members have to improve or, at the very least maintain, the long-term competitive position and economic performance of the firm. Perhaps in military history we might find a leader who has failed, a person who led members of his command in a gallant but futile action, but in both large corporations and small ventures we recognize that most members want more positive and longer-lasting results from their effort and commitment.

A leader in a business organization, then, is a person who brings members of that organization to contribute an extra effort to improve the long-term competitive position and economic performance of the company. These are valuable people to have. We need more of them, particularly in this period of technological advancement, market change, industrial competitiveness, and international complexity.

CRITERIA FOR A STRATEGIC VIEW OF LEADERSHIP

This chapter seeks to delineate a strategic view of leadership—one that is compatible with the diversity of factors that influence the economic and competitive position of a business firm. Such a notion of leadership also must provide a logical reason—a rational explanation—for the commitment by members of that organization to its success.

This would be a *strategic* definition of leadership, for it would by necessity include all the factors in the strategy formulation and implementation processes, plus the means of developing commitment. It would enable us to make recommendations on the proper role of a leader, rather than an administrator or manager, in the strategy formulation and implementation processes. The role of the manager often has been seen as being primarily analytical; selecting the *right* strategy, designing a *good* structure, making use of the *best* incentives. However, leadership involves something more—the achievement of commitment. This chapter seeks to identify that *something more.*

THEORIES OF LEADERSHIP

It seems peculiar that strategic management, which purports to center on the duties and responsibilities of the general manager,[4] has never considered that *something more,* nor has it focused on a notion that long has been understood to be a key determinant of organizational success—leadership.

This chapter proposes a strategic definition of leadership. First, however it is useful to review five alternative definitions. All are useful. All provide insight. None is *wrong.* Yet, none provides a logically complete and intuitively satisfactory explanation of the means by which an executive formulates a successful strategy for an organization, implements that strategy through structure and systems design, and builds commitment to the strategy, structure, and systems among members of that organization.

The five alternative definitions of leadership are:

1. Managerial: the belief that leadership is inherent in the proper practice of management that generates clear objectives and sets explicit policies for members of the organization.
2. Personal: the belief that leadership is inherent in the personal character of the individual that produces a charisma or attractiveness for members of the organization.
3. Political: the belief that leadership is inherent in the hierarchical power of an individual, through the ability to reward or penalize members of the organization.
4. Participative: the belief that leadership involves members of the organization in determining objectives and setting policies; it is assumed that people become committed to consensus decisions in which they have participated.
5. Directive: the belief that leadership pulls rather than pushes members of the organization by articulating a clear vision of the future and generating confidence that the vision can be achieved; it is assumed that people become committed to realistic objectives in which they believe.

The Managerial View of Leadership

It is possible that management and leadership are synonymous; that leadership is just a considerably improved or more sharply honed group of managerial skills, doubtless oriented more toward people than toward dollars but still part of the "planning, organizing, staffing, directing and controlling"[5] sequence of managerial responsibilities. During the 1950s and 1960s Sears Roebuck graded their catalogue products, with suitable price differentiations of course, into "good," "better," and "best" classifications; perhaps, after all the study and all the concern, leadership may turn out to be just the "better," not even necessarily the "best," method of management.

We have an authority who appears to support that view. No one can doubt Lee Iacocca's credentials as a manager who has improved the long-term competitive position and economic performance of his organization while apparently inspiring employee commitment. Iacocca was formally responsible for the dramatic recovery of Chrysler Corp. from a cumulative loss of $3,586 million over the five-year period from 1978 to 1982 to a profit of $2,430 million in 1984. Iacocca has written a best-selling book describing his experiences first at Ford Motor Company and then at Chrysler Corp.[6] Chapter 5 of that book has not received the attention it deserves, perhaps because most readers are more interested in the moral aspects of the author's firing from Ford and his subsequent vindication at Chrysler than in the managerial methods that led to that vindication. In Chapter 5, "The Key to Management," Lee Iacocca describes his philosophy and methods of management. As anyone might expect who has seen Iacocca on television or has heard accounts of his style, this description is explicit and direct. What are the keys to management, according to the person who revitalized Chrysler? There are three.

1. *Quarterly review and planning sessions.* The author describes his adoption of a modified management-by-objectives procedure that provides a review of each subordinate's accomplishments every 90 days and a determination of that person's plans for the next 90 days.

> Over the years, I've regularly asked my key people—and I've had them ask their key people, and so on down the line—a few basic questions. "What are your objectives for the next ninety days? What are your plans, your priorities, your hopes? And how do you intend to go about achieving them?"[7]

The objectives for the coming quarter are negotiated, with input from both the superior and the subordinate, and, after agreement is reached, the plans are put in written form. "The discipline of writing something down is the first step in making it happen."[8]

The author admits that the quarterly review seems overly simple but says that it forces dialogues between managers throughout the company, helps develop new ideas and new solutions to existing problems, keeps people from getting lost in a large corporation, and, of course, provides motivation. People who feel that they have set their own goals, according to Iacocca, will "go right through a brick wall to reach them"[9] because they will be able to follow their own plans and show others that those plans will work.

2. *Prompt and resolute decisions.* The second key, according to the author, is the willingness to make decisions with inadequate information:

> If I had to sum up in one word the qualities that make a good manager, I'd say that it all comes down to decisiveness. You can use the fanciest computers in the world and you can gather all the charts and numbers, but in the end you have to bring all your information together, and *act.*[10]

Lee Iacocca explains that he does not mean that prompt decisions should be rash or uninformed, rather that as much information as possible should be gathered. However, he feels, too many managers may have 95 percent of the information they need but will still wait another six months to get the last 5 percent. You never do have 100 percent of the data, he says, and, even if you did, the right decision would be wrong if it was made too late.

> That's why a certain amount of risk-taking is essential. I realize it's not for everyone. There are some people who won't leave home in the morning without an umbrella even if the sun is shining. Unfortunately, the world doesn't always wait for you while you try to anticipate your losses. Sometimes you just have to take a chance—and correct your mistakes as you go along.[11]

3. *Employee motivation.* The last key is motivation, but Iacocca means motivation

through personal encouragement and commendation more than by financial incentives and rewards:

In corporate life, you have to encourage all your people to make a contribution to the common good and to come up with better ways of doing things. You don't have to accept every single suggestion, but if you don't get back to the guy and say, "Hey, that idea was terrific," and pat him on the back, he'll never give you another one. That kind of communication lets people know they really count.[12]

Iacocca believes that communication is more than praising others on their personal accomplishments and more than speaking to others about corporate objectives. It is also listening to others so that it is possible to help them solve their problems. He admits, though, that "Money and a promotion are the tangible ways a company can say: most valuable player."[13] Nevertheless, he believes in motivation through communication about the need for teamwork.

I've always felt that a manager has achieved a great deal when he's able to motivate one other person. When it comes to making the place run, motivation is everything. You might be able to do the work of two people, but you can't be two people. Instead, you have to inspire the next guy down the line and get him to inspire his people.[14]

Lee Iacocca says that the keys to management, and the means he used to bring Chrysler back to economic vitality, are negotiated, short-term (three month!) objectives between each stage in the hierarchical structure, quick and explicit decisions despite the occasional lack of data and the continual presence of uncertainty, and motivation through communication, commendation, and promotion. Is this enough? The financial outcome in this instance says "yes" very emphatically. However, this is a single instance, and the evidence is anecdotal and without empirical support. I know of no research study that was designed to test the hypothesis that management by objectives, prompt decisions, and a friendly pat on the back lead to improved organizational performance and increased personal commitment. We all know of situations where these keys to management were not enough; where negotiated objectives, intuitive judgments, and appreciative comments led not to success but to failure. Iacocca's keys to management neglect company resources, industry characteristics, national conditions, international trends, structural relationships, control mechanisms, and so forth; in short, all the factors except short-term plans and individual incentives that we have agreed have an impact on organizational performance. Perhaps Iacocca believes that these factors are inputs into the second key—the decision-making process—but they are not emphasized in his book.

Are management techniques, practiced as successfully as by Lee Iacocca, the equivalent of leadership? I think that we all recognize that something more is needed, and the reason I have quoted so extensively from the autobiography is that the *something more* comes through in this instance. The words describe routine and mundane managerial methods; the style and spirit portray a very vital, energetic, and likable man. "Ah ha," you say, "we've found the secret of leadership; it is that combination of vitality, energy, and attractiveness that we call charisma, combined with excellent managerial methods." Is it? Let us look next at the personal theory of leadership.

The Personal View of Leadership

The personal theory of leadership is the belief that there is something in the character of the individual and something in the way in which that person interacts with others in both one-on-one and group situations that brings others within an organization to accept his or her directions. We are looking at ways to improve performance and build commitment. Lee Iacocca appears to say that you can build commitment by management through objectives (remember that comment about "going right through a brick wall" to achieve personally determined goals) and motivation through communication ("let people know they really count"). However, we know that these managerial methods do not work in all instances. Now we are looking at the possibility that their effectiveness depends on the characteristics and capabilities of the manager. The question is: can personal characteristics and interpersonal capabilities build commitment?[15]

1. *Personal characteristics.* There are numerous personal characteristics or traits that distinguish managers as individuals; that is, as people by themselves, not as people in interactions with others. These personal characteristics include factors such as intelligence level, technical knowledge, physical size, emotional stability, energy output, risk tolerance, and personal ambition. I think we can agree that none of these traits by itself is adequate to build commitment. Superior intelligence is doubtless useful but apparently not essential (judging from the well-known statistic that presidents of successful companies tend not to come from the upper grade-point ranges of their M.B.A. classes[16]) and physical size may be imposing but Napoleon was only 5′ 2″. The major exception may be technical knowledge; in a high-technology company, technical competence is generally respected, but technical competence does not seem to be enough, by itself, to move from respect to commitment.

2. *Interpersonal capabilities.* Numerous interpersonal characteristics and capabilities distinguish managers in the ways in which they interact with others, whether one-on-one or in small groups or large meetings. These include traits such as self-assurance (Are they at ease with others?), verbal facility (Do they express their opinions clearly and generally say the right thing?), humor (Are they self-depreciating at times, and do they understand the ridiculous nature of many human situations?) and courtesy (Do they listen to others and tend to respect the worth of individual human beings?). Again, I think that we can agree that none of these traits, by itself, is enough to build commitment. Admittedly, it is easy to like the person who is self-assured, clear-spoken, humorous, and courteous, but popularity is not the same as leadership. It would seem that likable interpersonal traits would not be enough, by themselves, to ensure commitment.

The personal concept of leadership goes beyond likable individual characteristics and interpersonal traits and focuses on the means of guiding managerial decisions and/or actions and facilitating small-group and even large-meeting processes. The thought is that people act by themselves and interact with others in small groups and large meetings in ways that will gain the approval of the senior executive. Why do people worry about managerial approval beyond the obvious reasons of promotion and pay? Approval, it is believed, is important for the self-esteem and sense of worth of the members of the organization, and that approval is particularly important if it comes from a person who is liked, respected, even admired.

Is it possible, then, for a manager to lead by personal appeal and approval, gaining commitment by building the self-esteem and sense of worth of members of the organization? Are the personal characteristics and interpersonal capabilities of a respected individual the equivalent of leadership? Intuitively, the idea does have some appeal. We have all known people whom we have liked, and we have all been pleased when those people have been appointed or elected to direct an organization of which we have been members. "He (or she) will be good to work with," we say (we never use the expression "work for" when discussing these likable individuals). Furthermore, we have all been pleased when these well-liked people have approved of our actions. However, we never consider whether being "good to work with" and approval of our decisions and actions are major criteria for organizational success. "It wasn't his (or her) fault" is more often our lenient judgment for these people when things go wrong, and things often do go wrong because, once again (as in the managerial definition of leadership), explicit consideration of company resources, industry conditions, national trends, and so forth has been either neglected or assumed.

What can we say about the impact on leadership of respected personal characteristics and likable interpersonal capabilities? I think that we can say that it is one possible way of building commitment but is probably not an essential input. I think that we have all known people who seemed dull and routine, whether in casual conversations or formal meetings, who were able to combine adaptation to external changes, integration of internal activities, and inspiration of organizational members. We also have known respected managers, some even admired within their organizations, who were unable to get certain members to go along with their directives. Perhaps we need to add the concept of political power—the ability to compel members of an organization to accept directives on adaptation and integration—as a substitute for or a supplement to personal appeal.

The Political View of Leadership

The political theory of leadership is based on power, on the ability to impose one's will on others. Managerial power, in the explicit definition proposed by John Kotter, is the "measure of a person's potential to get others to do what he or she wants them to do, as well as to avoid being forced by others to do what he or she does not want to do."[17]

Kotter argues that a manager depends on others to achieve organizational goals. The "others" have limited time and energy within the organization and other interests outside the organization, so they may or may not cooperate. Consequently, it is necessary to use power to gain cooperation. What are the sources of this managerial power? Kotter lists six, but they can be summarized in two classifications.

1. *Hierarchical.* A manager's superior position within the hierarchy and his or her ranking relative to others automatically provides the authority to approve or reject proposals and to allocate or withhold resources. A superior ranking also carries with it the right to recommend or endorse promotions and to review salaries. It even includes, in many organizations, the ability to terminate employment. In short, a senior position in the hierarchy enables an executive to assist or to frustrate others in the performance of their organizational assignments and the achievement of their individual ambitions.

2. *Reciprocal.* Power within organizations does not depend solely on a superior position within the hierarchy. Managers at the same level, or even at lower levels, can assist or frustrate others in the performance of organizational assignments and the achievement of individual ambitions by offering or withholding functional and/or technical information or expertise. Executives depend on information and expertise. Providing information and expertise is often done on a reciprocal basis, building up mutual interests or personal obligations on both sides of the exchange.

There are a number of negative aspects to managerial power, with which we are all familiar; its use obviously can harm the organization as well as people within the organization, particularly when used informally on the reciprocal basis. Despite these problems, Kotter believes that managerial power is central to the direction of organizations, and he quotes extensively from Abraham Zaleznik and Manfred Kets de Vries:

> "Power" is an ugly word. It connotes dominance and submission, control and acquiescence; one man's will at the expense of another man's self-esteem . . . . Yet it is power, the ability to control and influence others, that provides the basis for the direction of organizations and for the attainment of social goals. Leadership is the exercise of power.[18]

Is managerial power equivalent to leadership? Not as we have defined leadership: adjustment to change, integration of activities, and commitment to the organization. It is possible to compel adjustment to change and integration of activities, but it is impossible to compel commitment, particularly when that commitment means a willingness to contribute some extra effort for the success of the organization. Thus we have to look for something else, perhaps the sense of belonging that comes from participating in managerial decisions, to find the source of commitment.

The Participative View of Leadership

In 1960, Douglas McGregor proposed his two famous alternative philosophies of management: Theory X and Theory Y.[19] Theory X is based on three assumptions about human nature: (1) the average person has an inherent dislike for work; (2) the average person has to be coerced to work; and (3) the average person prefers to be directed in his or her work to avoid responsibility. Theory Y, on the other hand, is based on three very different assumptions: (1) the average person finds a source of satisfaction in work beyond monetary rewards; (2) the average person will work without coercion provided he or she can receive those satisfactions; and (3) the average person prefers to be self-directed and to take responsibility for his or her own performance to ensure receipt of those satisfactions.

From the Theory Y assumptions came the principle of participative management: that the senior executives of an organization should create conditions such that members of that organization would believe they could best satisfy their personal needs by directing their efforts toward organizational goals. This could be done by permitting, even encouraging, each employee to participate, in an interactive process, in the determination of his or her work performance standards that would jointly satisfy organizational objectives and personal goals. Participation in the managerial process would result, it was alleged, in a meaningful commitment to organizational objectives and a substantial increase in worker productivity.

McGregor stopped short of calling participative management "leadership"; that claim was left to others.[20] He did write that leadership was a combination of (1) the personal characteristics of the leader; (2) the attitudes and needs of the followers; (3) the characteristics of the organization, including its purpose, structure, and tasks that have to be performed; and (4) the characteristics of the social, economic, and political environment.[21] We can find little to quarrel with in this definition, except for the means by which the attitudes and needs of the followers are to be included in the managerial process.

Let us agree that participation in the managerial process does increase commitment. There seems to be adequate evidence from the Japanese industrial success to support this conclusion. The value of widespread participation leading to consensus in Japanese management decisions has been reported by William Ouchi and by Richard Pascale and Anthony Athos.[22] However, there seems to be an inherent contradiction in the management-by-objectives means of achieving this participation and consensus. As the attitudes and values of the members of the organization are brought up in interactive discussions between superior and subordinate at different levels of management and as the requirements and policies of the organization are brought down, there has to be a conflict at some point between the personal ambitions of the employee and the strategic plans of the organization. Management-by-objectives is a combination of top-down and bottom-up planning, with no means of reconciling conflicts between organization objectives and personal goals except through interactive negotiations between different managerial levels. This might work in a stable environment with limited technological and market change to alter organizational objectives, but stability is not a characteristic of current global competition. Something more is needed; perhaps that something comes from the employees accepting changes in the mission or direction of the organization and adjusting their personal goals to their understanding of that mission or direction.

The Directive View of Leadership

Directive is not a good word to describe this last alternative theory of leadership. I adopted it from a clear and concise statement in Harry

Levinson and Stuart Rosenthal's book, *CEO: Corporate Leadership in Action*. "The most significant difference between one organization and another is neither sociological nor economic. Rather, it lies in a leadership style that gives direction, evolves structure, and allocates power."[23] The problem is that *directive* can be confused with *authoritative*, which has a very different meaning from that intended by the authors; it oversimplifies the contributions of the executives described in the book, for these people not only provided a new direction in response to external change, but they convinced others of the need for the new direction and inspired commitment to it. They created a new vision of the future, but *visionary* seems too impractical. They inspired commitment throughout the company, but *inspirational* seems too vapid. *Directive* will have to serve as the descriptive word.

Levinson and Rosenthal studied the decisions and actions of six senior executives who had had a major, positive impact on their organizations. All of the organizations were in a period of change; the executives in the study pushed for a means of adaptation. The organizations all were complex; the executives emphasized the need for mutual support, for integration. The organizations all had a sense of collective purpose, of being "the best," of sharing ambitions and goals; the executives recognized the need for involvement, commitment, and that *extra effort*. Here, of course, we are getting very close to our original concept of the proper functions of leadership. How did the six executives achieve major, positive results? The authors examined, through extensive interviews with both the individuals concerned and selected colleagues, the personal characteristics, behavior patterns, and managerial methods of the executives. They then looked for consistencies among the six interviews. What did they find? They found five explicit patterns.

1. Each executive was able to take charge. Each one appeared to have a definite view of organizational problems and environmental changes. Each executive knew what he was going to do in response, generally by creating a new direction or shifting to a new focus for his organization, and he adhered to that direction or focus. Each one was able to describe this new direction or focus with clarity and conviction, so that understanding extended throughout the organization.

2. Each executive had a strong self-image. Each one had the self-confidence to emphasize his own philosophy and to abide by his own choices. Each one set high personal standards for himself, which in turn led to high personal standards for others. Each executive brought people together by urging the corporate traditions and values, and then enforced this integration through positive rewards and negative sanctions.

3. Each executive supported members of the organization. Each one could use negative sanctions, when necessary, but primarily offered support and helped people cope with problems and setbacks. Each used the word *we* more often than *I* and was accessible. Each one recognized the need for consent of the governed and obtained consensus from a sufficiently large number of subordinates to support movement toward the new direction or focus. Each one continually brought people along and dealt with those who either could not or did not want to be brought along.

4. Each executive encouraged effort and risk. Each one did not just exhort others to try new methods; he assumed part of the blame for failure. He "got in the boat." He used the organizational structure and controls for his support of autonomy and creativity. Each recognized that autonomy without accountability is chaos, and none of them was in the business of creating chaos. Each was in the business of building an enduring corporation through a continuous socialization process that reinforced the customs, values, and policies of his organization.

5. Each executive was a thinker as well as a doer. Each one conceptualized problems and looked for rational reasons to support his actions. Each one gave meaning to his organization by stressing a sense of collective purpose, a sense of shared ambitions and goals. Each one had a highly developed capacity for abstraction, for vision, and had the strength

to take charge and to *pull* the organization into the future.

This view is supported by Warren Bennis and Burt Nanus in their book *Leaders: The Strategies for Taking Charge.*[24] It is unusual that two major research projects on the same topic should be published within the space of a few months, and it is even more unusual that they should reach such similar conclusions. The first and most obvious similarity is that of *taking charge* as the primary characteristic of leadership in the first book and as the subtitle of the second. Another is the general observation, implied in Levinson and Rosenthal but clearly stated in Bennis and Nanus, that "most organizations are overmanaged and underled."[25] The authors of both books view leadership as critical for organizational success in a complex and changing environment.

What are the other similarities? Bennis and Nanus interviewed 90 people who had reputations as being leaders in business, government, education, arts, and so forth and found four major themes or patterns of behavior.

1. *Attention through vision.* Each executive provided a vision of the organization's future that created a new focus or direction and that portrayed a better existence for the organization than it currently enjoyed. A clear view of the future, the authors explain, is essential so that members can understand their role within the organization, and the means by which they can contribute to its achievement. A realistic view of the future is also essential; the members have to believe that they are capable of performing the necessary acts to reach that better existence.

2. *Meaning through communication.* Each executive was able to describe convincingly his or her vision of the future to the members of the organization. Often this was done not through verbal descriptions, for many of the executives were not personally articulate, but by means of comparisons, analogues, and symbols. The vision has to be shared to be effective, the authors state, and they quote John Young, the chairman of Hewlett-Packard in support: "Successful companies have a consensus from top to bottom on a set of overall goals. The most brilliant management strategy will fail if that consensus is missing."[26]

3. *Trust through positioning.* Each executive was able to establish an atmosphere of predictability, reliability, and trust by selecting a position that was "right for the times, right for the organization, and right for the people working in it,"[27] communicating that position, and then consistently staying with it. Members of the organizations in the study trusted their executives because they knew where the organizations were going and what was expected of them in consequence. Consistency is as important as consensus.

4. *Deployment of self through positive self-regard.* Each executive had a strong self-image and trusted his or her own judgment. Each knew his or her strengths, yet was able to develop and improve skills continually in areas of both strength and weakness. "Manage yourself, lead others" was the statement of the authors.[28] Why? Because they found that a positive self-regard extended to others, increasing their self-respect, their confidence, their ability to succeed.

What are the most obvious similarities between the conclusions of the two studies? The strong self-image; the new direction or focus; the ability to communicate that direction or focus, though not necessarily by words, to others; the trust and mutual support by members; the need for consensus; the need for consistency; the belief that leaders pull rather than push organizations into the future by creating a vision of the possible, communicating that vision to members of the organization, and then motivating them to achieve the vision.

What are the least obvious similarities between the conclusions of the two studies? Neither explains how to achieve a new focus or direction—let us start to call this focus or direction *strategy*—nor the analytical process that leads to a competitively successful focus or direction. Neither explains how to communicate this vision of the future to members beyond recommending the use of metaphors and symbols. Neither explains how to develop consensus among diverse groups beyond suggesting that people be "brought along." Neither states how to develop consistency beyond admitting that it is essential. Neither considers how to motivate members of these groups beyond inspiring them with a view of the future. In short, both describe the need to develop a sense of collective purpose, of shared ambitions and goals, of consistent policies and procedures, of mutual trust and respect, but neither then says, "and this is how you do it." This is where I think our concepts of strategic management—our analytical methods of formulating and implementing strategy—have something concrete to add.

A Strategic View of Leadership

The concepts of strategy formulation can help to develop the direction or focus for the firm. This is what the procedures of strategic planning are designed to do. The concepts of strategy implementation also can help to communicate that new direction or focus to members of the organization through clear statements of goals and objectives, policies and procedures, programs and plans, and so forth and to encourage adherence to the new direction or focus by design of the structure and systems. How do we build that consensus? How do we establish those attitudes of trust and mutual support? How, in short, do we gain commitment?

Let us start to answer these questions by examining what is meant by strategy formulation and implementation, and how these two processes fit together.

Strategy formulation is an iterative decision process. A range of strategic alternatives, representing different methods of competition open to the firm within a given industry, are compared to different forecasts of environmental conditions and different assessments of organizational resources. Then, a single strategy is selected based on criteria developed through a comparison of the corporate performance and position with the organizational mission or charter and the managerial values and attitudes. Strategic planing is conceptually simple though pragmatically complex; the intent is to determine what needs to be done to improve the competitive position and economic performance of the firm within the constraints imposed by its industry competitors, its environmental trends, and its organizational strengths.

The strategy implementation mechanism is a sequential decision process. The selected strategy of the firm is translated into a series of statements on the goals and objectives, policies and procedures, programs and plans, and immediate actions for each of the divisions within the corporation and then for the functional and technical units within the divisions. In turn, these objectives, policies, programs, and actions define the tasks that are required for strategic success. The organizational structure and the managerial systems for planning, control, and motivation are then designed to coordinate and integrate the performance of those tasks.

The concepts of strategic management can analytically develop the direction or focus for a firm, can rationally communicate that direction or focus to members of the firm, and can logically design the structure and systems, but something more than analysis, reason, and logic is needed. We need, according to Levinson and Rosenthal and Bennis and Nanus, feelings of trust and mutual support, of consensus and shared objectives. We need, according to our own definition of leadership, commitment or the willingness of members to contribute something more, to make an extra effort to achieve the strategy of the firm. How do we gain feelings of trust and mutual support, of consensus and shared objectives? How do we gain commitment? To understand we need to view the strategy formulation and implementation process not from the point of view of the corporation but from the point of view of people working within the corporation, as shown in Figure 4-1.

This is the reverse of the usual method of examining the content of the strategy, structure, and systems of an organization from the

Factors	Perception	Result
Organizational mission or charter Corporate performance and position Managerial values and attitudes Opportunities and risks within the industry Range of strategic alternatives open to the firm Strengths and weaknesses within the company	Individual perception of the "proper" organizational strategy	Individual sense of direction and purpose, resulting in improved corporate performance and competitive position
Departmental goals and objectives Departmental policies and procedures Departmental programs and plans Departmental immediate actions	Individual perception of the "present" organizational strategy	
Corporate organizational structure Corporate planning system Corporate control system Corporate motivational system	Individual perception of the "personal" organizational strategy	

Figure 4-1. Outline of the impact of strategy formulation and implementation on members of an organization (from L. T. Hosmer, *Strategic Management: Text and Cases on Business Policy,* Englewood Cliffs, N.J.: Prentice-Hall, 1982, 669).

impersonal and top-down perspective of the senior executives. Here, we are looking at the impact of the strategy, structure, and systems from the very personal, bottom-up viewpoint of the individual members.

Each individual can have three very different viewpoints toward the strategy, direction, or focus of the firm.

1. *The proper strategy of the organization.* Each member depending on his or her posi-

tion within the organization, is able to gauge the current performance and position of the business firm and has some understanding of the industry competitors, the environmental trends, and the organizational resources. Each member, therefore, has a concept of what the organization *ought* to be doing.

2. *The present strategy of the organization.* Each member, again depending on his or her position within the firm, has some knowledge of the goals and objectives, policies and procedures, programs and plans, and current actions of the product divisions and the functional and technical departments. Each member, therefore, has a concept of what the organization *is now* doing.

3. *The personal strategy of the organization.* Each member within the organization is directly affected by the hierarchical structure and by the managerial systems for planning, control, and motivation. The relationships imposed by the structure and the behavior encouraged by the systems give each person a concept of what he or she *should be doing* for personal self-interest.

Members of an organization will develop a sense of organizational direction and purpose if they believe that what the organization ought to be doing, what it is currently doing, and what it is in their own best interest to do are not only approximate or parallel but identical. That sense of direction and purpose, in turn, will lead to feelings of trust and mutual support. Why? Because each individual knows what is expected of him or her and of others and feels that he or she can rely on the control system to accurately measure that performance and the incentive system to reward it. The sense of direction and purpose will develop consensus and shared objectives. Why? Because each individual understands the reasons for strategic change and can anticipate the impact of that change on himself or herself. The sense of direction and purpose can build commitment. Why? Because members of the organization know where the organization is going and generally approve of that direction; it is easy to devote your efforts to an objective or cause of which you approve.

How is it possible to develop this consistency in the minds of the members of an organization from what the organization ought to be doing, what it currently is doing, and what is in the best interest of each of the members to do? Recognizing the true nature of strategic management is, of course, the answer. Strategic management—the combination of strategy formulation and implementation—has historically been said to be the responsibility of the general manager, but the task cannot be accomplished by a single person. The amount of information needed to evaluate each alternative, the number of alternatives to be considered, the detail in the statements on objectives, policies, programs and actions, the precision in the definition of the critical tasks, and the variations in the design of the organizational structure and managerial systems are simply too great. Strategic management has to be an organizational task, not an individual effort. Consequently, responsibility for the development of the strategy, structure, and systems of an organization has to be diffused throughout the organization. Members of an organization have to decide on the strategy, the structure, and the systems, yet they need some guidance, some direction, some person in charge.

The strategic definition of a leader, then, is the person in charge of the strategic management process. It is the person who is capable of managing this complex process through others, who is able to set the general direction of the organization with clarity yet leave specifics to members, who is willing to insist on participation at all levels to gain consensus, and to demand consistency between strategy, structure, and systems to gain commitment. The strategic definition of leadership is the ability to understand not only the rational content of the strategy, structure, and systems for an organization but the emotional impact of the strategy, structure, and systems on members of that organization. It is the ability to develop a strategy, structure, and systems for an organization that will adapt to external changes, integrate internal activities, and gain the commitment and support of members of the organization.

SUMMARY

Earlier in this chapter I defined the functions of leadership as the need to direct members of an organization as they adapt to rapid environmental changes, to encourage members as they integrate their diverse functional and technical activities in response to those changes, and to inspire members to contribute that *something extra,* that degree of commitment, that often seems to be the difference between the success and failure of an organization. Gaining that commitment, getting that effort is the mark of a leader, not of a manager. Finally, the purpose of leadership is to improve the long-term competitive position and economic performance of the organization.

Given those functions and that purpose, we then looked at five alternative definitions of leadership: managerial, personal, political, participative, and directive. Each provided some understanding. Each had some validity. The leader of an organization doubtless uses such managerial methods as quarterly reviews, prompt decisions, and individual motivations, such personal traits as force of character and ease of interaction, such authoritative powers as resource allocation and promotion and/or pay approval, such participative incentives as the integration of employee goals with group objectives, and such directive means as a new focus or vision of the future and a widespread consensus or agreement on that vision.

However, none of these methods by itself seems to be adequate for a complete definition of leadership. None, for example, explained how to achieve the new direction of focus. None explained how to inform others of that new focus or to combine activities within the firm to achieve it. For this, we must turn to strategic management and the techniques of strategy formulation and strategy implementation. Strategy formulation can provide the focus and the adjustment to external change that we need. Strategy implementation can inform members of the organization about the new direction and offer the integration of internal activities that is necessary. How, then, do we achieve consensus? How do we get commitment and gain that extra effort and dedication that is the hallmark of leadership rather than management?

We achieve consensus and gain commitment by recognizing the true nature of strategic management: that it is too complex to be the responsibility of a single individual or even a group of individuals and becomes a task for the full organization. Someone has to lead that process, providing direction, ensuring consensus, and insisting on consistency. That is the strategic definition of leadership.

REFERENCES

1. P. R. Lawrence and J. W. Lorsch, *Organization and Environment* (Homewood, Ill.: Richard D. Irwin, 1967).
2. R. T. Pascale and A. G. Athos, *The Art of Japanese Management* (New York: Warner Books, 1981).
3. Ibid., 252.
4. K. R. Andrews, *The Concept of Corporate Strategy* (Homewood, Ill.: Richard D. Irwin, 1980).
5. H. F. Koontz and C. O'Donnell, *Principles of Management* (New York: McGraw-Hill, 1955).
6. L. Iacocca, *Iacocca: An Autobiography* (New York: Bantam Books, 1984).
7. Ibid., 47.
8. Ibid.
9. Ibid., 48.
10. Ibid., 50, emphasis added.
11. Ibid., 52.
12. Ibid., 54.
13. Ibid., 55.
14. Ibid., 56.
15. For a good discussion of the impact of personal and interpersonal characteristics on leadership, see G. A. Yukl, *Leadership in Organizations* (Englewood Cliffs, N.J.: Prentice-Hall, 1981).
16. H. Levinson and S. Rosenthal, *CEO: Corporate Leadership in Action* (New York: Basic Books, 1984), 7.
17. J. P. Kotter, *Power in Management* (New York: AMACOM, 1979), iv.
18. A. Zaleznik and M. R. F. Kets de Vries, *Power and the Corporate Mind* (Boston: Houghton Mifflin, 1975), 3.

19. D. McGregor, *The Human Side of Enterprise* (New York: McGraw-Hill, 1960).
20. See, for example, R. Likert, *New Patterns of Management* (New York: McGraw-Hill, 1961).
21. McGregor, *Human Side of Enterprise,* 182.
22. W. G. Ouchi, *Theory Z* (New York: Avon Books, 1981); Pascale and Athos, *Art of Japanese Management.*
23. Levinson and Rosenthal, *CEO,* 4.
24. W. Bennis and B. Nanus, *Leaders: The Strategies for Taking Charge* (New York: Harper and Row, 1985).
25. Ibid., 21.
26. Ibid., 92.
27. Ibid., 107.
28. Ibid., 19.

II

Strategic Planning and Management Outputs

Part II deals with the products or outputs of strategic planning. The form of these outputs—the document that is called the plan—is relatively unimportant. It is the substance of the plan—the mission, strategy, and other strategic *choices* made in the strategic planning process—that constitute the fruits of the strategic planning process.

Chapter 5 by William R. King and David Cleland defines the choice elements of strategic management. The particular terminology used in various firms or contexts for these choice elements may differ, but the underlying constructs are of fundamental importance. Since the interrelationships among the selected mission, strategy, objectives, programs, and goals are critical, generic criteria on which the strategic choices must be made are also identified.

Chapter 6 deals with the most easily-slighted significant planning outcomes—the mission chosen for the organization. John A. Pearce II, Richard B. Robinson, Jr., and Kendall Roth provide a practical framework within which the organization can choose the mission that is most appropriate to its situation. They emphasize current trends toward multinationalization, concerns with social responsibility, and the potential impact of these trends on the mission that is chosen.

CHAPTER 5

The Choice Elements of Strategic Management

William R. King and David I. Cleland

William R. King is University Professor of business administration in the Graduate School of Business at the University of Pittsburgh. He is the author, co-author, or co-editor of more than a dozen books and 150 articles that have appeared in the leading journals in the fields of strategic management, information systems, and management science. He has served as senior editor of *Management Information Systems Quarterly* and serves as associate editor for *Management Science, Strategic Management Journal, OMEGA: The International Journal of Management Science* and various other journals. He frequently consults in these areas with corporations and government agencies in the United States and Europe.

Dr. David I. Cleland is currently a professor of engineering management in the Industrial Engineering Department at the University of Pittsburgh. He is the author or editor of 15 books and has published many articles appearing in leading national and international technological, business management, and educational periodicals. Dr. Cleland has had extensive experience in management consultation, lecturing, seminars, and research. He is the recipient of the Distinguished Contribution to Project Management award given by the Project Management Institute in 1983, and in May 1984 received (jointly with Dr. King) the 1983 Institute of Industrial Engineers-Joint Publishers Book-of-the-Year Award for the *Project Management Handbook*.

At the heart of strategic management is a series of choices that must be made. Because of the semantics jungle in the field, it is useful to define precisely these choice elements. This is not to suggest that there is only one set of terminology that is appropriate but rather that there are fundamental constructs that must be dealt with in strategic management.

The choice elements of corporate strategy—those choices that must be made explicitly or implicitly in the corporate strategic planning process—are the organization's

Mission: the business that the organization is in.
Objectives: desired future positions on roles for the organization.

Adapted with permission from W. R. King and D. I. Cleland, *Strategic Planning and Policy* (New York: Van Nostrand Reinhold, 1978) and W. R. King, "Implementing Strategic Plans through Strategic Program Evaluation," *OMEGA, The International Journal of Management Science* **8** (1980): 173-181.

Strategy: the general direction in which the objectives are to be pursued.
Goals: specific targets to be sought at specified points in time.
Programs and/or projects: resource-consuming sets of activities through which strategies are implemented and goals are pursued.
Resource allocations: allocations of funds, manpower, and so forth to various units, objectives, strategies, programs, and projects.

These informal definitions are meant to provide a common framework for communication rather than to define the "correct" terminology. Firms may use different terminology, but none can escape the need to make choices of each variety.

Most organizations conduct planning processes aimed at explicitly choosing all or some of these strategic choice elements. However, many firms fail to deal with all the choice elements in the detail and specificity each deserves.

Often, for instance, missions are dealt with implicitly, as in the case of the firm that responds to the mission concept by stating "We make widgets." Such a product-oriented view of the organization's business ignores new market opportunities and perhaps the firm's generic strengths. It is these opportunities and strengths that form the most likely areas for future success. Thus, these opportunities and strengths rather than the current product line should define the mission.

Strategies are almost always explicitly chosen by firms but are often thought of in output rather than input terms. In such instances, strategies may be described in terms of expected sales and profits rather than in terms of strategic directions such as product redesign, new products, or new markets.

Thus, the elements of strategic choice are inescapable in the sense that the avoidance of an explicit choice about any of the elements means that it is chosen implicitly. However, many firms make poor or inappropriate choices, both explicitly and implicitly, because they do not have a clear understanding of the relationships among the strategic choice elements and their innate interdependence.

RELATIONSHIPS AMONG THE STRATEGIC CHOICE ELEMENTS

One of the most important conditions for the effectiveness of plans has to do with the relationships among the strategic choice elements. If these relationships are well defined and carefully analyzed and conceived, the plan is likely to be successful. If they are not, the plan is likely to be a voluminous document that requires substantial time and energy to prepare but is filed on the shelf until the next planning cycle commences. Indeed, many plans are so treated, precisely because they do not carefully spell out relationships among various strategic choice elements and therefore do not provide the appropriate information necessary to guide the many decisions that must be made to implement the plan and to develop and manage the projects and programs that are the operational essence of the plan.

Figure 5-1 shows the elements of strategic choice in the form of a triangle that illustrates that the mission and objectives are the highest level elements. They are supported by the other elements—the strategies, goals, programs, and projects. The strategic resource allocations underlie each of these elements.

Figure 5-2 illustrates these concepts in terms of a business firm. The mission chosen is that of supplying system components to a worldwide nonresidential air-conditioning market. Note that although this mission statement superficially appears to be product-oriented, it specifically identifies the nature of the product (system components) and the market (worldwide nonresidential air conditioning). By exclusion, it guides managers in avoiding proposals for overall systems and strategies that would be directed toward residential markets. However, it does identify the world as the company's territory and (in an elaboration not shown here) defines air conditioning to include air heating, cooling, cleaning, humidity control, and movement.

Supporting the base of the triangle are strategies, goals,and programs. The firm's strategies are stated in terms of a three-phase

Figure 5-1. Relationship of strategic choice elements (from W. R. King and D. I. Cleland, *Strategic Planning and Policy,* New York: Van Nostrand Reinhold, 1978, 133).

approach. First, the company will concentrate on achieving its objectives through existing products and markets while maintaining its existing image. Then it will give attention to new markets for existing products, foreign and restricted, while improving the company's image. Restricted markets may be thought of as those that require product-safety certification before the product can be sold in that market. Finally, it will focus on new products in existing markets while *significantly* improving its image.

Clearly, this is a staged strategy; one that focuses attention first on one thing and then on another. This staging does not imply that the first strategy element is carried through to completion before the second is begun; it merely means that the first element is given primary and earliest attention, then the second and third in turn. In effect, the first element of the strategy is implemented first. This will be made more clear in terms of goals and programs.

At the right base of the triangle, a number of the firm's programs are identified. Each of these programs is made up of a variety of projects or activities. Each program serves as a focus for various activities having a common goal. For instance, in the case of the Product Cost Improvement Program, the associated projects and activities might be as follows:

Quality Control Project
Production Planning Improvement Project
Production Control System Development Project
Plant Layout Redesign Project
Employee Relations Project

All of these projects and activities are focused toward the *single* goal of product cost improvement.

In the case of the Working Capital Improvement Program, the various projects and activities might include a terms-and-conditions study aimed at revising the terms and conditions under which goods are sold, an inventory-reduction project, and so forth. Each of the other programs would have a similar collection of projects and activities focused on some single well-defined goal.

The goals are listed in the lower center portion of the triangle in Figure 5-2. Each goal is stated in specific and timely terms related to

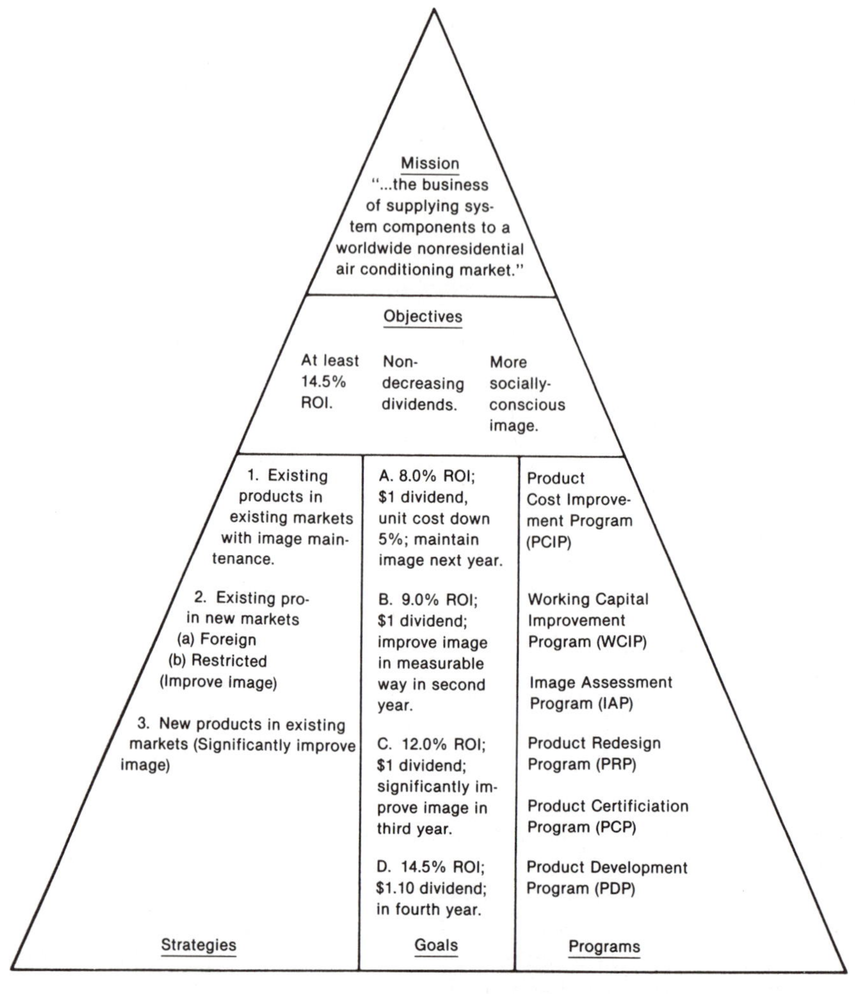

Figure 5-2. Illustrative strategic choice elements (from W. R. King and D. I. Cleland, *Strategic Planning and Policy,* New York: Van Nostrand Reinhold, 1978, 135).

the staged strategy and the various programs. These goals reflect the desire to attain 8.0 percent return on investment (ROI; a step along the way to the 14.5-percent objective) next year, along with a $1 dividend (the current level), a unit-cost improvement of 5 percent while maintaining image. For subsequent years, the goals reflect a climb to 14.5 percent ROI, a steady and then increasing dividend, and an increasing and measurable image consistent with the staged strategy that places image improvements later in the staged sequence. This is also consistent with the program structure, which includes an Image Assessment Program, a program designed to develop methods and measures for quantitatively assessing the company's image.

Figure 5-3 shows the same elements as does Figure 5-2 with each being indicated by number, letter, or acronym. For example, the circle labeled 1 in Figure 5-3 represents the first stage of the strategy in Figure 5-2, the letter *A* represents next year's goals, and so forth.

The arrows in Figure 5-3 represent some illustrative relationships among the various objectives, programs, strategy elements, and goals. For instance, the arrows *a, b,* and *c* reflect direct relationships between specific timely goals and broad timeless objectives:

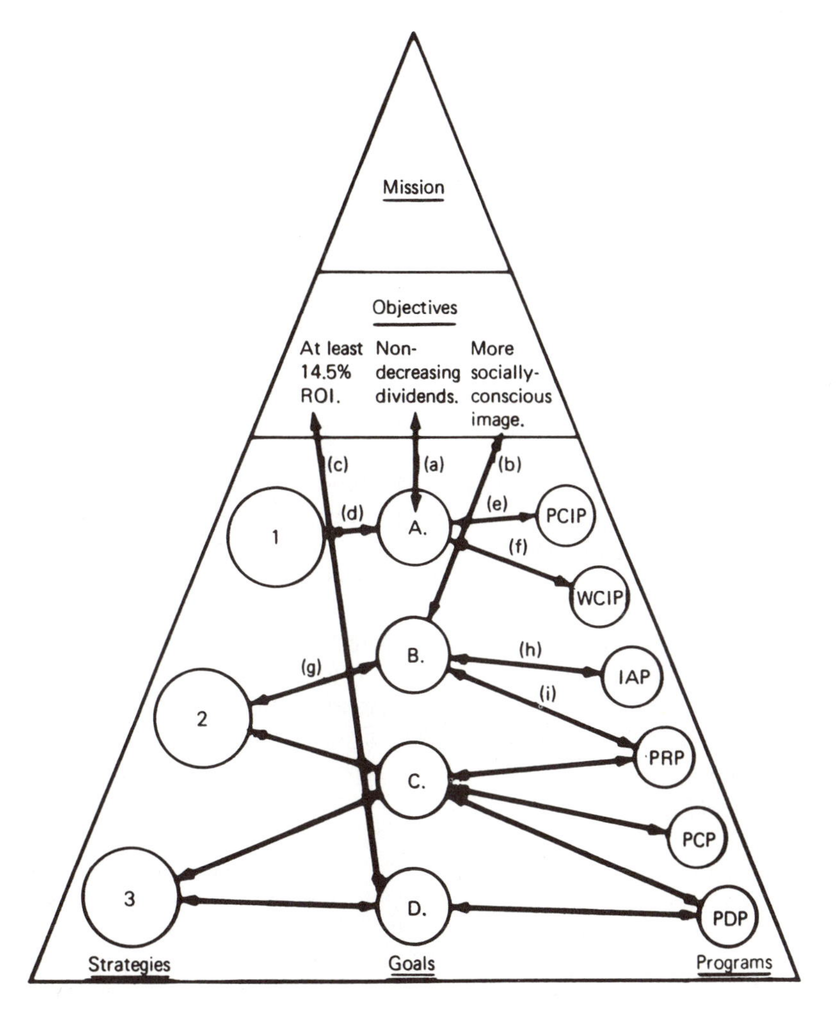

Figure 5-3. Relationships among illustrative choice elements (from W. R. King and D. I. Cleland, *Strategic Planning and Policy*, New York: Van Nostrand Reinhold, 1978, 137).

(a) *A*, next year's goals primarily relate to the objective of nondecreasing dividends.
(b) *B*, the second year's goals relate to the more socially conscious image objective.
(c) *D*, the quantitative ROI figure is incorporated as a goal in the fourth year.

Of course, each year's goals relate implicitly or explicitly to all objectives. However, these relationships are the most direct and obvious.

Similarly, arrow *d* in Figure 5-3 relates the first year's goals to the first element of the overall strategy in that these goals for next year are to be attained primarily through the strategy element involving existing products in existing markets with image maintenance. However, arrows *e* and *f* also show that the Product Cost Improvement Program (PCIP) and the Working Capital Improvement Program (WCIP) are also expected to contribute to the achievement of next year's goals.

The second year's goals will begin to reflect the impact of the second strategy element (Existing products in new markets) as indicated by arrow *g* in Figure 5-3. The effect of the Product Redesign Program (PRP) is also expected to contribute to the achievement of these goals (arrow *i*) as is the Image Assessment Program (IAP) expected to provide an ability to measure image by that time. The other arrows in Figure 5-3 depict various direct relationships whose interpretation is left to the reader.

From this figure, relationships among the various strategic decision elements can be seen:

1. Goals are specific steps along the way to the accomplishment of broad objectives.
2. Goals are established to reflect the expected outputs from strategies.
3. Goals are directly achieved through programs.
4. Strategies are implemented by programs.

Thus, the picture shown in Figure 5-3 is that of an interrelated set of strategic factors that demonstrate *what* the company wishes to accomplish in the long run, how it will do this in a sequenced and sensible way, and *what performance levels* it wishes to achieve at various points along the way.

CRITERIA FOR STRATEGIC CHOICE

Figures 5-1, 5-2, and 5-3 make it clear that the various elements of strategic choice are mutually supportive. However, there remains the question of how this high degree of interdependence can be effectively attained.

Certainly, an understanding of the logical relationships that are depicted in these figures should itself lead to better subjective choices in the planning process. However, although subjective judgment can be appropriately applied at the higher levels of strategic choice because these elements are tractable, it is an inadequate basis for choice at the lower levels of program and/or project selection and funding.

In other words, no formal techniques are needed in choosing among alternative missions and objectives because these choices must be made inherently on a primitive basis of the personal values and goals of management and other stakeholders. At this level, there are only a few viable options from which choices must be made.

At the level of programs, projects, and resource allocations, quite the opposite is the case. There are many contenders and combinations of contenders to be considered. Thus, some formal approach may be useful. Indeed, such an approach is not only practically useful, but it can form the integrating factor in the array of strategic choice elements.

The integrating factor is a strategic program evaluation approach *that directly uses the results of the higher-level strategic choices to evaluate alternative programs, projects, and funding levels.* Project selection approaches are well known and widely used in industry for the selection of engineering projects, R&D projects, and new product development projects. However, if program and/or project evaluation is to be the key link in unifying the array of organizational strategic choice elements, the evaluation framework and criteria must be an integral element of the strategic planning process.

Figure 5-4 indicates in general terms how a program and/or project evaluation process can serve as an integrating factor for the firm's array of strategic choices. It shows a wide variety of potential projects and programs being filtered through the application of strategic criteria that are based on the higher-level choices that have previously been made—the organization's mission, objectives, and strategies. The output of this filtering process is a set of rank-ordered project and program opportunities that can serve as a basis for the allocation of resources.

Other important criteria must come into play in implementing this evaluation process. These criteria are those that are *generic* to a good specification of the organization's mission, objectives, and strategy. However, they must be *specifically addressed* if programs and projects are to truly reflect corporate strategy. These criteria are:

1. Does the opportunity take advantage of a *strength* the company possesses?
2. Correspondingly, does it avoid a dependence on something that is a *weakness* of the firm?
3. Does it offer the opportunity to attain a *comparative advantage* in the marketplace?

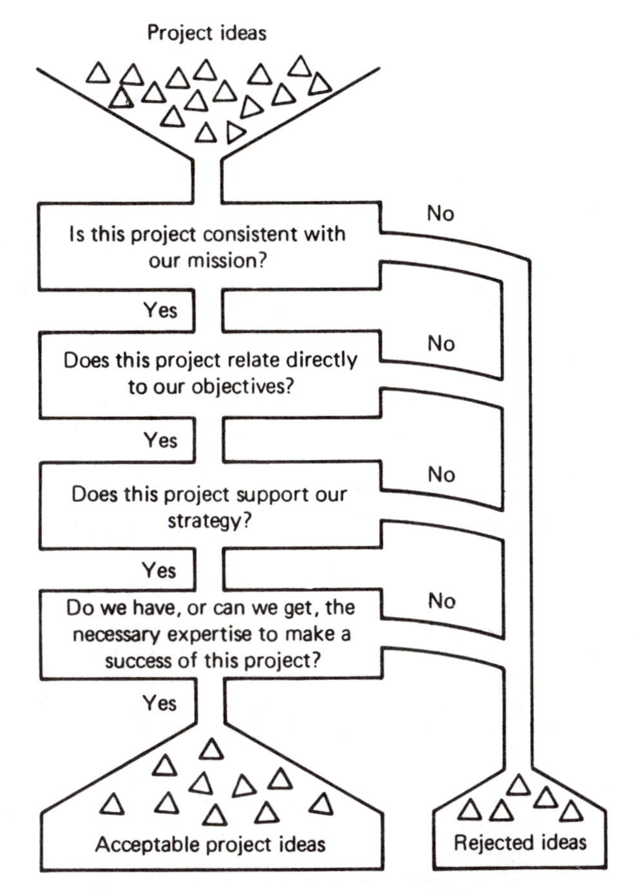

Figure 5-4. An evaluation of programs and/or projects (from W. R. King and D. I. Cleland, *Strategic Planning and Policy*, New York: Van Reinhold, 1978, 201).

4. Does it contribute to the *internal consistency* of existing strategies?
5. Does it address a *mission-related opportunity* that is presented by the evolving market environment?
6. Is the level of *risk* acceptable?
7. Is it consistent with established *policy guidelines?*

These criteria may be applied either judgmentally or by using some formal evaluation mechanism such as that described by King.[1]

SUMMARY

This chapter defines the choice elements of strategic management—the organization's mission, objectives, strategy, goals, strategic programs and resource allocations—in terms of the interrelationships. It also describes appropriate criteria to apply in making the choices. The terminology used is not universally accepted since there is a tendency for individual organizations to create unique terms. However, the fundamental concepts are universal and, regardless of which terminology is used, an organization that ignores any or all of these strategic choice elements does so at its peril.

REFERENCE

1. W. R. King, "Implementing Strategic Plans Through Strategic Program Evaluation," *OMEGA, The International Journal of Management Science* **8**(1980):173-181.

CHAPTER 6

The Company Mission as a Guide to Strategic Action

John A. Pearce II, Richard B. Robinson, Jr., and Kendall Roth

John A. Pearce II is LeRoy Eakin Professor of Strategic Management and chairman of the Management Department in the School of Business at George Mason University. He has published more than one hundred papers on strategic management, industry analysis, and long-range planning topics and is an active business consultant and trainer.

Richard B. Robinson, Jr., is an associate professor of strategic management and entrepreneurship at the University of South Carolina. He has authored over sixty papers on management topics and has co-authored four strategic management books with John Pearce, including *Formulation and Implementation of Competitive Strategy* (Richard D. Irwin, 1985).

Kendall Roth is an assistant professor of international business at the University of South Carolina. His main research activities have been in the area of strategic management as applied to the multinational context.

The corporate mission statement can be an invaluable tool for directing the strategy formulation and implementation processes. Given the diversity and dispersion of individuals comprising the modern corporation, the volatility and often hostility of operating environments, and the rapidity of technological change, strategy processes can easily become simply reactionary to these forces. Thus, it becomes increasingly essential for a company to establish a heightened sense of its overall purpose and ultimate aims to provide a consistent foundation and focal point for strategy formulation and implementation. It is in establishing this foundation that the properly conceived and effectively articulated mission statement becomes an invaluable strategic management tool. To assume this role, a mission cannot be conveyed in bland, publicity-oriented generalizations—often characteristic of mission statements. Rather, it must explicitly and clearly identify and communicate the

Consolidated from the following three publications: J. A. Pearce II, "The Company Mission as a Strategic Tool," *Sloan Management Review* **23** (1982):15-24; J. A. Pearce II and R. B. Robinson, Jr., *Strategic Management: Strategy Formulation and Implementation,* 2nd ed. (Homewood, Ill.: Richard D. Irwin, 1985), 77-97; J. A. Pearce II and K. Roth, "Multinationalization of the Company Mission," *Sloan Management Review,* in press, 1986.

long-term intentions of the firm so that the firm's goals can serve as a basis for shared expectations, planning, and performance evaluation.

This chapter provides a practical framework that can be used in defining a meaningful company mission. The framework offers both recommendations for mission content and a process for melding the diverse and often conflicting demands placed on strategic direction. In particular, this framework is applied to questions of multinationalization and social responsibility and their impact on the mission statement.

WHAT IS A COMPANY MISSION?

In order to develop a new business or to reformulate the direction of an ongoing company, strategic decision makers must determine the basic goals, characteristics, and philosophies that will shape the strategic posture of the firm. The outcome of this task, known as the company mission, provides the basis for a culture that will guide future executive action.

The company mission is a broadly defined but enduring statement of purpose that distinguishes a business from other firms of its type and identifies the scope of its operations in product and market terms. Not only does the company mission embody the strategic decision makers' business philosophy, but it also reveals the image the company seeks to project, reflects the firm's self-concept, and indicates the principal product or service areas and the primary customer needs the company will attempt to satisfy. In short, the company mission describes the firm's product, market, and technology in a way that reflects the values and priorities of the strategic decision makers. For example, the Zale Corporation's mission statement (presented in Table 6-1) effectively incorporates these major features and provides the broad framework needed for strategic decision making.

Formulating the Mission

The process of defining the mission for a specific business is perhaps best understood by thinking about a firm at its inception. The typical business organization begins with the beliefs, desires, and aspirations of a single entrepreneur. The sense of mission for such an owner-manager is usually based on several fundamental elements:

Belief that the company's product or service can provide benefits commensurate with its price.

Belief that the product or service can satisfy a customer need currently not adequately met for specific market segments.

Belief that the technology to be used in the product or service is cost and quality competitive.

Belief that with hard work and the support of others, the business can do better than just survive; it can grow and be profitable.

Belief that the management philosophy will result in a favorable public image and will provide financial and psychological rewards for those willing to invest their labor and money.

Belief that the self-concept the entrepreneur has of the business can be communicated to and adopted by employees and stockholders of the company.

As the business grows, the need may arise to redefine the company mission, but, if and when it does, the revised mission statement will reflect the same set of elements as did the original. It will state: the basic type of product or service to be offered; the primary markets or customer groups whose needs will be served; the technology to be used in the production or delivery of the product or service; the fundamental concern for survival through growth and profitability; the managerial philosophy of the firm; the public image that is sought; and the firm's self-concept—that is, the image of the firm that should be held by those affiliated with it.

The Need for an Explicit Mission

Defining the company mission is time-consuming, tedious, and not required by any

Table 6-1. Zale Corporation: Summary Statement of the Company Mission

Our business is specialty retailing. Retailing is a people-oriented business. We recognize that our business existence and continued success are dependent upon how well we meet our responsibilities to several critically important groups of people.

Our first responsibility is to our customers. Without them we would have no reason for being. We strive to appeal to a broad spectrum of consumers, catering in a professional manner to their needs. Our concept of value to the customer includes a wide selection of quality merchandise, competitively priced and delivered with courtesy and professionalism.

Our ultimate responsibility is to our shareholders. Our goal is to earn an optimum return on invested capital through steady profit growth and prudent, aggressive asset management. The attainment of this financial goal coupled with a record of sound management, represents our approach toward influencing the value place upon our common stock in the market.

We feel a deep, personal responsibility to our employees. As an equal opportunity employer, we seek to create and maintain an environment where every employee is provided the opportunity to develop to his or her maximum potential. We expect to reward employees commensurate with their contribution to the success of the company.

We are committed to honesty and integrity in all relationships with suppliers of goods and services. We are demanding but fair. We evaluate our suppliers on the basis of quality, price, and service.

We recognize community involvement as an important obligation and as a viable business objective. Support of worthwhile community projects in areas where we operate generally accrues to the health and well-being of the community. This makes the community a better place for our employees to live and a better place for us to operate.

We believe in the Free Enterpise System and in the American Democratic form of government under which this superior economic system has been permitted to flourish. We feel an incumbent responsibility to insure that our business operates at a reasonable profit. Profit provides opportunity for growth and job security. We believe growth is necessary to provide opportunities on an ever increasing scale for our people. Therefore, we are dedicated to profitable growth—growth as a company and growth as individuals.

This mission statement spells out the creed by which we live.

external body. The mission contains few specific directives, only broad outlines or implied objectives and strategies. Characteristically it is a statement of attitude, outlook, and orientation rather than of details and measurable targets.

What then is a company mission designed to accomplish? King and Cleland provide seven good answers:

1. To ensure unanimity of purpose within the organization.
2. To provide a basis for motivating the use of the organization's resources.
3. To develop a basis, or standard, for allocating organizational resources.
4. To establish a general tone or organizational climate, for example, to suggest a businesslike operation.
5. To serve as a focal point for those who can identify with the organization's purpose and direction; and to deter those who cannot from participating further in the organization's activities.
6. To facilitate the translation of objectives and goals into a work structure involving the assignment of tasks to responsible elements within the organization.
7. To specify organizational purposes and the translation of these purposes into goals in such

a way that cost, time, and performance parameters can be assessed and controlled.[1]

COMPONENTS OF THE MISSION STATEMENT

Product or Service, Market, and Technology

Three indispensable components of a mission statement are the company's basic product or service, the primary market, and the principal technology to be used in producing or delivering the product or service. In combination, these three components define the firm's present and potential business activity. A good example of these three mission components is found in the business plan of ITT Barton, a subsidiary of ITT. Under the heading of "Business Mission and Area Served," the company presents the following information:

> The Unit's mission is to serve the industry and government with quality instruments used for the primary measurement, analysis, and local control of fluid flow, level, pressure, temperature, and fluid properties. This instrumentation includes flow meters, electronic readouts, indicators, recorders, switches, liquid level systems, analytical instruments such as titrators, integrators, controllers, transmitters, and various instruments for the measurement of fluid properties (density, viscosity, gravity) used for process variable sensing, data collection, control, and transmission. The Unit's mission includes fundamental "loop-closing" control and display devices when economically justified, but excludes broadline central control room instrumentation, systems design, and turnkey responsibility.
>
> Markets served include instrumentation for oil and gas production, gas transportation, chemical and petrochemical processing, cryogenics, power generation, aerospace, government and marine, as well as other instrument and equipment manufacturers.

This segment of the mission statement, accomplished in 129 words, clearly indicates to all readers—from company employees to casual observers—the firm's basic products, primary markets, and principal technologies.

Often a company's most referenced public statement of selected products and markets is presented in "silver bullet" form in the mission statement. For example: "Dayton-Hudson Corporation is a diversified retailing company whose business is to serve the American consumer through the retailing of fashion-oriented quality merchandise." Such a statement serves as an abstract of company direction and is particularly helpful to outsiders who value condensed overviews.

Company Goals: Survival, Growth, and Profitability

Three economic goals guide the strategic direction of almost every viable business organization. Although it is not always explicitly stated, a company mission reflects the firm's intention to secure its survival through sustained growth and profitability.

Unless a firm is able to survive, it will be incapable of satisfying any of its stakeholders' aims. Unfortunately, like growth and profitability, survival is such an assumed goal of the firm that it is often neglected as a principal criterion in strategic decision making. When this happens, the firm may focus on terminal aims at the expense of the long run. Concerns for expediency—a *quick fix* or a *bargain*—can displace the need for assessing long-term impacts. Too often the result is near-term economic failure, owing to a lack of resource synergy and sound business practice.

Profitability is the main goal of a business organization. No matter how it is measured or defined, profit over the long term is accepted as the clearest indication of the firm's ability to satisfy the principal claims and desires of employees and stockholders. Clearly, the key phrase here is "over the long term," since the use of short-term profitability measures as a basis for strategic decision making would lead to a focus on terminal aims. For example, a firm might be misguided into overlooking the enduring concerns of customers, suppliers, creditors, ecologists, and regulatory agents.

Such a strategy could be profitable in the short run, but over time its financial consequences are likely to be seriously detrimental. The following excerpt from the Hewlett-Packard statement of corporate objectives (that is, mission) ably expresses the importance of an orientation toward long-term profit:

Objective: To achieve sufficient profit to finance our company growth and to provide the resources we need to achieve our other corporate objectives.

In our economic system, the profit we generate from our operations is the ultimate source of the funds we need to prosper and grow. It is the one absolutely essential measure of our corporate performance over the long term. Only if we continue to meet our profit objective can we achieve our other corporate objectives.

A firm's growth is inextricably tied to its survival and profitability. In this context, the meaning of growth must be broadly defined. Although market share growth has been shown by the PIMS studies to be strongly correlated with firm profitability, there are other important forms of growth. For example, growth in the number of markets served, in the variety of products offered, and in the technologies used to provide goods or services frequently leads to improvements in the company's competitive ability. Growth means change, and proactive change is a necessity in the dynamic business environment. Hewlett-Packard's mission statement provides an excellent example of corporate regard for growth:

Objective: To let our growth be limited only by our profits and our ability to develop and produce technical products that satisfy real customer needs.

We do not believe that large size is important for its own sake; however, for at least two basic reasons, continuous growth is essential for us to achieve our other objectives.

In the first place, we serve a rapidly growing and expanding segment of our technological society. To remain static would be to lose ground. We cannot maintain a position of strength and leadership in our field without growth.

In the second place, growth is important in order to attract and hold high caliber people. These individuals will align their future only with a company that offers them considerable opportunity for personal progress. Opportunities are greater and more challenging in a growing company.

The issue of growth raises a concern about the definition of a company mission. How can a business specify sufficiently its intention in product, market, and technology terms to provide direction without delimiting its unanticipated strategic options? How can a company define its mission to permit consideration of opportunistic diversification without removing valuable parameters that the mission should incorporate to guide growth decisions? Perhaps, such questions are best addressed when a firm outlines within its mission statement the conditions under which it would consider departures from its ongoing stream of operations. Dayton-Hudson is one company that has pursued such an approach, as is disclosed in its growth philosophy.

The stability and quality of the corporation's financial performance will be developed through the profitable execution of our existing businesses, as well as through the acquisition or development of new businesses. Our growth priorities, in order, are as follows:

1. Development of the profitable market preeminence of existing companies in existing markets through new store development or new strategies within existing stores;
2. Expansion of our companies to feasible new markets;
3. Acquisition of other retailing companies that are strategically and financially compatible with Dayton-Hudson;
4. Internal development of new retailing strategies.

Capital allocations to fund the expansion of existing operating companies will be based on each company's return on investment, in relationship to its ROI objective and its consistency in earnings growth, and on its management capability to perform up to forecasts contained in capital requests.

Expansion via acquisition or new venture will occur when the opportunity promises an acceptable rate of long-term growth and profitability, acceptable degree of risk, and compatibility with the corporation's long-term strategy.

Company Philosophy

The statement of a company's philosophy, or company creed as it is sometimes known, usually accompanies or appears as part of the mission. It reflects or explictly states the basic beliefs, values, aspirations, and philosophical priorities that the strategic decision makers are committed to emphasize in their management of the firm. Fortunately, the topical content of company philosophies vary little from one firm to another. This means that business owners and managers implicitly accept a general, unwritten, yet pervasive, code of behavior by which actions in a business setting can be governed and largely self-regulated. Unfortunately, statements of philosophy display so great a degree of similarity among firms and are stated in such platitudinous ways that they look and read more like public relations promotions than like the commitments to values they are intended to be.

But despite the similarities among company philosophies, cynicism toward strategic managers' intentions in developing these philosophies is rarely justified. In almost all cases, managers attempt, and often succeed, to provide a distinctive and accurate picture of the firm's managerial outlook. One such valuable statement is that of Zale Corporation, whose company mission was presented earlier. As shown in Table 6-2, Zale has subdivided its statement of management's operating philosophy into four key areas: marketing and customer services, management tasks, human resources, and finance and control. These subdivisions serve as a basis for even greater philosophical refinement than was provided through the mission statement itself. As a result, Zale has established especially clear directions for company decision making and action.

The mission statement of the Dayton-Hudson Corporation is at least equally specific in detailing the firm's management philosophy, as shown in Table 6-3. Perhaps the most noteworthy quality of the Dayton-Hudson statement is its delineation of responsibility at both the corporate and business levels. In many ways, it could serve as a prototype for the new three-tiered approach to strategic management, which argues that the mission statement needs to address strategic concerns at the corporate, business, and functional levels of the organization. To this end, Dayton-Hudson's management philosophy is a balance of operating autonomy and flexibility on the one hand, and corporate input and direction on the other.

Company Self-concept

A major determinant of any company's continued success is the extent to which it can relate functionally to its external environment. Finding its place in a competitive situation requires that the firm be able to evaluate realistically its own strengths and weaknesses as a competitor. The ideal—that the firm must know itself—is the essence of the company self-concept. The firm's ability to survive in a dynamic and highly competitive environment would be severely limited if it did not understand the impact that it has or could have on the environment, and vice versa. This notion per se is not commonly integrated into theories of strategic management; however, scholars have appreciated its importance to an individual: "Man has struggled to understand himself, for how he thinks of himself will influence both what he chooses to do and what he expects from life. Knowing his identity connects him both with his past and the potentiality of his future."[2]

There is a direct parallel between this view of the importance of an individual's self-concept and the self-concept of a business. Fundamentally, the need for each to know the self is crucial. The ability of a firm or an individual to survive in a dynamic and highly competitive environment would be severely limited if the impact on others and of others is not understood.

In some senses, then, the organization takes on a personality of its own. Hall stated that much behavior in organizations is organizationally based, that is, a business acts on its members in other than individual and interactive ways.[3] Thus, businesses are entities that act with a personality that transcends those of particular company members. As such, the firm can be seen as setting decision-making

Table 6-2. Zale Corporation's Operating Philosophy

I. Marketing & Customer Services:

We require that the entire organization be continuously customer oriented. Our future success is dependent more on meeting the customers' needs better than on our competition.

We expect to maintain a marketing concept and distribution capability to identify changing trends and emerging markets and to promote effectively our products.

We strive to provide our customers with continuous offerings of quality merchandise that is competitively priced, stressing value and service.

We plan to constantly maintain our facilities as modern, attractive, clean, and orderly stores that are pleasing and exciting places for customers to shop.

II. Management Tasks

We require profitable results from operations—activity does not necessarily equate with accomplishment. Results must be measurable.

We recognize there are always better ways to perform many functions. Continuous improvement in operating capability is a daily objective of the entire organization.

We expect all managers to demonstrate capabilities to plan objectives, delegate responsibilities, motivate people, control operations, and achieve results measured against planned objectives.

We must promote a spirit of teamwork. To succeed, a complex business such as ours requires good communication, clearly understood policies, effective controls, and, above all, a dedication to "make it happen."

We are highly competitive and dedicated to succeeding. However, as a human organization, we will make mistakes. We must openly acknowledge our mistakes, learn from them, and take corrective action.

III. Human Resources

We must develop and maintain a competent, highly, motivated, results-oriented organization.

We seek to attract, develop, and motivate people who demonstrate professional competence, courage, and integrity in performing their jobs.

We strive to identify individuals who are outstanding performers, to provide them with continuous challenges, and to search for new effective ways to compensate them by utilizing significant incentives.

Promotion from within is our goal. We must have the best talent available and, from time to time, will have to reach outside to meet our ever-improving standards. We heartily endorse and support development programs to prepare individuals for increased responsibility. In like manner, we must promptly advise those who are not geared to the pace, in order that they make the necessary adjustments without delay.

IV. Finance & Control

We will maintain a sound financial plan that provides capital for growth of the business and provides optimum return for our stockholders.

We must develop and maintain a system of controls that highlights potential significant failures early for positive corrective action.

Table 6-3. Management Philosophy of Dayton-Hudson Corporation

I. The corporation will:

Set standards for ROI and earnings growth.
Approve strategic plans;
Allocate capial;
Approve goals;
Monitor, measure, and audit results,
Reward performance;
Allocate management resources.

II. The operating companies will be accorded the freedom and responsibility:

To manage their own business;
To develop strategic plans and goals that will optimize their growth;
To develop an organization that can assure consistency of results and optimum growth;
To operate their business consistent with the corporation's statement of philosophy.

III. The corporate staff will provide only those services that are:

Essential to the protection of the corporation.
Needed for the growth of the corporation;
Wanted by operating companies and that provide a significant advantage in quality or cost.

IV. The corporation will insist on:

Uniform accounting practices by type of business;
Prompt disclosure of operating results;
A systematic approach to training and developing people;
Adherence to appropriately high standards of business conduct and civic responsibility in accordance with the corporation's statement of philosophy.

Source: W. Ouchi, *Theory Z* (Reading, Mass.: Addison-Wesley, 1981), pp. 204-205. The author presents more complete mission statements of three companies discussed in this chapter.

parameters based on aims different and distinct from the individual aims of its members. The effects of organizational considerations are pervasive.

Organizations do have policies, do and do not condone violence, and may or may not greet you with a smile. They also manufacture goods, administer policies, and protect the citizenry. These are organizational actions and involve properties of organizations, not individuals. They are carried out by individuals, even in the case of computer-produced letters, which are programmed by individuals—but the genesis of the actions remains in the organizations.[4]

The actual role of the corporate self-concept has been summarized as follows:

1. The corporate self-concept is based on management perception of the way others (society) will respond to the corporation.
2. The corporate self-concept will function to direct the behavior of people employed by the company.
3. The actual response of others to the company will in part determine the corporate self-concept.
4. The self-concept is incorporated in statements of corporate mission to be explicitly communicated to individuals inside and outside the company, that is, to be actualized.[5]

A second look at the company mission of the Zale Corporation in Table 6-1 reveals much about the business's self-concept. The strategic decision makers see the firm as socially responsive, prudent, and fiercely independent.

Characteristically, descriptions of self-concept per se do not appear in company mission statements. Yet, strong impressions of a firm's self-image are often evident. For example, material from Intel Corporation, as seen in Table 6-4, provides a solid basis for such an understanding of the company's self-concept.

Public Image

The issue of public image is important, particularly for a growing firm that is involved in redefining its company mission. Both present and potential customers attribute certain qualities to a particular business. Gerber and Johnson & Johnson make "safe" products; Cross Pen makes "professional" writing instruments; Aigner makes "stylish but affordable" leather products; Corvettes are "power" machines; and Izod signifies the "preppy" look. Thus, mission statements should reflect the anticipations of the public whenever the goals of the firm are likely to be achieved as a result. Gerber's mission should not open the possibility for diversification into pesticides, nor should Cross Pen's allow for the possibility of producing thirty-nine-cent, private-brand disposables.

On the other hand, a negative public image often prompts firms to reemphasize the beneficial aspects of their character as reflected in their missions. For example, as a result of what Dow Chemical saw as a "disturbing trend in public opinion," it undertook an aggressive promotional campaign to fortify its credibility, particularly among "employees and those who live and work in [Dow's] plant communities." Dow's approach was described in its 1980 annual report:

Table 6-4. Abstract of Intel's Mission-Related Information

I. Management Style: Intel is a company of individuals, each with his or her own personality and characteristics.

Management is self-critical. The leaders must be capable of recognizing and accepting their mistakes and learning from them.

Open (constructive) confrontation is encouraged at all levels of corporation and is viewed as a method of problem solving and conflict resolution.

Decision by consensus is the rule. Decisions once made are supported. Position in the organization is not the basis for quality of ideas.

A highly communicative/open management is part of the style.

A high degree of *organizational skills and discipline* are demanded.

Management must be ethical. Managing by telling the truth and treating all employees equitably have established credibility that is ethical.

II. Work Ethic/Environment

It is a general objective of Intel to line up individual work assignments with career objectives.

We strive to provide an *opportunity for rapid development.*

Intel is a *results-oriented* company. The focus is on *substance* vs. form, *quality* vs. quantity.

We believe in the principle that *hard work, high productivity* is something to be proud of.

The concept of *assumed responsibility* is accepted. (If a task needs to be done, assume you have the responsibility to get it done.)

Commitments are long term; if career problems occur at some point, reassignment is a better alternative that termination.

We desire to have *all employees involved and participative* in their relationship with Intel.

All around the world today, Dow people are speaking up. People who care deeply about their company, what it stands for, and how it is viewed by others. People who are immensely proud of their company's performance, yet realistic enough to realize it is the public's perception of that performance that counts in the long run.

A firm's concern for its public image is seldom addressed in an intermittent fashion. Although public agitation often stimulates a heightened corporate response, a corporation is concerned about its image even in the absence of expressed public concern. The following excerpt from the mission statement of Intel Corporation is exemplary of this attitude:

> We are sensitive to our image with our customers and the business community. Commitments to customers are considered sacred and we are upset with ourselves when we do not meet our commitments. We strive to demonstrate to the business world on a continuing basis that we are credible in describing the state of the corporation, and that we are well organized and in complete control of all things that determine the numbers.

THE ROLE OF CLAIMANTS IN DETERMINING THE MISSION

In defining or redefining the company mission, strategic managers must recognize and acknowledge the legitimate claims of other stakeholders of the firm. These stakeholders can be divided into two categories as indicated in Figure 6-1. *Inside claimants* are individuals and groups who are stockholders or are employed by the firm. *Outside claimants* are all other individuals and groups who are not insiders but who are affected by the actions of the firm as a producer and marketer of goods or services. Such outsiders commonly include customers, suppliers, governments, unions, competitors, local communities, and the general public. Each of these interest groups has justifiable reasons to expect, and often to demand, that the company act in a responsible manner toward the satisfaction of its claims. In general, stockholders claim appropriate returns on their investments; employees seek broadly defined job satisfaction; customers want what they pay for; suppliers seek dependable buyers; governments want adherence to legislated regulations; unions seek benefits for members in proportion to their contributions to company success; competitors want fair competition; local communities want companies that are responsible citizens; and the general public seeks some assurance that the quality of life will be improved as a result of the firm's existence.

However, when a specific business attempts to define its mission to incorporate the

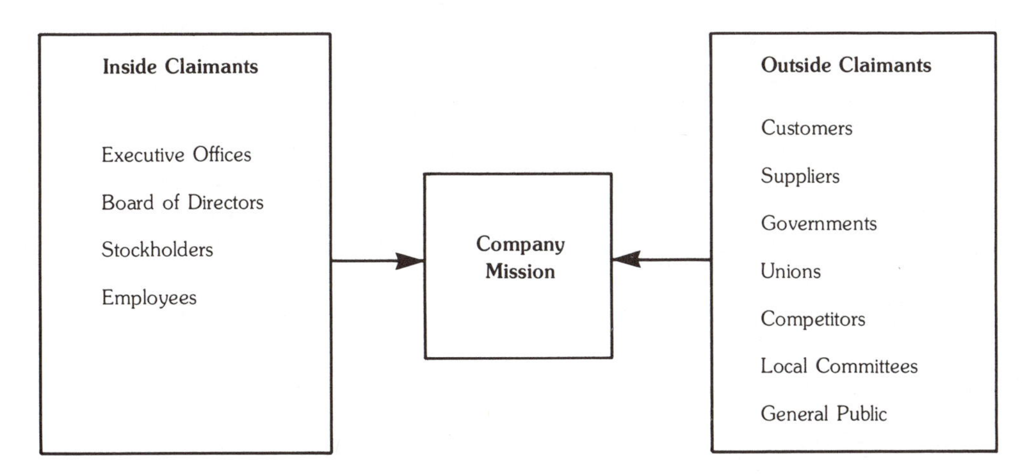

Figure 6-1. Inputs to the company mission.

interests of these various claimant groups, such broad generalizations are insufficient. Four steps must be taken: (1) identification of claimants; (2) understanding claimants' specific demands vis-à-vis the company; (3) reconciliation and prioritization of the claims; (4) coordination of claims with other elements of the mission.

Identifying Claimants

Every business faces a slightly different set of claimants who vary in number, size, influence, and importance. In defining the company mission, strategic managers need to identify each claimant group and to weigh its relative ability to affect the firm's success.

Understanding Claims

Although the concerns of principal claimants tend to center on general issues such as those described above, it is important that strategic decision makers understand more specifically the demands that each group will make on the firm. By so doing, strategic managers will be better able to appreciate the concerns of claimants and to respond to them effectively.

The importance of identifying claimants and understanding their claims is nowhere better illustrated than in examining the issue of a firm's social responsibility. Broadly stated, outsiders often demand that the claims of insiders be subordinated to the greater good of the society—that is, to the greater good of the outsiders. This extremely large and often amorphous set of outsiders insists that the company be socially responsible. They believe that issues such as the elimination of solid and liquid wastes, the contamination of air and water, and the exhaustibility of natural resources should be principal considerations in the strategic decision making of the firm. On the other hand, insiders tend to believe that the competing claims of the outsiders should be balanced against each other in a way that protects the mission of the company. For example, the consumers' need for a product must be balanced against the water pollution resulting from its production if the company cannot totally afford to eliminate the pollution and remain profitable. Furthermore, some insiders argue that if society were sufficiently concerned about unwanted business by-products, such as water pollution, it could allocate tax monies to eliminate them.

A representative overview of various claimant concerns is presented in Table 6-5. While far from exhaustive, this list provides some feel for the tremendous breadth of claims made on a company.

Reconciling Claims

Unfortunately, the concerns of various claimants often are in conflict. For example, the claims of governments and the general public tend to limit profitability, which is the central concern of most creditors and stockholders. Such conflicting claims must be reconciled. In order for strategic managers to prepare a unifying approach for the company, they must be able to define a mission that has resolved the competing, conflicting, and contradictory claims placed on the company's direction by claimants and claimant groups. Internally consistent and precisely focused objectives and strategies require mission statements that display such a single-minded, though multidimensional, approach to business aims.

Claims on any business number in the hundreds, if not thousands: high wages, pure air, job security, product quality, community service, taxes, OSHA, EEO, product variety, wide markets, career opportunities, company growth, investment security, high ROI, and so forth. Although most, if not all, of these claims are desirable ends for the company, they cannot be pursued with equal emphasis. Claims must be prioritized in a way that reflects the relative attention the firm will give to each major claim. Such emphasis is shown by the criteria used in strategic decision making, by the company's allocation of human, financial, and physical resources, and by the long-term objectives and strategies developed for the firm.

Table 6-5. A Claimant View of Company Responsibility

Claimant	*Nature of the claim*
Stockholders	Participation in distribution of profits, additional stock offerings, assets on liquidation; vote of stock, inspection of company books, transfer of stock, election of board of directors, and such additional rights as established in the contract with the corporation.
Creditors	Legal proportion of interest payments due and return of principal from the investment. Security of pledged assets; relative priority in event of liquidation. Participate in some management and owner prerogatives if certain conditions exist within the company (such as default of interest payments).
Employees	Economic, social, and psychological satisfaction in the place of employment. Freedom from arbitrary and capricious behavior on the part of company officials. Share in fringe benefits, freedom to join union and participate in collective bargaining, individual freedom in offering up their services through an employment contract. Adequate working conditions.
Customers	Service provided with the product: technical data to use the product; suitable warranties; spare parts to support the product during customer use; R & D leading to product improvements; facilitation of consumer credit.
Suppliers	Continuing source of business: timely consummation of trade credit obligations; professional relationship in contracting for, purchasing, and receiving goods and services.
Governments	Taxes (income, property, etc.), fair competition, and adherence to the letter and intent of public policy dealing with the requirements of fair and free competion. Legal obligation of businessmen (and business organizations); adherence to antitrust laws.
Unions	Recognition as the negotiating agent for employees. Opportunity to perpetuate the union as a participant in the business organization.
Competitors	Norms established by society and the industry for competitive conduct. Business statesmanship on the part of peers.
Local communities	Place of productive and healthful employment in the community. Participation of company officials in community affairs, regular employment, fair play, purchase of reasonable portion of product of the local community, interest in and support of local government; support of cultural and charity projects.
The general public	Participation in and contribution to society as a whole; creative communications between governmental and business units designed for reciprocal understanding; bear fair proportion of the burden of government and society. Fair price for products and advancement of state of the art technology which the product line involves.

Source: From W. R. King and D. I. Cleland, *Strategic Planning and Policy* (New York: Van Nostrand Reinhold, 1978), 153. © 1978 by Litton Educational Publishing Inc. Reprinted by permission of Van Nostrand Reinhold Company.

Coordinating Mission Components

Demands on a company for responsible action by claimant groups constitute only one set of inputs to the mission. Managerial operating philosophies and the determination of product-market segments are the other principal components that must be considered. These factors essentially pose a *reality test* that the distilled set of accepted claims must pass in order to serve as a basis for a sound company mission. The key question to be addressed is: How can the company simultaneously satisfy claimants and optimize its success in the marketplace?

Again, the question of a firm's social responsibility illustrates the challenge that faces those who must coordinate the numerous components of the mission statement. In this realm, the issues are so complex and the problems so situational as to defy rigid rules of conduct. Thus, each business must decide on its own approach in trying to satisfy its perceived social responsibility. Different approaches will reflect differences in competitive positions, industries, countries, environmental and ecological pressures, and a host of other factors. In other words, they will reflect both situational factors and differing priorities in the companies' acknowledgments of claims.

Despite differences in their approaches, most American companies now take steps to assure outsiders of their efforts to conduct the business in a socially responsible manner. Many firms, including Abt Associates, Eastern Gas and Fuel Associates, and the Bank of America, have gone to the effort of conducting and publishing annual social audits. For example, the social audit of Eastern Gas and Fuel, as published in its 1981 annual report, provides numerical evidence of the firm's success on social responsibility topics including health and safety, charitable giving, minority employment, and employee pensions. Such social audits attempt to evaluate the business from a social responsibility perspective. They are often conducted for the firm by private consultants who offer minimally biased evaluations on what are inherently highly subjective issues. Many other firms periodically report to both insiders and outsiders on their progress to reach self-set social goals. Primarily through their annual reports, companies discuss their efforts and achievements in acting in a socially responsible fashion.

MULTINATIONALIZATION OF THE CORPORATE MISSION

Few strategic decisions bring about a more radical departure from the existing direction and operations of a company than the decision to expand internationally. Multinationalization subjects a company to a radically redefined and challenging set of environmentally determined opportunities, constraints, and risks. To prevent the company's direction from being dictated by these externalities, it is essential for top management to reassess the corporation's fundamental purpose, philosophy, and strategic intentions prior to multinationalization to ensure that these basic values will continue as decision criteria in proactive planning.

A mission statement designed with a domestic perspective does not recognize the added dimensions inherent in multinationalization. Thus, in this section, the content of the domestic mission statement is analyzed to determine appropriate modifications for a multinational corporation (MNC) that can help to ensure that the resultant mission statement will continue to provide the necessary framework for strategic decision processes.

Maintaining MNC Mission Statement Relevance

Expanding across national borders to secure new market or production opportunities initially may be viewed as consistent with a company's growth objectives as outlined in its existing mission statement. However, as multinationalization occurs, the direction of a company is inherently altered. For example, as a company expands overseas, its operations are physically relocated in foreign operating environments. Since strategic decisions are made in the context of some perception or understanding of the environment, information from new sources will be absorbed into planning processes as the environment becomes pluralistic, with revised corporate directions as a probable and desirable result. Thus, prior to the reconsideration of the corporate strategic choice, top management must reassess the mission statement and institute changes as required so that the appropriate environmental information is defined, collected, analyzed, and integrated into existing data bases.

Management must also provide a mission statement that continues to serve as a basis for evaluating strategic alternatives as this additional information is incorporated into the organization's system of decision-making processes. Reconsider one component of the Zale Corporation's mission statement from this perspective:

> Our ultimate responsibilities are to our shareholders. Our goal is to earn an optimum return on invested capital through steady profit growth and prudent, aggressive asset management. The attainment of this financial goal, coupled with a record of sound management, represents our approach toward influencing the value placed upon our common stock in the market.

From a U.S. perspective this financial component seems quite reasonable. However, it could be unacceptable in a global context where financial goals are frequently divergent. Research has shown differing propensities for financial goals in various countries.[6] Executives from France, Japan, and the Netherlands have displayed a clear preference for maximizing growth in after-tax earnings. Norwegian executives place a higher priority on maximizing earnings before interest and taxes. In contrast, the maximization of stockholder wealth has received little acclaim in any of these four countries. Thus, a mission statement that specifies that a firm's ultimate responsibility is to its stockholders may or may not be appropriate as the basis for the company's financial operating philosophy when viewed from a global perspective. This responsibility illustrates the criticalness of reviewing and revising the corporate mission prior to international expansion so that the mission statement maintains its relevance by overarching divergent contingencies and environmental imperatives.

Components of the Corporate Mission Revisited

Each of the components of the mission statement remains relevant and fundamental to the MNC mission statement. However, because multinationalization will mandate changes in strategic decision making, corporate direction, and strategic alternatives, the content of each mission component must be revised to incorporate multinational contingencies. The additional strategic capabilities that will result from internationalizing operations must continue to be encompassed by the corporate mission statement. Therefore, each of the basic components needs to be analyzed in light of specific considerations that serve to multinationalize the corporate mission.

Product or Service, Market, and Technology. The definition of the basic market need the company aims to satisfy is likely to remain intact in the MNC context because the company has acquired domestic competences that can be exploited as competitive advantages when transferred internationally. However, confronted with a multiplicity of contexts, some degree of prioritizing and redefining of the primary market and customer is necessary.

The MNC could define a global market necessitating standardization in product and company responses or it could pursue a marketing-concept orientation by focusing on each national market's particular or unique demands. Thus, the mission statement must provide a basis for strategic decision making in this trade-off situation. For example, Hewlett Parkard's statement includes the directive that: "HP customers must feel that they are dealing with one company with common policies and services," implying a standardized approach designed to provide comparable service to all customers. In contrast, Holiday Inns, Inc.'s corporate mission reflects the marketing concept: "Basic to almost everything Holiday Inns, Inc. does is its interaction with its market, the consumer, and its consistent capacity to provide what the consumer wants, when, and where it is needed."

Subsequent to defining the company's target market, the mission statement should suggest the corresponding internal mechanisms required to support this definition. For example, if a company defines a global market the mission statement may establish centralized corporate functions and activities that promote standardization. Illustrative of this example, the mission statement of Drew Chemical Corporation (see Table 6-6) establishes three industry-defined divisions and one international division that represents and supports the others. Thus, the international market is viewed as a standardized extension of the domestic divisions. To support this view, the mission statement creates economies of scale and product standardization across the industry divisions by specifying centralized research, production, and administrative operations to support the marketing organizations.

Company Goals: Survival, Growth, and Profitability. In the United States, growth and profitability are considered essentials for survival. Similarly, in environments relatively supportive of the free enterprise system, these priorities are widely acceptable. However, following international expansion, the firm may operate in economies not unequivocally committed to the profit motive. A host country may view social welfare and development goals as taking precedence over free market capitalism. For example, in third-world countries, employment and income distribution goals often take precedence over rapid economic growth.

Opposition to profit goals may also come from a nonphilosophical perspective, that is, even in environments that accept the profit motive, MNC profits are often viewed as having a unidirectional flow. At the extreme, the multinational company is seen as a tool for exploiting the host country for the exclusive benefit of its parent's home country. Profits are the evidence of perceived corporate atrocities. Thus, in a multinational context, a corporate commitment to profits may not only fail to help secure survival, it may even increase the risk of failure.

Therefore, an MNC mission statement must reflect the firm's intention to secure its survival through dimensions that extend beyond growth and profitability. An MNC must develop a corporate philosophy that expresses the need for a bidirectional flow of benefits between the firm and its multiple environments. Such a view is expressed deftly by Gulf & Western Americas Corporation: "We believe that in a developing country, revenue is in-

Table 6-6. Drew Chemical Corporation Mission Statement

The Drew Chemical Corporation is a service-oriented company that consists of four basic marketing organizations. Drew's assets are in people—not land, plant or equipment.

Each of the operations is primarily concerned with the beneficiation of the world's most basic resource—water—through the use of performance.

The Water and Waste Treatment Division markets chemicals, equipment and services for boiler and cooling water treatment and for use in waste disposal plants.

The Marine Division markets chemicals, equipment and services for marine boilers and maintenance products and fuel oil additives for use aboard ships.

The Specialty Chemicals Division markets process additives and chemical specialties for the pulp and paper, paint and latex, textile and chemical industries.

International Operations consist of twelve wholly-owned subsidiaries. Eight subsidiaries are essentially sales agents for the Marine Division and licensees of the Water and Treatment and Specialty Chemicals Divisions. Drew Ameroid International is a tax shelter for Marine business outside of the United States. The other three subsidiaries function quite independently of the parent company, drawing on it mainly for technical support.

The marketing organizations are supported by centralized research, production, and administrative operations.

separable from mandatory social responsibility and that a company is an integral part of the local and national community in which its activities are based."[7] This statement illustrates a corporate attitude that it intends to make MNC contributions to the host country, yet clearly maintains a commitment to firm profitability.

In a broader context, a company's mission must recognize not only economic dimensions but also an effectiveness dimension—where effectiveness is the ability to meet the desires of all major organizational claimants. This implies that effectiveness is in part an external standard, relative to and defined by each MNC context. Therefore, an MNC must identify the coalitions in each environment that determine these external standards or, more importantly, the specific coalitions on which the MNC depends for its existence and prosperity. The organization must then include as an essential part of its corporate mission, a legitimization element to proactively create and sustain this coalition support, thereby securing the organization's survival. For example, management of Caterpillar Tractor Co. has stated:

> We affirm that Caterpillar investment must be compatible with social and economic priorities of host countries, and with local customs, tradition, and sovereignty. We intend to conduct our business in a way that will earn acceptance and respect for Caterpillar, and allay concerns—by host country governments and others—about multinational corporations.

The legitimization element ideally should exhibit a hierarchical nature, that is, to the degree that the subsidiary environments are heterogeneous with respect to their support coalitions, separate subsidiary-level mission statements are desirable. However, the subsidiary statements should derive their framework from the overarching legitimization of the corporation.

As the multinational corporation pursues its growth and profitability goals it is inevitably confronted with the conflicting imperatives of attempting to maximize resource deployments corporatewide, irrespective of location, while at the same time being responsive to local market demands and host governments. These conflicting imperatives create a condition of strategic ambiguity, where a clear strategic response is not easily distinguishable.

Resolution of this strategic ambiguity involves situational approaches in which strategic responses are contingent on issues. Management apportions strategic responsibility, then manages the overall resultant process of strategic decision making. This ensures that conflicting imperatives are considered and appropriately balanced. However, revisions in strategy formulation processes cannot be relied on exclusively for the resolution of strategic ambiguity, nor can management allow strategic decisions to be exclusively situational—particularly where geographical separation disallows close scrutiny of all strategy-related processes. Strategic decision makers need a guiding framework or philosophy external to the formulation process that will provide a basis for promoting consistency in strategic decision making across time and location. Providing this framework is a fundamental contribution of the MNC mission statement. Therefore, multinationalization of the growth and profitability components implies the establishment of corporate perspectives that reduce the ambiguities resulting from conflicting strategic imperatives.

Company Philosophy. Domestically, the corporate philosophy often is stated in relatively bland and equivocal terms. This lack of specificity is thought to be tolerable in that the domestic environment offers rather homogeneous expectations of a corporation that are readily conveyed among strategic observers through a journalistic form of business argot.[8]

Although an inclusive and detailed corporate philosophy may go unstated, the implicit understanding of the domestic environment results in a general uniformity of corporate values and behavior within the domestic setting. Few events occur domestically that directly challenge a company to self-examine its espoused or actual philosophy to ensure that it is both properly formulated and implemented. Multinationalization is, however, clearly such an event. A corporate philosophy developed from a singular perspective cannot be assumed to maintain its relevance in variant cultures. Corporate values and beliefs are primarily culturally defined, reflecting the general philosophical perspective of the society in which the company operates. Thus, when a company extends into another social structure it encounters a new set of accepted corporate values and preferences that must be assimilated and incorporated into its own.

The critical concern for the MNC is to decide which values and philosophical priorities will be attended, given multiple cultures. The MNC has the basic choice of adopting a company philosophy for each operating environment or of defining a supranational corporate philosophy. Although each subsidiary must verify that its philosophy is not in direct conflict with existing cultural norms, the preferable choice for the MNC is usually the latter approach—striving to define an acceptable overall corporate philosophy not contingent on specific environments. The communication capabilities of modern society seldom allow exclusively subsidiary-defined corporate philosophies. For example, numerous U.S. multinational corporations have been subjected to considerable criticism from U.S. special interest groups regarding the policies in their South Africa, Namibia, and Dominican Republic subsidiaries. The allegation of corporate social responsibility violations has, in general, not been instigated, motivated, or defined by coalitions within the host countries. Rather, values and beliefs regarding working standards and conditions have been directly transferred to each country by external coalitions from the United States, such as the Interfaith Center on Corporate Responsibility. Thus, though corporate philosophical preferences, values, and beliefs can be tailored to each host company, they cannot be fundamentally redefined because of the capability of concerned interest groups to selectively match or mismatch potentially conflicting positions. This results in a significantly enlarged external conscience to which the corporations must be responsive. Consequently, the MNC's accountability to this collective conscience must be recognized in its entirety when expressing a corporate value system.

The mission statement must also provide for operationalizing the corporate philosophy such that it can serve as a basis for strategic decision-making and operational activities. The acquisition of corporate resources as well as their allocation and utilization must be consistent with the corporate value objectives. This necessitates a formal mechanism by which contemplated corporate actions are assessed to ensure that value preferences will be implemented. For example, in responding to world criticism regarding the use of its infant formula, Nestle established an independent audit committee to monitor its market practices and its compliance with the international code of the World Health Organization regarding infant formula marketing activities. Unfortunately, the committee was viewed by some critics as merely a public relations ploy. This criticism may have been avoided had the corporate philosophy provided for the committee prior to the infant formula crisis. Nevertheless, a committee is an example of a mechanism by which the operationalization of the corporate philosophy can be monitored on an on-going basis.

Self-Concept. Multinationalizing the self-concept component of the corporate mission depends on management's understanding and evaluation of the company's strengths and weaknesses as a competitor in each of its operating arenas. The firm's ability to survive in multiple dynamic and highly competiitive environments is severely limited if strategic decision makers do not understand the impact the company has or could have on the environment and vice versa. They are in fact, partially responsible for determining the environment to which they must subsequently respond. That is, a company may engage in actions to proactively select and impact its environments or it may decide to take a reactive stance in which its posture is to respond and adapt to environmental changes once their impact is known with greater certainty. Activities by which management may proactively manage the environment include: (1) developing leadership and innovation in products, marketing, and technology; (2) securing monopolistic market or distribution positions; (3) coopting key environmental determining coalitions; (4) participating in environmental determining political processes; (5) joint venturing with environmental determining coalitions; and (6) selecting particular portions of the environment to engage in competitively.

The multinationalized mission must convey the overall corporate intentions and strategic orientation toward a proactive versus reactive choice in each operating context. Subsidiaries cannot be allowed to determine their own environmental posture if the MNC is to fully capitalize on the potential advantages inherent in internationalized operations.

Public Image. Domestically, the public image is often shaped from a marketing viewpoint. The firm's public image is considered a marketing tool that is managed with the objective of customer acceptance of the firm's product in the marketplace. Although this dimension remains a critical consideration in the multinational environment, it must be properly balanced with concern for organizational claimants other than the customer. The multinational firm is a major user of national resources and a major force in the socialization processes of many countries. Thus, the MNC must manage its image with respect to this larger context by clearly conveying its intentions to recognize the additional internal and external claimants resulting from multinationalization. The following excerpt from Hewlett-Packard's mission statement is exemplary of this broadened public image: "As a corporation operating in many different communities throughout the world, we must assure ourselves that each of these communities is better for our presence. . . . Each community has its particular set of social problems. Our company must help to solve these problems." Through this statement, Hewlett-Packard conveys to the public an image of responsiveness to claimants throughout the world.

SUMMARY

Undertaking the definition of a company mission is one of the most easily slighted tasks in the strategic management process. Emphasizing operational aspects of long-range management activities comes much more easily for most executives. But the critical role of the company mission as the basis of orchestrating managerial actions is repeatedly demonstrated by failing firms whose short-run actions are ultimately found to be counterproductive to their long-run purpose.

The principal value of a mission statement as a guide to strategic action is derived from its specification of the ultimate aims of the firm. It thus provides managers with a unity of direction that transcends individual, parochial, and transitory needs. It promotes a sense of shared expectations among all levels and generations of employees. It consolidates values over time and across individuals and interest groups. It projects a sense of worth and intent that can be identified and assimilated by company outsiders, that is, customers, suppliers, competitors, local committees, and the general public. Finally, it affirms the company's commitment to responsible action that is symbiotic with its needs to preserve and protect the essential claims of insiders for sustained survival, growth, and profitability of the firm.

REFERENCES

1. W. R. King and D. I. Cleland, *Strategic Planning and Policy* (New York: Van Nostrand Reinhold, 1978), 124.
2. J. Kelly, *Organizational Behavior* (Homewood, Ill.: Richard D. Irwin, 1974), 258.

3. R. H. Hall, *Organization—Structure and Process* (Englewood Cliffs, N.J.: Prentice-Hall, 1972), 11.
4. Ibid., 13.
5. E. J. Kelly, *Marketing Planning and Competitive Strategy* (Englewood Cliffs, N.J.: Prentice-Hall, 1972), 55.
6. A. Stonehill, T. Beekhuisen, R. Wright, L. Remmers, N. Toy, P. Pares, A. Shapiro, D. Egan, and T. Bates, "Financial Goals and Debt Ratio Determinants: A Survey of Practice in Five Countries," *Financial Management* **4**(1975):27-41.
7. O. Williams, "Who Cast the First Stone?" *Harvard Business Review* **62**(1984):151-160.
8. For examples see G. A. Steiner, *Top Management Planning* (New York: Macmillan, 1969), 145.

III

The Strategic Planning and Management Process

In Part III the organizational processes that may be used to effectively conduct strategic planning and management are identified and discussed.

In Chapter 7 John H. Grant ties in the strategic-choice elements and decision-making criteria of Part II to the strategic planning process. In Chapter 8 Robert L. Eskridge provides details on the desirable format of a plan and describes an appropriate planning review process.

Narendra K. Sethi uses the results of a survey of multinational companies to identify the planning processes that are used and are useful in a global context in Chapter 9. In Chapter 10 Archie B. Carroll and John Hall identify processes of particular relevance to the corporate social policy context. These latter two chapters carry through the theme—established first in Chapter 6—of first developing strategic planning and management ideas in a domestic corporate context and then extending those notions to deal with the special problems and opportunities that exist in the multi-national and social policy arenas.

CHAPTER 7

The Corporate Strategic Process: A Synthesizing Framework for General Management

John H. Grant

John H. Grant is a professor of business administration and director of the Strategic Management Institute at the University of Pittsburgh. Since receiving his doctorate from the Harvard Business School, he has pursued research involving the management of diversified firms and has published in *Strategic Management Journal, Academy of Management Review, Advances in Strategic Management* and has co-authored *The Logic of Strategic Planning* (Little, Brown, 1982). He has served as visiting professor at IMEDE in Lausanne, Switzerland, and as a consultant to many major corporations.

The world of the general manager is a complex and dynamic one. An increasing number of markets are developing global characteristics, the distinctions between private and public sector activities are becoming blurred, and rapid advances are occurring in information processing and transmission. As a result, it has become increasingly important that those men and women responsible for guiding our diverse organizations have useful frameworks and processes to establish relationships and priorities for the systematic resolution of problems and the pursuit of new opportunities. The conceptual framework of a corporate strategic process—or somewhat more broadly, organizational strategic process—is proposed as a means for linking the firm to its external environment. This managerial process of formulating and implementing strategy over time can be called strategic management.[1]

GENERAL MANAGERS—WHO ARE THEY?

The word *manager* has become one of the most ubiquitous of our day. Slogans, such as "every employee a manager," do little to add meaning to the word. For our purposes, however, we can respect the notion that all resource allocators are managers, while at the same time limiting the field markedly by adding a single adjective. *General managers*, then, can be viewed as individuals with multifunctional responsibilities for guiding operating units toward one or more objectives through the shifting of scarce resources within an environment posing various constraints and expectations. From this description, we can conclude that the term *general manager* need not be limited to corporate presidents, nor, on the other hand, can we infer that all vice-presidents are also general managers. In a typical large, diversified corporation, for example, we might find at least a few dozen individuals with titles such as president, group vice-president, division manager, and, in some instances, product manager who legitimately fit our description of a general manager. On the contrary, a vice-president of manufacturing or marketing is not apt to have general management responsibilities.

Identification of general managers is relevant only to the extent that it permits us to define a subset of people within an organization who face similar issues, and whose jobs are thus amenable to study as a group or population, not exclusively as unique positions. As is implied in the preceding paragraph, general management positions within a given firm vary in their degree of *generality,* or, in other words, their range of discretionary action is constrained in a cascading fashion at each lower level in the hierarchy. However, in spite of our generalists' organizational levels, we have posed for them some tasks typically not amenable to explicit optimization. The general manager must (1) set and/or interpret objectives, typically over several facets of the firms or subunit; (2) balance and time resource commitments to production, engineering, marketing, administration, personnel, and finance; (3) recognize internal and external limitations on actions; and (4) evaluate the importance of objectives and the progress being made toward attaining those objectives over time. In short, the general manager is responsible for determining both *what to do* and *how to do it,* that is, the formulation and implementation of plans for action.

SYNTHESIZING FACTORS FOR STRATEGY

As our outline of general managers' tasks implies, relatively few fixed or given variables exist within their sphere of activity. Some factors act as tentative boundaries or constraints for varying periods of time, but it is recognized that most such limits are subject to modification over the long run. In order to organize their decisions within such a setting, the general managers need a dynamic, open-system model that will help structure interrelationships among commitments over time. A concept of *corporate strategy* incorporating the logic for such a task is proposed as a model.[2]

Many books have been written about various notions of *strategy, business strategy,* and *strategic management.*[3] It is not our purpose here to review or analyze all these ideas; rather we want to sketch a composite concept of the corporate strategic process that will serve both to guide the readers' inquiries into related strategy literature and to facilitate application of the process to actual situations. The observant reader will recognize that the conceptual framework that follows is not limited in applicability to the business corporation; many elements of the model are useful in a wide variety of organizational contexts.[4]

The corporate strategic process has several characteristics including, but not limited to, the following:

1. It is comprised of two conceptually separable, but operationally interdependent processes—*formulation* and *implementation*. The phrase *strategic process* may usefully describe the pattern of interaction of the variables over time, that is, the sequential formulation, execution, feedback, and modification that occur continuously.
2. It is a *dynamic, open-system* model because it influences and is influenced by exchange transactions with the broader socioeconomic environment. It assumes proactive rather than reactive management behavior.
3. It is *multileveled* in the sense that substrategies exist at various levels of the corporate totality, and these may either complement or conflict with higher-level strategies.

We should begin by recognizing that organizations are continuously engaged in processes for formulating and implementing strategy,[5] but we must simplify our discussion by outlining the processes sequentially. Recognizing the futility in trying to specify all of the factors and relationships included within the strategic process in a given organization, it may be helpful to present some primary relationships as points of reference. Toward this end, the schematic representation in Figure 7-1 should be helpful for purposes of orientation, but remember, it is neither a map nor an answer. The most direct application of Figure 7-1 is to a single-product firm; other more complex systems for use in multidivisional companies will be discussed later.

The formulation process can begin with a preliminary definition of the business and an analysis of corporate resources (financial, technical, managerial, and so forth) in relationship to those of both existing and potential competitors.[6] During this process, particular attention should be directed toward those *strengths* and *weaknesses* that yield relative competencies or points of vulnerability in relationship to competitors. The result might be a list of product and/or market possibilities. A subsequent analysis of the forces at work in the external economic environment might constrain the possibilities to a more limited number of currently feasible options. On the other hand, the environmental analysis might reveal opportunities or threats not suggested by the internal evaluation. Such a corporate SWOT (strengths, weaknesses, opportunities, threats) analysis must then be tested against the personal perspectives of the top managers before the formulation process can yield an acceptable set of strategic options.[7] The importance of such personal values in the choice process seems to be particularly significant at the senior-executive decision levels.[8] The integrating or matching process noted in Figure 7-1 includes technical, economic, and creative steps, but the resulting strategy should include an internally consistent notion of the firm's business(es) and its mode of competition or way of doing business.

After formulating the concept or statement of strategy, the manager faces the process of implementation.[9] It is here that he or she must decide the size, objective, sequence, and timing of resource commitments. In order for needed action to occur, an organizational structure must be chosen so tasks can be divided and assigned and information can flow efficiently among subunits. Processes for developing standards or benchmarks of performance must be established, and a scheme of incentives should be devised to stimulate the movement of human and other resources toward the objectives. Additionally, the choice of a leadership style frequently seems to be a critical catalytic element in the implementation process.

The above paragraph strongly suggests that *strategy* should determine *structure and systems*. Some authors find that relationship a comfortable normative position, but there is considerable evidence to suggest that many managers assume their organizational systems are largely a given factor, and they limit their strategic product and/or market options to those which seem compatible with the structure.[10] In short, there is a widely acknowledged interdependence between the two concepts, but the direction of causation depends on the ways in which the top corporate officers define the flexible elements of strategy in a given firm at a particular point in time.[11]

The real test of a strategy does not occur until resources are committed and the firm's actions are manifest in the marketplace. As the economic and other results begin to appear, the *sequencing* and *conditionality* in the strategy become active. Do *good* results suggest we should commit more resources, or are we only going to attract further competition? Do *bad* results demand the acknowledgment of sunk costs, or rather do they signal the need for greater effort or more resources? Have circumstances changed so much since the initial strategic decision that basic assumptions and premises are no longer valid? The range of outcomes and their possible implications are virtually limitless,[12] but an important factor to remember is that *any* results can transmit signals both within the firm, as an incentive to reallocate resources, and to the external environmental forces, as a stimulus for adaptation.

We have thus far described the strategic process as a rather deliberate, straightforward activity. We should be forewarned, however,

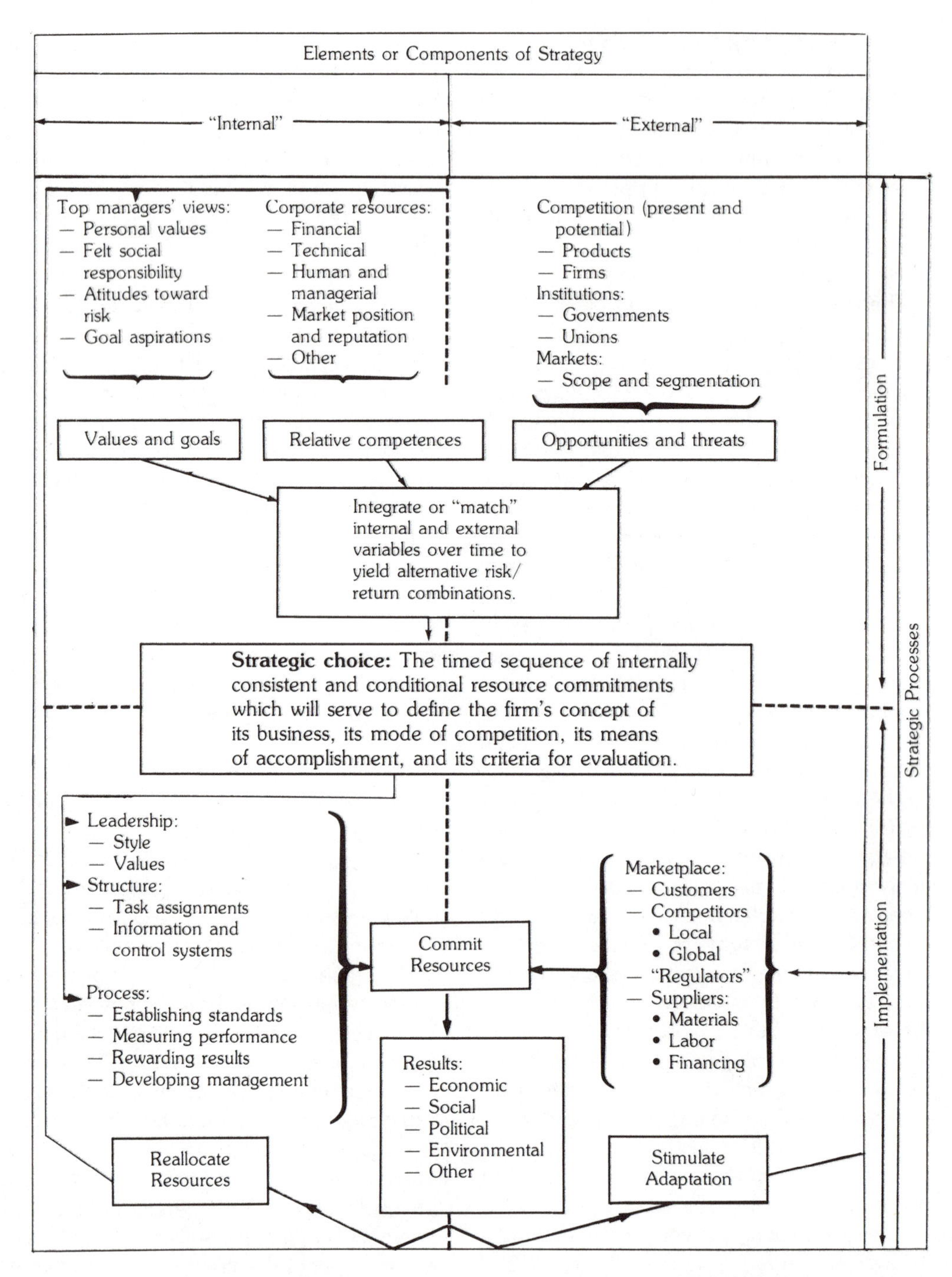

Figure 7-1. Corporate strategic process—a schematic representation. (Acknowledgment: I am indebted to J. L. Bower, "Strategy as a Problem Solving Theory of Business Planning," Harvard Business School, 1967, p. 27, for several of the relationships originally expressed herein.)

that a strategy may be either explicit or implicit; it may or may not be understood by more than a very limited number of the company's executives. Because the study of a written statement of a firm's strategy is neither a necessary nor sufficient condition for truly understanding its strategy, it behooves the analyst to study carefully the ways in which a firm deploys its resources as well as the way it describes its business.[13] The frequent absence of forthright, operational statements of a firm's strategy will become increasingly understandable as one explores the implications of a firm's strategy for its competitive position and for power relationships within the firm.[14] In short, a strategy may contain certain elements that should not be widely communicated in advance.

We have now reached another juncture in our discussion. We have thus far circumscribed the territory of general management and have proposed a corporate strategic process as a means of analyzing, synthesizing, and acting on many of the ambiguities or difficulties facing such generalists. We have not, however, devoted much attention to the actual processes of identifying, evaluating, or choosing corporate strategies. Hence, it is to these tasks that we must now direct our attention.

The identification of possible corporate strategies requires the analyst to understand the objectives a firm seeks to accomplish, the reasons why those objectives are being pursued, and the means management is expected to use to accomplish them. Among the more detailed issues to examine are the following:

1. What *objectives* are important to the firm?
   - Economic
     - ROI, EPS, profit margins
     - Sales growth, share of market
     - Market value of common stock
   - Prestige
     - Industry leadership in size, technology, and so forth
     - Positions or independence of executives
   - Social
     - Useful products
     - Employment security

The apparent set of objectives should be ordered in terms of importance and considered as being subject to change over time. Risks associated with the pursuit of the various objectives should not be overlooked.

2. What *reasons* justify these objectives?
   - Values of those stakeholders with power to influence
     - Executives
     - Stockholders
     - Unions, employees, and so forth
   - Regulatory expectations
   - Competitors' behaviors in markets for products and resources
3. What *means* can be employed for attaining the objectives?
   - Marketing posture
     - Products and services in terms of variety, quality, and so forth
     - Pricing patterns
     - Diversity of customers and geographic regions
     - Distribution channels
   - Operations and technology
     - R&D that leads or follows the industry
     - Technical inputs from highly skilled workers or automated production
     - Cost structures that are capital or labor intensive
     - Locational advantages
     - Minimum efficient scale (MES) operating units

When marketing and operating factors are combined, one begins to see alternative ways of doing business in an industry (for example, high-volume, low-cost mass producer; technology leader with small market niche) or what some have labeled *generic strategies.*[15]

   - Financial considerations
     - Capital structure and expenses

Dividend practices
Timing, sequencing, and risk analyses of alternatives
Flexibility and control of budgeted cash flows
Incentives and contingency payments
Taxation requirements
Stock value enhancement techniques
Organizational design
Leadership style
Formal hierarchy and task assignment
Information systems for planning and control
Performance standards and evaluation process
Personnel recruitment and development programs

A careful effort on the part of analysts to identify a firm's actual positions or actions with respect to the above dimensions should permit them to summarize a firm's strategy. Consideration should be given to both what a firm says it is doing and what it really seems to be doing; words and deeds sometimes interact in interesting ways!

Following the identification of a corporate strategy, one frequently is placed in a position of evaluating the alternative strategies. The evaluation might encompass the legitimacy of the objectives and the reasons for them, or it might be limited to an analysis of the probable effectiveness or efficiency of the means to attain the objectives. For our purposes at this point, we will restrict most evaluative considerations to the reasonableness of the objectives and the ability of the firm to reach them. Some of the tests or questions we might pose during the evaluation can be developed from prior work.[16]

1. *Comparative advantage:* Does the firm have a distinctive competence not found among competitors, for example, location, clientele, market position or reputation, outstanding R & D, low production costs? Does the strategy take consistent advantage of such positions or capabilities?
2. *Relation of opportunity to corporate resources:* Does the firm have the resources required to succeed (for example, money, facilities, personnel, reputation)? Can the essential elements be developed? When? Remember, it is possible to be so successful that a firm may not take full advantage of its success and thus become vulnerable to external takeover threats, which may or may not be consistent with long-term objectives.
3. *Risks and opportunities:* What can we gain? What can we lose? Are the trade-offs worthwhile in terms of the values and stakes of those who must make the decision? If the strategy fails or is delayed, will the company have enough reserve resources to pursue an attractive alternative?
4. *Internal consistency:* Are marketing and production efforts and abilities consistent with the requirements of market segments being sought? Will available resources carry the firm to the objective? Is the product and/or market posture consistent with the values of management? Are product and employment practices consistent with societal expectations? What responses can be expected from employees and customers?
5. *Sequencing commitments:* Does the firm recognize the order in which actions must be taken? What lead times are required for evaluation? What alternatives exist at each point? What are the conditions for their pursuit?
6. *Competitive response:* Does the proposed sequencing include the timing, magnitude, probability, and potential effects of competitors' responses?

Based on an assessment of these general criteria and others that might be relevant in a given situation, a general manager will be in a position to begin to judge the reasonableness of the strategy in terms of the specific firm and the particular environment in which it is operating. The data with which he or she must work may seem limited and the uncertainties may seem great, but the important questions to answer are these: Have we asked the right questions? Have we extended the analyses far enough?

After general managers have satisfied themselves that they have raised the critical issues and answered them to the best of their ability with the resources available, they are then in a position to choose future actions of the company. It is possible that the recommendation will simply be, "Full steam ahead!"

Such a statement presupposes that the manager is satisfied with the reasonableness of the firm's objectives and its means of accomplishing them. If, on the other hand, he or she is less enthusiastic about the firm's existing strategy, then recommendations must encompass the perceived weaknesses and include viable alternative actions. In the usual situation, many recommendations will be derived from the strategic issues[17] raised above in conjunction with the identification and evaluation of strategy; however, there are some additional facets one might want to consider. As with those preceding, the list that follows is intended to be illustrative, not exhaustive.

1. Do proposed objectives constitute an integrating concept, myth, or mission for important management personnel?
   Are the objectives operational, that is, do they provide useful guides to action and can they be measured adequately?
   Are the objectives feasible?
2. Are product and service offerings conceived in terms of function as well as form?
   Is the nature of market segmentation explicit?
   Are distinctions in offerings perceptible to customers?
3. Have present and probable future industry structures been considered?
   What is the flow of goods and services from raw material through final consumption? Who contributors value-added?
   What is the nature (size, reputation, and so forth) of competitors, both present and potential?
   Given existing relationships within the industry, have you considered relative bargaining power of various parties? What implications would arise from mergers or other restructuring?
   Are traditions (for example, price leadership, advertising restraint, and so forth) within the industry incorporated into the analysis?
4. Do recommendations consider the basic economic structure for the relevant industries?
   Cost-volume-profit relationships in production and marketing?
   Elasticity of demand?
   Barriers to entry and exit (for example, patents, technology, cost of plant or market position, liquidation expenses, and so forth)?
5. Are existing or available personnel willing and able to execute the proposed strategy?
   Can the unskilled be replaced or retrained?
   How much time is required to develop the needed skills?
6. Can the recommendations be integrated over an appropriate time horizon?
   What resources will be required?
   How will they be generated and when?
   How sensitive are the results to reasonable changes in the assumptions?

Although many formats and procedures can be useful in structuring the evaluation and choice phases, a set of *pro forma* financial statements for a relevant time period is frequently a convenient means for compelling analysts to make many of their assumptions explicit and quantifiable, where possible. When time and resources permit, the development of competitive scenarios can greatly aid the assessment of probable future outcomes.[18]

Our identification, evaluation, and choice phases of the strategic process have presupposed a thorough understanding of the competitive environment surrounding the firm. A brief elaboration of this system will highlight sources of competitive trends and reasons for governmental involvement in many such sectors.

COMPETITIVE ENVIRONMENT AND THE STRATEGIC PROCESS

Our description of the strategy formulation process accompanying Figure 7-1 identified the external environment as an important ingredient in that analytical process. In order to fully understand the potential consequences of a particular strategic choice, however, we must evaluate the sequence of relationships in the surrounding competitive environment. A value-added chain, as shown in Figure 7-2, can be valuable in assessing the stages of the overall production sequence at which the profit margins may be seriously squeezed and those where competition is much less severe.

If we consider, for example, one of several competitors in the manufacturing stage—Firm X—to be positioned between various suppliers and distributors, we can see that potential additional competition could emerge from any one of several sources. Not only can the existing manufacturers be expected to seek more effective ways to improve profits or other benefits, but both component suppliers and distributors may begin asking themselves, "Should we integrate forward (or backward) to increase our prospective profits from this industry?" In competitive and flexible economic systems, such ideas for restructuring particular industry sectors stimulate innovative ways to improve operations that often ex-

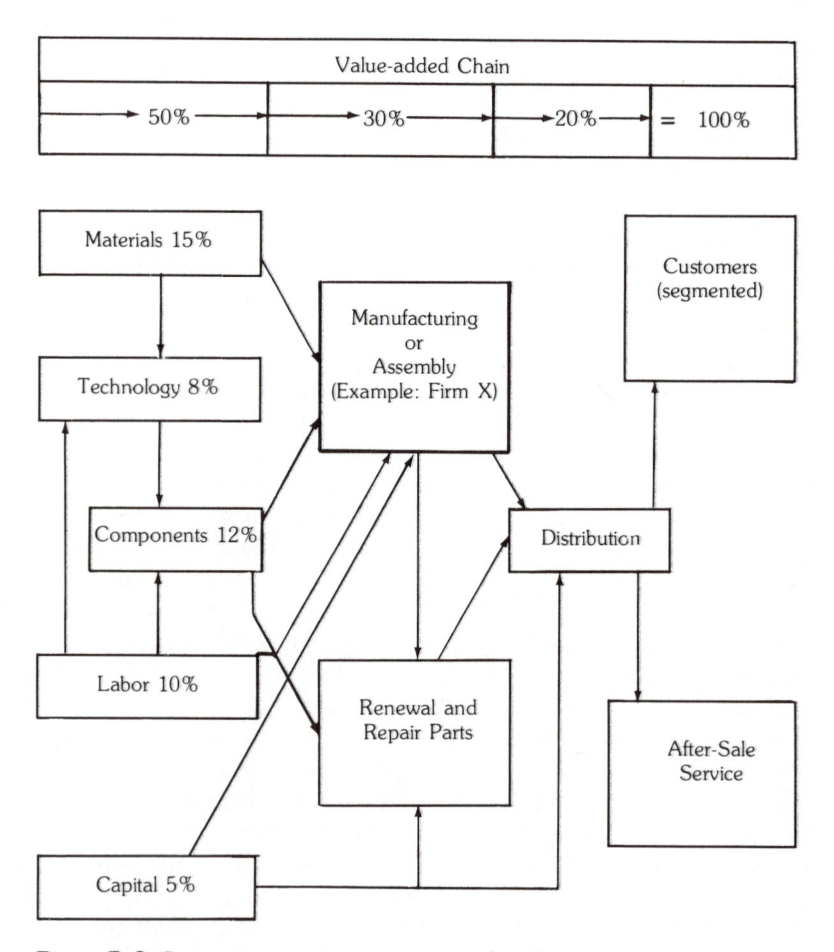

Figure 7-2. Competitive environment external to the strategic process.

tend well beyond the boundaries of an existing organization.[19]

The response of many organizations to severe competitive pressures in their existing markets or innovative ideas for other environments has been to diversify, either through R&D or acquisitions. Although the economic and related consequences of such actions continue to be debated,[20] considerable progress has been made in understanding how to effectively manage the resulting complexities.[21] Such diversification moves often simply serve as vehicles for accelerating the restructuring of industry sector relationships and for the movement of resources across the sectors. With rapidly changing technologies and national priorities, such restructuring of industries frequently will generate both threats and opportunities for the strategies of particular firms.[22]

The level of competitive pressures and the opportunities for strategic flexibility available to a given firm are, however, influenced by the sociopolitical characteristics of particular countries and industries, so general managers must be cognizant of such possible diversity if they are to focus their corporate strategic processes effectively.

DIVERSITY OF COMPETITIVE ARENAS

Most of our descriptions and analyses of the corporate strategic process have thus far implied the pursuit of such activities within a domestic or national market setting by a single or dominant product firm. The rapid internationalization of many markets[23] and the significant interests of governments in many strategic decisions[24] are but two trends that demand, however, that the analyst expand the spheres of analyses to include the appropriate geographic scope and the full range of competing organizational forms, whether private firms, not-for-profit organizations, or government agencies.[25]

The contextual framework for positioning a firm within its relevant competitive arena can be visualized in terms of three dimensions. Initially, the *geographic region* surrounding a given firm must be studied so the strategist will understand the varying channels through which particular products and services can be provided in the area. For example, does a fragmented industry serve specialized local needs, or, in the other extreme, are multinational firms offering widely recognized branded products throughout the region?

After developing an understanding of the appropriate geographic scope for analysis, the strategist must understand the relative roles of various *organizational forms,* including private firms, not-for-profit institutions, and government agencies in the industry sector of interest. Corporate strategists often view themselves as operating largely in a privately owned and economically competitive arena, but in many parts of the world, major portions of economic activity are undertaken by government agencies, trade associations, and other types of organizations. In addition, growing trends toward various collaborative ventures among firms offering different marketing, technological, and financial resources make assessments of a given firm's competitive posture complex.[26] Further, because most countries alter their extent of involvement in particular industries over time, general managers must monitor a third dimension to understand the *proportion* of each sector controlled by various owners. These trends must be thoroughly understood if managers are to determine whether they are competing against companies or countries.[27] In those situations where governmental influences may be significant, it is important to recognize the differences in objectives and decision processes among the various types of organizations.[28]

When the strategist has developed an understanding of the contextual framework within which he or she must conceptualize and implement a firm's strategy, the next step is one of selecting appropriate systems for executing the process.

PLANNING AND CONTROL SYSTEMS

As organizations grow and become more complex, it becomes increasingly important that various phases of the corporate strategic process become systematized for timely and efficient consideration of key issues.[29] Planning and control systems can be designed to support the strategic objectives of organizations and the managerial styles of diverse executives, but the purpose is always one of balancing the potential gains of such systems against the unavoidable costs of maintaining them.[30]

Some planning and control activities must be undertaken at regular intervals to assess customers' needs and streamline internal operations. On the other hand, consideration of major new corporate directions may only be appropriate every few years.[31] Likewise, as competitive pressures become more intense, general managers must be sure to evaluate their planning and control systems to see that they are continuing to support the primary corporate mission.[32] An important part of this process, as noted in Figure 7-1, is an examination of the relationship of managerial incentives and rewards to the strategic objectives.[33]

Another aspect of planning and control system evaluation involves the capacity to efficiently accommodate the complexity that typically arises from diversity. The strategic process we have been describing usually must be divided into corporate-level and division-level processes in order to assure both coordination and integration at the top of the organization and responsiveness within the divisions. By involving the appropriate people and permitting them to function according to different schedules or cycles, the strategist can maintain consistent division-level strategies while retaining desirable corporate-level flexibility.[34] Space here does not permit a full examination of diversity's consequences for administrative systems, however, strategic processes must occasionally be reexamined to see if they should be elaborated or streamlined to meet an organization's current requirements.

As organizations evolve and become more sophisticated, it remains important that managerial skills be enhanced along with refinements in the systems themselves.[35] Effective and timely strategic changes, whether aimed toward efficiency, innovation, or diversification, thus can become increasingly predictable and rewarding.[36] Because general managers can control most of the organizational factors that influence the quality of their administrative systems, it is important that they make full use of such mechanisms in learning about their environments and in pursuing their corporate strategy.[37]

SUMMARY

The purpose of this chapter is to provide an overview of the elements of a corporate strategic process as a framework for general managers' analyses and actions. As the pace of change accelerates and the scope of competition expands, organizations will find it increasingly important to be able to assess diverse changes and respond promptly to their implications. At the same time, researchers are constantly seeking to improve our understanding of the basic concepts and their relationships.[38]

The pursuit of simplified solutions may provide useful guidance to various operating personnel, but such approaches can yield a disservice to corporate strategists who must confront the external complexities that will influence greatly the long-term well-being of most organizations.[39] The strategic processes I have outlined require the *identification, evaluation, recommendation, implementation,* and *reevaluation* of corporate strategy through the assessment of organizational strengths and weaknesses and environmental opportunities and risks. Differences among competitors will be more relative than absolute, and their strategic importance is derived from the unique combinations of factors being considered. The process is not only *analytical* and *perceptual,* but, because it usually involves many people, it is also *social.*[40] Preoccupation with one set of skills to the exclusion of the others can lead to recommendations that may challenge observers' concepts of feasibility or reasonableness.

From the selection of an operating environment and corporate objectives through the formulation of a competitive strategy and supporting administrative systems and incentives, the strategic process can be useful to both general managers and their observers who are seeking a framework for synthesizing diverse theories and models for the improvement of organizational performance. Many facets of one's educational background are relevant to the study of this process. The question is not so much one of *what* is relevant, but rather, *when* is it relevant and *how* can it be applied?

REFERENCES

1. D. E. Schendel and C. W. Hofer, eds., *Strategic Management: A New View of Business Policy and Planning* (Boston: Little, Brown, 1979).
2. J. H. Grant and W. R. King, *The Logic of Strategic Planning* (Boston: Little, Brown, 1981).
3. R. Evered, "So What Is Strategy?" *Long Range Planning Journal* **16**(1983):3.
4. C. E. Summer, *Strategic Behavior in Business and Government* (Boston: Little, Brown, 1980); M. L. Hatten, "Strategic Management in Not-for-Profit Organizations," *Strategic Management Journal* **3**(1982):89-104.
5. K. Ohmae, *Mind of the Strategist* (New York: McGraw-Hill, 1982).
6. D. F. Abell, *Defining the Business: Starting Point of Strategic Planning* (Englewood Cliffs, N.J.: Prentice Hall, 1980).
7. A. A. Thompson, Jr. and A. J. Strickland III, *Strategy Formulation and Implementation,* rev. ed. (Dallas, Tex.: Business Publications, 1983).
8. L. L. Cummings, "The Logics of Management," *Academy of Management Review* **8**(1983):532-538.
9. L. G. Hrebiniak and W. F. Joyce, *Implementing Strategy* (New York: Macmillan, 1984).
10. D. I. Hall and M. A. Saias, "Strategy Follows Structure!" *Strategic Management Journal* (1980):149-163.
11. R. H. Hayes, "Strategic Planning—Forward in Reverse?" *Harvard Business Review* **63** (1985):111-119
12. R. T. Lenz, " 'Determinants' of Organizational Performance," *Strategic Management Journal* **2**(1981):131-154.
13. J. B. Quinn, *Strategies for Change: Logical Incrementalism* (Homewood, Ill.: Richard D. Irwin, 1980).
14. H. Mintzberg, *Power in and Around Organization* (Englewood Cliffs, N.J.: Prentice-Hall, 1983).
15. M. E. Porter, *Competitive Strategy* (New York: Free Press, 1980).
16. S. Tilles, "How to Evaluate Corporate Strategy," *Harvard Business Review* **41**(1963): 111-121; R. P. Rumelt, "Evaluation of Strategy: Theory and Models," in *Strategic Management,* ed. D. Schendel and C. Hofer (Boston: Little, Brown, 1979), 196-212; M. E. Porter, *Competitive Advantage,* 2nd ed. (New York: Free Press, 1985).
17. W. R. King, "Using Strategic Issue Analysis," *Long Range Planning Journal* **15**(1982):45-49.
18. P. Wack, "Scenarios: Shooting the Rapids," *Harvard Business Review* **85**(1985): 139-150.
19. Porter, *Competitive Strategy.*
20. R. P. Rumelt, "Diversification Strategy and Profitability," *Strategic Management Journal* **3**(1982):359-369.
21. R. G. Hamermesh and R. E. White, "Manage Beyond Portfolio Analysis," *Harvard Business Review* **84**(1984):103-109.
22. J. Diebold, *Making the Future Work* (New York: Simon and Schuster, 1984); S. Fraker, "High Speed Management for the High-Tech Age," *Fortune* **51**(1984):62-68.
23. T. Hout, M. E. Porter, and E. Rudden, "How Global Companies Win Out," *Harvard Business Review* **60**(1982):98-108.
24. D. P. Rutenberg, "Multinational Corporate Planning and National Economic Policies," in *Strategic Management Frontiers,* ed. J. H. Grant (Greenwich, Conn.: JAI Press, forthcoming).
25. C. Fombrun and W. Graham Astley, "Beyond Corporate Strategy," *Journal of Business Strategy* **3**(1983):47-54.
26. J. P. Killing, "How to Make a Global Joint Venture Work," *Harvard Business Review* **60**(1982):120-127
27. B. R. Scott, "National Strategy for Stronger U.S. Competitiveness," *Harvard Business Review* **62**(1984):77-91.
28. J. L. Bower, "Managing for Efficiency, Managing for Equity," *Harvard Business Review* **61**(1983):82-90.

29. W. R. King, "Using Strategic Issue Analysis," *Long Range Planning Journal* **15**(1982):45-49.
30. P. Lorange, *Corporate Planning: An Executive Viewpoint* (Englewood Cliffs, N.J.: Prentice-Hall, 1980).
31. J. C. Camillus, "Reconciling Logical Incrementalism and Synoptic Formalism," *Strategic Management Journal* **3**(1982):277-283.
32. W. R. King, "Evaluating Strategic Planning Systems," *Strategic Management Journal* **4**(1983):263-277.
33. "Rewarding Executives for Taking the Long View," *Business Week* (2 April 1984).
34. Grant and King, *Logic of Strategic Planning*.
35. F. W. Gluck, S. P. Kaufman, and A. S. Walleck, "Strategic Management for Competitive Advantage," *Harvard Business Review* **58**(1980):154-161.
36. D. F. Harrington, "Stock Prices, Beta and Strategic Planning," *Harvard Business Review* **61**(1983):157-164; P. J. Strebel, "The Stock Market and Competitive Analysis," *Strategic Management Journal* **4**(1983):279-291; I. M. Duhaime and J. H. Grant, "Factors Influencing Divestment Decision-making: Evidence from a Field Study," *Strategic Management Journal* **5**(1984):301-318.
37. I. I. Mitroff, "Archetypal Social Systems Analysis: On the Deeper Structure of Human Systems," *Academy of Management Review* **8**(1983):387-397; P. Shrivastava and J. H. Grant, "Empirically-derived Models of Strategic Decision-making Processes," *Strategic Management Journal* **6**(1985):97-114.
38. N. Venkatraman and J. H. Grant, "Construct Measurement in Organizational Strategy Research," *Academy of Management Review* **11**(1986):71-87.
39. T. J. Peters and R. H. Waterman, Jr., *In Search of Excellence* (New York: Harper and Row, 1982); D. T. Carroll, "A Disappointing Search for Excellence," *Harvard Business Review* **61**(1983):77-84, 88.
40. C. R. Anderson and F. T. Paine, "Managerial Perceptions and Strategic Behavior," *Academy of Management Journal* **18**(1975):811-823.

CHAPTER 8

The Strategic Planning Review Process

Robert L. Eskridge

Robert L. Eskridge holds a B.S. in economics and journalism from Northwestern University and an M.B.A. from Pepperdine University. He has served as director of new products, advertising, and brand management of the Grocery Division of Ralston Purina Company and was a member of the president's planning council and vice-president of the Seafood Division. He also has served as vice-president of marketing at Carnation International and was founder and chairman of Growth Management Center with offices in the United States and Paris, France. The Growth Management Center associates have developed and used their own proprietary Strategic Management Process with more than one hundred clients on five continents.

Strategic management is a relatively new discipline for which there is no common, precise language in general use. Therefore, terse definitions of terms, generally consistent with those provided in Chapter 5, are included here. The reader may want to substitute the terms used in his or her own company's process.

This chapter assumes that the strategic plan to be reviewed has the following basic elements, although not necessarily in this sequence.

1. A description of the relevant environment, including the external environment (that is, those things that cannot be controlled), the internal environment (that is, those things that can be controlled), and the relevant industry structure.
2. Conclusions about the company (including strengths, problems, threats, and opportunities), its competitors, and its consumers and/or customers, all drawn from the environmental description.
3. Resource portfolio analysis.
4. A mission or charter statement that tells where the company is, what it wants to become, when, through what products and/or services, to which consumers and/or customers.
5. Priority objectives (eight to ten macro things that must be done to achieve the mission) and strategies (how the company will use its finite resources to activate the priority objectives).
6. Specific programs and projects for implementing the strategic plan throughout the company, including pertinent communication to appropriate departmental personnel, calendar, individual commitments, and a clear review process.

These are the choice elements of the strategic plan that should be reviewed during the upcoming fiscal year.

SHOULD YOU BOTHER TO REVIEW YOUR STRATEGIC PLAN?

"Would you tell me, please, which way I ought to go from here?"

"That depends a good deal on where you want to get to," said the Cat.

"I don't much care where—" said Alice.

"Then it doesn't matter which way you go," said the Cat.[1]

Alice reviewed her simplistic strategic plan, the mission of which was to walk, not arrive anywhere at a specific time, but recognizing a change in the environment (the cat), she reviewed her plan. Meanwhile, the cat used a good technique for reviewing a strategic plan. He asked meaningful questions.

A few years ago Growth Management Center made a simple study centered around the question: "What has happened to the *Fortune 500* companies during the first 25 years of that publication's reporting of such companies?" One hundred fifty-two or 30 percent were no longer listed. A few were acquired (perhaps because they were in bad shape), but our conclusion was that these companies, with all their power and resources, did not anticipate and manage change well.

Strategic management is the art and/or science of anticipating and managing change for the purpose of building defensible strategies to ensure the future of the business. The company has a substantial management investment in the strategic plan that it has already built. It is worth thoughtful review because the return on its investment increases geometrically with each review. The plan gets better, but the time required diminishes. Yes, the review process is important.

A properly built strategic plan is a document made up of the time, mental sweat, words, thoughts, personal goals, and perhaps dreams of the planners. These planners have a stake in the success of the company and in the productivity of the strategic plan. However, there are others not involved in the planning process who also have a substantial *stake* in the productivity of the plan. Therefore, the plan also should embody the welfare and goals of the other *stakeholders* associated with the company, including all levels of current and future employees, owners, customers, and suppliers. These diverse people's *stakes* and interests in the company are intertwined into a maze of change taking place at different times and different rates in different areas. These changes should be considered when reviewing the strategic plan (see Chapter 13).

Like the ship captain and his crew crossing the Atlantic, in charge of the destiny of thousands of lives, so the CEO and his corporate planners, through the strategic plan, are plotting the direction of the company, changing the operational course with the various environmental changes, but always helping move the corporate ship toward its destination, its mission. Unlike the ship captain, however, the corporation leader need not check the course hourly or daily because the shifts in his or her business environment generally are more subtle, but check he or she must, as changes occur, to keep the company going toward its mission. This is the purpose of the strategic plan review.

WHEN TO REVIEW?

This is really two separate questions: How *often* to review? And *when* to review during the business cycle?

Frequency of Review

Strategic management is an iterative process. It should never stop once it is started anymore than the ship's captain should stop checking the charts and let the ship drift, since all major decisions are directed by the strategic plan's mission statement. The frequency (How often?) and depth of the review are determined by these factors:

1. Speed: Rapidity of change in the company's environment. For example, in the inexpensive toy or novelty market, things change quickly. A new item is sold to the distributors and/or wholesalers in July and is into retail stores in October, and by January the manufacturer knows whether it has a hot product on which it can build or whether it has to go back to the drawing boards. In the whiskey business things

change more slowly. It takes four years to age the product and established consumer tastes do not change quickly.

2. Change: Whenever there is a major change in the company's external or internal environment, for example, if the industry will soon be deregulated (external) or the company has a new CEO (internal).
3. Geography: Geographical location of planners. If the key planners live and work on five continents, the planning cycle might be quite different (even in the same industry) than if all the key planners work in the same building daily.
4. Tracking: How well is the plan tracking toward its mission? If it is frequently off course, more frequent reviews are necessary until it gets back on course.
5. Money: Whenever there are *major* financial considerations or expenditure approvals because those investments must support the company's mission in the strategic plan.

Generally, the strategic plan should be reviewed annually to determine the *cumulative* effect of seemingly small environmental changes that occurred during the year. This kind of a review may require a great deal of time or very little time, depending on the quality of the initial plan, the degree to which environmental changes affect the business, the skill of the planners, and the process that they use.

During the year, meaningful changes, or *anticipated* changes, in the environment should always trigger a review. When deregulation was forthcoming in the banking business, few financial institutions reviewed their strategic plan (if they had one) thoroughly enough or soon enough. Hence, banks and savings and loans, even large ones such as the Bank of America, failed or severely cut back their operations as the effects of deregulation began to have an impact. Citibank, on the other hand, immediately rethought their consumer business in light of this environmental change.

The strategic plan is a directive document that requires periodic monitoring and updating, either annually or when there is a major change in the external or internal environment.

What Month? At what time during the business cycle should the strategic plan be reviewed? Whenever the boss wants it? When there is a crisis? When everyone is tired so they rush through to get it over with and get back to *important* work? Concurrently with the budget? Of course, none of the above.

Generally, when a strategic planning process is part of the company calendar and multiechelon commitments are made by the planning participants, the review takes place at different times during the year for different portions of the plan by different echelons of the planning team, *but never near budget approval time.* However, the budget planning time does determine when the strategic planning should be reviewed.

The strategic plan, which drives the budgets, should be completed three months earlier than budget approval time in order for the budget work to be completed because many of the people working on the strategic plan also are in charge of operating budgets. Particularly in the United States, quarterly operating statements and year-end closings occupy so much of management's time and thinking that trying to review a strategic plan during this period is folly. In such cases, the strategic plan review always takes second place.

However, in companies where the personnel is well trained in the review process and there are not a large number of changes to consider, the *preliminary* annual budget can be added to the end of a strategic plan review meeting. For example, if senior managers review and approve the plan through the strategies section, they then may want to have operations personnel join the meeting to discuss the action programs that will activate the shorter-term strategies to get a first cut at the costs involved. Still, strategic plan reviews should never be conducted near budget *approval* time.

How? The major problems with the strategic planning review are:

1. The senior officers and/or planners try to do the complete review themselves instead of sequentially delegating portions of the plan to other echelons. In this scenario, the review becomes too time consuming for the manage-

ment group and they become discouraged with the process.

2. Using a complicated or laborious review process that involves many forms to fill out and demands extrapolated numbers or a process that begins with the question: "How much profit will your division or department make five years from now?" This can be done, of course, but such forecasts are a *result* of the strategic plan, not a foundation for it. In some case, using no process is a problem, exemplified by the CEO saying: "Here's how we'll (meaning I'll) change the strategic plan" or "Let's just get together and review the plan."
3. Inadequate or ill-conceived preparation.
4. Wrong people at the wrong time and/or place.
5. Inadequate *commitment*—psychological, emotional, financial—by the CEO and sometimes other senior planners to have the best possible current strategic plan and the willingness to pay the required price to get it.
6. Improper communication to the review participants, particularly regarding management's expectations and the individual's role in the process.

The experienced strategic planner can probably add more problems that he or she has encountered to this list. Think through your own experience and add such problems here. Once identified, these problems can be solved by adapting the review process to accommodate them.

There are four phases in the strategic planning review:

1. The environment: (a) Review and modification of the business environment assessment; (b) Conclusions about the effect of environmental changes on the planner's perception of customers and/or consumers, competitors, and the company itself; and (c) Resource analysis. This phase is prepared by second or third echelon personnel and department heads and is *approved* by management.
2. The direction: Mission and purpose, priority objectives, and the strategies for using resources to make the priority objectives become a reality. This is done by the CEO and senior management.
3. The action: Action programs and projects that accomplish the individual priority objectives and together, therefore, the *company's mission*. This generally is prepared by operations personnel.
4. The integration: Definitive programs for communicating the plan to appropriate personnel, making the plans work, blending in with the company's operations, working out the bugs, smoothing over interdepartmental relationships and coordinating budget implementation. This is recommended by department heads and operation officers and approved by senior management.

Content of the Review Process

Where to Begin? This section describes the content of the review and how to conduct it. A later section suggests *who* should do the review and a specific timetable.

Some companies begin the process with the senior planners (CEO and officers) reviewing the *macro crossroads issues*. These generally are 8 to 12 issues that should be addressed about the future of the company, expressed as questions, for example, "Since several of our senior managers are within five years of retirement, do we have an adequate succession plan in place?" Addressing these issues normally requires little time before moving to the first major phase in the review process.

The important starting place in the review, regardless of the structure of the plan, is a reexamination of the environment, because the external environment the company must live with and the internal environment it creates drive its business. In some cases, a description of assumed *changes* (assumptions) in the external and internal environments also are described.

The review begins with a brief *description* of the changes that have occurred since the last review in each of the relevant categories listed below. Categories can be added to or deleted from this list. Some categories will be much more important and therefore require more time and space than others, depending on the nature of the industry, the company's position in it, and the degree of change in each category.

Internal Categories
- Product lines and/or services offered
- Proprietary processes and/or technology
- Organization and human resources

Financial structure
Management style and internal culture
Marketing practices
Stated goals and objectives

External Categories
Government factors
Economic factors
Social trends
Technology
Distribution channels
Customers and/or markets
Competition

Clearly, technological changes in the computer field is a vital issue, whereas deregulation (legal) may be more important in the transportation industry. Therefore, for a computer company there might be few if any changes in the legal environment, but the technical factor would have a more thorough description of the technological environmental changes. In most cases, the length will range from a maximum of two pages to a simple statement such as "No Change."

There are many ways to conduct this environmental scan, but the three most commonly used that have proven to be effective are:

1. Treat each category separately and assign a segment of the environmental scan to appropriate departments or individuals in those departments (see the example in Figure 8-1).

The person's or department's job is to gather relevant information during the planning period and note any changes that have taken place in the assigned area of the environment. Then, when it is time to review the environmental portion of the strategic plan, he or she can study the environmental assessment from the current plan that relates to his or her assignment, examine the notes collected during the planning period relative to the changes that have taken place, and describe these changes. Each of those environmental revisions is then circulated to other people in the department or those who have the right to an opinion about them for review, criticism, critique, or modification.

The assigned person then pulls together comments into a terse document that de-

Environmental Category		Assigned to
External	**Department**	**Individuals**
Technical	R&D or production	John Jones
Legal	Legal	Henry Jackson
Customers	Marketing	Bob Smith
Competition	Sales	Jack Brown
Internal	**Department**	**Individuals**
Organization changes	Human resources	Olivia Harris
Marketing Practices	Sales-Advertising	Joe Steiner
Financial Structure	Finance	Fay King

Figure 8-1. Example of category assignment for environmental scan.

scribes the environmental changes that have taken place since the last review and the cumulative effects of these changes on the *conclusions* about the company's business situation (see Conclusions section). This document is then provided to the planning-coordinating department (or person) for consolidation and review by the proper senior planning personnel before the formal review takes place.

The advantage of this alternative is that it minimizes meeting time. It is most effective in companies where many of the planners have worked together through several plans.

2. Assign a multidepartmental group of second or third echelon personnel from the various departments involved to review all the relevant categories. For example, the group might consist of someone from sales, marketing, R&D, production, and human resources. This task force normally operates in meetings headed by a planning coordinator or a senior person in the planning department. Each of the individuals performs the same task as stated in alternative one above for review by the group before a final document is created. This cooperative document will describe the environmental changes that have taken place since the last plan had been approved and what these changes mean to the company (see Conclusions section).

The advantage of this process is that the ramifications of the environmental changes as they affect the various departments of the company can be worked out before submission of the final document to those who eventually approve the plan. For example, substantial change in media costs could have an impact on the advertising department's strategy for the coming planning period, but it also could affect the sales department, which would use the advertising plan with their dealers and/or distributors or other factors in the distribution system. Therefore, the conclusions that would be drawn by the advertising department would have to be weighed against the impact that these conclusions would have on the sales department, and the mutual ramifications would be expressed jointly in the Conclusions.

3. Each of the elements listed earlier that go into evaluating changes of the environment can be addressed by the senior planners who must eventually approve the plan. When the senior management review the environmental changes themselves, or as a group, the latter is normally done in an off-site, two-to-five day meeting. The disadvantage of this method is the amount of time required by the senior people. In addition, they normally are less knowledgeable about changes, particularly smaller ones, than the second and third echelon people who are closer to the situation. Of course, combinations of these alternatives can be used, as will be shown later in the Calendar section.

Conclusions

Regardless of the review planning alternative used, after the description of the environment(s) is completed, the information is used to draw conclusions about the company's current and anticipated situation in its environment. This is one of the weakest parts of most plans and, therefore, the weakest part of most strategic plan reviews. Too often, planners tend to talk about what has happened but avoid the discipline of answering the question: "So, what? What does this mean to us and our future?"

One of the ways of doing this is to conduct a strengths, problems, opportunity, and threats (SPOT) analysis for the customer and/or consumer, the competitor, and the company, in addition to judgmental conclusions about the descriptions of the relevant categories in the environment. As a part of this conclusion process, this information is used to conduct a SPOT review (sometimes called SWOT, in which case *weakness* is substituted for *problems*).

The terms used in the plan are unimportant. The concept is to consider the elements described in the environmental assessment to decide whether those elements are a strength that creates an opportunity, a problem worth solving, or a threat that can lead to a problem in the future and, therefore, should be squelched early. Such an exercise brings the environmental description elements into clearer focus regarding their effects on the company's direction. Figure 8-2 contains an example of a SPOT analysis. Deciding which of the two alternative conclusions given in Figure 8-2, or other conclusions, would be selected would require a more complete SPOT analysis than space here permits.

The preparation for modifying last year's plan by conducting the environmental scan and the conclusions drawn from it are by far the most important steps in the review process. Therefore, adequate time should be allowed for this endeavor to be completed, regardless of which of the alternative methods outlined earlier is used.

Once thorough, complete, concise descriptions have been made of the environmental changes and conclusions drawn from them, these descriptions and conclusions are used to review and assess the impact of the changed conclusions on the resource analysis, priority objectives and mission statements contained in the strategic plan being reviewed.

Resource Analysis

The resource portfolio analysis generally is used to divide the company's products and/or services into four quadrants, each of which indicates the individual products' and/or services' future destiny in the current strategic plan. Generally, the quadrant placements are related to growth potential. There are many terms for the use of these quadrants, but the most commonly used are: *Question Marks, Stars, Cash Cows,* and *Dogs.* This exercise most often is done on the basis of mathematical computation based on market share, market size, and market growth.

The quadrant management concept has

External Environment Category: Markets

Description: During the past 12 months our industry/category grew from ____ to ____. Our sales were flat so our share declined from ____ to ____.

Strengths	**Problems**
Industry increased ____ percent.	Our share declined from ____ to ____.
Opportunity	**Threat**
Invest more marketing dollars to participate more fully in industry growth	A continued loss of share could lead to loss of distribution.

Conclusion: "Our marketing plans and share of voice (our percent of advertising dollars share relative to the industry) need to be reviewed and possibly budget adjustments made to capitalize on the opportunity created by industry growth." Or the conclusion might be: "Our position in this category is weak so we should cut media expenditures, use only trade deals, and invest the savings in better growth areas for us."

Figure 8-2. Example of SPOT analysis (after George A. Steiner & John B. Mines *Management Policy & Strategy* New York: Macmillan, 1977, pp. 386-388).

been refined and its use expanded in recent years. It is now not only used for mathematical positioning of products, services, divisions, and business units but also to prioritize almost any aspect of the business, including customers, plants, personnel, markets segmentation, and so forth. In addition, the rigid mathematical criteria of market size, market growth, and market share have been replaced by predetermined judgmental criteria consisting of those factors important for the analysis and management of these other categories for which the quadrant management process is being used.

Regardless of which methods or terms are used, the strategic plan should have a resource-portfolio-analysis section that indicates how company resources are going to be used within the total product or service mix. Environmental changes could cause modifications to last year's resource analysis, which is why the portfolio should be analyzed after the environmental conclusions are drawn.

Senior Management's Involvement

In the strategic management review process, the mission and purpose, priority objectives, and strategies are the prime responsibility of the CEO and his officers. The description of the environmental changes, the company's strengths, problems, threats, and opportunities, and the conclusions drawn therefrom, balanced with the changes in the company's resource portfolio analysis, together form the basis for senior management's review of these factors to consider appropriate changes in the company's mission.

This review of the mission statement calls for the most precise wording. Read it carefully in the light of those things that sequentially preceded it in the plan. If it and the preceding steps are well done, it should not change substantially from year to year since the mission sets the direction for the future and directs the efforts of many employees. It should state clearly where the company is, where it wants to go, when, by providing what customers with what products or services. An example of a mission change is: "We will shift our growth emphasis from new products to acquisitions that have the following characteristics." The priority objectives should stay relatively constant from review to review if the mission does not change much. Generally, the changes would involve adding or subtracting a priority objective rather than rewriting all of them. In the mission change above, from new products emphasis to acquisition emphasis, the prior priority objective outlining the new product direction may be dropped, and one directing acquisition activity added.

The strategies that state how resources will be used to accomplish each priority objective, however, may change if the prior strategy was not working well. For any given priority objective, where this is the case, several alternative strategies should be considered before writing the new one. For example, if a priority objective directing acquisition activity is added, the review should examine all the alternatives, for example, "build our acquisition department, hire an investment banking company, consultants, and so forth."

The steps that follow the mission statement, priority objectives, and strategy reviews are reviews of action plans, programs, and projects. Once the action programs are reviewed and approved, the integration plan should be started.

Integration

The integration review should begin with an implementation analysis, which is an exercise designed to remove internal impediments that might hinder the completion of the programs. When the programs are reviewed, ask the person in charge of each to isolate those internal factors that helped him or her complete the program on time or that slowed down or hindered the completion. Examples of responses to the "help" question might be, "The marketing people really took the initiative and finished their assignment early, allowing me to complete the program on time," or "The computer department gave us more usable information than requested, which provided different insights for completing the program." Examples of responses to the "hinder" question might be, "I could never get the right marketing people together to make a decision," or "The computer messed up our financial reports."

When this approach is used for all programs, there sometimes are great similarities or repetition. This then allows management to examine these interdepartmental forces and cycle the observations back to the internal environment description or the SPOT analysis. If the computer performance or advertising-sales relationship shows up as hindering or helping the completion of a program, then that analysis becomes a strength to utilize more fully or is listed as a problem to solve.

In one company the packaging department failed to make a seasonal packaging-design change in time to distribute the product to the trade before the season began. They said marketing was a month late in approving the colors. The purchasing department said the marketing people were late in deciding how many of these new packages were to be sold, so the purchase order to the supplier was delayed. Although this sounds like everyone is blaming the marketing department, the fact was that there was a poor communication-responsibility process in the company that involved other departments as well. This was uncovered by the implementation analysis and corrective action was taken for all departments, not just marketing.

It is helpful to use this exercise not just for past events but also for planning the implementation of future programs by anticipating what should go well or not so well across several such programs.

Budget

Budget integration is a major factor to be reckoned with when reviewing the strategic plan, especially during the first review. This is true for several reasons. Annual budgets deal with short-term allocation of funds and the measurement of operational performance (doing things right) whereas the strategic plan review is concerned with long-term direction (doing the right thing). Therefore, the two are not naturally compatible, but they must be made compatible.

Budgets should show who will pay for the strategic planning review activity investments (sometimes called expense or even waste), including facilities, travel, rent, equipment, room and board at meetings, and, in some cases, outside consultants, resources, or experts.

Financial systems, forms, and procedures are firmly entrenched. Sometimes, the strategic plan will call for new products and the planners agree that, with the current method of allocating overhead, the new products would be costed out of the marketplace. The chances of changing the allocation method in many companies may range from difficult to impossible, but in this case should be addressed by a program of the new product priority objective. Even good strategic plans sometimes underestimate the cost of activating programs or overestimating income from such programs.

By the time everything is added up, the budgets will not allow for all programs to be activated at the planned rate and timetable, therefore, substantial time and planning effort has been wasted. This is not to say that the budget people are impediments to strategic planning, but their ways are set and the strategic planning process is relatively new and fluid. The two forces must mesh. These and other such budget integration issues should be addressed during the strategic planning review.

Plan Communication and Reinforcement

This is perhaps the single most important opportunity for improving plan effectiveness, productivity, and efficiency. In the past many companies went through undisciplined long-range planning efforts made up of either straightlined forecasts such as "15 percent compounded growth in sales and profits" or wish lists consisting of dreams of "things that we should do." The document was prepared, distributed, and filed in the bottom drawer until the next year and everyone went back to doing "important work."

Then, management by objectives came along and forced detailed follow-through. Too often, this tool was misused to deal with operating matters instead of longer-range strategies. It communicated deep into the organization the message of *doing things right* and on time. The misuse came by not adequately addressing broad issues in less detail, that is, *doing the right things,* which our present strategic plan stipulates.

The strategic plan review process should use all meaningful tools to establish a program that will reinforce the importance of the strategic planning assignments to individuals (or departments and/or task forces) who have such assignments. Some companies, for instance, have a program that says that each officer, once a week when in town, visits each of his or her key subordinates to briefly ask for a verbal update on some specific strategic planning program for which that individual has responsibility. Some companies start each department meeting with a brief review of the strategic planning programs pertinent to that department.

During the annual strategic planning review, the important thing is to search for ways to reinforce, during the year, the importance and progress of the strategic planning assignments.

Calendar

The key document in the strategic plan review process is the calendar delineating *who, what, where,* and *when.* This must be adhered to dogmatically. It should be distributed and its importance emphasized by the CEO. No one else! No one else, because he or she ultimately is responsible for the company's future well being. That is what strategic planning is all about.

The review process involves different people at different times, and the success and productivity of planning meetings depend on proper preparation. Therefore, assignments—what is expected of participants, their behavior patterns, the purpose, and the expected output of each of the meetings—should be clearly communicated well in advance.

The process, as it relates to the participation of several echelons in the review, can be visualized as an hour glass, with broad participation at the beginning and end but narrow, senior management participation in the center for the mission, priority objectives, and strategy reviews. When the review process is properly structured and adhered to, the entire process should require no more management time than a total of 15 days (or 6 percent of working time) during the year.

For purposes of a calendar example, assume that: (1) this company is on a fiscal year basis; (2) it has $100 million gross sales (United States only); (3) it has 500 employees; (4) it makes a line of 20 products sold through its own sales force to distributors who sell to retailers who sell to users; (5) its organization structure is vice-presidents of planning, production, R&D, human resources, finance, and administration reporting to the CEO; and (6) budgets for the following year must be approved by November 1.

Following is a typical strategic planning review calendar for such a company. Whether a month or two weeks is used between steps is unimportant. In some companies, the time frame can be shortened substantially. The schedule should be tailored to the company's need, work habits, travel schedules, local customs, and so forth, but enough time must be allowed to get assignments completed and reviewed.

It is important that the review process takes place in stages, and that the same people are not involved at every stage. This spreads the workload, saves management's time, and gets the appropriate people involved when they can make their best contribution. This involvement will provide "buy in" on the part of those people who must make the plan work. It also provides a multiechelon, multidepartmental check and balance system to improve the quality of the plan.

For this example, we will divide the planners into three groups: Group I consists of the CEO and his officers; Group II consists of key people who report to the officers; and Group III consists of technical experts, for example, an advertising agency key person or two, patent specialists, and/or lower echelon personnel with specific review assignments, for example, plant manager describing relationships with union.

Another key person, a coordinator should be a part of Group I, even though he or she may not be an officer. The coordinator's role is to coordinate the calendar, the people, the distribution (and sometimes the editing) of documents to be reviewed and to provide equipment and facilities for meetings. The coordinator should be selected carefully based on these suggested criteria:

- Objectivity
- No scars from prior confrontations in the company
- Diplomacy
- Good writer and/or editor
- Can make time available
- Can get things done through others
- Reports to a senior manager and can function with the CEO's authority

This person can be the head planner, head of administration, assistant to president, or an outside consultant. The person usually is not an operating officer.

The Strategic Planning Calendar

Building up, or back, from the close of the fiscal year and the annual budget approval date, Figure 8-3 gives an example of a strategic planning review that coordinates the planning calendar timetable with the budget cycle.

The purpose of the July 1 senior officer's meeting is to:

1. Review, discuss, and, if necessary, modify environment assessment, description, conclusions, and the resource analysis reports made by Group II. *These should be read before the meeting by Group I planners.*
2. Review and modify, if necessary, the mission and priority objectives required to accomplish the mission.
3. Review and modify, if necessary, the strategy and alternative strategies for activating the priority objectives.
4. Approve the strategic plan through the strategies sections. The strategies in this review process articulate briefly how the company will utilize its resources to accomplish each priority objective.

This is the key meeting. The time required for the meeting will be determined by the quality of the preparatory work, how thorough the Group I members have reviewed the material prepared by Group II, and the number and significance of the changes in the environment. Usually, the meeting will last two-to-five days. The preparation materials consist of the following provided by Group II.

1. Description of the changes in each category of the environment.
2. Conclusions drawn from these environmental changes and the SPOT analyses that support them.
3. Portfolio resource analysis changes the current strategic plan being reviewed, including an updated status report of the priority objectives and action program and projects.

All planning meetings for both Groups I and II should be held away from the company offices, even if just across the street or across town in a hotel. Generally, the farther away from phones and the normal work atmosphere, the better, and isolated facilities (away from distractions) are desirable. For this key meeting of senior management, the meeting must be off company premises.

Equipment for the meeting should consist of the usual meeting considerations, for example, charts, and so forth, plus, if possible, word-processing equipment and an operator

What	When	By Whom
PHASE I: The Environment		
Determine the method to be used to conduct the business environment assessment (described later) and make assignments	April 1	Group I
Conduct environmental assessment and draw any revised relevant conclusions regarding customers, consumers, competitors, and the company (see "how" section on p. 105) and distribute to Group I		
Write peer and interdepartmental consensus on environment assessment and conclusions	May 15	Group II
Resource analysis modifications	June 1	Group II with appropriate department heads
Phase III: The Action		
Approve Phase I and review the prior mission and purpose, priority objectives, and strategies at special strategic planning meeting	July 1	Group I sometimes with all or part of Group II attending for reference
Phase III: The Action		
Operational action plans, programs, projects. (one of these programs should address the strategic planning communication when, to whom, how it will be reinforced and tied to the budget for Phase IV integration	August 1	Department and division heads within their areas of expertise
Phase IV: The Integration Program	August	Coordinator
One of the Phase III programs		
Date Reference Only		
Submit budgets	October 1	
Budget approval	November 1	
Fiscal year ends	December 31	

Figure 8-3. Example of strategic planning review.

to make modifications in the plan as soon as the group makes such adjustments. The revised plan should then be disseminated to Group II and other appropriate personnel as soon as possible.

SUMMARY

1. Always keep in mind that the strategic plan review should focus on the company "Doing the right things, not just doing things right."
2. Strategic planning should be reviewed at least annually; in some companies, it is an on-going process.
3. The review should never take place in proximity to budget approval time.
4. A description of changes in the relevant business environment and the conclusions drawn from this description are the foundation for the review.
5. The planning review should be done sequentially by different people with different assignments at different times during the year.
6. The key documents are the current year's plan and the next year's planning calendar.
7. Thorough and well-organized preparation saves much time and improves immensely the productivity of the review.
8. A review process without a strong, committed, well-communicated strategic planning process reinforced by example from the CEO is not productive.
9. A good process, tailored to each company, is necessary for continuing strategic planning review success.

REFERENCE

1. L.B. Carroll, *Alice in Wonderland* (New York: Macmillan, 1966), p. 59

CHAPTER 9

Strategic Planning Processes in a Global Context

Narendra K. Sethi

Narendra K. Sethi is a professor of management at St. John's University in Jamaica, New York. He is the author of over twelve books and two hundred professional articles in administrative management and is a consultant to several large multinational firms in the areas of corporate planning and control. He has received several research awards, prizes, and citations for contributions made toward the advancement of knowledge in planning.

Contemporary research in multinational corporate strategy has focused on the development of international negotiated agreements, investment and divestment programs, organizational structure, locational decision, and technology transfer, to name the more pronounced research areas.[1] Although these issues are of extreme importance to long-range planning, there are other issues that also deserve studies on their role in the multinational corporation (MNC). The study of the process of planning in the global corporate matrix still remains a relatively unexplored area of inquiry.[2] It is indicative of inadequate administrative vision to believe that the only difference between the domestic planning process and the global planning process is one of locational consideration.[3] Many factors must be considered in multinational planning that need not be considered, or at least considered to a lesser degree, in domestic planning. Views of the various cultures toward business activities, philosophies of different political parties and how they relate to business, development of labor pools, and historical and ecological factors of each nation are only a few of the factors that differ on the multinational and domestic scenes. Global planning requires an approach to strategy and its development that is distinguishable from domestic corporate planning.[4]

This chapter reports on the results of a research study that illustrate major components of global planning in the context of modern administration. The study is based on data collected from a representative sample of MNCs who reported having a formal planning system in their total work organization. Both U.S. and foreign companies were included in the study.[5]

Adapted with permission from N. K. Sethi, "Strategic Planning System for Multinational Companies," *Long Range Planning Journal* **15**(1982):80-89.

ASSUMPTIONS

Following are the assumptions in the planning process of the selected group of MNCs.

1. Formal planning systems are emerging in the area of multinational corporate planning.[6]
2. Development of strategic planning parameters is still a function performed by the parent company's home office away from its operational locations.[7]
3. Tactical planning is delegated to branches and overseas locations.[8]
4. Integrated planning and control systems are emerging in home offices as well as branch locations.[9]
5. Environmental factors are recognized as being critical in MNC planning.[10]
6. Planning feedback systems are given top priority in MNC planning.[11]

Formal Planning Systems

Planning activity is a systematic, intellectual, and creative exploration of the future. In searching for the obvious strategic and comparative advantages that are the basic rationale for justified existence and growth of the MNCs, formality of the planning system is emerging in multinational administration. Planning groups develop planning premises, gather market intelligence, suggest resource allocation choices, and study industry trends in the regions served. Very often in large companies discussion rooms are set up and nicknamed the *War Room* by executives outside the planning department. The planning department is run by intellectuals who are more open minded than the other executives of the company who may suffer from managerial myopia. In the large MNC it is not rare to see a diversity of long-range planning systems, each addressing itself to a selected planning horizon and time frame.[12]

The number of formal planning departments in large MNCs has grown at an enormous rate in recent years due to the advantages of long-range as well as short-range planning. The prime merit of long-range and short-range planning does not lie in being able to out-guess others with regard to future events but rather in recognizing the human inability to forecast all tangible variants accurately.[13] The major benefits of using long-range and short-range plans are:

1. They aid in determining corporate goals and the policy procedures to be used to attain those goals. According to Terry, "Preciseness in goal identity is enhanced by long-range planning."[14] When setting goals for growth, account is taken of the precise growth sought, timing, risk involved, and people.
2. "Long-range and short-range planning help orient management to current conditions in that they supply a yardstick to measure achievement of goals to date. By comparing sought goals with present stature, areas of strength are revealed and areas of weakness requiring correction are disclosed."[15]
3. "An overall, integrated view-point is gained since long-range planning and short-range planning cut across functional lines. The problems that stand in the way of desired accomplishments are identified and brought into true focus."[16]
4. Strong managers are identified and weak managers are spotted by the formulation and implementation of planning, especially in long-range planning where foresight, creativity and genuine skill in planning is demonstrated.[17]
5. "Long-range planning and short-range planning provide pertinent answers to where, when, and how much financing is required for what purpose and what is the expected return. Such planning is inclusive in that it takes into account major matters of sales, production, manpower and research for projected activities, and relates them to the financial requirements."[18]
6. "The use of long-range planning and short-range planning brings attention to new techniques and developments either of a technological or of a managerial nature. Managers became aware of what is new whether it is taking place in marketing, engineering, production, or data processing."[19]

Strategic Planning

Strategic planning, which addresses itself to selection of corporate missions, goals, objectives, and subobjectives, remains an area of top managerial authority. It is developed in the home office at the top administrative level and uses both top-down and bottom-up planning procedures.[20] Strategic plans are routed from the home office to the MNC's different locations.

According to Dymsza,

> Strategic planning of the multinational company is far more complex than that for domestic business. Some of this complexity arises from the unique aspects of international business, which are:
>
> 1. The multinational company faces a multiplicity of political economic, legal, social and cultural environments as well as a differential rate of change in them.
> 2. There are complex interactions, which are difficult to analyze, between the multinational firm and such national environments because of national sovereignty, widely disparate economic and social conditions, and other factors.
> 3. Geographical distance, cultural and national differences, variations in business practices, and other differences make communications difficult between headquarters and the overseas organizations.
> 4. The degree of significant economic, marketing, and other information for planning purposes varies a great deal among countries in availability, depth and reliability. Furthermore, modern techniques for analyzing and developing data may not be highly developed.
> 5. Analysis of present and future competition may be more difficult to undertake in a number of countries because of differences in industrial structure and business practices.
> 6. The multinational company is confronted not only with different national environments but also with regional organizations.[21]
>
> All these differences make planning of the multinational corporation far more difficult than domestic planning. Even the most carefully developed plans can become obsolete very rapidly in some areas of the world. At the same time, the greater complexity makes planning even more essential for the profitability, growth, and its survival of the multinational company, its divisions and its affiliates in various countries.[22]

Steiner and Cannon state that "Strategic planning is a systematic way of dealing with the opportunities, risks, and problems that are important to a multinational company's future over an extended period of time."[23] Through strategic planning, the firm's top executives try to shape the corporation's future. The MNC uses strategic planning to (1) coordinate and integrate its future directions, objectives, and policies around the world; (2) anticipate and better prepare for change and to innovate programs to deal with developments around the world; (3) get managements of overseas affiliates more actively involved in setting goals and developing means to more effectively utilize the enterprise's total resources in a decentralized organization; and (4) establish basic international strategies for long-term and short-term decisions about major issues that will determine the firm's profitability, growth, and development in the future.[24]

Tactical Planning

Tactical planning, which addresses itself to operational implementation of strategic plans, is delegated to the individual branch locations of the MNCs. There is, however, a pattern of regional control, administrative feedback, and contingent plans that do in fact curb the authority of overseas MNCs.[25]

Tactical planning is more detailed than strategic planning and operates for short periods of time—about six months to two years. According to Dymsza,

> It is undertaken by many layers of management. It deals with more specific allocation of a company's resources and usually involves highly detailed annual budgets for overseas affiliates, divisions, and the entire enterprise. It provides a basis for the company's operations and day to day decision making. It deals more with current and immediate problems; it is more operational and more fully involves all levels of management. It calls for elaborate accounting data, for projections of operating budgets and cash requirements for each division and each overseas subsidiary.[26]

Integrated Planning and Control Systems

Because of the duality of reporting at strategical and tactical levels, MNCs are developing integrated planning-control modules. These modules assist the home offices in their selection of strategic choices as well as the overseas locations in their operational considerations.

Dymsza says that "The degree to which the multinational company decentralizes management and establishes profit centers is of major importance in control."[27] The problems that hamper setting up control systems in MNCs is a result of (1) various government regulations; (2) differences in the environments of each country; (3) differences in business practices; (4) differences in accounting procedures; and (5) difficulties in communications.[28]

Each of these factors can cause problems in developing an integrated planning and control system. Companies in the multinational scene, however, have been able to circumvent these problems by concentrating on different aspects of each factor that cause a problem.

Environmental Factors

Environmental factors are recognized as affecting the total planning process in both strategic and tactical plans. Although there was no uniformity in the listing of environmental considerations, it was suggested that several alternative plans be proposed, each using or deploying a new or different planning scenario based on several alternative environmental options and decision choices.[29]

Developing different planning scenarios for each environmental consideration is both costly and burdensome. It is costly because each MNC must deal with multiple political and economic systems, each with its attendant controls and risks, advantages and disadvantages, opportunities and dangers.[30] As the company expands into new countries, more variables arise that must be incorporated into plans, thus placing a heavier burden on management. Having numerous planning scenarios can also be a burdensome task because it requires too many reports and thus an overflow of paperwork, which can become burdensome to both the management of affiliates as well as headquarters management.

Multinational companies must operate within an environment that has numerous components. These components are:[31]

I. Government, laws, regulations and policies of home country (United States, for example).
  A. Monetary and fiscal policies and their effect on price trends, interest rates, economic growth, and stability
  B. Balance-of-payment policies
    1. Mandatory controls on direct investment
    2. Interest equalization tax and other policies
  C. Commercial policies, especially tariffs, quantitative import restrictions, and voluntary import controls
  D. Export controls and other restrictions on trade with Eastern European and other Communist nations
  E. Tax policies and their impact on overseas business
  F. Antitrust regulations and their administration and their impact on international business
  G. Investment guarantees, investment surveys, and other programs to encourage private investments in less-developed countries
  H. Export-import, and governmental export expansion programs
  I. Other changes in government policy that affect international business

II. Key political and legal parameters in foreign countries and their projection
  A. Type of political and economic system, political philosophy, national ideology
  B. Major political parties, their philosophies, and their policies
  C. Stability of the government
    1. Changes in political parties
    2. Changes in governments
  D. Assessment of nationalism and its possible impact on political environment and legislation

E. Assessment of political vulnerability
    1. Possibilities of expropriation
    2. Unfavorable and discriminatory national legislation and tax laws
    3. Labor laws and problems
F. Favorable political aspects
    1. Tax and other concessions to encourage foreign investments
    2. Credit and other guarantees
G. Differences in legal system and commercial law
H. Jurisdiction in legal disputes
I. Antitrust laws and rules of competition
J. Arbitration clauses and their enforcement
K. Protection of patents, trademarks, brand names, and other industrial property rights

III. Key economic parameters and their projection
A. Population and its distribution by age groups, density, annual percentage increase, percentage of working age, percentage of total in agriculture, percentage in urban centers
B. Level of economic development and industrialization
C. Gross national product, gross domestic product, or national income in real terms and also on per capita basis in recent years and projections over future planning period
D. Distribution of personal income
E. Measures of price stability and inflation, wholesale price index, consumer price index, other price indexes
F. Supply of labor, wage rates
G. Balance of payments equilibrium or disequilibrium, level of international monetary reserves, and balance-of-payments policies
H. Trends in exchange rates, currency stability, evaluation of possibility of depreciation of currency
I. Tariffs, quantitative restrictions, export controls, border taxes, exchange controls, state trading, and other entry barriers to foreign trade
J. Monetary, fiscal, and tax policies
K. Exchange controls and other restrictions on capital movements, repatriation of capital, and remission of earnings

IV. Business system and structure
A. Prevailing business philosophy: mixed capitalism, planned economy, state socialism
B. Major types of industry and economic activities
C. Numbers, size, and types of firms, including legal forms of business
D. Organization: proprietorships, partnerships, limited companies, corporations, cooperatives, state enterprises
E. Local ownership patterns: public and privately held corporations, family-owned enterprises
F. Domestic and foreign patterns of ownership in major industries
G. Business managers available: their education, training, experience, career patterns, attitudes, and reputations
H. Business associations and chambers of commerce and their influence
I. Business codes, both formal and informal
J. Marketing institutions: distributors, agents, wholesalers, retailers, advertising agencies, advertising media, marketing research and other consultants
K. Financial and other business institutions: commercial and investment banks, other financial institutions, capital markets, money markets, foreign exchange dealers, insurance firms, engineering companies
L. Managerial processes and practices with respect to planning, administration, operations, accounting, budgeting, control

V. Social and cultural parameters and their projections
A. Literacy and educational levels
B. Business, economic, technical, and other specialized education available

C. Language and cultural characteristics
D. Class structure and mobility
E. Religious, racial, and national characteristics
F. Degree of urbanization and rural-urban shifts
G. Strength of nationalistic sentiment
H. Rate of social change
I. Impact of nationalism on social and institutional change

Feedback Systems

The logistics of MNCs places a heavy premium on the effective development of feedback systems in the formal planning process. Information management appears as a major new consideration for the MNCs[32] and in developing an effective feedback system problems are encountered in communications as a result of (1) distance; (2) time; (3) cultural differences; (4) social differences; (5) nationalistic differences; and (6) managerial variations.[33]

To resolve these problems in communications, MNCs have incorporated elaborate feedback systems into their planning process to assist in reporting from affiliate managers to headquarter managers.

Figure 9-1 illustrates the integration of the aforementioned assumptions in the MNC planning process.

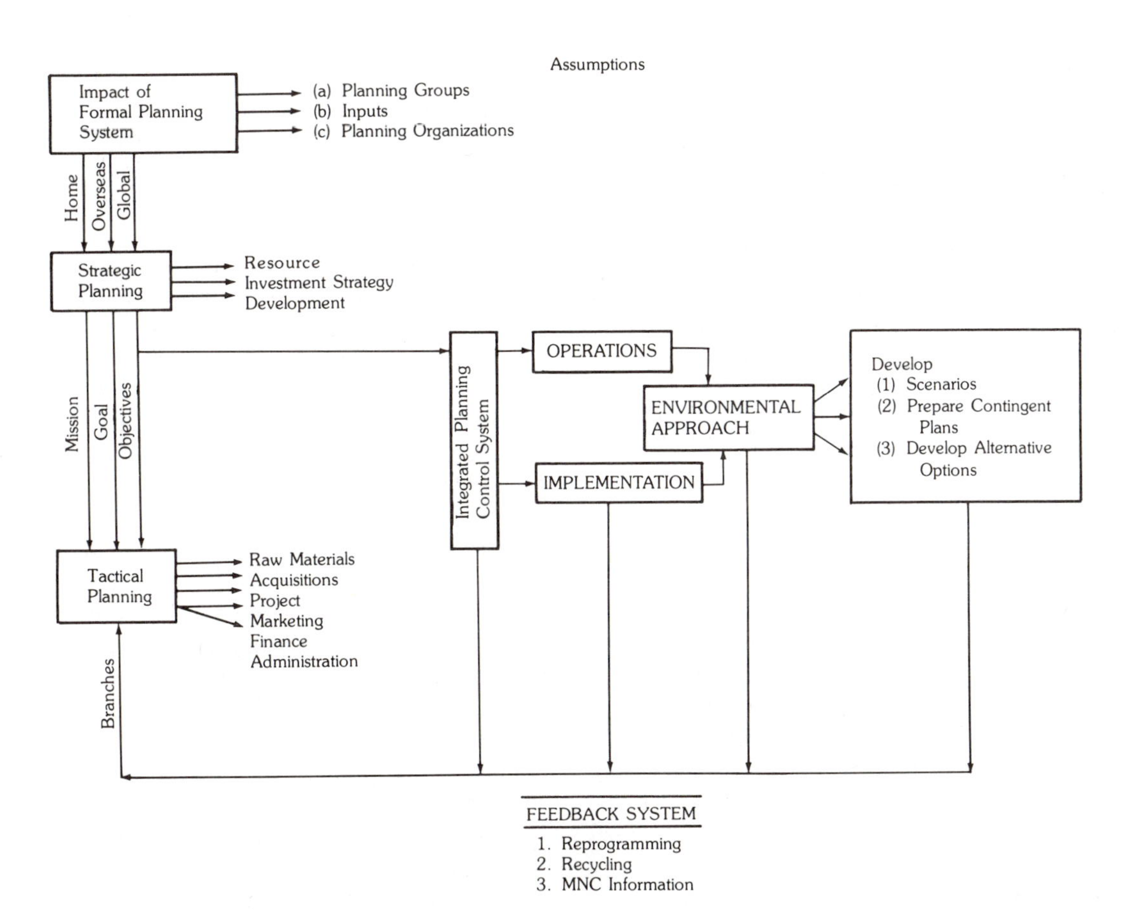

Figure 9-1. MNC planning process (from N. K. Sethi, "Strategic Planning System for Multinational Companies," *Long Range Planning Journal,* **15,** 1982:85. Copyright © 1982 by Pergamon Press).

OBJECTIVES OF PLANNING

Based on the assumptions noted above, responding multinational planners noted the following objectives characterizing their *total* planning process:

1. Continued existence in the regions served by the MNC.[34]
2. Anticipated response and development of action strategy for any environmental changes or modifications.[35]
3. Program of administrative action in such functional areas as investment, divestment, expansion, negotiations, and international personnel.[36]
4. Reassessment of the MNC's contribution to the total profitability and productivity of the enterprise.[37]

Continued Existence

One primary objective of MNC planning is the reassurance of survival in the regions served by the corporate enterprise. It has assumed major importance due to current situational factors developing in the sociopolitical and diplomatic context. Both before multinational commitment as well as during the existence of administrative work, search for resource optimization as a basis of continued business existence becomes an important planning objective.

Action Strategy

The development of strategy and counterstrategy to offset any anticipated changes in the environment is a major planning objective for the MNCs. Often the planners are required to include a nonmanagerial dimension in their planning effort: one that addresses itself to the deployment of strategy in such areas as governmental changes, shifts in public opinion, revival of social responsibility thinking, and corporate morality.

Functional Areas

Strategic planning objectives are transformed into a series of highly developed operational or functional objectives. These include such important areas of functional planning as (1) logistics; (2) marketing; (3) personnel; (4) investment; (5) production; (6) finance; and (7) management information systems (MIS). Each of these functional planning objectives receives attention at both home and overseas locations.

Total Profitability

Planning groups are constantly being asked to reassess the overall contributions made by the overseas branches in the total profit perspective of the MNC. Reprogramming of plans and recycling of planning efforts require sustained profitability of the MNC in as many overseas operations as is decided on by the home office groups.

Figure 9-2 illustrates a breakdown of these objectives of the total planning process into both corporate-level (home office) and divisional-level (overseas locations) areas of responsibility. As illustrated, home office objectives comprise a corporate strategic plan, including a corporate development plan, a divestment plan, a diversification plan, and a acquisitions and merger plan.

Overseas divisional objectives comprise a corporate operations and development plan, including a raw material plan, an acquisition plan, a project plan, a research and development growth plan, a technical support plan, a production plan, a marketing and sales plan, a financial plan,and an administrative plan.[38]

The objectives of the MNC are a very important part of the planning process. Objectives should first be set on the company level and then on departmental levels. The previous four objectives were company objectives, whereas the following are departmental objectives.[39]

1. Profitability: level of profits; return on investment, assets, and sales; growth of profits per year; growth of earnings per share.
2. Marketing: level of sales; market penetration—share of market; product diversification and mix; marketing of new product; improving distribution and promotion; improving marketing services.
3. Production: shifting to production in various national markets based on production and distribution economies and competitive conditions; developing production efficiencies in use of labor, materials, techniques, and other resources; manufacturing of new and improved products; cost reduction programs; quality control programs.
4. Finance: financing part of funds required for expansion, for example, 40 percent, from retained earnings; developing sources of local funds for expansion and acquisitions; protection of company's funds from possible currency depreciation and inflation. Specific aims with respect to transfer of part of profits to headquarters include optimum or well-balanced capital structure and minimizing taxes in various countries or globally.
5. Technology: developing new products; adapting products to specific national markets; improving techniques of production; product improvement.
6. Personnel: developing local managers, technical and professional personnel, and skilled labor force; aim of having local managers take over top executive positions of overseas affiliates; improving labor relations.
7. Acquisition: goal of acquiring companies overseas characterized by rapid growth and related to operations of company in selected countries; diversification of business by regions, countries, industries, or product lines.

PROCESS

Based on the assumptions and objectives described above, the MNC planning process consists of the following identified steps:

Step 1: Establish (or reiterate) overall enterprise objectives.
Step 2: Obtain planning inputs from MNC strategists representing their locational objectives.
Step 3: Determine present and potential environmental constraints affecting the MNC's overseas activity.
Step 4: Determine the area of the MNC's strengths and weaknesses in terms of the environment affecting it.
Step 5: Prepare working strategic alternatives for implementation at individual locations of the MNC.
Step 6: Choose the optimum strategy by linking it at both the parent and overseas levels.
Step 7: Prepare a detailed action plan to provide for the formulation, implementation, and reprogramming of functional plans that are derived from the chosen strategy.
Step 8: Develop policies, procedures, and rules for the fuller deployment of the functional plans.
Step 9: Prepare and retain an inventory of several contingent plans based on alternative environmental scenarios characterizing the MNC's overseas patterns.
Step 10: Develop an MIS type of feedback for recycling and readjusting both strategic and tactical plans.
Step 11: Develop an evaluation program for the formal planning process.

Figure 9-3 illustrates the typical flow chart and time phase for the planning process outlined above.

CONFLICT

Planning groups noted the following sources of conflict in the MNCs planning process:

1. Global plans may differ from regional or an individual country's plans.
2. The viewpoint of the government and the individual company may differ.
3. The viewpoint of the parent firm and the individual country's operations may differ.
4. Third-world countries may pose a threat to planning stability.
5. Considerations of international liquidity and exchange may force premature termination of plans.
6. Organizational restrictions may be imposed by host governments.

It was suggested that the following issues be incorporated in the total planning process to minimize the above-mentioned areas of potential danger:

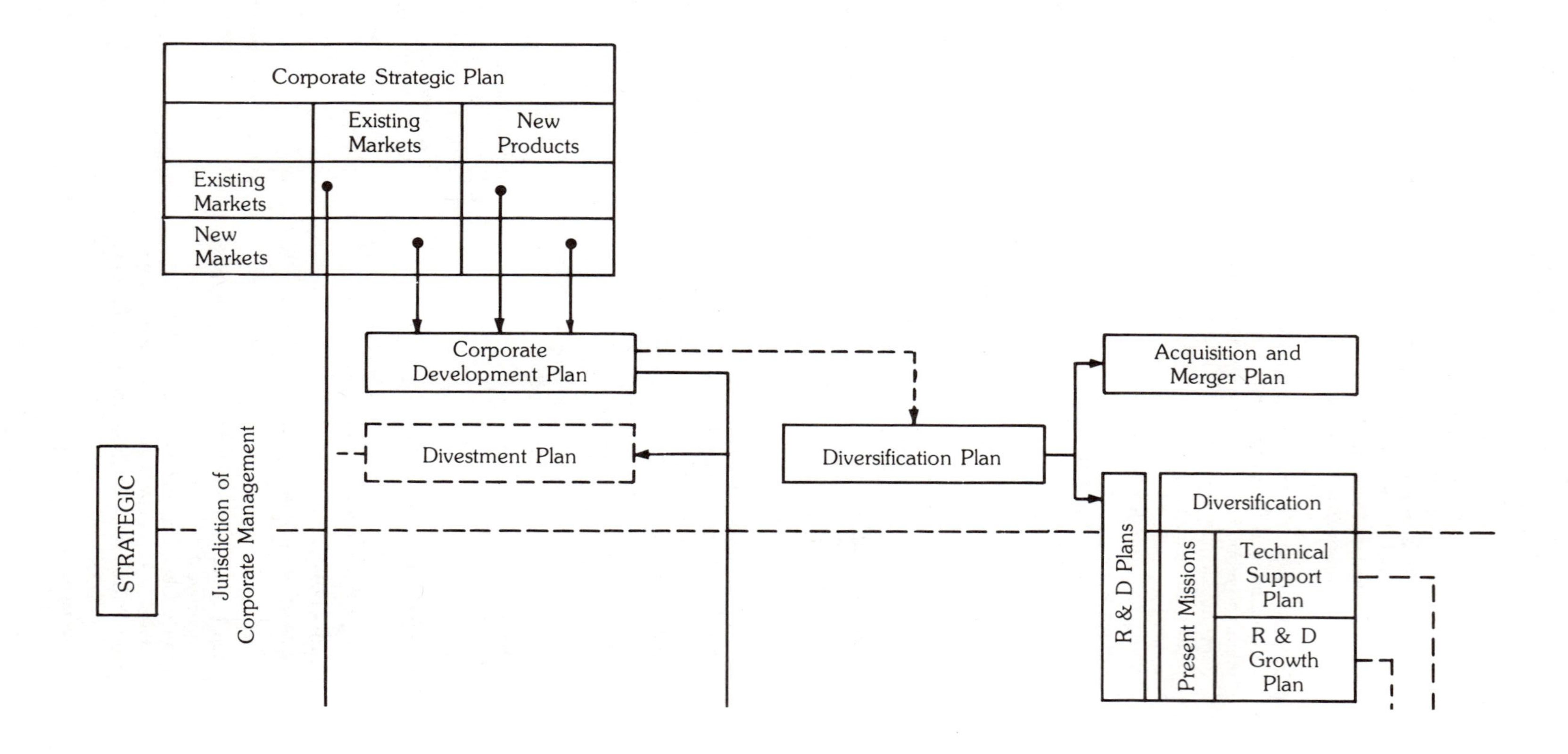
Corporate Strategic Plan
Existing Markets
New Products
Existing Markets
New Markets
Corporate Development Plan
Divestment Plan
Diversification Plan
Acquisition and Merger Plan
Diversification
R & D Plans
Present Missions
Technical Support Plan
R & D Growth Plan
STRATEGIC
Jurisdiction of Corporate Management

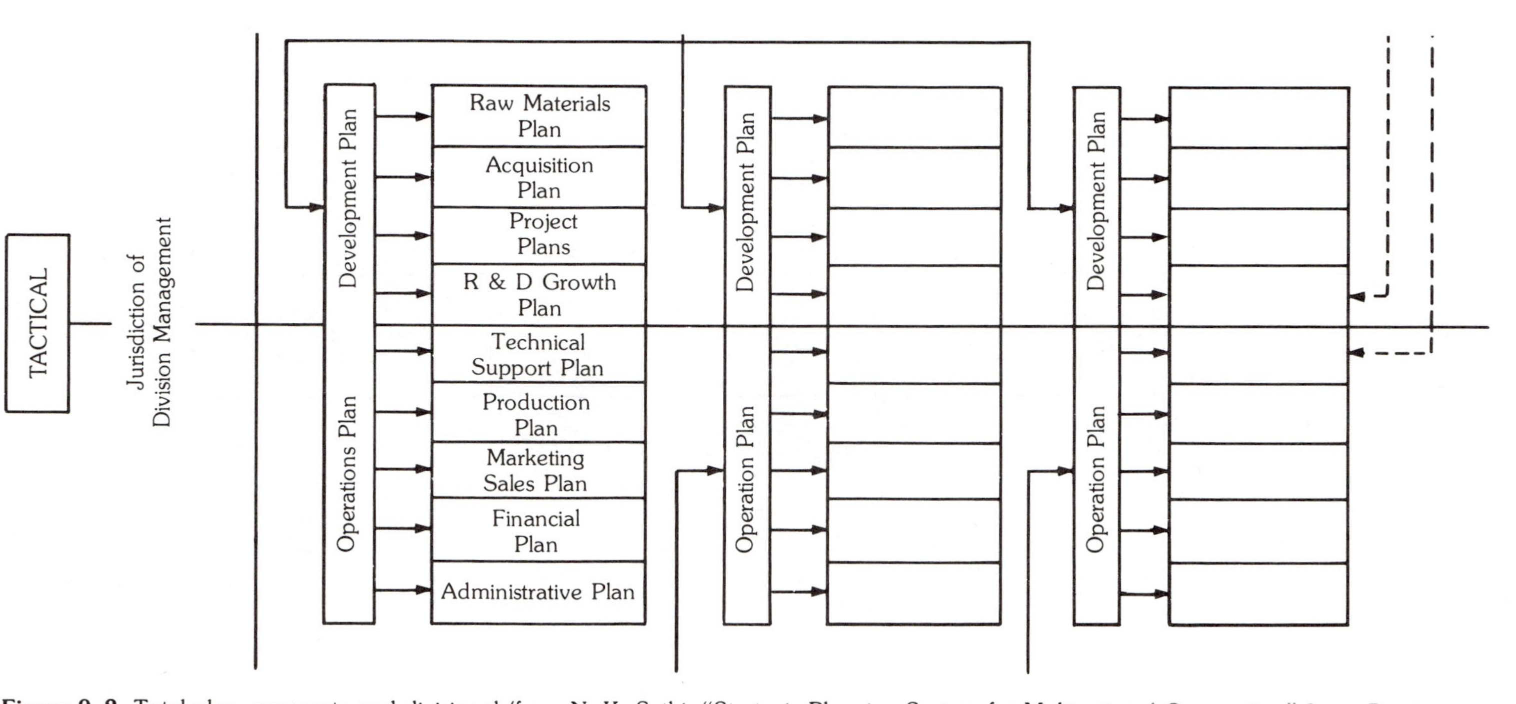

Figure 9-2. Total plan, corporate and divisional (from N. K. Sethi, "Strategic Planning System for Multinational Companies," *Long Range Planning Journal* **15,** 1982:86. Copyright © 1982 by Pergamon Press).

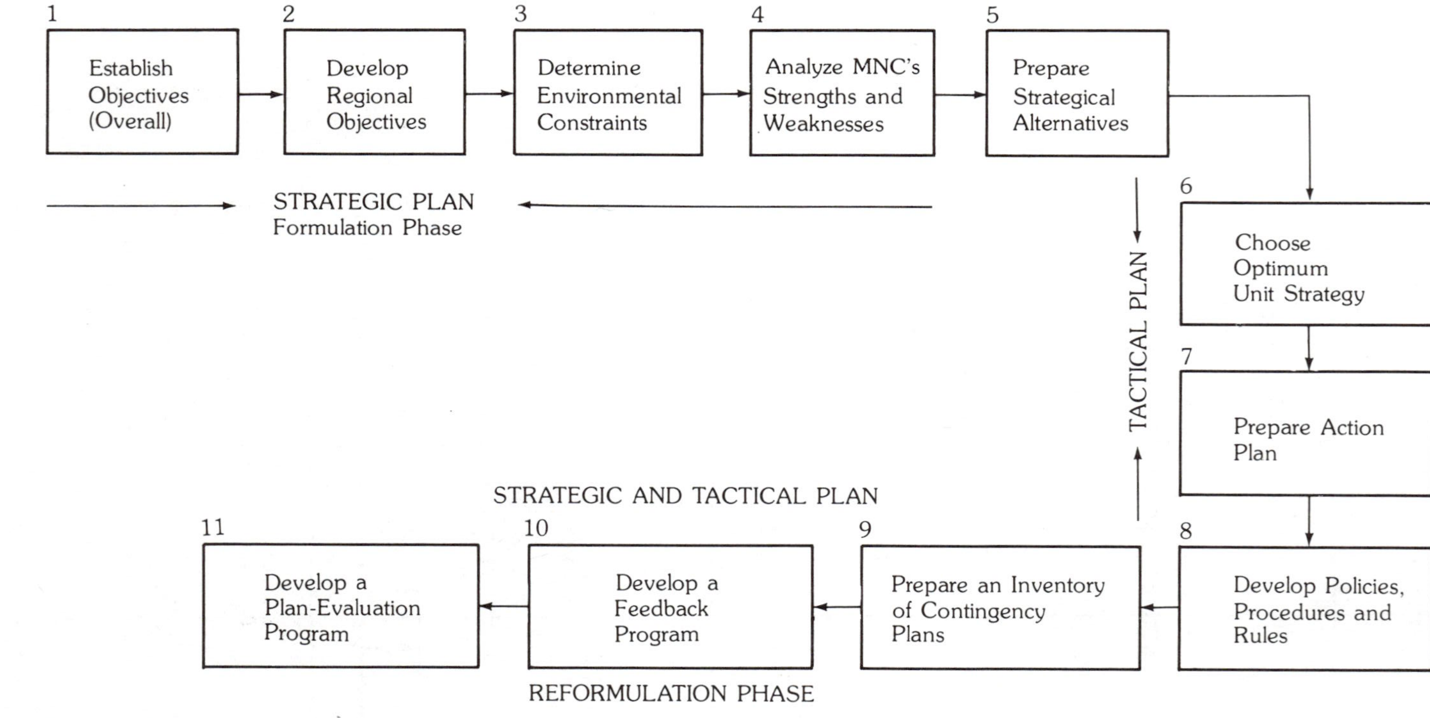

Figure 9-3. Flow chart of MNC plan (from N. K. Sethi, "Strategic Planning System for Multinational Companies," *Long Range Planning Journal* **15,** 1982:88. Copyright © 1982 by Pergamon Press).

1. The MNC should make several concurrent and congruent plans that will resolve the different approach requirements between global and regional plans.
2. Regional planning should be preceded by a vigorous and in-depth research-cum-negotiation phase so as to discount any unforeseen governmental outlook shifts.
3. A sharply defined planning horizon should be established covering the territory included in parent office domain and locational domain.
4. Emphasis should be on tactical as opposed to strategic planning framework as far as third-world operations are concerned.
5. Plans should be exposed to techniques such as risk analysis and/or sensitivity analysis prior to their multinational implementation.
6. Tactical planning, particularly its implementation aspects, may be regionalized.
7. Internationalization should not be a forced growth strategy. It should also not be a modular strategy happening in a nonsystematic manner.[40]

FUTURITY

MNC planning will become a blend of both administrative and institutional types of plans because it cuts across corporate lines and interacts with organized groups such as national and foreign governments, parent and home office plus overseas offices, and international agencies. Formalized planning models are being developed and are attaining widespread use by MNCs. As noted in this chapter, the planning process in the surveyed MNCs follows a path of strategy based on formal planning systems, fluid demarcation of authority between the parent organization's home office and the global offices, critical regard for environmental variables, and the development of integrated planning feedback methods. What needs to be studied in depth is the empirical data about actual planning case-studies reflecting MNC practices in the global context.

NOTES AND REFERENCES

1. A. Kapoor, *Planning for International Business Negotiations* (Cambridge, Mass.: Ballinger, 1975); L. Vignola, *Strategic Divestments* (New York: Amacom, 1974); Y. Zeira and E. Harari, "Genuine Multinational Staffing Policy—Expectations and Realities," *Academy of Management Journal* **20**(1977):327-333; B. M. Bass, D. W. McGregor, and J. L. Walters, "Selecting Foreign Plant Sites—Economic, Social and Political Considerations," *Academy of Management Journal* **20**(1977):535-546.
2. A few notable exceptions, although dated, are H. Schollhammer, "Long-range Planning in Multinational Firms," *Columbia Journal of World Business* **6**(1971); P. Lorange, "Formal Planning in Multinational Corporations," *Columbia Journal of World Business* **8**(1973); J. S. Schwendinan, *Strategic and Long-Range Planning for the Multinational Corporation* (New York: Praeger, 1973).
3. D. Hussey, *Corporate Planning—Theory and Practice* (New York: Pergamon Press, 1974).
4. Compare C. K. Prahalad, "Strategic Choices in Diversified Multinational Corporations," *Harvard Business Review* **54**(1976):67-78.
5. The method of research was a combination of questionnaire and interviews. The study was carried out during the fall of 1977.
6. Lorange, "Formal Planning in Multinational Corporations."
7. C. R. Christensen, K. R. Andrews, and J. L. Bower, *Business Policy* (Homewood, Ill.: Richard D. Irwin, 1978), 131-133.
8. J. Argenti, *Systematic Corporate Planning* (London: Thomas Nelson, 1976), 244-247.
9. Little research documentation is available on multinational corporate control from the planning perspective as opposed to functional or financial perspective. The respondents were enthusiastic about the future role of *integrated control* in their plans.
10. The literature on environmental factors in management is vast. The works of Farmer and Richman, Boddewyn, Negandhi, Prasod, Estofan, and Kapoor are established sources. For bibliography see N. K. Sethi, *Social Sciences in Management—An Environmental View* (The Hague: Martinus Nijhoff, 1972).
11. As in the field of multinational corporate control, literature on multinational planning feedback is virtually nonexistent. It suggests an area of future research inquiry.
12. Several respondents pointed out that there often are opposing and differing plans in the same corporate organization, each plan addressed to a particular geographical or divisional entity. A related source is R. M. Kinnunen, "Hypotheses Related to strategy Formulation in Large Divisionalized Companies," *Academy of Management Review* **1**(1976):7-14.

13. G. R. Terry, *Principles of Management,* 6th ed. (Homewood, Ill.: Richard D. Irwin, 1972), 254.
14. Ibid., 255.
15. Ibid.
16. Ibid.
17. Ibid.
18. Ibid., 256.
19. Ibid.
20. On this see R. C. Shirley, M. H. Peters, and A. I. El-Ansary, *Strategy and Policy Formulation—A Multifunctional Orientation* (New York: Wiley, 1976); T. J. McNichols, *Policymaking and Executive Action* (New York: McGraw-Hill, 1977); and W. F. Glueck, *Business Policy Strategy Formation and Management Action* (New York: McGraw-Hill, 1976).
21. W. A. Dymsza, *Multinational Business Strategy* (New York: McGraw-Hill, 1972), 50.
22. Ibid., 51.
23. G. A. Steiner and W. M. Cannon, eds., *Multinational Corporate Planning* (New York: Macmillan, 1966), 9.
24. Dymsza, *Multinational Business Strategy,* 57.
25. Several respondents expressed concern about this issue.
26. Dymsza, *Multinational Business Strategy,* 53.
27. Ibid., 222.
28. Ibid., 223.
29. R. E. Linneman and J. D. Kennell, "Shirt Sleeve Approach to Long Range Planning," *Harvard Business Review* **55**(1977):141-148. The authors develop a technique called "Multiple Scenario Analysis" containing ten steps.
30. Dymsza, *Multinational Business Strategy,* 221.
31. Ibid., 83-85.
32. Compare E. R. McLean and J. V. Soden, *Strategic Planning for MIS* (New York: Wiley, 1977).
33. Dymsza, *Multinational Business Strategy,* 223.
34. The recent problems of survival faced by such MNC giants as IBM, Coca Cola, and Exxon among others in India and in other parts of the world were repeatedly reported by the survey respondents.
35. Terry, *Principles of Management;* M. Z. Brooke and H. L. Rummers, *The Strategy of Multinational Enterprise* (New York: Elsevier, 1970); Dymsza, *Multinational Business Strategy.*
36. Prahalad, "Strategic Choices in Diversified Multinational Corporations."
37. See particularly F. T. Paine and W. Naumes, *Organizational Strategy and Policy* (Philadelphia: Saunders, 1978), 101-102.
38. The 11-step planning process developed here is conceptionalized on the basis of the assumptions, objectives, effective practices, and hypotheses mentioned by the respondents. To make it viable in the context of MNC action, the model has been further modified to incorporate current research orientation and thinking in the domestic planning area as well.
39. Dymsza, *Multinational Business Strategy,* 98-99.
40. Christensen et al., *Business Policy,* 131-133; Hussey, *Corporate Planning,* 148-150; Argenti, *Systematic Corporate Planning,* 246-247.

CHAPTER 10

Strategic Management Processes for Corporate Social Policy

Archie B. Carroll and John Hall

Archie B. Carroll is a professor of management at the University of Georgia where he has served since 1972. Dr. Carroll has published 8 books and 50 articles. His articles have appeared in publications such as *Sloan Management Review, Journal of Business Strategy, Public Affairs Review, California Management Review, Academy of Management Review, Academy of Management Journal.* He is a contributor to the recently published *Handbook for Professional Managers* (McGraw-Hill, 1985) and author of *The Social Responsibility of Management* (Science Research Associates, 1984).

John Hall is a former vice-president for a savings and loan association where he served for five years. He also served as an officer in the U.S. Navy Supply Corps. Currently he is a doctoral student in strategic management at the University of Georgia. He also teaches business policy on the staff of the Management Department.

The purpose of a business is to earn a profit. This maxim has been driven home for over two hundred years—from Adam Smith to Milton Friedman. It remains a fact of life: a business that does not earn a profit fails. However, the path to profit is no longer as straight-forward as it once was. Consider the following example:

One-third of the more than 350 U.S. companies conducting business in South Africa seek the approval of Leon Sullivan, a black Baptist minister from Philadelphia. In 1977, Sullivan promulgated a voluntary code of conduct for companies operating in South Africa and now supervises a complex system for grading compliance with his code. The companies that agree to follow the "Sullivan Principles" pay up to $7,000 to support the administration of the grading system and incur significant expenses in attempting to implement the six principles outlined in the code: from 1977 to 1984, the companies spent over $78 million on schools, housing, and other social programs in South Africa.[1]

From a strict economic view, the firms voluntarily following the code are not behaving rationally. However, these firms are responding to external issues, voluntarily participating in Sullivan's program to help head off legislative action, stockholder resolutions, and other public relations headaches associated with firms operating in the land of apartheid.

Examining and interacting with issues originating outside the firm has become commonplace in corporate America. The causes of this change in orientation, from a narrow internal view of the firm and its responsibilities to a macro view of the firm operating in the world at large, have been attributed to the increasing complexity of organizations, accelerating changes in social and political views of the firm's responsibilities, and the increasing interdependence of all organizations and societies in today's era of instant communications.

One theory is that the way firms view themselves has been in transition from a *production view* to a *managerial view* to a *world view*.[2] Not too many years ago, the majority of firms were run by owner-managers operating relatively simple, single-product businesses. The focus of these firms—both "Mom and Pop" stores and larger firms—was on the groups with the greatest impact: suppliers and customers. If the firm could acquire raw materials and produce a better mousetrap, the world would indeed beat a path to its door.

A number of factors caused a shift in orientation from the production view to the managerial view of the firm. New technologies and production processes made larger firms more economical; larger firms required amounts of capital beyond the capacity of most owner-managers and the addition of more outside employees. The eventual separation of ownership from management created a new group with which the firm had to deal. The managerial view of the firm encompassed the firm itself, customers, suppliers, employees, and owners.

The transition from a managerial view to a world view of the firm has occurred due to changes in the key players in the managerial view and to new players outside the managerial view model having an impact on the firm. The firm's relationship with owners has changed due to the rise of corporate takeovers and shareholder activism; terms such as *greenmail* and *poison bullets* have been coined to describe emerging concerns in this area. OPEC is the most frequently cited example of changes between the firm and its suppliers. Consumer advocates (such as Nader's Raiders), environmentalists (Friends of the Earth, Sierra Club), special interest groups (from narrowly focused groups such as Mothers Against Drunk Driving to more broadly focused ones such as The Moral Majority), increased government actions, and media coverage of businesses have combined to strain the ability of firms to operate within the framework of the managerial view.

Our purpose here is to discuss the extent to which external issues—those issues arising outside the managerial view of the firm—have become important to businesses, to review some of the processes for identifying external issues, to discuss why the firm must have a corporate social policy to identify and cope with external issues, and to provide a framework for integrating corporate social policy into the strategic planning process of the firm.

THE TURBULENT ENVIRONMENT

We live in a world of constant change. Some, such as Alvin Toffler with his "Third Wave," view the change as symptomatic of transition to a new era; others, such as Peter Drucker with his "Age of Discontinuity," see this period of change as a new era in itself; John Naisbett sees ten new directions ("Megatrends") transforming our lives. Business must cope with a multitude of changes in its external, or social, environment, and these changes are generally identified as originating in four major areas: society, government, technology, and the economy.

Society

Increasing affluence and higher education levels have created a revolution of rising expectations in our society. Consumer rights groups have sprung up, appointing themselves guardians of the public's safety and demanding better, safer, and less costly products. Environmentalists, concerned with both public health and safety per se and the broader issue of preservation of the environment, have had an impact on many industries: in agriculture in the debate over pesticides, with electric utilities over nuclear plants, in the steel and automobile industries on clean air standards.

The media has played a major role in enhancing the awareness of societal issues. In addition to the widely shared nightmare of arriving at work and finding the "60 Minutes" crew on your doorstep, the media help make societal issues both visible and immediate, as illustrated in this example:

> One week after the Union Carbide accident in Bhopal killed more than 2,000 people, the city of Akron, Ohio, passed a right-to-know law. The director of the Toxic Action Project for the Ohio Public Interest Campaign stated, "Bhopal made it difficult for anyone to argue against people needing to know [what substances they are being exposed to in the workplace].[3]

Activist cries for legislation requiring manufacturers to disclose the content and possible side effects of materials handled on the job and pollutants emitted have occurred since the early 1970s; the Bhopal incident, with graphic images telecast worldwide the day of the disaster, provided impetus to a movement no chemical manufacturer could ignore.

Societal values and trends that management must monitor are too numerous to catalog. For example, we are a maturing society. There are now more people 65 or over than teen-agers in the population. And, by 1990 the older population is expected to surpass 31 million, whereas the teen-age population will shrink to 23 million. The graying of society will pose knotty problems in the workplace and in consumer markets. The lure of the sunbelt is also an issue companies will have to watch closely as a dramatic population shift occurs. The rise of minorities and the changing role of women are critical factors, too. Minorities and women are gaining power and influence as their numbers grow.[4] In addition to shifts in demographics and values, John Naisbitt suggests that we are transforming from an industrial society to an information society, from a short-term perspective to a long-term perspective, from centralization to decentralization, from institutional help to self-help, from representative democracy to participatory democracy, and from hierarchical structures to networking.[5]

Government

For the most part, government's impact on business has been a result of pressure from society at large. Social concern over civil rights led to equal opportunity and affirmative action; consumer and environmental groups' actions have resulted in environmental and product-related legislation. Deregulation, regulatory reform, tax code alterations, government waste cutting, and industrial policy are current issues causing special uncertainty in the business-government relationship today. In the last decade deregulation has occurred in many areas, including finance, telecommunications, and transportation. Regulatory reform and tax code revisions are a constant agenda item in the 1980s. Statutes being examined closely include environmental regulation, job health and safety, surface transportation, energy prices, exports, banking, and marketing. Government waste-reduction programs not only cause problems for government departments affected but special interest groups, some of which are business-related, too. The President's Grace Commission in 1983 recommended $60 billion a year in cost cutting. Industrial policy continues as an important issue as the federal government strives to remedy declining U.S. competitiveness in world markets.

Complicating the difficulty in tracking and coping with government's actions are state and local laws. These are sometimes more stringent than the more easily forecasted federal laws, such as in the Akron right-to-know law or the automobile antipollution legislation in California. Businesses must track legislation at the federal level, at the state and local levels in the communities in which they operate, and at the state and local levels in bellwether areas (for example, Delaware on banking laws). Increasingly, state and local regulators are rushing in where Washington no longer treads.

Technology

Advances in technology can radically affect a business and are among the most difficult to forecast. Even after a technological advance has been announced, many fail to estimate the impact of the advance (for example, the development of semiconductors and their impact on consumer electronics and data processing equipment). One approach to thinking about technological change is to ask what-if and what-then questions about technological advances; for example, what if a safe alternative to petroleum-based energy becomes widely available? What effect will this have on your markets?[6]

Advances in computing seem to overshadow other technological changes insofar as business is concerned. As *U.S. News & World Report* put it, "It's not often that the engine of an economic and social revolution can be held inside the human fist."[7] Indeed, the declining price and expanded memory of today's computers are combining to put these devices into the hands of millions of individuals and businesses. They are dramatically changing business operations and affecting the workplace in the form of information collection, analysis and dissemination, product design, robotics, and control and flow of production lines.[8]

The Economy

Economic indicators were once fairly stable, or at least fairly predictable. This is no longer the case. The double-digit inflation of the 1970s changed businesses' and consumers' views on many economic issues. Pertinent indicators and trends that now must be monitored carefully include interest rates, growth of gross national product (GNP), the economic recovery, the interaction of supply and demand factors, consumer confidence and attitudes, labor force participation rates, fiscal policy, monetary policy, labor's strategy, capital spending, and, of course, the international sector. Though our concern in this chapter is not primarily with the economic aspects of strategic management, this sector is in much more flux than it once was and this must be acknowledged.

The Interaction of Issues: Business Ecology

Issues interact among society, government, technology and the economy, making it difficult to identify the source or true nature of an issue. For example, the conventional view is that pollution standards for automobiles were a social issue, became a matter for government action, required technological changes to be implemented, and had economic consequences (due to increased prices) for the automobile and related industries. Two alternative views are that automobile pollution was originally an economic issue, with the costs of installing antipollution devices outweighing the benefits, or that it was a technological issue, with the technology to significantly reduce automobile pollution unavailable. The benefit of identifying causes and effects among these forces is that it permits a longer lead time in formulating response to change.

Figure 10-1 illustrates the position of the firm in terms of its relationship to both traditional managerial concerns and social issues. The firm maintains an interactive relationship with owners, customers, suppliers, and employees. Social issues—societal, governmental, technological, and economic—can affect the firm both directly and indirectly. An example of an indirect effect would be a change in the tax rates for personal capital gains, which would not affect the firm's taxes but could impact on the ownership group, which would in turn affect the firm; an example of a direct effect would be a change in corporate tax rates.

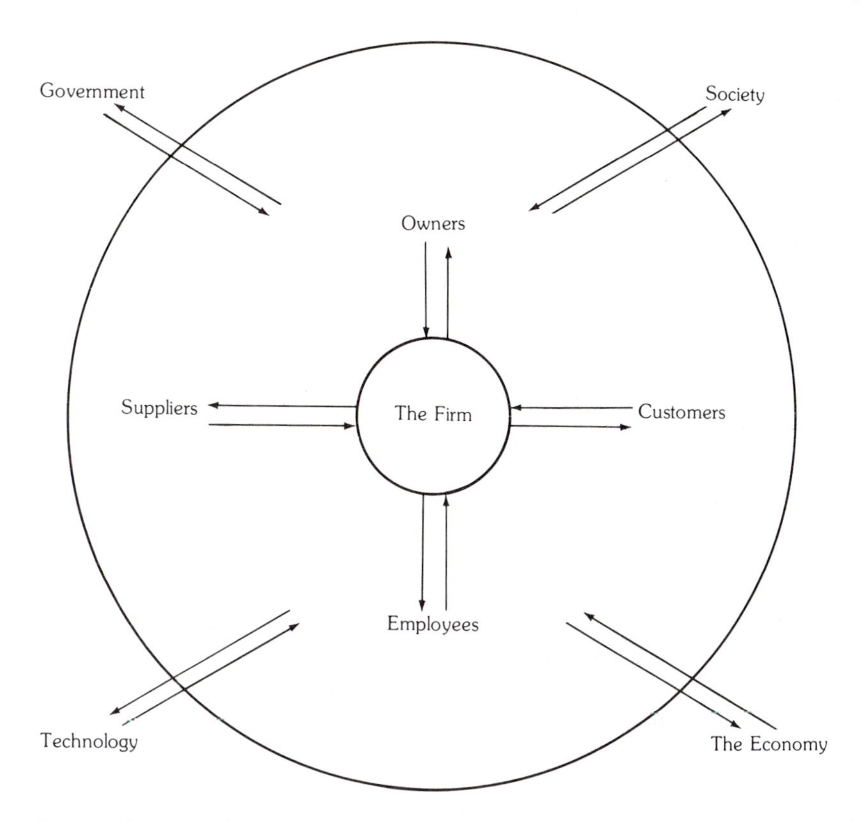

Figure 10-1. The business ecology model.

The actions of the firm can also affect the other players and forces in this model. The Chrysler loan and its effect on business-government relationships, the social changes brought about by the development and marketing of the birth control pill, and the technological revolution caused by advances in computer technology are all examples of firms having an impact on or creating social issues.

Viewed in these terms, the objective of the firm is to maintain a delicate balance between its goals and the goals and actions of the players and forces that can affect the achievement of its goals. As in the ecology of a marsh or pond, the unilateral actions of any group can upset the balance. Corporate social policy is an approach to help maintain that balance.

Although corporate social policy has no clearly accepted definition, it encompasses top management's commitment to the pursuit of social as well as economic goals. Corporate social policy requires that managers ask not only, "What do we owe our stockholders?" but also "What do we owe customers, consumer groups, government, environmentalists, and other groups that have a stake in the achievement of the firm's objectives?" To properly implement corporate social policy, the firm must integrate this externally oriented approach into its total strategic management program.

STRATEGIC MANAGEMENT AND CORPORATE SOCIAL POLICY

As a direct consequence of the turbulent environment in which business must operate, top managers must spend an increasing amount of their time with groups and issues outside the firm. A 1976 survey of *Fortune* 1,000 Chief Executive Officers indicated that the CEOs allocated an average of 20 percent of their time to dealing with external issues; by 1978 the average was 40 percent. Today, that figure is almost certainly over 50 percent.[9] From an awareness of social issues in the 1960s to the need for social response in the 1970s, managers in the 1980s must focus on corporate social policy. The essence of corporate social policy involves recognition of issues in which the firm has an interest, conceptualization of the firm's response to these issues, and the preparation of policy positions that adequately reflect management's philosophy on the issues. Top management, therefore, becomes the chief designer or architect of the firm's corporate social policy.[10]

HIERARCHY OF STRATEGY

One useful approach to viewing levels of involvement in corporate social policy is to examine the hierarchy of strategy. As suggested in Table 10-1, no single paradigm of the hierarchy of strategy is universally accepted. Examining several paradigms, however, permits drawing a line of demarcation between levels of involvement in corporate social policy.

The traditional view outlines three levels of strategy. *Corporate strategy,* encompassing the entire enterprise, involves the formulation of missions, goals, environmental analysis of risk and opportunity, assessment of the firm's distinctive competences, and the choice of products and/or services the firm wishes to pursue as its portfolio of businesses. Corporate strategy answers the question "What business are we in?" *Business strategy* is concerned with product-market choices at the division level or product-line choices in a diversified company and includes competitor and market analysis. A primary focus is on how the business will compete. *Functional strategy*—in marketing, finance, research and development, and manufacturing—has the shortest time frame and least risk of the three levels and deals with the policies that guide these various functions of the firm.[11]

Several authors have expanded the traditional hierarchy of strategy to include an overarching fourth level: *enterprise,* or *societal-level, strategy.* Focusing on social-legitimacy concerns, the purpose of enterprise strategy is to integrate the firm with its environment, examining the role of the business in society. Enterprise strategy does not require a particular mode of social response on the part of the firm. Instead, it poses questions concerning the nature of corporate governance and the manner in which the firm will relate to the society of which it is a part. Just as it has been necessary to separate and distinguish among corporate, business, and functional strategies, enterprise strategy must be treated as a dis-

Table 10-1. Hierarchy of Strategy

Traditional View of the Hierarchy of Strategy	*Expanded View of the Hierarchy of Strategy*	*Key Questions Posed by the Levels*
	Enterprise-level or societal-level strategy	What is the role of our business in society? What do we stand for? What should be our role?
Corporate-level strategy	Corporate level strategy	What is our business? What businesses should we be in?
Business-level strategy	Business-level strategy	How do we compete effectively in a given business?
Functional-level strategy	Functional-level strategy	What functional (marketing, sales, financial, production) policies should we employ to implement our strategies?

tinct level to ensure that proper attention is given to the role of the firm in society. The remaining levels in this expanded paradigm follow the traditional view. Corporate-level strategy focuses on what businesses the firm should be in and how these businesses can be integrated, taking into account the economic, social, political, and cultural factors. Business-level strategy is concerned with seeking competitive advantage over the firm's rivals and integrating the various functional areas that comprise a business. Functional-level strategies are designed to integrate the firm's subfunctional activities in order to relate functional policies with changes in the functional area environments.[12]

The concept of enterprise strategy appears at first glance to be abstract. One approach to making this comprehensive level of strategy of practical value is to define enterprise strategy as one writer has as the answer to the question, "What do we stand for?" The purpose of answering this question is to define the mission of the organization and to provide a broad statement of what the corporation and its employees represent to their customers, to society, and to themselves. The process for answering that question could include three interactive stages: stakeholder analysis, values analysis, and social issues analysis.[13]

Drawing on the literature of corporate planning, corporate social responsibility, systems theory and organizational theory, a stakeholder is defined as any group or individual who can affect or is affected by the achievement of the firm's objectives. This first stage entails: identifying the firm's stakeholders; recognizing what effects the firm has on each stakeholder in political, economic, and social terms; understanding the perceptions of the firm held by the stakeholders; and recognizing what effects the stakeholders can have on the firm. Stakeholder analysis provides a method of capturing a broad range of participants in the achievement of objectives that the organization might otherwise neglect.

The second stage in developing enterprise strategy—value analysis—requires the identification and understanding of the values of the organization, of the top executives in the organization, and of the major stakeholders. The key to this process involves the separation of intrinsic values—those values that are good in and of themselves—from instrumental values—those activities that have value because they contribute to intrinsic value. Once the values of the important players are separated, it is possible to understand the origin of conflicts and inconsistencies among these players.

The third stage, social-issues analysis, involves identifying the major social issues facing the firm today, forecasting the major social issues of the future, and examining how these issues affect both the organization and its stakeholders.[14] As shown in Figure 10-2, a practical example of this process could be IBM's respect for individual rights. Thomas Watson, Jr., said of his company, "IBM's philosophy is largely contained in three simple beliefs. I want to begin with what I think is the most important: *our respect for the individual.* This is a simple concept, but in IBM it occupies a major portion of management time."[15] The stakeholder groups with the most power (that is, those with the greatest potential impact on the achievement of the firm's objectives) could be defined as customers, who control the fate of both sales and profits, and employees, who are the keys to product innovation, sales, and service. Although respect for the individual is an intrinsic value, IBM uses a number of techniques that have instrumental value in contributing to respect for the individual: an extraordinary degree of customer service, intensive employee training programs, a broad range of employee benefit programs, and a sophisticated due-process mechanism. The issue of individual rights has been of increasing social significance over the past 20 years.

Although not specifically designed to formulate corporate social policy, the enterprise strategy formulation model provides a framework for understanding the firm in terms of its relationship to society or societal elements.

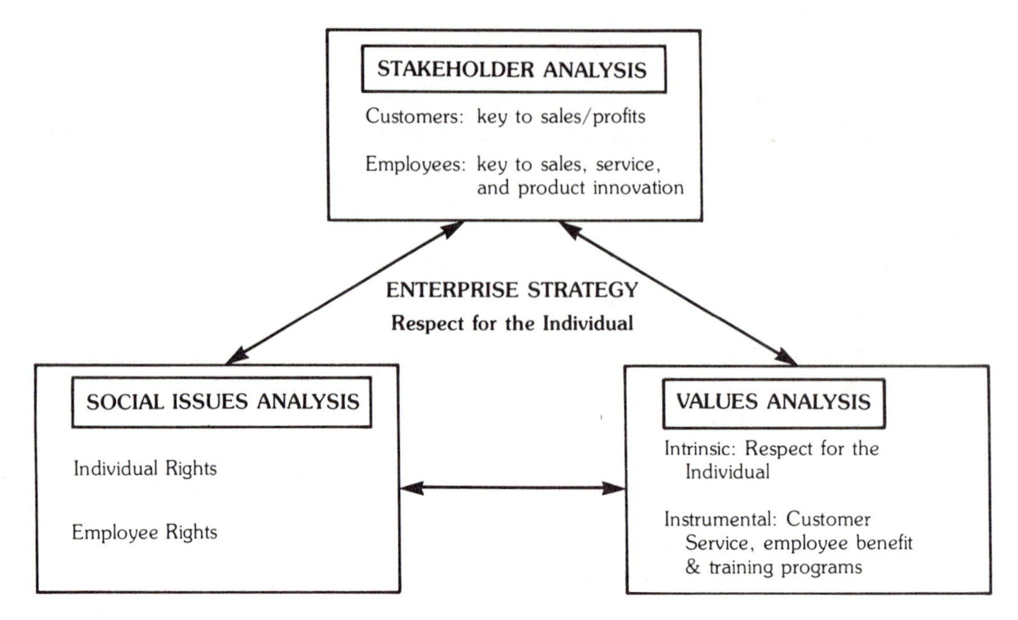

Figure 10-2. Formulating enterprise strategy: IBM.

The Strategic Management Process

Another approach to visualizing the integration of corporate social policy into strategic management is in the strategic management process itself. There are at least six major tasks in the strategic management process: (1) goal formulation; (2) strategy evaluation; (3) strategy formulation; (4) strategy implementation; (5) strategic control; and (6) environmental analysis.[16] As shown in Figure 10-3, elements of the corporate social policy process occur in each of these tasks. Although the tasks are discussed sequentially, they are in fact interactive and do not progress in a neatly diagrammed pattern.

Goal Formulation. The complex task of goal formulation involves both establishing goals and setting priorities for those goals. A politically charged process, goal formulation involves the personal values, perceptions, attitudes, and power of the managers and owners involved in the process. Social norms are one of the most general means of influence on the goal-formulation process. The social norms that affect the participants in the goal-formulation process come from a variety of sources: parents, teachers, friends, church, civic organizations, the media. These norms are in a constant state of transition and range from very broad ("Thou shalt not steal") to fairly specific guidelines (employees should be kept informed of corporate goals and objectives). Conflicts over social norms arise both because individuals have different social norms and because of conflicts in the norms: plant efficiency versus pollution, managerial effectiveness versus enlightened management, charitable obligations versus obligations to owners. At the goal-formulation stage, the general framework and attitude of corporate social policy is established.[17]

Strategy Evaluation. Strategy evaluation in an established organization involves evaluating the organization's current strategy and evaluating a proposed strategy. There are at least six criteria for evaluating strategy: (1) internal consistency with the organization's goals, objectives, and functional policies; (2) environmental consistency with current and projected social forces; (3) the appropriateness of the strategy in view of the organization's available resources; (4) the degree of

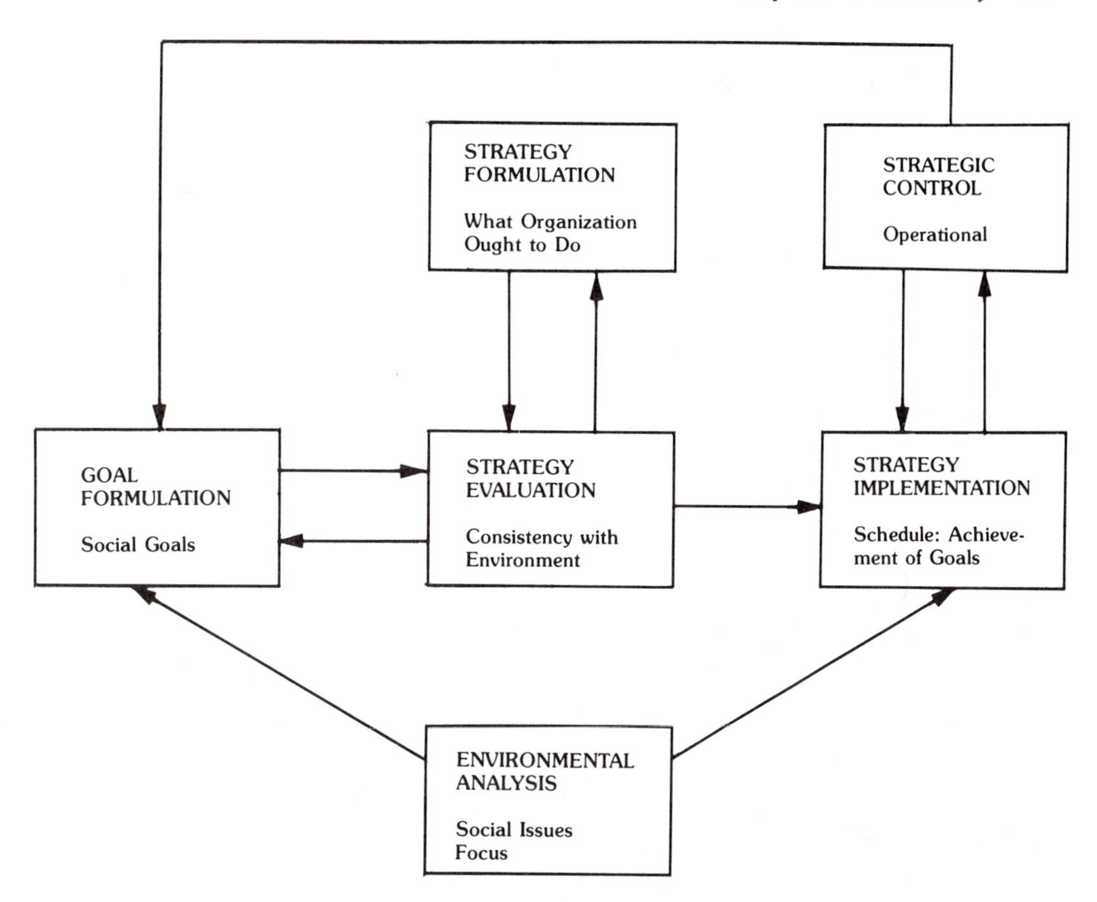

Figure 10-3. Strategic management process and corporate social policy.

risk involved in the proposed strategy; (5) the time horizon involved; and (6) the strategy's workability.[18] The second criterion, environmental consistency, requires at least an acknowledgment of social forces and issues.

Strategy Formulation. This stage involves environmental, industry, and competitor analysis and occurs after the proposed strategy has been evaluated and found wanting in some aspect of goal achievement or is used when a new goal has been formulated. As shown in Figure 10-4, there are four major determinants of strategy formulation: (1) appraisal of present and forecasted market opportunities and risks; (2) assessment of the firm's distinctive strengths and weaknesses; (3) the personal values and aspirations of managers; and (4) acknowledged obligations to society.[19]

The first two factors are basic to the success of every organization, dealing with what the organization is capable of accomplishing in terms of its strengths and weaknesses and with what the organization is able to do in terms of its market. The third factor—values and aspirations of top managers or owners—is what the top managers or owners want to do as a statement of the organization's strategy.

The fourth factor—acknowledged obligations to society—contains the core of corporate social policy. Because it concerns what the organization ought to do, it refers to the question of how social policy affects, interacts with, and helps determine overall strategic choice. Some, such as Milton Friedman, do not believe that the rational process of determining organizational direction and policy should be influenced by the value-laden con-

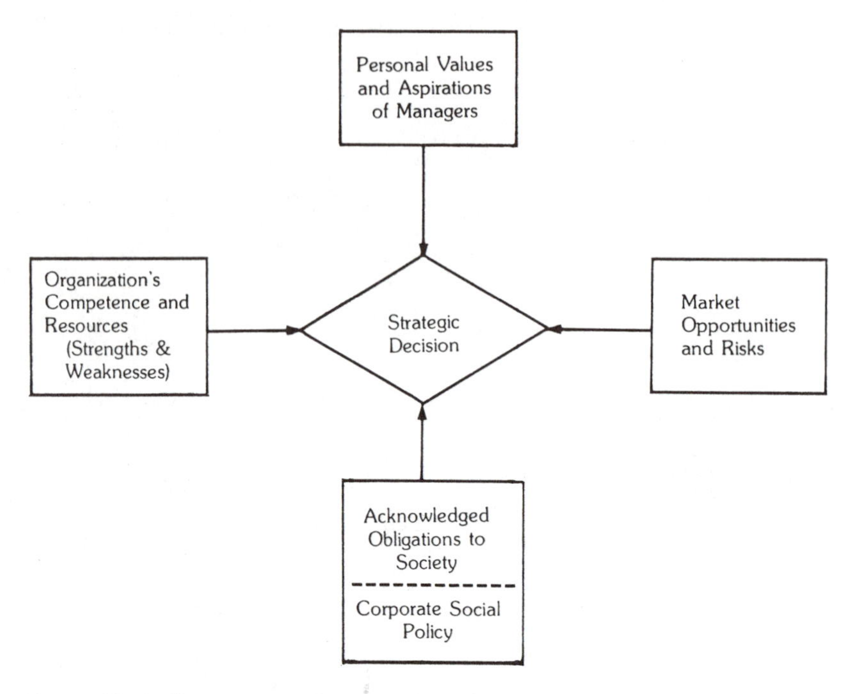

Figure 10-4. Components of strategy formulation.

siderations of organizational responsibilities that extend beyond economic and legal obligations. There are certainly problems and difficult decisions involved, but the business person who is concerned about social policy must closely examine the impact on the organization and on society of the policy alternatives chosen.

Strategy Implementation. No design, however grand, will aid the organization if it is left on the drawing board. Strategy implementation involves the fit between the organization's strategy, structure, and processes.[20] The purpose of this administrative task in regard to corporate social policy is to ensure that the social policy component of the organization's strategy is communicated to the appropriate levels in the organization and that the organization's structure is arranged to accommodate the duties and jobs required by the social policy. Specific examples of this will be discussed later. At this point, it is sufficient to note that, although strategy formulation contains the core of corporate social policy, it also in a sense captures the organization's intentions. A number of proverbs come to mind here; the most appropriate is that the road to hell is paved with good intentions. The strategy implementation stage translates social intentions into social actions.

Strategic Control. Strategic control focuses on the fit between formulation and implementation, ensuring that the strategy is both on schedule and on target.[21] For corporate social policy, at issue in this task is measuring the progress of implementation against established benchmarks and ensuring that the policies (for example, affirmative action, product safety, workplace safety, fairness in advertising) are achieving the desired results.

Environmental Analysis. The final task in the strategic process, environmental analysis, is sometimes listed solely as a subset of strategy formulation. It is listed here as a separate task because the other tasks in the strategic management process cannot be accomplished in an intraorganizational vacuum. A public utility might formulate the goal of bringing five nuclear plants on-line within the next ten years; to formulate such a goal, however, would require total ignorance of public attitudes toward nuclear power. Because an organization cannot control its environment, environmental analysis is used to examine the present and forecast the future for both opportunities and threats.

The degree of formalization of environmental analysis in an organization depends on the degree of uncertainty in the environment and the importance of external changes to the firm.[22] A shoe repair shop faces a relatively certain environment and is not greatly affected by environmental change; the degree of environmental analysis needed would be limited to competitor and market analysis. An international maker and marketer of pharmaceuticals is at the opposite extreme and would require a highly formalized method of scanning the environment.

Two major approaches to environmental analysis are environmental scanning and sociopolitical forecasting. Although these two are quite similar, there are some notable differences.

Scanning the environment has been classified into three models: irregular, regular, and continuous.[23] The irregular model consists of *ad hoc* studies of specific events that generally are a short-term response to a crisis, such as the introduction of a new product by a competitor or a sudden, unexpected shortage of a vital resource or material. The regular model is more comprehensive and involves periodic review of issues identified as important to the firm; examples of this model include annual surveys and forecasts of the economy and technological advances. The continuous scanning model is the most advanced, emphasizing constant monitoring of various environmental systems rather than specific events. This model is long term in orientation and typically uses relatively sophisticated forecasting techniques.

Social, or sociopolitical, forecasting techniques are the most long-term methods of analyzing the environment. Social forecasting is a process of identifying social trends, analyzing the opportunities and threats the trends present to the organization, and projecting the trends with other findings at least five years into the future.[24] There are three broad categories of social forecasting: authority methods, conjecture methods, and mathematical modeling.[25] Authority methods rely on the oldest method of getting advice—asking an expert. The sole-source method involves either hiring an expert or purchasing an expert's newsletter. Polling methods work on the old premise that two (or more) heads are better than one. In the Delphi technique, developed at the Rand Corporation, a group of experts are independently polled concerning the likelihood of future events. The results of the poll are tabulated and sent back to the experts for comments and revisions. As this procedure is repeated, the group tends to move towards a consensus opinion.

Conjecture methods are systematic attempts to identify alternative futures that may have an impact on a social issue. There are two categories of conjecture methods, qualitative and quantitative. The qualitative conjecture uses words to describe the future, so the content cannot be measured or counted. A scenario is a kind of qualitative conjecture in which two or more descriptions of a future attitude or issue are formulated. For example, a scenario describing employee rights in the year 2000 might forecast an increasing trend in employee rights, with the result that employees manage the corporation; an alternative scenario might project a backlash against employee rights. Scenario construction permits managers to widen their horizons by projecting a number of alternatives to their standard view of the future. The second type of qualitative conjecture is future history construction. In this technique, the course of events and developments are traced over time to explain how a set of circumstances developed out of preceding circumstances.

Two quantitative conjecture methods are cross-impact matrices and relevance trees. The cross-impact matrix relates the impact of one event on the likelihood of occurrence of subsequent events. Assuming that the occurrence of one event will affect other events, the cross-impact matrix can be used in conjunction with other methods of social forecasting, such as the Delphi technique. Relevance trees start with a future goal; in order to achieve this goal, two or more alternative policies are planned; finally, a number of steps are planned to implement those policies. Relevance values are assigned to the alternatives and can be compared to see which set of actions is most likely to lead to the desired objective.

Mathematical model building can be either deterministic, where knowledge of some of the parameters is assumed, or stochastic, where the parameters of the models take on a series of possible values. The Club of Rome's report, *The Limits to Growth*, is one of the best-known mathematical models in futures research.

Because of increased computer power, organizations now have the option of combining several social forecasting models to form more powerful analytic tools. Although the use of the more complex methods of social forecasting is still in the development phase, these techniques are being used by larger firms, particularly those in industries requiring sophisticated technology.[26]

BEYOND ANALYSIS: INFLUENCING THE ENVIRONMENT*

Earlier, it was stated that organizations cannot control their environment. Although this is technically true, organizations can move to influence their environment. As shown in Table 10-2 there are at least four possible stances to environmental or social issues. One writer views these as reactive, defensive, accommodative, and proactive. Another way of conceptualizing them is "Fight all the way," "Do only what is required," "Be progressive," and "Lead the industry." In certain situations involving specific issues and specific organizations, any one of these four stances may be best. In the Tylenol poisoning case, Johnson & Johnson had no reason to proact prior to the initial police reports; it could only react once the cause of the deaths became known. Many issues do not remotely relate to the activities of a given business, and the firm is afforded the luxury of inaction rather than reaction or some other posture.

However, from the broad view of dealing with all social issues, many firms are adopting a proactive stance, attempting to influence the nature and direction of change in order to maximize opportunities and minimize threats to the firm. The reasoning behind a proactive stance is that issues have a relatively predictable life cycle and that changes are possible at certain stages of the life cycle.

*The topic of this section is treated more fully in Chapter 15 entitled "Strategic Issue Management."

Issue Life Cycles

Interest and activism in social or public issues tend to follow an S-shaped curve over time. In most instances, the only initial indicators of the incipient issue are bits and pieces of data, such as comments in speeches by leading experts or minor news stories. Even if these disparate pieces of information were gathered together, the information at this point would appear to most observers as random and unrelated to any trend. Some issues, such as the thalidomide drug case or the Bhopal tragedy, come to life slightly ahead of this point due to a catalytic event.

As the bits and pieces of information increase, activist groups form around the issue. Frequently started by crusaders and gloom-and-doom peddlers, the groups attract both supporters and media attention. An example of how this stage develops is the issue of automobile safety and the efforts of activist groups, started by Ralph Nader, to bring the issue to the forefront during the mid-1960s.

As more and more attention is drawn to the issue, politicians jump on the bandwagon. One commentator has suggested that political statesmen and ideological leaders enter the process at a relatively early point, rallying

Table 10-2. Postures in Responding to Social and/or Environmental Issues

Negative Posture ◄			► *Positive Posture*	*Advocate*
Reactive	*Defensive*	*Accommodative*	*Proactive*	*Ian Wilson*
Fight all the way	*Do only what is required*	*Be progressive*	*Lead the Industry*	*T. W. McAdam*

Sources: I. Wilson, "What One Company is Doing About Today's Demands on Business," *UCLA Conference on Changing Business—Society Relationships*, July 30, 1974, p. 12; T. W. McAdam, "How to Put Corporate Responsibility into Practice." *Business and Society Review/Innovation* (Summer 1973)

public opinion around an issue. After 20 percent to 40 percent of the population begins to see the issue as important, legislators, reflecting the views of their constituents, begin to take up the cause and the issue becomes the subject of government studies, legislation, and ultimately regulation.[27]

The organization that waits to become involved in an issue until government studies or legislation has commenced can only react to the changes the issue will bring about; during the take-off period, it is possible for organizations to predict and prepare for the eventual changes. Identification and action during the earliest stages of the issue life cycle permits the organization to influence the nature and direction of change.

A number of methods are employed by organizations to spot emerging issues. John Naisbitt's firm, the Naisbitt Group, uses content analysis of over one hundred daily newspapers that have primarily local interests. Based on the principle that change begins at the grassroots and that reading local newspapers is a useful way of identifying social priorities, content analysis enabled Naisbitt to predict the end of mandatory retirement one year before it was legislated by Congress, the decline of nuclear power before the Three Mile Island incident, and the failure of the Equal Rights Amendment four years before it came to a vote. Other methods of identifying issues in the early stages of development include public-opinion surveys and analysis of trends in other countries (such as Sweden and Denmark) and in U.S. states (California, Florida) where social issues tend to increase in importance sooner than in other areas.[28]

Working with a shorter time frame than social forecasting, the issues life cycle and trend watching have given rise to a new kind of executive in organizations: the issues manager. The issues manager has become one of the principal mechanisms by which social, political, and environmental information has found its way into strategic decision making.

Issues Management

Issues management is a much more applied effort than social forecasting or environmental scanning. Whereas the latter focuses on a longer time frame and is more concerned with forecasting or prediction, issues management considers issues that are of more immediate interest (the next 1-3 years) and attempts to integrate these concerns into an effective response pattern for the firm.

Issues management is a process of identifying and evaluating social or public issues in terms of the impact these might have on the business and then energizing management toward integrating this knowledge into corporate decision making. Issues management, therefore, is an approach to seriously anticipating and integrating what heretofore has been considered the unmanageable environment into planning and operations.

One study of the important purposes of an issues-management program identified the following, listed in order of their significance:[29]

1. Allows *management of* vs. *reaction to* public issues.
2. Allows management to select issues that will have the greatest impact on the corporation.
3. Inserts relevant issues into the strategic management process.
4. Gives the company ability to act in tune with society.
5. Protects the credibility of business in the public mind.
6. Provides opportunities for leadership roles.

The underlying assumption in issues management is that we can no longer assume away the social environment as unpredictable. Rather, management must do its best to ascertain which public issues have the potential to greatly affect the company and then use this information to protect it or to exploit opportunities previously unrecognized.

An issues-management program entails a number of steps or stages. It is crucial to mention that an effective program assumes executive commitment and top management support. Assuming that commitment is there, the following steps outline the essential ingredients of a successful program.[30]

1. Identification of issues and trends in public expectations. Involves environmental scanning, tracking trends and developing issues, developing forecasts of issues and trends, identifying those of interest to the company.
2. Evaluation of issues' impact and setting priorities. Involves assessment of impact, probability of occurrence, corporate resources and response ability, and preparations for additional analysis.
3. Research and analysis. Involves categorizing issues, prioritizing for additional research, involving functional staff where applicable, using outside information sources and developing and analyzing optional positions the firm might take.
4. Strategy development. Involves analysis of options, managerial decisions on positions and strategy, and integration with overall corporate strategy.
5. Implementation. Involves dissemination of chosen positions and strategies, development of appropriate tactics and alliances with external organizations and linkage with both internal and external communication networks.
6. Evaluation. Involves assessment by staff of results, evaluation of management, modification of plans where needed, and additional research where needed.

These steps are fairly straightforward, but one might ask how to get an issues management program going in a company. One recommended plan by a corporate staff executive who had to implement an effort is as follows.[31]

Phase One:

1. Get to know the company as completely as possible: its mission, personality, structure, history, and needs.
2. Find out what is already being done, who is tracking what issues, what kind of information they have.
3. Make an initial judgment about what issues should be monitored.
4. Systematize the collection of information.

Phase Two:

1. Decide on a basic communication mechanism. What information ought to go to whom and how often? Possibilities include issue summaries, article or book abstracts, research findings, speculations of major futurists, international developments, trends.
2. Decide what structural or organizational arrangements are needed to carry out the program. This may dictate a top level management group that meets regularly to discuss issues and their implications. A process of issue identification that involves people with relevant expertise throughout the company should be established.
4. Personnel and budget considerations need to be determined and assessed.

Some payoffs that companies have experienced through issues management programs are worth briefly mentioning.

The S.C. Johnson Company removed fluorocarbons from its aerosol sprays three years before federal action forced others in the industry to do so.

Sears, Roebuck spotted the flammable-nightwear and Tris-treated fabric controversies early and removed the goods before government action.

Bank of America moved to change its lending policies two years before Congress required banks to disclose whether they were barring all loans in certain parts of a city ("redlining"). Early action cut the eventual cost of compliance and spared Bank of America public relations grief and antagonism from cities and activist groups.

Shortly after Ronald Reagan's election, Atlantic Richfield's issues management department accurately predicted that Reagan budget cuts would prompt states to compensate for lost federal money by taxing business. By lobbying early, they were able to head off tax proposals in a number of states.[32]

Organizationally, companies implement issues management in a variety of ways. Some companies assign the responsibility to the public affairs office. This is logical since public affairs staffs are increasing in numbers and stature in companies today. Other companies assign the function to corporate planning staffs. Also, a high echelon committee centering its attention on issues management is not uncommon. For example, Polaroid has a Public Issues Policy Committee as a center for its issues management efforts. The committee is made up of line and staff professionals from across the company. After gathering background information on a variety of issues and conducting appropriate analyses, the committee recommends actions that Polaroid ought to take. The company thinks that by identifying important issues early it can formulate policies to deal with them responsibly rather than waiting to respond to crises.[33]

The acid test of successful issues management is integrating it with strategic planning. Actually, issues management parallels and contributes to strategic planning. Issues management addresses public policy events and issues that shape or are shaped by the total business environment. Strategic planning must employ alternate scenarios about externalities and must include information on public issues that are crucial to the firm.[34] Strategic planning necessitates a thorough consideration of public issues if it is to be comprehensive.

CONCLUSIONS

The factors that affect business success today go far beyond the relatively simple factors of the past. Rapid, sometimes radical, change has become a fact of business life.

The business that hopes to survive into the twenty first century must move beyond merely being adroit at managing change. Corporate social policy is a way of thinking that provides a framework not only for coping with the turbulent environment but also with attempting to influence the environment in a positive way for the firm.

No specific moral or ethical stance is espoused in recommending that organizations adopt a corporate social policy and integrate that policy into their strategic management process. Rather, it is recommended because, by relating social issues to the organization's strategic decisions, business has a greater opportunity to succeed.

REFERENCES

1. S. P. Sherman, "Scoring Corporate Conduct in South Africa," *Fortune* (9 July 1984):168-172.
2. R. E. Freeman, *Strategic Management: A Stakeholder Approach* (Boston: Pitman, 1984).
3. M. Recio and V. Cahan, "Bhopal Has Americans Demanding the 'Right to Know'." *Business Week* (18 February 1985):36-37.
4. "Ten Forces Shaping America," *U.S. News & World Report* (August 1984):41-55.
5. J. Naisbitt, *Megatrends* (New York: Warner Books, 1982).
6. W. E. Rothschild, *Putting It All Together* (New York: AMACOM, 1976).
7. "Ten Forces Shaping America," p. 43.
8. Ibid.
9. G. A. Steiner, *The New CEO* (New York: Macmillan, 1983).
10. A. B. Carroll, *Business and Society: Managing Corporate Social Performance* (Boston: Little, Brown, 1981).
11. K. R. Andrews, *The Concept of Corporate Strategy* (Homewood, Ill.: Richard D. Irwin, 1980).
12. D. E. Schendel and C. W. Hofer, *Strategic Management* (Boston: Little, Brown, 1979).
13. Freeman, *Strategic Management.*
14. Ibid.
15. T. J. Peters and R. H. Waterman, Jr., *In Search of Excellence* (New York: Warner Books, 1982).
16. Schendel and Hofer, *Strategic Management.*
17. H. Mintzberg, "Organizational Power and

Goals: A Skeletal Theory," in *Strategic Management*, eds. D. E. Schendel and C. W. Hofer (Boston: Little, Brown, 1979), 64-80.
18. S. Tilles, "How to Evaluate Corporate Strategy," *Harvard Business Review* **41** (1963):111-121.
19. Andrews, *Concept of Corporate Strategy.*
20. J. R. Galbraith and D. A. Nathanson, *Strategic Implementation: The Role of Structure and Process* (St. Paul, Minn.: West Publishing, 1978).
21. Schendel and Hofer, *Strategic Management.*
22. J. M. Utterback, "Environmental Analysis and Forecasting," in *Strategic Management*, eds. D. E. Schendel and C. W. Hofer (Boston: Little, Brown, 1979), 134-144.
23. L. Fahey and W. R. King, "Environmental Scanning for Corporate Planning," *Business Horizons* (August 1977):61-71.
24. K. E. Newgren, "Social Forecasting: An Overview of Current Business Practices," in *Managing Corporate Social Responsibility*, ed. A. B. Carroll (Boston: Little, Brown, 1977), 170-190.
25. R. F. Lusch and Laczniak, "Futures Research for Managers," *Business* **29** (1979):41-45.
26. J. E. Fleming, "Public Issues Scanning," in L. E. Preston (ed) *Research in Corporate Social Performance and Policy* (Greenwich, Conn.: JAI Press 1981):155-173.
27. G. T. T. Monitor, "How to Anticipate Public-Policy Changes," *S.A.M. Advanced Management Journal* **42** (1977):4-13.
28. B. Abrams, "John Naisbitt Makes a Handsome Living Reading Newspapers for a Living," *Wall Street Journal* (30 September 1982):55.
29. R. A. Buchholz, "Education for Public Issues Management: Key Insights from a Survey of Top Practitioners," *Public Affairs Review* **111** (1982):70.
30. R. A. Buchholz, *Essentials of Public Policy for Management* (Englewood Cliffs, N.J.: Prentice-Hall, 1985), 187.
31. J. K. Brown, *Guidelines for Management Corporate Issues Programs* (The Conference Board, New York n.d.), 29.
32. E. C. Gottschalk, Jr., "Firms Hiring New Type of Manager to Study Issues, Emerging Troubles," *Wall Street Journal* (10 June 1982):33.
33. "Public Issues Policy Committee," *Ethics* **95** (1985):5.
34. C. B. Arrington, Jr. and R. N. Sawaya, "Issues Management in an Uncertain Environment," *California Management Review* 26 (1984):153-154.

IV

Strategic Environmental Assessment

Part IV covers the processes and techniques used to deal with one of the most fundamental aspects of strategic management—strategic assessment of the environment.

In Chapter 11 V. K. Narayanan and Liam Fahey describe strategic environmental analysis in terms of its basic goals, overall conceptions of the organization's environment, analytic processes, the linkage to strategy analysis, and the organizational prerequisites for managing the environmental analysis effort.

Stakeholder analysis, an approach to assessing the existing potential and desirable influence of the stakeholders of the organization—those individuals or groups who have a stake in or claim to it—is discussed in Chapter 12 by Aubrey Mendelow. He emphasizes the use of stakeholder assessments in the various stages of the strategic planning process.

Another key area of environmental assessment—competitive analysis—is discussed in Chapter 13 by Robert E. MacAvoy. He enumerates the factors that have contributed to the rapidly increasing importance of competitive analysis and distinguishes between it and the collection of competitor information. His focus is also on integrating competitive analysis into strategic planning through a process applicable to both mature and dynamic businesses.

The competitive analysis area is further explored in Chapter 14 by John E. Prescott who focuses on the techniques that may be employed in the various stages of the competitive analysis process.

In Chapter 15, William R. King discusses the emerging area of strategic issues management in terms of its rationale, objectives, and process. As with each of the other chapters in this section, strategic issue management is viewed as an integral element of an overall process of strategic management.

CHAPTER 11

Environmental Analysis for Strategy Formulation

V. K. Narayanan and Liam Fahey

V. K. Narayanan is an associate professor of strategic management and co-director of the Field Studies Program at the University of Kansas, Lawrence, Kansas. His primary research and consulting interests are in the areas of strategy formulation in technology-intensive industries, cognitive and political processes in strategy formulation, and organizational development for turbulent environments. He has published over a dozen articles in various journals such as *Strategic Management Journal, Journal of Applied Psychology*, and *Journal of Applied Behavioral Sciences*. His recent book *Macro Environmental Analysis for Strategic Management* is co-authored with Liam Fahey.

Liam Fahey is an associate professor of management policy at Boston University. His research has appeared in such publications as *Strategic Management Journal, Journal of Business Strategy*, and *Academy of Management Review*. He is co-author of the *The New Competition* and is founder and editor of *Strategic Planning Management*, a monthly publication for executives involved in strategy formulation.

Wrist watches used to be differentiated by accuracy. Now, however, in the wake of mass-produced, large-scale integrated (LSI) chips and frequency oscillators, accuracy is no longer a source of differentiation. Accordingly, the successful watchmakers have rather quickly shifted their emphasis. Elegance and high fashion have become the major source of differentiation.

In the 1970s, in the wake of the consumer movement, product liability became a central concern for automobile firms. Many of these firms were engaged in lawsuits brought by their customers for alleged product defects. These suits often resulted in multimillion dollar penalties for these corporations.

During the last decade, there has been a population migration from the northeast United States to the south and the west coast. In addition, there has been a movement of those with higher incomes from the cities to the suburbs. As a consequence, supermarket chains have closed some of their stores and opened new stores in other locations.

Each one of the above examples illustrates how trends and events in the environment affect industries and destinies of the firms within them. The environmental shifts described above generally took place outside the traditional boundaries of the industry surrounding specific firms. This chapter is concerned with the analysis of such environmental shifts and their implications for an organization's strategy. The conception of environment advanced here is that of the macroenvironment; the focus is not on industries or competitive environments traditionally highlighted in strategic planning.

The predominant characteristic of macroenvironment in the past decade has been an acceleration in rate of change: this is often referred to as turbulence or discontinuity.[1] Such discontinuities constitute the gist of writings of a number of recently popular thinkers.

For example, Toffler's description of the *third wave* is a presentation of the major discontinuities—some current and others speculative—posed by the emergence of postindustrial society.[2] Similarly, Naisbitt's description of ten *megatrends* is an attempt to synthesize the currently observed discontinuities into coherent, comprehensive patterns.[3]

Discontinuities and turbulence rapidly render the cognitive maps of key decision makers obsolete; they also necessitate adaptation by organizations at speeds faster than was considered possible in the early 1960s. In efforts to cope with discontinuity and turbulence, two approaches to managing organizations have emerged: one attempts to redesign an organization to enhance its adaptive capabilities and is primarily behavioral in its import;[4] the other attempts to enhance an organization's *cognitive* capacities for understanding environmental changes to provide sufficient lead time for adaptation as well as to formulate actions to influence and shape such changes. Although both approaches are significant, our concern here is with the latter.

There is widespread evidence of the prevalence of macroenvironmental analysis in organizations. First, in many organizations environmental analysis is conducted on an *ad hoc* basis by different organizational subunits. For example, legal departments typically monitor political and regulatory sectors of the macroenvironment, corporate planning departments address the economic environment, R & D functions focus on the technological environment, and public affairs departments monitor sociopolitical developments. Second, environmental analysis is practiced formally or informally in many organizations. This is often reflected in the emergence of subunit titles such as *environmental scanning unit* or job titles such as *director of environmental analysis*. Finally, we note the emergence of *issues management* in many organizations. As noted earlier, macroenvironments often create discontinuities, and many organizations have begun to address these discontinuities by engaging in issue analysis. Issues management is now so prevalent that the recently formed Issues Management Society has over five hundred members. (See chap. 15 for more detail.)

A number of recently published studies investigating the practice of environmental analysis in organizations suggest that the extent, scope, and degree of sophistication of environmental analysis efforts are highly variable across organizations.[5] Further, except in a limited number of organizations, these efforts are scattered and poorly integrated. In this chapter, we present a *comprehensive framework for environmental analysis*.

Three *basic goals* are postulated for environmental analysis in this discussion:

1. The analysis should provide an understanding of *current* and *potential* changes taking place in the macroenvironment. We emphasize current and potential in order to highlight that: (1) understanding of current changes is an important guide to anticipating the future and, hence, choosing strategic actions; and (2) environmental analysis should cover a time frame from short run to long run. The role of current changes is often emphasized (in practice) at the expense of potential changes, however, both are important.
2. Environmental analysis should provide *important intelligence* for strategic management. The analysis generally will provide information that is nice to know, but the primary goal is not generation of information for its own sake, but information that is useful in determining and managing the firm's strategies.
3. Environmental analysis should facilitate and foster *strategic thinking* in organizations. It typically is a rich source of ideas and understanding of the context within which a firm operates. It, therefore, often challenges the current wisdom by bringing fresh viewpoints into the organization, thereby serving as an antidote to obsolescence of cognitive maps.

The organization of the chapter is represented pictorially in Figure 11-1. The figure identifies four key considerations in environmental analysis: (1) conception of the macroenvironment; (2) process of analysis; (3) linkage to strategy analysis; and (4) organizational prerequisites. The central themes here are that one's vision of macroenvironment (content) and approaches to analyzing

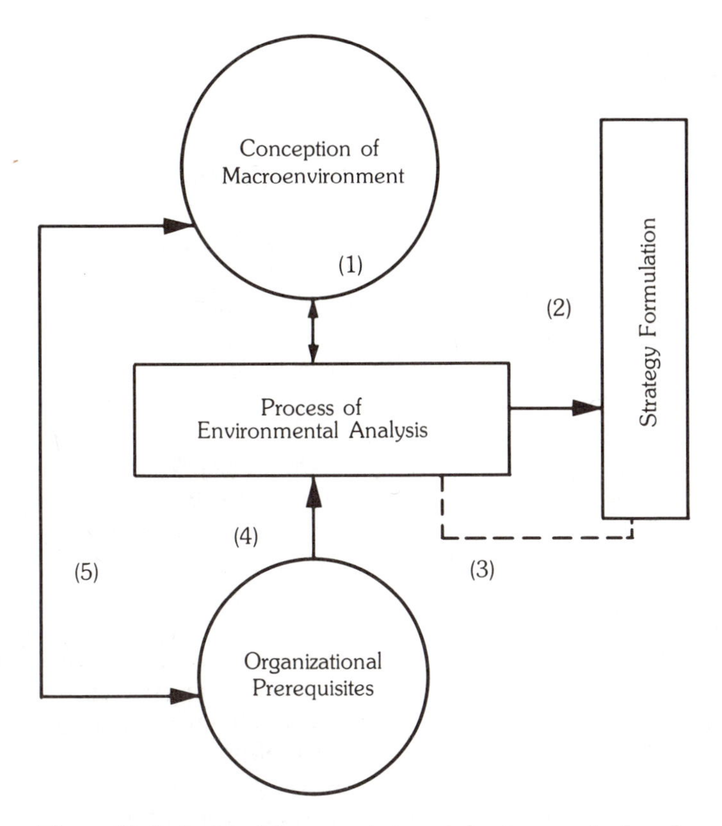

Figure 11-1. A pictorial representation of chapter organization. 1. Content-process interplay; 2. Precedence of environmental analysis to strategy analysis (outside-in approach); 3. Induced environmental analysis (inside-out approach); 4. Organizational support-strong link; 5. Structural differentiation in organizational prerequisites.

it (process) are inextricably *intertwined;* that environmental analysis should *precede* strategy analysis, though the latter may induce the former for short-run purposes; and that analysis, as carried out in organizations, needs to be *supported* by organizational prerequisites that should reflect the structural differentiation implicit in the conception of environment.

The scheme of the chapter is as follows: first, a selective review of research in this area is presented; second, we portray a conception of macroenvironment; third, the process of analysis is elaborated; fourth, links to strategy analysis are laid out; and fifth, the organizational prerequisites for managing environmental analysis are briefly indicated. The chapter concludes with a few observations about the value and limitations of analysis.

MODELS OF UNDERSTANDING THE ENVIRONMENT

A growing number of authors have addressed issues related to how organizations understand their environments. A selective review of these works is presented here to delineate the intellectual indebtedness of the model presented in this chapter as well as to anchor the discussion of environmental analysis.

Three strands of work of this genre deserve

mention: (1) enactment theorists; (2) scanning models; and (3) systemic models.

Enactment

Working from cognitive psychology, Karl Weick developed the notion of enactment as the active process organizational participants carry out in defining their environments.[6] Individual information processors in an organization enact the environment to which the system then adapts. The process of enactment introduces information into the system, which is dealt with according to organizational rules and routines. Weick emphasizes the importance of *attention* processes as they determine what kinds of information get selected and what kinds get neglected from the available cues. These processes are highly variable since they take into account a wide range of situational factors.

Weick's imaginative ideas cannot be done justice here; but four comments are in order to place the enactment process in perspective. First, the notion of enacted environment states that an organization or the key decision makers in it generate a socially shared way of envisioning the complex world of environment beyond their borders; Weick's model thus draws attention to the tendencies to develop cognitive distortions. A key implication here is that the image of environment, in which organizational strategies and structures are anchored, may be incomplete or distorted. Second, we suggest that systematic environment analysis is one mechanism of enactment that may reduce such distortions. Third, and anticipating our discussion in the next section, the notion of *relevant* environment is squarely anchored in this view since it highlights the judgmental processes involved in analysts' focusing on one or more segments of the environment rather than others. Finally, it must be noted that no one to date has searched for systematic tendencies of distortion in the enactment process; as such, specific prescriptions for enhancing the quality of environmental analysis are not yet forthcoming.

Scanning Models

Scanning models focus on the key modes of exposure and perception of information about environment. The classic description of these modes was provided by Aguilar as a result of his field study of information-gathering practices of managers.[7] Aguilar identifies four modes of information collection:

Undirected Viewing. Refers to the manager's exposure and perception of information that has no specific purpose. The source and substance of information are highly varied; and typically much information is dropped from attention, however, it can be valuable as a means of reducing our natural tendency to perceive and digest only information of immediate relevance.

Conditioned Viewing. This mode involves a degree of purposefulness by the manager as he or she receives information inputs. The individual is receptive to information and begins to assess its significance for him or her. There is a natural tendency to *react* or respond under this mode.

Informal Search. In this mode, the individual moves on to a proactive orientation toward searching for information. However, there is a limited and relatively unstructured effort to seek out information for a specific purpose.

Formal Search. This is a highly proactive mode deliberately undertaken to obtain information for specific purposes. Here, there is emphasis on formal procedures and methodologies for obtaining informational inputs.

Two comments are in order with respect to this stream of work. First, the focus is primarily on the kinds of analytical activities involved in understanding environment. Relatively little attention is paid to systems for environmental analysis. Second, given the key role played by individuals, the work offers a framework for studying analytical activities.

The activities involved in environmental analysis—detailed later—are anchored in this stream of research.

Systemic Models

In recent years, a growing number of studies have described the systems for environmental analysis found in organizations. Unlike the scanning models, the focus of the studies is not analysis-related activities but systems of management found in organizations. In an early study, Fahey and King provided a typology of such systems: irregular, periodic, and continuous.[8]

Irregular systems are found in organizations primarily on an *ad hoc* basis and tend to be crisis initiated. These systems focus on specific events, they deal with retrospective data for decisions in current and near-term future. At best the structure for analysis may consist of *ad hoc* teams; they engage in simplistic data and budgeting projections. The environmental analysis, in these cases, is not integrated into the mainstream of activity.

Periodic systems are guided by the need for environmental information for problem solving or decision making. They focus on selected events and deal with current, retrospective, and somewhat prospective data for decisions in the near-term future. They engage various staff agencies for periodically updated studies of economic and market environment. They allocate relatively low levels of resources for this and employ statistical forecasting-oriented techniques. These systems are partially integrated into the strategic planning process.

Continuous systems focus on finding opportunities and avoiding problems in addition to information for decision making. The breadth of environmental analysis is high; they collect data on social, political, technological, and economic trends of a current prospective nature. They are utilized for decisions of long-term nature, involve structured data collection and processing units and employ sophisticated futuristic methodologies. Scanning units with a relatively high level of resources are the norm. These systems are fully integrated into strategic planning systems as crucial for long-term growth of organizations.[9]

Table 11-1 presents a complete description of this taxonomy.

Studies of this genre point up three issues related to environmental analysis:[10] (1) managerial and/or organizational factors influence the analytical activities in the process; (2) there is great variability among organizations as to how analysis is performed; and (3) these systems are in flux; they will need to be developed and studied further. We must note here that these studies form the basis of our section on managing environmental analysis.

CONCEPTION OF MACROENVIRONMENT

This section presents the constructs necessary for understanding the various elements of macroenvironment, the linkages among these, and their evolution. It does not specify the kinds of linkages or paths of evolution; these are deemed to be the outputs of the process of analysis. In this sense, it is a framework rather than a theory.

Three key assumptions underlie this conception. First, macroenvironment needs to be analyzed in its own right irrespective of the immediate context of an organization, a position we share with Ansoff and Emery and Trist.[11] Second, specific implications of environmental evolution for a *firm* need to be discerned by the analysts in conjunction with the traditional industry-competitor analysis described in strategy formulation models.[12] Finally, action responses that are chosen should be anchored in these implications.

The concept of environment is composed of three constructs: (1) levels of environment; (2) a model of macroenvironment; and (3) constructs for describing environmental evolution. Each of these constructs provides the lenses through which the analysts can begin

Table 11-1. Typology of Environmental Systems

	Irregular	*Periodic*	*Continuous*
Impetus for scanning	Crisis-initiated	Problem-solving decision or issue-oriented	Opportunity finding and problem avoidance
Scope of scanning	Specific events	Selected events	Broad range of environmental systems
Temporal nature	Reactive	Proactive	Proactive
1. Time frame for data	Retrospective	Current and retrospective	Current and prospective
2. Time frame for decision impact	Current and near-term future	Near term	Long term
Types of forecasts	Budget-oriented	Economic- and sales-oriented	Marketing, social, legal, regulatory, cultural, and so on
Media scanning and forecasting	*Ad hoc* studies	Periodically updated studies	Structured data collection and processing systems
Organization structure	1. *Ad hoc* teams 2. Focus on reduction of perceived certainty	Various staff agencies	Scanning unit, focus on enhancing uncertainty-handling capacity
Resource allocation to activity	Not specific (perhaps periodic as fads arise)	Specific and continuous but relatively low	Specific, continuous, and relatively substantial
Methodological sophistication	Simplistic data analyses and budgetary projections	Statistical forecasting-oriented	Many futuristic forecasting methodologies
Cultural orientation	Not integrated into mainsteam of activity	Partially integrated as a stepchild	Fully integrated as crucial for long-range growth

Source: L. Fahey, W. R. King, and V. K. Narayanan, "Environmental Scanning and Forecasting in Strategic Planning: The State of the Art," *Long Range Planning Journal* (April 1981).

to understand the environment. This framework thus addresses the *content* of macroenvironmental analysis; the *process* of analysis is detailed in the next section.

Levels of Environment

The framework posits three levels of environment: task environment, competitive or industry environment, and general or macroenvironment.

Task environment refers to the set of customers, suppliers, competitors, and other environmental agencies such as trade associations directly related to the firm. Much of the day-to-day operations of a firm involves activities or concerns dealing with its task environment. Thus, a firm negotiating a loan with a bank, requesting supplies from a supplier, or dealing with customer service are examples of operating within a task environment. The task environment is more or less specific to a firm and is not necessarily shared by its competitors. The customers are often loyal to a firm's brand; suppliers may have granted preferred-customer status to a firm. Such factors as brand loyalty or preferred-customer status ensure that the task environment of a firm is generally distinct from that of its competitors.

Beyond the task environment, lies the *competitive or industry environment.* This comprises a firm or a business unit and its competitors functioning in the same industry. At this level, environmental factors directly affect all competitors in the same industry. Porter, for example, notes that new entrants, substitutes, suppliers, buyers, and rivalry among competitors influence what happens in an industry. The analyst here is concerned with the competitive environment in order to assess the attractiveness of an industry and the feasibility of alternative competitive strategies for a specific firm.

At the broadest level lies the *general environment or macroenvironment*. This is sometimes referred to as the political economy. Factors in the general environment influence all the industries functioning within it. Although the influence of these factors may be experienced differently by different industries, such issues as inflation or interest rates—factors in the general economic environment—affect a host of industries from automobiles to financial services to restaurants.

The various levels of environment can be presented schematically as in Figure 11-2. Here the general environment affects the specific industry and competitive environment, and the industry environment, in turn, affects the task environment of the firm. In addition, in the figure, we have noted a concept called *relevant* environment. Relevant environment refers to the boundaries of general environment drawn for analytical purposes. It must be noted that general environment, in its broadest sense, includes almost everything outside the firm's task and competitive-industry environments. However, such a broad definition makes the analysis task well nigh impossible because of the multiplicity of factors involved. So for the purpose of analysis, the analyst needs to focus on aspects of the environment deemed relevant. In some sense, all definitions of relevant environment require judgment and contain some degree of arbitrariness; yet such judgments are necessary for engaging in worthwhile analysis. The construct *relevant environment* is introduced to highlight the role of enactment processes

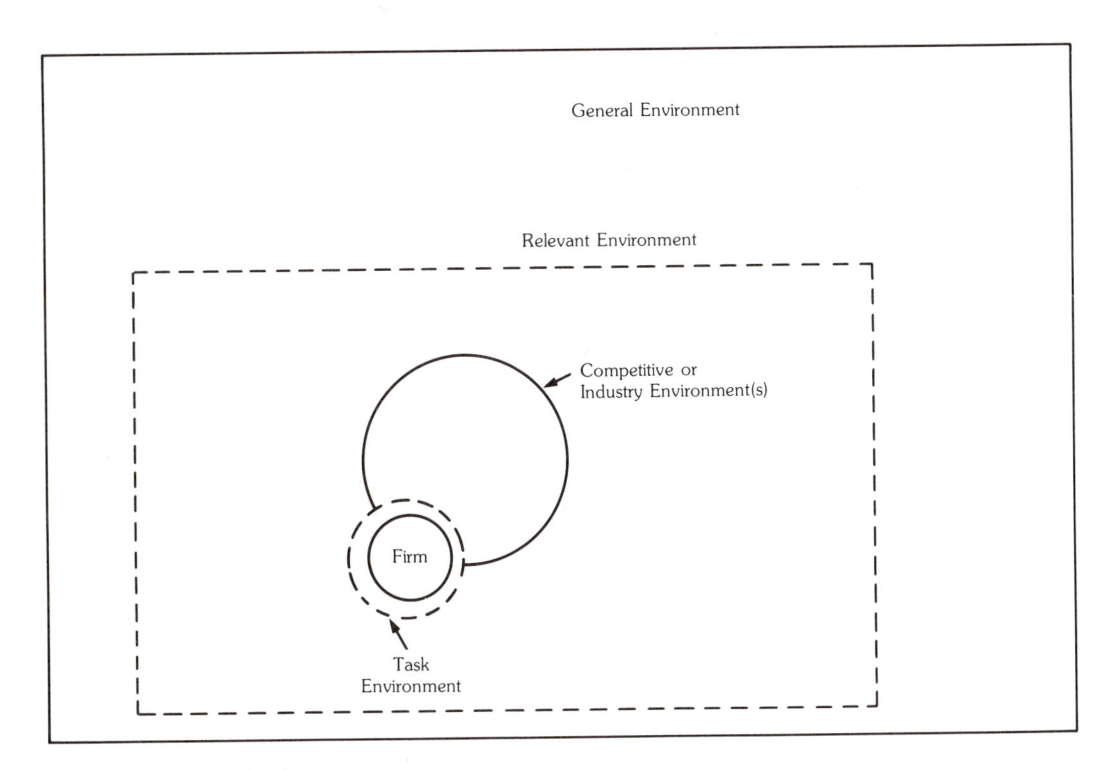

Figure 11-2. Levels of environment.

in organizations, especially as an analyst attempts proactively to understand macroenvironment.

A final comment is in order with respect to Figure 11-2: the levels as posited are applicable not only to single-business firms but to diversified firms with multiple-business units. The latter will face a number of competitive and industry environments corresponding to their business units. Further, their task environment becomes complex since different business units will have distinct elements in task environment in addition to sharing common ones at the corporate level. The concept of relevant environment, aspects of macroenvironment common across all industries, that the analyst pays attention to is relevant for both types of firms, single-business and diversified, although working out their implications is rendered more complex in the case of diversified firms.

Model of Macroenvironment

As an analytical construct, the model of macroenvironment decomposes *relevant* environment into four segments: social, economic, political, and technological.

The *social* segment focuses on demographics, life styles, and social values of a population in a society. The analyst, in this segment, is interested in understanding shifts in population characteristics or emergence of new life styles or social values.

The *economic* segment focuses on the general set of economic actors and conditions confronting all industries in a society. This has perhaps the most direct impact on all business organizations.

The *political* segment deals with political processes occurring in a society and regulatory institutions that shape the codes of conduct. This is perhaps the most turbulent segment in the model.

The *technological* segment is concerned with the technological progress or advancements taking place in a society. New products, processes, or materials; general level of scientific activity; and advances in fundamental science (for example, physics) are the key concern in this area.

The model is presented in Figure 11-3. As shown in the figure, the model also presents various linkages among these segments. This is to highlight the notion that macroenvironment can only be understood in a *systemic* fashion. Thus, the macroenvironment is presented as a system of interrelated segments: every segment is related to and affects every other segment. However, the model does not *specify* the types of linkages; these are deemed to be the output of a process of analysis. In other words, the linkages are to be discerned by the analysts during their efforts to scan, monitor, and forecast macroenvironmental trends and patterns.

Constructs for Describing Environmental Evolution

Three key constructs are useful for describing changes in environmental segments: (1) types of change; (2) forces driving change; and (3) type of future evolution.

Changes in macroenvironmental segments may be *systematic* or *discontinuous*. Gradual, continuous, and potentially predictable changes are termed *systematic,* whereas random, unpredictable, sudden changes are termed *discontinuous*. Despite theorists who disavow discontinuous change and attribute perception of such change to lack of conceptual tools for seeing systematic orderly change,[13] there are strong pragmatic reasons for retaining the notion of discontinuous change: given the cognitive limitations of the analyst, most organizations have at some time or other experienced discontinuity in environmental change. The potential for such discontinuities is present even in the most sophisticated environmental analysis efforts.

For analytic purposes, it is important to go beyond description of change to assess the *forces* driving it. In Figure 11-2, it was noted that various segments are linked together in a systemic manner. The forces driving change

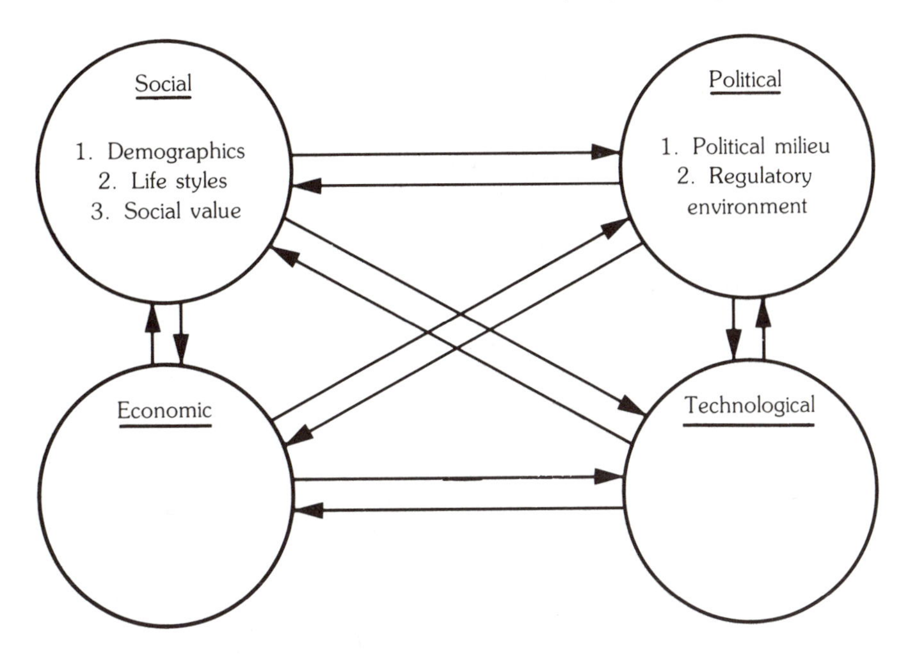

Figure 11-3. A macroenvironment model.

in one segment, therefore, often lie in changes in other segments. Thus, shifts in the social segment (for example, migration of population) may affect the political segment (for example, distribution of regional power). However, each segment can evolve quasi-autonomously, that is, the forces driving the change are located in the segment itself. Thus, social values may shift independently of technology. The existence of induced and autonomous evolution necessitates that changes in segments be analyzed *independent of each other and in conjunction* to identify the underlying forces.

Often driving forces interact with one another. These interactions may be *reinforcing, conflicting,* or *disjointed.* When the forces support one another in terms of their effect on changes in a third segment, the effect is reinforcing; when they dampen each other they are conflicting, and when they do not affect each other, they are disjointed. In addition, the effects of changes in one segment may have *primary* or *secondary consequences* for other segments. When the effects are direct, they are called primary consequences. In some cases, changes may not have a direct impact on other segments, however, consequences may ensue as a result of direct effects on a third segment. These are called secondary consequences: they often *lag* and take place only after the direct consequences are visible.

Finally, in charting the evolution of change in the future, it is important to characterize whether such evolution is completely predictable from the present trends or whether it is contingent on actions of the firm or other entities in the environment. This refers to *closed* and *open* versions of the future, respectively. This distinction is often crucial: unlike in the closed version, open versions should alert organizations to potential action domains that need further analysis or, where firm level responses may enable it, to shape the future evolution of the change.

As noted earlier, these constructs provide a cognitive map for analyzing macroenvironments and the changes taking place in them. Decomposition of the macroenvironment into four segments (social, economic, political, and technological) is presented to enable the analysis task to be manageable and to define the *scope* of such analysis. Three comments are in order regarding these segments. First, the scope of analysis—the number of segments attended to—will and should vary from organization to organization. Thus, we expect consumer goods firms to pay closer attention to socioeconomic sectors and technologically intensive firms to technoeconomic sectors. Second, irrespective of the nature of the firm, the longer the time horizon of analysis the broader should be the scope of environmental analysis. In the short run, the firm may focus on segments that critically affect its functioning; however, in the longer run, the systemic properties of the macroenvironment are likely to be clearly manifest. Third, analysis tasks differ somewhat among segments due to their unique properties. The process of analysis takes place against these properties and needs to be fine-tuned when a *specific* segment is considered. To this, we now turn.

PROCESS OF ENVIRONMENTAL ANALYSIS

Conceptually, the process of environmental analysis can be divided into four analytical stages: (1) scanning the environment to detect warning signals; (2) monitoring specific environmental trends; (3) forecasting the future direction of environmental changes; and (4) assessing current and future environmental changes for their organizational implications.[14] Table 11-2 provides an overview of these four activities, and Figure 11-4 shows a conceptual view of their interaction.

Scanning

In its *prospective mode,* scanning focuses on identifying precursors or indicators of potential environmental changes and issues. Environmental scanning is thus aimed at alerting the organization to potentially significant external impingements *before* they have fully formed or crystallized. Indeed, successful environmental scanning draws attention to possible changes and events well before they have revealed themselves in a discernible pattern.

Scanning, in the prospective sense, implicitly or explicitly feeds into monitoring early signals or indicators of potential environmental change. In this sense, scanning as an analytical activity becomes useful when environmental change takes time to unfold. This gives an organization time to work out implications for its actions. For example, social-value shifts do not occur in days or even in months; technology changes often take many years. Large-scale social movements may take years to develop from small pockets of disquiet. In these instances, it is possible during scanning to detect early indicators or precursors of these types of environmental change. For example, scanning might indicate signs of potential shifts toward political conservation or liberalism, the emergence of a pro-business judicial and administrative system, or changes in technology. In these cases, the *immediate* action implications for an organization may not be clear without further tracking and careful assessment of these trends.

Scanning, in the *current* and *retrospective* sense, identifies surprises or strategic issues requiring action on the part of an organization. In this case, the outputs may feed directly into assessment, and influence current and imminent strategic decisions of the organization.

Scanning frequently detects environmental change that is already in an *advanced* state, a change that has evolved to the point where it is actual or imminent rather than potential at some, as yet, unspecifiable date. A scan of demographic data might pick up population drifts or changes in household formation. A scan of the economic environment might reveal intermittent shortages in local energy or water supplies that have already taken place. Scanning frequently unearths actual or imminent environmental change because it explicitly focuses an organization's antennas on areas that previously may have been neglected, or it challenges the organization to rethink areas to which it had paid attention.

Table 11-2. Distinction among Scanning, Monitoring, Forecasting, and Measurement

	Scanning	*Monitoring*	*Forecasting*	*Assessment*
Focus	Open-end viewing of environment Identify early signals	Track specific trends and events	Project future patterns and events	Derive implications for organization
Goal	Detect change already underway	Confirm or disconfirm trends	Develop possible and plausible projections of future	Derive implications for organizations
Scope	Broad, general environment	Specific trends, issues, events	Limited to trends and issues deemed worthy of forecasting	Critical implications for organizations
Time horizon	Retrospective	Real time	Prospective	Prospective and current
Approach	Unconditioned viewing of heterogeneity of stimuli	Conditioned viewing of selective stimuli	Systematic and structural	Systematic, structured, and detailed
Data characteristics	Unboundable and imprecise, vague and ambiguous	Relatively boundable gains in precision	Quite specific	Very specific
Data interpretation	Acts of perception, intuitive reasoning	Weighing evidence, detailing patterns	Judgments about inferences	Judgments about inferences and/or implications
Data sources	Broad reading, consulting many types of experts inside and outside	Focused reading, selective use individuals, focus groups	Outputs of monitoring, collected via forecasting techniques	Forecasts, internal: strategies, competitive context, and so forth
Outputs	Signals of potential change, detection of change underway	Specification of trends, identification of scanning needs	Alternate forecasts; identification of scanning and monetary needs	Specific organization implications
Transition	Hunches regarding salience and importance	Judgments regarding relevance to specific organization	Inputs to decisions and decision processes	Action plans
Organizational outcomes	Awareness of general environment	Consideration and detailing of specific unfoldments, time for developing flexibility	Useful decision models and processes	Specific actions

Source: L. Fahey and V. K. Narayanan, *Environmental Analysis for Strategic Management* (St. Paul, Minn., 1985).

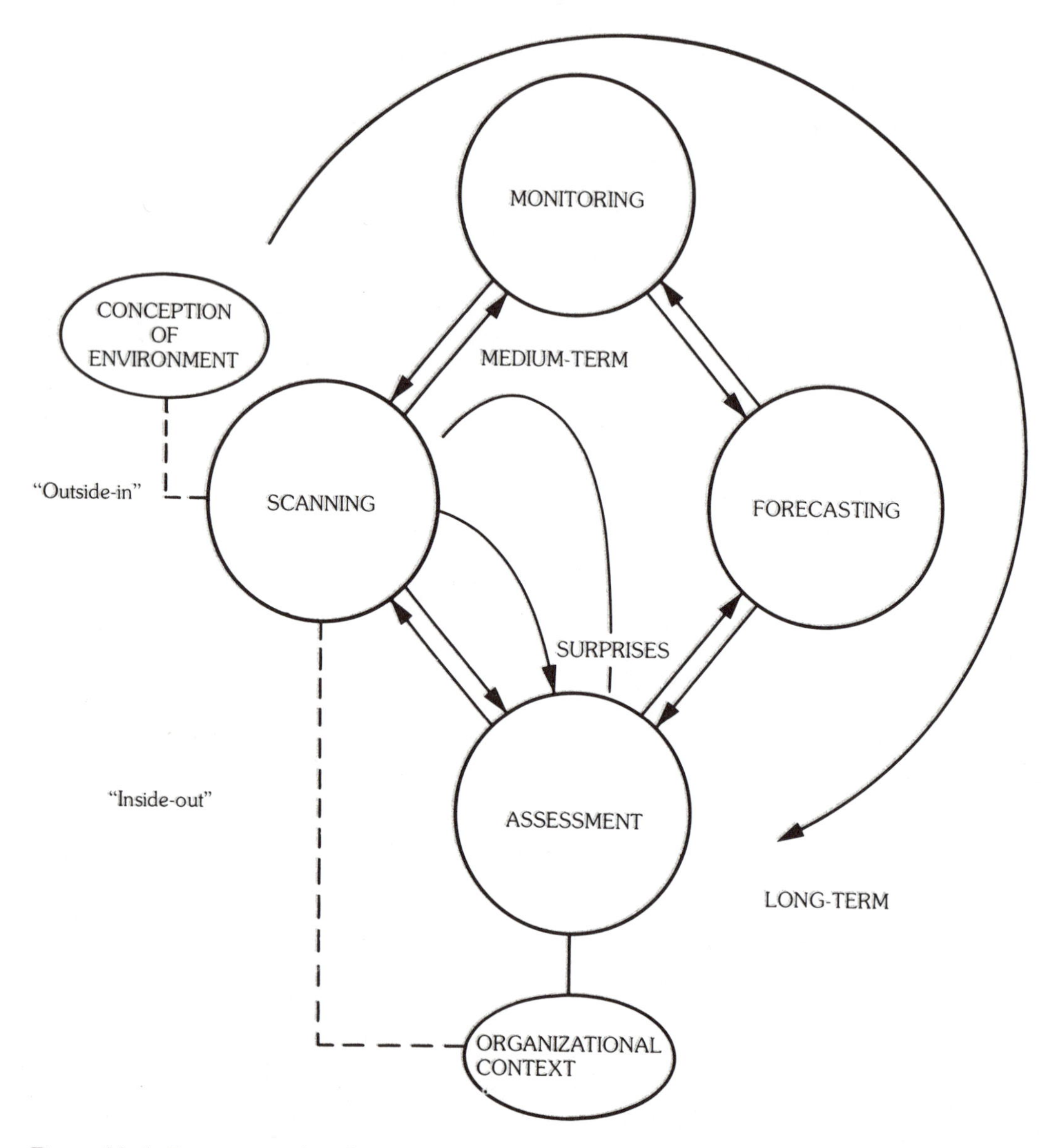

Figure 11-4. Environmental analysis activities.

It is important to recognize that scanning is the *most ill-structured* and ambiguous environmental analysis activity. There are essentially no limits on the potentially relevant data. The data are inherently scattered, vague, and imprecise. Data sources are also many and varied. Moreover, a common feature of scanning is that early signals often show up in unexpected places. Thus, *the purview of search must be broad,* but there are no guidelines as to where the search should be focused. In short, the noise level in scanning is likely to be high.

The fundamental challenge for the analyst in scanning is to make sense out of vague, ambiguous, and unconnected data. The analyst has to infuse meaning into the data; he or she has to make the connections among discordant data such that a signal of future events is created. This involves acts of perception and intuition on the part of the analyst. It requires the capacity to suspend one's beliefs, preconceptions, and judgments: these may inhibit connections being made among ambiguous and disconnected data.

Three critical decisions during scanning need highlighting: First, the scope and breadth of the data and data sources inevitably influence the analyst's perception. Second, the data do not speak by themselves; the analyst has to breathe life into them. Third, critical acts of judgment are required of the analyst in his or her choice of events and/or precursors to consider for monitoring, forecasting, and/or assessment. These decisions depend heavily on the skill and expertise of the analyst and are not easily formalized.

Monitoring

Monitoring involves tracking the evolution of environmental trends, sequences of events, or streams of activities. It frequently involves following signals or indicators unearthed during environmental scanning. Sometimes, it entails tracking trends, events, or activities that the organization accidentally becomes aware of or that are brought to the organization's attention by outsiders.

The purpose of monitoring is to assemble sufficient data to discern whether certain patterns are emerging. Two comments are in order with respect to these patterns. First, they are likely to be a complex of discrete trends. For example, an emergent life-style pattern may include changes in entertainment, education, consumption, work habits, and domicile-location preferences. In the initial stages of monitoring, the patterns are likely to be hazy because they are the outputs of scanning: the analyst has only a vague notion of what to look for. Second, highly formalized and quantified data bases usually (if) found in archives of organizations represent a characterization based on previously identified patterns and may have limited utility in tracking emergent patterns. The analyst has to breathe life into such data bases; he or she does this by arming himself or herself with a general sense of the pattern he or she is looking for and extending data searches to qualitative and quantitative outside sources.

Monitoring, viewed this way, almost always follows scanning; the activity ensures that the hunches and intuitive judgments about the weak signals made during scanning are tracked for confirmation, elaboration, modification, and (in) validation. In monitoring, the data search is *focused* and much more *systematic* than in scanning. By *focused,* we mean that the analyst is guided by *a priori* hunches (usually made during scanning or brought to the organization's attention by outsiders). *Systematic* refers to the notion that the analyst has a general sense of the pattern(s) he or she is looking for and collects data regarding the evolution of the pattern. The systematic character of the search renders data regarding trends cumulative. Thus, what were hazy outlines during scanning can now be imbued with details and clarity.

As monitoring progresses, the data frequently move from imprecise and unbounded to reasonably specific and focused. For example, in tracking the emergence of social issues, the first indicators (often picked up through scanning) are feelings of discontent or loosely distributed concerns expressed by a few individuals. These sentiments begin to attract the attention of others and gradually what is often referred to as a *social movement* begins to evolve. Such has been the evolution of the consumerist and environmentalist movements at the national level and community-interest movements at the local level.

A number of data interpretations or judgments are unavoidable in monitoring. As a sequence of events or a potential pattern is tracked, judgments must be made as to what data are relevant, what the valid and reliable data sources are, how the data fit together, how conflicts in the data can be reconciled, and when there is sufficient data to declare that a pattern is evident. It is tempting to suggest that these judgments are easier to make in monitoring than in scanning because the pattern is more evident. Yet, it may not be so. It is often difficult to judge when a pattern is evident: it may be simply a fad or a figment of the analysts' fancy. He may be imputing groundless linkages among disparate, unconnected data points. This problem is further confounded when different individuals within the organization make different and conflicting judgments—a not uncommon occurrence.

The outputs of monitoring are threefold: (1) a specific description of environmental patterns to be forecast; (2) the identification of trends for further monitoring; and (3) the identification of patterns requiring future scanning. Thus, the outputs of monitoring go beyond simply providing inputs to forecasting. Monitoring may identify trends or apparent trends that were not included in the scope of the original monitoring program. Also, it is not unusual for monitoring activities to indicate areas where further scanning may be desirable.

Forecasting

Scanning and monitoring provide a picture of what has already taken place and what is currently happening. However, strategic decision making requires a future orientation: it needs a picture of what is likely to take place. Thus, forecasting is an essential element in environmental analysis.

Forecasting is concerned with developing plausible projections of the direction, scope, speed, and intensity of environmental change. It tries to lay out the evolutionary path of anticipated change. How long will it take the new technology to reach the marketplace? Will the present social concern with some issue result in new legislation or administrative agency action? Are current life-style trends likely to continue? These kinds of questions provide the grist for forecasting efforts.

There are two conceptually separable, though integrally related activity elements in forecasting. The first concerns *projections* based on trends that are evident and can be expected with some margin of error to continue unabated in a given period of time in the future. For example, demographics (not anchored in birth rates) may be projected reasonably correctly for 5-10 years. The second relates to *alternative futures* that may come about not only on the basis of current trends but judgments regarding events that may take place or that may be made to happen by an organization or entities outside of it. These represent *possible* futures. The distinction corresponds to the one drawn in the previous section regarding closed and open futures. Forecasting, in the former sense (as projections) involves a closed perspective on the future, whereas in the latter sense (as alternative futures) it corresponds to the version of an open future.

There are a number of key analytic tasks and outputs involved in forecasting. The first concerns untangling the *forces* that drive the evolution of a trend. This is a necessary prerequisite to charting out the trend's evolutionary path. The second concerns understanding *the nature of the evolutionary path*, that is, whether the change is a fad, or of some duration, or cyclical or systematic in character. The third concerns more or less clearly *delineating the evolutionary path* or paths leading to projections and alternative futures. The critical outputs of forecasting are a specific understanding of the future implications of current and anticipated environmental changes and decision-relevant assumptions, projections, and information.

Forecasting typically is well focused in comparison with scanning and monitoring. Emphasis is placed on environmental changes of importance to the organization. Forecasting inherently requires that the organization identify fairly precisely what it is that it wishes to forecast. To forecast social-value change, the analyst needs to identify specific value changes to be forecast: for example, political, religious, or economic values; to forecast technology change the specific technologies to be forecasted have to be clearly identified.

Since the focus, scope, and goals of forecasting are more specific than in scanning and monitoring, forecasting is usually a much more *deductive* and rigorous activity. A wide variety of forecasting techniques are available ranging from simple extrapolation to methodologies involving multiple participants making forecasts in a number of iterations such as Delphi, to scenario development, which involves individuals laying out different paths of likely future developments. It is important to note that forecasting as an activity may take a good deal of time from initial commitment to do the forecast until actual outputs are acquired. For example, a Delphi-based forecast, depending on the number of iterations involved, may take six to nine months or more. Development of multiple scenarios may also consume many months.

Assessment

Scanning, monitoring, and forecasting are not ends in themselves. Unless their outputs are assessed for their implications for the organization's current and potential strategies, scanning, monitoring, and forecasting merely provide nice-to-know information. Assessment involves identifying and evaluating how and why current and projected environmental changes (will) affect strategic management of the organization. In assessment, the frame of reference moves from understanding the environment—the focus of scanning, monitoring, and forecasting—to identifying what that understanding of the environment means for the organization. Assessment thus endeavors to answer the question: What are the implications of our analysis of the environment for our organization?

Typically, environmental analysis generates a whole host of issues for the organization. However, any organization's capacity to identify and assess the patterns is limited: if there are too many issues to handle, the organization becomes bogged down in analysis with the result that organizational action becomes unlikely. On the other hand, it is considerably more difficult to identify the key issues that need to be tackled than to generate a whole host of issues.[15]

From the perspective of linking environmental analysis and strategic management, the critical question is: What are likely to be the positive or negative impacts of environmental patterns on the firm's strategies? This question compels the linking of environmental patterns and the organization's context. Those patterns judged to have already had an impact on the organization's strategies or to possess the potential to do so are deemed to be *issues* for the organization. (Also see chap. 15.)

Emphasis must be placed on the judgment required to identify issues, that is, to determine which patterns are affecting or will affect the organization. Judgment involves assessing and prioritizing patterns against specific criteria. These criteria should include the following:

1. How might the pattern have an impact on the organization?
2. What is the probability that the pattern will develop and become clearly recognizable?
3. How great will be the eventual impact on the organization?
4. When is the issue likely to peak? Near term? Medium term? Long term?

The intent of the first criterion is to determine whether the pattern has or will have an impact on the organization. This question forces members of the organization to make a preliminary assessment as to whether the pattern is likely to evolve into an issue. Does it have relevance given the organization's context? Those patterns assessed to have potential relevance to the organization are then subjected to the other three criteria.

Issues can then be conveniently arrayed on a probability-impact matrix, (Figure 11-5) with a separate matrix being prepared for each of the three planning periods: short-, medium-, and long-term. Although the scoring system for this assessment of probability and impact can be simple or complex, a general categorizing of high, medium, or low is usually sufficient. The merits of the matrix display are that it provides a comprehensive, at-a-glance array of issues, orders them in a manner that facilitates discussion and planning, and places them in time frames appropriate to the allocation of resources and management attention.[16] The issues arrayed on a

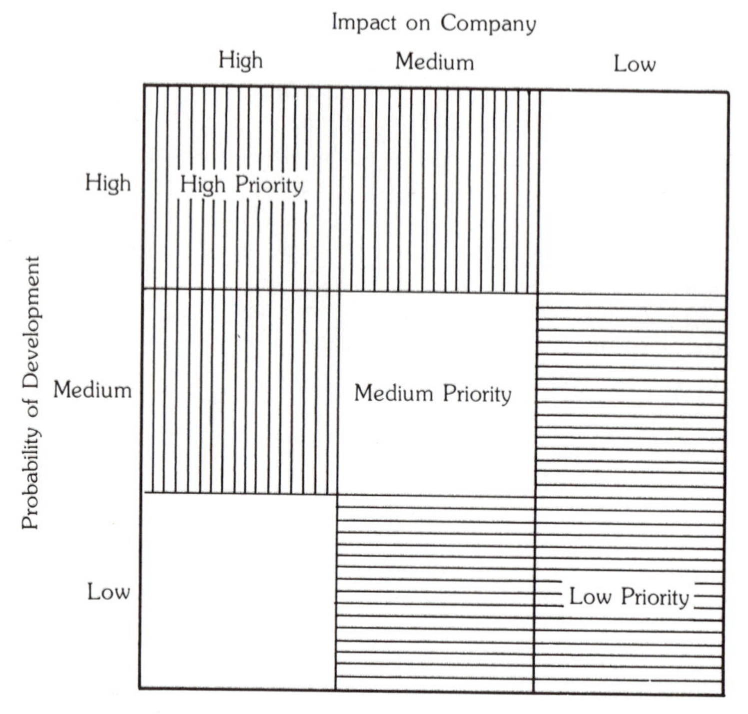

Figure 11-5. Probability-impact matrix (from I. H. Wilson, "The Benefits of Environmental Analysis," in The Strategic Management Handbook, ed. K. Albert, New York: McGraw-Hill, 1983).

temporal basis then should be fed into various strategic analyses; these linkages will be portrayed in detail in the next section.

The above discussion emphasized the distinctions among scanning, monitoring, forecasting, and assessment. Although they are conceptually separable activities within environmental analysis; scanning, monitoring, forecasting, and assessment are *inextricably intertwined*. As shown in Figure 11-4, each one can and should influence the others. For example, upon unearthing an emerging trend through scanning, one might quickly jump to potential implications for the organization (assessment) by implicitly forecasting the future path of the trend. If warranted by the potential impact, one may then continue scanning and monitoring. Also, forecasting often proves difficult, if not impossible, because of insufficient knowledge and data about the topic or trends being forecast, thus, forcing a return to scanning and monitoring efforts. Deriving implications (assessment) often alerts the organization to the need to conduct further scanning, monitoring, and forecasting. In short, environmental analysis is not as simple and as linear as moving from scanning to monitoring to forecasting to assessment.

Three key aspects need to be emphasized with respect to figure 11-4: (1) the approach to analysis adopted by a firm; (2) temporal cycles; and (3) differences among various environmental segments presented earlier regarding the nature of these activities.

Two Approaches to Environmental Analysis

It is possible to conceptualize and execute two distinct though related approaches to environmental analysis. One approach, which we shall call the *outside-in* or *macro* approach, adopts a broad view of the environment, focuses on longer-term trends, develops alternative views or scenarios of the future environment, and then, derives implications for the industry surrounding the firm and, of course, for the firm itself. An alternative approach, which we shall call the *inside-out* or *micro* approach, adopts a narrow view of the environment, develops a picture of what is currently happening through ongoing monitoring as a basis for forecasting the immediate future environment, and then derives implications for the industry and the firms within it. A characterization of these two approaches is presented in Table 11-3.

The two approaches are integrally related. The outside-in perspective ultimately leads to inside-out analysis: What are the short-run and medium-run implications of our broad environmental scan for the organization (for example, strategy issues)? The inside-out perspective can help sharpen points of departure for the outside-in approach. An example of the latter occurred when a bank adopting the narrow focus of the inside-out approach began to notice the importance of information technology in almost all phases of its operations. It then launched a major scan of bank-related technology developments to detect strategic and operational implications.

Indeed, it is impossible to do environmental analyses that focus on the broad environment and derive specific implications for the organization without adopting both approaches. Many organizations do both with varying degrees of integration. The point to be emphasized here is that organizations will find it productive to adopt both approaches to environmental analysis and to know which approach they are employing.

Temporal Cycles

Figure 11-4 portrays the multiple time-bound cycles in environmental analysis. Discovery of surprises or discrete issues during early scanning activities may require immediate action on the part of the organization: this is a short-term cycle. In a similar vein, monitoring activities may engender short- and medium-term actions, whereas a complete cycle typically is useful for long-term actions. Stated differently, the issues requiring organizational action are partly characterized by time spans for action implications. There are two considerations here. First, in environmental analysis, it is important to distinguish among these issues based on their time frame of impact and to consider implications *separately*. Second, depending on resources, time, and inclination, an organization may choose to analyze an emerging pattern in a segment or multiple segments. These decisions determine the breadth of focus of environmental analysis. As

Table 11-3. The Outside-In and Inside-Out Perspectives

	Outside-In	*Inside-Out*
Focus and scope	Unconstrained view of environment	View of environment constrained by conception of organization
Goal	Broad environmental analyses before considering the organization	Environmental analysis relevant to current organization
Time horizon	Typically 3-5 years, sometimes 5-10 years	Typically 1-3 years
Frequency	Periodic or *ad hoc*	Continuous or periodic
Strengths	Avoids organizational blinders Identifies broader array of trends Identifies trends earlier	Efficient, well-focused analysis Implications for covert operations

Source: L. Fahey and V. K. Narayanan, *Environmental Analysis for Strategic Management* (St. Paul, Minn., 1985).

noted earlier, broader analyses typically are necessary for longer-term cycles.

Differences among Environmental Segments

The activities presented in Figure 11-4 tend to assume different characteristics among environmental segments detailed earlier. Table 11-4 presents the key areas of differences among segments. As shown in the table, the segments present differential difficulties due to differences in availability of data, types of problems to be resolved, availability of indicators, and general accuracy of forecasts. Typically, the social segment is relatively slow to evolve, whereas the political segment is extremely turbulent and the technological segment is rendered obscure in the long run by considerations of secrecy. The table also shows the primary areas of impact on organizations. We now turn to a general framework for consideration of these impacts.

INTEGRATING ENVIRONMENTAL ANALYSIS INTO STRATEGIC ANALYSIS

The major purpose of integration is to derive from issues identified during environmental analysis their implications for strategic actions of a firm. Integration is, thus, primarily inside-out in character. Three points need to be kept in mind regarding integration. First, environmental analysis, although often intrinsically interesting, is useful only to the extent that it results in strategy-related insights and actions. Second, integration does not just happen, it is *made to happen*. The specific linkages to various kinds of actions need to be thought through and not left to evolve in a happenstance manner. Third, integration needs to take place for short-, medium-, and long-run horizons.

The implications of environmental issues need to be assessed for strategic planning at three levels: (1) corporate strategy where product-market decisions, that is, decisions to enter, withdraw, or remain in an industry, are usually the focus; (2) business strategy where the focus is on how to compete within an industry; and (3) functional-level strategies, where operational decisions are the concern. Given the pervasiveness of business strategy, we discuss this level first and in greater detail and then point out implications for the functional and corporate levels, respectively.

Business Strategy

Macroenvironmental analysis provides one set of critical intelligence inputs into the formulation of business strategy, that is, strategy at the level of a single-business firm or a business unit in a multibusiness firm. For a comprehensive strategy analysis, issues need to be assessed for their implications for the industry and competitor environment first and together with these for business strategy (see Figure 11-6 and chaps. 13 and 14).

Industry Structure. Changes in the macroenvironment may affect (1) the boundaries of the industry; (2) the forces shaping industry structure, such as suppliers, customers, rivalry and product substitution, and entry barriers; (3) strategic groups; (4) the key success factors; and (5) the general expectations within the industry. These elements provide the competitive context within which business strategy is formulated.

First, and of perhaps the greatest importance, is the impact of environmental change on the survival of an industry or specific industry segments. Environmental change can sometimes have more sudden and significant impact on *industry* (segment) *boundaries,* and thus survivability, than the structural and competitive forces within the industry. Economic progress can be viewed as one long history of technology change contributing to shifting industry boundaries. For example, technology advances underlying frozen foods and personal computers have irrevocably altered our conceptions of the food and computer industries.

Table 11-4. Environmental Segment

Unique Features of Analysis	*Demographics*	*Social Life Styles*	*Social Values*	*Political Milieu*	*Regulatory*	*Technological*	*Economic*
Activities							
Scanning							
Data availability	Abundance of quantitative data	Some quantitative data, qualitative data needs to be searched out	No standard measure available	Data available from personal sources, access may pose problems	Data available from primary and secondary sources	Data on national trends available, otherwise shrouded in secrecy	Abundant data on a weekly, monthly, and yearly basis
Key problem area	Organization and interpretation of data	Detection of new life styles	Values have to be inferred in many cases	Data need to be elicited on real-time basis	Relatively detectable; discontinuities at the judicial level	Detection is difficult because of secrecy issues	Structural shifts difficult to detect
Monitoring							
Availability of indicators	Plentiful; available on ongoing basis	Indicators sometimes unique to life style: market research techniques needed	None. Have to be created for specific purposes	None. Have to be created for specific issues; key is tracking events	No standard measures	No standard measures. Specific performance parameters for specific changes. No timely availability	Wide array of standardized indicators, availability on real-time basis
Forecasting							
Theories of change	Models of population growth and shifts	None available for prediction	None available for application	None available	No theory of regulation	No theory of invention; some for diffusion of innovation	Several available for cyclical and seasonal change (for example, Kordratiefs cycle for long term)

(continued)

Table 11-4. (*Continued*)

Unique Features of Analysis	*Demographics*	*Social Life Styles*	*Social Values*	*Political Milieu*	*Regulatory*	*Technological*	*Economic*
General accuracy	Fairly high in the short run	Low for new life style	Unknown	Variable; none for long term	Generally moderate	Generally moderate to low	Low in terms of magnitude of effects, high in terms of direction of effects
Assessment Primary impact[a]	Market potential, growth, regional concentration[b]	Market segmentation, product differentiation[c]	Creates discontinuities and unique pressures[d]	Risk, creates discontinuities, stakeholders[e]	Entry barriers, cost position diversification pattern, product requirements	Product substitution, product differentiation cost position, new industries for diversification	Demand, cost of doing business, availability of capital, acquisitions, divestitures
Techniques	Transition matrices, log linear models	Focus groups, life-style profiling diffusion matrices	Value profile (Wilson, 1977) social pressures priority analysis (Wilson, 1977), socio-political forecasting	Network analysis, political-risk analysis, socio-political forecasting	Network analysis, event-history analysis	Logistic curves, time independent comparison, Delphi, morphological method, relevance trees	Economic models, input-output matrices simulation models, industrial dynamics time series analysis, trend extrapolation

[a]Selectively presented
[b]Scenarios
[c]Cross Impact
[d]Matrices
[e]Alternative futures

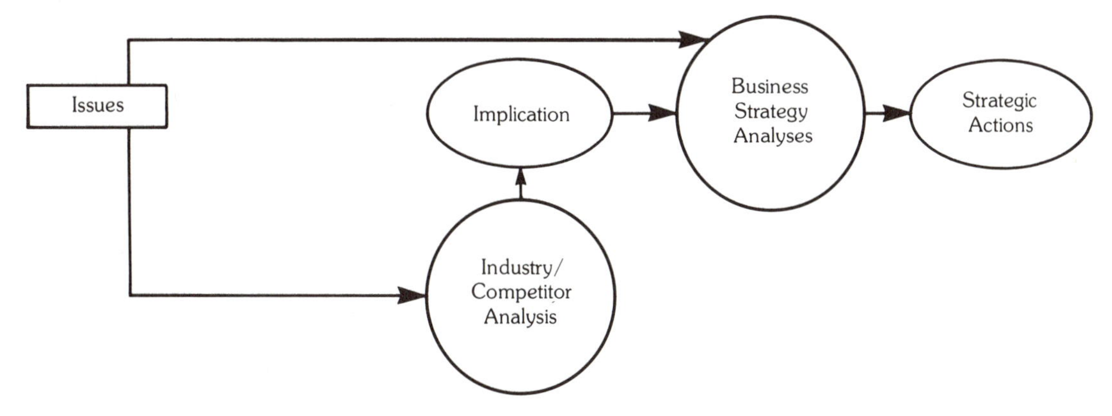

Figure 11-6. Issue assessment.

Second, environmental trends directly influence *the forces shaping the industry structure.* The forces may be conveniently summarized as suppliers, consumers, new entrants, and substitutes. Environmental changes can affect (1) the number, type, and location of suppliers, their products and supply costs and competitive dynamics of suppliers; (2) the size, characteristics, and behavior of a firm's customers; (3) the rate and trends in product substitution; and (4) a change in entry barriers.

Third, environmental changes have differential impact on various *strategic groups* within the industry. Changes, to the extent that they affect customers' preferences, suppliers' capabilities, substitute products, and so forth, could potentially enlarge or decimate the product-market arenas in which different strategic groups operate. Perhaps more importantly, environmental changes may afford opportunities for firms in a specific strategic group to overcome mobility barriers, that is, the barriers inhibiting a firm from moving from one strategic group to another. For example, deregulation of the airline industry in the late 1970s had an adverse impact on longer-haul firms relative to those with shorter hauls, thus facilitating restructuring of the routes of larger airlines and the capturing of these by smaller airlines.

Fourth, environmental change can potentially affect the *key success factors* in almost any industry or industry segment. At a minimum, environmental changes need to be assessed in terms of their impact on factors such as relative cost positions, product quality and functionality, image, reputation, and resource requirements for major product-market segments.

Fifth, the above discussion can perhaps be summarized by noting that environmental changes potentially affect *general expectations about an industry* and about the firms within it. These general expectations are important because they may have an impact on the level of investment funds into the industry and on the stock price behavior of the firms within it. Similarly, firms that have not had to contend with major competitive battles may be completely misled if they do not carefully assess implications of environmental change. Their prior general expectations may no longer be valid. For instance, since the deregulation of the telecommunications industry, many telephone companies are discovering that their assessments of competitors' responses shaped by past behavior are no longer valid.

In summary, forces in the macroenvironment can have an impact on industry structure and, therefore, on the choice and implementation of business-level strategy. The implications of macroenvironmental trends need to be assessed for each of the industry elements noted above. In this context, it is convenient to develop issue-impact matrices tracing the influence of environmental issues on each of the industry factors. These issue-impact matrices detail the effect of each one of the selected set of environmental issues on industry-level factors. Matrix displays of the type shown in Figure 11-7 facilitate assessments of these impacts. These assessments should include not only the general direction of change but its *timing* and *intensity*. Such assessments form much of the industry backdrop against which business strategies are formulated.

Impact on Business-unit Strategy. Together with industry and competitor analysis, environmental analysis outputs need to be assessed for their impact on business-unit strategy in terms of (1) business definition; (2) assumptions; and (3) general strategic thrust. These assessments are likely to be *medium-* and *short-term* oriented.

In broad terms, strategy formulation at the business or corporate levels includes a *definition of the business* and positioning of the business in an industry. Definition and positioning are inevitably affected by industry structure; macroenvironmental trends as they affect industry structure open up opportunities and threats for business strategy. Each of the three elements of business definition suggested by Abell (What customers does business serve? What customer needs are satisfied? What technologies are employed to satisfy these customer needs?) can be affected by environmental change.[17]

Thus, any given business definition needs to be evaluated in light of environmental change. Each of the three key elements in a business definition may be affected directly or indirectly by change in any one of the environmental segments discussed here.

Some pivotal *assumptions* underlie any strategy; for example, anticipated actions of suppliers, customers, competitors, or new entrants in the industry. These assumptions need to be anchored in a thorough analysis of the macroenvironment surrounding the industry if they are to be realistic and hence useful for strategy formulation.

The merits of identifying the pivotal environmental assumptions are threefold. First, it forces a thorough assessment of the macroenvironment, culminating in the major environmental assumptions that form the basis of strategy analysis. Second, it facilitates sensitivity analysis of key variables to which strategy options are vulnerable. Third, it frequently serves to heighten awareness of environmental change and its importance to strategic management.

Finally, the *general strategic thrust* of the firm or its business units such as market-share building or market-share maintaining are built around assumptions about industry. As we

INDUSTRY	TRENDS, EVENTS, DEVELOPMENTS IN THE MACROENVIRONMENT			
	Social	Economic	Political	Technological
Markets Customers Competitors Technology Materials and suppliers Production				

Figure 11-7. Issue-impact matrix.

have noted, these assumptions are influenced by changes in the environment. Environmental analysis often signals the need for changes in strategic thrust by opening up pathways to gain market share or rendering share-maintaining strategies obsolete.

Functional-level Strategy

Macroenvironmental change has implications for the functional-level strategies of an organization, over and beyond the business strategy. Consider the following examples:

Advances in telecommunications and computer applications have made it possible for marketing specialists to much more precisely segment and target customers. This is leading to changes in the implementation of marketing research, advertising, and distribution.

Changes in demographics, life styles, values, and technology have created a set of customers who are much more sophisticated than their predecessors in terms of the information they demand on products and how they use them. Partly as a consequence, some firms are retraining or revamping their sales forces to meet the *new* customer.

In the early seventies, as a result of general expectations regarding work hours, many organizations introduced flextime and alternative work schedules. In the late seventies, due to the increase in women in the work force, some organizations have created child-care centers.

As illustrated in the above examples, macroenvironmental change often necessitates changes in how various functional strategies are performed. Traditionally, these functional changes are regarded as operating issues within the context of strategic management. Environmental trends may open up avenues for enhancing operating capabilities or for rendering some obsolete.

Corporate-level Strategy

Corporate-level strategy focuses on questions related to business portfolio, including patterns of diversification and risk-return issues. Macroenvironmental impacts need to be assessed *separately* for each one of the corporation's business units; however, in addition, they have implications for the corporate-level strategy. The latter implications are likely to be medium- and long-term.

At the level of corporate strategy, environmental impacts on three key issues need to be considered: (1) patterns of diversifications; (2) portfolio planning; and (3) risk-return trade-offs.

Patterns of Diversification. There are at least three modes by which the environment influences patterns of diversification. First, firms differ in terms of the synergies they try to exploit across their businesses. These synergies could be upset or enhanced by macroenvironmental change. Second, different patterns of diversification manifest different vulnerabilities. Macroenvironmental change may amplify these vulnerabilities. Third, macroenvironmental trends may open up or close out existing patterns of diversification; this is particularly so when the pattern of diversification is not conglomerate. For example, regulatory changes in the 1950s rendered impossible vertical (forward) integration for the big "five" movie houses resulting in their divestiture of theaters. Similarly, technology and regulatory changes are opening up pathways for diversification in the financial-services industry.

Portfolio Planning. Macroenvironmental trends have important implications for the *bases* of portfolio planning. Typical portfolio approaches focus on business's competitive advantages within an existing industry, constrained by the internal financial resources of the firm. Ansoff notes that macroenvironmental trends may necessitate portfolio planning based on such bases as resources or technology.[18]

Environmental analyses are also particularly important for planning potential future portfolios. Product portfolio approaches are useful for portfolio planning *within* the existing set of businesses or at best pointing the direction of search for additional businesses. The specific businesses to be targeted need to be considered in the light of macroenvironmental forecasts and predictions.

Risk–Return Trade-Offs. Political, economic, technological, and demographic shifts have an impact on the returns and risks of existing and planned portfolios.

In a conglomerate firm, the accepted macroenvironmental trends suggested a persistant level of high inflation in the economy. One of the consequent considerations was that additional investment in any business unit should be justified not only by competitive position but by returns in excess of the forecasted rate of inflation.

In a technology-related firm, a technology study suggested obsolescence of some key technologies within the next decade. As a result, the firm is searching for methods of converting existing technology so as to retain competitive advantage in its existing markets.

It is important to consider environmental impacts on each of these characteristics of corporate-level strategy.

As was noted earlier, the primary utility of environmental analysis lies in its potential for enhancing effectiveness of strategic actions. However, these benefits are not likely to be realized unless analysis efforts are supported by some crucial organizational prerequisites.

ORGANIZATIONAL PREREQUISITES

Environmental analysis is an activity that requires people, resources, and time. Someone in the organization must spend the time to do the requisite analytical tasks involved in environmental analysis. Resources beyond people are often required: money to fund data collection, to buy outside analysis capability, or to support internal analysis efforts. Much managerial time is consumed in organizing to carry out the analytical tasks inherent in environmental analysis; deciding who should do what in collecting, analyzing, and interpreting data; establishing scanning and monitoring roles and overseeing their implementation; and creating the organizational processes such as task forces, *ad hoc* teams, or working groups required to effect these tasks.

The importance of the need to manage organizational prerequisites cannot be overemphasized. Almost all the studies dealing with environmental analysis in large organizations suggest that the management of environmental analysis often dominates the technical or analytical factors as a determinant of the quality and effectiveness of environmental analysis.[19] In this section, we will consider in brief some of the organizational prerequisites and approaches to managing environmental analysis in general. The discussion is organized around three prerequisites: (1) cultural-political; (2) structural; and (3) process factors.

Cultural-Political Prerequisites

There are some key cultural-political prerequisites to getting environmental analysis initiated, sustained, and integrated into the strategic management process in organizations. These are necessary for maintaining the legitimacy, visibility, and credibility of environmental analysis.

Top Management Commitment. Top management plays a key role in shaping the culture of an organization; without top management commitment, environmental analysis is not likely to be initiated or sustained. Commitment means more than the allocation of resources; it also means a propensity to highlight the importance of environmental analysis. Top management involvement sends a major signal to the organization that environmental analysis is important and legitimate.

An Environmental Analysis Champion. Environmental analysis, especially in its *early stages,* requires a credible champion to make it visible within the organization. The champion should ideally be able to wield *sufficient influence* in the organization. The role of the champion is important because not all members of top management can pay continual attention to environmental analysis issues, given the demands on their time. The champion's role is to *nurture* environmental analysis efforts; this includes political amnesty to those engaged in environmental analysis who are likely to be viewed with suspicion in the organization until that time when environmental analysis becomes ingrained as part of the culture.

The Sustenance of Ambiguity. A key role for top management and the environmental analysis champion is to create and sustain a sufficient level of ambiguity in the process of doing environmental analysis. This implies an ability to live with uncertain information and conflicting interpretations and to downplay expectations of certainty, often demanded by the operating sectors of the organization. This also involves legitimizing analyses that challenge the status quo, through such processes as devils' advocacy or institutionalizing *dialectical* processes.[20] Further, individuals engaged in these analyses should be shielded from punishments and norms of conformity and should be encouraged to think openly and creatively.

Interfaces into Strategic Management. Commitment to environmental analysis at the operating levels in the organization needs to be generated primarily by highlighting the role of environmental analysis in strategy analysis and particularly the need to weave environmental *assumptions* into strategy analysis. This could be accomplished by having senior line management ask environment-related questions in strategy presentations and making environmental analysis a *formal* part of analysis in strategy decisions, strategy development, and implementation.

Structural Prerequisites

Structural prerequisites are important because they delineate allocation of tasks and responsibilities for doing environmental analysis and have an impact on the quality of environmental analysis outputs and their usefulness for strategic analysis. Five key issues will be discussed.

Multiple Structural Mechanisms. A wide number of choices are available to an organization in terms of structural mechanisms to facilitate the doing of environmental analysis. At the simplest level, an organization could rely on outside agencies to provide it with the requisite analysis; at the other end of the spectrum, an organization may have a unit devoted entirely to doing environmental analysis. To some extent, the structural mechanism employed defines the scope of environmental analysis; the skills and biases represented in the structure determine which segments are attended to and which are omitted.

It is important to note that most organizations have multiple structural mechanisms for environmental analysis. What structure is used depends on the purpose of the analysis and the amount of resources available. Outside agencies very often are relied on for routine data provisions (for example, economic data) or quite sophisticated data analysis and forecasts (technological and political issues); *ad hoc* teams frequently are assembled to identify and assess strategic issues; task forces involved in such strategic activities as new product development or diversification may include environmental analysis representatives; more generally, environmental analysis units can serve multiple functions in ongoing strategic planning processes. Where the intent is to institutionalize environmental analysis, these structural mechanisms should be considered as complementary and not mutually exclusive.

Linkage Mechanisms. Without linkage mechanisms, environmental analysis is not likely to be integrated into strategic analysis. The driving forces behind the creation and maintenance of these linkages are the notions that involvement in the process of environmental analysis is an important learning experience for organizational members and that environmental factors should be considered in all stages of decision making.

Linkage problems are likely to be acute in the case of reports from outside agencies. Thus, sufficient time must be devoted to understanding the implications of outside agency reports. This may involve getting such reports on the agenda of *ad hoc* committee meetings, task force meetings, or regular management meetings. Linkage problems may also arise in the case of environmental analysis units at an advanced stage of expertise. For many reasons, the experts may have problems communicating with the rest of the organization. Thus, it is useful to consider rotation of individuals through environmental analysis units to partially alleviate these problems. In the case of *ad hoc* task forces, it may be necessary to have an environmental analysis representative during strategy analysis to help ensure that critical environmental factors are considered during all phases of decision making.

Differentiation between Corporate and Business-unit Levels. In large corporations, environmental analysis is done at both the corporate and business-unit levels. For example, many types of macroenvironmental forecasts (for example, economic and political forecasts) frequently are produced at the corporate level, whereas other types of environmental analysis (for example, demographic and lifestyle assessments) are done at the business-unit level. It is important that such differentiation be established on a rational basis. Environmental analysis at the corporate level is frequently germane to a majority or all of the business units, whereas that at the business-unit level may have little relevance to other parts of the organization. Thus, where the corporation is composed of very diverse business units, a certain degree of decentralization of environmental analysis efforts may be necessary.

The decisions with respect to the above structural mechanisms may be guided by two principles pertaining to the location of environmental analysis centers and the design of environmental analysis units or task forces.

Location of Environmental Analysis Center. We use the term *analysis center* to emphasize that environmental analysis may be conducted within many different structural mechanisms—environmental analysis units, *ad hoc* teams, task forces, outside agencies, or as a part of the activity of other organizational entities such as market research or research and development groups. Irrespective of the structural mechanism employed, in attempting to assure linkage it is important to locate environmental analysis *as proximate as possible* to strategy analysis so that its outputs are useful for and utilized in strategy analysis. Unless there is close interaction between the two streams of analysis, the value of both the process and the products of environmental analysis will not be fully realized. Such proximity in location enables those engaged in environmental analysis to (1) portray the outputs in terms meaningful to strategy analysts; (2) fine-tune the process and outputs of environmental analysis so that they correspond to the strategy analysis requirements, thus enhancing the chances that action implications are relevant and meaningful; and (3) help strategy analysts to think through the implications in specific action terms.

Design of Environmental Analysis Units or Task Forces. A final set of considerations involves design of environmental units or task forces. We had noted earlier that heterogeneous viewpoints need to be represented throughout environmental analysis and especially in the early stages so that premature convergence is avoided. This in turn necessitates that design should focus on creating groups with a *heterogeneity of perspectives.* Diversity should be maintained on specific projects over time. Diversity over time is particularly important for long-standing environmental analysis units so that group-think does not ensue. This may necessitate that new blood with fresh viewpoints be brought in from time to time to resuscitate what may be an ossified analysis unit.

Managerial Process Considerations

Process considerations involve the dynamics of managing environmental analysis. They influence how environmental analysis gets done and also how it is linked to strategy analysis. We will address four major considerations.

Managing the Design Process. It is critical to manage the process of setting the overarching premises of environmental analysis and how it gets done. First, the premises involve two key issues: that environmental analysis outputs should be included in strategy analysis and that environmental analysis is not a quick fix or panacea for organizational problems. Setting these premises is not an exercise in rational analysis but should be viewed as an *educational effort:* The key individuals who play a role in strategic analysis need to be educated about the role of environmental analysis. Managing these expectations about the role, scope, and nature of environmental analysis is a key to the success of environmental analysis.

Second, environmental analysis is a function: there are multiple ways of carrying out the functions that are contingent on the circumstances. For example, we noted that there are multiple structural mechanisms for effecting environmental analysis. The choice of these structural mechanisms should be a deliberative process intended to fulfill the function of environmental analysis.

Managing the Internal Process. Managing the internal dynamics of environmental analysis units or tasks groups is another important consideration. Issues germane to managing organizational processes are relevant here, however, two issues specific to environmental-analysis-related activities deserve mention. First, as previously noted, environmental analysis is characterized by divergent, retroductive, and recursive processes; the analyst is expected to exercise judgment, intuition, and sometimes speculation. This necessitates internal norms different from those in operating cultures. Stated differently, norms of decisiveness, specificity, and consensus are not always functional; in contrast, norms of reflection, ambiguity, and legitimacy of differences in viewpoints need to be nurtured. The tasks of managing are twofold: to reinforce to those engaged in the analysis that such norms, though possibly drastically different from the rest of the organization, are functional and to highlight their functionality to those not involved in environmental analysis. The first task ensures quality of environmental analysis, whereas the second serves to protect those engaged in environmental analysis from negative political pressures.

Second, it is critical to manage the time frame of those engaged in environmental analysis. It is always easy to focus on the current environment because data is plentiful and analysis is relatively straightforward, but as the time horizon lengthens, data becomes murky and interpretations are much more difficult. Where individuals are rewarded on short-term considerations, the focus will be predominantly on the short term. It is therefore necessary to manage the motivational and reward schemes so that individuals are focused not merely on the short term but also on the long term.

Managing Linkages. Like the structural factors, managing the process of linkage is important if environmental analysis is to be useful for strategy analysis. Four key issues need to be considered here. First, and perhaps most important, the environmental analysis outputs should have direct utility to the strategy problems at hand. Thus, the kind of outputs useful for corporate strategy are likely to be different from those at the business unit or functional levels. Loss of linkage results when the outputs are mismatched.

Second and related to the above, environmental analysis outputs should be timed to fit into different planning cycles. We have noted earlier that short-, medium- and long-term characterization of environmental issues is useful for different planning horizons.

Third, the form of environmental analysis outputs should be intelligible to those engaged in strategy analysis. Environmental analysis is facilitated by the use of several methodologies and its own jargon, however, the outputs are not necessarily intelligible to those unfamiliar with them and thus may be ignored. This is particularly likely to happen when there is a highly differentiated environmental analysis unit staffed with individuals of unique expertise.

Finally, environmental issues should be given visibility at all stages of strategy analysis. The role of the linking person is often to make sure environmental issues are raised at the various stages of decision making.

Evaluation of Environmental Analysis. Of particular importance to organizations where environmental analysis has been in existence as a formalized process is the need to evaluate periodically the performance of environmental analysis. Stated differently, the function of environmental analysis should be *audited* with a view to assessing its contribution to strategy analysis. For example, at the analytical level, as the environment changes, previously held definitions of the environment will have to be discarded. At the organizational level, structures and processes will have to be changed given the particular set of circumstances facing the organization. At the cultural-political level, leadership changes may necessitate that environmental analysis redefine itself and renew its legitimacy and utility.

CONCLUSION

The value of environmental analysis inheres in the process of doing it in addition to its outputs. The process reflects a philosophy of managing: It underscores the key notion that organizations are incessantly pervious to the influence of outside forces. Environmental analysis when conducted properly leads to enhanced capacity and commitment to understanding, anticipating, and responding to external changes on the part of the firm's key strategic managers. Responsiveness to environment is achieved by inducing managers to think beyond their task or industry environment and to consider the environment and changes within it as a source of opportunities to be exploited. Thus, environmental analysis often forces managers to reflect on their cognitive biases. It unearths unidentified assumptions and blindspots in the thinking of key managers. In short, at the level of process, it offers one basis for organizational learning.

Despite these potentials, environmental analysis should be placed in proper perspective: In and of itself it is not a sufficient guarantor of organizational effectiveness. It forms only one input, although that input is crucial for strategy development and testing. Further, it does not foretell the future, nor does it eliminate uncertainty for any organization. Thus, organizations that practice environmental analysis do sometimes confront unexpected events—events not anticipated by the firm, although the frequency of such events is reduced by systematic efforts at understanding environment.

REFERENCES

1. P. F. Drucker, *Innovation and Entrepreneurship: Practice and Principles* (New York: Harper & Row, 1985).
2. A. Toffler, *Third Wave* (New York: Morrow, 1980).
3. J. Naisbitt, *Megatrends* (New York: Warner Books, 1982).
4. For example, R. H. Kilmann, *Beyond the Quick Fix* (Jossey-Bass, 1984).
5. L. Fahey, W. R. King, and V. K. Narayanan, "Environmental Scanning and Forecasting in Strategic Planning: The State of the Art," *Long Range Planning Journal* (April 1981).J. Diffenbach, "Corporate Environmental Analysis in Large U.S. Corporations," *Long Range Planning Journal* **16**(1983):107-116; R. T. Lenz and J. L. Engledow, "Environmental Analysis Units and Strategic Decision Making: A Field Study of Selected Leading Edge Corporation," 1983; C. Stubbart, "Are Environmental Scanning Units Effective!" *Long Range Planning Journal* **15**(1982):139-145.
6. K. E. Weick, *The Social Psychology of Organizing* (Reading, Mass.: Addison-Wesley, 1969).
7. F. J. Aguilar, *Scanning the Business Environment* (New York: Macmillan, 1967).
8. L. Fahey and W. King, "Environmental Scanning for Corporate Planning," *Business Horizons* (August 1977):62-63.
9. Fahey et al., "Environmental Scanning and Forecasting."
10. Fahey et al., "Environmental Scanning and Forecasting"; Stubbart, "Are Environmental Scanning Units Effective!"; Diffenbach, "Corporate Environmental Analysis"; Lenz and Engledow, "Environmental Analysis Units."
11. I. Ansoff, *Strategic Management* (New York: Halstead Press, 1981); F. Emery and E. Trist, "The Causal Texture of Organizational Environments," *Human Relations* (1965).
12. For example, M. A. Porter, *Competitive Strategy* (New York: Free Press, 1981).
13. Ibid.
14. This section is drawn from L. Fahey and V. K. Narayanan, *Environmental Analysis for Strategic Management* (St. Paul, Minn.: 1985).

15. W. R. King and D. I. Cleland, "Information for More Effective Strategic Planning," *Long Range Planning* **10**(1978):59-64.
16. I. H. Wilson, "The Benefits of Environmental Analysis," in *The Strategic Management Handbook*, ed. K. Albert (New York: McGraw-Hill, 1983).
17. D. F. Abell, *Defining the Business: The Starting Point of Strategic Planning* (Englewood Cliffs, N.J.: Prentice-Hall, 1980).
18. I. Ansoff, "Managing Strategic Surprise by Response to Weak Signals," *California Management Review* **19**(1976).
19. Fahey et al., "Environmental Scanning and Forecasting"; Diffenbach, "Corporate Environmental Analysis"; Stubbardt, "Are Environmental Scanning Units Effective!"; Lenz and Engledow, "Environmental Analysis Units."
20. R. Mason and I. I. Mitroff, *Strategic Assumption Analysis* (New York: Wiley, 1981).

CHAPTER 12

Stakeholder Analysis for Strategic Planning and Implementation

Aubrey L. Mendelow

Aubrey L. Mendelow holds a doctorate in business leadership from the School of Business Leadership of the University of South Africa. He is on the faculty of Duquesne University where he leads courses in the areas of business policy and management of information resources. In addition to his consulting interests in policy and information systems, Dr. Mendelow is researching the manner in which information systems can be used as competitive devices.

Stakeholders—those groups or individuals who can affect or are affected by the achievement of the organization's objectives[1]—have gained increasing attention from practitioners and academics since they were first identified by Abrams.[2] This rise in attention paralleled the growing recognition that the environment of the organization was becoming increasingly dynamic and complex. As firms began to suffer more and more due to jolts from their environment, planning systems emerged to help managers include environmental trends in their plans. These plans were developed in an attempt to reduce the adverse effects of environmental impacts and to take advantage of emerging opportunities identified during the course of the environmental analysis.

However, with the increasing complexity of the environment, a major problem was soon encountered: that of identifying the relevant environmental components to be taken into consideration for planning purposes. King and Cleland proposed a process of claimant analysis to help ameliorate this problem.[3] Claimants—"those groups or institutions who have a demand for something due from the organization"[4]—are thus a subset of the stakeholders and represent a framework within which the environment can be reduced to a series of components. These components, by definition, could have an impact on the extent to which the firm could achieve its objectives.

Thus a key assumption of the stakeholder framework is that successful strategies can be developed and implemented only when strategic planners take the potential impacts of their organization's stakeholders into account. It will soon emerge that the stakeholder approach is particularly useful during strategy formulation at the business level because of the ease with which stakeholders can be identified at this level. Corporate stakeholders include not only the direct stakeholders of the corporation itself but also the stakeholders of all its business units. Stakeholder analyses at the corporate level thus are particularly useful in developing uniform corporate policies with respect to stakeholders who might interact with more than one business unit. Therefore, except where specifically necessary, the words *organization, firm,* and *company* will be used interchangeably.

This chapter examines methods for performing stakeholder analyses and the manner in which the results of these analyses may be included at each stage of the strategic management process. Accordingly, the chapter is divided into several distinct sections, each dealing with the application of stakeholder concepts to a stage of the strategic management process. Sections thus will be devoted to objective formulation, environmental scanning, the position audit, alternative generation, alternative selection, and finally to successful implementation. Prior to developing these applications, it is necessary to discuss more fully some of the concepts underlying the stakeholder approach and then to examine techniques for identifying the organization's stakeholders.

CONCEPTS UNDERLYING THE STAKEHOLDER APPROACH

Rise of the Concept of a Stakeholder

Organizations do not exist in isolation. Indeed, the existence of most organizations, irrespective of whether or not they are formed to make a profit, are legitimized by some mechanism of society. For instance, companies are legitimized by certificates of incorporation from the state, charities by registration as welfare organizations, and health-care bodies by both local authorities and national boards. On this basis, it is argued that organizations will be allowed to continue to exist only for as long as they provide benefits by way of output to their stakeholders and indeed to society as a whole. Clearly, such outputs are both tangible and intangible. Moreover, benefits arise only when the output of the organization exceeds the resources it consumes. Hence, managers need to understand fully the nature of the resources that the organization is using and the output it is providing.

In the past, the stockholders have occupied somewhat of a privileged position in organizations—their contribution of funds in exchange for equity was tangible proof of their interest in the organization. Moreover, the funds were often provided at a moment when the firm was financially strapped. As a result, both management and academics alike have viewed stockholders' needs as being of paramount importance. Indeed, a widely held view is that the primary task of management is to increase the wealth of the stockholders. However, Abrams was one of the first to realize that firms should "strive to maintain an equitable and working balance among the claims of various directly interested groups—stockholders, employees, customers and the public at large."[5] Freeman and Reed report that an internal memorandum written in 1963 at the Stanford Research Institute coined the word *stakeholder,* which was then defined as being those groups without whose support the firm would cease to exist.[6]

Some twenty-three years after Abrams, Drucker asserted that the social impacts and social responsibilities of a firm have to be managed.[7] At about the same time, the Business Environment Research Group at General Electric realized the need to examine issues. In their view "without a proper business response, societal expectations of today become the political issues of tomorrow, legislated requirements the next day and litigated penalties the day after that."[8] The stakeholder approach has a far broader utility than one that simply makes a plea for social responsibility. Indeed, the stakeholder perspective encompasses the societal responsibilities of business since it recognizes that society at large is but one of the organization's stakeholders.

Stakeholder management thus is aimed at proactive action—action aimed, on the one hand, at forestalling stakeholder activities that could adversely affect the organization and, on the other, at enabling the organization to take advantage of stakeholder opportunities.

The first aim is achieved by ensuring that the organization is effective vis-à-vis each category of stakeholder; stakeholder opportunities are used to advantage by developing an in-depth understanding of the forces that influence the relationship between the organization and its stakeholders.

This understanding can be achieved only through a conscious decision on the part of managers to adopt the stakeholder perspective as part of their strategy formulation process. However, this in itself is insufficient. Strategies relating to each of the stakeholders have to be implemented, and, indeed, an all-encompassing approach to stakeholder management requires organizations to adopt a series of integrated strategies capable of dealing with the entire stakeholder environment.[9]

There is yet a further justification for the adoption of the stakeholder perspective in strategy development. Organizations can be viewed as goal-seeking entities. The achievement of these goals requires the organization to execute a series of ongoing transactions with its stakeholders. Thus it can be argued that the extent to which an organization achieves its goals is influenced by the actions of its stakeholders: stakeholder cooperation enhances goal achievement, whereas non-cooperation hinders it. Management should therefore attempt to ensure that the organization's stakeholders are prone to behave in a manner that will enhance the organization's ability to achieve its objectives.

Identifying Stakeholders

The first step in applying the stakeholder approach to strategic management is to identify the organization's stakeholders. During this process, care must be taken to ensure that *specific* stakeholders are identified. Failure to make specific identification gives rise to a generic set of stakeholders typically comprising shareholders, lenders, suppliers, government, employees, society, customers, and competitors. These stakeholder categories can be applied to most organizations, hence, little insight is gained for the specific organization. Generic stakeholder categories may be of some use in developing generalized policies, but they are of little use in developing detailed strategies.

The organization's competitors have to be included in any list of stakeholders. In examining the competitors, attention traditionally is focused on those firms who compete against the focal firm in the marketplace. This is myopic. Competitors abound for the favors of each of the organization's stakeholders. Thus competitors that achieve a special relationship with a specific stakeholder of an organization could be given favorable terms of business, resulting in the competitor deriving a competitive advantage from the relationship.

Freeman points out that a historical analysis of an organization's interaction with its environment is a useful point of departure.[10] This method requires the construction of a list of the issues that have affected the organization over, say, the past five years. The stakeholders involved in these issues can be identified from this list. However, this process provides only a historical perspective. It therefore is necessary to identify issues the firm is currently facing (or are about to face) in order to include any new stakeholders on the list. Once the stakeholders are identified, it is useful to tabulate them (Table 12-1) and to set them out diagrammatically in the form of a map (Figure 12-1). The map highlights interactions among the stakeholders and is useful for analytical purposes and scenario building.

Porter's value chain is another useful basis for identifying stakeholders.[11] In his view, an organization adds value to the parts and raw materials it purchases by way of its primary and support activities. Primary activities relate directly to the production of the final goods or services, to their distribution to the customer, and to after-sale service. Support activities are those necessary to support the primary activities. Primary activities thus comprise five distinct categories: inbound logistics, operations, outbound logistics, marketing and distribution, and service. Each of the primary activi-

Table 12-1. Tabulation of Stakeholders of a Large Corporation

Generic	*Component*	*Details*[a]
Shareholders	Management (give names)	Shares held
	Board (give names)	Date acquired
	Institutions	Changes in holdings
Suppliers	Supplier 1	Extent of dependence
	Supplier 2	Special terms
	•	Alternative sources
	•	
	Supplier n	
Lenders	Bank 1	Value of loans
	Bank 2	Lending officer
	•	Date of first association
	•	
	Bank n	
Employees	Union 1	Name of negotiator
	•	Contract renewal date
	•	
	Union n	
Competitors	Competitor 1	Market share held
	•	Basis of competition
	•	CEO
	Competitor n	Relative size
Society	Interest Group 1	Nature of interest
	•	
	•	
	Interest Group n	
Government	Department 1	Include local, state, federal and foreign departments.
	•	
	•	
	Department n	
Customers	Segment 1	Market size
	•	Percent of unit revenues
	•	
	Segment n	
Industry	Trade Association 1	
	•	
	•	
	Trade Association n	

[a]Some examples of the data that can be listed are shown.

ties depend on the support activities that comprise the firm's infrastructure: human resource management systems, technology development procedures, and general procurement. The beauty of this model is that it forces managers to think in a structured and detailed fashion about the stakeholders who support each specific stage of the value-added chain, giving rise to insights that may be overlooked in the normal course of events.

For instance, the information system, which Porter groups under the heading of firm infrastructure, depends on the hardware vendors for the supply and service of the hardware necessary for the satisfactory operation of the information system. A further example

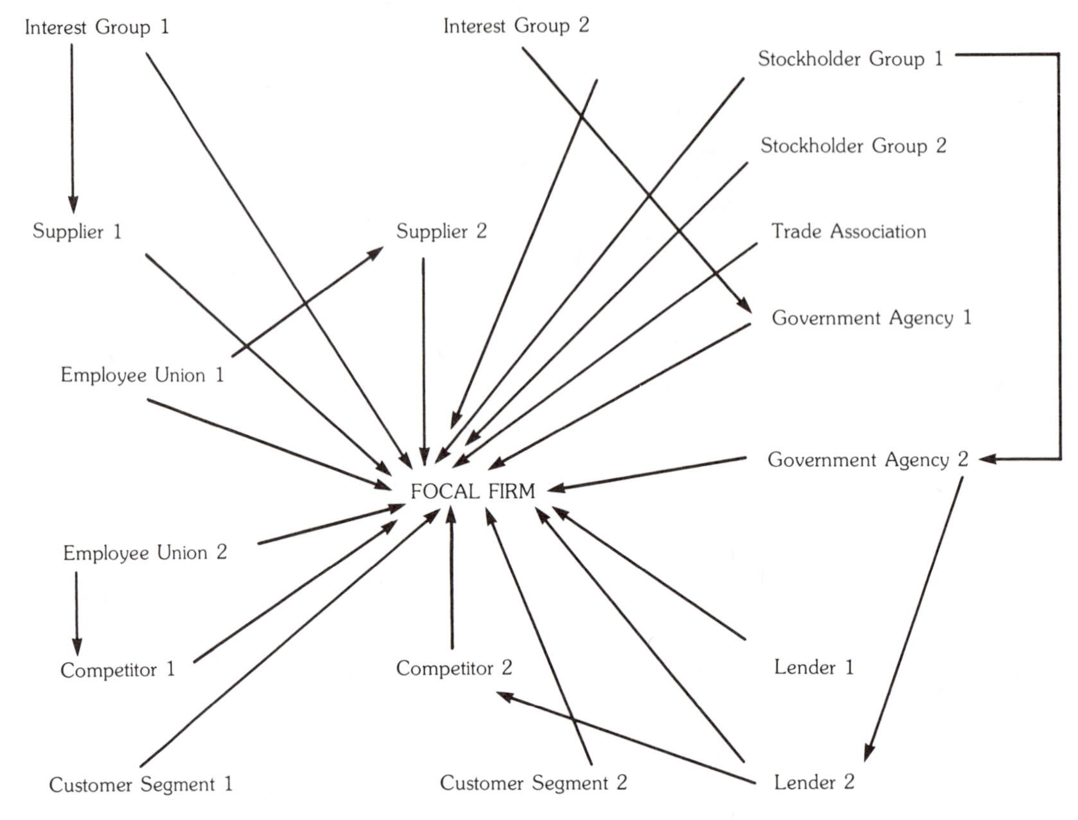

Figure 12-1. A Stakeholder map. Arrows represent the direction of influence.

of often overlooked stakeholders are those who interact with the outbound logistics system. Stakeholders of this component are the firm's distributors, who in turn may fall into several categories. Each of these categories have to be identified because of the differences in their expectations.

Stakeholder Attributes

Once the stakeholders have been identified, it is useful to examine them for several attributes. These attributes can serve as a basis for the development of a series of policies. Useful attributes for stakeholder analysis have been found to be orientation, opinion, and power.

Orientation. Stakeholders can be classified into one of two orientation categories, depending on whether they are in favor of the organization (pro) or against it (con). Stakeholders whose orientation favors the organization generally perceive that their own interests are best served if the organization achieves its goals. Stakeholders with this orientation thus tend to act to facilitate achievement of the organization's goals. In a dispute, pro-stakeholders would tend to side with the organization, generally going out of their way to assist it in any way possible should the need arise. Indeed, such stakeholders often are used by the organization to win over other stakeholders to its position on any given issue. From a policy perspective, it therefore is important for an organization to keep its pro-stakeholders well informed of its goals, the extent to which they will be achieved, and any influences that might inhibit their achievement.

Conversely, stakeholders who are against the organization perceive that their interests are best served if the organization does not achieve its goals. Their tactics tend to be disruptive. Clearly, care has to be exercised in dealing with this group of stakeholders. In fact, information released directly to this class of stakeholders has to be screened to ensure that it will not be used against the organization some time in the future. Despite this, it is important to keep communication channels open with this class of stakeholders since this is the most effective way of putting the company's point of view across. Failure to attend to these direct communication channels frequently drives these stakeholders to seek information from secondary sources—a process that often gives rise to inaccuracies and misunderstandings.

Opinion. An underlying tenet of stakeholder action is that stakeholders act in a manner that, in their opinion, will further their interests. Thus, from a strategic perspective, stakeholder opinions must be categorized with respect to issues concerning the organization. During this process, it is crucial to realize that the stakeholders may alter their opinion category according to the corporate issue with which they are confronted. Therefore, it is conceivable that a stakeholder could be classified into more than one opinion category! This means that managers have to identify the issues that may be important to the dual stakeholder and take such issues into account during strategy formulation. In addition to their orientation toward the organization, Gruning points out that publics, and thus stakeholders, may be categorized into three types: active, aware, and latent.[12]

Active stakeholders. These stakeholders actively pursue the advantages or actively attempt to reduce the disadvantages of their transactions with the corporation. Active stakeholders who are for the firm are a blessing but dare not be taken for granted. Should they be taken for granted over a lengthy period, these stakeholders could turn against the firm. Conversely, active stakeholders who are against the firm have to be handled with care. This group will use any pretext or scrap of information against the organization. As a result management has to adopt policies to reduce the adverse impact of this class of stakeholders.

Aware stakeholders. Aware stakeholders generally recognize the advantages or disadvantages associated with their entering into transactions with the organization, but are not sufficiently motivated to act to seize the advantages, on the one hand, or to act to reduce the disadvantages, on the other. As aware stakeholders, they tend actively to gather information about the corporation in an effort to update their perspectives on the corporation, to gauge where it stands in respect to the issues in which they—the stakeholders—are interested. While formulating strategy, it is crucial that managers give serious consideration to these stakeholders. In effect this is a make-or-break situation. Failure to win these stakeholders over to the organization's side could result in its facing a group of stakeholders who actively oppose it.

Latent stakeholders. Latent stakeholders are ambivalent in their attitude toward the organization—they perceive nothing particularly good or irksomely bad about the organization. In fact, they are happy as long as the organization continues to proceed on the basis on which it is currently operating. Indeed, they will remain in this state for as long as there is no issue that causes them to alter their opinion. Thus, these stakeholders have few problems with the manner in which the corporation is acting and perceive few current threats by the corporation to their freedom. As a result, these stakeholders have a low level of involvement with the organization and thus do not actively seek information relating to it. Hence, latent stakeholders tend to depend on information that comes to their notice without any effort on their part.

Although these stakeholders are welcome ones because superficially they are not very demanding on management's time, top management cannot afford to ignore them. They have to be identified and fed corporate information, because latent stakeholders will not actively seek it of their own accord.

Latent stakeholders can switch to the aware and/or active categories very quickly should they feel themselves threatened in any way. Moreover, latent stakeholders, switching to the active or aware state due to a perceived threat to their interests, could well orient themselves against the organization as it attempts to resolve the issue. This means that management has to act to defuse any issue that could jolt the stakeholder from a state of latency to one of awareness or even to one of activity with a con orientation.

Power. One of Porter's contributions to the field of strategy was to point out that the power of suppliers and customers of the firm has to be considered as a factor in strategic analysis.[13] Stakeholders tend to interest themselves actively in the affairs of the firm in two general instances: when they perceive themselves to be threatened or potentially benefitted by the organization's actions. Stakeholders' interest will give way to action only when they perceive that they have the power to make the desired changes. It is thus strategically necessary for top management to evaluate the power each group of stakeholders possesses. In this way managers will be able to anticipate possible moves by the stakeholders before the effects of these moves are felt by the organization.

Top management has to determine the circumstances under which stakeholders will rely on their power base. As alluded to above, stakeholders will attempt to use their power to influence the corporation to act in their favor when they perceive an incongruity between their goals and those of the corporation. In addition, stakeholders will resort to power tactics when they compete with the firm for a scarce resource; labor resorts to power tactics as they compete with the firm and stockholders for monetary reward. The possibility of resorting to power plays is enhanced when there is no other player in the domain who has the authority to dictate the action of the firm and its stakeholders. Thus the threat of government regulation limits the extent to which a firm and its specific stakeholders resort to actions based on power. Finally, power is used by a stakeholder in an attempt to implement a critical set of decisions. In recent years, corporations have threatened that their very existence was at stake in order to push through job reforms and cost-cutting measures.

Baird suggested that stakeholders derive power from three key sources: the resources they possess, their ability to influence the decision process, and, finally, their position in the industry.[14] In addition, MacMillan indicated that power was also derived from the ability on the part of, say, a stakeholder to dictate alternatives, use authority, and wield influence.[15]

Possession of resources. Porter points out that suppliers possess power if they are dominated by a few companies and if, at the same time, the supplier's industry is more concentrated than the industry to which it sells.[16] In addition, should a company depend on a specific resource that is not readily available, the power of the supplier is enhanced. This power is manifested in negotiations over price, quality, payment conditions, and delivery specifications.

Influence on the decision process. A stakeholder may single out a manager who it perceives to be favorably disposed toward it. It may then direct all its communication to him or her. In this way the stakeholder feels assured of a sympathetic hearing. In addition, part of this communication could be aimed at setting out the criteria by which a decision should be made. For instance, a supplier obtains a toehold in a company and offers to help a manager write the specifications for the goods or services to be purchased. It is difficult to imagine that the supplier will not set out a series of criteria that its competitors will find difficult to match. By so doing, the supplier attempts to influence the purchasing decision in its favor.

Position in the industry. Power in this instance arises from the stakeholder's dominance of its industry. For example, IBM is able to set standards for software and hardware in the computer industry on the basis of its market share.

Dictation of alternatives. Where management has the option of alternative courses of action, stakeholders do not possess the power to dictate their terms. This means that stakeholders, in an attempt to achieve their aims, often attempt to limit the choices open to management. For instance, unions attempt to dictate management's alternatives by using picket lines and threats of violence to prevent nonunion replacements from entering the firm's premises.

Use of authority. Authority is the right to enforce obedience. Clearly, the power of central government is derived from its authority.

Wielding influence. Strategic planners may decide to influence industry standards by appointing a manager to the industrial body governing the industry. This manager would then attempt to pass legislation that would favor his or her company's position.

Once the organization's stakeholders have been identified and classified using the variables of power, orientation, and opinions, the data can be used as input to the strategic planning process.

STAGES OF THE STRATEGIC PLANNING PROCESS

The strategic planning process comprises a series of distinct phases: objective formulation, environmental scanning, the position audit, alternative generation, alternative selection, and implementation.

Stakeholders and Objective Formulation

A prerequisite for the strategic management of a firm's stakeholders requires that managers formulate objectives for each of the stakeholders. These objectives must have the same standing as those traditionally aimed at satisfying the needs of shareholders. In effect, this reduces the long-standing, almost total dominance of the shareholders. Of course, profits are necessary, but they are not the alpha and omega of an organization's existence, just as breathing, a prerequisite to human survival, is not the sole purpose of human existence.

Clearly objectives should be set in an attempt to meet, as far as possible, the needs of each of the stakeholders. This is easier said than done because there may be conflicting interests. Where such conflicts are encountered, they must be resolved in terms of stakeholder priorities; these priorities have to be determined based on the power, opinion, and orientation of each stakeholder group. Now the extent to which the expectations of the stakeholders are met by the corporation is, conceptually, the same as the effectiveness of the corporation. Several approaches may be used to identify the needs of each stakeholder.

The most direct method is to approach the stakeholders themselves. This method was adopted by a company wanting to set a series of debt-to-equity objectives. They sent their financial statements to potential lenders and asked them how much they would be willing to lend the company, using the financial statements as a basis for the decision. Management then calculated the debt-equity ratios that would arise from the loans of each of these potential lenders. The average of these ratios represented the basis for the debt-to-equity objective.

The deductive approach is another method for determining the effectiveness criterion of each stakeholder. Typically a literature survey is conducted, experts consulted, or representatives of each stakeholder category (rather than each individual stakeholder) are polled to ascertain the needs of the relevant stakeholder. This approach tends to result in a generalized set of stakeholder needs from which the specific needs of each stakeholder group can be gauged.

Mendelow suggested a combination of the two preceding methods.[17] Once the generalized list of stakeholder needs has been completed, the individual stakeholders are approached for an assessment of the extent to which these needs ought to be met. At the same time, the individual stakeholders are asked to rank each need in terms of its relative importance. By combining these ratings (weighting techniques could be used) and averaging the extent to which the needs have to be met, managers can develop a series of strategic objectives for each group of stakeholders. In addition, the relative priorities that should be accorded each stakeholder group can be decided. Once this is done, an ongoing review system has to be developed to track environmental changes that could result in changing stakeholder expectations.

Stakeholders and Environmental Scanning

Environmental scanning is one of the key elements of the strategic planning process. Scanning initially is undertaken as part of the strategy formulation process to assemble data for the position audit. However, scanning has to be an ongoing process to enable managers to track any changes in the environment that could affect the conclusions drawn from the position audit. In addition, the forces giving rise to these changes have to be identified and tracked.

In the same way, changes in stakeholder needs have to be tracked because they could alter the perspective of the stakeholder vis-à-vis the corporation. Failure on the part of the corporation to keep abreast of the changing needs of the stakeholders may give rise to stakeholder dissatisfaction. In turn, a dissatisfied stakeholder may attempt to change the manner in which the company relates to it by, for instance, altering its business terms. In order to anticipate impending changes in stakeholder needs, it is necessary to monitor the issues causing these changes.

Issues. Brown suggests that an issue is a condition or pressure that, if it continues, will have a significant effect on the power of an organization's stakeholders.[18] In order to successfully identify issues it is necessary to analyze the bases from which stakeholders derive their power. Two examples of issues may be:

The progress of legislation aimed at regulating the company's industry. This issue would alter the basis of power of government and regulatory agencies.

Emerging technology resulting in the availability of substitute raw materials. This issue would reduce the power base of suppliers of the existing raw materials.

Issues may be categorized in various ways to further aid the scanning process. Brown reports that General Electric uses a framework in which issues progress from (1) societal expectations to (2) political issues to (3) legislated requirements and, finally, to (4) punitive action.[19] The consumer movement is a widely-known illustration of the use of this framework: the need for product safety was widely discussed, then politicians took up the cudgels of the consumer, making the need for product safety part of their election manifestoes. Once the politician was elected, legislation pertaining to product safety was passed, and, after an interim period during which manufacturers were given the opportunity to comply with legislated requirements, punitive action was taken against those who broke the law.

This framework has several requirements. At the outset it requires that management identify those issues that have to be tracked. This is achieved by recognizing the basis on which stakeholders attain their power in relation to the company. Then the issues that can alter the power base of the stakeholders are identified. In addition, those issues that are likely to alter the stakeholders' disposition toward the company, for example, from latent to aware or from pro-company to against it, have to be identified. This can be done using group techniques such as the nominal group technique. Furthermore, Klein and Newman's SPIRE approach can be used to obtain a grasp of the possible impact of these issues.[20] Using this process, it is possible to develop a cognitive map of the cross impacts that the stakeholders exert on each other and on the company. Those issues that seem to have the greatest potential impact on the organization and its stakeholders are then tracked. In this way, the company may receive at least some warning of the impending alteration of stakeholder expectations as a result of a set of developing issues.

The chances are that, even using this framework, managers will find themselves overburdened by the amount of effort and time required to scan the environment for the issues that have been identified. Refinement is necessary and is achieved by determining the extent to which the environment of each stakeholder is changing. Clearly, more attention has to be given to the more dynamic aspects of the environment.

Modes of Scanning. Several authors have attempted to identify specific methods to scan the environment. Aguilar recognized four modes of environmental scanning: undirected viewing, conditioned viewing, informal search, and formal search.[21] Although this represents a useful conceptual framework, it is not action oriented. In order to overcome this, Fahey et al. suggested an action-oriented framework comprising three scanning activities: irregular scanning, periodic scanning, and continuous scanning.[22] In their view, irregular scanning is applicable when the organization has to react to a crisis initiated by slow-changing events. Conversely, they suggest that periodic scanning is a systematic attempt to scan the environment on a regular basis. Finally, continuous scanning is an ongoing process aimed at recognizing opportunities.

Selecting a Mode. The variables of power, dynamism of the stakeholder environment, and stakeholder opinion may be combined to develop an indication of the scanning mode that has to be adopted with respect to each of the stakeholders and their respective issues. It should be noted that in the very act of scanning, new issues may arise whose progress will also have to be tracked. Thus, for example, much attention should be devoted to those stakeholders who are actively acting against the firm and who, in addition, are in a position of power with respect to the firm, irrespective of the dynamism of their environment. The firm's major competitors are an example of such a stakeholder category. Their actions, and indeed their environments, have to be scanned with much care and on a continuous basis in an attempt to find ways to reduce their power and to anticipate their moves.

Irregular scanning would best be applied to situations in a static environment wherein stakeholders act in the interests of the firm but are, at the same time, in a powerful position with respect to the firm. In this situation, the purpose of the scanning effort is to alert management to any changes that might cause the stakeholder to alter its favorable opinion toward the firm. However, because the environment has been identified as being slow changing, very little additional information could be expected from a more continuous scanning effort.

Periodic scanning is best suited to those instances where the environment has been identified as being dynamic but where the stakeholders have little power. In this instance, periodic scanning is necessary because the dynamic environment may result in an alteration of the basis from which the stakeholders derive their power. This could result in a previously low-power stakeholder acquiring a basis for power, a possibility the organization needs to be aware of.

Allocating Responsibilities. In many instances, once the stakeholders and the associated issues have been identified, the sources of data often become readily apparent. It is particularly important to allocate the responsibility for environmental scanning to the department that acts as a boundary spanner to the specific stakeholder. (Boundary spanners are people who interact most frequently with the given stakeholder group and who therefore have some specialized knowledge of the stakeholder and the attendant issues.) Where several people or departments interact with the same stakeholder, ultimate responsibility for the scanning effort should be allocated to a specific person, with the others who interact with the stakeholder designated as contributors. The person who is ultimately responsible for information relating to a specific stakeholder should synthesize the information collected and forward it to a centralized collection area, often the planning department or the firm's library.

Stakeholders and the Position Audit

In traditional strategic planning, this phase often takes the form of an analysis of the firm's strengths, weaknesses, threats, and opportunities. The position audit has two components: the external audit and the internal audit. The internal audit, being introspective, is often focused on a functional analysis to determine the firm's strengths and weaknesses. The results of this analysis are input to the alternative-generation phase. Traditionally, the external audit is focused on the firm's market and resources. This approach is a narrow one.

In general the stakeholder approach is able to provide another dimension to the position audit during both the internal and external phases. This is due to the symbiotic relationship that exists between the firm and its stakeholders: the firm cannot exist without its stakeholders and, conversely, the stakeholders rely to some extent on the firm for their existence. Stakeholders provide support in return for the organization's inducements. Thus a new perspective may be gained by identifying and understanding the reasons why the firm's stakeholders provide it rather than its competitors with their resources. Adequate inducements represent strengths, whereas inadequate inducements represent weaknesses. For instance, where a firm has a low debt-equity ratio, this would be considered to be a strength. Banks would be prepared to lend funds to the firm. Conversely, where customers need substantial technical support in order to use a product efficiently, a firm lacking this type of support would have identified a weakness.

The stakeholder approach can also be extended to the external phase of the position audit. Threats are viewed as those environmental factors that will increase the inducements anticipated by the stakeholders and, conversely, opportunities as those factors that reduce the stakeholders' expected inducements. Where the firm finds that it is providing inducements to its stakeholders in excess of their expectations, this represents an opportunity. Management would then be able to redeploy the firm's resources by reducing the extent to which it provides the inducements.

Stakeholders and Alternative Generation

The stakeholder approach represents an additional tool that can be used by management for identifying alternative corporate strategies. Porter's model of the forces that shape strategy formulation gains additional applicability and a wider impact when it is extended to incorporate the existence of stakeholders.[23] Whereas Porter's model is based on industry structure, Freeman extends it to stakeholder structure.[24] By doing this, he suggests that a series of generic stakeholder strategies can be developed based on the extent to which a specific stakeholder represents a threat to the organization and the willingness of the stakeholder in question to cooperate.

This perspective can be expanded further by incorporating orientation, opinion, and power to generate a set of generic strategies. Then, using these stakeholder descriptors, it is possible to select the generic strategy best able to take advantage of the stakeholder's position.

Big gun strategy. Stakeholders who are actively for the corporation but not in a powerful position relative to it may, however, be in a position of power relative to other stakeholders. This situation may be very useful to the corporation. The organization should attempt to influence the environment of these stakeholders to provide or to increase the base from which these stakeholders derive their power so that, in turn, the organization might improve its competitive stance. This line of action is even more appropriate when the corporation itself is in a powerful position in relation to the stakeholder group.

Defense strategy. To deal with stakeholders who are actively against the organization and who, at the same time, have a strong power base requires that the organization mobilize its defenses. Care has to be exercised to ensure that any transactions with these stakeholders take place flawlessly; any hitches will give this group more ammunition to be used against the organization. In addition, consideration has to be given to finding a way of switching the orientation of these stakeholders.

Alliance strategy. Stakeholders who are in a powerful position and are actively for the corporation can be used to great effect to further the organization's interests. This group of stakeholders is committed to the future of the organization. In this instance, the interests of the stakeholder and the organization are parallel. An alliance with the stakeholders in this category can be used to improve the firm's position with respect to other clusters of stakeholders. Hence, the organization should actively seek issues on which it can cooperate with this group to their mutual benefit.

Switch strategy. Management should adopt a strategy aimed at switching the orientation of those stakeholders who are actively against the corporation but who are in a low power position. Freeman points out that such strategies could include attempts to alter legislation in favor of the stakeholder in anticipation that the stakeholder will switch its orientation.[25] Alternatively, the manner in which business is transacted between this group and the corporation may be changed, again in an attempt to influence the stakeholders to switch to a more favorable orientation.

Hold strategy. The goal of hold strategy is to retain the status quo. As such, this strategy is most applicable to the latent stakeholder groups. Strategies aimed at these stakeholders should monitor the corporate-stakeholder relationship on an ongoing basis. In addition, attempts should be made to ensure a pro-corporate stance in the event that the stakeholder emerges from the latent category.

Build strategies. Build strategies are aimed at building stakeholder awareness in those groups that are oriented in favor of the organization. In this way management would hope to gain a valuable ally in its pursuit of strategies.

In addition to the above set of generic strategies, several alternative-generation techniques may be adapted to identify strategic alternatives using the stakeholder approach.

The stakeholder approach in effect encourages a systems perspective of the organization. This in itself can help to develop a series of creative alternatives that flow from both the careful identification of stakeholders and the realization of the forces at work among the stakeholder groups. In addition, a detailed appreciation of the inducements and expectations of each of the stakeholder groups will provide further impetus to the generation of alternatives. Stakeholders do not necessarily affect only the focal firm but may also affect other stakeholders. Strategies must therefore be developed to take this interaction into account.

Although it is recognized that strategy formulation is accomplished by both individuals and groups, comprehensive alternative strategies—strategies that have an impact on a broad spectrum of stakeholders—generally are better generated by a group of senior executives. This is because the individual manager in a large corporation often does not have an intimate working knowledge of all the corporation's stakeholders. A group of senior executives, if carefully constituted, could be sure to include individuals with the necessary stakeholder perspectives and experience.

Van de Ven and Delbecq compared the creativity arising from the use of three group techniques—the Nominal Group Technique, the Delphi Technique, and the Discussion Group Technique—frequently used for alternative generation.[26] Their findings indicate that the Nominal Group and the Delphi techniques resulted in more ideas and hence greater creativity than did the group discussion approach. For the purposes of developing strategic alternatives using the stakeholder perspective, the groups should be constituted to ensure that their members have an appreciation of and experience with a broad selection of the organization's stakeholders. The statement of the problems to be considered by the group should be couched in terms that encourage members to take a stakeholder perspective as they generate their alternatives.

Stakeholders and Alternative Selection

The stakeholder approach lends itself to use in the selection of strategic alternatives, once the alternatives have been developed. At the outset, the alternatives are examined to determine whether they will result in the breach of any effectiveness criteria adopted by the organization. If none are apparent, the usual alternative selection procedures may be applied. Conversely, where the proposed strategy shows the potential for breaches-of-effectiveness criteria, the possible outcome of such breaches has to be weighed before the alternative is considered further. Mason and Mitroff developed and applied an approach by which the underlying assumptions inherent in any existing or proposed strategy may be brought to the surface.[27] Their process involves several phases: group formation, assumption surfacing, assumption checking by each group, assumption defense against other groups, and a synthesis of the underlying assumptions and the resulting decision.

This is not the place to give a detailed description of each of the phases; however, they do bear some comment and expansion. Two or more homogeneous groups are formed. These groups should be small with no more than about a dozen individuals. Mason and Mitroff point out that two objectives that should be applied in selecting the groups are to minimize the interpersonal conflict, by selecting individuals with similar personalities, and to maximize the differences in experience and knowledge of each of the groups.[28] In the latter instance, it is conceivable to constitute groups with affinities for the interests of a group of similar stakeholders.

Each alternative is then considered by each group. The stakeholders affected by the alternative are identified, and assumptions are made as to the manner in which the stakeholders will react to the proposed set of actions. In addition, assumptions are made regarding the contributions the stakeholders will be expected to make, and the inducements that will have to be made by the organization to obtain these contributions.

Mason and Mitroff suggest that it is necessary to test the relevance of each assumption to the alternative. This is achieved by examining the converse of the assumption: "If the converse of the assumption were true, would it affect the favorable outcome of the alternative if it were selected?"[29] The assumption is discarded if there would be no effect on the outcome if its converse applied. Group members are then required to rate the relative certainty and importance of the remaining assumptions. Consensus is pursued and those assumptions with a low level of importance and low certainty are dropped from further discussion. The group then identifies the key assumption on which the ultimate successful execution of the alternative relies.

The groups then compare their assumptions for each alternative and support their position in a debate. At this point a series of assumptions often becomes untenable, and the associated alternative is then dropped from further consideration. The final phase involves all the members of the groups working to identify and to resolve the controversies still in place. This in turn will indicate the most viable alternatives, which should then be pursued.

Stakeholders and Implementation

Once the strategies have been selected the task then is to implement them. Any group that has had a hand in formulating the strategy and that supports it becomes an internal stakeholder with an orientation in favor of the strategy. Those that do not support it, however, can wreck its successful implementation and have to be won over before attempts are made to implement the strategy.

External stakeholders can also act as the

basis for strategic control. During the implementation phase, the assumptions made for the purposes of alternative selection have to be monitored. Although these assumptions may have seemed accurate at the time of strategy formulation, their accuracy has to be continuously checked as the strategy is implemented. Where an assumption is found to be faulty, jeopardizing the successful implementation of the strategy, the strategic planning process has to be reactivated to reformulate at least that particular aspect of strategy.

Strategic control is also used to check the relationship between the stakeholders and the organization. This process has to be ongoing because the forces and currents of the environment often affect the stakeholders of the organization. This necessitates that the stakeholders' relationships with the organization—their opinion, orientation, and power—have to be evaluated. Once again, such evaluations may result in minor trimmings to the strategy sail to take full advantage of alterations in the speed or direction of the wind of change sweeping the environment. Alternatively, the stakeholder audit might dictate the wholesale reformulation of the organization's strategy. In this case, the full strategic planning process has to be initiated.

Several additional factors can influence the extent to which an organization's stakeholder strategies can be successfully implemented.

Board Composition. Use of a stakeholder perspective can influence the composition of the board. It is relatively common that a representative of the corporation's bankers is given a seat on the board. Also, some of the recent innovative labor settlements have resulted in a labor union representative being given a seat on the board. Provision of a seat to a stakeholder representative provides several advantages to the corporation. At the outset, the stakeholder group is given an information conduit to the thinking of the highest echelons in the organization. Conversely, these stakeholders provide the corporation with a continuous update of stakeholder expectations. In this way, it becomes easier for the corporation to tailor its strategies to meet the stakeholders' expectations. Moreover, it could be argued that a stakeholder representative on the board would be likely to respond to corporate requests in a favorable manner. These are indeed advantages, but the disadvantage of composing the board of too many stakeholder representatives is that it could render the board ineffective, racked with dissension as each representative pursues his or her own special interests.

If the stakeholder perspective is used to influence the composition of the board, the power, opinion, and orientation attributes must be considered. There is a tendency to appoint to the board representatives of stakeholders who are powerful and actively for the corporation. This is the most comfortable situation for management also. However, recognizing that the alternative has shortcomings and difficulties, there could be greater benefit to the corporation in appointing stakeholder representatives who are to some extent against the company. These are the groups who would probably keep management on their toes!

Corporate Stakeholders. At the start of this chapter, it was pointed out that the stakeholder approach was particularly suited for business-level analysis. This does not mean that the corporate head office should ignore the stakeholder approach. It is important for the head office to gain an understanding of each of the stakeholders at the level of the business unit. Only then can they develop corporatewide policies for dealing with specific stakeholders who are able to affect the success of more than one business unit of the corporation.

In addition to standardized policies, it is possible that consolidation of specific stakeholder relationships at the corporate level could increase the power of the business unit. Clearly, this could be exploited to the advantage of the organization as a whole. Examples of such policies are centralizing purchase of common goods and services, standardizing relationships with unions, standardizing pension funds, and standardizing specifications for computer hardware needed by each of the business units.

Operationalization of Stakeholder Policies. The implementation of stakeholder policies is not without cost. Indeed, the cynics have stated that a firm has to be sufficiently well off financially before it can afford to have a comprehensive set of stakeholder policies. It should be clear from the preceding discussion that this is not necessarily so. However, an accounting statement of the costs of implementing the stakeholder policies can be developed in a generation and distribution of wealth statement. In terms of this statement, revenues can be thought of as wealth generation. Purchases are that portion of corporate wealth distributed to suppliers, as interest is to lenders, direct and indirect benefits to employees, taxes to government, charitable benefits to society, and dividends to stockholders. Depreciation represents that portion of corporate wealth that has been retained for self renewal, investment in additional facilities, and so forth.

Using this statement, it is possible to track the manner in which corporate wealth has been distributed and to think through strategic implications of these activities. A most striking aspect of the statement is the fact that the stockholders receive such a comparatively small portion of the generated wealth. This seems almost ludicrous when compared to the amount of power wielded by the stockholders in comparison to other stakeholders.

CONCLUSION

The preceding discussion has shown that stakeholder concepts can be usefully applied to the formulation and implementation of strategies at various levels of the organization. The fact that the stakeholder approach forces the strategic planner to take into account the needs and structure of the organization's environment can only result in a series of strategies closely geared to the demands of the environment. This should yield improved performance, provided the stakeholder-related policies formulated in this manner are implemented.

REFERENCES

1. R. E. Freeman, *Strategic Management—A Stakeholder Approach* (Boston: Pitman, 1984).
2. F. W. Abrams, "Management's Responsibilities in a Complex World," *Harvard Business Review* **29**(3)(1951):29-34.
3. W. R. King and D. I. Cleland, *Strategic Planning and Policy* (New York: Van Nostrand Reinhold, 1978).
4. Ibid., 152.
5. Abrams, "Management's Responsibilities," 32.
6. R. E. Freeman and D. Reed, "Stockholders and Stakeholders: A New Perspective on Corporate Governance," in *Corporate Governance: A Definitive Exploration of the Issues,* ed. C. Huizinga (Berkeley and Los Angeles: University of California Press, 1983):173-205.
7. P. F. Drucker, *Effective Management Performance* (London: British Institute of Management, 1974).
8. I. H. Wilson, "Socio-Political Forecasting: A New Dimension to Strategic Planning," *Michigan Business Review* **26**(3) (1974):15-25.
9. J. Emshoff and R. E. Freeman, "Who's Butting into Your Business," *Wharton Magazine* **4**(1) (1979):44-48, 58-59.
10. Freeman, *Strategic Management.*
11. M. E. Porter, *Competitive Advantage—Creating and Sustaining Superior Performance* (New York: Free Press, 1985).
12. J. E. Gruning, "A New Measure of Public Opinions on Corporate Social Responsibility," *Academy of Management Journal* **22**(1979): 738-764.
13. M. E. Porter, *Competitive Strategy: Techniques for Analyzing Industries and Competitors* (New York: Free Press, 1980).
14. I. S. Baird, "Power: The Theory Underlying Competitive Strategy," Working Paper, Department of Management Science, Ball State University, Muncie, In., 1984.
15. I. C. MacMillan, *Strategy Formulation: Political Concepts* (St. Paul, Minn.: West Publishing, 1978).
16. Porter, *Competitive Strategy.*
17. A. L. Mendelow, "Setting Corporate Goals and Measuring Organizational Effectiveness—A Practical Approach," *Long Range Planning Journal* **16**(1) (1983):70-76.
18. J. K. Brown, *This Business of Issues: Coping*

with the Company's Environments (New York: National Industrial Conference Board, 1979).
19. Ibid.
20. H. Klein and W. Newman, "How to Use SPIRE: A Systematic Procedure for Identifying Relevant Environments for Strategic Planning," *Journal of Business Strategy* **1**(1) (1980):30-45.
21. F. Aguilar, *Scanning the Business Environment* (New York: Macmillan, 1967).
22. L. Fahey, W. R. King, and V. K. Narayanan, "Environmental Scanning and Forecasting in Strategic Planning—The State of the Art," *Long Range Planning Journal* **16**(1) (1981):32-39.
23. Porter, *Competitive Strategy*.
24. Freeman, *Strategic Management*.
25. Ibid.
26. A. H. Van de Ven and A. L. Delbecq, "The Effectiveness of the Nominal, Delphi and Interacting Group Decision Making Processes," *Academy of Management Journal* **17**(4) (1974):605-617.
27. R. O. Mason and I. I. Mitroff, *Challenging Strategic Planning Assumptions* (New York: Wiley, 1982).
28. R. O. Mason and I. I. Mitroff, "Assumptions of Majestic Metals: Strategy through Dialectics," *California Management Review* **20**(2) (1979):80-88.
29. Ibid., 86.

CHAPTER 13

Establishing Superior Performance through Competitive Analysis

Robert E. MacAvoy

Robert E. MacAvoy is a certified management consultant and a founding partner of Easton Consultants, Inc. He has been doing original research and publishing in the field of competitive analysis and strategy development since the mid-1970s. He has co-chaired the North American Society for Corporate Planning's workshop on competitive analysis. In recent projects he has concentrated on analyzing the special factors that must be considered in competitive analysis and strategy development in industries under pressure to change.

Competitive analysis is in the ascendancy. It is a vital and growing influence on the manner in which many corporations position their businesses, spend their resources, and determine the level of performance that is reasonable to expect from their business units. It has garnered the attention of top management. Many corporations acknowledge its current or potential importance by appointing a director of competitive analysis or assigning it as a specific responsibility to an existing staff organization. In fact, in a recent survey of the *Fortune 500* corporations, competitive analysis was designated "a vitally important" determinant of future performance by 82 percent of respondents.

WHY DO COMPETITIVE ANALYSIS?

The high visibility and interest in competitive analysis primarily are results of secular changes in the U.S. business environment. Now, more than in earlier periods, making enough profit to finance growth and pay dividends threatens to become a zero-sum game for the competitors in a business. Today, one firm's superior performance may cause another firm's serious problems. Prior to the 1970s, companies sought to achieve their business goals through pursuing *positional* strategies, that is, strategies that *positioned* the firm in profitable businesses or profitable segments of a business. Businesses such as forest products and agricultural chemicals were considered unattractive; businesses such as specialty chemicals and computer peripherals were considered a better bet. Companies diversified out of unattractive businesses and into attractive ones, sometimes through acquisitions and sometimes through grassroots efforts. Companies recognized that in any given business some firms did better than others but felt that even an average or somewhat below average performance in an *attractive* industry was tolerable. Management's attitude was that this level of competitive performance was adequate to reward stockholders and finance growth.

This attitude has changed. Relative competitive position has become much more important. No longer is the most severe penalty for inferior competitive performance the embarrassment of being relegated to the bottom of the heap in *Fortune, Forbes,* and *Business Week's* surveys of industry performance. As the competitive environment has become harsher, all the players are under increasing stress. In many industries, below-average performance has resulted in serious business setback for the players. The experience of companies such as International Harvester, Eastern Airlines, Bank of America, and Storage Technology Corporation, to name just a few, bear this out. For these and the many others like them the need to improve competitive position seems essential.

A number of factors are responsible for this most strenuous level of competitive challenge, some of which are pervasive in their influence across many different industries. Others only affect specific industries. When these factors converge on a particular industry, all but the most superior firms are threatened. Some of the most crucial are:

1. *Slowing growth in many industries that forces companies to seek volume increases through greater market share.* The U.S. economy has been evolving into a period of slower growth. This has been masked somewhat by high inflation rates. A tire company with a growth objective of 10 percent, 12 percent, or 15 percent a year will find this difficult to achieve with a market growth of 2 percent or 3 percent per year in units. To achieve its objective, it will need a resurgence in inflation, increases in market share, or the force feeding of capital (either through acquisition or startup) into other fields. Given limited inflation, increase in market share is the seemingly easiest place to start. Seasoned managers know better, but frequently succumb to use of price to achieve increased market share, particularly in mature product areas. The misuse of the growth-share matrix, so beloved of consultants and business planners, has not helped.

2. *Corporate development programs that focus on resource rechanneling into adjunct businesses.* Major corporations have developed staffs who are constantly combing their business borders for better positional opportunities. Entering new businesses through grassroots efforts or strategic acquisitions is common. "Buying your way into the game" is an accepted practice. This creates discontinuities in competitive equilibrium as companies seek to expand their beachheads through market-share acquisition.

3. *Globalization of businesses.* There was a time, not too long ago, when Americans were considered the superior practitioners of business. Managements in other countries emulated American organizational structure, information systems, decision-making processes, and managerial training. Now, it is more apparent that the seeming supremacy of the American businessperson was really the phenomenon of a well-developed and stable business community ensconced in the most well-developed and stable market in the world. The market was well understood, and American companies found equilibrium in the competitive matrix. Managers wanted to make money, not necessarily increase market share. What was good for the country was good for General Motors (or Ford, or Chrysler, or even, once upon a time, American Motors). Foreign competitors lacked capital, a base market, and knowledge regarding the U.S. market. The halcyon period is now behind us.

Now, markets outside the United States are significant and can provide the base for development of new and highly competitive business enterprises. Technical and financial capital provided by the developed Western economies fused with natural resources and cheap labor in many foreign countries has produced a competitive infrastructure capable of prevailing in foreign markets and, in many cases, of carrying the battle to U.S. firms on their native turf. The days are behind us when steel from countries such as Brazil generally was considered lower grade because of the high variability of impurities or when continuous-manufacturing processes were considered highly risky in countries such as India because of the instability of the power supply. In the early 1980s many American businesspersons viewed China as an exciting potential market. Now, it is apparent that China can also be a source for competitive manufacture of goods not simply for domestic consumption but for sale in the world markets. For example, Macy's has begun to import high quality rugs from China. The Japanese, long an import threat to U.S. firms in many fabricated metal product areas, are phasing out certain of these businesses in the belief the Chinese can not be beaten in world markets.

In many industries, foreign competitors evolved from small companies clinging precariously to existence with a toehold in a market niche to powerful presences seeking new markets. Honda grew from humble beginnings to become a good motorcycle manufacturer and then a good car manufacturer. Along the way, it contributed to the demise of several U.S. motorcycle companies. Harley Davidson alone remains. The time to deal with foreign competition in that industry may be past. Other U.S. industries will surely follow the same path unless creative competitive strategies can be developed. Singapore may seem remote from U.S. markets, but with labor rates much lower and a reputation for quality, customers will find the distance less disquieting. Even when the product is stamped Made in America, the chances are increasing that its innards were made offshore.

4. *An explosion of information and a cheapening of information access that has created an ever-widening window on competitive opportunities.* Increased loyalty to professions and decreased loyalty to companies, the emergence of a highly mobile management class, and higher employee turnover at all levels brought about by a work force less concerned about job security than their parents have contributed to a freeing of proprietary company knowledge and skills. If Chrysler wishes to emulate Ford marketing skills, it can hire a raft of professional marketers from Ford by buying out their contracts. Knowledge also has become portable. The blueprints for a Boeing 747 weigh more than the plane itself. Twenty-five pounds of optical discs can contain the same blueprint information. Proprietary containment of such information is increasingly difficult. This diffusion of knowledge facilitates the ability of companies to broaden their product offerings and to enter adjacent businesses. With more information available, success is more likely.

Because of these factors, there is also a greater general insight regarding the heart of the profitability of a company (something competitors take great lengths to avoid disclosing in their line of business financial reporting to the SEC in the 10-K). This means that competitors can pick their shots more accurately.

5. *Maturation of businesses.* In many businesses, growth and change keep competitors busy minding the store. Growth requires new plants, recruitment and development, responsive product line expansion, and so forth.

Change requires coping with new raw materials, new processes, new product challenges, and new distribution paths. Meeting all these challenges satisfactorily requires intense management attention. This situation is typical in the early and middle stages of a business. During these periods, firms tend to evolve in unique ways, creating unique competitive positions. In the early stages of the fast-food franchise business, International House of Pancakes, Arby's, McDonald's, and Colonel Sanders were only marginally competitive with one another. Later, when change subsides, companies in an industry tend to drift toward similarity. Also, firms less caught up with the management of growth tend to focus on increasing market share. The invasion of McDonald's into the breakfast and chicken fast-food business is a classic example. This ordinarily causes eroding margins, as prices are reduced to maintain competitive position. McDonald's, of course, has emphasized such products as Chicken McNuggets® and Egg McMuffins®, which can be advertised as unique. It remains to be seen, however, whether such product differentiations will succeed.

6. *A dissipation in the potency of many old sources of competitive advantage that ordered industries, defined business boundaries, and freed winners from serious competitive worry.* Sources of competitive advantage such as entrenchment in traditional distribution channels decline in competitive importance as the channels themselves diffuse. The distributor of paper goods now wants to carry janitorial supplies. The food supply distributor now wants to carry paper goods. Retail supermarkets now carry T-shirts, hammers, lawn chairs, and garbage cans. Automobile parts jobbers now sell engine tune-up kits in blister packs.

The potential comparative advantage of manufacturing scale efficiencies is roughly similar for leading firms in many important industries. There is a size for a high-density polyethylene plant beyond which costs per unit can actually rise. If all competitors are building plants of the most efficient scale, none has an advantage. Large commodity businesses tend to fall in this category.

The trend toward buying systems rather than products reduces the market demand for a stand-alone superior product. For example, the key source of comparative advantage in the engine business may be in the design and durability of the product. In the past, truck purchasers bought engines as a decision separate from their choice of trucks. Truck fleets would buy White Motor truck bodies but Cummins engines. If OEM truck makers integrate backward into engines, however, this could become less important if buyers drift into a pattern of buying the system. Another example would be in the HVAC businesses. Making high quality pumps or motors or electronic controls for building heating systems will be less a key to success than selling an environmental control system that links them all together. This threat of a wrap-around competitive approach is evident in virtually all manufacturing product categories.

7. *Accelerating value drift, in which customers prefer new product and service attributes that may not be as efficiently provided by current suppliers.* Today, an increasingly sophisticated consumer spends more time evaluating his needs and reviewing his alternatives. Buying habits are less ingrained in industrial and commercial markets and the noncompetitive buy is increasingly rare. Remaining sensitive to the needs of the market is always a struggle, particularly for the larger firms. Firms tend to fall into a habit of selling what they can make, rather than what the customer wants. In the 1960s and 1970s automobile buyers complained about their choices, objecting to poor paint finishes, doors that did not fit properly, short car life, high prices of captive parts, low gas mileage, and uncomfortable seats. Automotive manufacturers insisted during this period that these stated product objections were superficial and that what the customer really wanted was low price, a wide variety of colors and styles, instant availability, and acceleration. The automotive manufacturers were simply extrapolating aspects of product value manifested in the 1950s. The organizational rigidity developed through years of success resisted change, even though there were powerful managers in the companies who could see what was happening and tried to react. A rift developed between the perception of value held by American car buyers and the capabilities of the manufacturers that has yet to be closed.

These lags between recognition of changing desired values by competitors and actual value changes are becoming more critical as some industries become vulnerable to upstart competitors offering new and more relevant value packages.

8. *Rapidly evolving technologies that have reduced the potency of scale in many business functions.* Many sources of competitive advantage in an industry have been based on volume. A fixed-cost component spread over a large number of copies means a base cost advantage to whomever enjoys the highest volume. However, increasing efficiencies in design and manufacture have reduced these

advantages. Mini-mills can now compete with large steel mills. In publishing, superior firms spread design and layout costs across large print runs. Now, design and layout costs are decreasing because of processing technology. Smaller publishing firms can compete in narrow segments. Recovery of product development costs becomes more difficult as the product cycle becomes shorter. Getting new products out quickly becomes more important than getting a large number of them out too late.

These are the major reasons for increased competitive pressures in many businesses. The effects are a general deterioration in the financial health of many industries and for individual companies less margin for strategic error. Winning over competition is, in some industries, a necessary rather than simply a desirable performance goal. Attaining this goal will require more than shrewd positional strategies or even superior execution. It will require getting ahead of competitors who are altogether a tougher bunch than they were in the past. This is not to say that trying harder will not be necessary, simply that it is not enough.

To win against competition, you must know what you are up against. In the late 1970s, however, most U.S. firms knew surprisingly little about competitors. They analyzed financial performance, performed product testing, and monitored new plants that were coming on stream. That was it. Companies did not know their competitors' costs by market segment, or even whether buyers thought their products were of greater or lesser value than the competition. Firms did no know where competitors were most vulnerable, nor did they know where their own flanks were seriously exposed. They had no approach to developing a prioritized plan for maintaining and improving competitive position in a logical fashion. People talked about the formula for success. In many cases, this meant doing what the company had done in the past. The formula was static, impervious to the changes in competition and management was managing looking backwards.

Top management recognized the need and since approximately 1980 has been devoting resources to upgrading competitive analysis. So far the results have been mixed. A recent survey of corporate and divisional planners indicates that on a scale of one to ten the planners felt competitive analysis as done by their companies rated a seven for living up to expectations.

However, a similar survey of top corporate management revealed a different picture. Top managers rated competitive analysis at three on a scale of one to ten regarding living up to expectations. The disappointed managers had several complaints. They thought that competitive analysis as done by their organizations was too vague, too financially oriented, too open-ended, and too nonspecific to be useful. Fewer than 5 percent were able to cite instances in which competitive analysis resulted in some material improvement in the competitive position of their business units or influenced a major decision the firm had to make. They did not care about learning such tidbits as the risk orientation of competitive managers, retirement dates, new plant openings, amount of rust in the rails, pollution fines, and whether their competitors were interested in getting into high-growth segments of the market. (They felt they already knew the answer to that one.) Top managers said that they wanted pertinent information that was concrete and actionable. That is, they wanted information that, when taken into consideration with other factors, allowed specific strategic actions to be taken in running their business.

INHIBITORS TO PERFORMING USEFUL COMPETITIVE ANALYSIS

Some observers believe the main shortcoming to competitive analysis lies in securing and disseminating data. The mechanics of collection, organization of activities, format decisions, frequency and distribution of reports, and so forth are important. However, from what I have seen, these problems are not central. Most companies have these types of skills well in hand. If not, there are many successful firms in other industries whose techniques can be emulated. If anything, these procedural aspects tend to suffer from over-organization. In many firms, the competitive analysis activity should be skinned down.

The problems are more deeply rooted. They deal with the relevance of the analysis: they fail at the "so what" stage. Who cares how much rust there is on competitor's railheads? The real issue is what change in your strategy should this dictate. I feel that the main problems are:

A lack of context for competitive analysis, which causes the exercise to fall into the category of fluff. In one memorable meeting the general manager challenged the usefulness of competitive analysis by asking what specific company activities would be hurt if the exercise was discontinued. Nobody had a clear and concise answer. Quite an indictment for a staff activity that was consuming several hundreds of thousands of dollars a year.

An attempt by corporations to homogenize the types of data gathered across very dissimilar businesses, resulting in a dilution of the potency of the exercise. In attempting to become all things to all people, it becomes meaningless to most. It confuses the very admirable notion that competitive analysis should be done equally for all with the idea that it should be done the same.

Failure of companies to recognize that all possibilities for competitive improvement suggested by competitive analysis are not real because of their own economic, cultural, and resource limitations.

An overly optimistic view that major changes in competitive position are possible within the span of two or three years. This is the exception rather than the rule.

COMPETITIVE ANALYSIS VERSUS COLLECTION OF COMPETITOR INFORMATION

Many companies, convinced of the value of competitive analysis, have moved to set up elaborate and all-inclusive files on a wide range of competitors. In many cases, competitors are those firms that the company currently meets in the marketplace. Others focus on identifying all information on only the major participants in the marketplace. In many cases, the emphasis is on magnitude of information and ease of access. Examples are annual reports, Dun & Bradstreet reports, newspaper clippings, product literature, trade journal articles, and uniform commercial code filings. These are perused and filed, with some firms moving toward elaborate computer-mounted data bases. A large telecommunications firm seeking to catch up with some of its faster moving competitors has set up an on-line data base of competitor information. The cost of getting the system into operation has been over $500,000. Maintaining and operating it will be $150,00 per year. The transactional efficiencies of a single point for collection and data access are the main argument for its development. The competitive analysis department tracks the inquiries and calculates a cost per inquiry. So far, the per-inquiry cost has been daunting. Users would apparently rather order their own competitors' annual reports for the cost of postage rather than sign up for the more elaborate program.

Much of this activity is wasted. Extraneous information is captured, filed, and never used. Financial information ordinarily is widely available from other sources. Information that might be useful to decision makers is not collected because its importance to a specific type of business has not been identified. In many instances, companies have simply created a centralized function of competitive librarian; the competitive analysis activity consists largely in the collection and cataloging of information. There is regular routing of information, but little interpretive analysis. In the case of a single-business company this is wasteful since it will not be incorporated in strategic planning. The manager likely will be as influenced by his readings in *Business Week* and *Forbes*. In diversified companies, its usefulness is questionable for other reasons. There are, I believe, serious questions regarding whether business-specific competitive data collection should be performed centrally by diversified companies in the first place.

Competitive analysis has greater promise

when it shadows the strategic development process. Competitive intelligence gathering reaches its most useful form when it is organized and guided by the needs of the strategic analysis. What follows is a concept for integrating competitive analysis into strategic planning that accommodates the differing needs of specific industries. Adjustments are probably necessary to the concept when applied by any individual corporation. The approach is meant as a guide. It does not work well as a cookbook formula. It does not seek to provide the unifying theory of business strategy development so ardently sought by serious students of business.

COMPETIVE ANALYSIS AND STRATEGY DEVELOPMENT

The orthodoxy of strategy development for business units includes analysis in two dimensions. In one plane lie the analytical elements necessary to distill the strategy possibilities. This plane relates the economic picture of the industry with the economic picture of the firm and is really a depiction of the possible. Analysis in the second dimension isolates the practically achievable, given the infrastructure, culture, latent resources, liabilities, and behavioral momentum of the competitors involved in the business at issue. Competitive analysis must deal with both dimensions or it will be viewed as of limited value.

The following outlines an approach to competitive analysis to isolate the possible. This, combined with the evaluation of the practically achievable, will yield answers to companies looking for ways to maintain and develop positions of competitive superiority.

Individual firm goals are typically many and complex. No two firms are the same. Usually, however, financial performance goals are important ones, and it can be argued that for the longer term they should be paramount. Two related aspects of financial performance typically considered to be key are profitability, as measured by return on equity, and profit growth. These are considered important in determining stock price-earnings rates and rate of growth of stock value. However, in a more competitive world, understanding the financial performance of a company relative to other competitors in its business is equally as important as these factors. In 1982 most firms in the home computer field did quite well from a return on equity standpoint. By 1985 it was clear that performance relative to competitors was much more important. IBM came to the conclusion it could not develop a satisfactorily defensible profit position and withdrew. Commodore and now perhaps Atari have a stronger relative position.

Industry Return on Equity Distribution Assessments

A good beginning point for reviewing the place of competitive analysis in strategy development is the documentation of the ways in which profit returns differ for competitors in a business. Two factors can be considered: the relative ranking of firms and the profile of profitability in a distribution analysis. In Figure 13-1, in which the rectangles represent a histogram of the profit distribution in an industry or sector of business, the vertical axis indicates the number of firms earning a particular level of return on equity. For example, suppose there are 50 companies in an industry. Twelve have achieved an average return on equity over the last five years of 10 percent to 12 percent. Four have achieved a return of 20 percent to 22 percent. This type of analysis can be represented in the abstract by the continuous curve that is drawn over the histogram in Figure 13-1. The points of interest in this distribution from a strategic standpoint are:

1. Median profitability of the business (in Figure 13-1 it is 14-16 percent).
2. The spread (which ordinarily is defined as the return on equity difference between the firms at the first and third interquartile). In Figure 13-1 it is 16 percent.
3. The actual shape of the curve.
4. Changes and rates of change in the position of

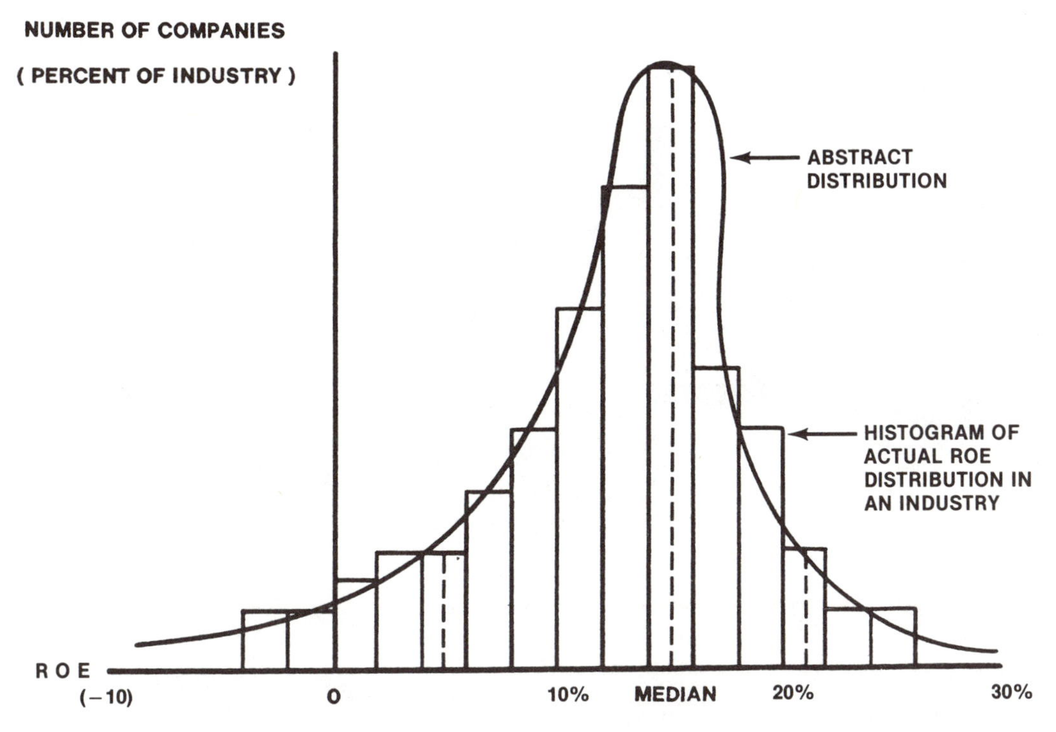

Figure 13-1. Industry profit profile.

> the median spread and shape. (In this case they are average values and the analysis is static.)

Different values along these four dimensions depict businesses of very different characteristics. Competitive positions when referenced in this way provide the beginning point for strategy development.

These four characteristics of an industry are determined by the characteristics of the business itself, the external factors influencing the business, and the types of activities necessary to serve the market (Figure 13-2). Analysis of competitor position is usually directed at influencing only one characteristic; that is, the relative position of one firm versus another. In the strictest sense, competitive analysis is successful if it improves relative standing, even if the entire industry curve in Figure 13-1 is trending to the left over time. However, in that case the business strategy itself would be a failure. Clearly, other strategy measures are not called for. The median profitability, the shape of the curve, and the trend over time have a lot to do with evaluating the possibilities for positional as opposed to performance strategies. Forgetting about the competitive position of the individual firm, the status of the overall curve signifies whether we are considering a good or bad business.

The spread, however, is important to consider. In some business areas, competitor variations around the mean have been quite narrow. In others, they have been extremely broad. In general, the wider the variation about the mean, the more promising the chances for differentiating the firms through competitive actions. The narrower the variation about the mean, the more the firm will be influenced by the positional component of strategies. In the days before deregulation, profitability in the banking industry went through cycles primarily determined by the degree of need to develop liquidity. When there was a need to improve liquidity ratios or when the economy was expanding, returns on equity for most banks went up. Banking was a good business to be positioned in. The only other concern was to *position* the bank in areas of economic growth. For example, if

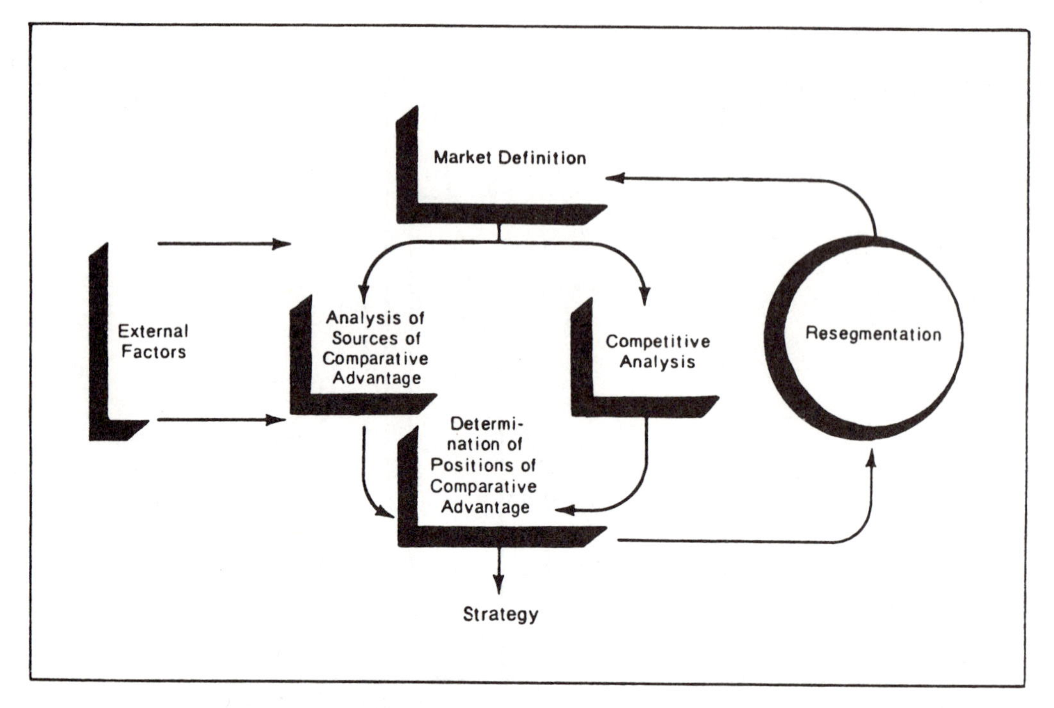

Figure 13-2. Elements of business analysis.

you wish to be in banking locate in Tampa and not in Detroit.

Most regulated industries have curves of relatively narrow spreads. For these types of businesses, strategic analysis was better concentrated on understanding and acting on the business cycles and searching out businesses that offered high average returns on equity.

Review and analysis of curves for various industries bring home some important points. For example, many so-called mundane businesses exhibit a wide spectrum of competitive profit positions, substantially overlapping profit possibilities in other more romantic industries. Everybody loved high technology stocks in the early 1980s while forest products companies were ignored. Yet as Figure 13-3 indicates, there has been significant overlap between the two curves since 1980. Positional strategies would have concentrated on getting out of the forest products business and into office products. For those that were in the forest products businesses, performance strategies, that is, strategies attuned to improving relative positions, may have been more worthwhile, but firms in the office products industry could have benefitted from this emphasis as well. Changes in office products and forest products from 1980 to 1984 are depicted in Figure 13-4. The spread in office products has narrowed considerably.

Where the company is in the profit spectrum of its business is also of interest to investment analysts, who are tending to look more closely at individual performance than they were ten or even five years ago. In the past, investment analysts tended to buy industries. If it was time to buy auto stocks or drugs, they bought them as a group. Now, they are increasingly likely to buy selectively within groups according to relative profit performance potential. Consequently, firms with superior competitive positions in their industry tend to enjoy superior price-earning stock ratios. This makes equity financing more palatable and diminishes the likelihood of an unfriendly tender offer. Also, not unimportantly, it rewards the existing shareholder.

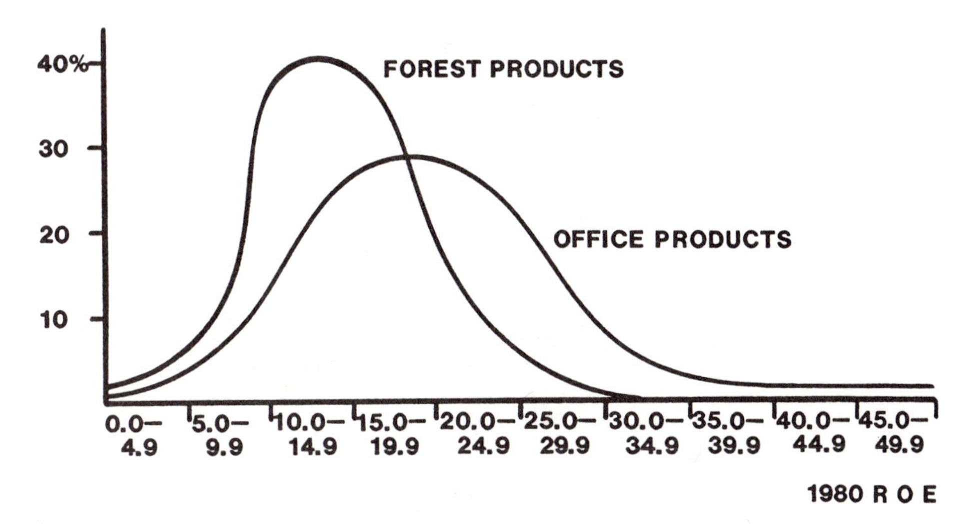

Figure 13-3. Overlapping profit profiles.

Relative profit differences are important, then, and competitive analysis of that which provides the key to creating them is valuable. Evaluation of competitive profit differences is a crucial first step. How wide or how narrow are they, and why? Financial data for at least five years should be examined. An analysis should be done of relative movements within the competitive rankings. High turbulence usually indicates firms of very different make-ups being unequally influenced by major industry events. Again the office products field is a good example. A relatively smooth trend of advancement or deterioration of competitive position by an individual firm may signal some important strategy information. Stability of relative ranking is the most straightforward data base for analysis but is increasingly rare. The competitive analysis should concentrate on identifying what factors are responsible for the profit difference between competitors. Where there is turbulence, it ordinarily signals changing sources of competitive advantage. Ordinarily, differences in profits are not simply volume or market-share related.

Market-share Analysis

Competitive market share has long been a popular way of explaining profit differences. In general, conventional wisdom has held that large market shares are desirable, should be protected, and should result in superior profits. One of the strong appeals of this concept was that it gave chief executives of diversified companies a factual base to compare the performance and the strategic potential of their various dissimilar business units. (Forecasts of financial performance do not qualify. They are rarely useful for anything except making sure that managers overpromise and work hard.) Conversely, small market shares came to be viewed as a problem. Some firms have a stated goal to be in the top three market-share positions in their business. Nothing less will suffice.

Although market-share analysis is essential in reviewing the competitive situation, it should be viewed as being a result rather than a causal factor in determining competitive superiority. Companies with large market shares

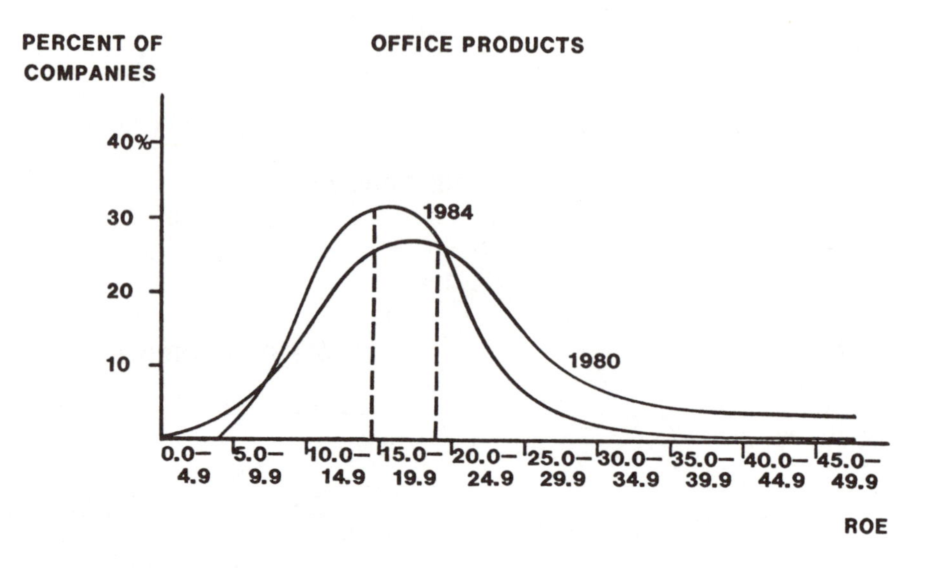

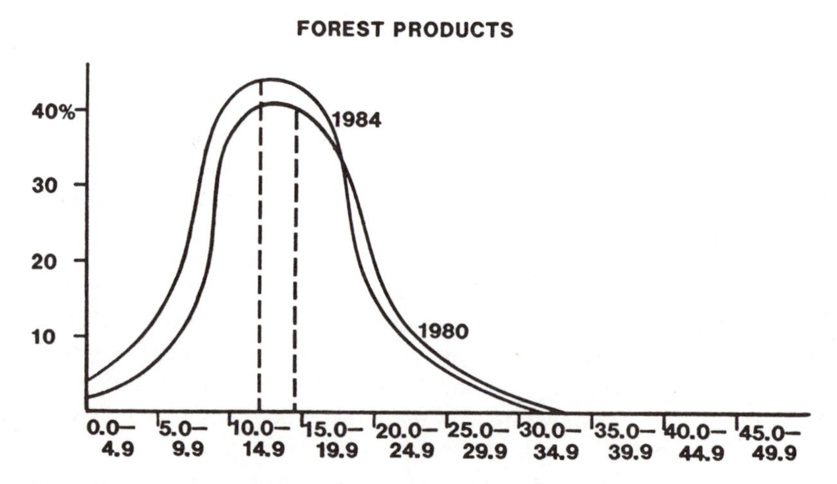

Figure 13-4. Changes in competitive profit distribution.

do not always enjoy higher profit returns than smaller competitors. Mercedes can be more profitable than Ford Motor Company. Deluxe Check Printers, although typically profitable, is not necessarily more profitable than John S. Harlan, a regional check printer. Maytag can be more profitable than White Consolidated's White Goods business.

The point of view that market share is something to be intensely sought, particularly in high-growth businesses, can be dangerous. Force-feeding capital into these businesses in a push to gain market share does not always work. All of the resources of Exxon Corp. could not turn Exxon Office Systems into a winner. In some cases, force-feeding of an entrant not only fails but can ruin the market for all competitors.

Bids for market share by companies in more mature markets can be even more dangerous. Such actions by individual companies may cause profit problems for the industry as a whole and particularly for the aggressors. For example, the wisdom of moves such as Miller Beer's reach for market share in the beer business or Maremount's bid for market leadership in the shock absorber market has been questioned. Whether the change in the U.S.

banking industry makes across the board market-share increases by major money center banks strategically sound is still to be resolved.

In many cases, the damage done to the business structure and general profit health of an industry that goes through market-share wars takes years to completely surface. Competitive differentiation is usually a much more successful strategy for an individual firm to follow, once a developing market-share war is discerned.

Another problem with competitive market-share analysis is in defining the relevant market. Certain businesses, when released from trade constraints, tend to be global. Others tend, for reasons such as transportation costs or varying regional product tastes, to be more regional in nature. Analyzing the national market share of a single-location salt company could possibly be useful, but a regional market-share analysis would probably be much more useful, at least, for most salt products.

The size of the conceptually appropriate product market is also difficult to determine. Product markets evolve over time, often drifting into new competitive grids. Glass bottles, for example, had to yield large parts of the soft drink and beer markets to cans. Cans, while benefitting from the new soft drink and beer markets, have found volume evaporating in other markets due to frozen foods, the pet bottle for soft drinks, retort pouches for off-the-shelf preservation of food, and the aseptic pack for noncarbonated beverages. Defining the appropriate market can be an elusive assignment. If you cannot get agreement as to what the appropriate product and market are, how can you analyze market share?

Finally, competitive market-share analysis does not indicate to a company what it should do to improve position, other than offering portfolio suggestions. Market-share analysis may be used to trigger investment in a high-growth business in which the company has a high share, or to precipitate divestment in a low-share, low-growth business. However, it does not help a company that wants to retain a business and improve its strategic position. In cases where the market-share analysis leads to a conclusion to pursue share gain, the avenue usually chosen is to cut price. Bausch and Lomb's position in soft lenses is a good example of this. In 1981, the company cut prices to gain share in extended-wear products. Share was gained, but the profitability of the business has yet to recover to 1981 levels. In 1984, the company changed financial reporting categories, muddying the picture of the true profitability of its vision-care business.

Value-added Analysis

Although the appeal of competitive market-share analysis is particularly strong in allowing corporations to compare dissimilar business units, its use for strategic purposes has declined. An evaluation of a much more specific approach to analysis of causal factors has taken its place. It centers on analysis of what determines value in a product and how resources are used to create it. By examining this relationship between value and resource expenditure, it is possible to identify with specificity the reasons for competitive success and what must be done to change relative competitive position. The reasons can be: (1) factor cost advantages; (2) advantages along very important and progressively leverageable economic relationships; or (3) actual superiority in execution. The first two reasons ordinarily are the most important for strategy reasons.

Value is measured in different ways in different businesses. Considering several different businesses will illustrate the point.

In the soap business, factors such as shelf availability and brand recognition are extremely important in creating a convenience value for customers.

In the machine tool business, availability of products ordinarily is not critical. Companies work against backlogs. Machine tools are not an impulse purchase, so brand recognition is not that important. However, the quality of the tool itself is an important determinant of value. The availability of spare parts may also be an important consideration. Once the tool is installed, it is expected to work consistently.

In the college textbook publishing business, users measure value by: (1) the quality of the published books; (2) the publisher's widespread recognition and acceptance in the academic community; and (3) the fit of the individual published books with each other. For example, a book titled *Heat Transfer Engineering Theory in Aerospace Applications* would probably be helped by a supporting publication entitled *Introduction to Heat Transfer Theory.*

In the computer peripheral business, dependability is important and is measured by factors such as the percent of time the machine is up, and the mean time between breakdowns. Compatibility with the users' host computer is also important.

Obviously, these are different values. For each business, the way value is added to the product in the minds of the people making the buying decision can be calibrated. In many cases, a qualitative understanding is sufficient. In others, a more precise quantitative understanding may be required. If necessary, the relative importance of all the different ways value can be added to the product can be established. Also, the quantitative tradeoffs in the buyer's mind between such things as price and availability can be determined.

Once determined, the way these particular values are created can be traced to causal expenditures of resources now or at sometime in the past. More importantly, the way these values can be created or enhanced in the future can be identified. Some hypotheses regarding how resources are spent to deliver value in the chosen examples are summarized in Figure 13-5.

1. *In the soap business,* the value of availability may be created through a number of interrelated factors, including a high level of entrenchment in existing outlets gained through intensive marketing activities in the past and through a strong current commitment to distribution (for warehouses and stocks) to assume minimal out-of-stock occurrences. These are past and current resource commitments that can be measured.
2. *In the machine tool business,* the value of parts availability can be created in a number of ways. Inventories of parts can be set at high levels, for example. A more efficient way may be to build into the design of the machine tool line as much parts standardization as possible.
3. *In the college textbook publishing business,* a strong relationship with recognized authorities in the various disciplines is required to guarantee a flow of quality manuscripts. A publisher's reputation in a particular discipline is also important. Colleges, although primarily concerned with the quality of the individual title, also tend to think of certain publishers as leaders in certain disciplines. This is often reflected in the quality and depth of the publisher's backlist. Also, a number of related texts facilitate the development of a program of study over time.
4. *In the computer peripheral business,* such factors as machine design and service availability can improve time up and lengthen mean times between breakdowns. Compatibility with the host computer is affected by the quality of the protocol software.

Comparative Advantage

As can be seen, the differences between businesses regarding what resources must be spent to create value are usually quite significant. The search among these relationships is for the limited few that determine competitive advantage and hence superior profitability. As *key sources of comparative advantage* in a business, they are the bond between the dependent variable that represents value (soap availability) and an independent variable that creates it and is controllable by business participants (strong distributing support). These key sources of comparative advantage are of two sorts: factor cost advantages, such as cheap labor in the case of a product being manufactured in various foreign locations or leverageable economic relationships, which provide efficiencies. This second type is the more critical, since it is more actionable by firms that cannot or do not necessarily want to move the location of their businesses. An example of a relatively straightforward key source of comparative advantage would be manufacturing cost per unit related to plant volume (Figure 13-6). To be a real source of comparative advantage these relationships

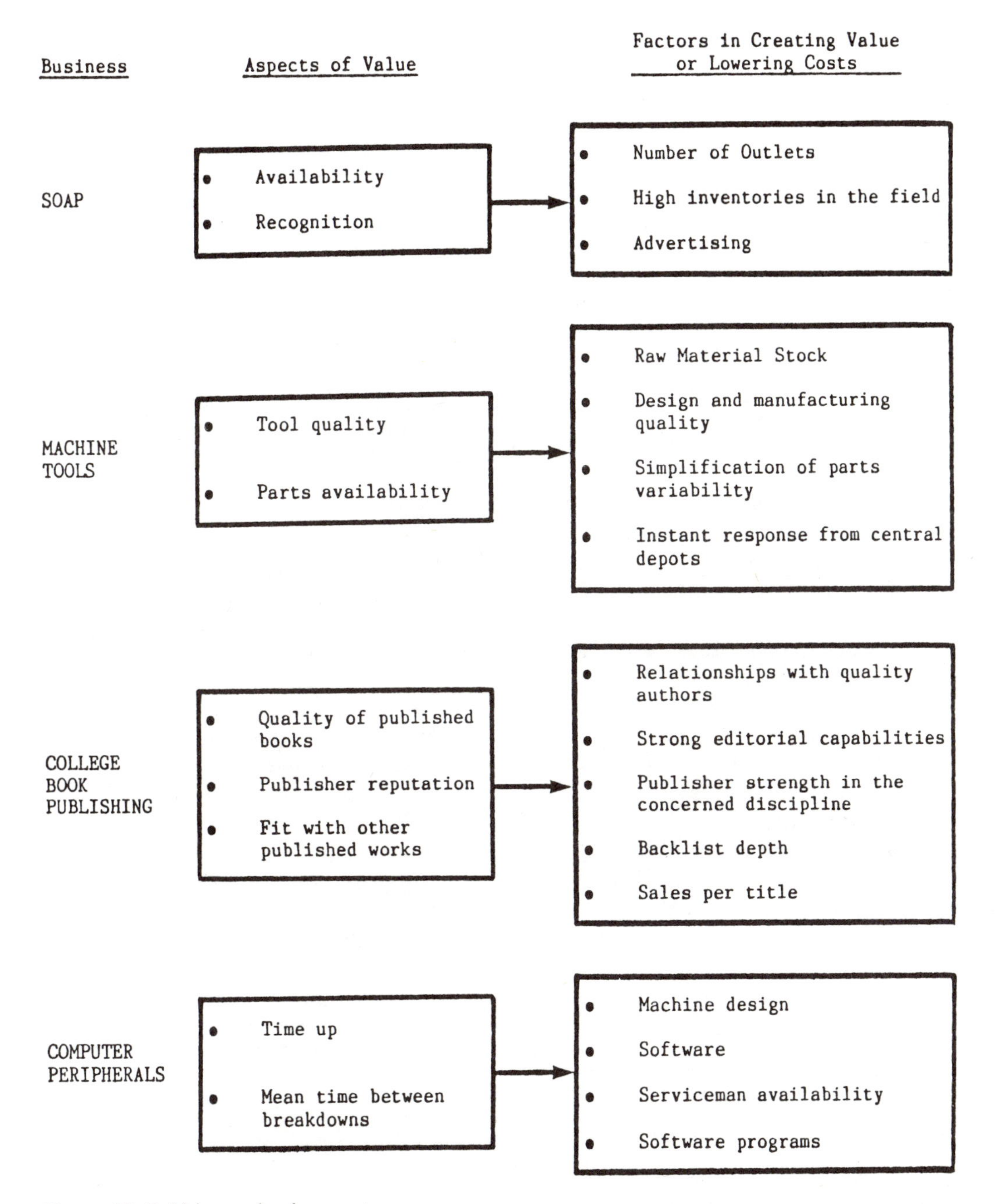

Figure 13-5. Value and value creators.

must be progressively leverageable along some dimension. In the case of the example in Figure 13-6, it is progressively leverageable by volume. Suggestions regarding key sources of comparative advantage in the four industries discussed earlier are listed in Table 13-1.

Sometimes these key sources of advantage are obvious, sometimes not. In cases where they are not obvious, a brief business analysis usually identifies them.

Where competitors sit on these curves determines to a great extent their relative profit performance. Competitors who have superior

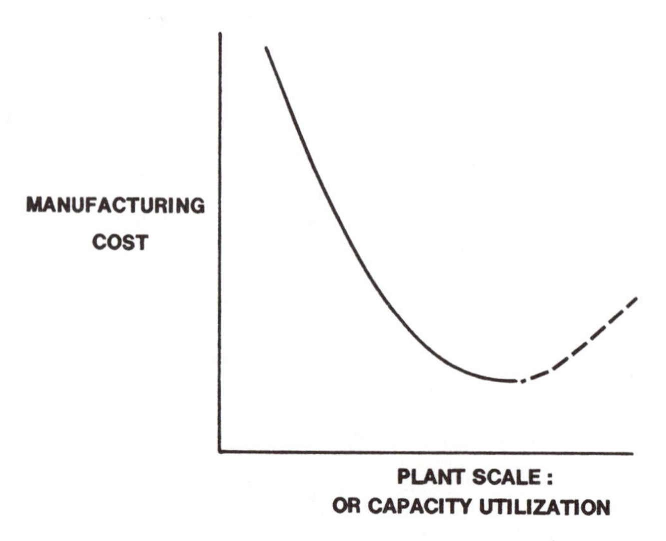

Figure 13-6. Manufacturing cost relationships.

relationships regarding these factors can defend themselves more easily from attack. They control the competitive environment.

Key sources of comparative advantage can take many forms, and many are not, in the strictest sense, cost related. All cannot be found on the income statement and balance sheet of the industry. The sources may have their roots in historical position in a business, or a reputation for quality, or an entrenched position in channels of distribution. None of these would appear on financial statements. Key sources may have to do with special knowledge of markets, adaptability to change, the management of minor costs, or a host of other similar factors. They may deal with economic relationships that have a very significant impact on total costs or total value, or they may not, but invariably they are economic relationships that are or have the potential to be progressively leverageable.

This then, identifies the necessary focus of competitive analysis typically missing from competitive analysis exercises. It is wrong to collect all available information on competitors. It is overkill, but, more importantly, it diminishes the visibility of truly important information by burying it. It is wrong to collect

Table 13-1. Key Sources of Comparative Advantage in Four Businesses

Business	*Key Sources of Competitive Advantage*
Soap	Number of products offered related to propensity of an outlet to carry soap Distribution costs per unit sold related to density of outlets Advertising costs per unit related to number of units sold
Machine tools	Manufacturing costs per unit related to number of copies of each model produced Design costs per unit related to number of copies of each model produced
College textbook publishing	Author remuneration per copy related to number sold Published books per editor related to number of published books Number of books in one backlist related to marketing costs per copy sold
Computer peripherals	Design costs per copy related to copies sold over design life Software costs per installed machine related to number of installations on the same host Service repair costs per machine related to number of machines installed

similar selected competitive information for all business units. Manufacturing costs per unit produced may be pertinent for one but not for most. Marketing-cost analysis may be important for some but not for all. The best shopping list for competitive information comes from an understanding of the economic relationships in a business. Information must be developed that will permit the relative positions of all competitors in the business to be determined.

When a company can find and document these factors, it can make very specific decisions regarding how it wishes to run its business vis-à-vis the relative position of competitors. It can gauge how much benefit it can expect from a move to improve its relative position.

Sources of comparative advantage can be found by methodically reviewing what constitutes value or need in any business. A suggested approach is to review value by main features of the physical product or apparent service, then by what the user expects in peripheral values. Table 13-2 lists the product features that were considered in a review of value for computer peripheral devices.

Following a determination of the values people are interested in when they buy the product, the next step is to consider the additional costs to deliver the value to the customer. This should include costs both internal and external to the firm. Also, as mentioned earlier, examining only current costs is a superficial way of analyzing how value is created. Value created in products sold today can be the result of resources expended over the last five, ten, or thirty years. Conversely, resources expended today can create value five, ten, or thirty years in the future. Accounting systems make some limited gestures toward acknowledging the concept but generally fall substantially short of what is necessary. For example, accounting systems may acknowledge some resource expenditures for the future in the form of R & D programs, however, they completely ignore the costs of building a superior position in distribution systems external to the company or the costs of training personnel.

As noted earlier, the search is for important and progressively leverageable sources of comparative advantage. It does no good simply to identify large chunks of value or costs that are largely the same for everyone in the industry.

Some additional examples may help to clarify the point. Figure 13-7 illustrates the relationship between sales and distribution costs per unit and the density of sales coverage. This relationship has proved to be an important source of comparative advantage in busi-

Table 13-2. Computer Peripheral Device Features Considered in a Review of Value

Product Features	*Support Features*
Price	Company guarantees
Performance (speed)	Availability of product
Performance (dependability)	Training in use
Durability (life)	Availability of parts
Economy of energy use	Speed and effectiveness of service
Economy of operator use	Compatability with existing equipment
Value retention	
Reassurance of brand image (recognition)	
Variety of use	
Perfection of output	
Size of equipment	
Simplicity of use	
Heat discharge	

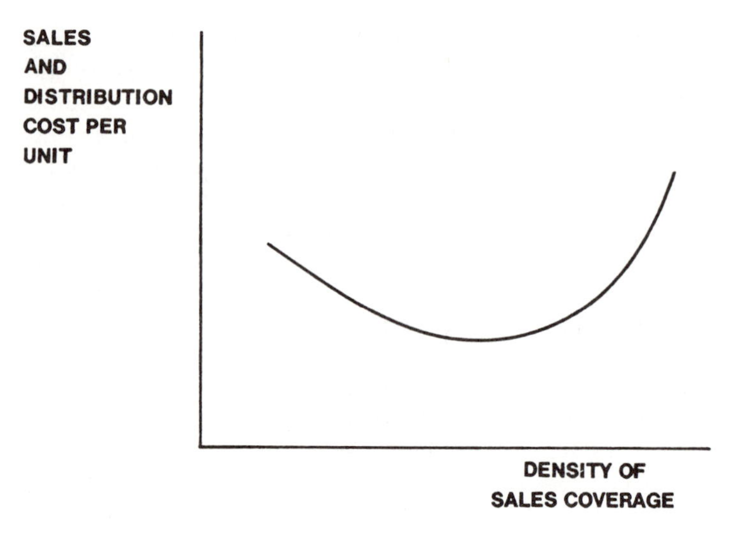

Figure 13-7. Sales-distribution costs and sales coverage.

ness requiring intensive selling and account maintenance activities. Good examples of this would be specialty chemical businesses such as warewashing compounds in which Economics Laboratories is preeminent or water treatment chemicals such as offered by Betz Laboratories. Sales and distribution costs per unit decrease as the density of sales coverage increases. Salespersons spend less time in transit from account to account. At some point, however, salesperson density becomes too high, an inefficiency that boosts costs.

A product-oriented business might offer a key source of comparative advantage in the relationship between product development costs and the number of copies to be sold (Figure 13-8). In many companies, product development costs tend to be controlled as a certain percentage of sales. Analysis of product development costs by specific product category generally is more promising.

It is much more relevant, however, to analyze these costs over the economic life of the product groups. In some businesses, a simplification of product lines, with fewer models having longer design lives or broader applicability, can result in longer runs for the product produced and lower costs. Lower costs

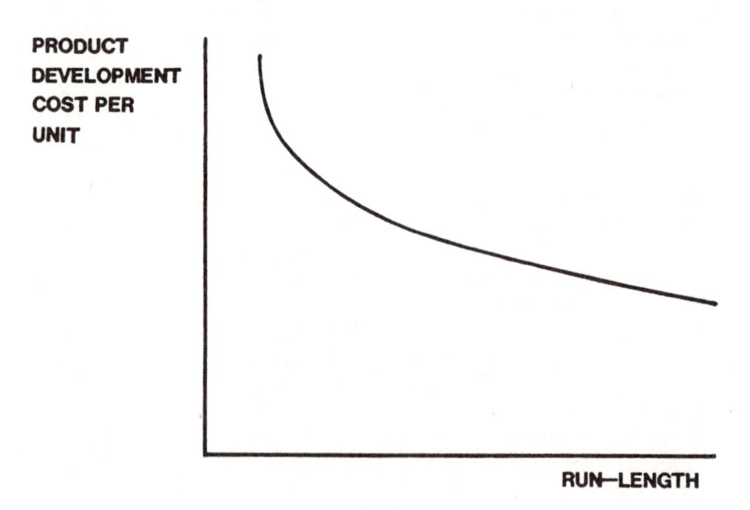

Figure 13-8. Developmental cost relationships.

can be used to price lower, which in turn can build volume. Beyond a certain point, however, a competitor will be operating out on the flat part of a curve, with little to gain by extending runs further. A firm that has significantly longer run lengths than competition can have important strategic advantages. Competitive analysis can measure the degree of superiority and use this superiority to manage pricing, new model introduction cycles, or other business factors controlling volume. For many years, General Motors' run-length advantages assured a very stable profit edge over Ford and Chrysler.

A competitor positioned favorably regarding product development costs per unit because of high volume can possibly move to increase the importance of the key source of comparative advantage. For example, if significant user differences in product need can be identified, model differentiation can be introduced. This will reduce run length advantages for the initiator but less than for other competitors who attempt to follow suit.

Spreading (Sharing) Costs. In performing analysis to identify key sources of comparative advantage in any industry, it is important to be alert for opportunities or threats of shared costs. For example, the costs of increasing product availability for tennis balls (signing up new accounts and maintaining relationships) may possibly be shared with the cost of similarly increasing availability of racket and squash balls or tennis rackets. Good news if you are in these other product lines; bad news if you are not. The salesperson-density advantage of a warewashing specialty chemical marketer may be leveraged by adding degreasers, lubricants, rug cleaners, glues, and washroom chemicals to its line.

However, a cautionary counterpoint should be added here. True opportunities for shared costs are unusual. The potential for shared costs initially may look bright but under closer inspection may prove to be illusory. Asking the sales force to handle additional products may simply result in all the products being sold poorly. In any event, the successful implementation of a shared-cost concept requires significantly more than simply putting another page in the sales catalog.

In the financial services industry, many participants are expending great effort and financial resources to cluster services that will most efficiently share costs and leverage existing relationships. Insurance companies, for example, have been interested in leveraging their marketing activities to sell stocks. The concept is appealing. Moving it from paper to reality has proven difficult. This will require an insurance salesperson who has built a career selling financial information for risk minimization to begin touting the attractiveness of risky stock investments. Will he or she want to do this? Will he or she be comfortable in doing this? Obviously, the transition can be made, but it will require cultural changes, organizational structure changes, and perhaps a new breed of sales managers and salespeople. There are training challenges. User resistance to the idea must be overcome. In many cases, the successful implementation of the concept will require superseding a relationship the user already has with his or her stock broker. Again, the second dimension of what is practically achievable cannot be ignored.

Positioning the Competitors. The identification of the key sources of comparative advantage in an industry sets the stage for truly effective competitive analysis. The goal, of course, is to analyze the competition's current positions and trends in position concerning these factors. Of course, the firm doing the analysis must also analyze its own position. The future plans of competitors that could result in any significant change in position will also be important to identify. This positioning of competitors will indicate:

1. *Who dominates and by how much.* The relative standing of competitors and the degree of strengths enjoyed by leaders.
2. *The nature of the dominance.* Whether leaders are strong in singular sources of comparative advantage or whether superior positions are enjoyed regarding several key relationships.
3. *Weak competitors.* Which competitors are most likely to yield market share to an aggressive and highly-focused competitive threat.
4. *How to defend against competitive moves by*

strong competitors. There may be sources of comparative advantage in which even the strongest competitor has a weak position. Aggressive competitors may be blocked by competing on these points. For example, a competitor may have a manufacturing cost advantage but poor after-market service. One response might be to lower price on service and upgrade the features and price of the hardware.
5. *How to most effectively improve competitive position*. This is really the bottom line of competitive analysis. Through understanding of the position of competitors concerning each key source of comparative advantage, a company can establish a plan for improving its own net competitive position with commensurate improvement in financial performance.

The identification of the priority and degree of resource investment starts with the identification of key sources of comparative advantage and proceeds with evaluating:

1. *The relative significance of the key sources of comparative advantage*. All key sources of comparative advantage are, by definition, important. However, some will be more critical than others. Understanding that the most important is twice or ten times as important as the second is critical. Understanding the possible relationships among them will also be important. For example, a higher rate of new product introduction may also influence shelf space position.
2. *Shape*. It is no great insight to observe that the steeper the curve, the more potentially powerful differences in competitor position can be. This would be true, for example, in a business where high initial design costs are spread over very few units sold. Another and perhaps more useful point is that in some cases exact curve shapes are not intuitively obvious. In fact, it is in these cases that the most promising chances for effective repositioning occur. An example would be a key source of comparative advantage in a check-clearing business. Customer loyalty was strongly dependent on how quickly information regarding account activity was posted. However, the customer perception of value increased only slightly when real-time information was offered. Costly investment in a system providing real-time information resulted in only marginal competitive improvement (Figure 13-9).
3. *Sensitivity*. The relative positions of competitors determine the degree of comparative advantage. The task is to improve position. To be effective, management must know how much improvement along each source of comparative advantage can be expected from some unit of investment. This can be difficult to assess. Also, in many cases, the potential for improvement per unit of investment may differ among competitors. For example, some competitors may have a strategy of being followers in new product introduction. These firms are unlikely to get as much per unit of investment as innovative companies in their business. Again, the culture and abilities of this type of firm will

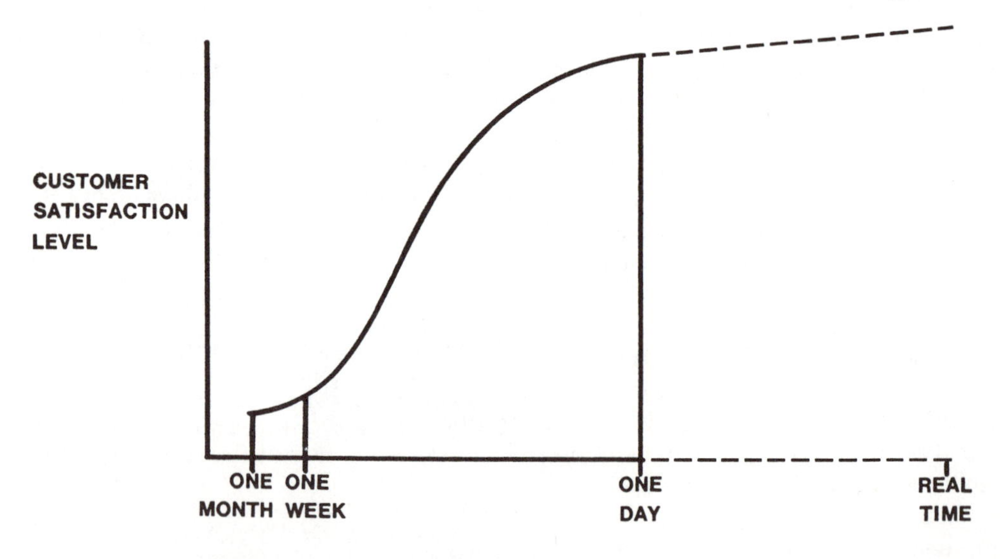

Figure 13-9. Cost-benefit analysis.

be a major determinant in how well this approach works.

4. *Shifts.* The analysis as reviewed so far has assumed a static world. This is an assumption made for the convenience of analysis and discussion. In fact, shifts in the relative importance of the sources of comparative advantage or the shapes of the curves are common. By far, the most critical factor to analyze is the potential for change, not simply of position but sometimes of the basic rules of the game. This change can be external and uncontrollable, in which case it must be anticipated; or it can be internal and controllable, in which case it must be managed to produce competitive superiority.
5. *Situation of competitors.* Positioning the competitors regarding the key sources of comparative advantage is the culmination of the competitive analysis.
6. *Sequence of investment.* From all these steps the most prudent sequence of investment in the business can be deduced. (Investment in terms of management time, monies, organizational resources, and so forth.) See Figure 13-10.

Changes can occur in the relative importance and shape of sources of comparative advantage. Changes also may occur in the ease with which competitors can revise positions regarding critical relationships. Gradual changes are particularly important to track in mature businesses, where managers tend to make decisions based on their perceptions of the business from long ago. Wrenching external discontinuities such as the oil embargo or the breakup of AT&T are usually needed. However, even these are often results of slowly gathering pressures that finally resulted in industry fracture.

In 1976 the profits profile of the trucking industry had a fairly wide spread with high shoulders (Figure 13-11). Subsequent to deregulation, the profile changed dramatically. Certain sources of comparative advantage became much more important, and selected competitors were able to retain attractive and defensible profit positions even though the median profit returns for the industry dropped. Deregulation changed the game significantly. Selected competitors were prepared and won. Why were these competitors prepared? Actually, the pressures to deregulate trucking had been building for some time. An analysis of trends in filing of special tariffs would have shown a significant increase of special deals. As trucking tonnage growth slowed, competitors were more inclined to file for special tariffs. The handwriting was on the wall for those who cared to look. Those who anticipated this external change in the indus-

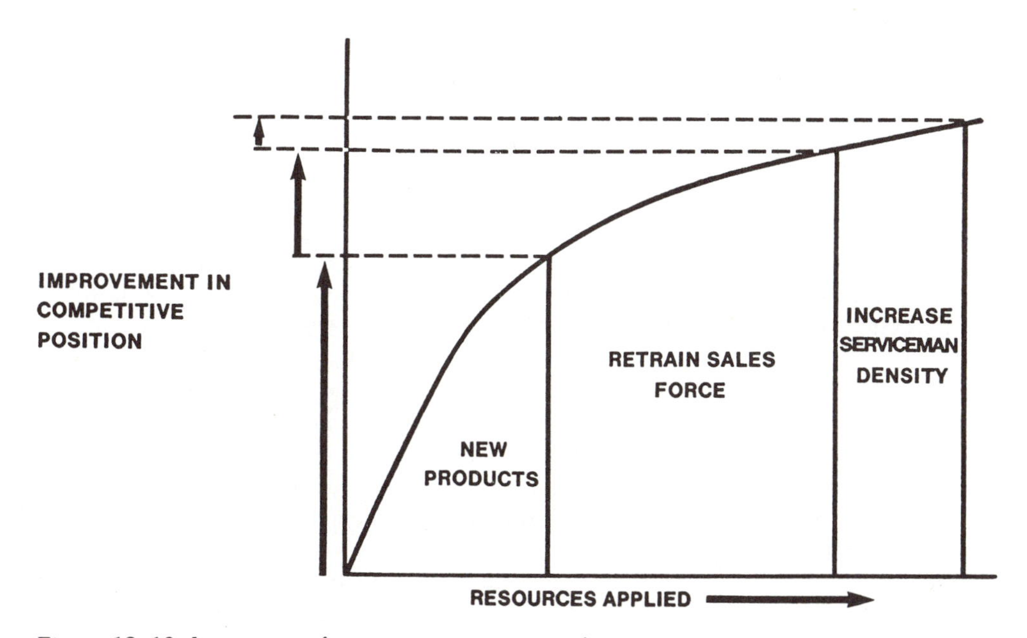

Figure 13-10. Investment of resources in strategy implications.

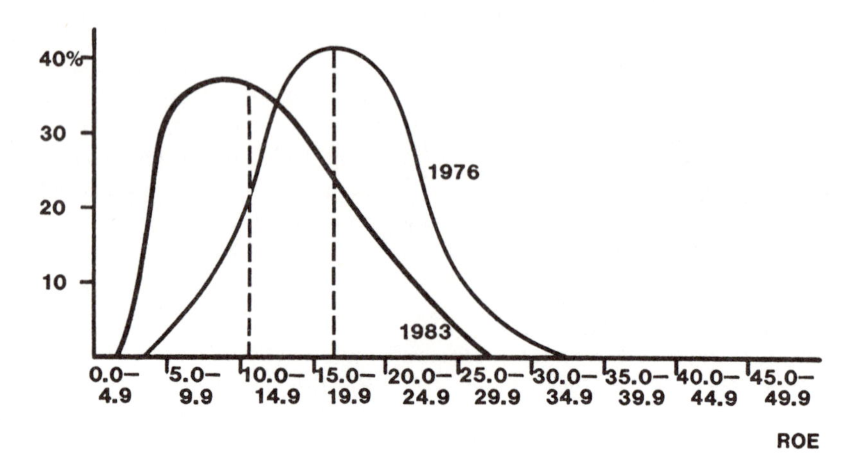

Figure 13-11. Trucking companies return on equity.

try analyzed the potential sources of comparative advantage in the trucking industry and took action accordingly. Yellow Freight, for example, has consistently been among the better performers in the business.

The analysis of the trucking industry raises another interesting point regarding profit distribution in various industries. In an industry in which prices tend to be fixed (airlines, health care, parking garages), the profit spreads tend to be wide. This is true even in regulated fields.

In regulated monopolies, the spreads are much more narrow. The electric utility industry is a good example of this (Figure 13-12). Competitive threats in the utilities business are limited in the conventional sense. The most dramatic competitive threat is that of fundamental change. Cogeneration, for example, is a competitive force that bears close watching.

Pressures for change are building in many other industries right now. In some cases, there are pressures altering the fundamental business structure of certain manufacturing businesses in which historical key sources of comparative advantage accrued from scale, specialized focus, or the ability to spread very large engineering costs over a long product life.

The advent of flexible automation (robots) and CAD/CAM systems threaten leverageability of these factors. Robots and CAD/CAM systems have been around for some time now, with some notable successes. The rate of advance, however, has recently seemed to accelerate.

The automotive industry is a prime example. Recently, very significant advances have been made in the applicability of robots for assembly line work. Roger Smith, chairman of General Motors, recently characterized old spot-welding robots as *junk* next to the newer sophisticated robots that seek out where to weld and adjust to different model designs. CAD/CAM systems continue to improve, with newer systems capable of creating (and storing for future instant alterations) detailed engineering drawings. Flexible automation should extend tool life over several model changes, weakening GM's advantage of spreading design and start-up costs over higher model volumes.

Another type of business ripe for a change in business gestalt is that in which information storage, manipulation, retrieval, and sale are significant activities. Costs of information manipulation are falling, and volume advantages will diminish. Examples would be banking, publishing, service bureaus, and data base

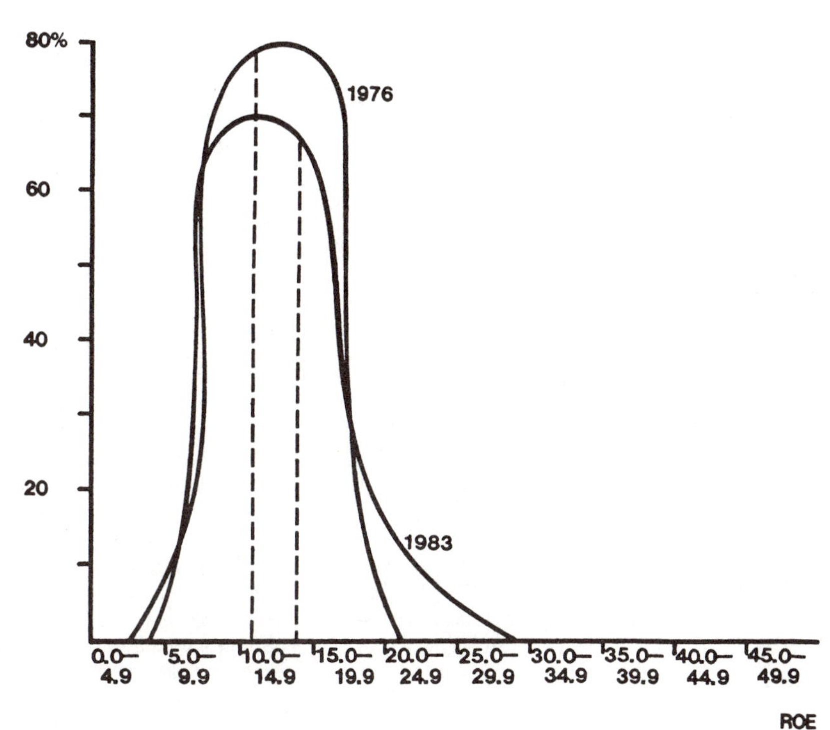

Figure 13-12. Electric utilities return on equity.

development integration and interpretation (such as Moodys and Value Line).

Retail banking, for instance, had long been a game of fixed prices (Regulation Q) and the management of availability (a branch on every corner). Bankers used to talk about the rule of three, meaning that standard operating procedure was to lend money at 3 percent more than it cost and to be on the golf course by three o'clock. Now, deregulation has essentially destroyed fixed prices, and availability can be accomplished through the mail, over the computer, or through the installation of a cash machine at the supermarket. Who is to say the cash machines will not end up being superseded by credit information systems hooked up to individual cash registers?

Transaction costs per customer related to total number of accounts seemed at one time to be a promising source of comparative advantage in banking. In some instances it still is, but trends in information-processing costs have reduced the relative advantage enjoyed by the biggest systems. The more likely new source of comparative advantage for retail banks will be the ability to generate and successfully deliver a constant flow of new products through a number of different channels. Innovation is a costly process that big banks may be able to shoulder more easily than smaller ones. On the other hand, some innovative banks may be willing to wholesale a flow of products to their correspondents.

Changes in the banking environment and the very significant alteration in the base and secondary resources necessary for delivering value to the user have expanded the product possibilities as well. Banking services are well on their way to being redefined. Banks now offer and process credit cards. They could also make payments for retail customers, send out retail bills for commercial customers, perform standard financial analyses, and store financial records for customers. They could perform payroll services, pay taxes, advise on data-processing systems, and do other service bureau functions.

The decreasing costs of handling data once it has been entered into a system argues that banks could look at their business as an integrated process of managing information flow from vendor to manufacturer to wholesaler to retail customer and back again.

New concepts for businesses become possible when significant changes occur regarding key sources of comparative advantage. Tracking them, as well as the positions of competitors, allows the firm to act quickly and seize the opportunity.

Process of Competitive Analysis

The actual process of competitive data collection and analysis can be very detailed. To be effective, it must be intertwined with other blocks of analysis already performed by companies (Figure 13-13).

The sequence of analysis ordinarily is as follows:

1. *Base Case Model.* The first step is an analysis of how resources are used by the firm to create the product. This is usually done by working with company financial, accounting, and marketing information and analyzing such things as shared costs between product lines, sensitivities to volumes, and trends in costs performance (labor costs tend to increase over time; information processing costs tend to decrease). It is usually possible to determine hypothetical sources of comparative advantage at this stage.
2. *Value Trade-offs.* The second step is the identification of customer indifference curves relative to the most important aspects of value. Homeowners may be willing to pay 20 percent more for a lawn mower that is 20 percent lighter. Data processing managers may be willing to pay twice as much for a printer that can print graphics as well as alphanumeric characters. In some cases, a qualitative assessment of these trade-offs is sufficient. Value trade-off analysis is usually performed through fairly extensive market research among the user base to calibrate, quantitatively where possible, how these trade-offs work.
3. *Competitive Modeling.* Starting with the cost model developed in step one and the value trade-offs developed in step two, competitive positions regarding the hypothetical sources of comparative advantage can be developed. Competitive information required usually centers on detailed information regarding products manufactured and marketed and highly selective information regarding purchasing, manufacturing, marketing, and management activities. If, for example, run length is a potentially important factor influencing cost per unit, then the concentration will be on developing competitive information in that area. In such a case, analysis of selling and marketing activities may be of minor importance.
4. *Competitive Performance Approaches.* Competitors in the same business may sell similar products yet manufacture and market in substantially different ways. For example, Xerox chooses to manufacture and market its line of nonimpact page printers. The printers use a laser technology, which has certain implications for design approaches, types of components used, and machine speeds. Delphax, a quasi-competitor, uses an ion-deposition technology that requires different components and runs at somewhat slower speeds. Furthermore, Delphax chooses to OEM the printer engine, rather than market directly to the end user. There are advantages and disadvantages to each approach. Clearly these differences must be dealt with in any competitive analysis.

Change

Most business analysis tends to view the market environment, the economics of the business, the position of competitors, and the resources of the firm itself as a stable system. Generally, only a cursory acknowledgment of change is made by some perfunctory comment or discussion regarding underlying trends.

To a certain extent, predicting future events is impossible. Catastrophic events are, by definition, unpredictable and ripples from these events can affect businesses and competitive positions in powerful ways. Lee Iacocca commented that the 1979 phenomenon of rising gas prices played right into the hands of the Japanese, who made a small car for their home market. Although the story is far from that simple, it is true that no amount of anal-

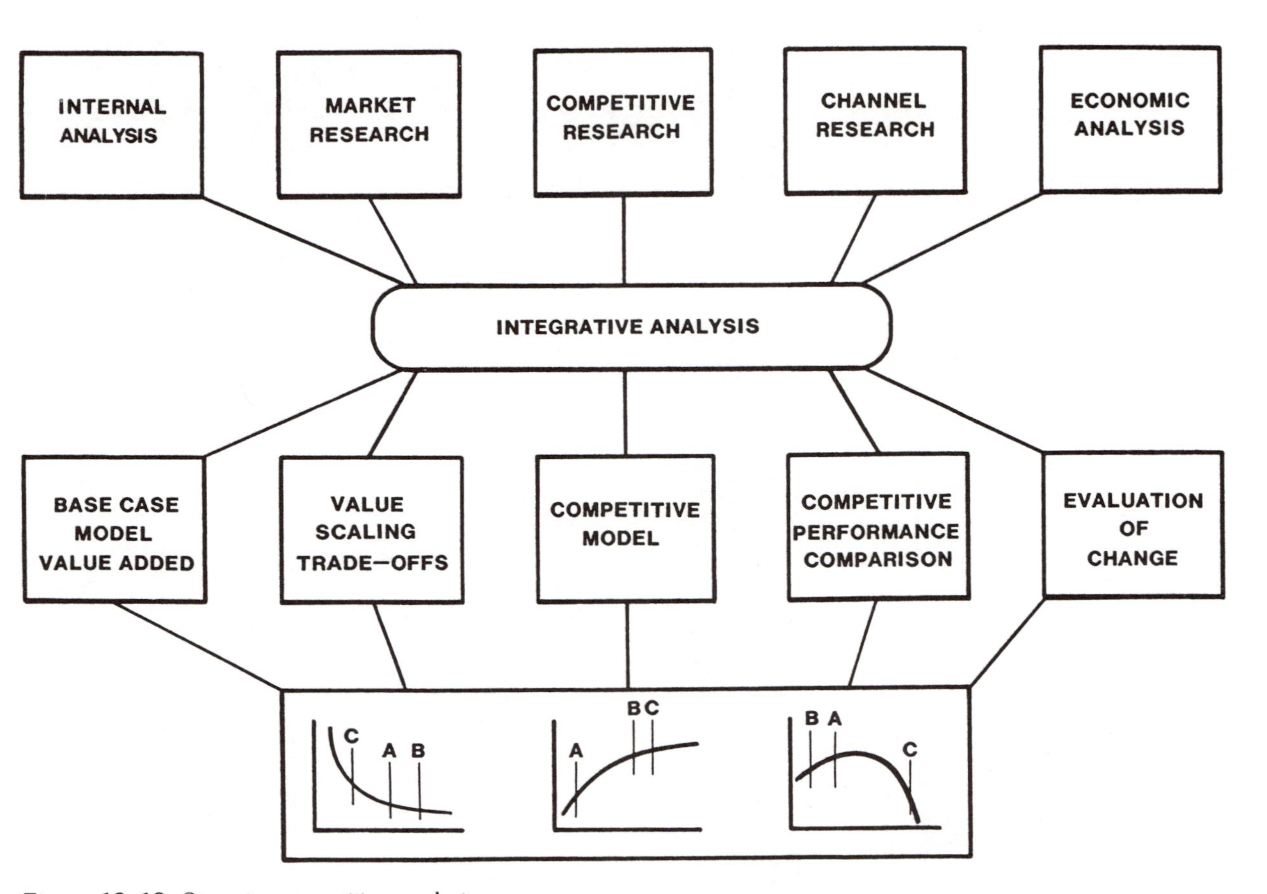

Figure 13-13. Steps in competitive analysis.

ysis would have yielded a significantly better forecast as to when such a catastrophic event might take place and to what degree.

But analysis of change in a business can be very useful, even disallowing the ability to predict these major discontinuities. Superior performance over competitors can be achieved through the anticipation of how changes in a number of areas might relate and what the net outcomes might be. In the car business, for example, the vulnerability of the U.S. car industry to high oil prices was clear from earlier experiences, as was the potential for instability. Although precisely when a price explosion might occur was unknown, a *Plan B* on the shelf would have helped. For example, the fabricated metals industry in this country is in the midst of short-term and more fundamental long-term changes that are creating threats and opportunities for U.S. firms. Improvement of competitive position is possible at this time. It requires analyzing how the universe of competitors will be affected by: (1) rate of capacity growth by country; (2) trends in energy costs; (3) declining labor content through automation; (4) declining ocean shipping rates; (5) shifting market locations; (6) long-term trends in currency exchange rates; (7) cost of capital by country; and (8) substitution of other materials for metals.

All these factors will have a profound affect on the relative position of competitors in the markets. Some areas will be lost to foreign competitors, others will not.

This type of structured approach to competitive analysis and the development of products in the form of recommended actions may seem abstract, mathematical, and inflexible. Do not be misled. The concepts as outlined serve as a departure point for the consideration of many other things that are less quantifiable and should not be dismissed.

Earlier the limiting (or releasing) factors involved in the rigorous approach were mentioned. No one devising strategy can ignore such factors as character and culture of competitors, organizational implications, and other important, yet hard to quantify, facets of the business. Iacocca mentions the fact that both GM and Ford management have lost their feel for the automobile business, perhaps because of the size of the companies. He seems to imply that GM's competitive advantages in the business will allow them to prevail in spite of the insularity and resultant inefficiency, whereas Ford may not be able to. Considerations such as these can hardly be ignored in developing competitive strategy. GM can guess wrong in new product development at tremendous costs time and time again and still prosper. Ford, on the other hand, has less margin for error. So would Chrysler, if it played that game, however Iacocca seems to suggest the company will have more of a specialty strategy in the future. The current culture of Chrysler makes pursuit of such a strategy possible.

The focus of appropriate competitive analysis varies significantly with the type of industry. Two examples will demonstrate the degree of difference that can exist.

Performing Competitive Analysis in the Mature or Maturing Business

Managers generally view a mature business as one in which market demand has peaked. Although this is true in many cases, it is not necessarily so. In a larger sense, mature businesses are those in which product values and the resources necessary for creating them have stopped changing, in other words, where key sources of comparative advantage have become very stable. In these situations, either a clear winner has emerged to dominate the business, in which case the profit environment is stable, or the general profitability of the business as it is presently constituted is either in decline or threatened with decline. Management becomes concerned that current and future investments may be caught in a cash trap. Signs of a mature business are: (1) similar products offered by competitors; (2) flattening or declining unit demand; (3) a fragility of pricing that results in extremely long-term effects of price cutting; (4) limited ways of segmenting the market and creating enriched and leverageable value packages for different customer groups; (5) product redesign to save on purchased materials, even at the expense of declining buyer loyalty; (6) use of an established brand name for remarketing related product lines; (7) stagnant distribution channels for the product; and (8) diminishing yield from research and development expenditures.

To survive in a corporate portfolio, a competitor in a mature business ordinarily must distance himself or herself significantly from other competitors. Either he or she remains or must become the dominant competitor since aggregate returns for the industry will be threatened. However, the possibilities of achieving such a position are diminished as firms tend to cluster close to one another in competitive position as the industry evolves.

What then are the aspects of competitive behavior that must be analyzed in a mature business, and what are the actions a firm can take to develop, expand, and maintain competitive advantage?

The first priority must be to monitor the business as it currently exists to detect increases in competitive intensity, which may result in a deterioration of the general profit character of the business. Things to watch for include:

Product line proliferation through superficial variations. This can occur in some industries for strategic reasons. For example, in the cigarette business, new brands mean more shelf space control. Generally, forced artificial differentiation of products means adding costs without adding value.

Change in ownership of assets and the terms of such arrangements. New ownership in a mature industry is usually negative. New owners may behave in ways that appear nonrational to longer-term players. Typically, new owners are more aggressive, less inclined to see the business as price-fragile.

Competitive programs to extend the life of existing assets rather than replacing them. This can possibly create manufacturing cost base differentials among competitors. Such an approach is less useful in more dynamic industries in that more efficient ways of manufacturing will make nurturing of old assets inefficient. In a mature industry however, there is a greater possibility of strategic advantage in the approach, if only as a way to gradually withdraw assets from the industry.

Significant new dedicated investment in the business, either in marketing or manufacturing. It is particularly critical to anticipate new plants coming into the business. To the extent facilities are dedicated (a typical trait of mature businesses), they represent a future pressure that will not be removed from the industry until the asset is destroyed. Sale of the asset will have little effect on the health of the overall business. It will simply be recapitalized and put to work again.

Competitors making arrangements to share costs with other businesses.

Competitors declaring bankruptcy as a first step in materially reducing their labor costs.

All of these elements, of course, are worth reviewing in any competitive analysis. In mature businesses, however, the chances are good that any significant shift in competitor position or any marked deterioration in the industry as a whole due to competitive action will be precipitated by these factors, hence they bear particularly close monitoring.

As in all industries, strategic decisions in mature industries tend to create a semipermanent reshaping of the use of resources in the business. Implementing strategies takes a raft of different types of decisions involving people, current expenditures, and investments. Some actions will have an impact life of months, others of years, and a very few of decades. In general, the strategy cycle is based on the length of time that important resources are committed to the business. For example, a decision to share costs by licensing defers action on other strategy options until expiration of the license (or its early repurchase). Building a dedicated manufacturing facility commits the firm to a strategy dominated by the useful life of the plant. Other strategy decisions are less obviously tied to time periods but, once made, cannot be undone overnight. For example, the decision to open up a high-priced-high-value product line to include moderate-priced products aimed at a larger and growing segment of the market will have a deep and lasting impression on users and middlemen. Using price as a weapon to gain market share can have an extremely lasting effect in a mature business. Reversal of such decisions is a costly and time-consuming process.

This time dimension to the strategy commitment is one of the fundamental deficiencies between businesses in the extended period of middle life and those in their end games. During the extended period of middle life, the firm can usually count on the business to remain intact long enough for asset-strategy decisions to work out. In mature businesses, the stability of the business as it is perceived can be threatened, and significant transformation may occur. Committed resources that cannot migrate may be lost forever. In end-game industries, therefore, well thought out strategies tend to make rescuable resource commitments, even at some penalty.

Competitive activities create stresses on the business environment. They can create pressures on the business that may or may not result in an improvement of position. For the aggressor, a competitor's pattern of resource investment must be matched to the key sources of competitive advantage. The resultant analysis may or may not indicate that a competitor has improved his or her relative position in the business. If the resource commitment has been particularly unwise (a new factory coming on stream in the middle of a chronic oversupply situation), then all competitors may suffer from a resultant deterioration in the profit levels. In a resource commitment of this type, it is important to know the impact life. For example, a competitor may mount an offensive to gain market share, using price as a primary weapon. In many cases, the competitor may intend for this competitive action to have a short natural life. It may last as long as a particular marketing activity. It may be a daily decision. The decision to build a new factory and the subsequent commitment of resources, however, is a different matter. This decision is much more costly and less reversible. Even if the company sells the factory, the new owners will simply operate it on a reduced capital base. This emphasizes the most important aspects of competitive analysis and the need for it in mature industries. The strategy cycle of competitors must be assessed. A competitor who has effectively shortened his strategy cycle by minimizing investments in new product R & D and leasing facilities will care less about the long-term profit health of the business. His or her strategy may be more opportunistic. To the extent the competitor succeeds at shortening the product cycle, he or she may be able to make a harvesting approach really work—a rarity. Thus, competitors must be able to direct their own actions in the light of what others are doing.

The Foreign Competitor

In mature businesses, the rate of technological development tends to be static. Because of this, manufacturing activities in mature businesses face particular challenges from foreign competitors. Many organizations are willing to put foreign firms in business for a price. Businesses in which manufacturing costs are significant in the overall delivered cost structure are particularly vulnerable to competitive dislocations.

In cases where shipping charges are an insignificant percent of delivered costs, the threat is greater. If such a situation is developing or already exists, it doesn't necessarily mean the end of the line for the U.S. manufacturer. If significant sources of comparative advantage exist outside of the manufacturing area, the U.S. competitor may still prevail. However, if the manufacturing disadvantages are severe, it may be prudent to move manufacturing offshore.

In this type of mature business, special emphasis should be placed on analysis of import statistics in volumes and prices of the products at issue. When activity is noted that is judged descriptive in the market, special competitive studies should be done to analyze the degree of factor-cost advantages enjoyed by the new competitors.

The competitive intentions of foreign firms

should be determined. Assessment of the level of their sophistication regarding the market should be done. Foreign competitors may have no understanding of market price points, or may be selling through inappropriate U.S. distributors. In some cases, these distributors may be viewing the product franchise opportunistically as found money, and therefore cut prices to get business with little regard as to what this may mean for the future of the business.

By definition, improvement in competitive position within a mature industry is difficult. Three basic concepts can be considered.

1. *Stop performing certain functions.* In general, firms are reluctant to stop performing marketing or manufacturing activities. In fact, in a mature business, there may be some aspects that can be done more efficiently by participants in other industries.
2. *Search for a factor-cost advantage.* Large companies are well aware of the manufacturing-cost advantages of foreign plant location and act accordingly. Smaller companies are sometimes intimidated by the thought of manufacturing in such areas as Hong Kong, Singapore, West Africa, or China. However, with careful planning, this can work very well. In many cases, costly subassemblies or components can be foreign made.
3. *Application of new technology.* In some cases, competitors in mature businesses become extremely rigid in the concepts of how the business should be run. In fact, the unchanging nature of the business attracts personnel who are comfortable in that type of environment. They become risk averse and are not interested in any concepts that might deviate from the formula they have become accustomed to applying. In some areas, innovation habitually comes from outside the business. Examples of this are machine tool automation from the machine tool manufacturer and information systems for cost control from the computer manufacturer. However, in areas that are more unique to the particular business, this may not happen. The application of new technology can break the historical business mold and create new profit opportunities. However, it is very difficult to achieve. General Motors is attempting to leverage use of computers and to increase potential technological contributions through the purchase of Electronic Data Systems and the Hughes Aircraft Company respectively. It should also be noted that these acquisitions, in their own right, diversify GM at least slightly (less than 10 percent) out of its vehicle business.

In general, competitive analysis and the development of competitive strategy in mature industries is much more difficult than for most businesses, and chances for achieving a dramatic competitive coup are less likely. However, firms cannot afford to ignore it. The very difficulty of it and the magnitude of the penalty for failure to at least remain abreast of the industry, suggests that more, rather than less, emphasis should be placed on it.

Technology—Dynamic Businesses Also Have Unique Areas of Focus for Competitive Analysis

In this instance, we are differentiating between highly complex and technical but static businesses and those in which rapidly changing technology is or can be an important aspect of how value is created in the product. For example, diesel locomotive manufacture is complex and certainly technical in nature but would not generally be called a technologically dynamic business. On the other hand, manufacture of steel plate could qualify as the technology moves to continuous casting. Other technologically dynamic industries, such as computer or genetic engineering, are more easily categorized.

Technologically dynamic industries are difficult to analyze from a competitive standpoint because they are vulnerable to significant changes in the competitive structure. In mature industries, great attention is placed on significant, irreversible, long-lived investments. In technologically dynamic industries, significant investments may be irreversible but may also be very short lived. Technological change tends to create discontinuities in values or costs or both and the degree and rate of such change is difficult to forecast very far in advance. Other types of change within an industry tend to be gradual. Market tastes shift, influenced by a number of factors. Generally speaking, the shifts are perceptible and can be analyzed. In other businesses, labor costs for some competitors tend to increase gradually as the economic activity in a particular region grows or shifts to production of more sophisticated goods. Some abrupt changes seem like discontinuities that are difficult to plan for but in fact are not. Pricing changes are a good example of this. In recent years, a number of industries have become deregulated. In fact, the forces for deregulation were evident and measurable many years before the changes came about.

Technologically dynamic industries typically are growth industries. As the range of ways in which value can be added to a product expand, new customer groups are discovered and the market grows. In some cases, the new customer groups may desire added features that require specialization. One of the characteristics of this type of industry is a high rate of bankruptcy and new business startup, which does not necessarily hurt the profit environment of the base business or alter the profit position of the original players. The computer industry is a good example of this. Key sources of comparative advantage may remain the same in the base business, but some of the new market offshoots may display additional ones. IBM remains a dominant force in the core mainframe market, yet other firms have carved out defensible profit positions. DEC has created a new concept utilizing minicomputers. Wang is a strong competitor in office systems. Commodore has a strong position in home computers. Newer concepts include desktop computers, and the incorporation of software for engineering computations, such as computer-aided design and computer-aided manufacture (CAD/CAM).

For competitive analysis and the development of competitive strategy in a technologically dynamic industry, the following factors must be considered.

1. *Competitive analysis must monitor the key sources of comparative advantage carefully.* Because of the dynamics of the basic technology, the significance and shape of the particular curves can change dramatically and unpredictably. Chip manufacturers have found certain types of chips vulnerable to new design approaches that basically wipe out old sources of comparative advantage.
2. *Competitive analysis must focus on potential crossovers.* In a technologically dynamic area, advances may cause one technology to supersede another, and the new one may negate the old sources of comparative advantage in the business. For example, manufacturers of car clocks and industrial timing devices were overwhelmed by alternative time-measurement devices using crystal, chip, and stepping motor components.
3. *Analysis of competitive capabilities regarding emerging sources of comparative advantage must be done.* In computer output peripheral devices, key sources of comparative advantage had to do with alphanumeric output speeds and dependability. Companies such as Printronics, Data Products, and Centronics had good positions regarding them. Later, interest developed in graphics output capability as data processing shops sought to expand their internal customer base. Manufacturing of machines for graphics reproduction requires new technologies, and entirely different key sources of comparative advantage emerged. It is still unclear whether the traditionally strong firms can adapt to the new requirements for competitive advantage.
4. *Monitoring of potential sources for technological upheaval.* Technological innovation in other associated businesses can have significant influence on the competitive position in an industry. These technological innovations can be forced by:
   Suppliers: When Alcoa Aluminum created an alterative to the steel can, it eventually forced can manufacturers to make significant investments. The advent of superstrength wood glues are making chip board a viable substitute for plywood, threatening investment in plywood mills.
   Commercial Research Laboratories: Innovations such as the transistor, high-temperature thermoplastics, or fiber-optics can have a significant impact on businesses.
   Customers: Industries are affected by changes in what their customers want to buy as when car manufacturers move to a new generation of engines, operating at higher temperatures and revolutions per minute. Such a technological change forces manufacturers of engine lubricants to create higher performance products.

Changes in these areas may result in significant changes in competitive position. Ordinarily, the effects of these changes are not equal among competitors.

When Competitive Analysis Works

Competitive analysis can result in the improvement of strategy, but the real test of its success is whether it is perceived as contributing to better and faster decisions in the management of the business. A general manager will feel that it is if it helps him or her decide such things as:

- The establishment of pricing practices.
- Whether or not to expand the product line and in which direction.
- Setting the R & D expenditure levels and determining the prudent focus of efforts.
- Whether or not investment in finished goods inventories is advisable and at what level.
- Whether or not to switch distribution channels.
- Whether or not to put substantial resources in back of new product introduction.
- When to become aggressively competitive.
- Whether or not to set up a strategy focusing on a particular segment of the business.

Dimensions of competitive analysis and strategy development are many and varied. The framework presented here is meant to provoke thought, not to provide all possible combinations and permutations. Competitive strategy development at its best is adaptive, and it can make a significant difference in the degree of success the firm enjoys.

As in many aspects of the conduct of business, it is not necessarily easy to do successfully—not all firms meet their objectives. As a matter of fact, most do not. But those that are successful will find the reward a significant improvement in the future of their firm, allowing fresh flexibility and greater promise of reward for the constituencies of the firm.

CHAPTER 14

A Process for Applying Analytic Models in Competitive Analysis

John E. Prescott

John E. Prescott is an assistant professor of business administration at the University of Pittsburgh. Since receiving his doctorate from The Pennsylvania State University, he has pursued research involving the relationship among industries, strategy, and performance and has published in *Academy of Management Journal* and *Strategic Management Journal*.

Virtually every manager acknowledges the importance of understanding his or her industry and competitors. As a result, there has been a growing interest in the use of competitive information as an essential input to strategy formulation and implementation. However, the effective and efficient generation of information pertinent to decision making has hindered the incorporation of a coordinated approach in most firms. The skepticism is partially a result of the lack of a framework and the associated analytical models for conducting competitive analysis. This chapter outlines such a framework for conducting a competitive analysis. The framework highlights the fact that there are many different types of competitive analysis objectives and activities.

The focus of the chapter is on identifying analytical techniques that can be used for competitive analysis and assisting managers in their selection of the analytic techniques that may be most appropriate for their needs. Many of these analytic techniques are discussed in more detail in other chapters. Here, the focus is on their application in the specialized context of competitive analysis.

THE PURPOSE OF COMPETITIVE ANALYSIS

The purpose of competitive analysis is to better understand one's industry and competitors in order to develop a strategy that will provide a sustainable competitive advantage, which will achieve continuing performance results that are superior relative to one's competitors. A fundamental understanding of one's industry and competitors should answer the following questions.

What are the characteristics of my industry?

The starting point of competitive analysis is to understand the dynamics of the industry. One should understand the attractiveness of the industry and how it affects the profit potential of the firms. Industries evolve and as a result change in their attractiveness to particular firms. It is also essential to understand the major issues currently facing the industry.

Who are my competitors?

A thorough understanding of one's industry should readily permit the identification of many of the competitors. However, competition often comes from other industries as well. It is necessary to identify the most relevant competitors as well as potential competitors. This will help bring focus to the following questions.

What are the current positions of my competitors?

Potential moves of a competitor and their impact on one's firm and the industry can only be assessed if the current position of the competitor is known. Since most industries comprise many firms, it is often desirable to develop profiles of each and make comparisons across firms. This will help identify the strengths as well as the vulnerable areas of a competitor. Further, it will shed light on whether a competitor is satisfied with his or her current position.

What moves are my competitors most likely to make?

Competitors can take two types of action: offensive and defensive. Offensive moves are aimed at solidifying or achieving a competitive advantage in some area. Defensive moves are reactions to offensive moves taken by others. A firm should understand the capabilities of its competitors in order to predict both types of moves and to assess their impact.

What moves can I make to achieve a competitive advantage?

Having a complete understanding of one's industry and competitors will permit the identification of opportunities to achieve a competitive advantage. Building on the information from the previous questions and a complete understanding of one's own business, one is now in a position to formulate and implement competitive moves.

The remaining sections of this chapter provide a framework and analytical techniques for answering these five questions.

A FRAMEWORK FOR COMPETITIVE ANALYSIS

As portrayed in Figure 14-1, the framework divides the competitive analysis process into five phases: objectives, data collection, data interpretation, implementation, and updating.[1] I describe briefly each phase and then use the framework for identifying techniques applicable for each phase of the competitive analysis process. The use of a circle in Figure 14-1 symbolizes the importance of the continual application of competitive analysis for a firm. Further, multiple competitive analysis assignments can be in progress and at different stages of completion. The multiple assignments can be independent or can be used as inputs for other assignments.

The first phase of any competitive analysis assignment involves the *establishment of the objectives*. The primary purposes are to ensure that everyone involved understands the type of assignment he or she will be undertaking and the constraints imposed on the project. Since most individuals conducting competitive analysis in firms also have other duties, this phase is critical for the successful implementation of the assignment.

The second phase involves the *collection of data*. In recent years a wealth of information and techniques has developed for acquiring information about competitors. Many firms have developed specialized skills in information collection, ranging from data base systems to interviewing to observation. The key of this stage is to gear collection techniques to the objectives of the assignment.

Data interpretation, the third phase, is probably the most well developed of all the phases of competitor analysis. This will become more evident later when I discuss the variety of techniques for analyzing information. Data interpretation not only involves the

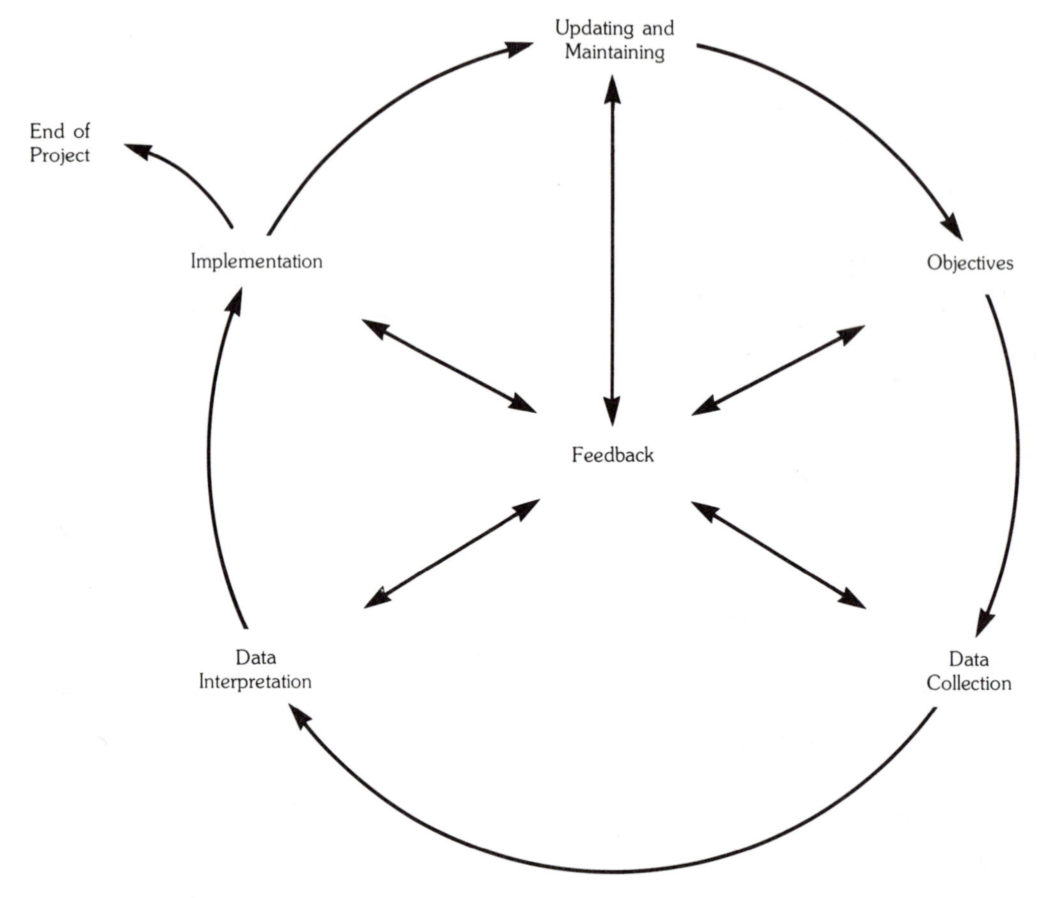

Figure 14-1. A framework for competitive analysis.

manipulation of techniques but also their implications and comparisons.

Probably the most neglected area of competitive analysis has been the *implementation of the finding.* Although the analysis is important, the communication of the implications of the study to the key individual who must make decisions is critical. Unless the information obtained from the assignment is incorporated into the strategic decision-making process, there is no reason to undertake the assignment.

The final phase is *updating and maintaining the system.* As indicated in Figure 14-1, some assignments come to a conclusion and no updating is required. I will discuss this more completely below. In continuous systems, decisions concerning the degree of formality and frequency of updating must be made. The objectives of the system should play a primary role in this decision. Having briefly outlined a framework for conducting a competitive analysis, I now use the framework to identify and describe techniques for each phase.

COMPETITIVE ANALYSIS OBJECTIVES

The first phase of any competitive analysis assignment is to determine the objectives. The objectives will influence every other phase of the assignment and should be well defined. Figure 14-2 illustrates a classification system for competitive analysis objectives that aids in clarifying the depth and purpose of the particular analysis. The two dimensions are the

		Comprehensiveness of Assignment			
		One Fact or Figure	Partial Company or Industry Profile(s)	Complete Company or Industry Profile(s)	Comprehensive System
Type of Assignment	Offensive				
	Informational				
	Defensive				

Figure 14-2. A classification of competitive analysis objectives.

comprehensiveness of the assignment and the type of assignment.

Competitive analysis systems come in many forms. Most managers are concerned about the costs of a system. The system can be designed around a project format or can be comprehensive and on-going. The comprehensiveness dimension addresses this issue and identifies four categories for competitive analysis systems. The first three categories can be run from a project format, whereas the fourth is a comprehensive on-going system. The four categories are:

1. *One Fact or Figure.* The objective is to find one piece of information. For example, what is the size of the fast-food market? How many distribution outlets does a drug retailer have? How many R & D personnel are employed by a particular steel firm? What is the age of a CEO?
2. *Partial Company or Industry Profile.* The objective here is to understand one particular area of an industry or company in depth. For example, what are the barriers to entry in the watch industry? How have service dealerships for stereos changed over the last five years? What is the financial position of the firm? What is the relationship between corporate headquarters and the business? What is the marketing strategy of the business?
3. *Complete Industry or Company Profile.* The objective is to understand, in depth, an industry or company. This assignment would cover every aspect of a company or industry. For example, a company profile would encompass all the functional areas, the administrative systems, management's goals and profiles, the organization's culture, a statement of strategy, its position in the industry, and so forth.
4. *Comprehensive System.* A comprehensive system would include complete profiles for every competitor, potential competitors, and the in-

dustry. Ideally, the system would be an interactive, computerized network in which decision makers could conduct in-depth strategic analysis. Alternatives could be analyzed and compared in real time. The on-line capabilities would also permit continual updating of information. An important distinction between a comprehensive system and the other three is that the comprehensive system is on-going, whereas the other three can be projects that have a definite end.

The comprehensiveness of the assignment dimension concerns the depth of the analysis and whether the system is continuous or project based. The second dimension, type of analysis, focuses on the purpose that the analysis will serve for the firm. Michael Kaiser has identified three types of assignments: offensive, informational, and defensive.[2] Offensive assignments are conducted to evaluate the impact of a strategic move on the industry and competitors. A second aspect of offensive assignment is to determine potential competitor responses that might influence the success of the move. The development of a new distribution system, a new product introduction, the attempt to develop a product that would be the standard for the industry, or a new warranty policy are all examples of where an offensive assignment would be appropriate.

Informational assignments are neutral in the sense that no apparent action is being taken by the firm conducting the analysis to gain an immediate competitive advantage. The results of the assignment, however, may lead to actions that may be either offensive or defensive. Informational assignments primarily have a purpose of gaining a better understanding of the industry or competitors. These types of analyses often arise as a result of a manager wanting to check out a rumor. They also are a good starting point for organizations that have little experience with conducting competitive analysis.

Defensive assignments are oriented at understanding the potential move that a competitor can make that would threaten the competitive position of the firm. Once the threat is understood, a second purpose is to develop responses to minimize or neutralize the threat. Examples of situations requiring a defensive analysis include two competitors discussing a merger to gain a dominant position in the industry, a technological breakthrough in another industry that results in a substitute product having a very favorable price-performance relationship, a weak competitor whose parent has disclosed a commitment to market leadership.

The three types of assignments are not mutually exclusive. Typically, an assignment will contain aspects of all three types, and the outputs of one often can serve as a starting point for another. When describing techniques for other phases of the competitive analysis process, I will return to the classification of offensive, defensive, and informational assignments as a vehicle for identifying the most applicable techniques for the assignment.

CONSTRAINTS ON THE ASSIGNMENT

A clear understanding of time, financial, organizational, informational, legal, and competitive constraints will assist the competitive analysis team in determining a realistic approach to the assignment. Further, the successful implementation of both the assignment and the recommendation of the assignment is elevated when competitive analysis constraints are known in advance. For example, the president of a major basic metals company wanted to know how a variety of automated technologies would influence the competitive dynamics in the industry over the next two years. A project group was formed to answer the question. The group quickly identified its key constraint as a lack of understanding concerning automated technologies and decided to hire an individual with expertise in the area before beginning the project.

Although project constraints should be identified during the objective phase of the assignment, additional constraints may arise during the implementation of the assignment.

For example, an airline company built an information system that could model the competitive moves of the other airlines. The information necessary for the model was available from federal government sources that required airlines to report financial and operating data. After deregulation, however, the governmental agencies were dismantled. The airline now faces the problem of securing comparable information to continue the model.

Constraints often limit the flexibility of the competitive analysis team by restricting the range and extent of actions that may occur. Further, they establish boundaries that help determine the competitive analysis tasks. For example, during an intense acquisition attempt, a firm needed to know the market potential of the acquiree's product and its competitors in Europe. However, because of time constraints imposed by the firm being acquired (due to other offers), the acquiring firm decided to concentrate on the financial and operating characteristics of the acquiree. After the acquisition was completed, the firm discovered to its dismay that a competitor offered the same product for 50 percent less.

The cost of acquiring the necessary information is a constraint that cannot be overlooked. Trade-offs have to be made concerning the comprehensiveness and quality of the information.

METHODS OF DATA COLLECTION

The data-collection phase of competitive analysis is probably the second most neglected aspect. Most of the analytic models for competitive analysis assume that the information is readily available. How does one find the market share of its competitors or identify the timing of a new product introduction? The goal of data collection is to gather pertinent, timely, and sometimes highly specific information about an industry or a competitor. The two key questions for the data collection phase of competitive analysis are: (1) What are the available methods of data collection? and (2) How do I choose the methods given my assignment?

Table 14-1 addresses the first question. The methods for collecting data can be grouped into two types of sites where the information is collected: library and field. Library sites refer to any public or private facility in which written, video, or audio materials may be accessed. The primary data-collection methodologies include the compilation and content analysis of information that has been documented by other individuals. This information often is readily available but it must be found. Often, however, the information needed for a competitive analysis assignment has not been previously documented. One must then turn to the second site—field research. Table 14-1 identifies four methods of field research with a description and example for each.[3] The competitive analysis team (or a firm hired to do the data collection) can use one or several of the methods to generate the required information.

The wide variety of methods for data collection poses a problem of choosing the method(s) most appropriate for the assignment. One approach for organizing data-collection efforts can be illustrated by modifying Leonard Fuld's description of how information travels in the real world (Table 14-2).[4] Fuld notes that for every competitive action that occurs, such as the purchase of property for the building of a new plant, there are a group of actors who are involved in the transaction. As a result, a vast amount of information is generated through the discussions of the actors and the public filings and documentation of the action. These sources can then be tapped to reveal the answer to the assignment. By understanding the actors that are involved in a competitive transaction, one should have a starting point in identifying the sources to tap and the methods that can be used to tap them.

A source can be defined as anything (for example, person, product, written material) from which information is obtained. The sources that one would pursue depend on

Table 14-1. Methods for Collecting Competitive Analysis Information

Site	*Methods*	*Description*	*Examples*
Library	1. Analysis of historical records of individuals, firms, industry: letters, speeches, diaries, and so forth.	The first four methods are oriented to identify published and nonpublished written, visual, and audio material. The purpose is to compile notes and statistics and to conduct content analysis of the material. Computerized information searches can identify sources for each of the methods.	Rather than provide an example, as will be done in the following parts of the table, following are typical information sources for written, visual, and audio materials.[a]
	2. Analysis of documents: statistical and nonstatistical records of firms and formal agencies.		Investment manuals Government industry reports Trade magazines Security exchange commission filings State corporate filings Industry directories Trade shows Classified ads Management biographies Annual reports Books on specific industries, companies, or people Credit services Wall Street transcripts
	3. Literature search for theory and previous research in books, journals, monographs, and case studies.	Small group of individuals is interviewed simultaneously; often referred to as focus group. Multiple groups often are interviewed.	A firm is considering a major design change in its consumer product. A group of typical consumers is brought together to assess the benefits and drawbacks of the design.
	4. Data bases.		
	5. Group interview.		
	6. Telephone interview or survey.	Same as a mailed questionnaire; often used as a follow-up for a mailed questionnaire.	A potential strike is pending with a group of suppliers. A telephone survey of other suppliers is conducted to identify potential alternative sources.

7. Case study.	An intensive, in-depth collection of data concerning a person, event, company, industry, and so forth. Often the case study is longitudinal, but it can be cross-sectional.	A starting point of many competitive analysis efforts is to conduct a case study of the industry in which the business competes.
8. Nonparticipant direct observation.	Observation of an event, people, meeting without direct participation—often referred to as *unobtrusive measurement*. Photographic recording can be a powerful medium to collect data.	Every year the industry has a trade show when all of the executives from the competing firms are in one location. During the events, one can trace the exchanges between the individual executives to gain insights into potential mergers, acquisitions, and other business deals.
9. Participant observation.	The observer is also a participant in the event, meeting, and so forth. Often various forms of recording the information are possible and desirable.	The sales force of a firm can accumulate a great deal of information concerning competitors, new developments in the industry, and so forth from the customers that it serves.
10. Mass observation.	Recording mass or collective behavior in public places. Can be done with both interviews and observation.	A fast-food chain observed that during peak periods many people would leave the restaurant rather than wait in line. As a result of the observations and interviews, they added more lines and developed express lines that had one type of meal comprised of the same entree, side order, and drink.
11. Purchase of competitor's product.	The purchase of a competitor's product will allow examination of the product characteristics and the costs of producing it (reverse engineering).	To gain insights into the production costs of a competitor, a product can be reverse-engineered to identify the costs of each component and the assembly costs.
12. Informants.	An informant is one who provides information. Informants can be within a firm or independent of the firm.	A housewife living near a competitor's facility was hired to track the number of deliveries and departures of trucks to help determine the sales volume in the plant.

(continued)

Table 14-1 (*Continued*)

Site	*Methods*	*Description*	*Examples*
Field	1. Mail questionnaires.	List of questions to obtain information or opinions from a selected group of individuals. Questions may be open-ended or have predesignated categories.	Consumer product firms often include a brief questionnaire in the package of their product that can be mailed in as part of the customer's warranty. The questions typically revolve around socioeconomic characteristics, buying patterns, and usage patterns.
	2. Personal interview.	Detailed set of questions administered face-to-face. Questions may be either open- or close-ended. Often the interview will cover a wide range of topics.	A new employee hired from a competing firm can be interviewed concerning the type of people in the firm, its culture, reward systems, and so forth.
	3. Focused interview.	Interviewer focuses attention on a specific event or experience and its implications; the interviewer knows in detail what will be concentrated on.	A new packaging material promises to be a substitute product for a firm's customers. A focused interview may be conducted with customers to assess the impact of the substitute material.
	4. Free story interview.	The focal person is urged to discuss freely subjects of interest to the interviewer.	A relatively unknown person is elected to head a union that operates in the firm's plants. Management sends a representative to discuss a wide variety of topics with the newly elected union president to help identify potential problem areas as well as nonconflicting areas.

[a]For a detailed list, refer to Washington Researchers, *Company Information: A Model Investigation* (Washington, D.C.: Washington Researchers, 1983), 45-98; L. M. Fuld, *Competitor Intelligence* (New York: Wiley, 1985), 85-458.

Table 14-2. A Method for Identifying Potential Sources

Action	*Actors*	*Potential Sources*	*Potential Information Revealed*
Purchase of property for building a new plant	Company management Seller Lawyers Industrial realtors Notary publics City officials	Filing with town assessor Bank filing Uniform commercial codes Site visit Interviews with actors	Location of proposed facility Details on capacity, costs, technology Expansion plans Personnel to be hired

Source: Adapted from L. M. Fuld, *Competitor Intelligence* (New York: Wiley, 1985), 12.

the assignment, the actors, and the constraints of the assignment. There are two primary types of sources: learning-curve sources and target sources.[5] Learning-curve sources are those with general rather than specific knowledge. They are used when time is not critical and when preparation is necessary before one goes to the target sources. Industry studies and books are representative of this source. Target sources typically harbor more specific information and are most capable of providing the greatest volume of pertinent information in the shortest period of time. Trade associations and company and competition personnel are typical sources in this category. Often these are one-shot sources that cannot be continually tapped.

Each of the methods of data collection can tap many different sources. Thus, the answer to which method(s) to use primarily depends on the constraints of the assignment. Given the assignment and an identification of the actors, the constraints of time and other organizational resources will strongly dictate the methods and thus sources for the project. The example of the purchase of property for the building of a new plant (Table 14-2) can be used to illustrate the role of constraints. Suppose that a manager hears a rumor that such a purchase was made. He or she wants a verification of the rumor in three hours. This is an informational assignment, attempting to identify one piece of information (Figure 14-2), that must tap target sources. The person does not have the luxury of a site visit but must either make some phone calls or visit the town assessor, if possible, within the three-hour limit.

Using the same example but changing the constraints, the manager, upon hearing the rumor, wants not only to verify it but also to know how this will influence his or her facility plans for the upcoming business strategy review with the corporate office in one month. The assignment's objective, although still trying to acquire one piece of information, now becomes a defensive assignment (Figure 14-2). Because there are few constraints on the person, he or she can tap both learning-curve as well as target sources. The added time could possibly allow for other information, such as industry growth rates, estimates of the costs for the new facility, what technologies will be in the facility, to be collected and incorporated into the analysis.

Ingenuity is often a key ingredient for collecting information. For example, the fleet manager of the U.S. Postal Service wanted to find out why United Parcel Service trucks lasted much longer than the Postal Service trucks. He collected information by actually walking up to their trucks and "poking at the body and paintwork."[6] He also talked to their drivers and even was permitted by one driver to enter the truck and check out the inside. Finally, he talked with their manufacturers. Using this knowledge and additional information, the Postal Service is studying the feasibility of building a truck body that can last for 24 years and an engine and transmission that can last for 12 years.[7]

DATA INTERPRETATION

Once data have been collected, they must be analyzed and interpreted before the results can be implemented. This phase of competitive analysis has benefitted primarily from developments in the disciplines of strategic management, marketing, and economics. An attempt has been made to classify the techniques according to the types of questions they seek to answer (Table 14-3). Those techniques that have their primary application to industrywide questions are grouped together. A second group of techniques addresses both industry and business-unit questions. Techniques that focus on competitive analysis at the business-unit level form a third group. The classification is somewhat arbitrary but it serves as a useful vehicle for organizing the discussion of a wide variety of competitive analysis techniques. The discussion of techniques begins with the industry level then moves to the dual level and finally the business level.

Table 14-3. A Classification of Competitive Analysis Techniques

Industry-level techniques

- Industrial Economists' Model of Industry Attractiveness
- Boston Consulting Group Industry Matrix
- Industry Segmentation
- Industry Scenarios
- Political and Country Risk Analysis

Dual-level Techniques

- Multipoint Competition
- PIMS
- Experience Curves
- Strategic Group Analysis
- Value-chain Analysis and Field Maps
- Critical Success Factors
- Technological Assessment

Business-level Techniques

- Management Profiles
- Value-based Planning
- Synergy Analysis
- Financial Analysis
- Stakeholder Analysis
- Market Signalling
- Portfolio Analysis

INDUSTRY-LEVEL TECHNIQUES

Industrial Economists' Model of Industry Attractiveness

A starting place for any competitive analysis is a thorough understanding of the industry in which one competes. Michael Porter has developed a model that identifies five forces driving industry competition: the bargaining power of suppliers, the bargaining power of customers, the threat of substitute products or services, the threat of entry, and the rivalry among existing firms in the industry (see Figure 14-3).[8] The bargaining power of suppliers is the extent to which the suppliers of the industry's raw materials can dictate higher prices or reduced services. The bargaining power of the customers is the extent to which purchasers of the product or service can influence a decrease in price or an increase in features. Increased power of either the suppliers or customers will erode the profits of the firms in the industry. The threat of substitution is the degree to which other products or services can perform the same function as those delivered by the industry. Substitute products often put a price-ceiling on the industry's products and become severe threats as their price-performance trade-off increases relative to the industry's product. The threat of entry is the degree to which potential new entrants are likely to establish profitable positions in the industry. New entrants typically bring substantial resources and a desire to gain market share and sometimes new forms of competition. The degree of rivalry concerns the in-

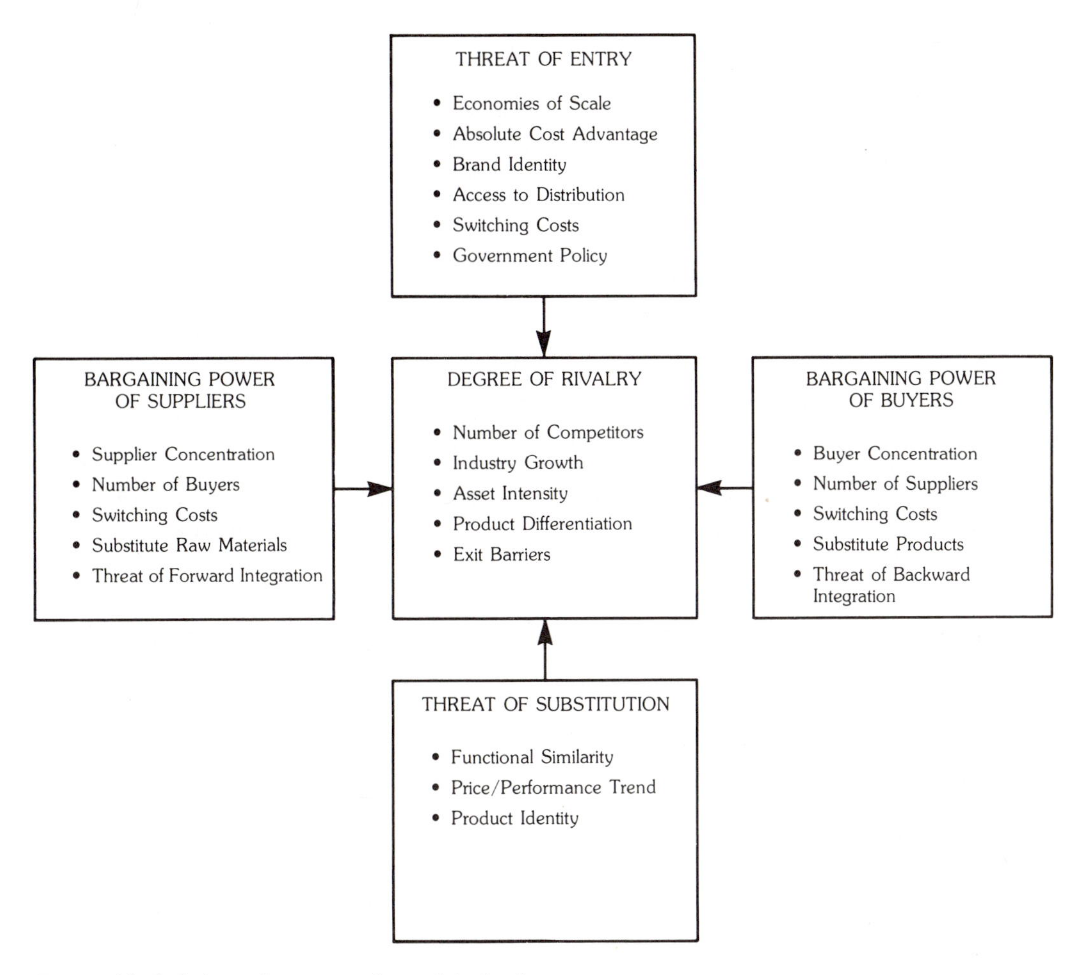

Figure 14-3. Industrial economist's model of industry attractiveness.

tensity of competition. Intense rivalry typically results in lower profits because of such factors as high advertising expenses, low prices, and increased cost of services that cannot be passed on to the consumer.

The basic assumption underlying the model is that the economic structure of the industry is the root of competition. Further, the combined influence of the five forces determines the long-run profit potential and thus the attractiveness of the industry. The strength of Porter's model is that it provides a structured approach for examining an industry. Porter has laid out in detail the characteristics of each of the five forces. Others such as Michael Kaiser have developed extensive checklists and detailed procedures for conducting an industry analysis.[9] Conducting an industry analysis is a time-consuming process and is complicated by how the boundaries of an industry are drawn. One of the most difficult tasks is determining which firms should be included and which excluded. This is also a strength of Porter's approach because it recognizes that competition comes from within one's industry as well as from other industries.

The implementation of Porter's model is an example of an informational competitive analysis assignment that can provide a starting point for both offensive and defensive assignments. The output typically identifies the economic bases of competition; the current and future profit attractiveness of the industry identifies a firm's competitors and raises several questions for the strategy formation and implementation tasks of planning.

The Boston Consulting Group's Industry Matrix

The Boston Consulting Group has developed a method for assessing the attractiveness of an industry or segment of an industry based on two industry characteristics (Figure 14-4).[10] The first characteristic identifies the number of potential sources for achieving a competitive advantage. In industries where there are multiple sources for achieving a competitive advantage, a wide variety of strategies can be successfully pursued. The second characteristic examines the size of the competitive advantage that a leading business can achieve. Industries in which the leading firms can achieve a large advantage typically have a higher profitability potential. The matrix identifies four industry segments that can be used for developing strategies. Volume segments are profitable for the low-cost producers but typically have low levels of profitability. Specialized segments offer significant opportunities for differentiating products and for small competitors. Fragmented segments differ primarily from specialized segments in that the size of the advantage is not influenced by either the size of the business or its strategy.

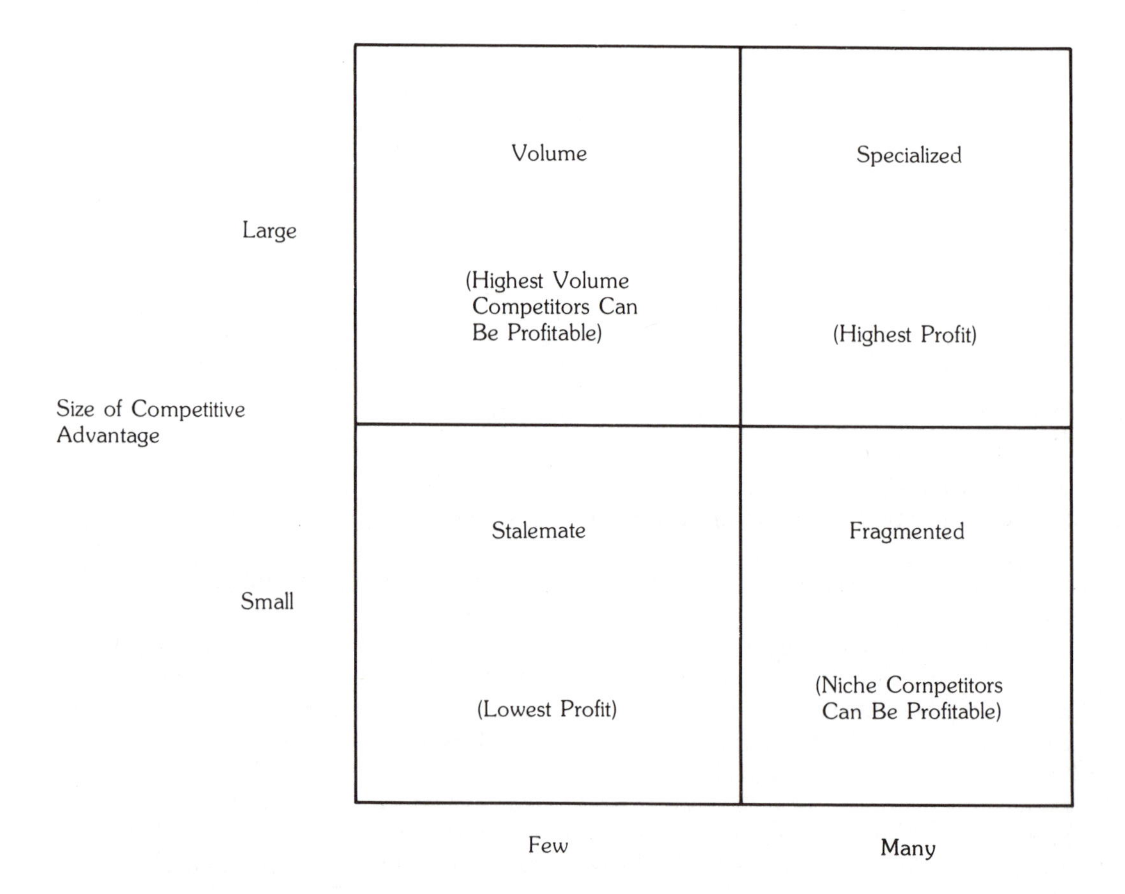

Figure 14-4. The Boston Consulting Group industry matrix.

In most fragmented segments, the advantages of differentiation are minimized because other firms can easily imitate them. Businesses in a stalemate segment are limited in strategic options. All the businesses compete in basically the same way.

Application of this matrix would require a fundamental understanding of an industry and its critical success factors and could probably not be used without other types of analysis. Its use appears to be primarily diagnostic in that it identifies the potential profitability of a segment. It can also be used to identify segments to avoid or segments to pursue. (Also discussed in chaps. 17 and 18.)

Industry Segmentation

Although Porter's model of industry analysis provides a good starting point for understanding an industry, the analyst will quickly notice that an industry is not homogeneous. Most industries are comprised of segments or discrete pockets of competition. The purpose of industry segmentation is to assess the competitive scope of one's business in the industry and to target which segments it could and should serve and how to serve them.

Industries can be segmented on the basis of product variety, buyer characteristics, channels of distribution, and the geographical location of the buyers.[11] A tool for analyzing industry segments is the industry segmentation matrix.[12] A simple matrix would comprise the selection of two segmentation variables and a number of categories for each variable (for example, domestic and international markets for geographical location and wholesalers, distributors, retail outlets for distribution channels). Many different matrices can be constructed and can be combined to provide a comprehensive picture of industry segments. The matrix can then be used to address questions such as: What is the attractiveness of each segment? Is a segment large or profitable enough to warrant a unique strategy? Are there segments currently not being exploited? In which segments do my competitors compete? Are different strategies required in different segments? Are there segments in which different businesses can share raw materials, manufacturing processes, distribution, and so forth?

The strength of industry segmentation is that by examining the industry from many different angles, a manager may identify pockets within the industry that currently are not being satisfied. A further strength is that it can identify which segments are most likely to sustain high levels of profits and which are susceptible to substitute products, technologies, and entry that would erode profits.

Industry Scenarios

Industries evolve and as they do they change in their attractiveness.[13] It often is impossible to predict with confidence the evolution of an industry. When there is a considerable degree of uncertainty as to the future structure of an industry, scenario planning is a way to prepare for that future. A scenario is a detailed, internally consistent description of what the future may be like and is based on a set of assumptions that are critical for the evolution of the industry. The output of a scenario is one possible configuration for the industry. A set of scenarios should cover a wide range of possible structures. The set can then be used to develop and evaluate strategic moves.

Scenario planning for industries is relatively new.[14] Scenarios have been used primarily in examining the macroenvironment[15] or in worldwide planning.[16] Zentner summarizes over thirty studies that describe and use scenario planning in a wide variety of roles.[17] The study of alternative futures is similar to scenario planning.[18]

Scenarios can be used to determine the sources of competitive advantage or critical success factors as an industry evolves. The consequences of the scenario for each competitor can be used to predict future offensive and defensive moves. One of the most valuable aspects of scenario development often is the sensitizing of management to the critical importance of adapting to industry evolution.

Since change tends to be slow and incremental, in most organizations scenarios can highlight the importance of recognizing the need to change well in advance.

Political and Country Risk Analyses

The techniques for analyzing a business's position and industry are appropriate for global as well as domestic markets. However, when an industry is global and a business is studying the possibility of competing globally, the process of competitive analysis becomes more complicated. Countries vary in terms of their views of business, culture, governmental structures, and so forth. Therefore, a global competitor needs to adopt a slightly different perspective than a domestic competitor. Political and country risk analysis techniques attempt to assess the types and extent of risks of competing in foreign countries.[19]

Two types of risk that a business must assess are asset protection and operational profitability.[20] Asset protection risk is associated with the loss of a business investment through confiscation, expropriation, damage to property and people and loss of the freedom to transfer people, money, and ownership. Operational profitability risk is associated with events such as currency instability, breaches of contract, discriminating taxes, and operating restrictions. The sources of the risks are either from host governmental intervention or the instability of the host country's environment.[21]

One of the major problems associated with political risk assessment is that U.S. managers often interpret the information in terms of U.S. norms rather than in terms of the host country's norms. A variety of techniques have been developed to overcome this problem.[22] Five of the more popular techniques are:

1. *Grand Tours.* Senior executive teams visit local business leaders, government officials, and so forth in the country under review.
2. *Old Hands.* The firm relies on the experiences of individuals who have lived, studied, or conducted business in the country.
3. *Delphi Techniques.* A two-step process of identifying the most important risk variables and then having a panel of experts rank or weigh the importance of each variable to arrive at an overall assessment. There are many possible variations to this technique.
4. *Quantitative Techniques.* A variety of quantitative techniques such as rank ordering, decision-tree analysis, and multiple regression analysis can be used to assess the risk in a country.
5. *Integrated Approach.* An integrated approach employs both qualitative and quantitative techniques to arrive at a story line of the risk situation.

The use of competitive analysis techniques and political and country risk analyses should address four questions: What are the risks of competing in the country? How are the roles of competition different from those in the base country? Who are the competitors in the host country? What are their strategies and competitive positions?

DUAL-LEVEL TECHNIQUES

Multipoint Competition

Multipoint competition occurs when diversified firms compete against each other in several markets. The markets can be segments of an industry, related industries, or unrelated industries. Multipoint competition is a special case of synergy because a firm operating under these conditions must consider the interrelationship of the group of industries when developing strategies. Although multipoint competition is a common occurrence, little theoretical or empirical work has been conducted. The research is primarily anecdotal with the exception of Shakun's research on advertising in coupled markets[23] and Karnani and Wernerfelt's and Porter's theoretical work.[24]

A simple tool for diagramming multipoint competition is a matrix that identifies competitors on one axis and the firm's businesses on the other axis. The matrix is completed by identifying the competitors that compete against multiple businesses. The value of this analysis is primarily to develop offensive moves. Part of an offensive analysis should identify competitor defensive moves that limit the success of the proposed offensive move. A multipoint analysis will assess the likelihood of a competitor retaliating not in the focal business but in other businesses in which both firms compete. This is referred to as a cross-parry.[25] Retaliation of this sort is less likely with firms that have diversified into unrelated industries because the costs of coordination are often perceived as outweighing the benefits.

PIMS

Probably the most comprehensive data base for strategic analysis is the Profit Impact of Marketing Strategy (PIMS) of the Strategic Planning Institute.[26] The data base is an ongoing project that collects data from member firms that pay a fee in return for several services. The purpose of the data is to assist in the planning efforts of participating businesses by addressing the following questions:

What profit rate is *normal* for a given business, considering its particular market, competitive position, technology, cost structure, and so forth?

If the business continues on its *current track*, what will its future operating results be?

What *strategic changes* in the business have promise of improving these results?

Given a *specific* contemplated future strategy for the business, how will profitability or cash flow change, short term and long term?[27]

The unit of analysis is a potentially free-standing individual-business unit for which a complete strategic plan can be developed. Currently there are data for over two thousand business units from over two hundred corporations over a four-year period. PIMS collects data on the business units' operating activities, their industry and competitors, their products and customers, and a description of the businesses themselves. The PIMS data base can be accessed through a computer terminal to conduct a variety of empirical analyses. The most well-known model developed by PIMS is a 37-variable equation that explains over 80 percent of the variance in return on investment (ROI). Member firms can conduct their own analyses and receive reports from PIMS. Four of the major reports are: (1) PAR Reports that indicate normal ROI and cash flows for look-alike businesses; (2) Strategic Sensitivity Reports that do a "what if" analysis for strategic changes; (3) Optimum Strategy Reports that predict which combination of strategic moves would maximize ROI and cash flow; and (4) Limited Information Model Reports that combine the PAR and Strategic Sensitivity reports but in a less comprehensive manner. The Strategic Planning Institute also publishes a newsletter that reports on empirical findings for specific topics such as the relationship between R & D expenditures and ROI.

The PIMS data base has been analyzed extensively by academic researchers.[28] The data base can assist managers in the strategy formulation process in that it can help identify why a business has not been performing well and how a variety of offensive moves might improve performance. Because the performance measures have a short-run bias (ROI, cash flow) and since there is limited time series data, the data base appears to be more appropriate for adjusting strategies for near-term improvements. In sum, the data base can provide excellent insights for a business unit's strategic analysis but is of limited use for examining competitor responses of either an offensive or defensive nature.

Experience Curves

Managers and workers alike realized that the time involved in completing many tasks decreased with practice. This idea evolved into the learning-curve concept, which states that the direct labor cost of a product decreases in a regular manner as the experience at producing it increases. The Boston Consulting

Group extended the learning curve to all the costs of producing a product.[29] They identified the experience-curve effect as the decrease in costs that occur over the total life of a product. In many industries a 20-25 percent decrease in costs has been observed with every doubling of accumulated production.

The experience curve has received considerable empirical analysis[30] as well as conceptual analysis.[31] A study that has significant implications for the use of experience curves in competitive analysis was done by Hall and Howell.[32] They built a convincing argument that the supposed effects of the experience can be attributed to current economies of scale rather than accumulated experience. The study is significant because experience-curve concepts have been used as a basis for individual firm and industry cost analysis and pricing decisions. Traditional views would suggest that a firm needs to understand its cost position relative to the market leader and calculate the number of years that it would take to achieve cost parity with that leader. During that time the market leader would have the most flexibility with respect to pricing decisions because it would have the lowest costs. If Hall and Howell's argument is correct, a firm should compare its current economies of scale across the value chain with respect to its competitors rather than from a historical perspective. Businesses pursuing experience-curve effects trade flexibility for low costs. The reduced flexibility can result in problems if the industry evolves in ways that the experience is not a critical success factor or if a new technology with considerable cost savings or differentiating capabilities replaces the existing technology.

Strategic Group Analysis

Strategic group analysis is a technique that serves a middle ground between analyzing an industry as a whole and examining individual businesses. Strategic groups are groups of businesses in an industry following similar strategies. Academic researchers originally developed the concept to explain profitability differences between businesses within an industry.[33] The concept has evolved into a tool for strategic group mapping and as an explanation of why some businesses continue to follow unsuccessful strategies.

Strategic groups form for a variety of reasons, including differences in business strengths and weaknesses, the entry and exiting of businesses from an industry, and historical accidents. Once strategic groups are formed, the businesses within the group tend to have similar characteristics in terms of their strategies and administrative systems. They also tend to be affected by and respond to competitor moves and external events in similar ways. The groups develop what Caves and Porter labeled mobility barriers.[34] Mobility barriers hinder the movement of a business from one strategic group to another. Thus, they protect groups from other businesses that attempt to enter or imitate their strategy. Mobility barriers help explain why some firms continue to follow average or unprofitable strategies.

Porter developed the tool of strategic group mapping, which is a two-dimensional graphic display of competition.[35] The analyst must choose the strategic dimensions for the axes of the map. Porter suggests that the choice of dimensions should be those that determine key mobility barriers and do not move together. The drawing of multiple maps allows the analysis of competition from different perspectives.

The concept of strategic groups is most helpful in organizing the many firms in an industry into meaningful groups. Managers can then identify the multiple forms of competition in the industry. By analyzing the mobility barriers of each group and charting strategic movement and trends over time, a firm can better predict the type of offensive moves that another group may make and its defensive reactions to those offensive moves. The maps also may identify opportunities for creating new groups and provide insights as to how the mobility barriers for a group can be strengthened.

Value-chain Analysis and Field Maps

An understanding of the competitive position of a business requires knowledge of the many discrete activities that a business performs. A tool for analyzing these activities is the value chain.[36] The value chain is an extension of the value-added concept, which can be defined as the difference between the selling price of a product or service and the costs of bringing the product to market.[37] The primary objective of value-chain analysis is to gain insights into a business's margins through a picture of its costs and differentiation through the chain. Porter's model is comprised of nine generic activities that every business performs (Figure 14-5).[38] The five *primary* activities involve the physical creation of the product, its marketing, and its service. The four *support* activities occur at every phase of the primary activities. For example, the support activity of procurement is spread throughout a business. Raw materials must be obtained, machinery must be bought, paper, pens, and pencils are necessary for the sales force, and laboratory supplies needed for R & D are examples of why the support activities must be viewed from a businesswide perspective.

A value-chain analysis is both detailed and time consuming. However, the details provide insights into how a competitor is trying to achieve a competitive advantage through cost reductions and/or differentiation. The analysis further identifies the weaknesses and inflexibilities in a competitor's activities. The value chain is probably the best technique for understanding the operating details of a business.

The analysis tends to separate the activities, but the analyst should search for linkages within the activities. The way that one activity is performed often has implications for the costs and performance of other activities. Linkages require management coordination and lead to trade-offs throughout the chain. Linkages not only exist within the business but extend to suppliers and customers as well. Thus, the value-chain analysis can be extended to the complete value system of a product, which may involve multiple industries. Another extension for a diversified firm is to examine linkages among its businesses or product types. Often linkages can be developed that lower costs or help to differen-

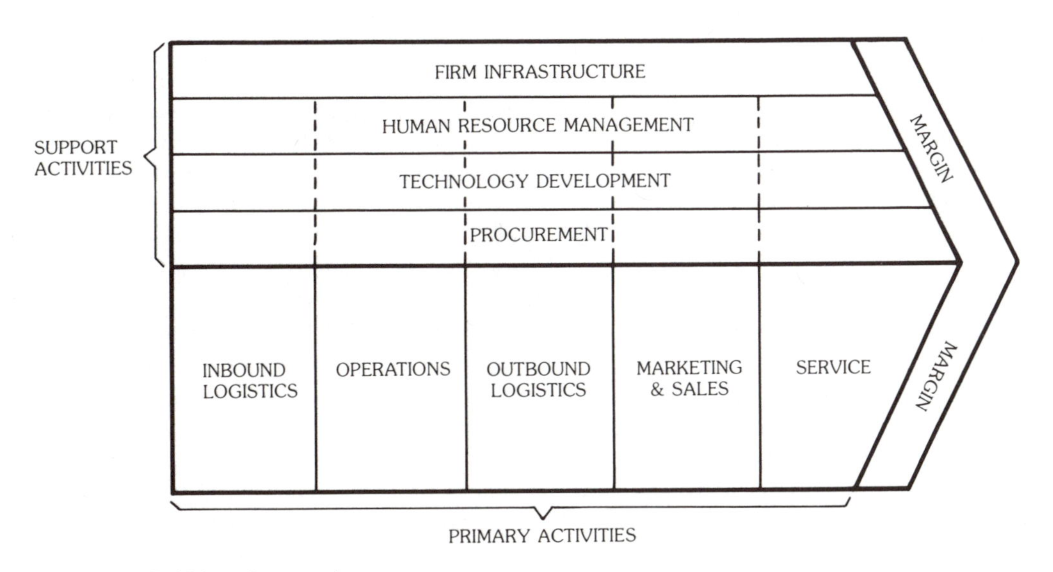

Figure 14-5. Value-chain analysis.

tiate the businesses. These linkages are referred to as synergies. A technique for examining these linkages is the *field map* developed by Strategic Planning Associates. The vertical axis of the map consists of the primary activities of the value chain. The horizontal axis contains a competitor's products or business. The content within the map involves the identification of shared activities across products or businesses.

Critical Success Factors

Most of the techniques that have been discussed in this section are oriented one way or another at identifying critical success factors (CSFs). CSFs are the few areas in which a business must do adequately in order to be successful.[39] John Rockart and his associates at M.I.T. have done the most extensive work on CSFs and have identified five sources: the industry, the competitive position of the business in the industry, broad environmental forces, temporal factors, and the managerial job requirements. They have further found that CSFs have a hierarchical nature. There are CSFs for one's industry, the corporation as a whole, business units within the corporation, and individuals.

Leidecker and Bruno have developed a framework for identifying and using CSFs.[40] They identify three primary uses of CSFs for strategic planning. First, they can be used to identify threats and opportunities in the industry and from competitors. A second use is to assess internal strengths and weaknesses relative to developing a competitive advantage. The third use is a means for setting priorities for the businesses resource-allocation process.

A matrix with competitors on one axis and the CSFs on the other is useful for organizing the information and addressing the three uses that Leidecker and Bruno identified. The CSFs can be weighted and each competitor can be assessed as to his or her position on each factor (also discussed in chap. 20).

Technological Assessment

Few events happen within an industry that can match the impact of a technological change. Every activity in the value chain of a business involves technology and a change can have an impact on the business in five ways: (1) it can change the costs of performing the activity; (2) it can lead to differentiation; (3) it can create new market segments; (4) it can bring previously separate activities together; and (5) it can replace or serve as a substitute method for performing the activity. The most significant threat of a technological change is when a firm is surprised and the change alters the basis of competition.

There are a variety of tools for conducting a technological assessment of a business. The technological matrix plots the attractiveness and relative position of each technology for a business.[41] A technological-share matrix divides businesses into four categories (Start-up, Follower, Leader, and Develop-Decline) based on technological share and market share.[42] A product-process matrix examines the evolution in manufacturing technologies as a product moves through stages in the product life cycle.[43] Morphology mixes scenario development and brainstorming techniques to generate possible new technologies.[44] Along with these techniques, there are many how-to articles that describe the importance of technology and how it can be woven into the strategic planning process.[45] All the techniques are oriented to help managers allocate resources to the technologies most appropriate for their competitive position and industry. The choice of technologies will be reflected in a competitor's R & D deportments, investment strategies, market development tactics, and lining practices for its personnel. A competitors' technological profile will help identify those competitors most likely to pose a threat and where the threat most likely will occur. Many technological changes occur outside the focal industry; businesses that can recognize the change and apply it seem to perform better than those that try to fight change.[46]

BUSINESS-LEVEL TECHNIQUES

Management Profiles

Whereas most of the techniques discussed in this section concern the economic analysis of a business, management profiles examine the goals, backgrounds, and personalities of the individuals making strategic decisions. Management profiles are a key aspect of competitor analysis because management does not always act in a rational, economic manner, nor should they. Management profiles should be developed for the board of directors and all top-level managers. Profiles for the board of directors should include their current positions in other companies, their ages, the number of years on the board, other boards on which they serve, decision-making styles, and any other background information available.

Profiles for the current management team should contain much the same information as for the board with particular focus on their employment histories, professional associations, hobbies, decision-making patterns, education, and other background information. Using this information, a management profile chart can be developed that traces the movement of management personnel through the organizational hierarchy. The charts help identify who is next in line for various positions, weak positions in the organization, if the organization is dominated by a particular group of individuals (for example, engineers or lawyers), and if decision making is confined to a single dominant individual, a small team, or a committee.

A technique for gaining insight into the decision-making styles of managers is to study their reactions and decisions to critical incidents—events that do not occur often but have a significant impact on the firm. For example, how did management respond to a strike, how did they handle a plant closing, scandal, or the dismissal of an executive? Insights gained from this analysis will help in evaluating how competitors might respond to an offensive move. It is more important to understand what managers will do rather than what they can do.

Value-based Planning Models

A recent thrust in the area of strategic planning and competitive analysis has been to evaluate strategies and strategic moves in light of a firm's stock market performance. The basic assumption underlying this trend is that the primary goal of every corporation is to create value for its shareholders. The variety of techniques that have been developed trace the relationship of a firm's capital market performance (typically its market-to-book ratio) to its financial performance (typically a ratio of the firm's return on equity to its cost of equity) over a period of years.[47] This analysis can determine if the firm has been creating, maintaining, or destroying value for its shareholders.

A value-based analysis of a competitor can help answer several questions.[48] Is the primary goal of a competitor growth or profits? For example, firms that have high profits and high stock valuation and are investing little in new assets are more concerned with profits than growth. Is the firm likely to change its strategy in the near future? A firm that has a low stock valuation is more likely to attempt a change in its strategy than one with a high valuation. Can the firm raise new capital from equity or debt? It is easier for a firm to issue new equity if it has a high market to book value. Is the firm vulnerable to takeovers? Firms that are poorly valued or undervalued may be targets for firms that can infuse capital or use their assets more efficiently.

The models are easily implemented for corporations as a whole, but are more difficult to use for individual businesses in a diversified firm. These individual businesses typically do not have stock that is traded, and the determination of the cost of equity for an individual business in a diversified firm becomes extremely difficult. This approach needs to be used, in my opinion, with much judgment because it overlooks several questions of interest to managers. For example, the approach does not address entry, exit, and mobility barriers. It does not take into account the political process of strategy. There are other goals in addition to maximizing shareholder value for many managers. Finally, the approach assumes that a firm should not reinvest in unprofitable businesses without consideration of their strategic importance to the corporation (also discussed in chap. 16).

Synergy Analysis

One of the arguments for diversification is that multiple businesses within a single corporation can obtain benefits exceeding those of the sum of the businesses if they were independent. The two basic types of benefits are tangible and intangible.[49] The tangible benefits arise from sharing of activities such as purchasing of raw materials, common production technologies, and common distribution channels among the business units. Tangible benefits from sharing lower the cost of competing in an industry or enhance the ability of a business to differentiate itself from its competitors.[50]

Intangible benefits involve the sharing of management know-how or reputation such as a brand name. The development of synergies between businesses is not costless. Three types of costs are the coordination of activities, compromise in terms of decisions that are probably not optimal for the businesses involved, and inflexibility in terms of responding to competitor moves and exit barriers.[51]

Synergy analysis is most useful for identifying competitors most likely to have staying power and who may act in seemingly irrational ways because of exit barriers. From an offensive perspective, synergy analysis can identify competitors that may be slow in responding to innovative technologies, substitute products, or the development of a new industry segment.

Financial Statement Analysis

Probably the most widely available information on all public firms is their financial statements. Financial statement analysis can be used to assess both the short-term health of a firm and its long-term financial resources. The objective of understanding the short-term health is to determine if the firm has potential cash flow, bankruptcy, and other short-term weaknesses.[52] The objective of assessing long-term financial resources is to determine intermediate and long-term financial strengths. One tool that is useful for long-term analysis is the sustainable growth formula developed by the Boston Consulting Group. The formula predicts the potential growth rate of the firm, given a set of financial policies.

Financial tools such as ratio analysis, funds-flow analysis, and financial modeling can be readily used by virtually anyone with the development of computerized spreadsheets. The spreadsheets can analyze and compare a large set of competitors in a very short period of time. Although financial analysis should be used in competitive analysis, its usefulness is sometimes limited to the corporation as a whole. Rarely do diversified firms publicly disclose comprehensive financial statements for their individual businesses. Firms that do not trade stock on the open market are not required to make their financial statements public, and foreign firms do not have the same reporting requirements as those in the United States. Financial analysis often provides a quick method of identifying weaknesses and strengths that can be further investigated using other tools.

Stakeholder Analysis and Assumption Surfacing and Testing

All competitors have stakeholders who influence and constrain their decision making. A stakeholder is any individual or group that affects or is affected by the realization of a business's goals.[53] Examples of stakeholders include employees, suppliers, management, competitors, customers, local community groups, and the media. An analysis of a competitor's stakeholders must also identify the assumptions made about each stakeholder.

The techniques of assumption surfacing can be used to uncover the assumptions driving the behavior of each stakeholder.[54] The technique typically would involve a group of individuals asking themselves the following question: What are the most plausible assumptions we must make about each stakeholder in order for our competitive strategy to be successful? Stakeholder assumptions can then be classified into two categories:[55]

1. Supporting assumptions, which give rise to strategic opportunities, strengths, and favorable conditions.
2. Resisting assumptions, which give rise to threats, weaknesses, and dangerous conditions.

The assumptions can be weighted and compared to assess their overall impact.

These techniques primarily have been used by firms to assess themselves. However, in competitive analysis it is equally important to understand the stakeholders and their assumptions of the firm's competitors. Competitive moves should exploit the resisting assumptions of a competitor since these will be the areas in which the competitor will be less likely to retaliate because of stakeholder resistance. Insights into the assumptions of competitors further indicate what they will do rather than what they can do (also discussed in chap. 12).

Market Signaling

Virtually all of the techniques in this section are based on gathering data for actions that have already occurred and evaluating their impacts (for example, the introduction of a new product). Another class of actions that is an indirect means of communicating are market signals, which are any actions by a competitor that provide a direct or indirect indication of his or her intentions, motives, goals, or internal situation.[56] Porter identifies two types of market signaling: One is the truthful indication of an intention, motive, or goal, and the second is a bluff intended to confuse or mislead the competition.

The use of market signaling in competitive analysis depends on the ability of the signaler to communicate a convincing message that is received and interpreted by the appropriate competitors. Tracking of a competitor's past signals and their ultimate outcomes can assist with the interpretation of current and future signals. If tendencies or patterns in the competitor's signaling can be uncovered, then the response time to those signals can be shortened.

Portfolio Management Techniques

The most well-known set of techniques for identifying and evaluating a business's competitive position are portfolio management techniques. Extending the work of financial theory, a variety of techniques have evolved that essentially locate a business, its product lines, or a group of businesses on a matrix that has on one axis a measure of industry attractiveness and on the other axis a measure of competitive position. The most commonly used matrices include the Boston Consulting Group growth-share matrix, the General Electric market-share matrix, the General Electric market-attractiveness-competitive-position matrix, the Royal Dutch Shell directional-policy matrix, and the Arthur D. Little product-life-cycle matrix.[57] Although the techniques are popular, there have been few attempts to test the propositions of these models.[58] (Also see chaps. 18 and 19.)

Figure 14-6 is representative of portfolio management techniques. The main purposes of the techniques are to assist managers in resource-allocation decisions and to evaluate future cash flows and profitability from the

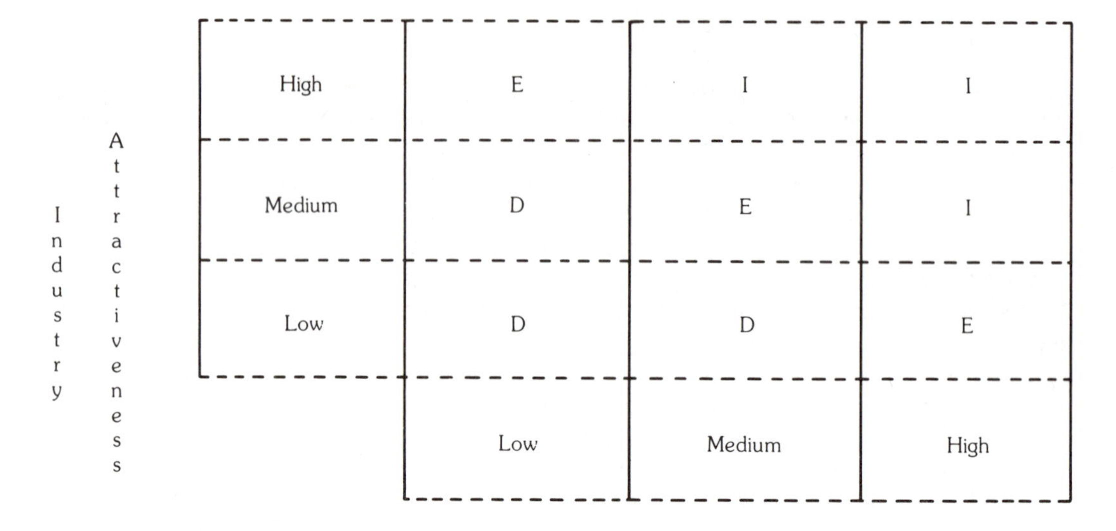

Figure 14-6. Business planning matrix.

portfolio. Whether applied at the corporate level or business-unit level, the techniques suggest which businesses (or product lines) to invest in and which to harvest or divest. A related benefit occurs when the managers are required to think systematically about the industries in which they compete and their business strengths and weaknesses. This may lead to discussions and decisions concerning the restructuring of the portfolio.

Understanding a competitor's position in the matrices can help answer several questions, such as: What is the position of the competitor in the corporate portfolio? What does this position imply about investment decisions for the competitor as a business unit? What does this imply for near-term strategic moves and responses? How does the position of the business fit with the corporation's goals concerning growth, profitability, and cash flow? The same set of questions can be asked about the product lines within a competitor's business. Portfolio models probably have their greatest uses as a heuristic method of decision making and visual summary for a host of more detailed analyses.

The wide variety of techniques described in this section illustrates the development of this phase of competitive analysis. Although new techniques will emerge, the prudent application of the techniques will be the test of their usefulness. Application of techniques is useful only when the raw data are valid and the results are implemented. We now turn to the implementation of competitive analysis results.

IMPLEMENTATION

The most common shortcoming of competitive analysis programs is the failure to implement their findings. Implementation involves the actual use of the output by managers to improve decision making and ultimately organizational performance. Since the individuals who conduct competitive analysis typically are not the ones who must implement the findings, the results must be conveyed to the appropriate managers. The techniques of implementation involve communicating and linking the analyses and their implications to the strategic management process. I have categorized these techniques according to their level of sophistication (Figure 14-7). The type and comprehensiveness of the competitive analysis assignment will determine what level of sophistication is necessary. As the comprehensiveness of the assignment increases, one would expect an increase in the number and sophistication of communication techniques. Each of the techniques, in order of increasing sophistication, are described below.

News Articles. The clipping of news articles and their circulation is an excellent way to convey recent developments in the industry or of competitors. News articles often provide credibility that a memo alone lacks. News articles are one of the most frequent methods of market signaling by competitors. However, they may also be "late news."

Bulletin Boards. Bulletin boards provide a central location to convey a wide variety of information to many people. Bulletin boards can contain information on a competitor's product lines, recent patent information, advertising, promotional brochures, and so forth. Bulletin boards should be placed in high-traffic areas to maximize exposure. Dated material should be renewed, and creativity in the layout of the material will heighten interest.

Demo Rooms. A demo room is a place where competitors' products can actually be seen and used by employees. A modification of the demo room is a strip-down room. In strip-down rooms a competitor's product is reverse-engineered. Various stages of stripping down can be displayed to illustrate production processes, costs, materials used, and so forth. Demo rooms add a sense of reality that promotional material or news articles lack.

Newsletters. Newsletters are periodic updates on industry trends and competitor activities. They are formal in the sense that an individual has responsibility to publish the newsletter at regular intervals. Newsletters typically are short, one-to-three pages, that highlight major events and help maintain current awareness. A newsletter can be published in stand-alone issues or can be designed to provide a continuing expanding picture of competition. If the latter is used, an

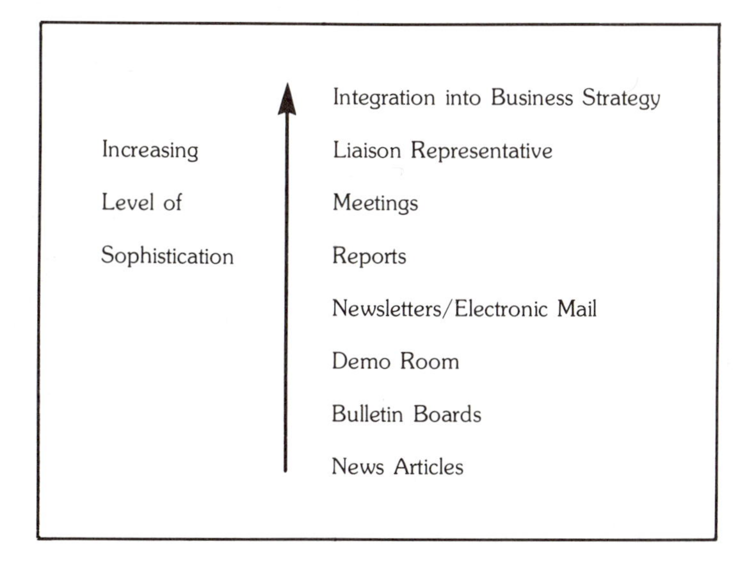

Figure 14-7. Techniques for implementing competitive analysis findings.

index and booklet should be provided to the individuals so they can organize the material. A sophisticated use of newsletters is electronic mail.

Reports. A practical way to organize and communicate the results of a competitive analysis assignment is a report. Sammon et al. developed a classification of reports (Figure 14-8).[59] The most basic report is spot intelligence, which is typically a request from sender management about a narrow topic of immediate interest. The danger of these reports is that the whole process of competitive analysis can turn into a disjointed, unfocused, reactive form of intelligence gathering.

Base case reports are comprehensive case studies of competitors and industries. The purpose of these reports is to provide information concerning all aspects of a competitor's business. A reader of the report should have a full understanding of the competitor's strategy, strengths and weaknesses, and potential impact on his or her firm. The base case is updated by the periodic report, which serves both an updating and monitoring function. Although the reports typically are distributed on a quarterly or semiannual basis, the time schedule should be one that makes sense for the company. The reports should be concise and descriptive of current developments. Periodic reports are more comprehensive and focused than newsletters.

Net estimate reports are simple, concise summaries of the most vital information concerning a competitor or industry. Similar to executive summaries, they should highlight the critical information that management requires. A net estimate of a competitor should

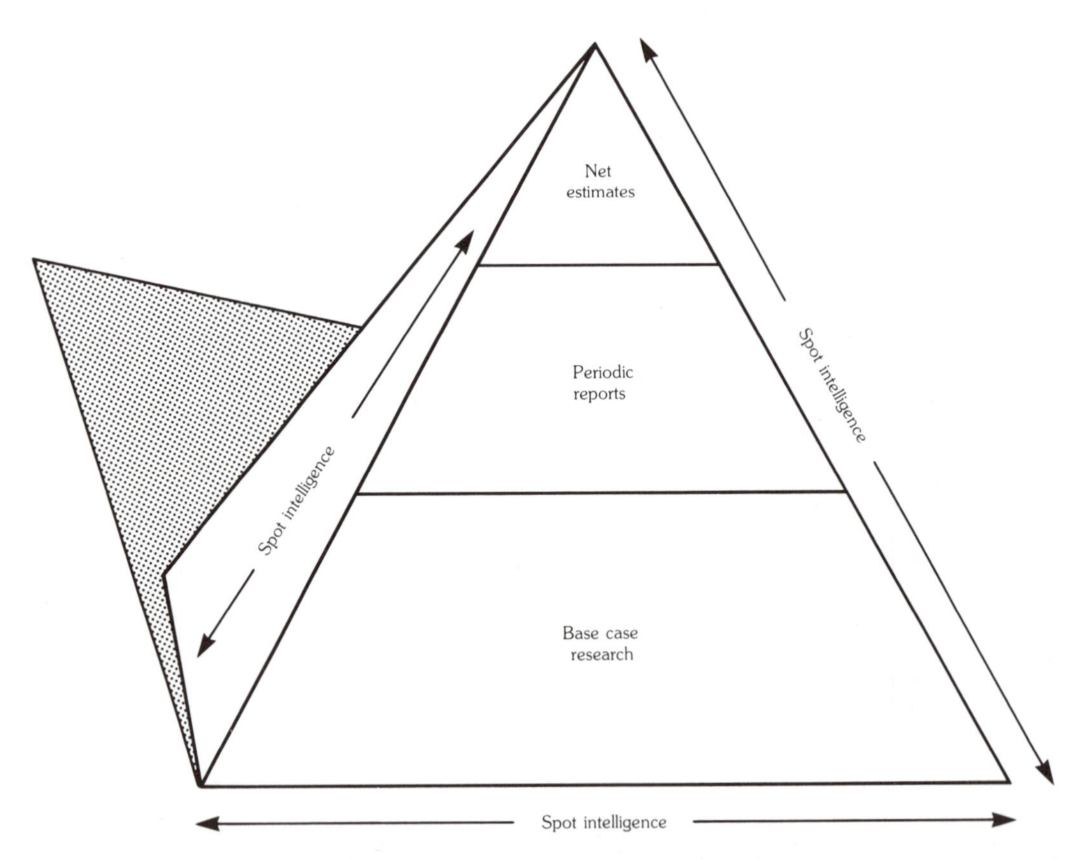

Figure 14-8. Classification of competitive analysis reports (after W. L. Sammon, M. A. Kurland, and R. Spitainic, *Business Competitor Intelligence,* New York: Wiley, 1984, 133).

identify his or her strategic position, assumptions concerning likely moves and the likelihood of their success. For any of the reports to be meaningful, they must be in the hands of those who make decisions, and the temptation to file them should be resisted.

Meetings. Meetings provide face-to-face communications between the individuals who conducted the competitive analysis assignment and those who will be using the findings. They provide the advantages of reports yet can be secretive if sensitive material is involved. Meetings can also be used to clarify issues, serve as showcases for the assignment, and help to guide future competitive analysis assignments. Meetings can be especially useful if all of the managers that may be affected by the assignment are present.

Liaison Representative. In most organizations several individuals would be involved in competitive analysis. Often these individuals are in different departments, divisions, or even hierarchical levels. Liaison representatives are individuals who represent a particular unit and interact with other liaisons. Typical interactions involve discussions concerning the methods of competitive analysis used throughout the organization, the possibilities of joint activity, and a fuller understanding of other units' strategies. These discussions are particularly useful when there is multipoint competition or when the organization is highly decentralized. The information can then be conveyed back to the key decision makers.

Integration into Business Strategic Plan. Many competitive analysis assignments and their implementation are tactical in nature, that is, they primarily emphasize how to make the current strategy work better or smoother. Another use of competitive analysis is to aid in the development of competitive strategy. One method for accomplishing this is to make competitive analysis a standard part of strategic planning, as discussed in chapter 13, so that a business unit's strategy is viewed in light of competitors' strategies and the dynamics of the industry.

These eight techniques for implementing the outputs of a competitive analysis assignment vary in terms of their sophistication. Further, the manner in which any of the techniques are implemented can take many forms. A business should develop the implementation methods most applicable to its needs. Thus, the delivery of current and usable information on a timely basis to those who need it should be the goal of implementation.

UPDATING AND MAINTAINING

The final phase in the cycle of competitive analysis depicted in Figure 14-1 is updating and maintaining the information. Before we move to a discussion of that topic, it is necessary to recognize that some competitive analysis assignments have an end. Figure 14-1 depicts this as an alternative to the updating and maintaining phase. Many of the assignments that fall into the first three columns of Figure 14-2 are projects that have a definite beginning and end. Although their results should be kept for reference purposes, these projects typically are not amenable to updating. Examples of these include an industry profile for a proposed new product introduction, a company profile for a possible acquisition, the analysis of a new technological innovation, and the analysis of the size of a market. Assuming that these assignments were conducted for a specific purpose within time constraints for a particular decision, they do not require updating. The updating and maintaining of information become more relevant as the comprehensiveness and permanence of the assignment increases.

When a business decides to institute a permanent comprehensive system, the updating and maintaining of the system become critical to its success. Questions that need to be addressed include:

1. How frequently should we update our information?
2. Who should have responsibility for the updating?
3. How do we determine the reliability and validity of the information that we have?

4. Are our assumptions concerning the process of competitive analysis realistic and/or changing?
5. How do we decide what and when information should be thrown away?

These questions need to be addressed because the amount of information can easily become overwhelming. Think of the information that can be accumulated with the advent of data bases and computerized storage systems. Even a less ambitious assignment can become unwieldy. For example, a newsletter service can, after a year, result in a large notebook of information. A method must be developed to purge the outdated material yet keep important information that is not outdated.

A related issue is how to keep the momentum of competitive analysis once it has been introduced into the business. Competitive analysis must provide added value to its users and the best way to ensure that it does is to make it customer oriented. Those involved with competitor analysis must constantly seek and receive feedback from those using the information and modify their approach as needed.

CONCLUSION

The practice of competitive analysis is in its infancy. Few firms have adequately addressed how competitive analysis should be conducted in their organization. The purpose of this chapter is to describe the techniques of competitive analysis that have been developed and dispersed among many disciplines. A framework comprised of five phases (objectives, data collection, data interpretation, implementation, and updating and maintaining) is used to organize the techniques of competitive analysis. An appropriate way of concluding this chapter is to highlight observations concerning each phase of competitive analysis.

The objectives phase illustrates that there are many different types of competitive analysis assignments. All firms need not approach competitive analysis in the same manner. Careful consideration as to the needs of competitive analysis will reduce the tendencies to build a big machine for its own sake. The techniques of data collection have made great strides in recent years. Several firms now specialize in the collection of information. Some concentrate on field research, whereas others emphasize data bases. Future developments are likely to come from the field of marketing because it is most closely aligned with customers.

The data interpretation phase currently comprises the broadest array of techniques. However, the usefulness of those techniques is limited by the availability of data. Few advances in this phase are likely to have an impact until better methods of collecting the relevant data are developed.

The implementation of the results of a competitive analysis assignment is the reason for undertaking the effort. However, unless the decision makers are provided with timely information, implementation efforts will fall short of expectations. Further, competitive analysis must become a daily way of thinking for managers. Decisions must be made with competitors and the industry in mind. Businesses will continue to struggle with the means of implementing competitive analysis findings for the foreseeable future.

The updating and maintaining phase involves keeping the ball rolling. Momentum is the key. For competitive analysis to be successful it must be used and challenged by managers. The practice of competitive analysis has a bright future. It is hoped that this chapter will help managers by providing a framework to organize their aspirations for competitive analysis in their business.

NOTES AND REFERENCES

1. I cannot take full credit for the development of this framework. It was developed based on the work of many individuals, several of the more influential being Michael Porter of the Harvard Business School, William R. King and David I. Cleland of the University of Pitts-

burgh, Michael M. Kaiser of Michael M. Kaiser Associates, Leila K. Kight of Washington Researchers, William Sammon, Mark Kurland, and Robert Spitalnic from their book *Business Competitor Intelligence,* and Leonard Fuld from his book *Competitor Intelligence.*

2. M. M. Kaiser, *Understanding the Competition: A Practical Guide to Competitive Analysis* (Washington, D.C.: Michael M. Kaiser Associates, 1984).
3. It is beyond the scope of this chapter to discuss the details, variations, strengths and weaknesses of each method. For additional information interested readers can refer to D. C. Miller, *Handbook of Research Design and Social Measurement,* 4th ed. (New York: Longman, 1982) or F. N. Kerlinger, *Foundations of Behavioral Research,* 2nd ed. (New York: Holt, Reinhart and Winston, 1973).
4. L. M. Fuld, *Competitor Intelligence* (New York: Wiley, 1985).
5. Washington Researchers, *Company Information: A Model Investigation* (Washington, D.C.: Washington Researchers, 1983).
6. L. M. Apcar, "Who Was That Mysterious Man Snooping around UPS's Trucks?" *Wall Street Journal* (18 March 1985):37.
7. Apcar, "Who Was That Mysterious Man?"
8. M. E. Porter, *Competitive Strategy* (New York: Free Press, 1980).
9. Kaiser, *Understanding the Competition.*
10. Boston Consulting Group, *Perspectives on Experience* (Boston: Boston Consulting Group, 1970).
11. T. V. Bonoma and B. P. Shapiro, *Segmenting the Industrial Market* (Lexington, Mass.: Lexington Books, 1983); J. P. Kotler, *Marketing Management: Analysis, Planning and Control,* 4th ed. (Englewood Cliffs, N.J.: Prentice-Hall, 1980); M. E. Porter, *Competitive Advantage: Creating and Sustaining Superior Performance* (New York: Free Press, 1985).
12. Kotler, *Marketing Management;* Porter, *Competitive Advantage.*
13. See Porter, *Competitive Strategy,* for a description of why industries evolve.
14. Porter, *Competitive Advantage.*
15. P. Wack, "Scenarios: Shooting the Rapids," *Harvard Business Review* **63** (1985):139-150.
16. H. Kahn and A. Wiener, *The Year 2000* (New York: Macmillan, 1967).
17. R. D. Zentner, "Scenarios in Forecasting," *C & E News* (6 October 1975).
18. B. Nanus, *Annual Report 1976* (Los Angeles: Center for Futures Research, University of Southern California, 1976).
19. S. J. Kobrin, *Managing Political Risk Assessment* (Berkeley and Los Angeles: University of California Press, 1982); A. Desta, "Assessing Political Risk in Less Developed Countries," *Journal of Business Strategy* **5** (1985):40-53.
20. C. W. Hofer and T. Haller, "Globescan: A Way to Better International Risk Assessment," *Journal of Business Strategy* **1**(1980):41-55.
21. Desta, "Assessing Political Risk."
22. For a review of the techniques, see Hofer and Haller, "Globescan"; Desta, "Assessing Political Risk"; R. J. Rumelt and D. A. Heenan, "How Multinationals Analyze Political Risk," *Harvard Business Review* **56** (1978):67-76; R. R. Davis, "Alternative Techniques for Country Risk Evaluation," *Business Economics* **17** (1981):36.
23. M. F. Shakun, "Advertising Expenditures in Coupled Markets—A Game-Theory Approach," *Management Science* **11**(1965):42-47; M. F. Shakun, "A Dynamic Model for Competitive Marketing in Coupled Markets," *Management Science* **12**(1966):525-530.
24. A. Karnani and B. Wernerfelt, "Multiple Point Competition," *Strategic Management Journal* **6** (1985):87-96; Porter, *Competitive Advantage.*
25. Porter, *Competitive Strategy.*
26. S. Schoeffler, R. D. Buzzell, and D. F. Heany, "Impact of Strategic Planning on Profit Performance," *Harvard Business Review* **52**(1974):137-145.
27. Strategic Planning Institute, *The PIMS Program* (Cambridge, Mass.: Strategic Planning Institute, 1980).
28. For a discussion of these studies and their findings, see D. Abell and J. S. Hammond, *Strategic Marketing Planning: Problems and Analytical Approaches* (Englewood Cliffs, N.J.: Prentice-Hall, 1979); V. Ramanujam and N. Venkatraman, "An Inventory and Critique of Strategy Research Using the PIMS Data Base," *Academy of Management Review* **9**(1984):138-151.
29. Boston Consulting Group, *Perspectives on Experience.*
30. K. M. Woolley, "Experience Curves and Their Use in Planning," Ph.D. diss., Stanford University, 1972; R. R. Cole, "Increasing Utilization of the Cost Quantity Relationship in Manufacturing," *Journal of Industrial Engineering* **10**(1958):173-177; W. Z. Hirsch, "Firm Progress Ratios," *Econometric* **24**(1956):136-143; G. Hall and S. Howell, "The Experience Curve from the Economist's Perspective," *Strategic Management Journal* **6**(1985):197-212.
31. A. C. Hax and N. S. Majluf, *Strategic Management: An Integrative Perspective* (Englewood Cliffs, N.J.: Prentice-Hall, 1984).

32. Hall and Howell, "Experience Curve from Economist's Perspective."
33. M. S. Hunt, "Competition in the Major Home Appliance Industry," Ph.D. diss., Harvard University, 1972; H. H. Newman, "Strategic Groups and the Structure Performance Relationships: A Study with Respect to the Chemical Process Industries," Ph.D. diss., Harvard University, 1973; K. J. Hatten and D. E. Schendel, "Heterogeneity within an Industry: Firm Conduct in the U.S. Brewing Industry," *Journal of Industrial Economics* **26**(1977):92-113; M. E. Porter, "The Structure within Industries' and Companies' Performance," *Review of Economics and Statistics* **61** (1979):214-219.
34. R. D. Caves and M. E. Porter, "From Entry Barriers to Mobility Barriers," *Quarterly Journal of Economics* **91**(1977):241-261.
35. Porter, *Competitive Strategy.*
36. Porter, *Competitive Advantage.*
37. J. L. Bower, *Simple Economic Tools for Strategic Analysis,* Case Study #9-373-094 (Cambridge, Mass.: Harvard Business School, 1972).
38. Porter, *Competitive Advantage.*
39. J. F. Rockart, "Chief Executives Define Their Own Data Needs," *Harvard Business Review* **57**(1979):81-92.
40. J. K. Leidecker and A. V. Bruno, "Identifying and Using Critical Success Factors," *Long Range Planning Journal* **17**(1984):23-32; see also Chapter 20 in this book.
41. B. Petrov, "The Advent of the Technology Portfolio," *Journal of Business Strategy* **3**(1982):70-75.
42. P. H. Nowill, "The Technology Share Matrix." Paper presented at the International Congress on Technology and Technology Exchange, Pittsburgh, 1982.
43. R. H. Hayes and S. C. Wheelwright, "Link Manufacturing Process and Product Life Cycles," *Harvard Business Review* **57** (1979):133-140; R. H. Hayes and S. C. Wheelwright, "The Dynamics of Process-Product Life Cycles," *Harvard Business Review* **57**(1979):127-136.
44. R. Shurig, "Morphology: a Tool for Exploring New Technology," *Long Range Planning Journal* **17**(1984):129-140.
45. Porter, *Competitive Advantage;* A. L. Frohman, "Technology as a Competitive Weapon," *Harvard Business Review* **60**(1982):97-104; A. L. Frohman, "Putting Technology into Strategy," *Journal of Business Strategy* **5**(1985):54-65; J. W. Williams, "Technological Evolution and Competitive Response," *Strategic Management Journal* **4**(1983):55-65; P. Lasserre, "Selecting a Foreign Partner for Technology Transfer," *Long Range Planning Journal* **17**(1984):43-49; G. Fairtlough, "Can We Plan for New Technology?" *Long Range Planning Journal* **17**(1984):14-23.
46. A. Cooper and D. Schendel, "Strategic Responses to Technological Threats," *Business Horizons* **19**(1976):61-69.
47. W. E. Fruhan, Jr., *Financial Strategy: Studies in the Creation, Transfer and Destruction of Shareholder Value* (Homewood, Ill.: Richard D. Irwin, 1979); C. Y. Woo, "An Empirical Test of Value-based Planning Models and Implications," *Management Science* **30** (1984):1031-1050; A. Rappaport, "Selecting Strategies That Create Shareholder Value," *Harvard Business Review* **59**(1981);139-149; P. J. Strebel, "The Stock Market and Competitive Analysis," *Strategic Management Journal* **4** (1983):279-291.
48. Kaiser, *Understanding the Corporation.*
49. Porter, *Competitive Advantage.*
50. E. E. Bailey and A. F. Frielaender, "Market Structure and Multi-product Industries," *Journal of Economic Literature* **20**(1982):1024-1048; J. C. Panzer and R. D. Willig, "Economies of Scope," *American Economic Review* **71**(1981): 268-272; W. J. Baumol, "Contestable Markets: An Uprising in the Theory of Industry Structure," *American Economic Review* **72**(1982):1-15; J. R. Wells, "In Search of Synergy," Ph.D. diss., Harvard Business School, 1984; R. B. Chinta, "Competitive Advantage from Pooling Divisional Activities: Theory and Empirical Tests for Strategy Implications," Ph.D. diss., University of Pittsburgh, 1985.
51. Porter, *Competitive Advantage.*
52. W. H. Beaver, "Alternative Accounting Measures as Predictors of Failure," *Accounting Review* **43** (1968):113-122; J. W. Wilcox, "A Gambler's Ruin Approach to Business Risk," *Sloan Management Review* **18** (1986):33-46; Hax and Majluf, *Strategic Management;* C. W. Hofer and D. Schendel, *Strategy Formulation: Analytical Concepts* (St. Paul, Minn.: West Publishing, 1978).
53. R. E. Freeman, *Strategic Management: A Stakeholder Approach* (Boston: Pitman, 1984); A. J. Rowe, R. O. Mason, and K. E. Dickel, *Strategic Management and Business Policy,* 2nd ed. (Reading, Mass.: Addison-Wesley, 1985); A. Mendelow, "Setting Corporate Goals and Measuring Organizational Effectiveness—A Practical Approach," *Long Range Planning Journal* **16**(1983):70-75.
54. R. O. Mason and I. Mitroff, *Challenging Strategic Planning Assumptions* (New York: Wiley, 1982).

55. Rowe et al., *Strategic Management and Business Policy.*
56. Porter, *Competitive Strategy,* 75.
57. For details of these and other portfolio models, see J. H. Grant and W. R. King, *The Logic of Strategic Planning* (Boston: Little, Brown, 1982); Hofer and Schendel, *Strategy Formulation;* Hax and Majluf, *Strategic Management.*
58. For an exception, see D. C. Hambrick, I. C. MacMillan, and D. L. Day, "Strategic Attributes and Performance in the BCG Matrix—A PIMS-based Analysis of Industrial Product Businesses," *Academy of Management Journal* **25** (1982):510-531.
59. W. L. Sammon, M. A. Kurland, and R. Spitalnic, *Business Competitor Intelligence* (New York: Wiley, 1984).

CHAPTER 15

Strategic Issue Management

William R. King

William R. King is University Professor of Business Administration in the Graduate School of Business at the University of Pittsburgh. He is the author, co-author, or co-editor of more than a dozen books and 150 articles that have appeared in the leading journals in the fields of strategic management, information systems, and management science. He has served as senior editor of the *Management Information Systems Quarterly* and serves as associate editor of *Management Science, Strategic Management Journal, OMEGA: The International Journal of Management Science,* and various other journals. He frequently consults in these areas with corporations and government agencies in the United States and Europe.

The notion of a strategic issue is pervasive in the field of planning in both conceptual and practical contexts.[1] Truly strategic issues have a self-evident importance to an organization. This chapter describes a practical and comprehensive process of strategic issue management (SIM) that facilitates the systematic identification, assessment, and analysis of such issues. Such a process can ensure that issue-related factors are fully integrated into the formulation and implementation of strategy. As such, SIM is a process whereby strategic issues may become an integral element of strategic management.

STRATEGIC ISSUES

Energy shortages, changes in government policy, and major societal trends all generally are recognized as potential strategic issues for a business firm.[2] Such events and trends can radically alter the future demand for a firm's various products, its costs, or its ability to perform its chosen mission.

The practical problem facing the strategic manager, who somehow must factor these issues into strategy deliberations, is to identify issues that are not yet commonly recognized and to separate truly strategic issues from those that are less so.

The resolution of this problem requires both a concept of what constitutes a strategic issue and a process for dealing with issues as integral elements of the process by which the organization's strategy is formulated.

Brown provides a definition that can be useful. A strategic issue is "a condition or pressure, either internal or external . . . that will have a significant effect on the functioning of the organization or its future interests."[3] Although broad in scope, this definition, with its

Adapted from W. R. King, "Integrating Strategic Issues into Strategic Management," *OMEGA: The International Journal of Management Science,* **12** (1984):529-538.

use of terms such as "internal or external" and "significant," provides a useful concept of a strategic issue. It defines a broad domain and therefore specifically takes the notion of a strategic issue out of the *social issue* or *political issue* context that it connotes to many people. Moreover, it identifies *significance* as a criterion of definition, a point which I subsequently show to be important practically as well as useful.

Issues meeting such a definition should arise as questions stated as: "What will be the effect of *X?*" where *X* is an issue such as the threat of entry of a new competitor or some changing pressure within or outside the firm.

However, in practice, issues may be buried in questions, such as "How can we cope with *X?*" It is important to ferret out the underlying "What?" question from the stated "How?" question if issues are to be dealt with in an integrated and comprehensive manner.

SOURCES OF ISSUES

Table 15-1 shows some common sources of issues. They may arise because someone in the organization (the boss, board of directors, staff groups, and so forth) becomes aware of a new or evolving force, or as a result of recent inabilities to resolve conflicts in the organization, or as a consequence of elements of an ongoing planning process such as the environmental scan or strength-weakness analysis.[4]

Table 15-1. Common Sources of Strategic Issues

The boss
Boards of directors (for example, public policy or social responsibility committees)
Staff groups who interact with clients (for example, government relations, stockholder relations)
Controversies in the organization
Unanswered questions from prior planning cycles
Assumptions made in plans that have been developed by subunits
Strength-weakness analyses
Environmental scanning

Source: W. R. King, "Integrating Strategic Issues into Strategic Management," OMEGA: The International Journal of Management Science, **12** (1984).

Figure 15-1 shows a way of organizing these various sources in terms of the internal, operating, and general environments. It emphasizes the three levels of environmental subsystems, the diversity of stakeholders who may be sources of or involved in strategic issues, and the multiple roles of certain stakeholders[5]—such as the direct jurisdiction of some regulatory agencies as contrasted with their broader role in the general environment.

When issues are thought of in this context, the importance of having a management system to deal with them is apparent; since any nonsystematic approach to coping with such complexity is unlikely to be fruitful on anything but a fortuitous basis.

RATIONALE FOR MANAGING STRATEGIC ISSUES

The management of strategic issues admittedly has been a nebulous concept. Issues are often thought to be uncontrollable, so that the best that may be hoped for is to have the firm react quickly and appropriately to them. Under this view, perhaps they may be influenced to some minor degree, but they certainly cannot be managed.

Although it is true that issues generally cannot be managed in the same way that a department can be managed, this view reflects a naive concept of management in the sense that it implicitly views management as deterministic. *In the sense of fully controlling with certainty, literally nothing can be managed.*

Management, in a more sophisticated and meaningful sense, means identifying the issue, dealing with it rationally in terms of the way in which it may have an impact on the interests of the organization, and attempting to influence it to the maximum degree of cost effectiveness. Thought of in this way, not all

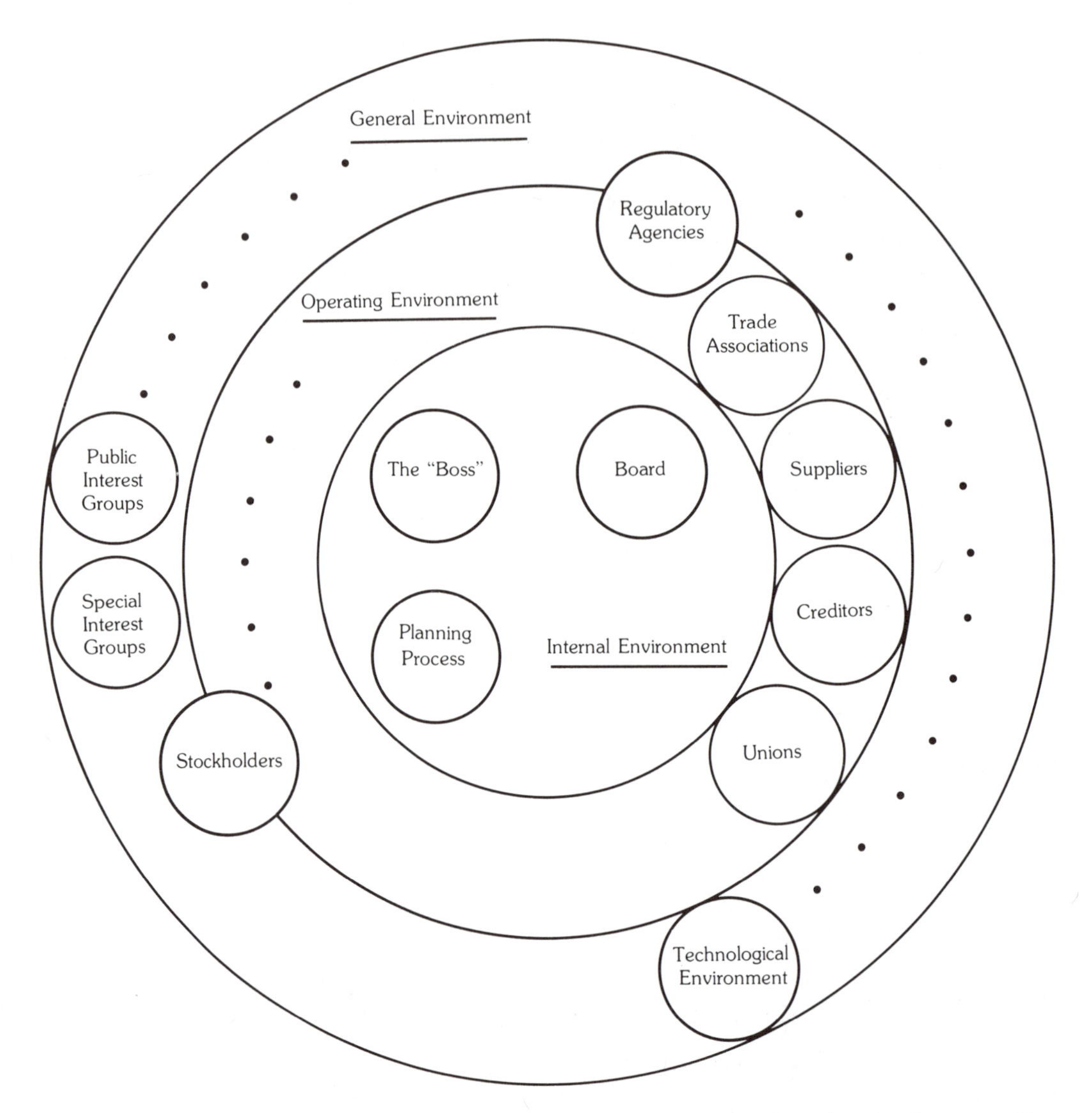

Figure 15-1. Environmental subsystems (after P. Thomas, "Environmental Analysis for Corporate Planning," *Business Horizons*, **17** 1974:27).

important issues should be dealt with because, although some are very significant, they may not be subject to much influence. On the other hand, some issues may be so potentially important that they will merit the expenditure of significant resources when only a modest degree of influence is even possible.

This is certainly understood by firms that establish public affairs and governmental relations functions specifically to identify and monitor public issues. However, such units do not necessarily fulfill, or even address, the broad role of issue management because they often are not integrally involved in the organization's strategic management process.

The reason for this is made apparent when one considers the life cycle of a public issue in terms of four phases: conversation, contention, legislation, and regulation.[6] According to General Electric's Business Environmental Research Group, "without a proper business response, societal expectations of today become the political issues of tomorrow, legislated requirements the next day, and litigated penalties the day after that."[7]

The issue life cycle, although it directly applies only to the special sort of public issues arising from the general environment, is applicable to all issues in the sense that the symptoms of an issue usually appear well before the issue becomes readily apparent. A production cost increase caused by poor equipment maintenance or the deferral of purchase of replacements will be reflected in the maintenance budget or reports substantially prior to being reflected in product cost reports. So too will competitive strategy changes be reflected by actions of the competitor's personnel before they are formally announced.[8] In both instances—one from the internal and one from the operating environment—the analogy to the life cycle of a public issue in the general environment is clear.

The implications to issue management are also direct. The *early identification* of issues is an important part of issue management because *information is available that will permit issues to be dealt with before they become fully defined, solidified, and apparent to all.* Since, in general, issues are more readily dealt with at early stages in their life cycle than they are at later stages, this means that issues must be evaluated and monitored in a fashion that facilitates action. This usually dictates that issue management be integrally related to or a part of a strategic management process.

OBJECTIVES AND CRITERIA FOR STRATEGIC ISSUE MANAGEMENT

The objectives of strategic issue management are therefore: (1) the *early identification* of potential strategic issues; and (2) the *assessment* of issues to facilitate *action*.

This must be done in a cost-effective manner that recognizes that all issues do not merit management and that some warrant much more attention than others. Accordingly, criteria must be established for selecting issues that are to be advanced to successive stages of the issue management process. Since these criteria give emphasis to the definition and meaning of a strategic issue, they are described here before the various stages of the SIM process are identified.

The proposed criteria for an issue to be identified as strategic and/or moved along to subsequent stages of the SIM process are: (1) strategy relevance; (2) actionability; (3) criticality; and (4) urgency.

The *strategy relevance* of an issue relates to its potential for having an impact on existing or contemplated strategies. In effect, this criterion addresses the questions: Is it likely to affect us? Is it likely to influence our choice of strategy or the likely consequences of the strategies that we are not following? If it is not strategy relevant, an issue may be important to society as a whole, but it probably does not warrant management.

The *actionability* of an issue is the other side of the coin. Can we affect the issue? Can we have influence over how the issue will proceed through its life cycle? If an issue is strategy relevant but not actionable, it warrants attention but at some relatively low level such as by monitoring to enable the organization to best predict the *timing* and the *degree of impact* that it will have.

The *criticality* of an issue is its adjudged importance or degree of impact. Clearly, if importance or impact is not judged to be great, it is not truly strategic.[9] However, since these assessments are made on a prospective basis early in the life cycle of an issue, issues that are initially judged to be noncritical should be monitored and periodically assessed to ascertain if the updated assessments are different from earlier ones.

The *urgency* of an issue has to do with the time period in which it merits or demands action. All else being equal, an issue that can or must be dealt with immediately takes precedence over one in which action may or must be deferred.

Issues that are strategy relevant, actionable, critical, and urgent clearly merit the highest level of management. Others may be less intensively managed.

STRATEGIC ISSUE MANAGEMENT PROCESS

The SIM process that logically evolves from these basic ideas is described in terms of a series of phases in Figure 15-2. The various stages are discussed below.

Issue Identification

Some sources of strategic issues identify them naturally. For instance, if the boss or the Board identifies an issue, it will undoubtedly at least begin its trek through the SIM process of Figure 15-2.

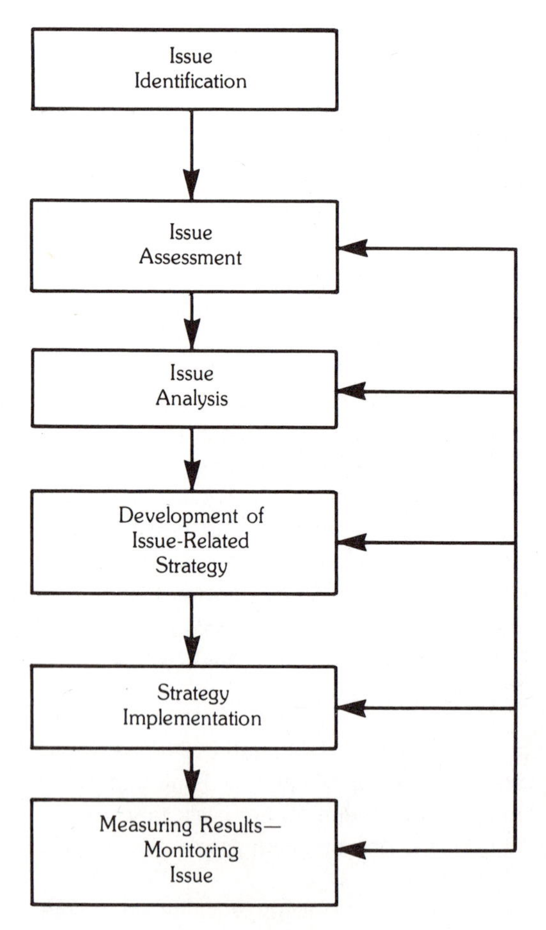

Figure 15-2. SIM process.

However, some systematic and proactive identification process is also needed. This process should be a part of or interface with the organization's environmental scanning activity.[10] The process may be as straightforward as the solicitation of the firm's managers to identify issues. Firms such as PPG, Inc. have used such a process for issue identification.[11]

A relatively informal process for identifying issues may rely on techniques such as the nominal group technique (NGT).[12] Other group processes such as brainstorming[13] are not as useful for this purpose because they do not necessarily facilitate the capture of a wide domain of issues. I have used the NGT very successfully in a number of firms for this purpose.

A more formal starting point may be desirable to enhance the group process. For instance, Molitor suggests that the media through which conventional scanning may be used to identify issues are leading events, authorities and advocates, literature, organizations, and political jurisdictions.[14] He identifies such leading entities, for example, states and cities, that tend to lead the nation in social trends. The tracking of such leading cities by a scanning activity can serve as the underlying issue identification mechanism.

In many instances, some framework in terms of which issues may be identified and assessed is a valuable aid to issue identification. One such useful framework is that of stakeholder groups, for example, each of the sets represented by circles in Figure 15-1. For each group, the following questions may be addressed:

1. What is the *nature of their claim on* or stake in our organization?
2. What are their *objectives*, and how do they affect their claims on us?
3. *How* might they see their objectives?
4. What is the likely *impact?*
5. How can we *measure the degree of activity and the impact?*
6. What are the *issues* that will be fostered?

Another formal approach to issue identification that may have merit is a computer program called SPIRE. As described by Klein and Newman, it is a technique that identifies clusters of interdependent environmental factors and organizational components that are commonly affected by environmental factor clusters.[15] While it does not directly identify issues, the logical groupings thus derived could be a powerful tool for the application of expert and managerial judgment to the identification of emerging issues and to the preliminary assessment of their impact points on the organization.

Assessing the Relevance of Issues

Once issues have been identified, they must be assessed. The assessment process itself may involve several phases, for example, an initial screen or filter followed by a more detailed prioritization. The purpose of the assessment phase is displayed in Figure 15-3, which shows a large number of unevaluated issues going through a mechanism that ranks them using the criteria of strategy relevance, actionability, criticality, and urgency.

In effect, the problem facing the planner or manager at this juncture is one of separating the truly strategic issues from those that are less so. It is not unlike the problem of separating critical *information* from less-relevant *data* in any managerial context.[16]

The filtering process shown in Figure 15-3 may be accomplished judgmentally by separating the issues into a number of groups.[17] An illustration of such groups is shown in Table 15-2. These groups are identified in priority classes that are described in terms of broad

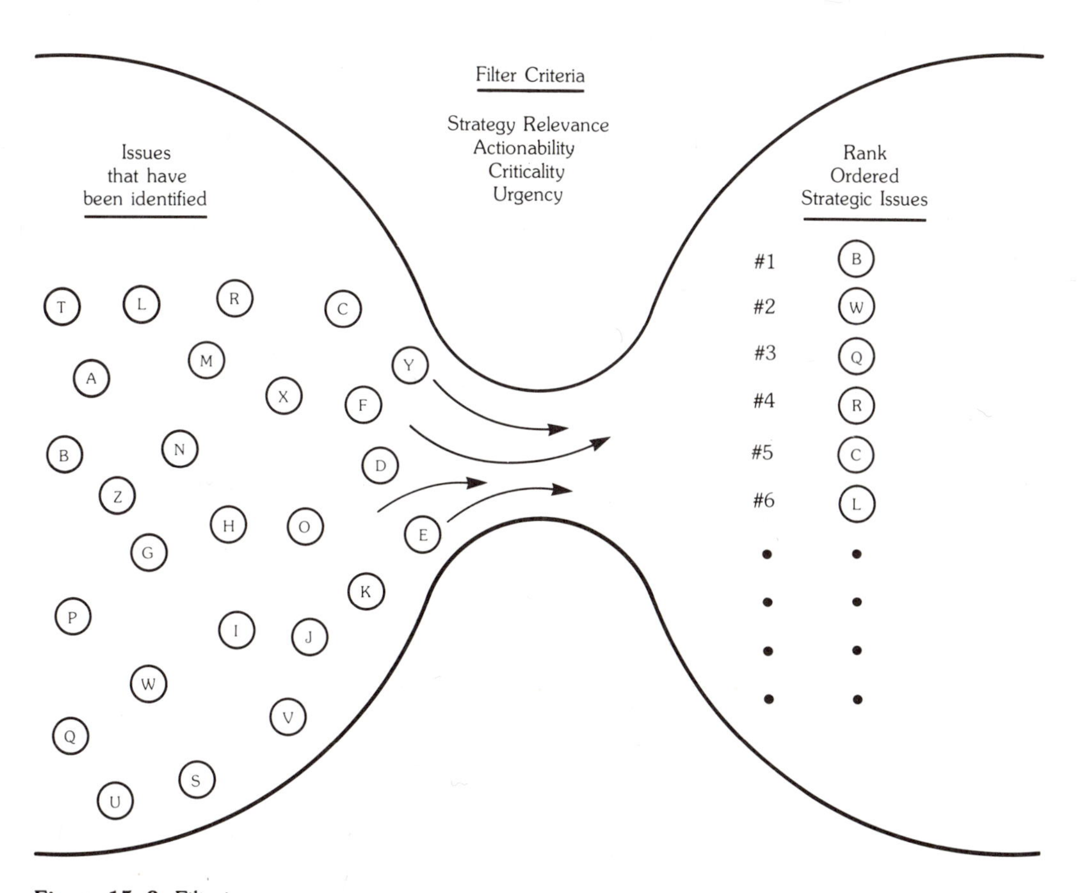

Figure 15-3. Filtering process.

Table 15-2. Priority Classes for Strategic Issues

Priority Class	*Action Identified*
Highest	Analyze in detail for corporate-level consideration
High	Analyze in detail for business-level consideration
Nice to know	Not sufficiently strategic to assess in detail (monitor)
Questionable	Still unframed on ill-defined issues (monitor the stakeholder)

Source: W. R. King, "Integrating Strategic Issues into Strategic Management," OMEGA: The International Journal of Management Science, **12** (1984).

categories of *action consequences.* In the two highest categories, judgments are made that the strategy relevance, actionability, criticality, and urgency are sufficiently great to warrant further analysis for consideration at either the corporate or business-unit level. The third or nice-to-know category of issue is to be monitored but not analyzed in detail. The fourth category, ill-defined issues, cannot be readily monitored, therefore, the *stakeholder groups* are identified for monitoring.

Such categorization may be made judgmentally by an individual, but it is generally much better to have a diverse group of knowledgeable people make the judgments. King and Cleland have described a task force process of developing "strategic data bases" to support an organization's planning process that may be applied.[18] In this approach, a task force is charged with applying their judgment to assess a broad set of issues in terms of the stated criteria. The *product* of the task force is an *issues strategic data base* that is sufficiently rich and concise to be used in the planning process.

A formal scoring technique may also be used to develop a unique priority scheme. Table 15-3 shows a simple version of such a scheme. In Table 15-3 a simple issue is rated in terms of the *likelihood* that it will turn out to be very highly strategy relevant, highly strategy relevant, moderately critical, and so on. The issue that is illustrated has been rated to have an 80 percent chance of being very highly strategy relevant and a 20 percent chance of being only highly strategy relevant, a 50-50 percent chance of being only highly or moderately actionable, and so forth.

These probabilities are then multiplied by the arbitrary but logical and consistent numbers used to represent the various levels of each criterion (in this case, 10 for very high, 8 for high, and so on), and the products are added in the fashion of calculating an average of expected value.

These expected values are then summed over the four criteria to obtain an overall score—in this case 30.8. Although this number means little in the abstract, it can be used as a logical basis for ranking issues, that is, one with a score of 40 would rank higher than this one since the score reflects the combined likelihood judgments of the issue in terms of the four criteria.

Often, more sophisticated scoring ap-

Table 15-3. An Issue-Prioritization Scoring Technique

	Very High (10)	*High (8)*	*Moderate (6)*	*Low (4)*	*Very Low (2)*	*Probability × Score*
Strategy relevance	0.8	0.2	—	—	—	9.6
Actionability	—	0.5	0.5	—	—	7.0
Criticality	1.0	—	—	—	—	10.0
Urgency	—	—	0.4	0.3	0.3	4.2
Overall score						30.8

Source: W. R. King, "Integrating Strategic Issues into Strategic Management," OMEGA: The International Journal of Management Science, **12** (1984).

Table 15-4. Life-cycle Analysis of Strategic Issues

	Who started the ball rolling?	*Who is now involved?*	*Who will get involved?*
Who?			
How?			
When?			
Why?			
Methods Used?			
Results?			

Source: W. R. King, "Integrating Strategic Issues into Strategic Management," OMEGA: The International Journal of Management Science, **12** (1984).

proaches may also be useful.[19] The Delphi technique may be used in obtaining judgments from a group if they need to be solicited. However, Table 15-3 captures the flavor of a priority scoring approach.

Issue Analysis

Once the set of issues has been prioritized, individual issues must be analyzed. Useful ways to do this analysis involve (1) backcasting on the relevant stakeholders, perhaps using a life-cycle approach; (2) a process of literal analysis (decomposition) called strategic issue analysis (SIA); and (3) strategic assumption analysis.

Backcasting on Stakeholders to Analyze Issues. In backcasting on stakeholders in order to preliminarily analyze the issue, an attempt is made to identify *all* relevant stakeholders, not just the small number who may have served as focal points for the identification of the issue. This may be done through a series of questions such as:

Who is affected by the issue?
Who has an interest in the issue?
Who is in a position to exert influence on the issue?
Who has expressed an opinion?
Who ought to care?

Note that these questions are at the level of the issue, whereas the former set of questions are at the level of the organization. Thus, these questions are useful to address comprehensively a specific issue, whereas the former questions are useful in identifying issues.

One related approach is to use the notion of the life cycle of an issue. Table 15-4 shows a simple life cycle in which an issue is treated in past, present, and future frames in terms of a series of attributes. The construction of such a life-cycle analysis is a good way of comprehensively describing issues such as those emanating from the public sector, for example, acid rain.

Strategic Issue Analysis to Analyze Issues. The second approach to the analysis of strategic issues is an SIA process developed by King.[20] Using SIA, an issue literally is analyzed—decomposed into its consistent parts in hierarchical fashion—to a level where data may be gathered to support judgments about the parts, how they relate to one another, and how they will affect the organization.

For instance, Figure 15-4 shows how an issue of the threat of entry of new competitors into a market, emanating from the operating environment, has been analyzed based on work by Porter.[21] The threat-of-entry issue is decomposed into subissues of economies of scale, product differentiation, capital requirements, and so forth. Each of these is successively decomposed to levels at which the sub-subissue is clearly defined and at which data may be collected. For instance, at the lowest level shown in Figure 15-4, the specific dollar

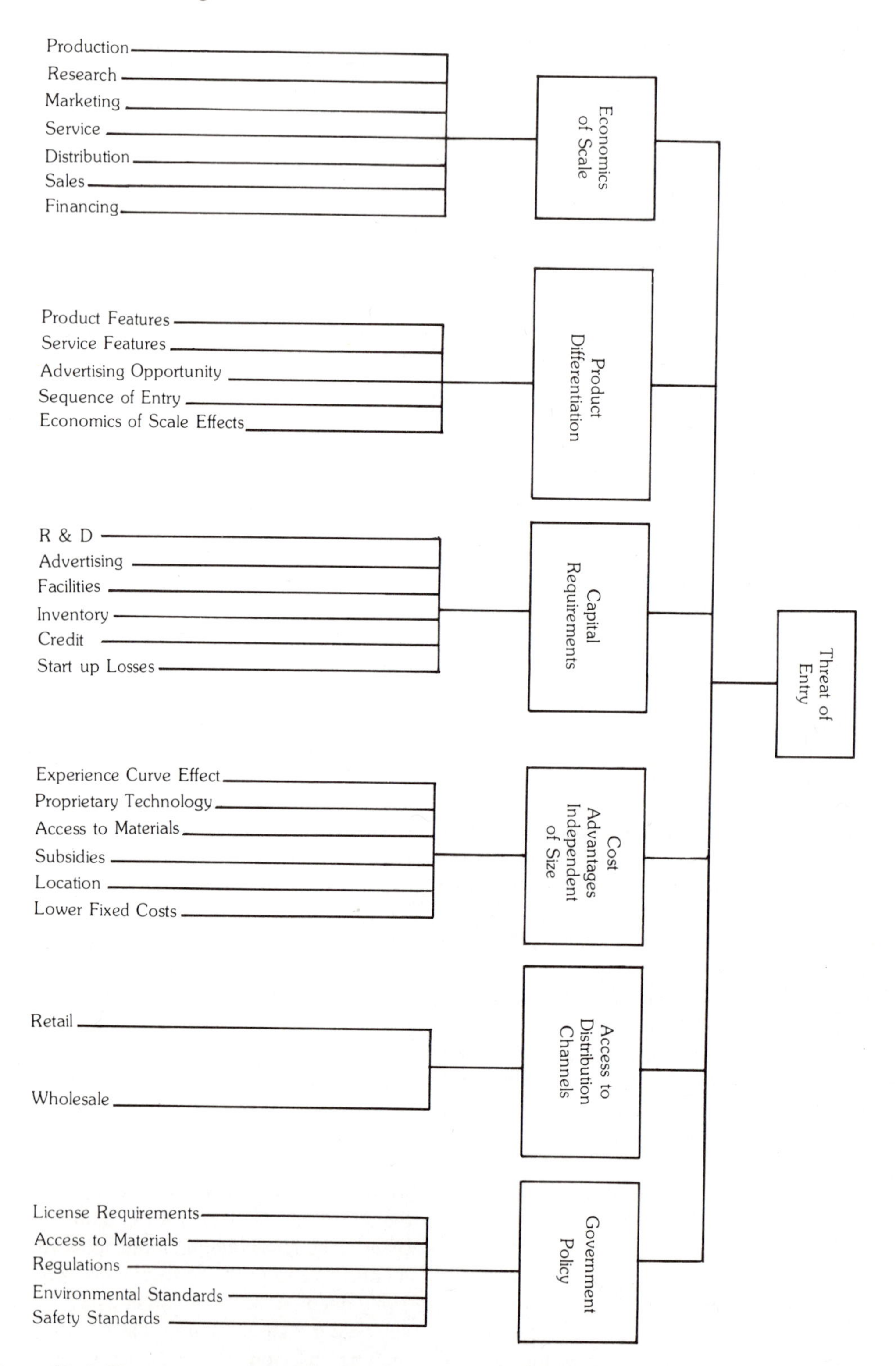

Figure 15-4. Threat-of-entry strategic issue tree (initially developed from material in M. E. Porter, "How Competitive Forces Shape Strategy," *Harvard Business Review* **57,**1979:137-145).

amounts of R & D capital requirements can be identified, the specific license requirements may be listed, and so forth.

Strategic Assumption Analysis to Analyze Issues. Mason and Mitroff have proposed an assumption analysis process that can be used to analyze issues.[22] As adapted to this domain, a previously accomplished stakeholder assessment may be used as the basis for

1. Eliciting assumptions about each stakeholder relative to the issue.
2. Identifying counterassumptions for each assumption.
3. Plotting the assumptions on an importance-certainty graph.

Assumptions about each claimant group are readily elicited from groups of relevant managers. One good approach is to have each one ask: "What assumptions about this group must be true if my personal ideas about them and the issue are to be correct?" In group sessions in which I have done this, assumptions concerning such things as the functionalities of the products that customers will value and the importance of foreign governments have been easy to identify.

The second step involves the identification of counterassumptions (the *deadly enemies* of the previously stated assumptions). As Mason and Mitroff suggest, these are not the logical opposites, but rather some reasonable contrary assumptions.[23] (The role of the counterassumptions will be discussed in the next section.)

Once the assumptions concerning the issue have been identified, each one should be located on an importance-certainty graph such as that in Figure 15-5. On such a graph, each assumption is rated on a subjective scale from very important to not very important and from very certain to very uncertain. This graph is a good device for focusing attention on the most critical assumptions that are implicit in an issue as well as a good way of comprehensively displaying the results of the assumption-analysis process. However, its most important role is in the development of issue-related strategies. (The letters depicted on the graph will be explained in the next section.)

Development of Issue-related Strategy

Once an issue has been analyzed, it must be synthesized in terms of its projected impact on the organization. This may be thought of in terms of answering the questions:

1. How do the issue elements fit together in terms of their likely implications?
2. What are the alternatives available for dealing with these implications?
3. What is the relative cost effectiveness of each alternative?
4. How does each alternative fit into overall strategic plans?

The alternative of doing nothing should always be formally considered because, despite the prior assessment of the issue in terms of the four criteria, after they are assessed in detail, many issues are perceived to be of a nature where little can be done that is, in fact, cost effective.

The fourth question of fit is particularly important in SIM since it is easy to get carried away with strategies that may be appropriate for an issue but that are not particularly *appropriate for the organization and its other strategic interests*. Grant and King discuss a number of criteria such as internal consistency, consistency with policy guidelines, and strength-weakness compatibility that may be used to make this assessment.[24]

If the strategic-assumption analysis process has been used to assess the issue, the importance-certainty graph of Figure 15-5 offers a direct device for developing strategy.

Assumptions that are portrayed to be in the upper-right quadrant of Figure 15-5 (labeled *A*)—those that are relatively important and highly certain—must be the bedrock assumptions of strategies that are related to and developed from the issue. This is so because these assumptions are so important and be-

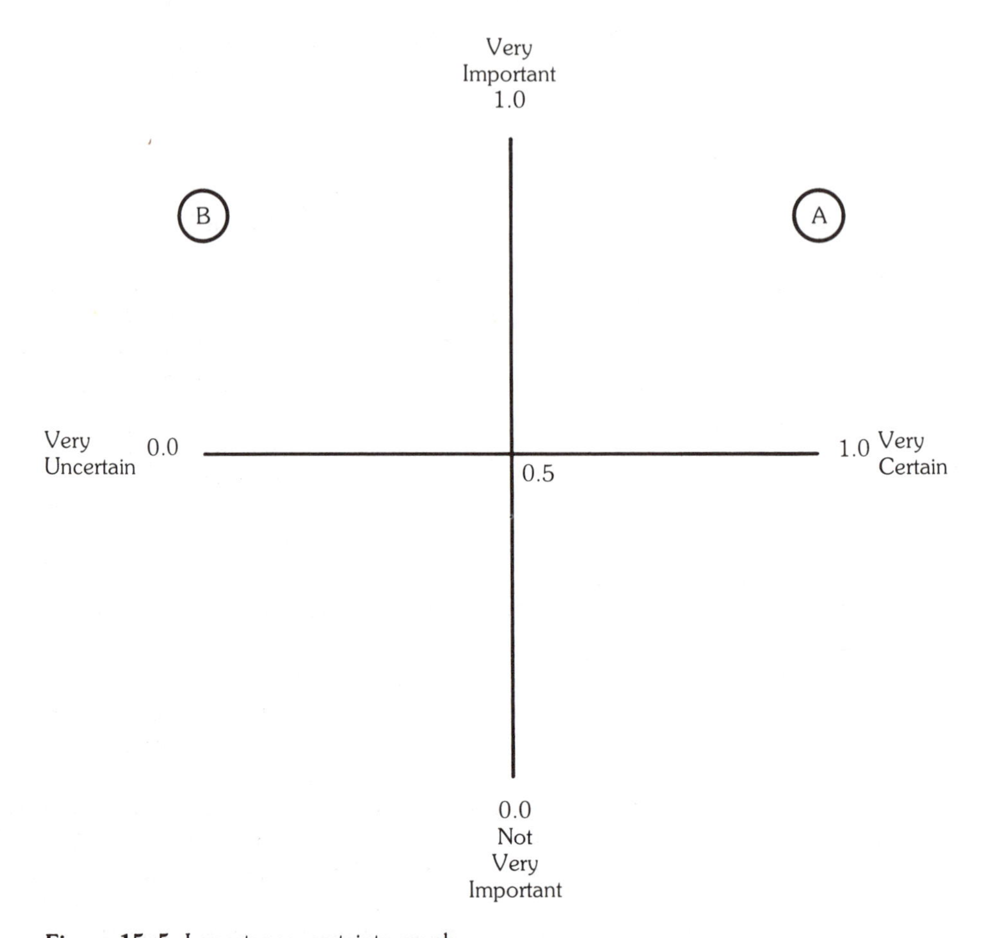

Figure 15-5. Importance-certainty graph.

cause we feel quite confident that they are valid.

However, those assumptions in the upper-left quadrant of Figure 15-5 (labeled *B*)—those very important ones about which we are rather uncertain—must play another role. They are the ones that must be carefully assessed with respect to alternative strategies.

At least two approaches may be used to do this: a contingency approach and a dialectic approach.

Contingency Approach to Issue-related Strategy. If a contingency approach is used, all of the relatively important assumptions (those labeled *A* and *B* in Figure 15-5) should form the foundation of the basic strategy.

Then, the *B* assumptions may be treated as contingency factors since they are the most uncertain of the important assumptions. In this way, the strategy is based on all important assumptions, but those about which there is relative uncertainty are used as the basis for contingency strategies that will be considered should one or more of the *B* assumptions prove invalid.

Dialectic Approach to Issue-related Strategy. The dialectic approach involves the development of a feasible counterstrategy to the basic strategy. Such a counterstrategy is most usefully based on the bedrock assumptions (A) and the counterassumptions to the B assumptions, that is, a counterstrategy is developed in which the fundamental foundations are the bedrock assumptions *and* the counterassumptions for the *B* assumptions that previously have been developed.

Table 15-5. Summary Table of Issue-related Strategies Derived from Assumption Analysis

Strategy	*Based on*
Basic	Assumptions *A* and *B*
Contingency	*B* Assumptions
Counter	*A* and *B* counter-assumptions

Source: W. R. King, "Integrating Strategic Issues into Strategic Management," OMEGA: The International Journal of Management Science, **12** (1984).

The dialectic notion presumes the existence of a *plan* and a *counterplan* that can form the basis for a dialectic debate. The purpose of the debate is not actually to choose between the two but to illuminate them both. In many instances it turns out that the basic strategy and the counterstrategy are not, in fact, mutually exclusive, so that some combined strategy can evolve from the debate.

Table 15-5 sums up the various strategies that may be developed based on an assumption analysis approach.

Implementation, Measuring Results, Monitoring, Feedback

The latter phases of the SIM process are no different for issue-related strategies than for any other strategy.

Because of the detailed assessments that precede the creation and implementation of issue-related strategies, the processes of strategic control are rather clear.[25] It is relatively easy to assess the impact of an issue-related strategy because the issue on which it will have an impact has been rather precisely defined, described, and measured. Thus, the basis for the strategy—the issue—becomes the standard by which it is assessed.

SUMMARY

The SIM process, as defined here, obviates the vagueness and passivity that is associated with the management of issues.

Often, a verbally described issue is given a "So what else is new?" reaction precisely because, although it may be generally recognized, its elements and their potential impacts are not sufficiently well understood for individuals to have developed an appropriate appreciation of the issue.

The SIM process deals with issues in a proactive way that is more likely to result in action—strategy change—than is a more passive approach to issue management.

The assessment phase of the process ensures that the organization will wisely allocate its resources to issues that are strategy-relevant, actionable, critical, and urgent.

The analysis phase serves to define precisely the issue to ensure that important aspects are not omitted and the relevant supporting data is identified. The analysis and strategy-development phases combine formal analysis and judgment in a structured fashion that enhances the understanding of participants. Participative approaches, such as the use of strategic data base task forces and group processes also will enhance the likelihood of implementation of the strategies that are developed using the SIM process.[26]

The final strategic-control phase is not unique to issue-related strategies. However, it is nonetheless enhanced by the SIM process because the issue that has been identified, assessed, and measured through the process becomes the standard by which strategy effectiveness is evaluated.

REFERENCES

1. J. K. Brown, *This Business of Issues: Coping with the Company's Environments* (New York: The Conference Board, 1979); W. R. King, "Environmental Analysis and Forecasting: The Importance of Strategic Issues," *Journal of Business Strategy* **2**(1981):74.
2. Brown, *This Business of Issues.*
3. Ibid., 1.
4. W. R. King and D. I. Cleland, *Strategic Planning and Policy* (New York: Van Nostrand Reinhold, 1978).
5. I. I. Mitroff and J. R. Emshoff, "On Strategic Assumption Making," *Academy of Management Review* **4**(1979):1-12.

6. Brown, *This Business of Issues.*
7. Ibid, 12.
8. Mitroff and Emshoff, "On Strategic Assumption Making."
9. Ibid.; R. O. Mason and I. C. MacMillan, *Challenging Strategic Planning Assumptions* (New York: Wiley, 1981).
10. L. Fahey and W. R. King, "Environmental Scanning for Corporate Planning," *Business Horizons* **14**(1977):61; L. Fahey, W. R. King and V. Narayanan, "Environmental Scanning and Forecasting in Strategic Planning: The State of the Art," *Long Range Planning Journal* **14**(1981):32; P. Thomas, "Environmental Scanning: The State of the Art," *Long Range Planning Journal* **13**(1980):20.
11. Brown, *This Business of Issues.*
12. A. H. Van de Ven and A. L. Delbecq, "Nominal Versus Interacting Group Processes for Committee Decision Making," *Academy of Management Journal* **14**(1971):605-632.
13. A. F. Osborn, *Applied Imagination* (1957).
14. G. Molitor, "How to Anticipate Public Policy Changes," *SAM Advanced Management Journal* **42**(1977):4.
15. H. Kline and W. Newman, "How to use SPIRE: A Systematic Procedure for Identifying Relevant Environments for Strategic Planning," *Journal of Business Strategy* **1** (1980):32.
16. W. R. King and D. I. Cleland, "Information for More Effective Strategic Planning," *Long Range Planning Journal* **10**(1977):59.
17. Brown, *This Business of Issues.*
18. King and Cleland, "Information for More Effective Strategic Planning."
19. For example, W. R. King, "Implementing Strategic Plans through Strategic Program Evaluation," *OMEGA* **8**(1980):173-181.
20. W. R. King, "Using Strategic Issue Analysis," *Long Range Planning* **15**(1982):45-49.
21. M. E. Porter, Jr., "How Competitive Forces Shape Strategy," *Harvard Business Review* **57**(1979):137-145.
22. Mason and MacMillan, *Challenging Strategic Planning.*
23. Ibid.
24. J. H. Grant and W. R. King, *The Logic of Strategic Planning* (Boston: Little, Brown, 1982).
25. P. Lorange, *Corporate Planning: An Executive Viewpoint* (Englewood Cliffs, N.J.: Prentice-Hall, 1980).
26. King, "Environmental Analysis and Forecasting."

V

Strategic Analysis

The biggest growth area in strategic management since the early 1970s has been in the development of tools and techniques for strategic analysis. Despite the fact that the rapid acceptance of many of these techniques has led to a possible overemphasis on techniques (as opposed to creative thinking) in the planning practices of some firms, these proven tools are at the core of modern strategic planning.

In Chapter 16 Alfred Rappaport discusses a criterion for strategy assessment that has become important in the modern era of strategic management—the ability of a strategy to create shareholder value. Some large corporations are committed to this criterion as that which guides their strategic decision making, so that it has come to be of fundamental importance in strategic analysis.

In Chapter 17, George S. Day describes the evolving role of strategy analysis methods and the new strategic planning and management environment in which they play a role today. In discussing implications for the use of strategy analysis methods, Day emphasizes that they must be reviewed as providing a useful but limited input to strategic management.

Chapter 17 serves as an introduction to various strategic analysis methods, but it can also serve to facilitate one's thinking about which methods are most appropriate for any given situation. As a result, the reader may wish to skim this chapter on first reading and then return to it once the other chapters in the section have been perused.

Chapters 18 and 19 deal with portfolio planning—one of the most used and most criticized methods of strategic analysis. In Chapter 18, Malcolm B. Coate compares various portfolio planning models. He identifies the critical assumptions of such models from the viewpoint of the relevant economic theories. The limitations and advantages of portfolio models are then summarized.

In Chapter 19, Philippe Haspeslagh presents the results of a survey of *Fortune 1000* industrial companies that shows the role of portfolio models in achieving improved strategic management in these firms. He presents factors that appear to be key determinants of the success of portfolio planning as well as advice concerning the do's and don'ts of introducing the methods into an organization.

Critical success factor (CSF) analysis is described in Chapter 20 by Joel K. Leidecker and Albert V. Bruno. They trace the development and use of the CSF notion and describe the role that CSF analysis can play in the various stages of strategy development.

One of the most fundamental ideas in management, that of the product life cycle (PLC), is described in Chapter 21 by David R. Rink and John E. Swan as an analytic device that can be used to determine the strategies appropriate at the various stages of the life cycle. They enumerate the different strategies in the context of each of the functional areas of management.

In Chapter 22 William R. King and David I. Cleland discuss strategic strength-weakness analysis as having the objective of providing a useful and succinct informational input to the strategic planning process. This chapter provides the transition to the next section, which deals with information systems for strategic management, in demonstrating that analytic tools are merely ways of processing data to obtain more useful and more highly valued strategic information.

Other discussions of strategic analysis methods also appear in Part IV (particularly in Chapter 14) where they are presented in the context of performing environmental, competitor, and industry analysis.

CHAPTER 16

Selecting Strategies That Create Shareholder Value

Alfred Rappaport

Alfred Rappaport is the Leonard Spacek Professor of Accounting and Information Systems at the Kellogg Graduate School of Management, Northwestern University. He also serves as chairman of The Alcar Group, which provides management education, consulting, and microcomputer software for companies interested in implementing the shareholder value approach to business analysis.

The analyses in this article were prepared using Alcar's microcomputer software The Value Planner® and The Strategy Planner™.

In today's fast-changing, often bewildering business environment, formal systems for strategic planning have become one of top management's principal tools for evaluating and coping with uncertainty. Corporate board members also are showing increasing interest in ensuring that the company has adequate strategies and that these are tested against actual results. Although organizational dynamics and the sophistication of the strategic planning process vary widely among companies, the process almost invariably culminates in projected (commonly five-year) financial statements.

This accounting format enables top managers and the board to review and approve strategic plans in the same terms that the company reports its performance to shareholders and the financial community. Under current practice the projected financial statements, particularly projected earnings per share (EPS) performance, commonly serve as the basis for judging the attractiveness of the strategic or long-term corporate plan.

The conventional accounting-oriented approach for evaluating the strategic plan does not, however, provide reliable answers to such basic questions as:

- Will the corporate plan create value for shareholders? If so, how much?
- Which business units are creating value and which are not?
- How would alternative strategic plans affect shareholder value?

My chief objective here is to provide top management and board members with a theoretically sound, practical approach for assessing the contributions of strategic business unit (SBU) plans and overall corporate strategic plans toward creating economic value for shareholders.

Revised and updated version of A. Rappaport, "Selecting Strategies That Create Shareholder Value," *Harvard Business Review* 59(1981):149-159. Used with permission.

LIMITATIONS OF EPS

A principal objective of corporate strategic planning is to create value for shareholders. By systematically focusing on strategic decision making, such planning helps management allocate corporate resources to their most productive and profitable use. It is commonly assumed that if the strategic plan provides for satisfactory growth in EPS, then the market value of the company's shares will increase as the plan materializes, thus creating value for shareholders. Unfortunately, EPS growth does not necessarily lead to an increase in the market value of the stock. This phenomenon can be observed empirically and explained on theoretical grounds as well.

Of the Standard and Poor's 400 industrial companies, 172 achieved compounded EPS growth rates of 15 percent or better during 1974-1979. In 27, or 16 percent, of these companies, stockholders realized negative rates of return from dividends plus capital losses. For 60, or 35 percent, of the 172 companies, stockholders' returns were inadequate to compensate them just for inflation. The returns provided no compensation for risk. Table 16-1 gives a more complete set of statistics. Additional evidence of the uncertain relationship between EPS growth and returns to shareholders is offered by the 1980 *Fortune 500* survey of the largest industrial corporations. Forty-eight, or almost 10 percent, of the companies achieved positive EPS growth rates for the 1969-1979 period, whereas their stockholders realized negative rates of return. Thirteen of these companies had EPS growth rates in excess of 10 percent during this period.

EPS and related accounting ratios, such as return on investment and return on equity, have shortcomings as financial standards by

Table 16-1. EPS Growth Rates Versus Rates of Return to Shareholders for Standard and Poor's 400 Industrial Companies

	1976-1979	*1975-1979*[a]	*1974-1979*
Companies with annual EPS growth of 10% or greater[b]	259 (100%)	268 (100%)	232 (100%)
Negative rates of return to shareholders	32 (12%)	7 (3%)	39 (17%)
Rates of return inadequate to compensate shareholders for inflation	65 (25%)	36 (13%)	89 (38%)
Companies with annual EPS growth of 15% or greater.[b]	191 (100%)	205 (100%)	172 (100%)
Negative rates of return to shareholders	14 (7%)	2 (1%)	27 (16%)
Rates of return inadequate to compensate shareholders for inflation	33 (17%)	20 (10%)	60 (35%)

Note: EPS growth and rate-of-return calculations prepared by CompuServe, Inc. using Standard and Poor's Compustat data base.

[a]The small number of companies with negative rates of return to shareholders for this period is due to low level of market at the end of 1974. Standard and Poor's stock index at the close of 1974 was 76.47 and, in subsequent years, 100.88, 119.46, 104.71, 107.21, and 121.02.

[b]Restated primary EPS excluding extraordinary items and discontinued operations. The annual growth rates in the consumer price index for 1976-1979, 1975-1979, and 1974-1979 are 7.7%, 7.6%, and 8%, respectively.

which to evaluate corporate strategy for the following six reasons:

1. Alternative and equally acceptable determinations are possible for the EPS figure. Prominent examples are the differences that arise from LIFO and FIFO approaches to computing cost of sales and various methods of computing depreciation.
2. Earnings figures do not reflect differences in risk among strategies and SBUs. Risk is conditioned both by the nature of the business investment and by the relative proportions of debt and equity used to finance investments.
3. Earnings do not take into account the working capital and fixed investment needed for anticipated sales growth.
4. Although projected earnings incorporate estimates of future revenues and expenses, they ignore potential changes in a company's cost of capital both because of inflation and because of shifting business and financial risk.
5. The EPS approach to strategy ignores dividend policy. If the objective were to maximize EPS, one could argue that the company should never pay any dividends as long as it expected to achieve a positive return on new investment. However, we know that if the company invested shareholders' funds at below the minimum acceptable market rate, the value of the company would be bound to decrease.
6. The EPS approach does not specify a time preference rate for the EPS stream, that is, it does not establish the value of a dollar of EPS this year compared with a year from now, two years from now, and so on.

SHAREHOLDER VALUE APPROACH

The economic value of any investment simply is the anticipated cash flow discounted by the cost of capital. An essential feature of the discounted cash flow (DCF) technique, of course, is that it takes into account that a dollar of cash received today is worth more than a dollar received a year from now, because today's dollar can be invested to earn a return during the intervening time.

Although many companies employ the shareholder value approach using DCF analysis in capital budgeting, they use it more often at the project level than at the corporate strategy level. Thus, we sometimes see a situation where capital projects regularly exceed the minimum acceptable rate of return, whereas the business unit itself is a problem and creates little or no value for shareholders. The DCF criterion not only can be applied to internal investments such as additions to existing capacity but also is useful in analysis of opportunities for external growth such as corporate mergers and acquisitions.

Companies can usefully extend this approach from piecemeal applications to the entire strategic plan. An SBU is commonly defined as the smallest organizational unit for which integrated strategic planning, related to a distinct product that serves a well-defined market, is feasible. A strategy for an SBU may then be seen as a collection of product-market related investments and the company itself may be characterized as a portfolio of these investment-requiring strategies. By estimating the future cash flows associated with each strategy, a company can assess the economic value to shareholders of alternative strategies at the business unit and corporate levels.

Steps in Analysis

The analysis for a shareholder value approach to strategic planning involves the following steps:

Estimation for each business unit and the corporation of the minimum pretax operating return on incremental sales (the Incremental Threshold Margin) needed to create value for shareholders.

Comparison of minimum acceptable rates of return on incremental sales with rates realized during the past five years and initial projections for the next year and the five-year plan.

Estimation of the contribution to shareholder value of alternative strategies at the business unit and corporate levels.

Evaluation of the corporate plan to determine whether the projected growth is financially feasible in light of anticipated return on sales, investment requirements per dollar of sales, target capital structure, and dividend policy.

A financial self-evaluation at the business unit and corporate levels.

Before proceeding to the case illustration in the next section, the reader may wish to refer to Appendix 16A to examine the basis for estimating the minimum pretax operating return on incremental sales needed to increase shareholder value as well as the calculation of the absolute shareholder value contributed by various strategies.

CASE OF ECONOVAL

Econoval, a diversified manufacturing company, divides its operations into three lines of business: semiconductors, energy, and automotive parts (see Table 16-2).

Before beginning their detailed analysis, Econoval managers must choose appropriate time horizons for calculating the value contributed by each business unit's strategy. The product life-cycle stages of the various units will ordinarily determine the length of the forecast period. If we were to measure value creation for all businesses arbitrarily in a common time horizon, say five years, then embryonic businesses with large capital requirements in early years and large payoffs in later years would be viewed as poor prospects even if they were expected to yield exceptional value over the life cycle. Therefore, in this case, I have extended the projections for the semiconductor unit to ten years and have limited projections for the energy and auto parts units to five years in the company's long-term financial plan.

Step 1. Estimation of the incremental threshold margins (minimum return on incremental sales needed to create value for shareholders).

The basis for calculating the incremental threshold margins appears as equation (4) in Appendix 16A. For each business unit, four parameters need to be estimated: capital expenditures per dollar of sales increase (incremental fixed-capital investment or F), cash required for working capital per dollar of sales increase (incremental working capital investment or W), the income tax rate, and the cost of capital (the discount rate or weighted average costs of debt and equity). Table 16-3 summarizes the results.

To estimate the recent values for incremental fixed-capital investment (capital expenditures required per dollar of sales increase), one simply takes the sum of all capital expenditures less depreciation over the preceding five or ten years and divides this amount by the sales increase during the period. Note that if a business continues to replace existing facilities in kind and if the prices of these facilities remain constant, then the numerator (that is, capital expenditures less depreciation) approximates the cost of real growth in productive capacity.

Table 16-2. Strategic Overview of Econoval's Lines of Business

Business Unit	*Product Life Cycle Stage*	*Strategy*	*Risk*	*Current year's sales ($ millions)*
Semiconductors	embryonic	invest aggressively to achieve dominant market position	high	50
Energy	expanding	invest to improve market position	medium	75
Automotive parts	mature	maintain market position	low	125

Table 16-3. Incremental Threshold Margins Based on the Initial Forecast

Investment Requirements per Dollar of Sales Increase

Business Unit	*Incremental Fixed Capital Investment (F)*	*Incremental Working Capital Investment (W)*	*Cost of Capital (K)*	*Incremental Threshold Margin (minimum) return on incremental sales)*
Semiconductors	.45	.20	.15	.157
Energy	.20	.20	.14	.091
Automotive parts	.15	.20	.13	.075

However, the costs for capital expenditures usually rise each year owing to inflationary forces and regulatory requirements such as environmental control. These cost increases may be partially offset by advances in technology. Thus the numerator reflects not only the cost of real growth but price changes in facilities as well as the impact of product mix changes, regulation, and technological improvements. Whether the historical value of this variable is a reasonable basis for the forecast period depends significantly on how quickly and to what extent the company can offset increased fixed-capital costs by higher future selling prices, given the competitive structure of the industry.

The increase in required working capital should reflect the cash flow consequences of changes in (1) minimum required cash balance; (2) accounts receivable; (3) inventory; and (4) accounts payable and accruals.

The appropriate rate for discounting the company's cash flow stream is the weighted average of the costs of debt and equity capital. For example, suppose a company's after-tax cost of debt is 6 percent and its estimated cost of equity 16 percent. Further, it plans to raise capital in the following proportion: 20 percent by way of debt and 80 percent by way of equity. It computes the cost of capital at 14 percent as shown in Table 16-4.

Table 16-4. Cost of Capital Calculation

	Weight	*Cost*	*Weighted Cost*
Debt	.20	.06	.012
Equity	.80	.16	.128
Cost of Capital			.140

Is the company's cost of capital the appropriate rate for discounting the cash-flow projections of individual business units? The use of a single discount rate for all parts of the company is valid only in the unlikely event that they are identically risky.

Executives who use a single discount rate companywide are likely to have a consistent bias in favor of funding higher-risk businesses at the expense of less risky businesses. To provide a consistent framework for dealing with different investment risks and thereby increasing shareholder value, management should allocate funds to business units on a risk-adjusted return basis.

The process of estimating a business unit's cost of capital inevitably involves a substantial degree of executive judgment. Unlike the company as a whole, ordinarily the business unit has no posted market price that would enable the analyst to estimate systematic or market-related risk. Moreover, it is often difficult to assign future financing (debt and equity) weights to individual business units.

One approach to estimating a business unit's cost of equity is to identify publicly traded stocks in the same line of business that might be expected to have about the same degree of systematic or market risk as the business unit. After establishing the cost of equity and cost of debt, the analyst can calculate a weighted-average cost of capital for the business unit in the same fashion as for the company.

The cost of equity, or minimum return expected by investors, is the risk-free rate (including the expected long-term rate of inflation) as reflected in current yields available in long-term government bonds plus a premium for accepting equity risk. The overall market risk premium for the last 40 years has averaged 6.1 percent.[1] The risk premium for an individual security can be estimated as the product of the market risk premium and the individual security's systematic risk as measured by its beta.[2]

Following is the estimate for Econoval's semiconductor unit's cost of equity:

$$\text{risk-free rate} + (\text{Beta} \times \text{market risk premium}) = \text{cost of equity}$$
$$9.25\% + (1.4 \times 6.1\%) = 17.8\%$$

Assuming an after-tax cost of debt of 6.5 percent and financing proportions of 25 percent debt and 75 percent equity, the semiconductor unit's risk-adjusted cost of capital is estimated to be 15 percent. Risk-adjusted rates for the energy and auto parts units are 14 percent and 13 percent, respectively.

Step 2. Comparison of the incremental threshold margins (minimum acceptable rates of return on incremental sales) with recently realized rates and initial planning projections.

Having developed some preliminary estimates of minimum return on incremental sales, Econoval now wishes to compare those rates with past and initially projected incremental profit margins for each business unit's planning period. This comparison (Table 16-5) provides both a reasonable check on the projections and insights into the potential of the various business units for creating shareholder value.

From Table 16-5, we can determine that the semiconductor unit is projecting substantial improvement over historical margins on the basis of a continuing product mix shift toward higher-margin proprietary items and substantial R & D expenditures to maintain competitiveness in the learning curve race.

If the forecasted margins materialize, the semiconductor unit will contribute to shareholder value. At this initial stage, the company is concerned with the reasonableness of the projections and the small distance between projected incremental profit margins and the incremental threshold margins (the minimum acceptable margins). The energy unit is projecting a rate of return on incremental sales in line with its recent experience, and this 11-percent rate is comfortably over the 9.1 percent incremental threshold margin.

The problem business unit is the automo-

Table 16-5. Econoval's Incremental Profit Margins (rates of return on incremental sales)

Business Unit	*Last Year*	*Past Five Years*	*Incremental Threshold Margin (minimum) acceptable)*	*Incremental Profit Margin (initial forecast)*
Semiconductors	.115	.110	.157	.155
Energy	.100	.120	.091	.110
Automotive parts	.070	.080	.075	.080

tive parts division. Margins have been eroding steadily, and the forecasted five-year margin is just above the acceptable minimum. Econoval managers are thus committed to investigating a full range of strategic alternatives for the automotive unit.

Step 3. Estimation of shareholder value contributed by alternative strategies at the business unit and corporate levels.

Once the company has developed and analyzed its initial planning projections, SBU managers and the corporate planning group can prepare more detailed analyses for evaluating alternative planning scenarios. Table 16-6 shows the semiconductor unit's planning parameters for conservative, most likely, and optimistic scenarios.

The worst case or conservative scenario assumes significant market penetration by Japanese producers via major technological advances coupled with aggressive price cutting. The most likely scenario assumes the semiconductor group's continued dominance in the metal-oxide-semiconductor (MOS) market, substantial R & D expenditures to enable

Table 16-6. Semiconductor Unit's Forecasts for Various Scenarios

	Year									
	1	*2*	*3*	*4*	*5*	*6*	*7*	*8*	*9*	*10*
Conservative										
Sales growth rate (*G*)	.25	.25	.20	.20	.18	.18	.18	.18	.18	.18
Operating profit margin (*P*)	.115	.12	.125	.13	.135	.135	.135	.135	.135	.135
Incremental working capital investment (*W*)	.20	.20	.20	.20	.20	.20	.20	.20	.20	.20
Incremental fixed-capital investment (*F*)	.42	.42	.42	.40	.40	.35	.35	.35	.35	.35
Cash income tax rate (*Tc*)	.41	.41	.41	.41	.41	.41	.41	.41	.41	.41
Most likely										
Sales growth rate (*G*)	.30	.28	.25	.22	.20	.20	.20	.20	.20	.20
Operating profit margin (*P*)	.12	.125	.13	.135	.14	.145	.15	.15	.15	.145
Incremental working capital investment (*W*)	.20	.20	.20	.20	.20	.20	.20	.20	.20	.20
Incremental fixed-capital investment (*F*)	.45	.45	.44	.42	.42	.40	.38	.38	.35	.35
Cash income tax rate (*Tc*)	.40	.40	.40	.40	.40	.40	.40	.40	.40	.40
Optimistic										
Sales growth rate (*G*)	.32	.30	.30	.25	.25	.25	.25	.25	.25	.25
Operating profit margin (*P*)	.125	.13	.14	.145	.15	.15	.15	.15	.15	.15
Incremental working capital investment (*W*)	.18	.18	.18	.18	.18	.18	.18	.18	.18	.18
Incremental fixed-capital investment (*F*)	.40	.38	.38	.36	.36	.35	.35	.35	.35	.35
Cash income tax rate (*Tc*)	.39	.39	.39	.39	.39	.39	.39	.39	.39	.39

the semiconductor group to maintain its competitiveness in the learning curve race, and gradual Japanese technological parity, which will place pressure on sales margins. The optimistic scenario projects more rapid industry growth and great success in the unit's effort to carve out high-margin proprietary niches.

Table 16-7 presents the shareholder value contribution for each of these three scenarios and for a range of discount rates.

Econoval expects the semiconductor unit's ten-year plan for the most likely scenario to contribute $10.60 million to shareholder value. The range of shareholder values from conservative to optimistic scenarios is from $4.87 million to $29.93 million for the estimated cost of capital or discount rate of 15 percent.

An assessment of the likelihood of each scenario will provide further insight into the relative riskiness of business unit investment strategies. For example, if all three scenarios are equally likely, the situation would be riskier than if the most likely scenario is 60 percent probable and the other two are each 20 percent probable.

Econoval performed similar analyses for the energy and automotive parts units. Table 16-8 summarizes the results for most likely scenarios. To ensure consistency in comparing or consolidating scenarios of various business units, it is important that the corporate planning group establish that such scenarios share common assumptions about critical environmental factors such as inflation and energy prices.

On closer inspection, we see that the analysis in Table 16-8 provides support for management's concern about the automotive unit's performance. Whereas the unit now accounts for 50 percent of Econoval's sales, the company expects it to contribute only $3.57

Table 16-7. Shareholder Value Contributed by the Semiconductor Unit under Different Scenarios and Discount Rates ($ millions)

	Cost of Capital (Discount Rate)				
Scenario	*.140*	*.145*	*.150*	*.155*	*.160*
Conservative	9.30	6.96	4.87	2.99	1.30
Most likely	16.92	13.59	10.60	7.91	5.48
Optimistic	39.64	34.53	29.93	25.79	22.05

Table 16-8. Shareholder Value, Sales Growth, and Earnings Growth Rates by Business Unit for Most Likely Scenarios

	Shareholder Value increase				*Growth rates*	
Business Unit	*Years in Plan*	*$ millions*	*Per Discounted $ of sales increase (VROS)*	*Per Discounted $ of investment (VROI)*	*Sales*	*Earnings*
Semiconductors	10	10.60	.077	.128	22.4%	26.1%
Energy	5	8.79	.175	.438	15	17.7
Automotive parts	5	3.57	.068	.194	10	11.9
Consolidated		22.96				

million, or about 15 percent of the total increase in shareholder value.

On the basis of traditional criteria such as sales and earnings growth rates, the semiconductor unit clearly emerges as the star performer. However, its high investment requirements and risk vis-à-vis its sales margins combine to limit its value-creating potential. Despite the fact that the semiconductor unit's sales and earnings growth rates are substantially greater than those of the energy unit, the semiconductor unit is expected to contribute only marginally more shareholder value in ten years than the energy unit in five years.

The shareholder value increase per discounted dollar of investment provides management with important information about where it is realizing the greatest benefits per dollar of investment. Indeed, this value return on investment (VROI), rather than the traditional accounting ROI, enables management to rank various business units on the basis of a substantive economic criterion.

The numerator of the VROI is simply the shareholder value contributed by a strategy and the denominator, the present cost or investment. When the VROI ratio is equal to zero, the strategy yields exactly the risk-adjusted cost of the capital, and when VROI is positive, the strategy yields a rate greater than its cost of capital. Note that the semiconductor unit ranks last, even behind the auto parts unit, in this all-important performance measure.

Ranking units on the basis of VROI can be particularly helpful to corporate headquarters in capital-rationing situations where the various parts of the business are competing for scarce funds. In the final analysis, however, corporate resources should be allocated to units so as to maximize the shareholder value of the company's total product-market portfolio.

Step 4. Evaluation of the financial feasibility of the strategic plan.

Once the company has established a preliminary plan, it should test its financial feasibility and whether it is fundable. This involves integrating the company's planned investment growth strategies with its dividend and financing policies. A particularly effective starting point is to estimate the company's maximum affordable dividend payout rate and its sensitivity to varying assumptions underlying the strategic plan.

To illustrate, Econoval calculates the maximum dividend payout for the first year of the five-year plan. On a consolidated basis, Econoval projects a sales growth rate (G) of 15.5 percent, an operating profit margin of 9.56 percent, investment requirements ($F + W$) of $.481 per dollar of incremental sales, a cash income tax rate of 42.2 percent, and a current debt equity ratio of 45.2 percent and a target debt equity ratio of 44.3 percent. Econoval can pay out no more than 6.3 percent of its net income as dividends. At the 6.3-percent payout rate, the earnings retained, plus added debt capacity, are just equal to the investment dollars required to support the 15.5 percent growth in sales from $250 million to $288.75 million.

It is easy to demonstrate this result. At $.481 per dollar of incremental sales, investment requirements (net of depreciation) on the projected $38.75 million sales increase will total $18.63 million. This amount will be financed as shown in Table 16-9. The maximum affordable dividend payout rate table (Table 16-10) shows how sensitive this rate is to changes in growth, profitability, investment intensity, and financial leverage. Note, for example, that if the sales growth rate (G) is increased from 15.5 percent to 16.5 percent,

Table 16-9. Investment Requirement Calculation ($ million)

Aftertax earnings on sales of $288.75 million	13.34
Less 6.3 percent dividend payout	.84
Earnings retained, that is, increase in equity	12.50
Increase in unused debt capacity	5.08
Increase in deferred taxes	1.05
	18.63

Table 16-10. Maximum Affordable Dividend Payout Rate Analysis

Investment requirements per dollar of sales increase (Incremental Fixed and Working Capital Investment, F + W)		.431			.481			.531		
			Debt/ Equity			Debt/ Equity			Debt/ Equity	
Sales growth rate (*G*)		.393	.443	.493	.393	.443	.493	.393	.443	.493
145										
Operating profit margin (*P*)	.086	− 9.6%	10.1%	28.5%	−20.8%	−0.7%	18.1%	−31.9%	−11.5%	7.6%
	.096	3.8	21.2	37.4	−6.1	11.7	28.2	−15.9	2.2	19.0
	.106	14.4	29.9	44.4	5.6	21.4	36.2	−3.2	12.9	28.0
.155										
Operating profit margin (*P*)	.086	−15.1	4.7	23.1	−26.9	−6.7	12.1	−38.7	−18.1	1.1
	.096	− 1.0	16.4	32.6	−11.4	6.3	22.9	−21.9	−3.7	13.2
	.106	10.1	25.7	40.2	0.7	16.6	31.5	−8.6	7.6	22.8
.165										
Operating profit margin (*P*)	.086	−20.4	−0.6	17.8	−32.8	−12.6	6.2	−45.3	−24.7	−5.4
	.096	− 5.7	11.7	28.0	−16.7	1.1	17.7	−27.7	−9.5	7.5
	.106	5.8	21.4	36.0	−4.0	11.9	26.8	−13.8	2.4	17.6

Notes: Book income tax rate = .460, cash income tax rate = .422, current debt/equity = .452, current market value of equity = $53.550 million.

the maximum affordable dividend payout rate decreases from 6.3 percent to 1.1 percent, whereas a 1-percent increase in the operating profit margin (*P*) raises the maximum affordable rate from 6.3 percent to 16.6 percent.

Table 16-11 presents Econoval's strategic funds statement for its five-year planning period. The cash required for working capital and fixed-capital investment exceeds the cash sources from operations in each year. This difference is reflected in the net-cash-provided line. Another source of funds is, of course, debt financing.

The increase in unused debt capacity is established by reference to the target debt/equity ratios of Econoval's three principal businesses. Adding the increase in unused debt capacity to the net cash provided provides the maximum affordable dividend, which, as seen earlier, is $.84 million or 6.3 percent in the first year and rises annually to 22.3 percent in the fifth year.

In Table 16-12, strategic funds statements for each of Econoval's main lines of business provide improved insights into product portfolio balancing opportunities. The semiconductor group places a substantial burden on corporate funds. Over the next five years it will require more than $26 million of cash, whereas the energy and auto parts units will throw off about $7 million in cash. Even after taking into account the increase in unused debt capacity that would be contributed by the semiconductors, corporate headquarters will still have to transfer $11 million to the unit.

After further analysis, Econoval managers concluded that the strategic plan was financially feasible. The analysis did, however, raise two concerns. First, Econoval had a low affordable dividend payout rate and was vulnerable to sales margins lower than those projected. Of immediate concern was that the current year's dividend is larger than next year's forecasted affordable dividend.

Also, the strategic funds statement underscored the risk associated with the semiconductor group's aggressive competitive positioning and the related high level of

Table 16-11. Econoval Strategic Funds Statement for Five-Year Forecast Period ($ million)

	Year					
	1	*2*	*3*	*4*	*5*	*Total*
Net income	13.34	15.74	18.54	21.75	25.44	94.81
Depreciation	3.84	4.74	5.82	7.03	8.32	29.74
Increase in deferred income taxes	1.05	1.29	1.56	1.88	2.23	8.02
Sources of funds	18.23	21.77	25.92	30.66	35.99	132.57
Fixed capital investment	14.71	17.58	20.22	22.55	25.66	100.72
Working capital investment	7.76	8.97	10.16	11.33	12.66	50.88
Uses of funds	22.47	26.55	30.38	33.88	38.32	151.60
Net cash provided	(4.24)	(4.78)	(4.46)	(3.22)	(2.33)	(19.03)
Increase in unused debt capacity	5.08	5.89	6.60	7.20	8.02	32.79
Maximum affordable dividend	0.84	1.11	2.14	3.98	5.69	13.76
Maximum affordable dividend payout rate	6.3%	7.1%	11.5%	18.3%	22.3%	14.5%

Table 16-12. Strategic Funds Statement for Five-Year Forecast Period by Business Units (in $ millions)

	Semiconductors	Automotive Energy	Parts	Consolidated
Net income	34.53	30.92	29.36	94.81
Depreciation	17.95	6.07	5.72	29.74
Increase in deferred income taxes	4.21	2.55	1.26	8.02
	56.69	39.54	36.34	132.57
Fixed capital investment	62.31	21.24	17.17	100.72
Working capital investment	20.45	15.17	15.26	50.88
	82.76	36.41	32.43	151.60
Net cash provided	(26.07)	3.13	3.91	(19.03)
Increase in unused debt capacity	15.05	9.26	8.48	32.79
Maximum affordable dividend	(11.02)	12.39	12.39	13.76

investment requirements. This group's large cash requirements, coupled with its modest VROI, prompted Econoval managers to launch a study of alternative product portfolio strategies.

Step 5. A financial self-evaluation at the business unit and corporate levels.

Increasingly, companies are adding financial self-evaluation to their strategic financial planning process. A financial evaluation poses two fundamental questions: How much are the company and each of its major lines of business worth? How much would each of several plausible scenarios involving various combinations of future environments and management strategies affect the value of the company and its business units?

The following types of companies would especially benefit from conducting a financial evaluation:

- Companies that wish to sell and need to establish a minimum acceptable selling price for their shares.
- Companies that are potential takeover targets.
- Companies considering selective divestments.
- Companies evaluating the attractiveness of repurchasing their own shares.
- Private companies wanting to establish the proper price at which to go public.
- Acquisition-minded companies wanting to assess the advantages of a cash versus a stock offer.

The present equity value or shareholder value of any business unit, or the entire company, is the sum of the estimated shareholder value contributed by its strategic plan and the current cash-flow level discounted at the risk-adjusted cost of capital less the market value of outstanding debt. Table 16-13 summarizes these values for Econoval and its three major business units. For example, the semiconductor unit's current cash-flow perpetuity level is $2.97 million, which, when discounted at its risk-adjusted cost of capital of 15 percent, produces a value of $19.8 million. Subtracting the $5 million of debt outstanding provides the $14.8 million prestrategy shareholder value. To obtain the total equity value or shareholder value of $25.40 million for the semiconductor unit, simply add the $10.60 million value contributed by the strategic plan.

The sum of the three business unit values is $83.79 million. Combining the cash flows

Table 16-13. Business Unit and Corporaate Financial Evaluation Summary for Most Likely Scenario (in $ millions)

	Semi-conductors	*Energy*	*Automotive Parts*	*Consolidated*
Pre-strategy shareholder value	14.80	20.93	25.10	60.83
Shareholder value contributed by strategy (see Table 16-8)	10.60	8.79	3.57	22.96
Shareholder Value	25.40	29.72	28.67	83.79
Percent of total shareholder value	30.3%	35.5%	34.2%	
Econoval equity value at corporate cost of capital at 14%				87.57

of the individual businesses and discounting them at the 14 percent risk-adjusted corporate cost of capital yields a value of $87.57 million. In this case, the difference between the value of the whole and the sum of the parts is minor. However, this may not always be true.

Aggregating the values of the company's business units is consistent with the assumption that the riskiness of each unit must be considered separately. If, however, the company's entry into unrelated businesses reduces the overall variability of its cash flows, then the lower expected probability of bankruptcy can decrease its cost of debt and increase its unused debt capacity.

What happens to the company's overall cost of capital naturally depends on any changes in the cost of equity as well as on the cost of debt. Analysis of the impact of business units on the total risk of the company is at best extremely difficult and subjective.

A more attractive alternative is to (1) assume risk independence in establishing the cost of capital for business units and (2) interpret the difference between the value of the company and the aggregate value of its individual businesses as a broad approximation of the benefits or costs associated with the company's product portfolio balancing activities.

Econoval's corporate financial evaluation gave management not only an improved understanding of the relative shareholder value contributed by each business but also the basis for structuring the purchase of an acquisition currently being negotiated. Econoval's market value was then about 25 percent less than its own estimate of value. Because the cash and exchange-of-shares price demanded by the selling shareholders was not materially different, Econoval management decided to offer cash rather than what it believed to be its undervalued shares.

MEETING THE FIDUCIARY DUTY

A fundamental fiduciary responsibility of corporate managers and boards of directors is to create economic value for their shareholders. Despite increasing sophistication in strategic planning applications, companies almost invariably evaluate the final product, the strategic plan, in terms of earnings per share or other accounting ratios such as return on investment or return on equity.

Surprisingly, the conventional accounting-oriented approach persists despite compelling theoretical and empirical evidence of the failings of accounting numbers as a reliable index for estimating changes in economic value. How should the board member of a company that has reported a decade of 15 percent annual EPS growth and no increase in its stock price respond when asked to approve yet another five-year business plan with projected EPS growth of 15 percent? The shareholder value approach to strategic planning would enable the board to recognize that, despite impressive earnings growth projections, the company's increasing cost of capital, rising investment requirements per dollar of sales, and lower margins on sales are clear signs of value erosion.

A number of major companies are now using the shareholder value approach to strategic planning. The method requires virtually no data not already developed under current financial planning systems; moreover, microcomputer programs such as The Value Planner™ and The Strategy Planner™ by Alcar (used in preparing the numerical illustrations) can help implement all of the steps I have outlined. Use of this approach should improve companies' prospects of creating value for their shareholders and thereby contribute to the long-run interest of the companies and of the economy.

ACKNOWLEDGMENTS

I wish to thank Carl M. Noble, Jr. and Robert C. Statius Muller, both of Alcar, Inc. for their many helpful suggestions.

APPENDIX 16A

CALCULATION OF VALUE CONTRIBUTED BY STRATEGY

The present value of a business is defined simply as the anticipated after-tax operating cash flows discounted by the weighted average cost of capital. The present value of the equity claims or shareholder value is then the value of the company (or business unit) less the market value of currently outstanding debt. The shareholder value for a business that expects no further real sales growth and also expects annual cost increases to be offset by selling price increases is given by the following formula:

$$\text{pre-strategy shareholder value} = \text{pre-strategy corporate value} - \text{debt} \quad (1)$$

where

$$\text{pre-strategy corporate value} = \frac{\text{perpetuity operating profit after tax}}{\text{cost of capital}} = \frac{PS\,(1-T,)}{K},$$

where P = operating profit margin; T = income tax rate; S = sales; K = weighted average cost of capital; and debt = market value of long-term debt and other long-term obligations (preferred stock, unfunded pension liabilities, etc.).

The change in shareholder value for a given level of sales increase, or the value contributed by strategy, is then

value contributed by strategy = present value of incremental cash inflows − present value of incremental cash outflows, (2)

where

$$\text{present value of incremental cash inflows} = \frac{(IPM)\,(\text{sales increase})\,(1-T)}{K},$$

$$\text{present value of incremental cash outflows} = \frac{(F+W)\,(\text{sales increase})}{(1+K)},$$

and where IPM = incremental profit margin (that is, operating profit margin on incremental sales); F = incremental fixed capital investment (capital expenditures minus depreciation per dollar of sales increase); and W = incremental working capital investment (cash required for net working capital per dollar of sales increase).

The value contributed by strategy (change in equity or shareholder value) is the difference between the after-tax perpetuity operating profit and the required investment outlay for fixed and working capital. Since all cash flows are assumed to occur at the end of the period, the outlays for working capital and fixed assets are discounted by $(1 + K)$ to obtain the present value. There is neither an increase nor a decrease in shareholder value for a specified sales increase whenever the value of the inflows and outflows is identical. Specifically, when

$$\frac{IPM(1-T)}{K} = \frac{(F+W)}{(1+K)}. \quad (3)$$

From equation (3), incremental threshold margin, (the break-even operating return on sales or the minimum pretax operating return on incremental sales needed to create value for shareholders) is derived as:

$$ITM = \frac{(F+W)\,K}{(1-T(1+K)} \quad (4)$$

The incremental threshold margins for a range of investment requirements per dollar of sales and costs of capital are presented in Table 16A-1.

The shareholder value contributed by any strategy can be estimated by taking the capitalized value of the incremental threshold spread, or ITS (the difference between the projected incremental profit margin and the incremental threshold margin). More specifi-

Table 16A-1. Incremental Threshold Margins (minimum pretax operating return on incremental sales to create value for shareholders)

	Incremental Fixed and Working Capital Investment (F + W)						
Cost of Capital	*.20*	*.30*	*.40*	*.50*	*.60*	*.70*	*.80*
.12	.040	.059	.079	.099	.119	.139	.159
.14	.045	.068	.091	.114	.136	.159	.182
.16	.051	.077	.102	.128	.153	.179	.204
.18	.056	.085	.113	.141	.169	.198	.226
.20	.062	.093	.123	.154	.185	.216	.247

Note: Assumed income tax rate = 46%

cally, the change in shareholder value for time *t* is given by the following equation, which assumes book and cash income tax rates are identical. If they are not, another term must be added.

$$\text{value contributed by strategy} = \frac{\text{sales increase}_t ITS_t\ (1-T)}{K(1 + K)^{t-1}}, \quad (5)$$

where *ITS* or incremental threshold spread = incremental profit margin − incremental threshold margin.

To illustrate, consider a business with sales of $50 million for its most recent year and the following assumptions for its five-year plan:

Sales growth rate (*G*) = 15 percent
Incremental profit margins (years 1-2) = 13.5 percent
Incremental profit margins (years 3-5) = 14.5 percent
Book and cash income rates = 46 percent
Incremental fixed-capital investment (*F*) = 35 percent
Incremental working capital investment (*W*) = 20 percent
Cost of capital (*K*) = 14 percent

Applying equation (4) for the incremental threshold margin (*ITS*), we obtain 12.5 per-

Table 16A-2. Shareholder Value Contributed by Five-year Plan ($ millions)

	Years					
	1	*2*	*3*	*4*	*5*	*Total*
Sales	57.50	66.12	76.04	87.45	100.57	387.68
Sales growth (G)	7.50	8.62	9.92	11.41	13.12	50.57
Incremental threshold spread (projected Incremental profit margin minus incremental threshold margin)	.01	.01	.02	.02	.02	
Value contributed by strategy (present value of the increase shareholder value)[a]	.29	.29	.59	.59	.60	2.36

Note: The present value of the five-year plan is $2.36 million.
[a]Computed by using equation (5).

cent. A summary of the shareholder value contributed by the five-year plan is presented in Table 16A-2.

NOTES AND REFERENCES

1. This average is derived from the geometric mean for the period 1926 to 1981; see R. G. Ibbotson and R. A. Sinquefield, "Stocks, Bonds, Bills and Inflation: The Past and the Future," *Research Foundation Monograph* **15**(1982).
2. For a method of predicting beta, see B. Rosenberg and J. Guy, "Prediction of Beta from Investment Fundamentals," *Financial Analysis Journal* (May-June 1976):60; B. Rosenberg and J. Guy, "Prediction of Beta from Investment Fundamentals," *Financial Analysis Journal* (July-August 1976):62.

CHAPTER 17

The Evolving Role of Strategy Analysis Methods

George S. Day

George S. Day is Magna International Professor of Business Strategy at the University of Toronto. He received the Ph.D. from Columbia University and previously taught at Stanford University and IMEDE in Lausanne, Switzerland. Recent publications include *Strategic Market Planning: The Pursuit of Competitive Advantage* (West Publishing, 1984). Much of the material in this chapter came from strategy consulting assignments with major manufacturing and financial services firms in Canada and the United States.

The past decade has seen widespread adoption of such strategy analysis methods as portfolio classification models and PIMS analyses of profitability determinants. Users were attracted by the logic and relevance of these methods to perplexing questions of strategic direction and resource allocation. A spate of success stories from Mead, General Electric, Borg-Warner, Norton, and other companies helped fuel the initial acceptance.[1] A new breed of strategic planning consultants including the Boston Consulting Group, Strategic Planning Associates, Braxton Associates, and Bain and Company made these methods an integral part of their practice and aggressively promoted them as solutions to management problems.

The new strategy analysis methods were well suited to the emerging planning concerns of the time. During the mid-seventies the orientation of strategic planning shifted from managing predictable growth to conserving scarce resources in a period of sharp discontinuities. Sorting out winning businesses from losers that were draining cash and then consolidating strong competitive positions became the primary concerns. The dominant features of the planning systems developed in response to these changes were

- The grouping of related businesses and products into strategic business units (SBUs) or organizational entities large enough and homogeneous enough to exercise control over most strategic factors affecting their performance.
- Strategic direction provided by strategic business plans that specify the product-market scope and focus of the SBU and strategic thrust and performance objectives for the SBU within the context of the overall corporate mission and strategy.
- Explicit consideration of distinct strategy alternatives, varying in terms of risk-reward profile or the importance of different objectives, such as market share gains versus short-term profitability.

Portions of this chapter are adapted from G. S. Day, *Analysis for Strategic Market Decisions* (St. Paul, Minn.: West Publishing, 1985).

Objectives for different SBUs are tailored to reflect differences in their strategic position and competitive environment that will influence long-run growth and profit potential.

The use of a portfolio logic to allocate resources in recognition of differences in the contributions of different SBUs to the achievement of corporate objectives for growth and profitability. A key is whether the SBU is designated a net cash generator or cash user.

By 1979 a survey of *Fortune 1000* companies found that 36 percent had adopted a portfolio approach, and this proportion was growing at 15 percent to 20 percent a year.[2] Close to half the adopters were using their portfolio as the basis for a management system, especially to guide the negotiation of explicit strategic missions (or investment strategies), between business unit managers and corporate management. The continued growth in acceptance was confirmed four years later in a replication of the survey with diversified Canadian companies that found 51 percent were using a portfolio approach, and a further 19 percent were considering its use. (Also see chapters 18 and 19.)

A NEW ENVIRONMENT FOR PLANNING METHODS

The challenges of the eighties are different from those that originally spawned strategy analysis methods. Competitive pressures have become even more acute as companies recognize that the maturity of most industries requires them to actively seek new opportunities in order to grow or even hold their position. At the same time, technological advances, deregulation, global markets, changing demographics, a reduced government role, and innumerable other factors are presenting new challenges. As a result, patterns of competition are becoming more complex, market boundaries are becoming fuzzy, and competitive advantages are increasingly short-lived.

There is also a growing recognition that strategic planning systems emphasizing centralized decisions are unbalanced and ineffective. There is a pressing need for greater attention to the integration of strategic and operational planning. Too many conceptually elegant strategic plans have failed from lack of commitment by operating management. As further encouragement to decentralization there is usually adequate funding available for attractive projects. Although corporate-level resource allocation problems remain important, they are no longer paramount.

Just as these new requirements for analysis of fast-moving and highly competitive markets emerged, there was growing skepticism of strategy analysis methods in the face of unrealized expectations. Critics faulted these methods for abdicating management imagination to quantitative factors and thereby suppressing creative alternatives, depersonalizing the resource allocation process, and prescribing strategies that were simplistic, doctrinaire, and possibly misleading.[3]

These criticisms cannot be ignored, but in the main they reflect a misunderstanding of the proper role of these analytical methods. Certainly there is an unavoidable adjustment period in the life of any management concept or method during which experience is gained and the limitations are appreciated. This is a necessary condition to informed use as well as a useful antidote to earlier overselling. What must be realized is that these methods can facilitate the strategic planning process and serve as a rich source of ideas about possible strategic options. However, on their own these methods can neither prescribe the appropriate strategy nor predict the consequences of a specific change in strategy.

For the remainder of this chapter I develop this theme further by putting the concepts and methods of strategy into perspective in a supporting role in the strategic planning process. Then I look at some of the important distinctions and relations among the various methods as a prelude to five specific proposals for enhancing the contributions of the methods to strategic thinking.

STRATEGY ANALYSIS METHODS AND CONCEPTS

The various strategy analysis methods share many features and often draw from a common conceptual source. An understanding of the benefits and limitations of one method will thus be helpful in illuminating the other methods. There is a natural hierarchical relationship among the methods that is shown schematically in Figure 17-1.

The foundation of the hierarchy provides the inputs to higher-level analyses, but these inputs generally have limited decision relevance. The higher one goes in the hierarchy the more directly the methods speak to the specific decisions to be made by the business unit until the pinnacle is reached, at which point we have the most generic and widely applicable method, which is the basic planning process. The decision relevance of the higher-level methods is achieved through greater complexity and integration of external and internal information.

Two methods are shown in a flanking position because they are by nature more flexible. In both cases they are a fundamental part of the foundation analysis but also provide significant inputs to higher-level analyses. For example, the starting point for a market analysis is a preliminary identification of the product class boundaries and market segments.

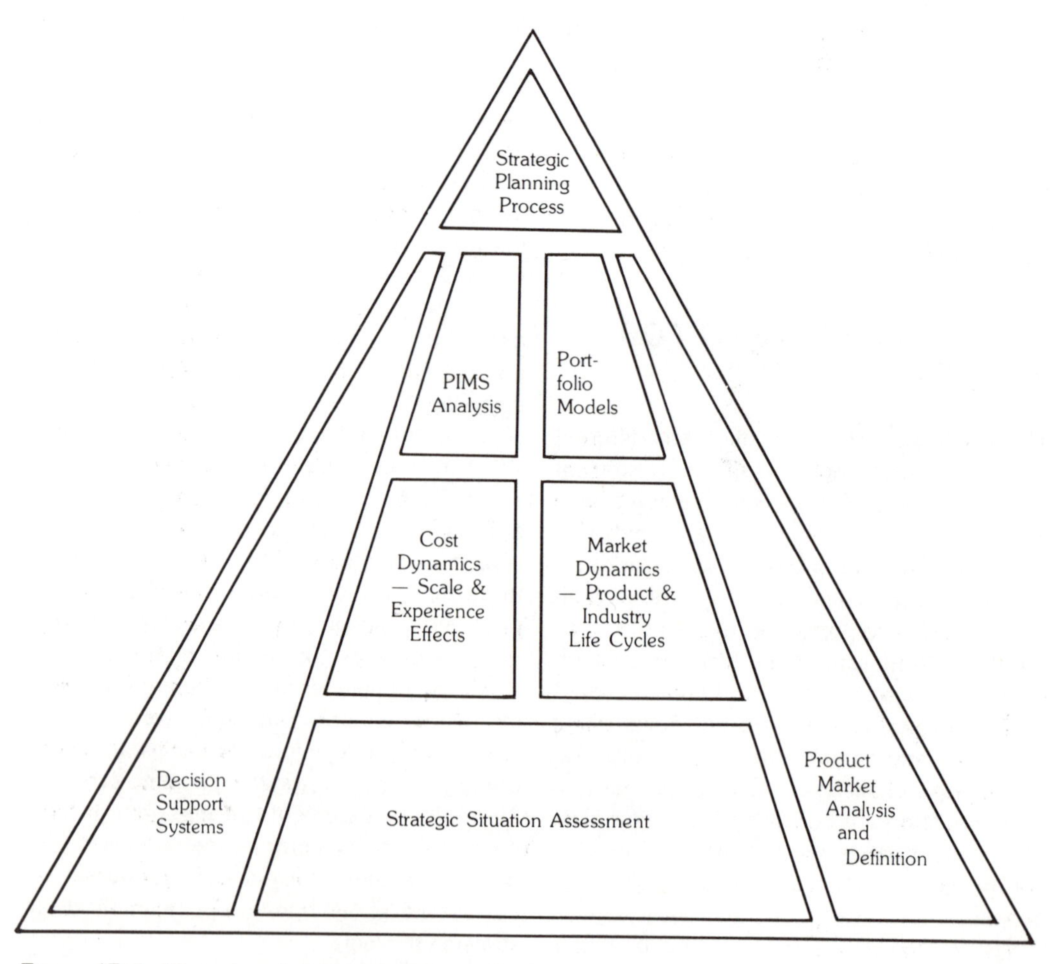

Figure 17-1. Hierarchy of strategy analysis methods.

However, part of the market analysis task is to identify and forecast exploitable discontinuities in the market. Also each of the models in this hierarchy—life cycle, experience curve, portfolio, or PIMS—relies on the proper definition of the market. Decision support systems play an analogous role in turning lower-level inputs into higher-level outputs.

Undergirding all the methods is the basic assessment of the situation. This entails the systematic assembly and analysis of past, present, and future data about the environment of the business and its competitive position. The trends, forces, and conditions identified during this situation assessment are the raw material for more refined analysis using the higher-level concepts and methods in the hierarchy.

Planning Concepts and Techniques

The second level of the hierarchy of strategy analysis methods is occupied by experience curves and product and industry life cycles. Each of these concepts and techniques enhances strategic thinking by providing frameworks for organizing the mass of strategic information unearthed in the first level of analysis. They play their role by separating important issues and relationships from those that are unimportant. The methods at this level do double duty by contributing to the conceptual framework for the models found at the next level.

Analysis of Cost Dynamics: Scale and Experience Effects. The organizing framework for dynamic analyses of costs and price is the experience curve.[4] Few strategy concepts have gained wider acceptance than the basic premise of the experience curve, that value-added costs—adjusted for inflation—decline systematically with increases in cumulative volume. These declines in real costs are usually attributable to some combination of learning by doing, advances in technology, or increased efficiency from economies of scale.

Three types of experience curves have been used to support strategy analyses:

The company cost compression curve relates changes in the total value-added costs, as well as individual cost elements, to the accumulated experience of the company with a particular product.

Competitive cost comparison curves are cross-sectional relationships of the relative cost positions of the competitors in an industry. With this curve it is possible to estimate the relative profitability of each of the competitors at the prevailing price.

Industry price experience curves relate the industry average price to industry cumulative experience.

The most compelling strategic message from the experience curve is that if costs per unit decline predictably with cumulative output then the largest competitor in the market has the potential for the lowest unit costs and highest profits. Only if the dominant player is being outsegmented by more focused competitors or is doing a poor job of cost control will the profit potential not be realized. However, the overall logic leads inexorably to the pursuit of dominance as a strategic imperative. A further implication is that gains in experience are most easily achieved in fast-growing markets by taking a disproportionate share of new sales. In practice, these strategic signals are likely to be misleading since they depend on a number of assumptions that are not easily satisfied. For example, all competitors must follow the same experience curve, realize the same price per unit, define their business the same way, share resources with other SBUs to the same extent, and operate at the same level of capacity utilization.

The strategic relevance of the experience curve also depends on the source of the observed cost reductions. Where the cost reductions are being achieved primarily from economies of scale through more efficient, automated facilities and vertical integration, then cumulative experience may be unimportant to the relative cost position. In these situations a new entrant may be more efficient than more experienced producers. If, however, experience effects are due to learning by doing or using and can be kept proprietary, the leaders will maintain a cost advantage.

Analysis of Market Dynamics. Market dynamics analysis is virtually synonymous with the product life-cycle concept. The notion of distinct stages in the sales history of a product category from birth → growth → maturity → decline has enduring appeal because of the biological analogy. What makes the life-cycle concept useful as an organizing framework for strategy analysis is the capacity to reflect the outcome of market, technology, and competitive forces. However, the capacity for prediction is much more restricted, and attempts to use it to prescribe appropriate strategies are dubious.

There is a strong temptation to view life cycles as immutable and inevitable. Although the stage in the life cycle can be a significant determinant of strategy, it is also possible that strategic choices can shape the growth rate. Ohmae describes what happened when American firms incorrectly assumed that radios had a natural life.

> Many financially oriented, diversified U.S. electrical manufacturers treated radios as species of hardware; in other words, essentially as a dot on a product portfolio matrix. Chronically starved of investment funds and management talent, a product line called "radio" accordingly lived out its natural life.[5]

This constraining—and invalid—assumption was not made by the Japanese. They redefined radios as a form of audio entertainment, and created a wealth of composite products such as radio cassettes and the Sony Walkman. (Also see chapter 21.)

Flanking Methods

Flanking methods are encountered during the most basic of preliminary analyses but are so broad in their potential utility that they also contribute to resolving higher-level strategic issues.

Product–Market Analysis and Definition. A strategy reflects a pattern of decisions as to how a business unit can best compete in the markets it elects to serve. The chosen market may be defined very broadly or comparatively narrowly, depending on the choice of customer segments, and the treatment of such issues as substitute techniques, geographic boundaries, and the number of stages in the value-added chain. These judgments in turn are pivotal when making assessments of the current performance and the establishment of a defensible competitive position.

A common feature of all strategy analysis methods is their reliance on the choice of product-market boundaries. This is the unit of analysis that is the basic context for the other measures. The choice is complicated by the need to adapt the definition of the product-market to the decision to be made and the reality that there is an element of arbitrariness in all such definitions.[6]

Decision support systems. Decision support systems (discussed further in Chapter 25) is a generic term for a combination of strategic intelligence and analytical models that are tailored to the requirements of the business.[7] The positioning of decision support systems as a flanking method reflects their versatility and potential contributions to the data base from which all strategy analysis methods draw for their inputs. Their greatest contributions to date have come from the enhancements to the quality of strategic information gained from the strategic intelligence system. This part of a decision support system directs intelligence gathering, collects the data, and then transforms it into information. Less progress has been made with the development of the models component, but trends in communications and computer technologies are rapidly enhancing their feasibility.

Analytical Planning Models

The most distinctive feature of these models is their capacity to synthesize a variety of judgments on the interaction of the capabilities of the business with the threats and opportunities in the environment. They are the most integrative and complex of all strategy analysis methods and certainly the most controversial.

Analysis of Pooled Business Experience: The PIMS Program. This approach to strategy analysis seeks guidance from the collective experience of a diverse sample of successful and unsuccessful businesses. By 1985 there were over twenty-seven hundred businesses in the sample.

The basic premise of the PIMS program is that an individual business can learn as much from the experiences of strategy peers, conducting a large series of strategy experiments from a similar position, as it can learn from industry peers who participate in the same industry but face different strategic situations.[8]

The search for strategically relevant insights is directed toward identifying the controllable and uncontrollable variables that do the best job of explaining the observed variance in profitability and cash flow in the sample of businesses. Three sets of influential variables have been identified: (1) the competitive position of the business, as measured by market share and relative product quality; (2) the production structure, including investment intensity and productivity of operations; and (3) the relative attractiveness of the served market, comprising the growth and customer characteristics. Overall, these variables account for 65 percent to 70 percent of the variability in profitability. These empirical insights from the data base are used to address a series of strategic questions for each business,

What rate of profit and cash flow is normal for this type of business?
If the business continues on its present track, what operating performance could be expected in the future?
How will this performance be affected by a change in strategy?

The strategy prescriptions derived from PIMS results have to be treated with great caution. The first problem is the difficulty of assessing the extent to which evidence of higher profits is compensation for taking higher risks. By only observing outcomes we do not know whether risks are reflected in the profitability of the survivors or the costs incurred by the failures. Furthermore, the outcomes cannot necessarily be interpreted as representing the objectives of the competitors. Those who ended up with low shares probably did not intend to do so. In the absence of data on goals and objectives this cannot be assessed, and prescriptions based on observed outcomes become correspondingly difficult.

Portfolio Classification Models. The essence of portfolio classification models (discussed further in Chapters 18 and 19) is the classification and visual display of the present and prospective positions of businesses and products according to the attractiveness of the market and the ability to compete within that market. In the original portfolio model, developed by the Boston Consulting Group, the corresponding dimensions were limited to market growth and relative market share.[9] This simple formulation yields powerful insights, so long as the premises on which it is based are reasonably well satisfied.

In many settings the key premises of the growth-share matrix are found invalid, for example, market share is not necessarily a determinant of profitability. More often share is a consequence of previous profits that in turn may have been generated by lucky or insightful moves into defensible positions. Similarly, growth markets are not always attractive if one takes into account the risks inherent in these markets. Another premise that is often questionable holds that the basic objective of portfolio strategies is to maintain the firm in cash-flow balance. Many of these deficiencies of the growth-share matrix are traceable to a reliance on two variables. This problem can be partially overcome with alternative portfolio models that use a composite of weighted factors to represent the two underlying dimensions.

Portfolio models are used widely as diagnostic aids for communicating strategic judgments about the position of a business or for understanding the behavior of competitors. They also may be used to guide the generation of strategic options and facilitate negotiations between corporate-level and business management on critical questions of feasible objectives and investment strategies. Seldom are portfolio models used as prescriptive guides to the choice of the appropriate strategic option. Although all portfolio models can be used to derive natural or generic strategies for a business according to its position in the portfolio matrix, in reality these possibilities are invariably dominated by the specifics of the strategic situation.

Integrating the Methods into the Planning Process

The experience of 3M illustrates how an already effective organization adapts its planning system to take best advantage of the developments in strategic planning methods. In 1981, when 3M's management began to revamp their planning system, it had achieved sales of $6.5 billion with over forty-five thousand products. Most of these products are rooted in the original coated-abrasives or pressure-sensitive tape technologies. The subsequent developments have taken the company into markets as diverse as recording equipment, medical products, and pharmaceuticals.

The initial challenge for 3M management was to cope with almost two hundred organizational units worldwide: divisions, departments, projects, and subsidiaries. Most of these units were accustomed to directing their business autonomously on the basis of their own perceived requirements and opportunities. However, there was a pressing need to coordinate these diverse capabilities to enhance competitive advantage in their markets. The planning systems that emerged in 1982 in response to this need were consistent with the history and culture of 3M in keeping responsibility for strategic decision making with the key operating unit managers. Some of the aspects of 3M's planning system are described in Appendix 17A.

ROLE OF STRATEGY ANALYSIS METHODS IN THE PLANNING PROCESS

Few generalizations about strategy methods are meaningful in the face of the variety of roles they can potentially play during the planning process. Still it is safe to say that these methods are limited to supporting and enhancing the effectiveness of the basic process and should never serve as a substitute for strategic thinking.

Possible Roles

The contributions of strategy analysis methods are realized primarily during the early stages of the planning process. They play a very modest role during the later stages of implementation and monitoring. Among the specific applications where they have been found to be particularly useful are:

Situation assessment

1. Provide a *structure for analysis* that enables management teams to separate important issues from unimportant issues and identify critical subproblems.
2. Isolate *information gaps* in order to assign priorities to marketing research and information system activities. One of the most valuable functions of the strategy analysis methods is to provide a logical and comprehensive checklist to ensure key variables are not overlooked.
3. Provide a common language and mutually acporation. This analysis helped develop a feel use to *communicate judgments and assumptions* about strategic issues.
4. *Evaluate the current strategy and competitive position* in terms of key performance measures such as cash flow and return on investment.

Strategy development, evaluation, and implementation

1. Facilitate the *identification of strategy options* to be given detailed consideration.
2. *Identify performance trade-offs* involved in choosing among the feasible strategic options.
3. Provide a logical basis for *allocating resources* across businesses or business segments.
4. *Test the validity* of forecasts of performance of a recommended strategy.
5. Facilitate the *monitoring* of actual performance of the chosen strategy and the isolation of reasons for variance from the objectives.

Although each method may be used for almost all purposes, in practice most methods are much better suited to some applications than others. Table 17-1 shows schematically the preferred patterns of use as indicated by double and single asterisks. Several generalizations are possible from this table even though the broad categories of analysis methods obscure some important distinctions (notably among types of decision support systems and varieties of portfolio models). First, we see that the more complex, higher-level methods can assume a greater variety of roles. Complexity here is a function of the number of variables and relationships that are incorporated into the method. This is why PIMS appears to be the most widely applicable. We will find, however, that the usefulness of a method also depends on how readily it can be adapted to a specific setting. Some concepts, such as the experience curve, offer powerful insights within certain industries but are downright misleading in other settings. PIMS and some portfolio models also suffer in this regard because they are constrained by rigidity in the specification of their variables and relationships.

Creative Combination of Methods. From Table 17-1 it seems that several methods serve the same purpose. This does not

Table 17-1. Potential Applications of Strategy Analysis Methods

	Product Life-Cycle Analysis	*Experience Curve Analysis*	*Decision Support Systems*	*Portfolio Models*	*PIMS*
Situation assessment					
1. Structure for analysis	**	**	**	**	**
2. Isolate information gaps	*	*	**	*	**
3. Communicate judgments and assumptions	*	*	*	**	**
4. Evaluate current strategy		*	*	*	*
Strategy development, evaluation and implementation					
1. Facilitate identification of options	*	*	*	**	*
2. Identify performance trade-offs			**	*	*
3. Guide resource allocations across businesses			*	*	*
4. Test validity of performance forecasts			*		*
5. Monitor performance against objectives			*		

mean they are substitutes. In most cases, it is more appropriate to treat them as complements. Each strategy analysis method has a distinctive profile of strengths and weaknesses stemming from differences in orientation, premises, and measures. This is an advantage to the strategy analyst because the differences in methods permit the examination of the same problem or issue from several perspectives. It is hoped they will give the same signals. If the signals are contradictory then further analysis must be undertaken to resolve the conflict. In this role, the methods are provocative devices for raising important questions, although the resolution usually lies outside the methods because it is embedded in the details of the particular competitive situation.

The experience of a strategic business unit within a medium-technology manufacturer of alarm and fire security systems illustrates the pay-off from vigorous pursuit of contradictions highlighted by the analysis methods. Early in the planning process each of the six business segments within this business unit were classified in a market attractiveness-business strength portfolio matrix. By all the usual signals this was a mature manage-for-current-earnings-while-holding-market-position type of business. But the forecast of the financial and market share consequences of the current strategy was wildly at variance with this portfolio interpretation. The problem was traced to a serious cash-flow problem created by an ill-advised decision to carry extensive inventories to buy coverage of a previously unused distribution channel. This situation had been obscured by an unusual one-time improvement in cash generated by the business during the year the new distribution strategy was implemented. Only when management had to reconcile the conflicting signals from two different ways of analyzing their strategic situation did they reconsider their new distribution strategy. The process of reconciliation began with a review of the assumptions made in the performance forecasts, followed by a search for explanations. By the end of the planning process the management team was committed to a major redirection of their strategy.

Assessing Competitive Position. A multiple-methods approach is particularly helpful when coping with incomplete, inaccurate, and contradictory data about competitors. One business overcame the paralyzing consequences of these data problems by using several approaches to estimating its costs and profitability relative to its two principal competitors. The business's problem was complicated by the fact that its competitors used different distribution systems, process technologies, and component sourcing strategies. The analysis was initiated by an evaluation of costs and profitability based on

1. A cross-section experience curve estimate of its relative cost position as a consequence of differences in accumulated experience and strategies. This required assumptions about each competitor's individual costs and ability to go down the experience curve at the industry rate for each major cost element. Adjustments were made for probable shared experience with other business units and access to cost reductions achieved by outside suppliers.
2. A PIMS estimate of the probable ROI and ROS given the market share, relative quality position, investment intensity, and so forth of each of the competitors. The critical assumption here was that the competitors would perform in a manner similar to other businesses with the same characteristics. The results were then used to estimate the cost structures for representative items in the product line.
3. Portfolio matrices of its respective parent corporation. This analysis helped develop a feel for how the business was viewed relative to the other businesses competing for corporate resources, and whether its parents would support an aggressive response to any strategic moves that would erode the position of these competitive businesses.

Not surprisingly, the three approaches yielded different insights as well as several contradictory results. The direct approach to assessing relative costs via the cross-section experience curve analysis suggested much lower costs than did the indirect PIMS approach to the same question. This required further analysis to reduce the uncertainty surrounding pivotal assumptions. For example, the preliminary cost analyses indicated that one of the competitors had a much more productive sales force, and indeed only employed a third as many salespeople per equivalent sales volume. What was not evident were the possible hidden costs of this competitor's marketing strategy, including significant distributor discounts.

The strategic planning process would have surfaced the issue of distributor discounts sooner or later. However, it is unlikely that it would have been assigned such a high priority among the issues to be resolved had it not been the major reason for the disparity in the cost estimates derived from different analytical approaches.

Appropriate Level of Analysis. This is a further dimension for distinguishing among the methods of analysis. It is not entirely independent of the relevant role of the method in the planning process and helps us appreciate why some methods are constrained in their application. The judgments of which methods can be used in which circumstances reflect the inherent limitations in the methods and their capacity to provide useful insights.

Although the initial impression from Table 17-2 is that strategy analysis methods are widely applicable, a closer look shows that they have very little to say about such critical issues as:

Charting new business directions. Although the diagnosis may reveal the need for an expanded business definition or diversification into a new business, none of the models says much about where to look for such opportunities or how to assess them.

Managing shared resource units (such as pooled sales forces and R & D facilities) or taking advantage of synergies that arise when two business units serve related markets or a common technology.

Directing the specific functional actions and programs necessary to implement a strategy.

IMPLICATIONS FOR THE USE OF STRATEGY ANALYSIS METHODS

These methods are no substitute for insightful strategic thinking. Indeed, their greatest weakness is the superficial appeal of the generic prescriptions that may override the careful analysis of fundamentals that are the basis for competitive advantage. The details of the situation will always dominate the facile gen-

Table 17-2. Finding the Appropriate Level in the Organization to Apply Strategy Analysis Methods

Level	*Product Life-Cycle Analysis*	*Experience Curve Analysis*	*Decision Support Systems*	*Portfolio Models*	*PIMS*
Corporate			*	**	*
Strategic business unit	*	*	**	*	**
Shared resource unit		*	*		
Business segment (product market unit)	*	*	*	*	*
Product category	**	**	*		
Value-adding component		**	*		

eralizations of the methods. Yet without the structure provided by the analysis methods, it is often difficult to know which details require attention or ensure that managers have a common language for discussing and questioning assumptions.

Many assessments of the value of strategy analysis methods seem to miss this point by focusing solely on the prescriptions. A typical commentary concluded that

> clearly, the quantitative formula-matrix approaches to strategic planning developed by BCG in the 1960s are out of favor. . . . These overly quantitative techniques caused companies to place a great deal of emphasis on market-share growth. As a result, companies were devoting too much time to corporate portfolio planning and too little to hammering out strategies to turn sick operations into healthy ones or ensure that strong businesses remained strong. In too many instances, strategy planning degenerated into acquiring growth businesses that the buyers did not know how to manage and selling or milking to death mature businesses.[10]

In the same article a well-known planner characterized "formula" planning as "a search for shortcuts. . . . It took the thinking out of what you had to do to be competitively successful in the future."[11]

Blanket criticisms can be just as misleading as the simplistic formulas they decry when they confuse the potential contributions of the analysis methods with the undeniable problems of interpreting and applying their signals. What is needed is a change in emphasis rather than a complete break with the past. These changes will be evolutionary, reflecting what has been learned through trial and error to get the most value from the methods. Five directions for change are especially important.

Integrate the Methods into the Planning Process

Strategic decisions are shaped during an ongoing dialogue between functions and levels. The challenge is to blend the top-down corporate concerns with resource allocations and long-run industry position with the bottom-up understanding of specific product-market opportunities. There need not be full consensus on the decisions taken. There seldom is. But it is critical that all operating managers understand why the strategic direction was chosen and have a substantial commitment to changing their functional activities in line with the strategy.

Strategy analysis methods are no replacement for top-down and bottom-up dialogue but can facilitate it by providing frameworks for sharing and challenging strategic insights and assumptions. From shared understanding comes the commitment to action, which is an essential ingredient of effective implementation.

Watch for Misleading Signals

All methods and concepts impose discipline on strategic thinking through some degree of quantification. This is both a source of strength and a potentially crippling weakness. The problem is the dependence of the methods on input measures and assumptions that may themselves be inaccurate, invalid, or inappropriate. Bad measures will invariably lead to poor results. The measurement of market boundaries and structure, competitive positions, and costs are notorious problems. The problems may be more fundamental, for example, when the basic premises of the analytical models are violated or inappropriate. Informed use requires a good understanding of these pitfalls and may sensibly lead to the conclusion that the concept or model should not be used.

Avoid the Planning Priesthood Barrier

Resistance, if not outright sabotage, by line managers is assured when planners are viewed as an elite group with their own dogma and arcane language of matrices, PIMS models, and SBUs. Whether planners construct this barrier inadvertently in an effort to show their analytical prowess, or deliberately use independent data and mysterious models to second-guess line managers, their effectiveness can be markedly reduced. To avoid this situation, planners and their methods need to be mainstreamed into the early stages of plan formulation. The models and concepts should only be used when they are understood and accepted by the line managers. This requires a commitment to education in strategic thinking and the capabilities of the methods by the planners. Then planners should act as internal consultants, offering guidance and answering questions.

Focus on Exploitable Advantages

The source of a competitive advantage for a business is often found in the unique complexities of a market and the skills and resources that have been assembled to exploit that environment. Yet by design, strategy analysis methods are myopic about these specific possibilities. The narrowing of vision comes about from the need to put order into complex situations. This means using measures that can be generalized across many different businesses. When these simplications are not recognized, there is a strong temptation to derive equally generalized strategy statements that are not relevant to the business. Such statements as the following provide little useful guidance:

> Dominate the market for zippers through aggressive capacity expansion.
> Gain market share through development of a position as the golf ball industry's low-cost producer.
> Manage the business for cash as it proceeds from maturity to decline over the planning horizon.

Such generic strategy prescriptions have a very limited role to play in the planning process and should be avoided by all users of planning concepts and methods. The argument against formula-based planning that yields these prescriptions has been made forcefully by Carroll in a critique of planning manuals that force all strategies into a common mold.

> The author of the manual has attempted to reduce all the possible courses of action for any company (in any business, in any competitive situation) into a very limited number of options. Each is given a name. A simple decision rule then guides the company into the appropriate category. [However] suppose customers rate service important . . . and installed base has a pronounced effect on the ability to supply service at a low delivered cost. In fact, it may be the *density*, not the *size* of the installed base that is predictably related to cost of service calls. Part of the firm's position and its strategy should reflect this relationship. The measure, "geographic density of installed base" is not permissible because it has no relevance in other businesses. The specific exploration of this relationship and its clear articulation are not encouraged.[12]

Although this is a parody of planning in multidivisional companies the warning should be heeded. The methods and concepts of strategy analysis can only serve as a means to an end, which is clear thinking about the possibilities for gaining and sustaining advantage.

Challenge the Boundaries

Planning methods help structure situations by placing boundaries around markets and organizational units. Although this is an essential step, it can also cause myopia unless it is recognized that these are arbitrary lines drawn for convenience. Some of the biggest strategic pay-offs come from stretching these boundaries by entering new products or markets or spanning boundaries to share skills and resources between business units. Indeed, diversified firms are only greater than the sum of their parts when these interdependencies are nourished.

SUMMARY

Strategy analysis methods and concepts have many roles to play in support of the planning process. However, their benefits will only be obtained through informed use that is sensitive to their possibilities and limitations. These methods are not for every situation; part of the art of strategy lies in knowing when to use them to advantage.

APPENDIX 17A

THE PRINCIPLES OF THE 3M PLANNING SYSTEM

Characterizing the Industry: The first step of the planning process in any SBC (SBU) or operating unit is to characterize the industry. Managers need to identify factors that will influence the strategies they select—for instance, the size and growth rate of the industry. And what about outside influences—new technology, regulatory changes, and competitive conditions? Who are the key competitors? What share of the market do they hold? What strategies have they adopted?

Determining Key Success Factors: In order to formulate effective strategy 3M stresses accurate evaluation of the basis of competition. This means understanding the key success factors of a particular industry—the factors that determine ultimate success or failure. For each 3M business, managers need to ask, "Where do I get the competitive edge for long-term success in my industry?" One way of getting these answers is to rank a 3M business against its major competitors in an industry to develop a measure of its relative competitive position. Reflecting the corporate strategy, 3M occupies a strong position in many of its industries, while in others a secure product or market niche provides a favorable position, defensible against competitors with broader product lines.

Selecting Business Strategy: The strategic condition of an SBC strongly influences appropriate strategies. For example, in embryonic industries 3M businesses emphasize market or product-oriented strategies to improve their competitive position. As their industries mature, these businesses will very likely emphasize integration, efficiency, and rejuvenation through innovation. In the late stages of maturity, they'll move forward to consolidation and disinvestment. The overall strategic position of each 3M SBC is typically displayed in the form of a maturity/competitive position exhibit. For a particular SBC or operating unit the product portfolio (Figure Four) or market portfolio (Figure Five) indicate different strategic positions.

Maturity Stage

Competitive Position	Embryonic	Growth	Mature	Aging
Leading				
Strong			○	Rare Earth X-ray
Favorable				
Tenable	○	Digital Processing Enhancement		○
Weak				

Diagnostic Imaging Example

Figure Four. Product Portfolio (Illustrative).

Reprinted with permission from M. Tita and R. Allio, "3M's Strategy System—Planning in an Innovative Organization," *Planning Review* **12**(1984):10-15.

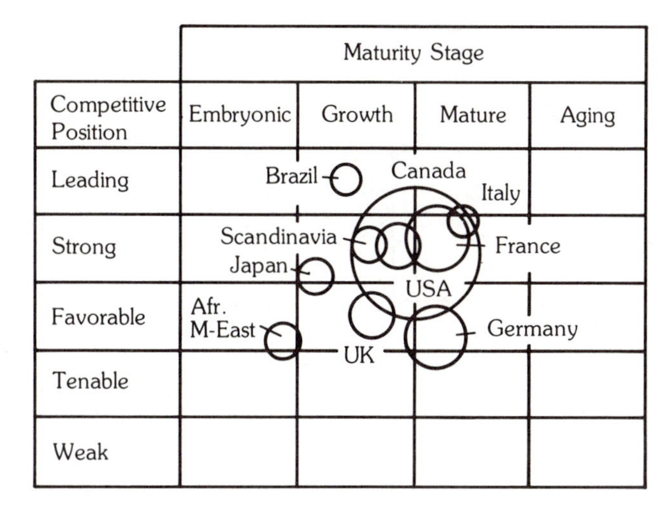

Figure Five. Strategic Condition of 3M Business by Geographic Market (Illustrative).

At this point in the planning process, implementation receives priority. 3M believes that the most eloquent strategy statement isn't much use if it's not part of the day-to-day operations of the business. This means that each SBC strategy must be backed up by appropriate action programs that identify the necessary costs, specify expected results, and establish schedules and responsibilities.

Every SBC is made up of individual operating units. These are usually divisions or departments responsible for specific products or market segments. Some SBCs cover two or more divisions—for example, Tape comprises several divisions, departments, and projects; and Memory Media contains Data Recording Products and Magnetic A/V Products. Other SBCs, such as Pharmaceutical (Riker Laboratories), are a single operating unit. Some of the larger SBCs contain operating units whose main function is to manufacture or market products for other divisions within that SBC or other 3M units.

The planning task for each operating unit is to develop strategic plans that support the overall strategic approach of the SBC. They detail how the SBC strategies will be carried out over the next several years. The responsibility for developing each operating unit's plan rests with the division general manager or department manager and subsidiary management.

Because some organizations, such as the Specialty Chemical and Film Division, provide a critical support function to other SBCs, their plans are necessarily finalized after the other SBC plans. Corporate staff plans (such as R & D and Human Resources) are also completed after SBC plans, although a considerable amount of discussion and integration takes place throughout the planning cycle.

Measuring Strategic Performance: In determining the expected financial performance of an SBC, 3M managers assess the maturity of an industry, competitive position, and strategy. Although regular operating reviews scrutinize a number of financial health indicators, the key financial parameters for an SBC include return on capital, operating income, and cash flow. Figure Seven shows how management can estimate the financial consequences of various strategies.

Effective monitoring of performance is based on financial and nonfinancial strategy measurements. For example, market share is a strong indicator of success for a business attempting to penetrate new markets, as are product quality, distribution expansion, and

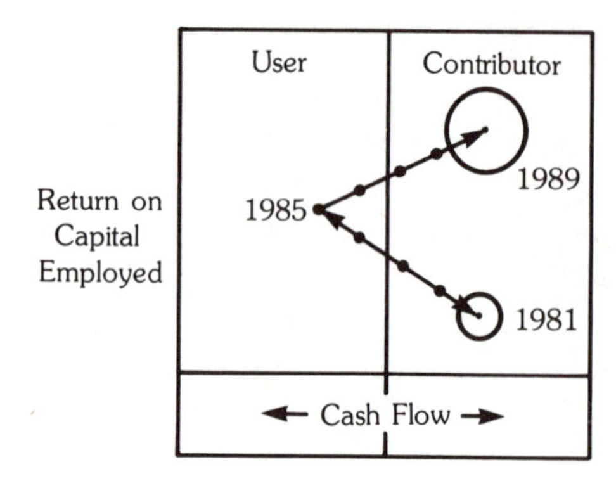

Figure Seven. Business Unit Financial Profile (Illustrative).

new products. As part of its planning process, each 3M SBC (and its component operating units) identifies a set of strategic measurements appropriate to its strategy. For many 3M businesses, this means that innovation and new product introduction are given high visibility.

Until recently, the corporation's rules demanded that every SBC grow at the same rate each year, that every SBC have 25 percent new products, and that every SBC target the same profit. Although the results were easy to monitor, needless to say, not every business could meet these demands. Now, the old mandate is being replaced by a measurement and reward system consistent with the strategy of each SBC, although overall corporate targets have not changed.

The final step in preparing the strategic plan is assessing risk—what's the probability of the plan's being successfully implemented? A number of factors come into play here, including the characterization of the industry and the competitive position of the business within that industry.

REFERENCES

1. Good examples of the enthusiastic literature of this era are "The New Planning," *Business Week* (19 December 1978):78-83; R. Cushman, "Corporate Strategy: Planning for the Future," Paper presented to the North American Society of Corporate Planners, Boston, October 1978; W. Loomis, "Strategic Planning in Uncertain Times," *Chief Executive* (Winter 1980-1981).
2. P. Haspeslagh, "Portfolio Planning: Uses and Limits," *Harvard Business Review* **60**(1982):59-73.
3. Representative criticisms can be found in W. Kiechel, "The Decline of the Experience Curve," *Fortune* (5 October 1981):139-146; S. Palesy, "Motivating Line Management Using the Planning Process," *Planning Review* **8**(March 1980):3-10; W. K. Hall, "Survival Strategies in a Hostile Environment," *Harvard Business Review* **58**(1980):75-85.
4. G. S. Day and D. B. Montgomery, "Diagnosing the Experience Curve," *Journal of Marketing* **47**(1983):44-58.
5. K. Ohmae, *The Mind of the Strategist* (New York: McGraw-Hill, 1982), 151.
6. G. S. Day, A. B. Shocker, and R. Srivastava, "Customer-oriented Approaches to Identifying Product-Markets," *Journal of Marketing* **43**(1979):8-19.
7. P. G. W. Keen and M. S. S. Morton, *Decision Support Systems: An Organizational Perspective* (Reading, Mass.: Addison-Wesley, 1978).
8. B. T. Gale, "Planning for Profit," *Planning Review* **6**(1978):4-7, 30-32.
9. B. Hedley, "Strategy and the Business Portfolio," *Long Range Planning Journal* **10**(1977):9-15.
10. "The New Breed of Strategic Planner," *Business Week* (17 September 1984):63.
11. Ibid.
12. P. Carroll, *Business-Specific Strategy* (Chicago: Hayes/Hill, 1984), 24.

CHAPTER 18

Portfolio Models and Profitability: An Economic Analysis

Malcolm B. Coate

Malcolm B. Coate is currently an economist with the Bureau of Economics of the Federal Trade Commission. He received the Ph.D. in economics from Duke University after completing his dissertation entitled "The Boston Consulting Group's Strategic Portfolio Planning Model: An Economic Analysis." Dr. Coate has published works on portfolio models, competitive analysis, and industry profitability.

The goal of any strategic planning system should be to define the set of investment proposals that maximize stockholder wealth.[1] Such a planning system usually requires two levels of strategy. First, planning requires that the opportunities facing a business be analyzed and the investment strategies be identified. Second, the planning system must contain a method for comparing all the alternatives and choosing the optimal strategy. Both tasks are difficult for a single business firm.

For a diversified firm, the strategic planning problem is even more complex because the firm must simultaneously identify strategies for all its business units, subject to the firm's particular environmental constraints. Portfolio models represent an approach that attempts to solve the planning problem with a structured method to compare alternative investments in different business units.[2] The portfolio system is based on an analysis of the competitive position and industry attractiveness of the various units that comprise the diversified firm. As Wensley notes, these models "are decision rules or heuristics for detecting areas of sustainable competitive advantage whereby the firm can realize economic profits."[3] Portfolio models define the appropriate strategies if their assumptions correctly approximate the environment of the firm. However, inappropriate assumptions can lead to suboptimal investment plans. Thus, a strategic planning system based on the portfolio methodology must identify the instances where the assumptions of a portfolio model are invalid and adjust the relevant strategies to ensure that the firm implements the opti-

Occasional paragraphs are adapted with permission from M. B. Coate, "Pitfalls in Portfolio Planning," *Long Range Planning Journal* **16**(1983):47-56. The analyses and conclusions set forth in this chapter are those of the author and do not necessarily reflect the views of other members of the Bureau of Economics, other Commission staff, or the Federal Trade Commission.

mal investment plans. In this analysis, I discuss various problems with the portfolio methodology to assist firms in implementing an optimal planning system.

The first section identifies the key assumptions of the various portfolio models and presents a brief overview of the general planning methodologies. Then, the problems with the individual portfolio model assumptions are discussed in detail and the relevant economic theories noted. I conclude by summarizing the limitations and advantages of the portfolio construct.

COMPARISON OF THE PORTFOLIO MODELS

A number of portfolio planning models are available to structure a diversified firm's search for the optimal set of business strategies. These models all incorporate assumptions on the independence of the business unit, the cost of capital, the value of dominating a market and the attractiveness of investing in a particular industry. The various models present a firm's portfolio of business units in matrix form and suggest general strategies for the particular business units. However, the models differ with respect to the actual measures of dominance and attractiveness. Following is a brief summary of the key points of the particular models and a comparison of their strategic implications for a diversified firm.

The Boston Consulting Group (BCG) defines a straightforward two-by-two matrix with dominance measured by relative market share and attractiveness approximated by the industry growth rate.[4] Then the matrix position of each unit is used to highlight the appropriate strategy for the business. The business-screen models of McKinsey & Co. (MC), General Electric (GE), and Shell Oil Company (Shell) all utilize similar three-by-three matrices.[5] Dominance is usually measured by share, firm growth, and technological, marketing, production or other competitive advantages. The attractiveness of an industry depends on market size, growth, structure, profits, and environmental characteristics. The various models use different weights to aggregate the measures of dominance and attractiveness into indexes for positioning the businesses in a portfolio matrix. The portfolio matrix is then used to suggest the appropriate allocation of resources in the corporation. Arthur D. Little (ADL) presents a twenty-cell matrix identified by the competitive position of a business and its industry maturity.[6] Competitive position is approximated by market share, share movement, technology, breadth of the product line, and special market advantages, and industry maturity is measured by considering industry growth, rate of technical change, stability of share, and customer switching. Again, weights must be defined to calculate the matrix position of a particular business. The matrix location of each unit can then be used to formulate a natural strategy to accomplish the business goals of the firm.

Each portfolio model requires the firm to calculate a single measure of market dominance and industry attractiveness to position business units in a portfolio matrix. The weighting scheme can be explicit as in the case of the BCG and Shell matrices or left to the user as in the case of the other models. We can even consider the BCG model to be a special case of the general models in which the user defines weights of one for relative share and industry growth and zero for all other variables. Once the matrix position has been assigned, all the models suggest strategies for the particular business units.

The investment strategies projected by each model are similar.[7] A firm should usually invest in its strong units in attractive industries, maintain its strong units in less attractive markets, and minimize investment (or even divest) its weak units in unattractive industries. Either an investment or divestment strategy is suggested for weak units in attractive markets. The individual investment strategies may have to be adjusted due to the lack of investment funds or the actions of competitors. However, the models are not usually clear concerning when and how the strategies should be changed. Overall, the portfolio models attempt to balance the corporate portfolio between growth and stable business units to ensure both current and future profits. Although the general investment plans of the portfolio models are comparable, the particular strategies for an individual unit are not necessarily the same.

The strategies, defined by the various models, can differ for three reasons.[8] First, the portfolio position of a unit in the relevant matrix can differ because different characteristics are used in the three classification schemes. The business strategy projected by each model will then vary with the unit's matrix position. Next, the ADL and business-screen models contain more potential classifications than the BCG model so they can generate different potential strategies for a particular unit. For example, a strong business in a marginally attractive industry could be chosen for heavy investment in the BCG model and selective investment in the ADL or MC models. Finally, the models differ with respect to their treatment of risk. The business-screen and ADL models both try to incorporate some concept of risk in the matrix position of the unit, whereas the BCG model downplays risk. Thus, the investment strategy for a risky unit can differ among the models. We can conclude that a one-to-one relationship will not exist for all the strategies advocated for a set of business units by the various portfolio models.

A recent empirical study confirmed the importance of the different characteristics used in the model's portfolio matrices.[9] For comparison purposes, the matrix position of each unit was classified into one of the four BCG segments. The BCG, Shell, and ADL models gave the same strategic classification for only two of the fifteen business units in the sample. On the other hand, four business units were characterized as strong in attractive markets by one model, whereas the same business was characterized as weak in an unattractive industry by another model. Moreover, in three of these four cases, each of the three portfolio models managed to characterize the business unit in a different sector of the matrix. Wind et al. conclude that portfolio planning methodologies should incorporate tests of the sensitivity of the strategies to changes in the underlying portfolio model.[10]

Thus, the strategic implications of the portfolio methodology appear to be sensitive to the choice of the particular portfolio matrix. Since there is only one profit-maximizing strategy, a firm using a portfolio planning system must consider the sensitivity of an investment strategy to the choice of a particular portfolio model and attempt to choose the appropriate portfolio approach when the investment implications differ.

THEORETICAL PROBLEMS WITH PORTFOLIO MODELS

The portfolio models have achieved their widespread popularity without sufficient analysis of their underlying assumptions. Important questions can be raised concerning each assumption, therefore the portfolio models may not be generally applicable to a firm's business units.[11] In particular, the assumption of independent business units ignores significant theoretical problems with the divisibility of the firm into business units; the assumption for the cost of capital appears to represent capital rationing and fails to consider risk; the assumption of dominance tends to give excessive consideration to the relative competitive position of a business while minimizing the influence of other characteristics that affect profitability; and the assumption of industry attractiveness ignores special investment opportunities and downplays the importance of competitors' strategies. Any of these problems can make a particular business unit strategy suboptimal. I analyze the four problems and present the applicable economic theories in the following subsections. By considering how the assumptions fail to represent the competitive environment, it may be possible for a firm to adjust the portfolio-planning construct to fit its particular planning needs.

Business Unit Assumption

The portfolio concept requires the firm to be divisible into independent business units. This would require only that the diversified firm identify all the markets in which it competes as if it is simply a collection of stand-alone business units. However, most diversified firms attempt to create and manage synergy between business units and to provide them with valuable resources at attractive terms. Also, some production technologies generate more than one product. Therefore, line of business divisions may be difficult. I discuss these problems with the independent-business-unit assumption and comment on a possible solution that minimizes the impact of the interrelationships on the planning process. Then, I discuss how economic theory can assist the firm in defining markets for its business units.

The potential for synergy clearly complicates a planning system because the strategies of the synergistic business units are interdependent. Investments in an individual unit benefit the related units and these benefits should be recognized in the planning process. For example, digital clocks, digital clock radios, and electronic timers can all use the same basic technology. In fact, any product that uses similar components can benefit from cost reductions caused by shared experience. Yavitz and Newman note synergies can evolve from vertical relationships, full utilization of raw materials, and combined services.[12] These synergies should be incorporated in the planning system.

The corporate provision of a particular resource does not make line of business divisions more difficult unless the resource is subject to economies of scope with respect to some business units. Economies of scope evolve when a resource is shareable across activities because use in one activity does not preclude use in another activity.[13] For example, a research lab can be used by different businesses if the relevant research technologies need similar equipment. Economies of scope among a subgroup of the firm's business units serve to link the investment strategies of these units because the optimal level of the shared resource depends on the strategies of the particular units. Yavitz and Newman list shared resources such as low-cost capital, outstanding executives, corporate research, and centralized marketing.[14] These resources will not complicate the business definition process if they are shareable across all units. However, if one of these factors has a significant fixed-cost component that only applies to some of the firm's businesses, the economies of scope should be considered in the planning process.

Finally, some production processes generate two or more products. For example, a petroleum refinery can produce gasoline, kerosene, fuel oil, and residue from a barrel of crude. Although some substitution is possible, a refiner must coordinate the strategies of its various businesses. Also, many mining operations produce a number of different minerals. Investments to increase capacity in one product and not the others are not likely to be profitable. Thus, independent strategies cannot be defined for all possible business units.

Heany and Weiss suggest that related business units be aggregated into clusters for planning purposes.[15] They note clusters are not always obvious but do represent an "improved strategic concept with an attractive profit potential."[16] The adjustment of the planning process to consider clusters of business units represents one approach to integrating synergy, economies of scope, and joint production technology into a portfolio model. The cluster takes the place of the business unit as the basic planning unit in the portfolio matrix. Of course, single business clusters are possible if an independent strategy is feasible for the particular unit.

Any portfolio planning system requires a definition of the relevant markets in which the business units compete. An overly broad market definition will bias down the competitive position of a unit, whereas a narrow market definition will artificially enhance the position of a business. Day suggests that a firm employ several alternative definitions and observes that the process of resolving the differences among the various markets will improve a firm's understanding of its competitive environment.[17]

Economic theory can offer insight into the problem of market definition. Traditionally, economists have delineated markets based on gaps in the chain of substitutes. The cross-elasticities of demand and supply have evolved as tools to approximate the degree of substitutability between various products. Since the data necessary to measure the cross-elasticities rarely exists, a checklist of characteristics has been defined to approximate the demand and supply cross-elasticities. The factors that indicate nonsubstitutability (and separate markets) include the persistence of sizable price disparities, the presence of sufficiently distinctive characteristics that render a product suitable for a specialized use, the preferences of particular purchasers for a product, and the judgment of purchasers or sellers that the products do not compete. Unfortunately, this procedure is *ad-hoc* so the analyst could influence the market definition process. Recently, two new concepts of a market have been advanced. They are the optimal cartel approach of Boyer[18] and the 5-percent test of the Department of Justice (DOJ).[19] Boyer defines a market "as an ideal collusive group."[20] The collusive group contains only those firms whose acquiescence to the collusion generates significant profits for the cartel members. This definition may be too broad for planning purposes because all the firms in the market may not be close competitors. The DOJ defines a market as a group of products "such that a hypothetical firm that was the only present and future seller of those products could profitably impose a small but significant and nontransitory increase in price."[21] The test usually considers a price increase of 5 percent for one year sufficient. The DOJ rule may be too narrow for planning purposes because it seems to exclude small firms from the market when the firms cannot render a price increase unprofitable. Planning should take a long-run view of competition and consider these small firms as rivals. These new definitions could be used to create upper and lower bounds for the relevant market and then the firm could undertake an analysis of all the relevant information to determine the actual market.

Henderson, of the BCG, rejects all conventional market definitions because he believes that they lead to an erroneous concept of competition.[22] He suggests that each firm has a comparative advantage in dealing with specific customers so competition can only occur at the boundaries of the various customer groups. Thus, Henderson posits each firm will dominate the customer group or market for which it holds an advantage. However, it is not easy to predict exactly which customers will prefer a particular firm. Moreover, it is not clear how many customers will switch in response to changes in product characteristics. Since competing firms are trying to maximize their sales by offering customers the best product possible, the particular group boundaries can change significantly in response to a competitive move. In addition, the desires of particular customers can evolve over time. Thus, the traditional concept of a market (a gap in the chain of substitutes) appears to capture the notion of competition better than the Henderson definition because it allows a firm to compete for the business of all the customers in a market.

Cost of Capital Assumption

The portfolio models assume that the firm allocates the available investment funds among its business units primarily on the basis of their portfolio position. This assumption suggests that the cost of capital is infinite once the fixed investment resources are exhausted. In addition, the portfolio models seem to imply that the liquidation value of a business is irrelevant to the investment decision and that detailed consideration of the risk of a project is unnecessary. However, it is theoretically possible for a firm to obtain additional investment resources by selling stock in the equity market or by borrowing in the financial markets. Moreover, a firm can obtain capital by liquidating its ownership interest in a particular business unit. Finally, risk can affect the profitability of an investment so it should be considered in a planning model. Thus, the portfolio models do not seem to incorporate the appropriate capital market assumptions in the planning analysis. I evaluate the problems with the portfolio investment strategies caused by inappropriate assumptions on the limitation of investment funds, the liquidation value of a business, and the risk of an investment. In addition, I discuss alternative approaches that may be able to correct for the problems with the assumption on cost of capital.

Portfolio models appear to accept the standard capital-rationing model in which a firm can only obtain a fixed amount of investment capital.[23] Further investment funds are not available at any price. This assumption generates interdependence among the strategies of the firm's business units because heavy investment in one unit reduces the funds available to the other units. This restriction on investment funds could be caused by either an institutional constraint in the capital market or a cost of capital schedule that eventually rises very sharply. These justifications do not seem plausible in the long run because the existence of profitable investment opportunities should allow the firm to acquire additional funds at reasonable cost. Ellsworth notes "it is not the financial community that imposes capital constraints on corporations, but managements' own policies.[24] Thus, the investment limitation is difficult to justify for long-run corporate planning. If a firm had unlimited access to the capital market at a competitive price, a portfolio model only highlights the firm's most profitable investments. However, other investments, not suggested by a portfolio model, could still be profitable.

As noted above, the investment strategies accepted by the firm should maximize the return to stockholders. A critical factor in analyzing the profitability of an investment is the cost of capital. Investment strategies that generate returns above the cost of capital create value for the stockholders and should be accepted. On the other hand, projects that fail to cover the cost of capital destroy the value of the firm and should be rejected. The concept of net present value allows the firm to evaluate investment strategies with returns in the future. The net present value calculation discounts both cash outflows and inflows by the cost of capital to determine the value of the project at the present time. Any present value greater than zero means the investment generates returns above the cost of capital and a negative value implies the project fails to cover the cost of capital. This general net present value model may be able to identify profitable strategies rejected by the portfolio planning system due to the assumption of fixed investment resources.

Williams presents a competitive strategy valuation method to integrate the net present value system with the portfolio approach.[25] The basic idea requires the firm to calculate the net present value of each business unit under the strategy suggested by the portfolio model. Then, the sum of the calculated present values for each unit is compared to the stock market's value of the entire firm and the differences analyzed. In addition, the firm can calculate the net present value for any radical change in a business unit's strategy and incorporate the results in the decision-making process. Williams discusses four limitations of this approach: (1) the estimated matrix position of a unit may not be reliable if the market is not well identified; (2) the quality of information may be too poor for reliable profitability calculations; (3) accounting assumptions can distort the cash-flow estimates; and (4) the full economic value of investment in new business units may not be able to be calculated with the net present value method. Williams concludes "that by emphasizing the economic value of corporate strategy, managers can be especially selective as to where they choose to build and defend competitive advantage."[26] The net present value methodology should allow the portfolio models to drop the assumption on the limitation of investment funds because the profitability of any particular investment project can be evaluated without any reference to the profitability of other investment projects.

The portfolio models also seem to ignore the importance of the opportunity cost of a unit's assets because they do not consider the liquidation value of a business in the investment allocation decision.[27] The cost of capital on the liquidation value of a business defines the opportunity cost to the firm of maintaining the business unit. The cost basis of a unit's assets is irrelevant except to the extent that it approximates the market value of the business. This insight suggests that the firm needs to supplement the portfolio display with information on liquidation values. For example, a firm should consider liquidating any unit if its salvage value is high enough. Moreover, the firm should consider retaining any unit with a low liquidation value. However, any unit with a negative discounted cash flow should be abandoned as soon as it is feasible.[28] Thus, the firm's strategic decisions must consider the liquidation value of each business. The net present value method can be used in the analysis if the liquidation value is used for the value of the existing assets of the business.

Finally, the portfolio models do not give risk the appropriate attention. The BCG portfolio model only attempts to balance cash flow to sustain stable profits in the long run. This approach completely ignores the effect of risk on the cost of capital. On the other hand, the ADL model incorporates a more detailed consideration of risk. Wright notes that the matrix position of a unit influences risk, with weak units in the early stages of the life cycle as risky and dominant units in the late stages of the life cycle as safe.[29] However, this approach does not allow the risk of a unit to be affected by market conditions. Moreover, the firm is not able to incorporate the effect of risk diversification in the analysis. Thus, the ADL model does not consider the appropriate cost of capital. The business-screen models can give risk some attention in the choice of the industry attractiveness measures but share the same shortcomings as the ADL model. In conclusion, none of the portfolio models adequately deal with the question of risk because they do not explicitly consider the effect of risk on the business unit's cost of capital.

Economic theory postulates two approaches to incorporate risk in investment analysis by adding a risk premium to the cost of capital. The first approach uses the capital asset pricing model (CAPM) to model risk as the covariance of a firm's return with the market return.[30] Pure random risk is assumed to be diversified away by the firm's stockholders in the equity market. In addition, a conglomerate firm could diversify random risk if it contained enough units. Thus, the firm only requires a risk premium on the variability of returns caused by market changes.

The second approach to risk considers the overall variance of a firm's return.[31] In this model, the firm demands a risk premium related to expected variability of the unit's return with respect to the firm's average return. This model only ignores the random risk actually diversified away by the firm so the risk premium should be larger than the CAPM premium.

The choice between the CAPM and firm-risk measures depends on the ability of a firm's stockholders to diversify risk. A large publicly held firm should probably use the CAPM approach because the stockholders are not concerned with nonsystematic risk. On the other hand, a small conglomerate with a core of owner-managers could prefer the firm-risk measure because the stockholders cannot fully diversify their risk. However, note that both methods tend to generate the same results for large conglomerates because the corporation diversifies the random risk through its choice of business units. Once an approach is chosen, the risk measure can be used to define a risk premium for the cost of capital. In a portfolio analysis, high-risk strategies should be identified and carefully evaluated before they are accepted.

Derkinderen and Crum present a different approach to adjusting the portfolio methodology for risk. They construct a model that considers the risk of a strategy and the endurance of the firm.[32] Risk is defined as the variability of a firm's cash flow to contingent setbacks, and endurance is characterized as the capability of a company to marshal resources to ensure survival if a setback occurs. These two factors determine the ability of a firm to handle adverse situations arising from its investment strategies. Derkinderen and Crum discuss a number of factors that indicate the firm faces risk or endurance problems. Also, they present a number of strategy recommendations designed to improve the strategic position of a firm with either risk or endurance problems.

The Derkinderen-Crum model may not appropriately correct for risk because it does not seem to consider the stockholders' ability to diversify risk within the portfolio. If the CAPM assumptions can be applied to the firm, a definition of risk as the variance of a business unit's return with respect to the market return would correct the problem.

In conclusion, the assumptions on the cost of capital implicit in a portfolio model are not always valid. This suggests that some investment strategies will not be optimal. In particular, the firm must attempt to identify business units where investment plans were cancelled due to the hypothetical lack of funds, units with liquidation values not closely related to expected earnings, and units where the consideration of risk can change the optimal strategies. Then the net present value model, with the appropriate cost of capital, can be used to evaluate the relevant investment strategies to ensure that the firm earns a return at least equal to the cost of capital.

Value of Dominance

The portfolio models generally suggest that profitability is directly related to the dominance of a business in a market. The inference is strongest with the BCG model, whereas the general portfolio models tend to consider other characteristics of the competitive environment. However, even these general models seem to measure particular production, marketing, or technical advantages relative to the competition. Therefore, a significant number of the determinants of business strengths depend on a unit's dominance of a market.

The basic concept of the value of dominance is not universally accepted in economic theory. Most commentators believe that the profitability of a business unit depends on the entire industry environment.[33] Also, profitability can be affected by the particular strategy implemented by the firm. Moreover, the optimal level of investment in a unit depends on the marginal, not average, return. I discuss criticisms of the value of dominance assumption and present alternative theories of profitability.

The simple dominance assumption seems to imply that an industry's technology offers continual increasing returns to scale. This type of cost structure defines a *natural monopoly* so only one firm should exist in each industry. A casual analysis of the structure of any representative market in the economy shows that this conclusion is clearly incorrect. Various firms compete to supply similar products to consumers with no evidence that one firm is approaching a monopoly position. Thus, the value of dominance must be qualified to explain the continued existence of a number of firms in individual markets.

One possible explanation is that the value of dominance relationship is only relevant for the long-run cost associated with a particular technology. In the short run, a firm's costs could increase with output and limit the optimal size of the firm. In the long run, the low-cost firm would tend to grow and dominate the market. However, technical change could allow a new firm to assume the dominant cost position before all the existing competitors were driven out of business.

A more reasonable theory would assume that the profitability of a business depends on its entire competitive environment. Porter lists five factors that impact on competition in any industry.[34] They are the rivalry among existing firms, the threat of substitute products, the bargaining power of suppliers, the bargaining power of buyers, and the threat of new entrants. Regulatory restrictions, the elasticity of demand, and the firm's underlying cost structure could be added to this list. A complete competitive analysis should be able to estimate the potential profitability of a business.

A competitive analysis would be designed to identify the sources of comparative advantage in particular industries. These advantages can allow a firm to earn higher profits than its competitors in the long run or allow all the firms in an industry to earn relatively high profits in the short run. Carroll takes a long-run view of comparative advantage by defining it as a "factor or effect which allows one firm to offer a product or service more effectively than its competitors."[35] He lists four areas in which comparative advantages can be found: barriers, product differentiation, organization, and cost. Only cost appears to be compatible with the notion of a value of dominance. A differentiation strategy requires the firm to offer a product with superior technology, features, service, distribution channels, or other desirable characteristics. Consumers are willing to pay higher prices for differentiated products because they believe they are receiving a higher-value product. For the differentiation strategy to offer long-run advantages, barriers must exist to prevent competitors from imitating successful differentiation strategies. Porter observes a firm may also be able to obtain high returns by implementing a focus strategy that attempts to serve a particular niche or segment of the market.[36] The focus strategy hinges on the assumption that a firm is able to serve a narrow segment of the market more effectively or efficiently than its competitors who compete for the entire market. Thus, a firm can earn high profits by providing a better product at a higher price or producing the same product at a lower cost. Again, barriers must exist to block competitors from copying the focus strategy. The organization strategy suggests the firm concentrate on creating a superior management team to ensure its business units minimize cost. For example, management must choose the location of factories to minimize the combination of production and transportation costs. Once defined, the sources of comparative advantage can be used in a competitive analysis of the firm's business units. Carroll concludes that "the application of this form of competitive analysis results directly in a strategy."[37]

It also is important to consider factors that can affect the profitability of the entire industry in the short run.[38] For example, the market demand can shift in response to changes in the number of consumers, tastes, income, and the prices of substitutes or complements. Also, market supply is influenced by various factors that change the incentives of firms to produce. The appropriate changes in either demand or supply conditions will make participation in a market profitable for all firms in the short run. Eventually, the expansion of existing firms and the entry of new firms will return the industry profitability to a normal level.

Economists posit that all the firms in a market can earn high profits if they recognize their mutual dependence and refrain from active competition.[39] The supranormal profits will attract entry, but the structure of the market may delay entry long enough for the firms to earn superior returns. If the firms can obtain government protection to block entry, they can earn the supranormal profits in the long run.

On balance, an overall competitive analysis of an industry is necessary to project the profitability of a firm in an industry. In many industries, the dominant firm will tend to be profitable. However, in other industries, differentiation or focus strategies will allow a small firm to earn significant profits. Also, some unprofitable industries can be expected to offer minimal returns to even the dominant firm, whereas profitable industries can offer substantial returns to firms with a mediocre position. Thus, the profitability of a strategy or an industry may dominate the profitability of a firm's position in the industry and render the value of dominance concept invalid for some interindustry comparisons of business unit profitability.

One can argue that the business-screen model can capture all the relevant information on a business unit's environment in the portfolio position of the unit. However, Wensley notes "the grave danger of leading the analyst to the tautological position of recommending preferential investment in those areas of highest market attractiveness and strongest business position."[40] This implies that a broad interpretation of the business-screen model simply reduces the model to a search for areas of comparative advantage.

The value of dominance assumption can also generate suboptional investment strategies since it keys on average, not marginal, relationships.[41] The final investment project of a strategy in a profitable business can generate a low marginal return but appear profitable on average. Moreover, a small investment project in a relatively unprofitable business can yield a high marginal return. The transfer of investment resources from the low marginal return uses to high return uses would increase the firm's profitability. Therefore, the strategic implications of a portfolio model may have to be adjusted slightly to avoid overinvesting in some units and underinvesting in others.

Econometric analysis can be used to evaluate the value of dominance assumption. Phillips et al. note cost position was a significant determinant of business performance in five of six types of businesses in their study.[42] Product differentiation, as measured by product quality, was significant in all six types of businesses. The PIMS study also uncovers a relationship between both cost position (measured by relative market share) and differentiation (approximated by perceived quality) and return on investment.[43] Similar results were found with the Federal Trade Commission line of business data.[44] These findings support the hypothesis that both the dominance and differentiation strategies can lead to competitive success.

In addition to the econometric evidence, case studies suggest that some low-share business units are profitable. Hamermesh et al. studied Burroughs, Union Camp, and Crown Cork and Seal, all successful low-share businesses.[45] They found that the three firms only competed in selected segments of their markets, used research efficiently, avoided unrestrained growth, and had effective management. Woo and Cooper undertook a more ambitious study of 84 low-share businesses associated with the PIMS project.[46] They found 40 low-share firms that were profitable in stable business environments. The successful firms seemed to follow a focus strategy tailored to the competitive environment. Also, the profitable businesses usually maintained a high-quality product for sale at a moderate or low price. This led Woo and Cooper to suggest that successful low-share businesses tended to have low costs probably as a result of their narrow product line and conservative marketing and research expenditures. In conclusion, the profitable experience of the 40 low-share business units "demonstrates that success is possible for well-positioned and well-managed small share businesses."[47]

The value of dominance assumption probably represents a good first approximation for defining profitable business units. However, the differences in the competitive environment must be incorporated in a strategic planning system. In particular, the factors that make a particular industry profitable should be reviewed. Also, differentiation and niche strategies deserve consideration.[48] Finally, the distinction between average return and marginal return should be reflected in the planning system.

Industry Attractiveness

The various portfolio models all attempt to identify characteristics that proxy the investment opportunities in particular industries. The BCG model concentrates on the industry growth rate, whereas the ADL approach considers other factors that have an impact on the life-cycle position of the industry. The business-screen models consider additional factors beyond the industry's life-cycle position. For example, market size, growth, industry concentration, capital intensity, and the state of technology are used to approximate market attractiveness in one analysis.[49] Thus, the business-screen models come closest to defining a general attractiveness measure. However, even these models tend to overlook a few important determinants of industry attractiveness. In particular, special industry characteristics, the opportunities for new product introductions, and possibilities for process innovation are all given minimal attention in defining the attractiveness of a market. Also, attractiveness itself may depend on the strategies of a firm's competitors and may be difficult to project. These problems are described below.

All the portfolio models ignore certain special characteristics of a market that can generate attractive investment opportunities. For example, a low-growth cyclical market may be attractive if a firm can expand its position to meet the peak demand and compete on equal terms with the existing firms when the sales fall to normal levels. Also, a market may contain niches that are not being served by the current competitors. Thus, special investment strategies may be viable in relatively unattractive markets.

Second, R & D may form the basis of a viable investment strategy in any market. A firm with an innovative product can invest profitably in a business because it will offer consumers a superior quality product in comparison to the existing competitors. This implies that a research-based strategy can be used to build share in certain markets, regardless of the underlying industry attractiveness.

Third, research into the production process in a particular market may create investment opportunities. A firm with a low-cost technology can build share in any business because it holds a competitive advantage over its rivals. Thus, a viable process innovation strategy can convert an unattractive market into an attractive one.

Case study evidence shows firms can succeed in some unattractive industries. Hamermesh and Silk note three strategies that can be successful in stagnant industries.[50] First, a firm can try to identify and exploit growth segments of the market. Second, the firm can emphasize product quality and innovation. Third, the firm can try to improve the efficiency of its production and distribution system. Hamermesh and Silk also stress the importance of implementing the appropriate management system for a low-growth business. They conclude that competing in stagnant industries "can be profitable and slow growth can become an ally."[51]

The standard analysis of industry attractiveness incorporates minimal consideration of the strategies of competitors. Williams notes that an attractive industry is likely to invite competition, so these industries can require heavy investment to maintain competitive position.[52] It logically follows that even more investment is needed to gain share. Thus, the attractiveness of an industry can be affected by the strategies of the competitors in the market. For example, if four firms are already trying to build share in a market, the fifth firm should be wary of investment. On the other hand, if a firm's competitors are trying to exit a market, it may be possible to profitably build market share.

Day also recognizes the crucial importance of the plans of a firm's competitors in his discussion of industry attractiveness.[53] He notes that competitors may respond aggressively if their expectations are not met regardless of the industry growth rate. Day concludes that the attractiveness of a market depends on "the competitors: who they are, how they intend to compete and whether they have the capabilities to compete aggressively."[54]

Even if industry attractiveness approximates the investment opportunities in a market, one is left with the problem of projecting the future values of the attractiveness variables. For example, deregulation has drastically improved the attractiveness of the long-distance telephone business. Also, increases in foreign competition can change the future opportunities in any market. Thus, the firm will need a sophisticated data system to identify investment opportunities with a portfolio model.

In conclusion, portfolio models do not give sufficient attention to the question of market attractiveness. The portfolio approach may fail to identify all investment opportunities, making analysis to detect special investment opportunities necessary. In addition, the need to project competitor strategies and future attractiveness can affect every portfolio strategy. A detailed system of competitor and market intelligence is required to estimate the necessary information so the appropriate strategic responses can be formulated.[55]

SUMMARY

This chapter discusses the problems with using the basic portfolio concept as a guide for strategic planning. The analysis starts by comparing the three basic types of portfolio models. The BCG growth-share strategic planning system requires minimal underlying information (relative market share and the industry growth rate) for each business unit. Other portfolio models require more data to structure a general planning system. All the models use portfolio matrices to project the most profitable investment strategies for a firm's business units. However, the actual strategic implications can differ because the models use different variables to approximate the environment, incorporate a different degree of detail, and consider different concepts of risk.

The bulk of the chapter analyzes the fundamental assumptions of the portfolio construct. The assumptions for the divisibility of the firm into business units, the cost of capital, the value of dominance, and the attractiveness of investment in a particular industry are all subject to criticism. A conglomerate firm may not be divisible into independent business units if synergies exist between units, economies of scope allow some units to share costs, or joint production technologies exist. The cost of capital theory implicit in the portfolio models assumes capital rationing exists, ignores the liquidation value of a business, and minimizes the importance of risk. The value of dominance assumption downplays a general competitive analysis of a market, tends to ignore differentiation or niche strategies, and fails to consider the marginal return on investment. Finally, the industry attractiveness assumption rejects the possibility of special market characteristics, new product introductions and process innovation changing the attractiveness of an industry, and fails to consider fully the implications of competitor's strategies for a market.

Even with its limitations, the portfolio construct represents a useful planning tool. First, the portfolio models utilize the concept of the business unit, which highlights the need for the diversified firm to plan strategies for each of its markets. Second, the cost of capital assumption forces the firm to address the issue of capital rationing and can lead it to consider the foregone profits from rejecting investment plans instead of acquiring the necessary investment resources. Third, the value of dominance and industry attractiveness assumptions focus the planning process on the economic value of a strategy to the firm. However, as Wensley notes "it is essential that any major project is assessed independent of its box classification."[56] Such an assessment should consider both the competitive environment of the market and the comparative advantages of the particular business unit. Moreover, the analysis should consider the effects of the strategies employed by a firm's competitors on industry attractiveness. Day notes that "portfolio models structure and simplify complex situations, and focus on important rather than unimportant issues."[57] Thus, the careful use of a portfolio model may improve the planning process.

In conclusion, the major shortcoming of the portfolio framework is its failure to apply to all business units. This problem can be offset partially by focusing on the competitive environment facing individual units. The integration of competitive analysis into a portfolio planning system should prevent a system from becoming a substitute for strategic thinking.

NOTES AND REFERENCES

1. A. Rappaport, "Selecting Strategies That Create Shareholder Value," *Harvard Business Review* **59**(1981):139-149.
2. The portfolio models implicitly address the issue of investment strategies because the portfolio position of a unit influences the viability of particular strategies.
3. R. Wensley, "Strategic Marketing: Betas, Boxes or Basics," *Journal of Marketing* **2**(1981):174.
4. D. Abel and J. Hammond, *Strategic Market Planning* (Englewood Cliffs, N.J.: Prentice-Hall, 1979).
5. Ibid.; M. Coate, "Pitfalls in Portfolio Planning," *Long Range Planning Journal* **16**(1983):47-56; S. Robinson, R. Hichens, and D. Wade, "The Directional Policy Matrix—Tool for Strategic Planning," *Long Range Planning Journal* **11**(1978):8-15; Y. Wind and V. Mahajan, "Designing Product and Business Portfolios," *Harvard Business Review* **59**(1981):155-165.
6. R. Wright, *A System for Managing Diversity* (Arthur D. Little, 1974); P. Patel and M. Younger, "A Frame of Reference for Strategy Development," *Long Range Planning Journal* **11**(1978):6-12.
7. Coate, "Pitfalls in Portfolio Planning."
8. Ibid., 52.
9. Y. Wind, V. Mahajan, and D. Swire, "An Empirical Comparison of Standardized Portfolio Models," *Journal of Marketing* **47**(1983):88-99.
10. Ibid.
11. Coate, "Pitfalls in Portfolio Planning."
12. B. Yavitz and W. Newman, "What the Corporation Should Provide Its Business Units," *Journal of Business Strategy* **3**(1982)14-19.
13. J. Waldron, "Strategic Management: An Application of the Theory of Multiproduct Firm Behavior," Ph.D. diss., Duke University, 1983.
14. Yavitz and Newman, "What the Corporation Should Provide."
15. D. Heany and G. Weiss, "Integrating Strategies for Clusters of Businesses," *Journal of Business Strategy* **4**(1983):3-11.
16. Ibid., 11.
17. G. Day, "Diagnosing the Product Portfolio," *Journal of Marketing* **41**(1977):29-38.
18. K. Boyer, "Is There a Principle for Defining Industries," *Southern Economic Journal* **50**(1984):761-770.
19. U.S. Department of Justice, *Merger Guidelines* (Washington, D.C.: Government Printing Office, 1984).
20. Boyer, "Is There a Principle for Defining Industries," 763.
21. U.S. Department of Justice, *Merger Guidelines,* 5.
22. B. Henderson, "The Application and Misapplication of the Experience Curve," *Journal of Business Strategy* **4**(1984):3-9.
23. Coate, "Pitfalls in Portfolio Planning."
24. R. Ellsworth, "Subordinate Financial Policy to Corporate Strategy," *Harvard Business Review* **61**(1983):174.
25. J. Williams, "Competitive Strategy Valuation," *Journal of Business Strategy* **4**(1984):36-46.
26. Ibid., 46.
27. Coate, "Pitfalls in Portfolio Planning."
28. This conclusion assumes that the unprofitable business unit does not create significant synergy for the firm.
29. Wright, *System for Managing Diversity.*
30. W. Sharpe, "Capital Asset Prices: A Theory of Market Equilibrium under Conditions of Risk," *Journal of Finance* **19**(1964):425-442.
31. This approach generalizes the simple project risk analysis to allow for diversification within the firm.
32. F. Derkinderen and R. Crum, "Pitfalls in Using Portfolio Techniques—Assessing Risk and Potential," *Long Range Planning Journal* **17**(1984):129-136.
33. M. Porter, *Competitive Strategy: Techniques for Analyzing Industries and Competitors* (New York: Free Press, 1980); Coate, "Pitfalls in Portfolio Planning,"; Wensley, "Strategic Marketing."
34. Porter, *Competitive Strategy;* M. Porter, "How Competitive Forces Shape Strategy," *Harvard Business Review* **57**(1979):137-145.
35. P. Carroll, "The Link between Performance and Strategy," *Journal of Business Strategy* **2**(1982):10.
36. Porter, *Competitive Strategy.*
37. Carroll, "Link between Performance and Strategy," 11.
38. T. Naylor, ed., *Portfolio Planning and Corporate Strategy* (Oxford, Ohio: Planning Executive Institute, 1983), 109-125.
39. Ibid., 123.
40. Wensley, "Strategic Marketing," 177.
41. Coate, "Pitfalls in Portfolio Planning."

42. L. Phillips, D. Chang, and R. Buzzell, "Product Quality, Cost Position and Business Performance: A Test of Some Key Hypotheses," *Journal of Marketing* **47**(1983):41.
43. S. Schoeffler, R. Buzzell, and D. Heany, "Impact of Strategic Planning on Profit Performance," *Harvard Business Review* **52**(1974):137-145.
44. D. Ravenscraft, "Structure-Profit Relationships at the Line of Business and Industry Level," *Review of Economics and Statistics* **65**(1983):22-31.
45. R. Hamermesh, J. Anderson, and J. Harris, "Strategies for Low Market Share Business," *Harvard Business Review* **56**(1978):95-102.
46. C. Woo and A. Cooper, "The Surprising Case for Low Market Share," *Harvard Business Review* **60**(1982):106-113.
47. Ibid., 113.
48. The firm could try to define the segment that it serves with a differentiation or niche strategy as a market, but this policy runs the risk of creating markets without economic rationale.
49. Abel and Hammond, *Strategic Market Planning*.
50. R. Hamermesh and S. Silk, "How to Compete in Stagnant Industries," *Harvard Business Review* **57**(1979):161-168.
51. Ibid., 168.
52. Williams, "Competitive Strategy Valuation."
53. G. Day, "Gaining Insight through Strategy Analysis," *Journal of Business Strategy* **4**(1983):51-58.
54. Ibid., 55.
55. I. MacMillan, "Seizing Competitive Initiative," *Journal of Business Strategy* **2**(1982):43-57.
56. Wensley, "Strategic Marketing," 181.
57. Day, "Gaining Insight through Strategy Analysis," 57

CHAPTER 19

Portfolio Planning: Uses, Limits, and Guidelines

Philippe Haspeslagh

Philippe Haspeslagh is assistant professor of business policy at the European Institute of Business Administration (INSEAD) in Fontainebleau, France, and director of its executive program on strategic issues in mergers and acquisitions. This chapter is based in part on his dissertation on portfolio planning approaches and the strategic management process in diversified industrial companies.

The diversity of large industrial—and mostly multinational—corporations can be at once their greatest source of competitive advantage and the wellspring of their most fundamental difficulties. Diversity provides an opportunity for these companies to use cash flow generated by their mature basic businesses to gain new leadership positions. Internally, however, this same diversity also creates a managerial gap between the corporate level, which has the power to commit resources but often only a superficial knowledge of each business, and the business level, where managers have the substantive knowledge required to make resource allocation decisions but lack the big corporate picture. Corporate managers may often feel they are too far away to see the trees yet standing too close by to take in the forest.

In the late 1970s, a new generation of strategic planning approaches called portfolio planning spread across a wide range of companies in response to the problems and prospects of managing diversity. On the basis of my survey, I estimate that, as of 1979, 36 percent of the *Fortune 1000* and 45 percent of the *Fortune 500* industrial companies had introduced the approach to some extent. Each year during the previous five years another 25 to 30 organizations joined the ranks.

Advocated by consulting firms such as the Boston Consulting Group, McKinsey, and Arthur D. Little and touted by organizations such as General Electric, Mead, and Olin, portfolio planning has struck the minds of many corporate executives. They speak a new strategic language and set up scores of bubble charts to explain their enthusiasm in corporate boardrooms. Most important, however, portfolio planning seems to have profoundly affected the way executives think about the management of their companies.

But what is all the fuss about? What is portfolio planning—as preached—and as practiced? How widespread is its application? What are the problems with its implementation? Does it really work? Or is it just another

Adapted with permission from P. Haspeslagh, "Portfolio Planning: Uses and Limits," *Harvard Business Review* **60**(1982):59-73.

set of words that consultants have sold to top management—words that must be learned but that are then easily forgotten?

I set out to investigate the impact of portfolio planning and its implications for corporate administration in a survey of *Fortune 1000* companies sponsored by the *Harvard Business Review* (see Table 19-1 for list of companies that agreed to be identified as participants). From subsequent conversations and interviews with planners, financial officers, and CEOs, I found that portfolio planning approaches are widespread among large diversified industrial companies and being increasingly introduced.

There seem to be some limits to the practice of portfolio planning as well as tremendous—and sometimes latent—opportunities. Among the more serious limits are that

- The road to portfolio planning is a long one; therefore, companies often get stuck trying to implement it and cannot realize the full potential of the approach.
- If a company looks on portfolio planning as merely an analytic planning tool, it will not realize its benefits.
- In implementing portfolio planning, companies often write in biases that block its usefulness, including the tendency to focus on capital investment rather than strategic expense allocation—or human resource management.
- Portfolio planning seems unable to successfully address the issue of new business generation.

Despite these difficulties, the corporate managers surveyed want to press ahead with portfolio planning, largely because the approach

- Promotes substantial improvement in the quality of strategies developed at both the business and the corporate level.
- Produces selective resource allocation.
- Provides a framework for adapting their overall management process to the needs of each business.
- Furnishes companies with a greatly improved capacity for strategic control when portfolio planning is applied intelligently and with attention to its pitfalls.

Before I explain the findings in detail, I briefly review the challenge facing diversified companies, the reason behind the widespread application of the portfolio planning approaches, and the characteristics they have in common. Then I delve closely into the findings before I offer some advice on the introduction of the technique and speculate on its future.

THE CHALLENGE: HOW BEST TO MANAGE DIVERSITY

The basic challenge for the modern corporation lies in the sheer number of businesses over which it holds sway. Managers of large companies in the 1980s cannot possibly be familiar with all the relevant strategic aspects of each business.

Faced with this challenge, companies react in two ways. They may seek a substantive solution and simplify the problem by limiting their activities to businesses that are easy to comprehend or that share a common strategic logic, or, to avoid the complexity of managing interrelatedness, they may treat their businesses as stand-alone units.

Usually, however, companies tackle the problem by trying to develop more administrative capability. The typical organization creates intermediate organizational levels (groups or sectors) and uses intermediate managers and administrative systems to measure, evaluate, and reward performance. Yet for all their sophistication, modern companies still experience difficulty in managing diversity.

Often top management is aware only of the short-term financial performance of its businesses (and even that is buried in the fragmentation of profit centers and the aggregation of reporting structures). Senior executives often end up delegating major decisions, which then become based on individual track records and managerial influence and heavily weighted by short-term career risks. Corporate top management becomes actively involved when dramatic across-the-board

Table 19-1. Participating Companies That Agreed to Be Identified as Participants

Out of 57 participating companies in Fortune "1-100"		*Out of 45 participating companies in Fortune "101-200"*	
Exxon	Armco	Eaton	Hercules
Mobil	Greyhound	American Cyanamid	Gould
Gulf Oil	Colgate Palmolive	NCR	Owens-Corning
Atlantic Richfield	W.R. Grace	Celanese	Control Data
Shell Oil	PepsiCo.	American Motors	Allis-Chalmers
E.I.du Pont de Nemours	Deere	Texas Instruments	Martin Marietta
Union Carbide	Aluminum Company of	Crown Zellerbach	Scott Paper
Phillips Petroleum	America	Borg-Warner	Pillsbury
Dow Chemical	Weyerhaeuser	Mead	Levi Strauss
Westinghouse Electric	TRW	Frehauf	Johns-Manville
United Technologies	Sperry Rand	General Tire & Rubber	Koppers
Rockwell International	Republic Steel	Whirlpool	Del Monte
Kraft	Allied Chemical	Avon Products	Olin
Monsanto	Inland Steel	Charter	Land O'Lakes
R.J. Reynolds Industries	General Mills		Studebaker-Worthington
Firestone Tire & Rubber	CPC International		
Cities Service			
Out of 45 participating companies in Fortune "201-300"			*Out of 30 participating companies in Fortune "301-400"*
U.S. Industries	Joseph E. Seagram & Sons	Certain-Teed	Consolidated Aluminum
International Minerals &	Diamond International	Blue Bell	General Host
Chemical	Air Products & Chemicals	Rexnord	Joy Manufacturing
Emhart	Baxter Travenol	Avnet	Pitney-Bowes
Stauffer Chemical	Zenith Radio	AMP	Outboard Marine
Rohm and Haas	Clark Oil & Refining	St. Joe Minerals	Parker-Hannifin
Corning Glass Works	Reliance Electric	Potlatch	Newmount Mining
Armstong Cork	Norton		AM International
Murphy Oil	Black & Decker		Hughes Tool
Meublein	Pennwalt		Crown Central Petroleum
Lear Siegler	Interlake		Cincinnati Milacron

(continued)

Table 19-1 (*Continued*)

Out of 30 participating companies in Fortune "301-400"	*Out of 35 participating companies in Fortune "401-500"*	*Out of 26 participating companies in Fortune "501-600"*	*Out of 29 participating companies in Fortune "601-700"*
Memorex	GATX	Armstrong Rubber	Shaklee
Signope	Dan River	Kohler	Universal Foods
Hart Schaffner & Marx	Louisiana Land & Exploration	Barnes Group	Graniteville
Johnson Controls	Washington Post	Wyman-Gordon	Franklin Mint
Norin	Wallach Murray	Keystone Consolidate	Alton Box Board
General Cinema	Ball	Sonoco Products	Prentice-Hall
Southwest Forest industries	American Bakeries	Beckman Instruments	Kaiser Cement & Gypsum
Saxon Industries	Envirotech	Maremont	Mississippi Chemical
Cluett, Peabody	New York Times	Houston Oil & Minerals	Sealed Power
Bell & Howell	Scott & Fetzer	Maytag	Media General
ConAgra	Smith International	Ametek	Mine Safety Appliances
Fairchild Industries	Wm. Wrigley Jr.	Dorsey	Raychem
	Bausch & Lomb	Allied Products	Condec
	Mattel	American Greetings	Toro
	Coca-Cola Bottling Company of New York	Snap-on Tools	Mitchell Energy & Development
	Maryland Cup	National Cooperative Refinery Assn.	H.B. Guller
	Arcata	Nucor	Crompton & Knowles
	General Refractories	Coachmen Industries	Alaska Interstate
	Butler Manufacturing	Freeport Minerals	Standard Register
	Varian Associates	Robertshaw Controls	
	Tyler	Albany International	
	Royal Crown Companies	Longview Fibre	

Out of 27 participating companies in Fortune "701-800"	*Out of 27 participating companies in Fortune "801-900"*	*Out of 24 participating companies in Fortune "901-1000"*
Commercial Shearing	Wolverine Worldwide	Safeguard Industries
Chamberlain Manufacturing	Lenox	Buckeye International
Management Assistance Corp.	Omark Industries	American Business Products
Mark Controls	Richardson	William Carter
Giddings & Lewis	Clow	Conrac
Norlin Industries	Reliance Universal	Cooper Laboratories
Chelsea Industries	Hesston	Marlene Industries
Western Gear	Mesa Petroleum	Imperial Sugar
Stanley Home Products	Datapoint	Robin Industries
Ludlow	Pope & Talbot	Stride Rite
Scott Foresman	Gullford Mills	Martin Processing
RTE	Affiliated Publications	P.H. Glatfeleter
Gould Pumps	Park-Ohio Industries	Courier
Quanex	American Sterilizer	Medalist Industries
Wean United	Wynn's International	Marion Laboratories
Medtronic	Ocean Spray Cranberries	Motch & Merryweather
Barber-Greene	Combustion Equipment	Machinery
Commerce Clearing House	Associates	Facet Enterprises
Chesapeake Corp. of Virginia	Farmer Brothers	Harper & Row
Atlantic Steel	Mohawk Data Sciences	Clevepak
Huffy	Bobbie Brooks	Polychrome
Texfi Industries	Union	Multimedia
	Cubic	
	Russ Togs	
	Elcor	
	Cross	

moves are called for. The uniformity of administrative systems indeed makes it very difficult to escape uniform pressures across all businesses.

As a result, a range of conflicts buffets almost any company. Even supposedly well-run organizations oscillate between periods of uniform emphasis on profits and emphasis on growth—often coinciding with the tenure of a particular CEO. What is needed to counter these problems is a management system that provides (1) corporate-level visibility of performance on both strategic and financial terms; (2) selectivity in resource allocation; and (3) differentiation in administrative attention among businesses.

The Essence of Portfolio Planning: How It Helps

Portfolio planning recognizes that diversified companies are a collection of businesses, each of which makes a distinct contribution to the overall corporate performance and that should be managed accordingly. Putting the portfolio planning philosophy into place takes three steps as the typical company

1. Redefines businesses for strategic planning purposes as strategic business units (SBUs), which may or may not differ from operating units.
2. Classifies these SBUs on a portfolio grid according to the competitive position and attractiveness of the particular product market.
3. Uses this framework to assign each a *strategic mission* with respect to its growth and financial objectives and allocates resources accordingly.

The approach, then, allows management to see business performance as largely determined by the company's position within the industry. Companies theoretically can assess the strategic position of each of their enterprises and compare these positions using cash flow as the common variable. A verbal and graphic language facilitates communication across organizational levels. Finally, the approach helps build a framework for allocating resources directly and selectively and for differentiating strategic influence.

Focusing Debate on the Real Issues

Given the attractiveness of portfolio planning theory and its rapid acceptance by major companies, it is not surprising that the approach has stirred up much debate. Most of it has been ill focused, however, for proponents and critics alike sometimes seem to be more interested in a dialogue about analytic techniques than in solving the practical problems inherent in implementation. So they argue about which portfolio and technology a company should choose—the Boston Consulting Group growth-share matrix or the General Electric-McKinsey, industry attractiveness-business position grid, the Arthur D. Little industry maturity-competitive position grid or the Shell Directional Policy matrix.

That discussion is sterile; the question of which grid to use and where to place a business on it is least important. The real issue is how a company can best define an SBU and assign a strategic mission to it. In short, what is a company to do with each of its businesses?

The decision on a strategic mission always requires a broad analysis of industry characteristics, competitive positions, expected competitive responses, financial resources, and the opportunities of other businesses in the portfolio. Whatever grid it chooses, a corporation's assessment comes down to a judgment heavily influenced by administrative considerations.

In selling a company on the "what," the consultants sometimes forget the "how." They make it appear that portfolio planning will emerge like a *deus ex machina* out on the corporate landscape—that its administration will pose no difficulties as long as top management has the will to implement it. A senior partner in a consulting firm explained to us his position as follows: "The challenge of portfolio planning really is analytical, not administrative. The way I see implementation is what I call the rule of the prince: once the analysis is done, a strong CEO should see to it that the portfolio strategy gets implemented."

If the experience with previous generations of planning approaches furnishes any lesson, the usefulness of portfolio planning is determined, of course, by the success a company has with its implementation. In formulating the basis for my survey, I made a number of assumptions. The first is most basic—that administrative rather than analytical problems create the greatest difficulties for companies implementing a portfolio planning approach.

My assumption was subsequently borne out as respondents reported that administrative problems loomed the largest. In fact, managers found the labels commonly used with the grid technologies to be largely irrelevant and often the source of psychological problems during the introduction of the technique. The better portfolio planning companies tend to avoid their use and focus on what to do with the business.

Any theory will mean different things to different practitioners, so I devised categories to help distinguish among the various forms portfolio planning has taken.

Developing Categories of Portfolio Planning

My survey was designed to allow me to make distinctions among companies (see Table 19-2 for the various types) with

No portfolio planning—No intention to introduce the technique.
Analytic portfolio planning—Use confined to a planning tool at the corporate level, no intention to negotiate explicit strategic missions with managers, and business strategies influenced by traditional administrative tools and profit pressures.
Process portfolio planning—Portfolio planning as a central part of the ongoing management process, as evidenced by the explicit negotiation of strategic missions with SBU managers.

As could be expected, getting to process portfolio planning is a long road, so I had to develop subcategories to take into account the various stages of introduction, such as companies with

New portfolio planning—Having just introduced the approach, the companies are still in the process of constructing the portfolio.
Undecided portfolio planning—The initial grid analysis completed, the companies have not yet agreed at the corporate level which strategic missions they want to assign to each of their SBUs.
Unassigned portfolio planning—Corporate-level decisions are reached on the strategic mission for each business unit, but not all these missions have been explicitly negotiated yet with the unit.
Process portfolio planning

THE USE OF PORTFOLIO PLANNING

Portfolio planning has indeed been widely accepted among diversified companies, particularly the larger ones. As of 1979 approximately 35 percent of the *Fortune 1000* companies (and even 55 percent of the *Fortune 250* companies) were introducing the approach at least to some extent. Over 90 percent of these companies use portfolio planning as a tool in their ongoing planning process, rather than as a one-time strategic analysis effort.

Nevertheless it is a long road to implementation. We estimate that only 13 percent of *Fortune 1000* companies (20 percent of *Fortune 250* companies) have reached the stage where missions are clearly decided and agreed upon by the business unit managers (see Fig. 19-1).

Application by Industry Sector

Though portfolio planning companies do not distinguish themselves in terms of their activ-

Table 19-2. Creating Portfolio Planning Categories (Sample section from questionnaire)

11. The following questions seek to determine how far your company has carried portfolio planning.	*Yes*	*No, not yet*	*No, will not*	*No, hasn't thought about it*
Has your company categorized its businesses on a portfolio grid?	☐	☐	☐	☐
Has your company made a corporate decision to have a strategic mission for each of its businesses?	☐	☐	☐	☐
Does your company explicitly label each of its businesses with a strategic category?	☐	☐	☐	☐
Has your company explicitly assigned each of its businesses a strategic mission?	☐	☐	☐	☐

Evaluating the responses

Does your company use portfolio planning?	Does it employ a grid analysis?	Has it made a corporate decision about missions?	Does it use labels?	Does it assign each business	Type of portfolio planning	Companies in sample	Years since introduction
Yes	Yes			No	Analytic	13	4.8
Yes	Yes	Yes	Yes	Yes	Process portfolio planning	78	5.5
Yes	Yes	Yes		Not yet	Unassigned portfolio planning	36	4.2
Yes	Yes	Not yet		Not yet	Undecided about portfolio planning	36	3.2
Yes	Not yet	Not yet		Not yet	New portfolio planning	13	1.5
No					No portfolio planning	106[a]	

[a]Excludes 67 respondents whose companies are not diversified.

ity at the general level of industrial products, consumer durables and consumer non-durables, a closer examination reveals that portfolio planning is more frequently practiced in capital intensive process industries such as chemicals, petroleum products, and paper products, and technology intensive but rather mature industrial products and consumer durables such as appliances, abrasives and industrial equipment. Such industries' competitive advantage, indeed, rests largely on relative costs, and relative costs closely relate to market dominance.

Technology intensive industries such as

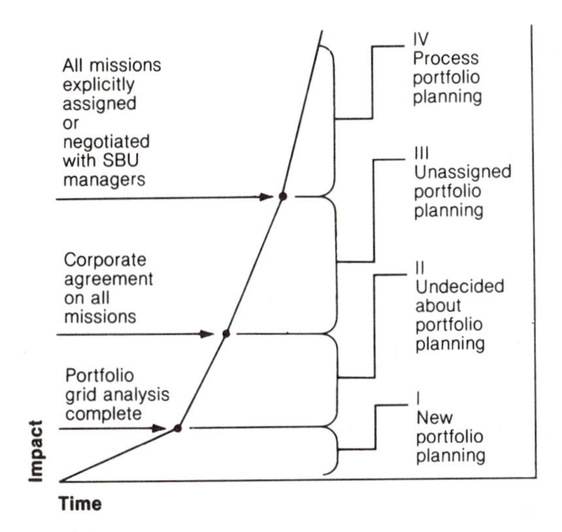

Figure 19-1. Process portfolio planning: stages of introduction.

computers, pharmaceuticals and scientific equipment, and marketing intensive industries such as beverages or food, are characterized by occasional use. Indeed, whereas market share may affect the cost position in such an industry, cost based dominance is not the only road to competitive advantage. Technological innovation or marketing based differentiation is equally important.

Resource based industries such as mining or oil exploration, or fragmented ones such as furniture, leather or jewelry, see little or no use. Indeed, in such industries, little association between market share and profitability is to be expected.

Application by Type of Company

Since they face the greatest challenges, the bigger and more diverse among *Fortune 1000* companies have introduced the technique. However, among diversified industries, conglomerates rarely use portfolio planning, whereas diversified industrials often do.

Two-thirds of portfolio planning companies oversee businesses that are related in some way. The majority not only overlap along one dimension (such as technology or market) but also along multiple dimensions (for example, technology, market, and raw materials). Moreover, companies usually attempt to integrate the management of these related businesses through the use of shared resources and staff at the corporate and group levels. (See Figure 19-2.)

It is because of the difficulty they have in assessing the strategic performance of each of their businesses and allocating resources selectively that diversified industrials need a formal tool such as portfolio planning. Conglomerates, on the other hand, speak portfolio planning prose like Monsieur Jourdain in Molière's *Le Bourgeois Gentilhomme*—sans savoir. How well companies can incorporate these interdependencies in applying portfolio planning will be crucial to the success of the approach.

Portfolio planning companies also tend to be international and likely to manage through complex organization structures. Again, the ease with which the planning approach incorporates both a product and a market dimension will prove crucial to the company's success.

STRUCTURING STRATEGIC BUSINESS UNITS

Before the introduction of portfolio planning, most companies divide up the corporate whole into organizational units (such as divisions) on the basis of operating control considerations. Often these units lack the necessary autonomy appropriate for strategic planning and resource allocation.

Defining what constitutes a business unit is the first step in all strategic—not just portfolio—planning. In the case of portfolio planning, two theoretical principles underlie the definition:

1. An organization must identify its various business units so that they can be regarded as independent for strategic purposes.
2. Companies then should allocate resources directly to these SBUs to support whatever strategies are chosen.

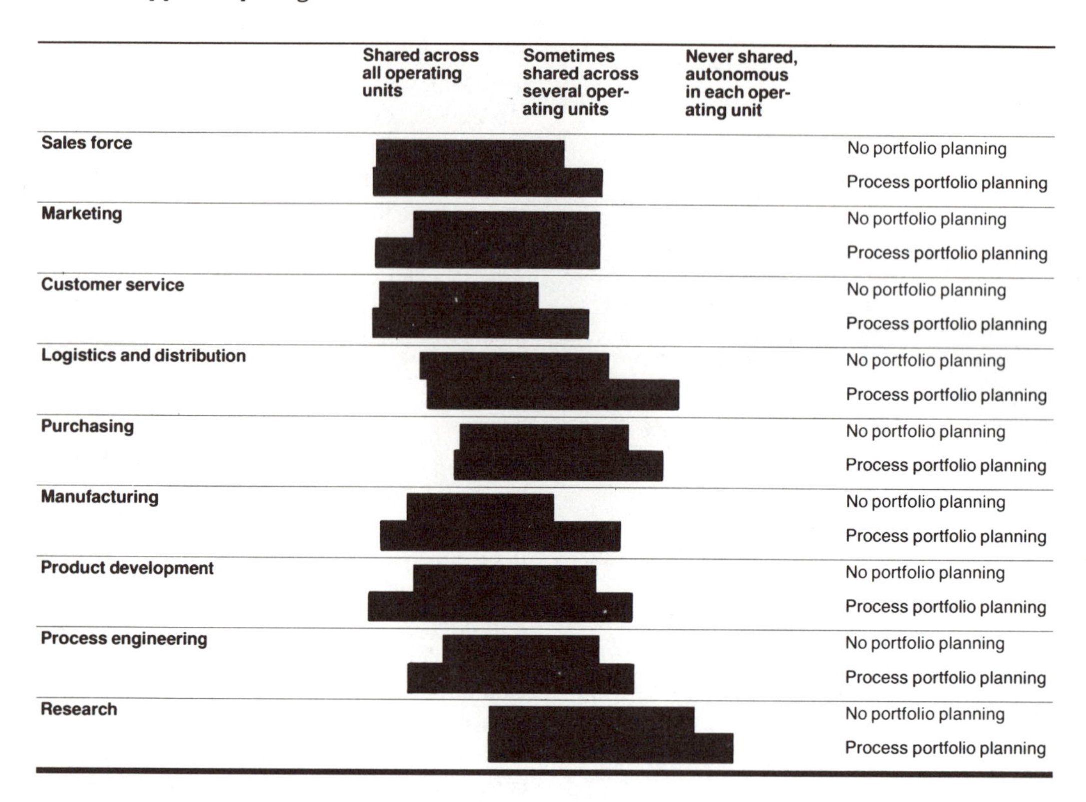

Figure 19-2. Use of shared resources.

The first principle is an attempt to solve the problem of inappropriate planning units by arriving at a good business definition on the basis of industry economics. Based on the experience curve, this definition sees a business as strategically independent if its value-added structure is such that market leadership in that business alone permits successful performance. To put it another way, a company looks at each market segment to see whether it can survive if it competes only in that segment.

The second principle, tying resource allocation directly to each SBU's strategic mission, attempts to solve the problems companies face with typical administrative systems. It hypothesizes that the practice of allocating resources on a project-by-project basis and the step-by-step aggregation of corporate operating results tend to shorten the focus of corporate management and create uniform rather than selective policies. To forge a dynamic strategy throughout the company, the theory states, a company must allocate its resources selectively and directly to strategically independent businesses.

SBUs in Practice

That is the theory, it is the application of the theory of portfolio planning to the realities of multinational diversified industrial companies that is most difficult.

In practice, of course, SBUs are not—and cannot be—strategically autonomous units rooted in an industry's structure. A company can only determine the size, shape, and number of SBUs in the light of prior organizational constraints and history, the limits of how many units corporate managers can handle and the multiple interdependencies among businesses.

According to my survey results, companies try to apply the theory; 70 percent of all organizations started off their introduction of portfolio planning by comprehensively reexamining the definition of each of their businesses. In 75 percent of these organizations, the reevaluation led to the classification of SBUs as sometimes or usually different from operating units. The larger the company, the more likely it is to have SBUs that do not coincide with operating units.

However, let us not assume that the companies started an administrative revolution. Despite the fact that they do not coincide with operating units per se, the resulting SBUs are generally aggregations of existing operating units or segments of single units; they clearly cut across organizational lines in only 7 percent of the cases. A close examination reveals that careful strategic analysis is rarely followed by an alignment of organizational units; rather, there is strong pressure to define SBUs quickly and good reason, in practice, to put the units clearly within the boundaries of existing organizational structure.

The Administrative Reality

Lest the theoreticians judge too harshly, the degree of diversification of most large companies obviates the possibility of simultaneous consideration of each relevant product and/or market segment at the corporate level. If the theory were strictly applied, a resulting grid would in most portfolio planning companies have over one hundred bubbles and in some over five hundred. It is no mere coincidence that portfolio planning companies—small and large alike—have ended up with, on the average, only 30 SBUs.

The end result is that instead of single, homogeneous units, most companies (58 percent of those in my survey, 72 percent of those among the *Fortune 1000*) consider each SBU as a portfolio itself—not of different businesses but rather of product and/or market segments that often may have quite diverse grid positions and strategic missions. In fact, the more experienced companies are with portfolio planning, the more likely they are to treat the exercise as a multilevel operation.

The impact of this redefinition comes alive when you realize the degree to which related businesses of each company share resources along different dimensions. Companies create SBUs at the organizational level, where shared resources can be managed. Top management should therefore see nothing inherently wrong with business units that cut across different market segments and that have widely different positions on the grid or a variety of strategic missions. To the contrary, forcing uniform strategic missions onto the managers of business units leads to either rejection of the mission or buildup of inappropriate strategies.

Those companies most advanced in the art of portfolio planning structure it on two levels. When analyzing the whole corporate portfolio and making trade-offs among businesses, they look at the company in the aggregate. In this larger picture, the companies look at SBUs that are in most cases organizational units and assign them strategic missions that reflect the expected cash-flow contribution to the whole company.

During the corporate plan review process, however, companies take a disaggregate view and look within those SBUs at the relevant strategic segments. These strategic segments are more likely to result from an analysis of the industry than from an accommodation to existing organizational structure. Their missions reflect the particular strategy the company wants a business unit manager to follow in each of his or her competitive arenas.

The definition of SBUs and strategic segments evolves throughout the introduction of the process. As the companies go through more planning cycles, the SBUs become more an organizational reality and the segment definition becomes finer. Often the segments are revised but the original SBU definition stays the same.

The definition of SBUs in each company raises two critical issues: how the company should define SBUs so as to accommodate the interdependencies and how a company can achieve the strategic aggregation and disaggregation that portfolio planning requires.

The question of interdependence, for example, creates nasty problems of the chicken-egg variety. One product-market segment may share manufacturing facilities with a few others, basic technology with an even broader group, and a sales organization with a different set of product-market segments. With all these differing dimensions, trade-offs must always be made. The guiding principle is that a company must define the SBU to incorporate control over those resources that will be the key strategic variables in the future. But how can the company do that before it knows the SBU's strategic mission?

Looking closer at the way in which the surveyed companies make these trade-offs in practice, you detect a bias toward cost efficiency in relation to responsiveness. Indeed, the business economics orientation of portfolio planning pushes cost structure as the only basis for business definition. In most industrial products and consumer durables companies, the market-based part of costs is small. Moreover, things like an SBU's responsiveness to local market conditions or governments are not quantifiable. As a result, these companies define SBUs along technological and manufacturing rather than market lines. Particularly in the international arena, many companies find that their worldwide SBUs are less responsive to local issues and cooler toward international activities than they were under the former international division structure and country plans.

PORTFOLIO PLANNING AND THE ADMINISTRATIVE CONTEXT

As seen from the corporate level of a large diversified company, business strategy is the outcome of a process of administrative influence. Although no one could expect a solar cell producer and an independent electrical wire manufacturer to perform under a similar set of administrative systems, as divisions of a hypothetical—and of course typical—U.S. diversified industrial company they are subject to fairly uniform sets of pressure and patterns of influence.

A theorist reasonably expects organizations to alter administrative systems to make way for portfolio planning, so planners would encourage this company to change its administrative systems—to be selective, not uniform, in its strategic decision making. In that way, new patterns of influence arise that correspond to the nature of the business, its competitive position, and strategic mission.

If we assume that the solar cell business is an SBU that our theoretical company has decided to grow while harvesting the electrical wire business, then you might expect the company to put an entrepreneur in charge, make him or her subject to frequent review and evaluation on the basis of market share and technological development, and encourage him or her to take some risks. In the case of the electrical wire business, the company gives the reins of that SBU to a penny-pincher and reviews his or her performance mainly on the numbers. It evaluates return and cash flow and discourages all investment other than cost reduction and maintenance.

In constructing the survey questionnaire, I tried to see if companies altered administrative processes to fit strategic missions. I asked whether they adjusted the financial planning system, the capital investment approval process, the incentive compensation system, or the strategic planning system itself.

Few Formal Administrative Changes

I found that in practice (except for the capital investment review system and, of course, the strategic planning system itself) companies do not alter formal administrative systems in accordance with the strategic missions of SBUs. On one level, this attitude simply reflects the time needed to implement the planning system. For example, one company (acknowledged as a leader in portfolio planning) finally brought its management compensation system into line five years after it had introduced the portfolio planning process.

In addition, the reluctance to modify administrative systems across businesses is a good indication of the perceived benefits of administrative simplicity. Controllers have excellent arguments for keeping administrative procedures uniform.

More profoundly, however, the basis for this reluctance may lie within the nature of the SBUs themselves. As I have said, diversified industrials look at SBUs as portfolios of various segments; tying the formal systems to portfolio planning would mean going beyond the business units and would require the company to gear itself to the specific strategic mission of each segment.

Differentiating the Informal Process

Though successful companies did not change administrative systems to accommodate portfolio planning, their managers did informally adapt systems to fit the various businesses. Time and time again in the survey, this informal differentiation seemed to make the difference between portfolio planning as an isolated exercise and as an integral part of the management process. In fact, implementation depends on how well the CEO and other top managers can tailor their attention to each SBU, especially how they monitor strategic plans, how they weight the financial numbers in light of the planning process, and how and where they promote managers.

Not all companies are sophisticated about the process. For example, in one company, the CEO had enthusiastically endorsed portfolio planning as the wave of the future. But two years' worth of effort virtually went down the drain when every business manager found a telegram on his desk one morning from that same CEO requesting a 5 percent across-the-board cut in manpower.

The impact of portfolio planning is most profound on the corporate review process and the capital investment appraisal process. Corporate review of business plans becomes more intense and focuses on different variables than in other companies. Generally, portfolio planning companies separate their strategic plan review from their financial review, so the planning process remains as meaningful as possible. As companies gain experience, the review goes into more and more of the detail of each segment within an SBU. My evidence indicates that the review process does shift from emphasis on short-term profits and sales objectives to long-term profits and sales targets and competitive analysis. (See Table 19-3.)

Tying Resource Allocation to Strategy

In a diversified company, strategy is essentially about resource allocation across businesses. In most companies, however, formal strategic planning is one thing and the capital investment appraisal process quite another.

My investigations show that companies engaging in process portfolio planning try to correct this inherent contradiction by tying the capital investment process closely to strategic planning (see Table 19-4). Not that many allocate resources primarily on the basis of strategies (only 14 percent do), but at least the business plan becomes an explicit element in the evaluation process for investment projects. Unfortunately, very few organizations tackle the allocation of strategic expenses (that is, investments that are expensed rather than capitalized, such as R & D, marketing, applications engineering) in the same way.

HOW MANAGERS PERCEIVE THE BENEFITS

Despite the difficulties associated with implementation, almost all the managers surveyed indicated that portfolio planning had a positive impact on the management process. The clearest evidence may be that only one of the 176 respondents that were introducing port-

Table 19-3. Ranking of the Most Important Issues in the Planning Review

	Rank under no portfolio planning	*Rank under process portfolio planning*
Next year's profit objectives	1	6
Long-range profit objective	2	1
Next year's sales objectives	4	8
Long-range sales objectives	5	3
Long-range resource allocation	6	2
Competitive analysis	7	5
Milestone for implementation	8	7
Contingency plans	9	9

Note: 1 = highest in importance to 9 = lowest in importance.

folio planning said it would hold less importance for the organization in the years to come. In fact, managers credit the approach with an array of benefits, with respect to not only the quality of strategy generation but also the commitment of resources and implementation. (See Figure 19-3 for a listing of the benefits.)

In the first place, companies gain a better understanding of each of their businesses. In turn, this allows them to make appropriate strategic decisions. One reason is the approach's emphasis on a company's ability to decipher industry logic and assess its competitive position. Another is the introduction of verbal and graphic language that facilitates improved communication in strategic, not just financial, terms.

According to the comments of managers, all of these benefits lead to improving resource commitments, for they improve resource allocation and facilitate strategic reorientation as well as entry and divestment decisions. Managers even credit the approach

Table 19-4. Link between Strategic Planning Process and Capital Investment Appraisal

Which statement, in your opinion, best reflects the relation between business plans and capital investment approval in your company?	*Percentage of companies in sample that use process portfolio planning*	*Percentage of companies in sample that do not use portfolio planning*
We explicitly allocate capital to our businesses as part of our business plans. Subsequent projects are expected to fall within this allocation guideline.	24	18
Fit with the business plan is an important and explicit element in our evaluation of capital investment projects.	57	22
The business plan serves primarily as general background information when we are considering capital investment projects on an individual basis	20	60

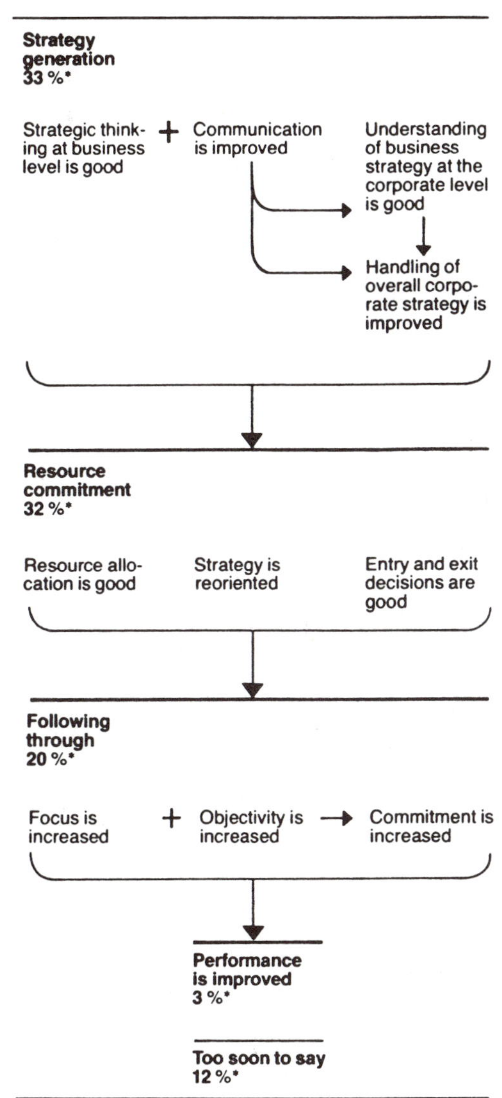

Figure 19-3. Managers' remarks about benefits.

with improved operations since it encourages focus, objectivity, and commitment.

IMPACT ON RESOURCE ALLOCATION

If resource allocation is what strategy is all about, then a fundamental question is whether portfolio planning actually affects the allocation of resources in the companies that adopt it. Since my survey format did not allow me to measure the actual shifts in allotments, I measured the impact as perceived by managers. I asked them how serious the following list of common allocation problems were for their companies on a scale from one (no problem) to five (severe problems):

1. We waste resources by continually subsidizing marginal businesses.
2. Our high-return businesses tend to underinvest.
3. Our low-return businesses tend to overinvest.
4. We do not fund our existing growth opportunities adequately.
5. We have a hard time generating new growth opportunities internally.

To reduce subjectivity, I did not ask for an assessment of improvement but rather for separate descriptions of the current situation as well as that before the introduction of portfolio planning. The results are striking. In general, the introduction of portfolio planning coincides with a perceived improvement in the allocation process. Both process and analytic portfolio planning help the company face the problem of marginal businesses. Changing the investment behavior of basic businesses or their attitudes toward risk that lead to inadequate funding of existing growth opportunities, however, requires a process approach to portfolio planning.

The one problem the approach does not address is the difficulty of generating new internal growth opportunities. I would add that on the basis of interviews I have conducted, the impact—if there is one at all—is rather negative. In theory, portfolio planning is about the allocation of all resources. In practice, however, companies focus on capital investment. The generation of new business requires explicit emphasis on human resource decisions and strategic expenses such as R & D and market research, only later to be followed by capital allocation.

On the most fundamental level, however, it appears that the impact of portfolio planning on resource allocation is a function of the degree and quality of its introduction. We can

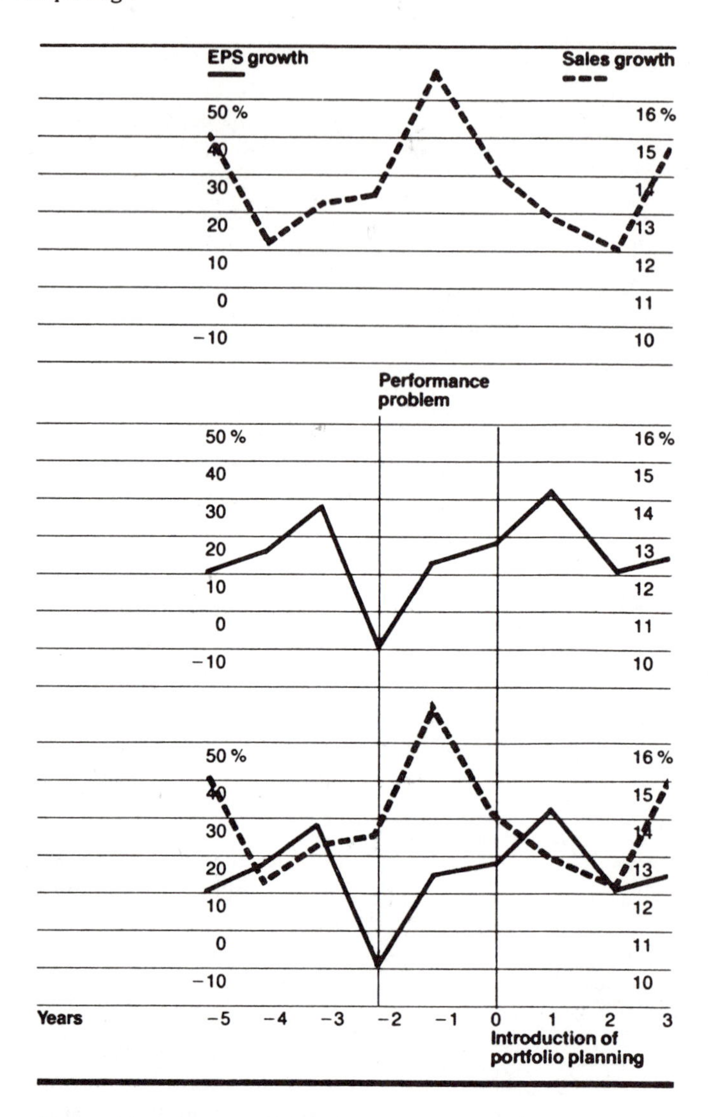

Figure 19-4. How portfolio planning is triggered.

draw a road map of the potential benefits according to the stages of introduction. First, companies face up to those businesses with untenably weak market positions, and make divestments. Next, businesses with growth opportunities feel liberated from short-term performance pressure and propose major growth programs from which the corporate level may not yet be ready to select.

In most cases, the investment inclination of base businesses changes slowly and requires the full commitment of top management behind certain power shifts. If a company has been overinvesting, all these resource demands may result in a resource crunch. The way that crunch is handled, in fact, gives a good indication of the degree to which portfolio planning has taken hold. If the company takes into account the various strategic missions of SBUs and counters the resource crunch selectively, then it has firmly established portfolio priorities. If, however, it institutes an across-the-board cut, you have a good sign that nothing has changed from the old days.

SUCCESS—OR FAILURE

Some companies in my sample reached the process portfolio planning stage very quickly, whereas others had not quite gotten there even though they had gone through five planning cycles. The difference between such fast and slow introduction coincided well with how successful or unsuccessful portfolio planning was.

In a set of telephone interviews, I compared the experience of 27 companies that had reached the process portfolio planning stage in three years with the experience of 24 organizations that, after five years or more, still had not negotiated strategic missions explicitly. I found that in these cases five factors determined the ease with which portfolio planning could be implemented. Most are endemic to the corporate situation; planners can do little about them. Yet they allow the corporate planner to gauge the magnitude of challenges that lie ahead.

The Performance Problem: Shock

Portfolio planning reallocates resources and thus implies a redistribution of power. As with all such shifts, a performance crisis—often, according to our data, arising from a profit crunch when fast growth goes out of control—triggers the initial decision to introduce the approach and greatly reduces the resistance (Figure 19-4 shows how a dip in EPS triggers the institution of portfolio planning).

CEO Role: Commitment

In all cases, a strong and continuous commitment from the CEO is the key to a fast introduction, even though in many companies the initial impetus comes from someone else—often the corporate planner. He signals with more than words, that is, with executive appointments, project approval decisions, and personal time spent on strategic review. Lack of real commitment was the source of major problems that companies in my sample ultimately faced.

Resource Imbalance Level: Low Resistance

Shifts in resource allocation can range anywhere from the improvement of segment strategies in divisions that can stand on their own two feet to major shifts across group, and even sector, levels. The degree of inertia and political resistance encountered with the corporate structure is a function of how important (or how high up in the structure) the shifts need to be.

Previous Planning Experience: Capability

Though the approach usually provides a dramatic jump in the quality of strategic analysis, its introduction as an ongoing process is based on the previous planning experience. Outside consultants cannot offer a good substitute for the strategic thinking of line managers and the experience acquired in earlier planning exercises. Skillful corporate planners often elicit support for portfolio planning by avoiding what line managers disliked about previous planning formats (for example, by reducing the amount of paperwork and financial information required).

Previous MBO Experience: Focus

As I have pointed out, good portfolio planning often requires the ability to treat SBUs as portfolios of segments. Companies with a tradition of management by objectives easily introduce the appropriate set of objectives specific to each segment and thereby mold the review process so that it incorporates both an aggregate and a detailed view.

SOME IMPORTANT ADVICE

Unfortunately, these five factors are largely out of the planner's control. My research turned up some valuable information about other variables that the planner *can* control and that will help him or her set the right priorities in introducing portfolio planning.

The following advice deals with the broad issues involved and applies to most companies independent of their specific situations:

1. Move quickly. Companies introduce portfolio planning initially on a wave, the strength of which depends on the extent of CEO support or the existence of a dramatic performance problem. It is important to establish the legitimacy of portfolio planning by pushing through some resource allocation decisions before their immediacy in the corporate atmosphere disappears. Many planners get bogged down because corporate attention shifts elsewhere.

If that happens, advocates may have to carry their selling job upstairs as well as downstairs. The best advice is to demonstrate the strength of the approach through successfully implementing it in one part of the company rather than pushing for across-the-board improvements.

2. Educate line managers. Portfolio planning is not a planner's exercise. It depends on improved strategic thinking at all levels of line management. From that perspective, many organizations—even though they invite consultants to analyze myriad divisions—have not taken advantage of the opportunity presented by the approach to improve the quality of management.

Successful companies usually conceive their planning process as a learning exercise for line management. They also invest heavily in education to allow managers to become familiar with both the basic thinking behind the approach and how best to use its tools.

3. Redefine strategic business units explicitly. Defining SBUs is the genesis—and nemesis—of portfolio planning. The units reflect and constrain corporate strategy in the most fundamental way. The period during which the company initially defines their scope gives astute managers the most leverage they will ever have to change strategic focus. This time should not be cut short and will prove to be time well spent if the right questions are asked and the practical problems and organization biases taken into account.

4. Avoid labels: focus on missions. The successful portfolio planning company shuns labels and does not haggle over grids. It focuses on a fundamental discussion of practical strategies for each SBU. Unhappily, many companies still leave fundamental decisions unmade and allocate resources continually by default. If portfolio planning is focused well, it can help by forcing the issue.

5. Acknowledge SBUs as portfolios to be managed. In practice, companies often determine what is a manageable number of SBUs around technological or market-based resources. These SBUs consist of many product-market segments. Successful companies, then, tend to accept portfolio planning as a multilevel approach. On the one hand, they develop the capability to take a detailed view of each segment when a strategic issue arises and, on the other, an aggregate view of the SBU when they discuss overall portfolio balance and resource commitments.

6. Invest corporate management time in the review process. Portfolio planning is a facilitating framework that allows substantive discussion between the corporate and the business levels. As such, the quality and extent of corporate review are crucial. Successful companies set aside time for reviewing strategic as well as financial plans, but they emphasize strategic planning. When analyzed at the same time, the financial numbers tend to drive out fundamental discussion of the strategic issues.

Good companies also maximize the time their top managers spend on strategic plan reviews. Though the theory of portfolio planning would call for reviewing all SBUs simultaneously, many companies spread out their review during the year in order to allow senior executives the most exposure possible.

7. Avoid across-the-board treatment. Portfolio planning companies do keep formal administrative systems uniform. However, they rely on a flexible, informal management process to differentiate influence patterns at the SBU level. Every corporate manager should remember an important caveat: have no across-the-board treatment. At a certain point the introduction of portfolio planning is likely to place demands on corporate resources. Those demands should be met squarely without equalizing the pain through uniform reductions. Any other method will dilute the real impact portfolio planning has.

8. Tie resource allocation to the business plan. The approach has no teeth without formal links to the resource allocation system. Forcing congruence between resource allocation, on the one hand, and the nature of the business and its strategic mission, on the other, is vital. In successful companies this link takes the form of setting an asset growth rate objective as well as approving a quota for various types of spending. Subsequent requests are evaluated first for their fit within this agreed-on spending pattern.

Such an approach actually alleviates the review burden because it allows decentralization of nonstrategic investments and facilitates approval of those projects that fall within agreed-on strategic priorities.

9. Consider strategic expenses and human resources as explicitly as capital investment. Portfolio planning is about the reallocation of all resources. Many companies subvert the process by focusing solely on the allocation of capital investment. That can work in industries that are highly capacity oriented, such as paper or steel, especially when the main problem is a deficient selection process between competing investments.

In most industries, however, strategic expenses in R & D, marketing, applications engineering, and recruiting hold the key to portfolio planning, especially if the company's main problem is in creation of growth. Companies that formally attempt to monitor the allocation of these resources tend to be the most successful.

10. Plan explicitly for new business development. Though it addresses the resource allocation imbalance in existing businesses, portfolio planning does not address the issue of new business generation. As a matter of fact, the way companies implement the system tends to inhibit innovative behavior at the business level. Companies with excellent track records of internal business development generally introduce some way to focus attention on the issue in their planning formats.

11. Make a clear strategic commitment to a few selected technologies and/or markets early. Corporate management can make specific portfolio choices only on the basis of strategy inputs from SBUs. Business unit managers, on the other hand, need guidance from the top to develop specific strategic proposals. When many companies in the survey could not initially decide on strategic missions, for example, business managers felt as if they were in limbo.

The best companies will announce firm commitments to a certain set of technologies and markets as early as possible. Setting up workable frames of reference to guide business level planning allows them to make better product-market decisions later on. Such commitments not only facilitate the introduction of portfolio planning, but in a planning approach that regards essentially interdependent businesses as stand-alone units, they also inject a dimension of commonality in a way that provides leverage for individual business strategies.

CURRENT FAD OR BASIC BREAKTHROUGH?

Does portfolio planning constitute a step forward in the management of diversity or is it simply a passing phenomenon? Often an administrative change, such as the introduction of a new planning system, is the way a CEO can address an organizational imbalance that might be at the root of a performance problem. Successful introduction may be self-defeating. The new system removes the problem and its own raison d'être.

It is true that in most companies the introduction of portfolio planning is triggered by a performance crisis and the need to allocate resources selectively in a capital-constrained environment.

Also, portfolio planning is not the discovery of the wheel. As I have defined it—the explicit recognition that a diversified company is a portfolio of businesses, each of which should make a distinct contribution to the overall corporate performance and should be managed accordingly—portfolio planning was practiced de facto by many companies before the development of formal technology.

Yet along with most managers, I feel that, in contrast to previous generations of planning approaches, portfolio planning is here to stay and represents an important improvement in management practice. After the initial portfolio imbalance is redressed, the approach can give companies a permanent added capacity for strategic control because it provides a framework within which the management process can be adapted to the evolving needs of the business. It also helps companies out of the dilemma between stifling centralization and dangerous decentralization. It allows them to reassert the primacy of the center in creating profit potential yet leave their strategic business units maximum operation autonomy in realizing that potential.

Portfolio planning can deliver on three fronts. The first is in the generation of good strategies, by promoting competitive analysis at the business level, more substantive discussion across levels, and strategy that capitalizes on the benefits of diversity at the corporate level. The second contribution is the promotion of more selective resource allocation trade-offs, not by solving the problems or eradicating the power game but by providing a focus for the issues and a vehicle for negotiation.

The most important contribution that portfolio planning can add is to the management process. The essence of managing diversity is the creation in each business of a pattern of influence that corresponds to the nature of the business, its competitive position, and its strategic mission. The benefit a company gets out of portfolio planning depends on its ability to create such a differentiated management process. Putting the approach into practice presents the company with some of its greatest challenges.

Success is based more on coping with administrative issues than on developing sophisticated analytic techniques. It requires a real commitment to good management and demands that an elegant theory be stretched to fit a complex reality.

CHAPTER 20

Critical Success Factor Analysis and the Strategy Development Process

Joel K. Leidecker and Albert V. Bruno

Joel K. Leidecker holds the B.A., M.B.A. and Ph.D. degrees, all from the University of Washington. He is currently an associate professor of business at the Leavey School of Business at Santa Clara University. At Santa Clara he teaches in the business policy area. Dr. Leidecker's research interests include the strategy development process and strategic planning systems in high-tech firms. He has published articles in the *Journal of Business Strategy, Long Range Planning, The Academy of Management Review, Mergers and Acquisitions* and the *Training and Development Journal.*

Albert V. Bruno holds the B.S. (with honors), M.B.A. and Ph.D. degrees, all from Purdue University. He is currently the Glenn Klimek Professor of Business at the Leavey School of Business at Santa Clara University. At Santa Clara, he teaches courses in entrepreneurship and business policy. Professor Bruno conducts research in marketing decision making, entrepreneurship, and strategic planning. He has published more than 50 articles, research notes, and book chapters in such diverse publications as the *Harvard Business Review, Management Science, Journal of Marketing Research, Journal of Consumer Research, Business Horizons, Sloan Management Review,* and the *Journal of Business Venturing.* His book on venture capital was published by North-Holland in 1986.

Critical success factors (CSFs) are those characteristics, conditions, or variables that, when properly sustained, maintained, or managed, can have a significant impact on the success of a firm competing in a particular industry. A CSF can be a characteristic such as price advantage, a condition such as capital structure or advantageous customer mix, or an industry structural characteristic such as vertical integration (see Tables 20-1 and 20-2 for some examples and chapter 24 for more applications).

The concept of CSFs can be applied at three levels of analysis: firm-specific, industry, and economic or sociopolitical environment. Analysis at each level provides a source of potential CSFs. Firm-specific analysis uses an internal focus to provide linkage to possible factors. Industry-level analysis focuses on certain factors in the basic structure of the industry that have a significant impact on the performance of any company operating in that industry. A third level of analysis goes beyond industry boundaries for sources of CSFs. This school of thought argues that one needs perpetually to scan the environment (economic,

Table 20-1. Critical Success Factors: Industry

Automobile	*Semiconductor*	*Food processing*	*Life Insurance*
Styling	Manufacturing process: Cost efficient, innovative, cumulative experience	New product development	Development of agency personnel
Strong dealer network	Technological competence: Adequate technical capability and personnel	Good distribution	Effective control of clerical personnel
Manufacturing cost control	Capital availability	Effective advertising	Innovation in policy development
Ability to meet EPA standards	Product development		Innovative advertising marketing strategy

Source: Automobile industry CSFs adapted from R. D. Daniel, "Management Information Crisis," *Harvard Business Review* 39(1961):37.

Table 20-2. Critical Success Factors: Firms in Semiconductor Industry

National Semiconductor	*Intel*	*AMD*	*Avantek*
Broad product line	Innovator and leader in technology	Proprietary innovative products. Does not compete in price-sensitive markets	Strong transistor product line
Large efficient production capacity	Strong product development and customer service capability	Effective location of fabrication and assembly. Operations strong	Solid customer range
Vertically integrated	High-margin proprietary devices	Technical marketing capabilities	High-yield manufacturing
Innovative packaging and assembly operations			

sociopolitical) to provide sources that will be the determinants of a firm's and/or industry's success.

EARLY CSF USE

Daniel first discussed CSFs in an article in the early 1960s.[1] The concept received little attention until a decade later, when Anthony et al. used the concept in the design of a management control system.[2] Anthony and Dearden pointed out that the management control system, in addition to measuring profitability, identifies certain key variables (also strategic factors, key success factors, key result areas and pulse points) that have a significant impact on profitability. These authors suggest, among other things, that there are usually six different variables that are important determinants of organizational success and failure; they are subject to change, which is not always predictable.[3] Rockart has used the concept of CSFs to assist in defining the CEO's and general manager's information needs.[4] This approach forces the key decision maker to identify those information needs that are critical or important to the success of the business. The factors identified become the basis for the company's management information system and provide the standards for subsequent performance measurement and control systems. Rockart and Anthony et al. believe the CSF approach can be an important tool in the two management areas cited above, however, another beneficial application of the concept is in the strategic planning process as shown by King and Cleland.[5] The identification of CSFs provides a means by which an organization can assess the threats and opportunities in its environment. CSFs also provide a set of criteria with which to evaluate (or assess) the strengths and weaknesses of the firm. These two elements (assessment of environmental threats and opportunities and specific firm resource analysis) are cornerstones of the strategic planning and strategy development processes.[6]

CSF AND THE STRATEGY LITERATURE

Recently, several authors have begun to recognize the value of the use of CSFs in strategic planning. Ohmae suggests that the key factors for success in an industry can be identified either by analyzing the industry's perceived market segment or by contrasting industry winners with industry losers.[7] Ohmae highlights the winner-versus-loser approach by contrasting the success of a Japanese forklift truck manufacturer with the failure of a truck manufacturer attempting to enter the forklift truck market. The forklift company's more efficient use of its sales force leads to the identification of sales force efficiency as a CSF in the forklift truck industry. In his forklift truck example, Ohmae indicates the importance of CSF identification by pointing out that success or failure of the firm is directly linked to the success factor.

Similarly, Ohmae examines a Japanese shipbuilding company that identifies its CSFs by recognizing that each market segment in the industry places different value weights on each ship type available. This approach aids the shipbuilding company in the identification of a CSF; that is in identifying which segments are of strategic importance to the firm. This information can, in turn, provide economic justifications in planning the use of resources. In this chapter we go beyond Ohmae and examine a number of other methods for identifying CSFs.

Andrews addresses the "requirements for success" in an industry.[8] These, he says, are the "critical tasks [that] must be performed particularly well to ensure survival."[9] Again, the emphasis here is on concentrating resources into areas or functions critical to the success of a given firm in a particular industry. He indicates the analysis does not stop there. CSFs are variables that can present new opportunities or threats over time. For those who are first to recognize a new requirement for success in an ever-changing competitive environment, the reward can be industry leadership. Andrews cites the case in which local radio station buyers paid much less for stations when they were the first to adopt new strategies in response to the rise of television.

In his search for competitor information, Goldenberg tells us that the business strategist would do well to start with determining the CSFs for his industry.[10] Although different CSFs can be found for each industry, in every case the strategist is called on to use a multidisciplinary approach. Goldenberg also recognizes the changing nature of CSFs. He offers as an example the textile industry, which converted from using only cotton, wool, and silk to using synthetic fibers. At the same time the industry underwent a parallel transition in CSFs. In this case, originally the only major CSF was purchasing. After the introduction of synthetic fibers as well as organized commodities markets, manufacturing and marketing emerged as CSFs equal to purchasing in importance. All three examples cited above indicate the emergence of CSF analysis in the strategic planning process and the importance of identification for a firm and industry. However, it is our contention that identifying industry CSFs is only the first step in the process. The firm must also determine the relative importance of the CSFs and continue to monitor factor importance over time. This not only provides vital information on current CSFs, but the trends that emerge may give strategists clues concerning CSFs of the future. We consider some parameters for determining factor importance and present a scheme for rating CSFs according to their relative importance.

Ferguson and Dickinson[11] recommend the use of CSFs by boards of directors to assist board members in functioning effectively. Whereas a CEO might be more concerned with planning and implementation, the board is more likely to use CSFs for monitoring the corporation's state of development and wellbeing and as a guide in evaluating the recommendations of the CEO. The need to assess whether or not the corporation has an adequate strategy for its own development is an area Ferguson and Dickinson believe will deserve the special attention of board members in the 1980s. This area considers such matters as management development, organizational development, research and development, and diversification efforts. We consider CSFs within the framework of this kind of an integrative strategic approach. In an article entitled "Identifying and Using Critical Success Factors," we suggested the use of CSFs in strategic planning.[12] In the following section we expand on this earlier work and focus more on the application of CSF framework in the strategy development process.

STRATEGY DEVELOPMENT USING CSFs

Hofer and Schendel, after reviewing a majority of strategy formulation models, indicate that strategy development is a seven-step process (strategy identification, environmental analysis, resource analysis, gap analysis, strategic alternatives, strategy evaluation, strategic choice).[13] We use these seven steps as a generally accepted representation of the strategy development process as well as the basis for our discussion of fit with CSFs.

CSF analysis can aid the strategy development process at three different junctures: environmental analysis, resource analysis, and strategy evaluation. Environmental analysis includes an assessment of the social, political, economic, and technological climates and their general impact on an industry and/or firm. In addition, this analysis usually focuses on the firm's competitive environment. Environmental analysis is used to identify the significant threats and opportunities facing a firm. CSF analysis, specifically at the macro and industry level, aids in the determination of threats and opportunities. CSF analysis provides a means by which a strategist can identify the essential competences, resources and skills necessary to be successful in a particular industry or specific economic climate. This type of information is particularly useful to the analyst responsible for identifying threats and opportunities. For example, if analysis indicates that vertical integration is a CSF for the soft drink industry, any firm in that industry or any firm contemplating entry into that industry will evaluate vertical integration either as a threat to or an opportunity for itself.

Resource analysis involves an inventory of a firm's strengths and weaknesses. Firm-level CSF analysis goes beyond the creation of an inventory. It identifies those variables that have been instrumental to a firm's success in a particular industry. This approach leads to a level of sophistication that provides greater depth and insight than a mere listing of a firm's strengths and weaknesses. This level of input provides more useful information for assessing a firm's competitive advantage (a firm's competencies vs. its competitors). In addition, firm-specific CSFs can be compared with threats and opportunities to aid in the identification of strategic options.

The third element in the strategy development process is strategy evaluation. Strategy evaluation involves comparing strategic alternatives with the specific goals and objectives of the firm and its various constituencies as well as any other evaluation criteria deemed pertinent. One other evaluation criterion could be the CSFs for an industry. For example, one available strategic option may be entry into the soft drink industry, and, as before, vertical integration may be a CSF. Now that this fact is known, the evaluation becomes whether the firm can negate this factor (neutralize or at least minimize its impact) or replicate it (do we have the resources financial and otherwise to become vertically integrated?). Obviously, the viability of the alternative is influenced by the firm's wherewithall relative to the CSF.

When a strategic alternative is tested as to its CSF fit, the strategy evaluation process becomes more rigorous and comprehensive.

In this section the strategy development process was outlined and the linkages to CSF analysis were set forth. Figure 20-1 shows the linkages between the three levels of CSF analysis (macro, industry, and firm) discussed earlier and the strategy development process. This linkage completes the discussion on CSF importance and use.

The focus now shifts to the third and most important objective: techniques for the identification of a CSF.

IDENTIFICATION TECHNIQUES FOR CSFs

The identification of CSFs can be an important element in the eventual development of a firm's strategy as well as an integral part of the strategic planning process. Eight techniques for identifying CSFs are discussed below. In addition, we present their respective advantages and disadvantages, discuss ways of applying them, and present examples of their use (see Table 20-3).

Environmental Analysis

This broad category includes a variety of approaches that identify the economic, political, and social forces that will have and are having an impact on an industry and/or firm's performance.

Environmental scanning, econometric models (such as the Chase Econometric Service or the Wharton models based on key environmental and/or economic variables), sociopolitical consulting services, and govern-

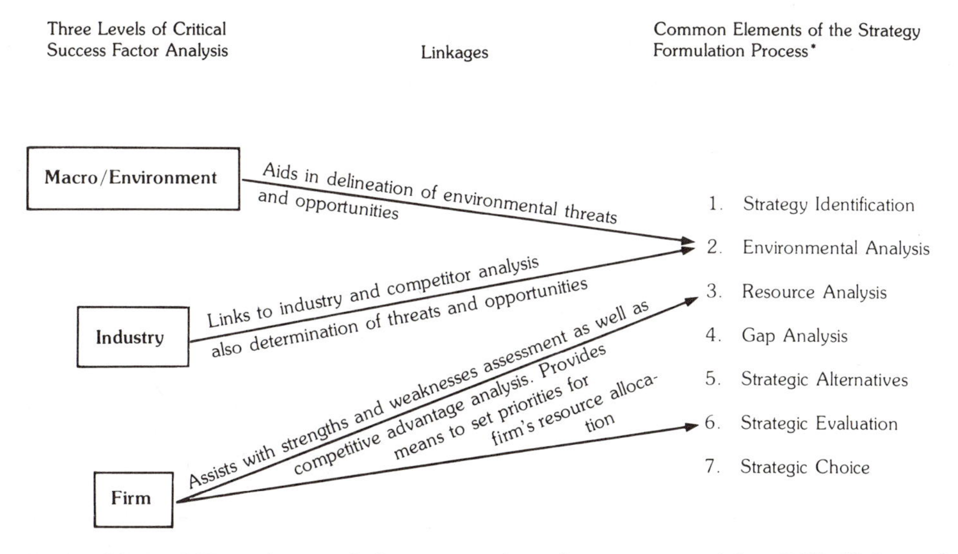

Figure 20-1. CSF analysis and the strategy formulation process (after C.W. Hofer and D. E. Schendel, *Strategy Formulation: Analytical Concepts,* West Publishing, St. Paul, Minn., 1978, 47). Strategic alternatives can be evaluated against CSFs and the relative importance of CSFs may influence prioritization of strategic alternatives.

Table 20-3. Critical Success Factors: Identification Techniques

Technique	*Focus*	*Sources*	*Advantages*	*Disadvantages*
Environmental Analysis	Macro	Environmental scanning (Corporate staff)	Future orientation	More difficult to operationalize into specific industry or firm CSFs
		Econometric models	Macro orientation: analysis goes beyond industry-firm focus	Results may not lend themselves to incorporate use in current time frame (today's CSFs).
		Sociopolitical consulting services	Can be linked to threats and/or opportunity evaluation	
Analysis of Industry Structure	Industry, Macro	A variety of industry structure frameworks	Specific focus is on industry	Excellent source for industrywide CSFs but not as useful in determining firm-specific CSFs
			Frameworks allow user to understand interrelationships between industry structural components	
			Can force more macrolevel focus (beyond industry boundaries)	
Industry and/or Business Experts	Industry, Micro	Industry association executives	Means of soliciting conventional wisdom about industry and firms	Lack of objectivity often leads to questions of verifiability and/or justification
		Financial analysts specializing in industry	Subjective information very often not discovered with more objective, formal, and analytical approaches	
		Outsider familiar with firms in industry		
		Knowledgeable insiders who work in industry		
Analysis of Competition (focus is limited to the competitive environment, how firms compete)	Industry, Micro	Staff specialists	Narrowness of focus, offers advantage of detailed, specific data	Narrowness of focus, CSF development limited to competitive arena (as opposed to industry structure approach)
		Line managers	Depth of analysis leads to better means of justification	
		Internal consultants		
		External consultants		

Table 20-3 (*Continued*)

Technique	*Focus*	*Sources*	*Advantages*	*Disadvantages*
Analysis of Dominant Firm in Industry	Industry, Micro	Staff specialists	Dominant competitor may in fact set industry CSFs	Narrow focus may preclude seeking alternative explanations of success
		Line managers	Understanding of #1 may assist in corroborating firm specific CSFs	May limit individual firm's strategic response and focus
		Internal consultants External consultants		
Company Assessment (Comprehensive Firm-Specific)	Micro	Internal staff Line organizations (Detailed analyses by organization function—checklist approach)	A thorough functional area screening reveals internal strengths and weaknesses that may assist CSF development	Narrow focus of analysis precludes inputs of more macro approaches
				Checklist approach can be very time consuming and become data bound
Temporal and/or Intuitive Factors (Firm-Specific)	Micro	Internal staff	More subjective and not limited to functional analysis approach	Difficulty in justifying as CSF if of short-term duration
		Brainstorming	Leads to identification of important short run CSFs that may go unnoticed in more formal reviews	Importance may be overstated, if in fact a short-run phenomena
		CEO/general manager observation		
PIMS Results	Industry, Micro	Articles on PIMS Project results	Empirically based	General nature
			Excellent starting point	Applicability to specific firm or industry Determination of relative importance

mental affairs departments are but a few of the diverse sources and approaches used to monitor and assess the environmental impact on an industry and the firms comprising that industry. The analysis is macro in approach and the data obtained do not always provide a clear linkage to the determination of industry CSFs, let alone firm-specific CSFs. The major advantage is the breadth of analysis, that is, the scope goes well beyond the industry-firm interface. This is particularly important to those industries whose survival depends on forces outside the control of the industry competitive environment.

Analysis of Industry Structure

Much of the strategic planning literature offers techniques to analyze the structure of an industry. The framework of analysis set forth by Porter provides an excellent example of this approach.[14] This framework consists of five components (barriers to entry, substitutable products, suppliers, buyers, and intefirm competition). The evaluation of each element and the interrelationships among them provide the analyst with considerable data to assist in the identification and justification of industry CSFs. One advantage of the analysis of industry structure approach is the thoroughness that the classification scheme provides. Another positive characteristic is the facility to depict schematically the industry's structural components as well as the critical interrelationships among elements (Figures 20-2 and 20-3 are examples of this technique).

Industry and/or Business Experts

This category includes inputs from people who have an excellent working knowledge of the industry and/or business. Although this technique may not be as objective and thorough as others, it does offer the advantage of obtaining information or a perspective not always available or discernible using the more standard analytic techniques. The conventional wisdom, insight, or intuitive feel of an industry insider often is an excellent source of CSFs and, coupled with more objective techniques, provides the analyst with a twofold data source to substantiate other CSF identification.

This approach is being used by the Center for Information Systems Research at MIT to identify firm CSFs that ultimately will be incorporated into a management information system.[15] We believe this technique is an equally rich, though subjective, source of CSFs to be used in the strategy development process. The disadvantages are obvious. The inputs may be no more than biased opinion and, therefore, result in a tenuous base for strategy development. Application is relatively straightforward but not simple; all the analyst must do is ask the right questions of the right knowledgeable sources and make the right interpretations.

Analysis of Competition

The focus here is a narrow one. It is limited to the competitive environment (or how firms compete) as opposed to the industry structure approach, which includes analysis of competition as one of the five structural elements to analyze.[16] The rationale for this approach is one of homing in on the target. The proponents of this approach argue that competitor analysis is one of the most important, if not the most important, source of CSFs. By concentrating analysis on competition—how firms compete—one does not dilute effort and possibly underanalyze competitive forces as users of broader industry structure approaches may do. The advantage of this approach relates to the specific nature of the firm; that is the thorough understanding of the competitive environment, and each firm's competitive posture allows a firm using this approach the facility to incorporate readily this information into the strategy development process. The major disadvantage is the inability to identify CSFs not linked to the analysis of how firms compete. Discussion of the importance of competitor analysis and approaches to competitor analysis can be found in the literature.[17]

Analysis of the Dominant Firm in the Industry

Often the way the leading firm in the industry conducts itself can provide significant insights into an industry's CSFs. This method is very useful in industries dominated by one or a few firms. The careful analysis of firms such as IBM, in the mainframe computer industry; H & R Block, in the tax preparation business; and Boeing, in the commercial jet aircraft business, would provide valuable information to identify and justify specific industry CSFs. Figure 20-4 is an example of this type of

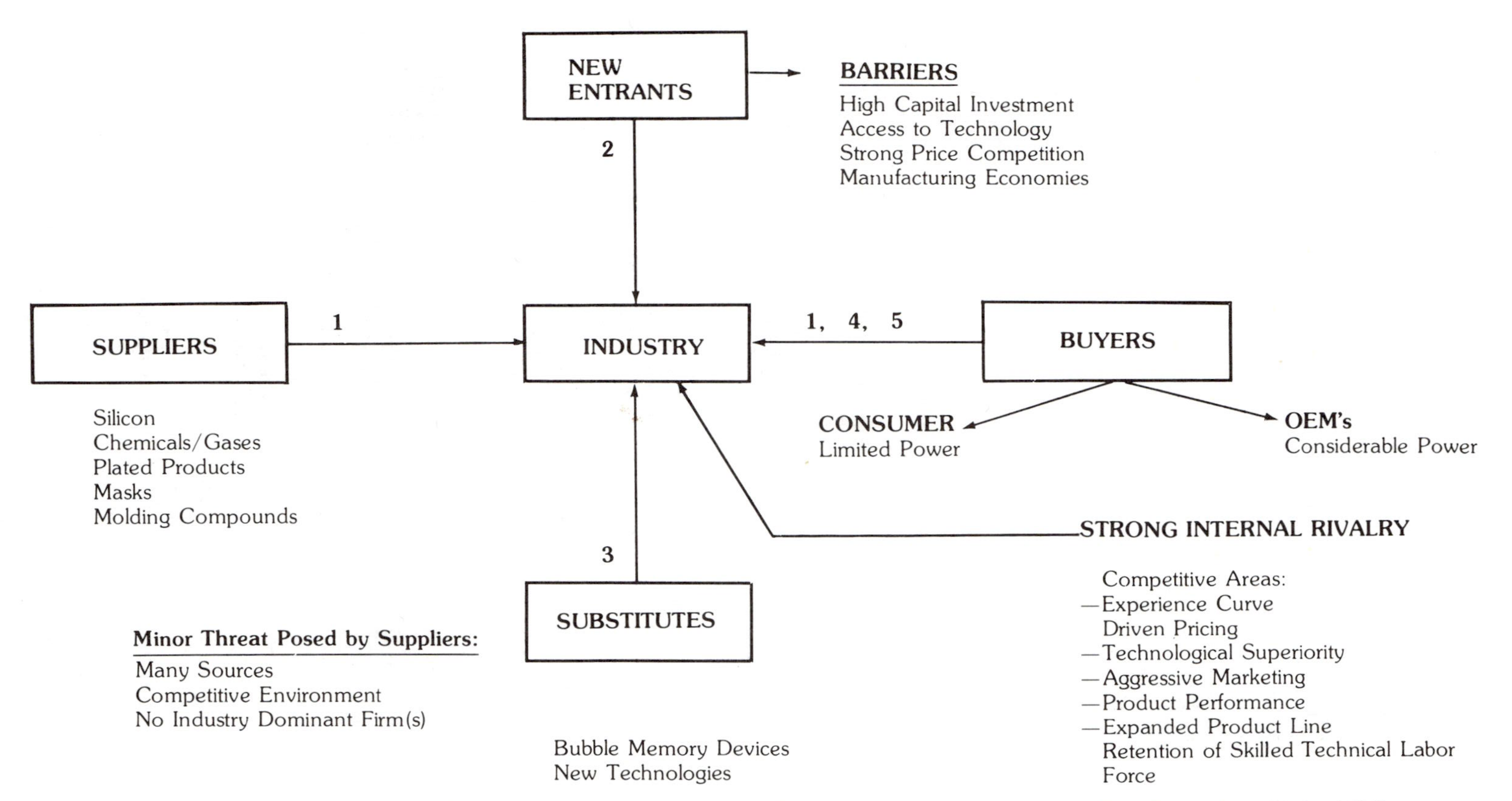

Figure 20-2. Industry structure analysis: Semiconductor industry (after M. E. Porter, *Competitive Strategy*, Free Press, New York, 1980). Interrelationships: (1) Increasing emphasis on forward and backward integration; (2) In recent years high barriers have been bridged by foreign competitors (European through purchase of U.S. companies, Japanese through government support and/or sponsorship and technology adaptation); (3) Development of new technology is a constant threat in this industry; (4) Pressure by both buyer groups for backward integration; (5) Manufacturers can create demand through marketing and product demand.

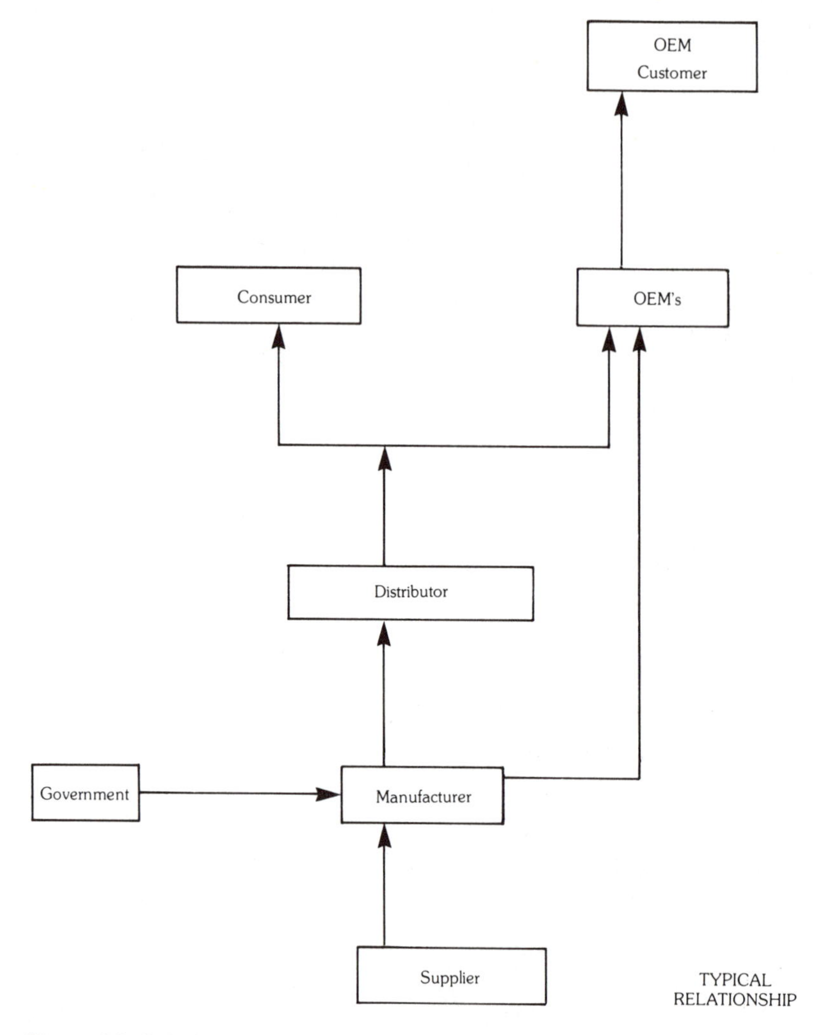

Figure 20-3. Industry structure analysis: Semiconductor industry.

analysis for a leading firm in the electronic components distribution (ECD) industry. The advantage of this approach is that if the dominant firm establishes the traditional success pattern for an industry, a thorough understanding of what the firm does successfully would aid in one's own internal analysis as well as in determining strategic posture. The major disadvantage is the narrow focus of this type of analysis. The strategic decision to emulate the dominant firm is fraught with danger. There are many examples of how firms have found ways of neutralizing or avoiding CSFs dictated by the dominant firm in the industry (for example, by-passing existing distributor networks by direct selling in the microwave components industry, the leasing of aircraft to offset capital intensiveness in the jet air freight business). However, the industry leader should be analyzed before deviations from its success pattern can be formulated.

Company Assessment

This approach is firm specific. The purpose of the analysis is to identify the CSFs for a particular firm. Although this is a worthwhile analytic exercise in itself, the results should be analyzed in light of industry and competitor

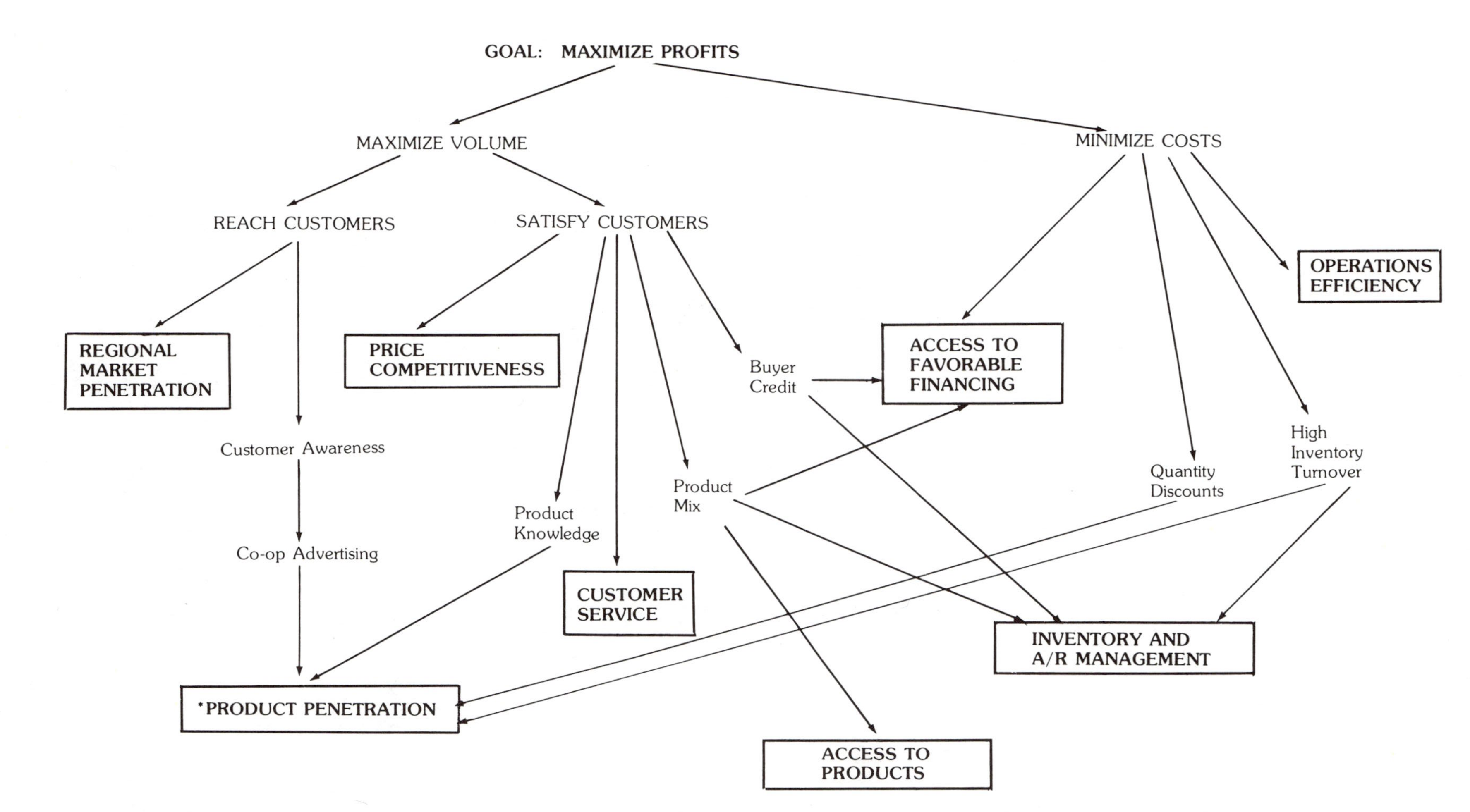

Figure 20-4. ECD Industry: CSFs. Product penetration = all the sales and/or marketing activities that make the distributor the preferred supplier for the customer by creating awareness of the firm's product mix, depth, and potential to benefit the customer through dissemination of product knowledge, value-added, and purchase discounts.

CSFs. A variety of approaches have been articulated: strength and weakness assessment, resource profiles, strategic audits, and strategic capabilities. All have one characteristic in common; the analyst must thoroughly explore what the firm does well and not so well. The positive aspects of the firm's operation may provide the means to determine its CSFs. Any of the approaches above, if applied by a good analyst, will most likely result in a useful set of CSFs. One specific method of application is through the use of a comprehensive checklist (see Appendix 20A). This is a series of questions, by functional area, designed so that the answers will provide the information necessary to determine a firm's CSFs. The specific questions and the weight given to each question will vary by firm. An evaluative prioritization scheme for the firm is a necessity. Also the set of questions must be tailored to each firm's specific requirements. Starting points would include checklists provided by CPAs, consulting firms, and the business literature.[18]

The step-by-step approach is methodical and time consuming but will result in a comprehensive analysis of each functional area. The narrow focus and the time to complete are the two disadvantages of this approach; however, the major disadvantage is the possible exclusion of obvious industry level CSFs, not uncovered with this approach. Moreover, the analyst or planner is constrained by the set of topics and/or questions that are available or that can be developed.

Temporal and/or Intuitive Factors

This is another approach with a firm-specific outlook. This, like its industry-level counterpart (Analysis of Competition), focuses on the intuition and insight of individuals very familiar with the firm. Although subjective, this approach often uncovers subtleties about CSFs that the more conventional and objective techniques overlook. The temporal issue deals with firm-specific occurrences, that in the short run may have a significant impact on performance and hence constitute a CSF (for example, key management people leave for a new company; purchasing fails to order enough of a critical material; in a noncapital intensive industry a new technological breakthrough in equipment design forces a financially weak firm into a disadvantaged position).

An example of how this approach can result in the determination of an important CSF is as follows: the founder and president of a local solar energy firm, discussing the key ingredients that make his firm successful, identified two important factors from his point of view. An earlier analysis using a more conventional objective approach failed to discern these two important aspects, machine design and manufacturing process. Both were people specific; each gave the firm a competitive advantage that was not available in the open market. The more conventional approach did not identify the major importance of either factor.

PIMS Results

In recent years the PIMS (Profit Impact of Market Strategy) project data indicates, among other things, that relative market share, degree of vertical integration, new product activity, capital intensity, and ratios of R & D and marketing to sales play a major role in determining profitability.[19] Profitability is certainly one of an industry's or firm's measures of success, if not the only one. If the PIMS results identify the key determinants of profitability, then these inputs provide a starting point for CSF analysis (other techniques may be used to substantiate).

The major advantage appears to be the empirical basis of the project results. The major disadvantage, as an identification technique, is the very general nature of the factors. The PIMS results do not provide a method of analysis to indicate whether the data are directly applicable to a specific firm or industry and/or what their relative importance may be.

The relative importance of a CSF is something that must be determined and periodically evaluated. Some CSFs are obviously more important than others. In the next section we discuss an identification and priority scheme for critical success factors.

DETERMINATION OF FACTOR IMPORTANCE

The profit impact of an activity or condition is usually the most significant factor for CSF identification as well as determination of factor importance. Although other areas may be important, we suggest four starting points for the profit impact analysis that will assist in the determination of factor importance.

Major Activity of Business

Usually CSFs are found in major areas of the business. For example, if a wholesaler were under examination, many of the factors that influence overall performance should be found in and around the inventory and warehousing function as opposed to, say, the advertising activity. If the opposite were true and most firms in a particular industry were concerned with marketing to consumers, then the advertising function might merit closer scrutiny.

Large Dollars Involved

Major CSFs probably have relatively large dollar amounts associated with them. For example, in a manufacturing firm direct labor may be a large dollar amount and the productivity of the workforce might be a CSF. Improving workforce productivity might lead to improved bottom-line performance. This compares to the wholesaling activity described above, where improved workforce productivity might not significantly improve bottom-line performance since most wholesale operations are not labor intensive.

Major Profit Impact

Another way of looking at a business is to assess the sensitivity of overall results to changes in certain activities. For example, in some circumstances, a small change in price might have enormous bottom-line impact, whereas doubling the advertising effort might have little impact. Value-added analysis would be an excellent tool to use for this level of analysis.

Major Changes in Performance

Sometimes it is a good idea to follow up on significant changes in the company's performance: for example, a dramatic drop in sales, a major profit reversal in a segment of the operation, a sizeable increase in margins. A significant change often eventually will be linked to a major CSF. Whether it will be of short- or long-term duration is to be determined by the analyst.

Another aid to the determination of relative factor importance and subsequent CSF use would be a priority scheme. The basis for establishing CSF priorities should be linked to the industry's and/or firm's success criteria (economic or otherwise). When classifying industry CSFs, the factor importance and profit impact analysis discussion above is relevant; the classification of firm-specific CSFs can be influenced by other, noneconomic, firm objectives. The discussion above provides the basis for the priorities. The decision as to how to prioritize the CSFs is, of course, up to the individual analyst. In the examples in Tables 20-4 and 20-5 we have used a four-factor scheme.

In most cases, the type of company or the nature of the industry will determine which CSFs are important. For example, the success of a retail business is heavily influenced by factors such as store location, effectiveness of the merchandising, and inventory control. The wholesaler(s) selling to this same retailer normally would not expect a CSF to be location-oriented. Tables 20-4 and 20-5 provide examples of how the prioritization

Table 20-4. Importance of CSFs: Industry Comparison

	Soft Drink (Bottler's)	*Semiconductor (Manufacturers)*	*Ferrous Metals Distribution*	*Tax Preparation*
Basic research and design	PF[a]	MF[b]	NF[c]	NF
New product development	SF[d]	MF	PF	NF
Manufacturing	MF	MF	SF	NF
Distribution	MF	SF	MF	PF
Customer service	MF	SF	MF	MF
Advertising	SF	NF	NF	SF
Postsales service	SF	PF	PF	MF

[a]Possible Factor: An activity or condition that could influence a company or industry results but it is not likely to. This category could also be used when analysis does not turn up sufficient data to warrant classification in MF or SF.
[b]Major Factor: An activity or condition that has a significant impact on company or industry results. This factor is usually linked directly to profit performance, but other success measures may be employed.
[c]Nonfactor: This activity or condition has very little impact on the company or industry results. Analysis indicates that there are no historical data or future expectations that link this activity with profit performance or other success criteria.
[d]Secondary Factor: An activity or condition that has, in most cases, modest impact on a company's results. Analysis shows this factor has an indirect link to profit performance or other success criteria. The magnitude of the impact is obviously less than a major factor.

Table 20-5. Importance of CSFs: Firms in Semiconductor Industry

	National Semiconductor	*Intel*	*Signetics*	*Texas Instruments*
Basic research and design	PF	MF	PF	SF
New product development	SF	MF	SF	MF
Manufacturing	MF	MF	MF	MF
Distribution	MF	SF	SF	MF
Customer service	MF	SF	SF	MF
Advertising	PF	PF	PF	SF
Postsales service	NF	NF	NF	SF

Note: See Table 20-4 for explanation of factor classifications.

scheme can be used with a set of potential CSFs. For comparative purpose, table column heads list various industries or firms and at the intersection of each factor category with the firm or industry the analyst would assess the relative importance of that factor for the respective firm or industry. This method of focusing the search provides more comprehensive CSF analysis and will assist with subsequent application.

All eight methods of identifying CSFs have weaknesses when used alone, but by using more than one method, business strategists not only reduce the probability of omitting important industry determinants of success, but they may even find that two or more methods concur on the validity of a particular CSF and provide a more integrative approach to strategic planning.

APPLICATION OF CSF ANALYSIS

IBM's successful PC project provides a good illustration of a large firm integrating the concept of CSFs into the strategy development process.[20] In formulating its strategy for entry into the PC market in 1981, IBM identified five key elements of success; namely, advanced design, widespread licensing of software developers, multichannel distribution, low-cost manufacturing, and aggressive pricing. The knowledge of these CSFs as well as its own resource strengths and weaknesses enabled IBM, through strategic evaluation, to make some wide-ranging strategic decisions. For instance, although IBM had the resources with which to automate its factories so as to meet the low-cost manufacturing CSF, it also displayed flexibility in being able to use outside software companies to meet software requirements as well as to acquire technology in the form of a 16-bit microprocessor, made by Intel Corporation, to satisfy the need for advanced design. Clearly, IBM's stunning success in grabbing 26 percent of the annual market for personal computers in only two years (1981-1982) can be traced to its recognition of the industry CSFs rather than being merely a matter of good fortune.

Dickinson et al. indicate that the use of CSF analysis is also becoming vitally important to small business, an area characterized by tight resource constraints and high failure rates.[21] Whether an entrepreneur is contemplating a new business venture or is seeking to improve his or her current operation, the use of CSFs in a planning framework can provide a comprehensive approach to problem solving. For instance, Dickinson et al. suggest that one of the most common CSFs at the small-business level is whether the firm's product or service offerings are unique, thereby creating an advantage over the competition. However, the planning process does not stop with the discovery of a CSF. Environmental analysis continues with management constantly watching for threats to its proprietary advantage or the threat that its particular uniqueness may no longer be a CSF. Likewise, through resource analysis, the firm monitors its strengths (its ability to protect and to nourish this CSF) and weaknesses (its inflexibility in response to changes in CSFs). In the way of strategic evaluation, the small-business manager can, for instance, design the structure of his or her organization either to fit the current CSFs or to respond in a timely way to changes in CSFs. We support this contention. CSFs used in conjunction with the strategy development model can provide the small-business manager with a comprehensive and ongoing perspective with which to pursue strategic planning.

Whether at the level of a giant IBM or that of a small-business entrepreneur, the application of the CSF concept is beginning to find its way into the strategy development process. These actions indicate that strategic planning can no longer remain a patchwork process. Rather, strategic planning must be approached in an integrative way and, as shown earlier, the identification of CSFs can be useful at various stages of the strategy development process.

SUMMARY

The purpose of this chapter is twofold. The first focus is to justify the existence and usefulness of CSF analysis in the strategy development process. Examples from the strategic planning literature and reports on company use were cited to support our contentions. The second focus is the identification, determination of importance, and use of CSFs.

This portion of the chapter concerns itself with a discussion of the application of CSFs in the strategic planning process. We argue that CSFs can be particularly instrumental in three areas of the strategy development process: environmental analysis, resource analysis, and strategy evaluation. For the practitioner, eight identification techniques are discussed and examples of use presented. In addition, a means of establishing the relative importance of a critical success factor is set forth.

APPENDIX 20A

FIRM ANALYSIS: Cost Accounting and Financial Management Checklist

Are clear pricing policies adhered to?

Has the relationship between price, volume, and profitability been studied?

Is pricing structure in line with the quality of the product or service?

Are products packaged in different forms to serve different segments of the market at different prices?

Are pricing decisions made by individuals with appropriate knowledge and responsibilities within the organization?

Where pricing discretion is delegated, are there clear policies to be followed and are discretionary decisions monitored?

Is zone pricing used to offset higher distribution costs in more distant markets?

Is the incentive scheme (commissions) for salespersons such that they are encouraged to quote at reduced prices?

Is there an attempt to develop information regarding bids lost (for example, price differentials involved)?

Is there sufficiently accurate cost information to develop sound estimates for quotation purposes? To what extent are sales lost because price quotations are too high or profits lost because bids are too low?

Are cost estimates compared subsequent to actual results?

Are quotations made at less than full markup carefully controlled?

Are pricing decisions so irregular that it is difficult to quote consistent prices to the same or similar customer?

Are trade discounts to customers in line with competitive practices or do they result in an unnecessary erosion of margins? Are they consistent from customer to customer?

Do volume discounts make sense in terms of the practices of the market? Do they make sense in terms of the actual savings through processing of large orders? (Has the cost of processing an order been established?)

Are certain customers granted volume discounts in spite of the fact that they order in small lots? Are volume discounts granted in advance or retroactively? How are volumes controlled?

Is the cash discount policy strictly enforced? Can cash discounts be eliminated? Is interest charged for late payments?

Are legal considerations taken into account when pricing policies are established?

How serious are bad debt write-offs? (As trend, as a percentage of sales, as a percentage of accounts receivable?)

Are there clear procedures to cover the approval of new accounts? Do these procedures result in delays of orders for new customers?

Is credit control based on appropriate criteria? (Dollar limits, period outstanding.)

Is order processing delayed because each order is given credit approval? Would exception procedures be more appropriate?

Are there clear policies regarding cutting off credit?

Have routine procedures been established for progressive collection steps?

Are there clear policies when collection agencies or legal action is to be used and what approval is required before such steps are taken?

Is the credit function supervised by a competent and knowledgeable individual? Is such an individual organizationally separated from the sales function?

Are suitable mechanisms available to balance credit control requirements with sales requirements?

Is there an adequate system for forecasting cash requirements? (For example, projected receipts and disbursements by month, forecasted schedules of bank borrowings, capital expenditure plans by month.)

Are cash needs effectively monitored and controlled? (Does someone in a senior position consciously monitor and control the movement and use of cash in the company?)

Is cash flow for operations adequate in relation to cash needs for working capital requirements, debt repayment, and expansion?

Are funds tied up in unproductive facilities?

Have sources and methods of financing been reviewed recently? (For example, seasonal borrowings, lines of credit, leasing, bonds, debentures.)

Does the company maximize the use of services offered by banks and other financial institutions (for example, payroll services, cash flow studies)?

Is information produced by the accounts payable and purchasing systems, as well as by the capital budget system, of use in cash forecasting?

Are suitable banking or other facilities used for speedy transfer of cash received?

Are policies regarding granting of cash discounts and the taking of cash discounts relevant in the light of interest rates?

Have payment delays caused by cash shortages resulted in supplier complaints, losses of cash discounts, higher prices, C.O.D. terms or refusal to supply?

Is an adequate line of credit maintained? Has contact been maintained with competing financial institutions? Have other financing possibilities been considered to improve liquidity?

Is utilization of credit lines reduced to the minimum on a day-to-day basis?

Are excess funds invested? Have policies or guidelines been established for such investment?

Is there unnecessary transcription of information from order to invoice? Could errors and cost be reduced through introduction of a more efficient system?

Are calculation-checking procedures appropriate taking into consideration the value of orders? Would a visual scan be more efficient? If invoices are subsequently computer processed, would after-the-fact checking of gross margins be sufficient?

Are there substantial billing delays that have an impact on cash flow and/or posting to inventory records?

Are there effective controls to ensure that all shipped orders have been invoiced?

Do customers pay on the basis of statements? If so, do delays in accounts receivable processing have an impact on cash flow?

Is current accounts receivable information required because of stringent credit control requirements? If so, are credit approvals on orders delayed because of backlogs in accounts receivable posting?

Are statements mailed promptly at month end?

Are the methods used to process accounts receivable appropriate in the light of needs, average value of account, credit control requirements, and problems experienced? Should alternatives be considered (one-write system, accounting machine, computer service bureau)?

Are there well-defined procedures for following up on outstanding accounts?

Is an appropriate amount of interest charged on overdue accounts?

Are invoices approved based on confirmation of receipt of goods or services? Are difficulties experienced because of delays in receiving such documentation or because invoices for services must be rerouted for approval?

Are satisfactory controls established to ensure that correct prices are used?

Are there unnecessary approval steps including checks to ensure that budgets will not be exceeded? (N.B., errors can lead to inventory differences.)

Are there satisfactory controls to ensure that transactions are posted both to general ledger and inventory records?

Are the methods used to process accounts payable appropriate, considering volume and circumstances?

Have clear policies been laid down regarding the taking of cash discounts? Are incidents of lost discounts investigated and recorded? Are the policies sound in the light of the costs of money? Is the system flexible enough to use credit to the maximum?

Are there significant complaints by vendors regarding slow payments. Have these led to higher prices, refusal to supply, or insistence on C.O.D. or prepaid shipments?

Are there suitable final checks before payment by the signing officers or some other party?

REFERENCES

1. R. D. Daniel, "Management Information Crisis," *Harvard Business Review* **39**(1961):110-119.
2. R. N. Anthony, J. Dearden, and R. F. Vancil, "Key Economic Variables," in *Management Control Systems,* ed. R. N. Anthony and J. Dearden (Homewood, Ill.: Richard D. Irwin, 1972), 138-143.
3. Anthony et al., "Key Economic Variables"; R. N. Anthony and J. Dearden, *Management Control Systems: Text and Cases* (Homewood, Ill.: Richard D. Irwin, 1980).
4. J. F. Rockart, "Chief Executives Define Their Own Data Needs," *Harvard Business Review* **57**(1979):81-92.
5. W. R. King and D. I. Cleland, "Information for More Effective Strategic Planning," *Long Range Planning* 10(1977):59-64.
6. I. Ansoff, *Corporate Strategy* (New York: McGraw-Hill, 1965); W. Glueck, *Business Policy and Strategic Management,* 3rd ed. (New York: McGraw-Hill, 1980); A. Thompson and A. J. Strickland, *Strategy Formulation and Implementation* (Dallas, Tex.: Business Publications, 1980).
7. K. Ohmae, *The Mind of the Strategist* (New York: McGraw-Hill, 1982).
8. K. R. Andrews, *The Concept of Corporate Strategy* (Homewood, Ill.: Richard D. Irwin, 1980).
9. Ibid., 21.
10. D. Goldenberg, "Search for Competitor Information" in W. L. Sammon, M. A. Kurland,

and R. Spitalnic (eds), *Business Competitor Intelligence* (Boston: Ronald Press, 1984), p. 47.

11. C. R. Ferguson and R. Dickinson, "Critical Success Factors for Directors in the Eighties," *Business Horizons* **26**(1982):14-18.
12. J. K. Leidecker and A. V. Bruno, "Identifying and Using Critical Success Factors," *Long Range Planning Journal* **17**(1984):22-32.
13. C. W. Hofer and D. E. Schendel, *Strategy Formulation: Analytical Concepts* (St. Paul, Minn.: West Publishing, 1978).
14. M. E. Porter, *Competitive Strategy* (New York: Free Press, 1980).
15. Rockart, "Chief Executives Define Their Own Data Needs."
16. Porter, *Competitive Strategy.*
17. S. C. South, "Competitive Advantage: The Cornerstone of Strategic Thinking," *Journal of Business Strategy* **1**(1983):15-25; J. L. Wall and B.-G. Shin, "Seeking Competitive Information," in *Business Policy and Strategic Management,* 3rd ed., ed. W. Glueck (New York: McGraw-Hill, 1980), 144-153; J. L. Wall, "What the Competition Is Doing: You Need to Know," *Harvard Business Review* **52**(1974):227.
18. South, "Competitive Advantage"; R. R. Buchele, "How to Evaluate a Firm," *California Management Review* (IV 1962):5-17; R. T. Lentz, "Strategic Capability: A Concept and Framework for Analysis," *Academy of Management Journal* **5**(1980):223-234; H. H. Stevenson, "Defining Strengths and Weaknesses," *Sloan Management Review* **17**(1976):51-68.
19. S. Schoeffler, R. Buzzell, and D. Heaney, "Impact of Strategic Planning on Profit Performance," *Harvard Business Review* **52**(1974):137-145; R. Buzzell, B. Sale, and R. Sultan, "Market Share: A Key to Profitability," *Harvard Business Review* **53**(1975):19-106.
20. "Personal Computers: And the Winner is IBM," *Business Week* (3 October 1983): 76-79.
21. R. Dickinson, C. R. Ferguson, and S. Sircar, "Critical Success Factors and Small Business," *American Journal of Small Business* (January-March 1984):49-57.

CHAPTER 21

Fitting Business Strategic and Tactical Planning to the Product Life Cycle

David R. Rink and John E. Swan

David R. Rink holds a Ph.D. from the University of Arkansas and is professor of marketing at Northern Illinois University, DeKalb. Dr. Rink has contributed articles to *Journal of Marketing, Journal of Purchasing and Materials Management, Business Horizons, Journal of Business Research, Transportation Journal,* and other publications. He has written papers for proceedings of the American Marketing Association, American Institute for Decision Sciences, Southern Marketing Association, and Southwestern Marketing Association. Dr. Rink has co-authored a textbook entitled *Marketing Research* (Merrill, 1982).

John E. Swan holds a D.B.A. from Indiana University and is the Birmingham Business Associates Professor of Marketing at the University of Alabama, Birmingham. Dr. Swan has contributed articles to *Journal of Marketing, Journal of Marketing Research, Journal of Personal Selling and Sales Management, Journal of Retailing, Journal of Applied Psychology, Business Horizons, Journal of Advertising Research,* and other publications. He has written papers for proceedings of the American Marketing Association, American Institute for Decision Sciences, Association for Consumer Research, and American Psychological Association. Dr. Swan serves on the editorial boards of *Journal of Marketing Research, Journal of Business Research,* and *Journal of Personal Selling and Sales Management.*

The purpose of this chapter is to provide a rich source of actionable suggestions for the busy executive. All the ideas presented here flow from the product life cycle (PLC) concept—the notion that products are conceived, born, grow, mature, decline, and die—and that management action can be made more effective by taking the life cycle into account. Just as the nutritional requirements of an infant are different from those of a teen-ager, the product has varying needs as it progresses through its life cycle. The product's life is sustained by the essential functions of the business: production, purchasing, finance, and marketing as directed by management.

Because product requirements change over the life cycle, management strategy and tactics must change as well. In this chapter, strategy and tactics across the PLC will be presented for the following functions: (1) general marketing plus personal selling, advertising, physical distribution, and transportation; (2) purchasing; and (3) finance. Production will not be treated in a separate section; however, production strategies will be mentioned within the other activities.

The terms *strategy* and *tactics* do not have uniform definitions in business, so it is important for us to define what we have in mind. We consider strategy to be relatively broad business actions, such as seeking to stimulate primary demand in the introduction stage of the PLC. Tactics are more specific actions that are relatively narrow in scope compared to strategy. A tactical recommendation is to introduce annual models with an emphasis on styling during the maturity stage. We have not tried to differentiate sharply between strategy and tactics. In short, this chapter will be concerned with alternative courses of actions, some more general than others.

The first major section of this chapter is a general orientation to the PLC—what it is as well as how and why it can have an impact on business activities. It is a useful background for those who are unfamiliar with the PLC. Next, separate sections for each functional area are presented, starting with general marketing management. Each functional section is self-contained, so a reader interested only in transportation could read that section. Some overlap exists between functions, and a transportation executive would probably benefit by also covering the section on physical distribution. However, some redundancies will be found across sections, which is unavoidable since each section was written to stand alone.

This chapter provides a number of concrete, practical suggestions. Of course, the applicability of any particular idea will depend on your situation, so you will need to be judicious in selecting your menu from the cafeteria of suggestions presented here. This is not a conceptual or theoretical treatment of the subject. The reader who is interested in that sort of approach can turn to any number of excellent textbooks on business management. We envision this chapter serving as a brainstorming session for you. By turning to material that is congruent with your area of responsibility, you can find numerous ideas. Most of the ideas represent common sense applied to the PLC. Some of our ideas may not be new to you, but we hope that by presenting a large set of possibilities, some will be new. Just a few good ideas would, we are sure, be worth the time and money invested in this book.

PRODUCT LIFE CYCLE CONCEPT

The product life cycle (PLC) is a generalized model depicting the unit sales trend of some narrowly defined product from the time of the market entry decision until withdrawal.[1] The PLC can be approximated by a bell-shaped curve (Figure 21-1), which is divided into several segments or stages.[2] For the sake of simplicity, we will adopt a five-stage PLC: (1) design (or precommercialization); (2) introduction; (3) growth; (4) maturity; and (5) decline. The major characteristics of each phase are summarized below.

Characteristics of PLC Stages

Design (or Precommercialization) Stage. This stage involves the development and test marketing of some product or service the company has never attempted to sell with full-scale efforts. The executive seeks flexibility because the project is uncertain. Operating decisions should be tentative or reversible. Experimentation with determining the appropriate levels for the marketing variables may involve a test market in this stage. Typical operational activities include coordination of R & D with other functions, product design, process planning, preparation of a marketing plan, and recruitment of personnel.

Introduction Stage. This stage commences with the full-scale marketing of the product or service in its intended market or in a large region. Sales are low but rising. This stage is also characterized by losses or low profits, uncertainty of ultimate sales level, product is vulnerable to attack from competing items or services, relatively few distributors, inexperienced personnel, middlemen

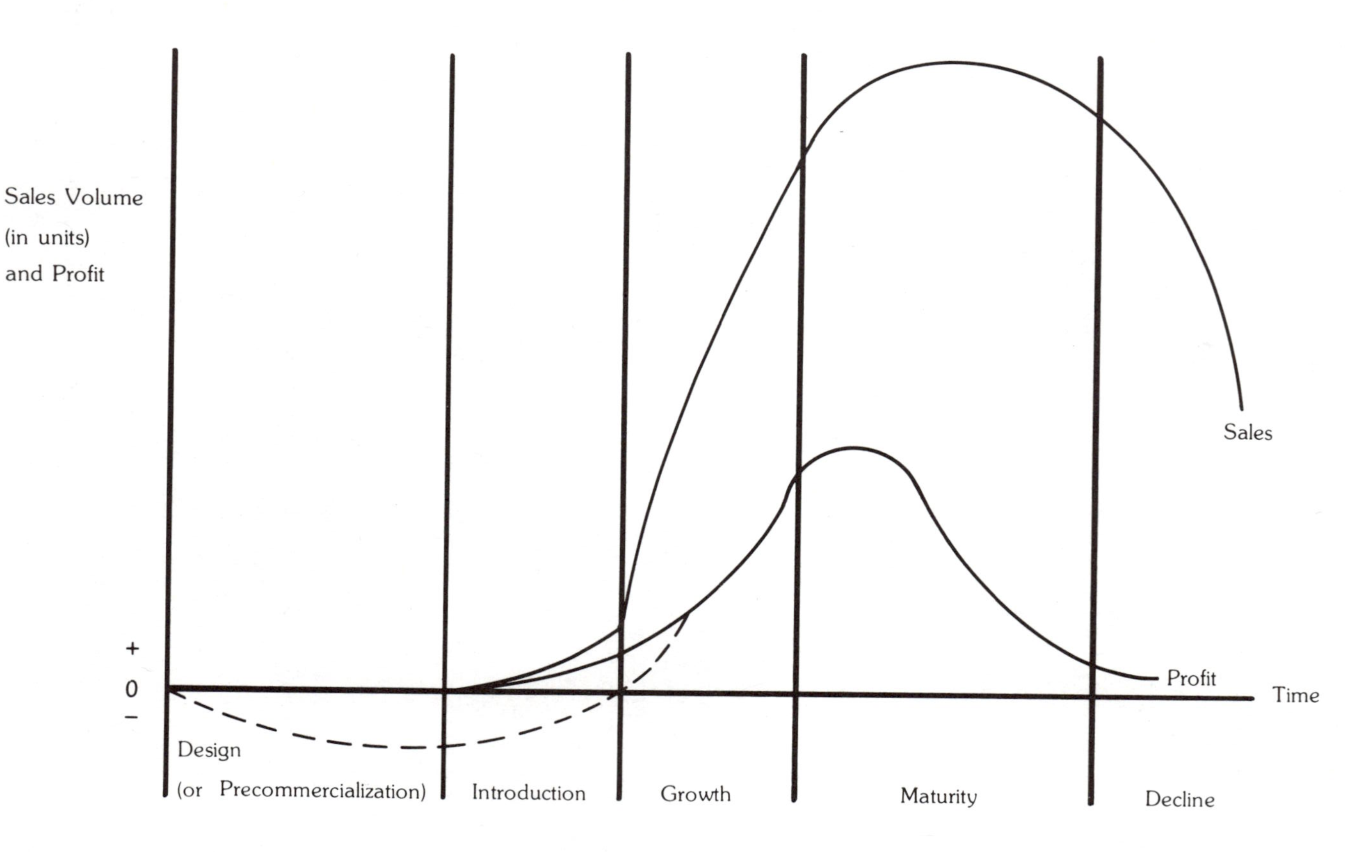

Figure 21-1. Generalized five-stage product life cycle.

buying base inventories, and the product is often manufactured in pilot plants. The uncertainty of the new product's ultimate success or failure creates a desire to maintain flexibility in functional activities. Systems are redesigned and modified to smooth out problems as they are discovered. Close monitoring of internal reports is critical both to identify these problems and to discover at the earliest possible moment indications of the ultimate destiny of the product. Product availability to the market is a crucial factor at this stage in the product's life. Toward the end of this phase, top management will decide either to withdraw the product as a failure or to continue to support the product if its sales are increasing. If the decision to support is confirmed, then performance standards and controls will be developed. Typical activities include product debugging, determining the ultimate scale of the plant, and introductory promotions.

Growth Stage. This stage begins when unit sales start increasing at an increasing rate. Trial sales have been largely completed. This phase is also epitomized by substantial profits, less product vulnerability, and use or development of full-scale production lines. Operating decisions reflect rapid expansion in demand, not buying of market share at the expense of competitors. Sales forecasting becomes critical to effective marketing of the product. Performance standards are implemented and maintained. Marketing strategy focuses first on encouraging potential new customers to try the product and later on developing brand loyalty with customers. As competitors enter the marketplace, customer service becomes critical as a competitive tool. Cost is secondary in importance. Typical activities include incurring a heavy amount of manufacturing overtime, adding new models to the product line, and increasing the number of distribution outlets handling the product.

Maturity Stage. This stage occurs when sales volume continues increasing but at a decreasing rate and eventually levels off or declines slightly. This stage is also represented by profits reaching a plateau and declining rapidly, existence of many aggressive competitors, declining prices, production facilities and/or processes in need of heavy repair, and a cost-price squeeze. Extremely vigorous competition yields to price-cutting tactics and large increases in promotion to maintain sales levels. To minimize the impact on profits, a companywide pursuit of efficiency is initiated. Also, new marketing opportunities in the form of product modification and repositioning are sought. Product improvements are encouraged in quality, style, and accessorial features to become more competitive in existing markets and to enter new markets. However, the pursuit of efficiency is paramount. Typical operations include development of new markets or new models or sizes, short production runs, and special sales inducements or concessions to customers.

Decline Stage. This stage occurs when unit sales decrease at an increasing rate. Other attributes of this stage are declining profits, distributors forsaking the product for other items, and sales and profit declines cannot be curtailed by the firm except in the very short run. Executives try to revitalize this product. If unsuccessful, the product is eliminated. The objective in this stage is to minimize risk and maintain flexibility. Business strategy is one of retrenchment—pulling back from declining markets but at the same time watching for opportunities both in those market segments abandoned by competitors and in those segments composed of significant groups of brand-loyal customers. The number of product varieties and promotional expenditures are reduced in anticipation of abandonment. The executive seeks to focus on true cost eliminations, not cost reassignments such as selling manufacturing equipment and cancelling R & D budget.

Different PLCs

The PLC curve as defined above is a general representation. The time length of any stage as well as the shape of the overall PLC may vary for different products and industries.[3] Furthermore, life cycles can be shortened or prolonged by deliberate managerial action. Excluding commodities (for example, wheat, steel) and/or premature intervention by the firm, most products pursue some type of PLC.[4]

The PLC notion is keyed to three assumptions:

1. Products move at varying rates of speed through the stages in sequence starting with market introduction and ending with decline.
2. Profits rise sharply in the growth stage and begin to decline in maturity due to competitive pressures even as volume continues to rise.
3. Marketing emphasis required for successful product exploitation varies from stage to stage as changes occur in competitive behavior.[5]

The PLC concept is especially appropriate for companies in which particular products dominate managerial thinking or actual volume. However, its importance varies by stage. The early periods are more volatile, more consequential in their impact on functional operations than the later ones. Seasonal patterns override the PLC in a protracted, stable maturity period. During decline, management's attention passes to successor products. This concept does not apply to organizations whose output has a steady long-term trend (for example, unbranded hardware) or has balanced diversity such that no one product line's sales pattern justifies special attention.

It is important to note that the PLC does *not* just happen. Instead, the PLC is the result of the interaction of a number of variables. It is shaped by both market demand factors and other external conditions—usually beyond the firm's control—as well as by the firm's marketing efforts.[6] Hence, the PLC represents a dependent variable that is the result of many factors—not just the passage of time per se.[7]

Rationale for PLC

Now that the PLC has been described, we discuss briefly why it occurs. The PLC is rooted in unfulfilled human wants and innovation. Innovation triggers the birth and death of products. As an example, before the development and commercialization of air conditioning, people wanted relief from summer heat, and various products helped to meet that want. The electrical ceiling fan was one such product. The fan did not completely solve the need, and air conditioning was invented. The ceiling fan was a mature product that was partially displaced by air conditioning, resulting in the decline stage for the fan. Decline is attributed to displacement by a product that better satisfies a need.

The introduction stage is triggered by innovation. During introduction, the new product has no direct competition so the firm has wide latitude in pricing, subject to customer sensitivity to price. The low sales volume keeps per-unit fixed costs high, and start-up costs of all types (for example, production, distribution, and promotion) are extensive. The result is a loss or low profits. Extensive promotion is needed to inform prospects of the innovation and to convert them into buyers as well as to gain distributors for the new product. During this stage, the product may appeal to a narrow segment of customers, and only specialized distributors would be interested in handling it.

The relatively long introduction stage is rooted in the process of adoption—the individual's decision to accept or reject a new product—and diffusion—the transmission of a new concept through a social system. The paradox of adoption is that person-to-person influence is important in directing the individual towards adoption. However, in introduction, few adopters are available to start the ball rolling. Personal influence is further restricted by the diffusion process. The first persons to adopt a new product tend to be oddballs who are not in the mainstream of their social system. The lack of social contact between the first adopters and everyone else further slows the rate of adoption.

The growth stage is driven by rapidly increasing customer acceptance of the product. The barriers to adoption fall as the product becomes acceptable to large numbers of people and visible to even more prospective customers. Distribution is broadened as less specialized wholesalers and retailers find their customers are buyers and as producers seek to tap a broader market. Costs fall due to economies of scale and also skill as producers learn how to reduce costs. Prices will probably be lowered as marketers seek new segments. Even so, prices will more than cover costs and profits will be high. Unless barriers to entry are considerable, the rapidly expanding market will attract competitors. However, increasing demand will moderate competitive pressures. Since buyers are aware of the product and its benefits, promotion will switch from persuading prospects to accept the basic innovation to brand promotion.

The main characteristic of the maturity stage is that the per capita rate of sales growth decreases as the pool of new buyers dries up. Essentially, every customer that wants the product has purchased it. The rate of sales for a durable good will depend on its replacement rate and on the growth, if any, of the market segment(s) that use the product. The replacement rate will be sensitive to the physical wear out of the product as well as sources of obsolescence, such as style changes or new product features. Sales of a nondurable will be a function of its consumption rate. Direct competition will become intense. The industry may suffer from overcapacity. Profitability can be attractive if overcapacity is not a problem. Profits will vary with the business cycle. Pricing will be competitive as will promotion. Producers strive to discover new market segments and add variations on the basic product. The product will be widely distributed. Since the maturity stage is the longest stage for most products, it represents normal marketing, production, purchasing, and financial activities.

The decline stage is reached when per capita sales of the product begin to decrease. The basic reason for the decline is that a new, innovative product has displaced the established item. Overcapacity is bound to occur and price competition will squeeze profits. Some firms may have already abandoned the industry during periods of overcapacity in the preceding maturity stage. Exit from the industry will increase. Some marketers may choose to milk their brand prior to exit, so relatively little will be spent on promotion and other demand-oriented marketing tools. The item will not be distributed as widely as it once was.

The PLC has been described and why it occurs has been explained. In the remaining sections of this chapter, PLC strategy and tactics models for selected functional areas of the business will be presented.

GENERAL MARKETING STRATEGY AND THE PLC

Given the large number of PLC marketing strategy models,[8] we selected the most complete model and filled it in by using portions of some of the remaining models. The most comprehensive PLC marketing strategy model is by Staudt and Taylor.[9] It will be supplemented by portions of the models from McCarthy,[10] Kotler,[11] and Wasson.[12] We have organized the general marketing strategies by five PLC stages: introduction, growth, maturity, saturation, and decline.

In the introduction stage, keen marketing judgment is essential. Consequently, the recommended strategies include:

1. Establish a high price.
2. Direct marketing effort at stimulating primary demand.
3. Eliminate as many product imperfections as possible.
4. Use job-shop methods of production, which makes use of general purpose tools rather than line or mass production methods.
5. Avoid heavy fixed investment in plant and equipment, especially that which is production oriented.
6. Advertising, personal selling, and other marketing expenditures should be high.
7. Distribute the product on a limited basis.
8. Manufacture and sell a limited product line, preferably consisting of one product model or version.[13]
9. Build up channels of distribution.

10. Promotion should be informative, that is, tell customers about the existence, advantages, and uses of the new product.
11. Use both mass and personal selling.[14]
12. High trade discounts and sampling devices are advisable.
13. Exclusive or selective distribution policy, with distributor margins high enough to justify heavy promotional expenses.[15]

When the product moves into the growth stage and encounters widespread buyer approval, the suggested strategies are:

1. Replace job-shop production methods with more efficient line or mass production methods.
2. Expand distribution by increasing the number of outlets and dealers handling the product.
3. Stimulate selective demand.
4. Drop price slightly.
5. Expand the product line by adding new models.
6. Make large fixed investments in plant and equipment.
7. In promotional messages, emphasize the advantages of the firm's product brand over those of competitors.[16]
8. Concentrate on using mass selling, although personal selling is still needed.[17]
9. Develop modular product design, if possible, to facilitate flexible addition of variants to appeal to every new segment and new use-system, as they appear.
10. Offer customary trade discounts.
11. Intensive and extensive distribution policy, with margins adequate to make product interesting to dealer, but no more than this.
12. Heavy emphasis on rapid resupply of distributor stocks is necessary, with heavy inventories at all levels.[18]

The maturity stage represents a period of extreme competitive volatility. Therefore, these strategies are prescribed:

1. Make a conscious effort to reach the mass market.
2. Introduce annual product models with emphasis on styling.
3. Carefully monitor and manage the costs associated with providing product service and parts, especially those related to labor, inventory carrying, storage, and personnel.
4. Stress the availability of good service and adequate supplies of spare parts in promotion.
5. Broaden the product line to more effectively reach the various available market segments,[19] but simultaneously tighten up product line to eliminate unnecessary specialties with little market appeal.
6. Concentrate on using mass selling, although dealer promotions and personal selling are still needed.
7. Intensive and extensive distribution policy and a strong emphasis on keeping dealers well supplied but with minimum inventory cost to them.[20]
8. Decrease price until a competitive parity is reached.[21]
9. Imitate competing features of the more successful products.
10. Promotion should be both persuasive and aggressive. It should emphasize the advantages of the firm's product brands, whether minor or psychological, as compared to those of competitors.[22]

During the saturation stage,[23] replacement sales tend to dominate the firm's sales volume. In order to survive this stage, the following strategies are suggested:

1. Manufacture and sell products that closely match and satisfy the buyer's self-image.
2. Achieve a cost structure that is comparable to competitors.
3. Strengthen distributors by having enough carefully selected, properly located, and sufficiently well-trained distribution outlets.
4. Try to achieve a large fraction of the dealer's total sales volume.
5. Experiment with different kinds of channels.
6. Monitor logistics costs.
7. Price competitively.[24]
8. Use markdowns and off-list pricing.
9. Spend about the same ratio on promotion to sales as competitors. Do not deviate widely from this level.
10. Emphasize consumer and trade deals.
11. Discover new uses for the product.[25]
12. Concentrate on mass selling although dealer-oriented promotions are still needed.
13. Introduce flanker products.
14. Reexamine necessity for design compromises.
15. Intensive and extensive distribution policy and a strong emphasis on keeping dealers well supplied but with minimum inventory cost to them.[26]

In the fifth and final stage of the PLC, decline, innovations from competing industries make the product obsolete, and the market declines. Confronted with this dilemma, it is recommended that the firm adopt these strategies:

1. Diversify into different products and markets.
2. Develop innovations to replace current products or make technological improvements in the old products.
3. Simplify product lines.
4. Rely on product differentiation to capture what little remains of the market.
5. Restimulate primary demand.
6. Reduce significantly both marketing and nonmarketing expenditures.
7. Use selective methods in both promotion and distribution.
8. Reduce price in the earlier half of this stage and increase it in the latter half.
9. Focus efforts on the core market.[27] Withdraw from smaller market segments.[28]
10. Prune product line constantly to eliminate any items not returning a direct profit.
11. Phase out distribution outlets as they become marginal.[29]

PLC FINANCIAL STRATEGY MODEL

In this model, the PLC notion is applied to the area of financial administration. For each stage of a five-stage PLC—precommercialization, introduction, growth, maturity, and decline—Fox suggests several strategies to financial executives for improving their administration of operating and financial factors.[30]

In the precommercialization stage, Fox recommends the following financial strategies:

1. Determine whether to lease or purchase manufacturing equipment for the future product.
2. Conduct business analysis of proposed new product.
3. Maintain financial control of R & D.
4. Maintain high internal security in R & D; this is the period of peak danger of theft of proprietary information.
5. Use full costs (except sunk cost).
6. Develop financial plans for prospective product.
7. Consider acquisition versus in-house R & D.
8. Consider treatment of interest on construction.
9. Determine method of depreciation for future use.
10. Consider inventory valuation method such as FIFO or LIFO for future use.
11. Plan for progressively higher cash outflows as R & D advances.
12. Budget no revenues, except on research contracts.
13. Budget fiscal losses.
14. Budget new personnel needs.
15. Analyze financial strength of future vendors.
16. Analyze prospective middlemen.
17. Develop estimated standards for manufacturing costs.

Within the introduction stage, Fox suggests these financial strategies:

1. Determine whether initially to price low or high.
2. Use subcontractors as much as possible, thereby minimizing investment in plant and equipment.
3. Obtain cancellable lease with option to buy.
4. Aggressive distributors or salespersons should be on commission.
5. Design product reporting system.
6. Develop marketing aids, such as salespersons' portfolios showing financial benefits to buyers.
7. Plan for continued net cash outflow for production and marketing.
8. Streamline credit and collections to process new accounts and give prompt feedback to marketing.
9. Determine ultimate scale of plant(s).
10. Determine location(s) of plant(s) for existing product.
11. Give stock options or other deferred inducements to executives.
12. Arrange for accounting and control on new finished goods.
13. Be prepared to extend credit to customers and/or ship on consignment.
14. Production should be labor intensive, if feasible.
15. Control engineering changes.
16. Budget heavy scrap in manufacturing.

In the growth stage, Fox emphasizes the adoption of these financial strategies:

1. When buying equipment for existing product, decide accounting treatment of investment credit.
2. Incur heavy amount of overtime in manufacturing.

3. Produce as much as possible in own plant, thereby minimizing subcontracting.
4. Salespersons should be salaried.
5. Switch production from labor intensive to capital intensive.
6. Budget receivables turnover improvement.
7. Budget for rapidly rising receivables.
8. Budget heaviest pressure on current and acid-test ratios.
9. Begin committing money to R & D for successor product(s).
10. Examine existing product's return on the basis of scarce resources.
11. Caution optimistic executives against overbuilding of specialized facilities.
12. Most propitious time to sell common stock.
13. Avert liquidity crisis.
14. Budget and control for highest profit of all stages.
15. Enter into foreign markets after considerable domestic experience.

With the advent of the maturity stage, Fox suggests these financial strategies:

1. Purchase parts and/or components from abroad.
2. Develop incentives for production efficiency.
3. Perform cost reduction analysis, such as learning curve and value analyses.
4. Place salespersons on salary plus commission basis.
5. Use direct costing for internal purposes.
6. Analyze logistics (or transportation) system for existing product.
7. Float bonds or incur intermediate debt instead of issuing stock.
8. Production should be capital intensive.
9. Develop engineered standards for manufacturing costs.
10. If capacity is excessive, liberalize credit terms.
11. If market share is very high, be extra careful to avoid actions that could trigger antitrust suits.
12. Budget receivables turnover stability.
13. Budget for normal current and acid-test ratios.
14. Budget declining profit margin.
15. Consider diversification, merger, forward integration, and so forth.
16. Prepare for heavy expenditures for extensive and complex inventories.
17. Computerize clerical and other routine jobs, especially reports on inventories of wholesalers and other resellers.
18. Conduct financial analyses for each product and market to assist the marketing manager relative to the existing product.
19. Evaluate financial strengths and weaknesses of existing product's competitors.
20. Authorize price reductions as forced competitively.
21. Stabilize net cash inflows.
22. When deploying field inventories, consider property taxes.
23. Prevent any proclivity for price collusion.
24. Cope with cost-price squeeze.
25. Give many *deals* (concessions to customers) and other special sales inducements, not initiatory.
26. Evaluate financial aspects of long-term procurement contracts.

In the last stage, decline, Fox recommends the following financial strategies:

1. Consider temporary price increase.
2. Simplify product line because many models have low revenues.
3. Sell manufacturing equipment.
4. If revitalization is hopeless, withdraw promotional support.
5. Cancel this product's R & D budget.
6. Determine escapable costs.
7. Maintain tight controls to keep finished goods inventory(ies) low.
8. Borrowing, if necessary, should be short term.
9. Dispose of functionable specialized facilities.
10. Convert plant to other products.
11. Write off unsalable inventory and equipment.
12. Capture temporary high net cash inflows.
13. Exert special efforts to prevent bad debts.
14. Administer systematic retrenchment.
15. Budget receivables turnover deterioration.
16. Budget substantial stock of replacement parts.
17. Realize temporary profit surge.

PLC TRANSPORTATION STRATEGY AND TACTICS MODEL

In this model, Kaminski and Rink[31] relate the PLC theory to the transportation aspect of physical distribution. As an aid to transportation managers in the formulation and implementation of effective and timely transportation strategies, Kaminski and Rink classify 78 transportation strategies and tactics according to various transportation functions (for example, carrier selection and customer service) and five PLC stages: planning, introduction, growth, maturity, and decline.

In the planning stage, Kaminski and Rink suggest the following transportation strategies and tactics:

1. Participate in generation of new product ideas.
2. Cooperate fully in providing transportation-related information to the new product developers.
3. Monitor the development of the new product.
4. Identify product transportability factors critical to the ultimate success of the new product.
5. Develop and monitor close interdepartmental and intercompany relationships.
6. Request company product-design engineers to consider the impact of alternative product designs on the company's transportation system.
7. Identify middlemen who may be integrated into the company's transportation network with minimum disruption.
8. Assess the abilities of the company's existing private transportation capability, if any, concerning existing routes, load factors, and delivery schedules.
9. Initiate contacts with for-hire carriers.
10. Identify preferred transportation modes and carriers.
11. Resolve the new product's common carrier rate classification.
12. Establish the availability of small-volume rates and economy rates on large-sized shipments.
13. Establish protective packaging requirements in line with carrier specifications and desired customer service standards.
14. Scrutinize transportation-related regulatory requirements.
15. Develop delivery schedules.
16. Monitor the newly designed transportation system to resolve potential interruptions and conflicts before the system comes on line.

Within the introduction stage, Kaminski and Rink recommend these transportation strategies and tactics:

1. Cooperate closely with middlemen to discover transportation difficulties and shortcomings.
2. Monitor carrier performance for conformance with desired customer service standards.
3. Replace carriers unwilling or unable to provide required service levels.
4. Continue carrier contacts to develop good business relationships with carrier personnel.
5. Delegate the routine task of monitoring carrier rate-increase proposals to lower management echelons.
6. Use direct shipment by air and/or motor carrier when possible to avoid unnecessary fixed investment.
7. Avoid longer-term commitments for transportation equipment, facilities, and services.
8. Consider consolidation programs to take advantage of existing transportation systems only if customer service levels can be maintained.
9. Rely heavily on air and/or motor carriers to maintain customer service on lower-volume routes.
10. Modify performance standards using feedback from the new transportation system.
11. Commence negotiations with for-hire carriers for additional necessary service adjustments.
12. Institute expediting and tracing services on important routes.
13. Assess the need for the return movement of the new product for servicing, recycling, or recall.
14. Monitor sales research reports to anticipate how soon the expected jump in sales will occur.

In the growth stage, Kaminski and Rink emphasize the adoption of these transportation strategies and tactics:

1. Monitor system performance to minimize the effects of rapid increases in demand.
2. Expand the transportation staff to handle the additional workload created by the increasing volume of orders.
3. Establish temporary market priorities in response to unexpected surges in demand.
4. Recognize the need for rapid and frequent deliveries.
5. Require compliance with established delivery schedules to maintain customer service standards.
6. Continue premium transportation where necessary to maintain service standards.
7. Gear carrier relations toward solving mutual problems.
8. Encourage carrier personnel in the development of expanded carrier services.
9. Review the product's classification and rating as the product undergoes minor modifications.
10. Initiate aggressive negotiations by company rate specialists for lower commodity rates.
11. Negotiate with competing carriers to secure lower rates or improved service.
12. Formulate procedures for setting up an internal freight-bill auditing system.

13. Develop a more stringent loss and damage prevention program.
14. Deemphasize consolidation programs on high-volume routes.
15. Substitute more efficient forms of consolidation for others already in use.
16. Institute variable route delivery schedules.
17. Give serious consideration to complete private carrier operations on high-volume routes.
18. Engage in longer-term commitments for transportation equipment, facilities, and services.
19. Initiate planning for transportation system expansion as new opportunities are identified.
20. In disputes with carriers, make concessions to obtain crucial services quickly but do not waive company's rights in future negotiations.

With the advent of the maturity stage, Kaminski and Rink suggest these transportation strategies and tactics:

1. Anticipate the impact of changes in the company's marketing strategy and of continued product modification.
2. Improve departmental operating efficiency.
3. Control transportation costs closely.
4. Reevaluate customer service policies.
5. Phase out carriers offering marginal or high-cost service.
6. Become increasingly diligent concerning favorable mode and carrier cost-service trade-offs.
7. Rely increasingly on rail transportation with its lower-cost volume rates.
8. Consider additional for-hire transportation service on a contractual basis.
9. Evaluate the cost-service trade-offs of shippers' associations, containerized freight shipments, distribution centers, and freight consolidation programs.
10. Identify shipments and routes that provide beneficial outbound and/or inbound consolidations.
11. Change to fixed-route delivery systems.
12. Monitor freight bills closely.
13. Consider using a preshipment audit procedure.
14. Delegate routine cost-monitoring activities to first-line supervisors and dispatchers.
15. Expedite shipments on a request basis only; bill customers for added costs.
16. Coordinate product and packaging modifications with the product manager.
17. Expand the private carrier operation using specialized vehicles and equipment.
18. Investigate additional for-hire carrier services that improve system operating efficiency.
19. Reevaluate the routing of shipments.
20. In rate-service disputes with carriers, enlist the help of the company's legal department where necessary to defend the company's rights.

In the fifth and final stage, decline, Kaminski and Rink recommend the following transportation strategies and tactics:

1. Monitor sales research reports for anticipated decreases in sales.
2. Simplify the transportation system to maintain flexibility.
3. Commit unneeded private transportation capacity to other burgeoning products.
4. Aggressively pursue freight consolidation programs.
5. Pool shipments into selected markets on selected days.
6. Employ an outside agency to audit freight bills.
7. Plan and implement a system for spare parts and product servicing.
8. Establish a contingency plan for transporting future product recalls.

PLC PERSONAL SELLING STRATEGY AND TACTICS MODEL

In this model, Dodge and Rink[32] apply the PLC concept to the area of personal selling. Using the PLC as a gauge of changing market conditions, Dodge and Rink formulate a list of personal selling strategies and tactics for each stage of a four-stage PLC—introduction, growth, maturity, and decline—which they feel will assist sales managers in the implementation of more effective and timely personal selling strategies and tactics.

Within the introduction stage, Dodge and Rink recommend these personal selling strategies and tactics:

1. Identify best prospects.
2. Distribute sales literature to salespersons.
3. Hold sales orientation and training meetings (emphasis on product knowledge and demonstration).
4. Recruit and select salespersons.
5. Develop sales and market forecasts (potentials).

6. Establish performance standards other than the traditional sales quota (demonstrations, calls on new customers, and so forth).
7. Make assignments and reassignments of salespersons to accounts, territories, and products.
8. Keep abreast of sales by salesperson, territory, and customer.
9. Form special sales teams to introduce product in designated market segments.
10. Provide and ensure adequate technical support.
11. Provide factory with feedback on customer reactions to product (bugs, adjustments for customers, and so forth).
12. Call on new customers.
13. Initiate sales contests.
14. Use merchandise promotions.
15. Use advertising to get leads.
16. Offer special credit terms.
17. Develop initial relationships with new customers and middlemen.
18. Involve sales manager in personal selling.
19. Involve sales manager in joint sales calls with salespersons and perhaps technical representation.
20. Organize demonstrations to groups of buyers and influencers (customer's business, plant, trade show, and so forth).

In the growth stage, Dodge and Rink prescribe these personal selling strategies and tactics:

1. Coordinate with production and distribution executives to gear production rates and inventories as closely as possible to actual needs.
2. Assign salespersons to different accounts as sales grow.
3. Adjust sales quotas to reflect sales production.
4. Analyze sales, particularly by customer and territory.
5. Hire and train additional salespersons.
6. Reduce time interval between sales reports (for example, requiring weekly rather than biweekly reports).
7. Increase number of sales meetings.
8. Set up new sales branches.
9. Allocate or ration product among customers, present and potential.
10. Stress developing customer loyalty.
11. Increase salespersons' workloads in terms of more sales calls.
12. Divide up sales territories as they increase in dollar sales.
13. Assign priorities in exploiting market (for example, which market segments are to be called on first).

With the advent of the maturity stage, Dodge and Rink suggest these personal selling strategies and tactics:

1. Reallocate or rebudget salespersons' expense monies.
2. Retrain salespersons in the use of different sales presentations (for example, shift from informational to persuasive format and interaction with customers).
3. Reassign salespersons.
4. Adopt a skimming philosophy with present customers (for example, place special emphasis on profitable, key accounts; set up national accounts).
5. Eliminate marginal accounts
6. Use package or systems approach in selling.
7. Have manager direct greater attention toward competitors, for example, competitors' actions have more effect on decisions, and then knowing this, the sales manager consults frequently with both salespersons and the marketing research department on competitive activity.
8. Search for and locate new profitable prospects in face of increasing competition.
9. Develop cold call quotas.
10. Have salespersons develop sales plans for individual customers.
11. Classify and make calls on accounts according to profitability and/or potential.
12. Organize selling efforts to customers of customers.
13. Develop merchandise programs that are persuasive in nature.
14. Develop programs whereby more liberal credit terms are available (for example, greater cash discounts and extending pay period).
15. Show customers how to routinize the whole buying process (for example, use automatic or computer reordering procedures).
16. Reinitiate sales contests.
17. Request brand and/or product manager to make minor modifications in product and/or product design.
18. Pay more attention to how salespersons cover their territories (routings).
19. Divide customers into classes or divisions based on corresponding sales or profit.
20. Have sales manager spend more time in supervision of salespersons.
21. Conduct more frequent reviews of salespersons' efforts.

In the decline stage, Dodge and Rink emphasize the adoption of the following personal selling strategies and tactics:

1. Project sales personnel needs in terms of possible terminations, transfers, and attrition.
2. Combine and/or redesign sales territories as sales decrease.
3. Set up inventory clearance sales.
4. Shift salespersons' efforts to other more promising products.
5. Forewarn customers of possible withdrawal from the market.
6. Assist customers in developing an effective program for disposing of surplus product inventory.
7. Develop a formalized feedback system to aid management in planning and timing withdrawal from the market.
8. Seek out customers who specialize in handling clearance sales.

PLC PURCHASING STRATEGY MODEL

In this model, the PLC concept is related to the purchasing function. The dependence of organizational buying operations on a product's unit sales trend was initially recognized by Berenson.[33] He formulated a set of normative procurement strategies for each stage of the PLC in an effort to aid the purchasing executive to "make decisions with greater accuracy, relevance and foresight."[34] Independent of this effort, other prescriptive models of organizational buying also used PLC as their integrating concept.[35]

On the basis of this past research and experience as well as discussions with many organizational buyers, Fox and Rink[36] substantially modified and enriched earlier PLC organizational procurement strategy models. The end result was a cross-classification of 83 procurement strategies according to nine departmental functions (for example, production, R & D, finance, and legal) and five PLC stages: precommercialization, introduction, growth, maturity, and decline, which "represents an integrated framework for formulating, organizing, and implementing both timely and effective purchasing strategies."[37]

In the precommercialization stage, Fox and Rink suggest the following purchasing strategies:

1. Evaluate custom shops and supply houses as prospective vendors of items and services not bought before or bought in small quantities.
2. Urge design department for new product specifications that will reduce future supply problems.
3. Try to have new product design incorporate materials and components already in satisfactory use on existing products.
4. Suggest in-house manual assembly of critical parts to prevent product disclosure to competitors.
5. Anticipate that in future months and years, requirements will change. Specifically, push for these options in early supply contracts: shipment acceleration or delay; cancellation or return; performance bonds or liquidated damages; right to change specifications; and a schedule of declining unit costs reflecting the supplier's learning curve.
6. Formulate joint plans with vendors for sudden or erratic production of quantities of recently tested materials.
7. Obtain samples for research and testing purposes.
8. Perform make-or-buy analysis for forthcoming product.
9. Relegate unit costs because vendors' cooperation in furnishing small quantities for trial orders is more important.
10. Delegate to junior buyers the purchasing of many different new items in small quantities.
11. Assist in screening forthcoming product for compliance with regulations on ecology, health, safety, and so forth.
12. Participate in evaluating proposals for equipment for the forthcoming product.
13. Approve specifications of new product on the basis of prospective materials availability.
14. Cooperate with marketing department by finding exotic materials that can be romanced in promotional efforts.
15. Participate in the design of materials management plan.
16. Make sure that company has required licenses or permits for use of toxic or other dangerous materials.
17. Participate in financial planning for the prospective product.
18. Purchase a wide variety of special parts, tailor-made components, and technical services—each in very small quantities.

Within the introduction stage, Fox and Rink recommend these purchasing strategies:

1. Develop list of preferred and standby suppliers, which reflects performance to date and

the capacity to deliver during expected sales growth.
2. Work closely with vendors concerning product defects and engineering changes.
3. Be prepared to handle avalanche of engineering changes.
4. Use subcontractors or rent facilities until the product's market acceptance has been demonstrated.
5. Monitor sales research reports to anticipate how soon a new spurt of sales orders can be expected.
6. Along with engineering and accounting departments, develop standards for cost, quality, and other factors relating to the product.
7. Near end of period, plan an orderly shift over from subcontractors to owned facilities.
8. Consolidate small-quantity orders with other incoming materials.
9. Assign senior procurement executives to culling vendors and establishing long-term relationships.
10. Keep flexible through leasing with option to buy.

In the growth stage, Fox and Rink emphasize the adoption of these purchasing strategies:

1. Selectively widen supply sources without disrupting established arrangements.
2. Maintain strict quality standards on purchased materials and components despite pressure for speedy deliveries.
3. Expedite vendors' shipments.
4. Build substantial inventories of raw materials and goods-in-process.
5. Shift to large-volume suppliers.
6. Phase out some subcontractors.
7. Use brokers, if necessary, to find scarce items for immediate delivery.
8. In disputes with vendors, give concessions under expedient policy of obtaining crucial materials quickly but do not waive purchaser's rights for future occasions.
9. Expand purchasing staff to handle increasing volume of requisitions, expediting, and so forth.
10. Avoid overbuying when some managers extrapolate or exaggerate this period's steep sales increase.
11. Have purchasing executive use high-level external contacts to get shipments promptly.
12. Urge the treasurer's office to pay vendors' invoices promptly to help maintain the flow of shipments.

With the advent of the maturity stage, Fox and Rink suggest these purchasing strategies:

1. Arrange for suitable mix of geographically dispersed vendors.
2. Conduct research on substitute materials.
3. Cancel overdue purchase orders.
4. Stabilize materials commitments.
5. Investigate feasibility of long-term contracts with fewer sources.
6. Because much of the buying is routine, replenishment of products in this period is suitable for on-the-job training of new buyers.
7. Along with designers and other staff personnel, conduct value analysis and other efficiency studies.
8. Participate in installation of economic order quantity or materials requirement planning.
9. Delegate buying to junior executives.
10. Participate in production decisions on the existing product.
11. Administer internal suggestion system for the improvement of purchasing operations.
12. Try to shift inventories to vendors.
13. Conduct cost-reduction efforts with purchasing department.
14. Preserve quality standards despite pressure for lower costs.
15. Adjust quality standards to conform with customers' buying criteria.
16. Urge the use of interchangeable parts on new sizes, attachments, and other modifications of the product.
17. Explore importing of labor-intensive parts or acquisition of suppliers, and notify the personnel and public relations departments accordingly.
18. If production decentralizes its manufacturing operations for each model variant, the purchasing manager may split up buying assignments the same way.
19. Perform make-or-buy analysis for the product with stable demand.
20. Use new buying techniques that will reduce lead times or eliminate the need to stock maintenance items.
21. Improve techniques of negotiation that will result in lower prices.
22. Evaluate usefulness of reports issued by purchasing both for its own internal control and for the purpose of advising management about purchasing operations.
23. Centralize some procurement at headquarters, for example, national contracts.
24. In disputes with vendors, hold out for protection of company's interests.
25. Due to the increase in models, sizes, and other

variants of the products, purchasing requirements fractionate.

26. Support shippers' association efforts for lower freight rates on incoming shipments.
27. Use trade associations as clearing houses for excess or needed materials.
28. Use trade association to compile industry statistics.
29. Get systematic price reductions based on vendors' experience.
30. Give preference to vendors who also are customers.
31. Get help of legal department if necessary to defend company's rights against vendors.

In the final stage, decline, Fox and Rink recommend the following purchasing strategies:

1. Accurate forecasts are very important, since excess specialized materials may lack alternative use.
2. Be sure sources for spare parts are adequate to serve customers still using the old product.
3. Reduce inventories and services for the existing product.
4. Revert to subcontracting if firm's equipment can be converted to burgeoning items.
5. Buy foreign or competitors' finished goods.
6. Economic order quantities are not serviceable.
7. Transfer specialized personnel to other duties or newer products.
8. Near end of this stage, forewarn suppliers that the existing product will be dropped.
9. Plan for disposal of surplus materials.
10. Have cautious controller-type of purchasing executive screen requisitions for the faltering product.
11. As cut-throat competition for declining volume makes some vendors desperate, enforce quality standards strictly.
12. If the company licenses another firm to take over the manufacture of the faltering product, ask the legal department to review and transfer purchasing commitments.

PLC PHYSICAL DISTRIBUTION STRATEGY AND TACTICS MODEL

In terms of physical distribution and transportation, writers originally applied the PLC concept to the larger area of physical distribution or logistics management.[38] However, these initial attempts were too broad and general to be useful to the practitioner. Later writers have partially overcome this problem by developing more specific and detailed strategic physical distribution and transportation planning models. Lynagh and Poist formulated a prescriptive guideline of 32 physical distribution strategies across four PLC stages.[39] In one of the most comprehensive efforts to date, Kaminski and Rink[40] developed a normative framework of 78 transportation strategies and tactics across five PLC stages, which was previously reviewed in this chapter.

In the following model, Kaminski and Rink[41] apply the PLC concept to the general area of physical distribution. The end result is 91 physical distribution strategies and tactics classified according to six physical distribution functions (for example, transportation, materials handling, inventory, and facility location) and five PLC stages: design, introduction, growth, maturity, and decline.

In the design stage, Kaminski and Rink suggest the following physical distribution strategies and tactics:

1. Cooperate fully in providing transportation-related information to the new product developers.
2. Request company product-design engineers to consider the impact of alternative product designs on the company's transportation system.
3. Assess the abilities of the company's existing private transportation capability, if any, concerning existing routes, load factors, and delivery schedules.
4. Monitor the newly designed transportation system to resolve potential interruptions and conflicts before the system comes on line.
5. Establish a no-stockout policy to support the new product.
6. Begin building inventory levels at central locations.
7. Establish small minimum order quantities to encourage middlemen to stock the new product.
8. Use centralized warehouse locations to minimize storage and inventory costs.
9. Identify target markets and major channels of distribution requiring a warehouse.
10. Avoid fixed investment by assessing unused capacity in existing warehouses, using public

warehouses whenever possible, and leasing short-term where necessary.
11. Coordinate with product design engineers to develop necessary packaging specifications (for example, shape, size and type of materials).
12. Assess the abilities of the company's existing packaging operation to accommodate the new product packaging requirement.
13. Make use of the professional expertise available from packaging suppliers.
14. Assess the abilities of the company's materials handling system to accommodate the handling and storage requirements of the new product.
15. Maintain flexibility by using multipurpose or standard materials-handling equipment instead of purchasing specialized equipment.
16. Avoid complex stock location systems; use deep storage bays, high stacks, and narrow aisles.
17. Monitor the development of the new product.
18. Identify middlemen who may be integrated into the company's existing distribution system with a minimum of disruption.
19. Perform credit checks on middlemen.
20. Establish standard order transmittal procedures.
21. Establish internal report generation and routing procedures.
22. Establish a customer complaint monitoring system.

Within the introduction stage, Kaminski and Rink recommend these physical distribution strategies and tactics:

1. Initiate and promote close carrier contacts to develop good business relationships with carrier personnel.
2. Monitor carrier performance for conformation with desired customer service standards; replace carriers unwilling or unable to provide required service.
3. Use direct shipment by air or other premium mode to avoid unnecessary fixed investment in transportation equipment and facilities.
4. Upgrade and improve the productivity and efficiency of manual inventory control systems.
5. Establish inventory requirement planning with the production and purchasing functions.
6. Assume the major share of the inventory burden in the channel.
7. Ensure system compatibility between warehouse location and the transportation system.
8. Use standard packaging materials until volume justifies tailoring these materials to the needs of the distribution system.
9. Be alert and able to react to changes in packaging specifications and problems in machine and/or operator performance.
10. Coordinate the technical development plan to include compatibility, quality control, and distribution testing criteria.
11. Unitize orders to avoid repetitive handling.
12. Minimize the number of handling decisions and alternatives to be considered (for example, pallet patterns).
13. Plan storage facilities for easy accessibility
14. Use expendable pallets and containers to eliminate the problems of return trips.
15. Evaluate order-filling accuracy as specified by customer service policy.
16. Expect frequent, erratic ordering.
17. Monitor sales reports for indications of product success.
18. Coordinate with middlemen to rectify start-up distribution system problems.

In the growth stage, Kaminski and Rink emphasize the adoption of these physical distribution strategies and tactics:

1. Expand the transportation staff to handle the additional workload created by the increasing volume of orders.
2. Require carrier compliance with established delivery schedules to maintain customer service standards and where necessary establish temporary market priorities in response to unexpected surges in demand.
3. Initiate planning for transportation system expansion as new opportunities are identified.
4. Anticipate and deal with various symptoms of rapid growth (for example, mix-ups, delays and late orders).
5. Implement a security system to control pilferage, deterioration, and damage.
6. Establish procedures to identify quickly customer service failures.
7. Continue to avoid stockouts but begin aiming for a balance between customer service levels and inventory costs.
8. Shift toward greater investment in fixed facilities.
9. Consider direct selling in lieu of intermediate inventory locations.
10. Shift the physical distribution burden toward middlemen.
11. Consider contract packaging service as demand outstrips capacity.
12. Investigate excessive customer and/or middleman complaints and product damage to determine if packaging is at fault.
13. Adopt packaging procedures designed to con-

trol pilferage and reduce deterioration (for example, shrink wrap; improved closures).
14. As mechanization and integration of the materials-handling system increases, recognize individual segment failure is critical to the system; build in redundancy and contingency plans.
15. Consider short-term leasing of additional equipment during peaks in demand.
16. Establish regional customer service centers.
17. Improve customer service through greater order-filling accuracy, reduction of backorders, and increased order transmittal speed.

With the advent of the maturity stage, Kaminski and Rink suggest these physical distribution strategies and tactics:

1. Anticipate the impact of changes in the company's marketing strategy and of continued product modification.
2. Evaluate the cost-service trade-offs of shippers' associations, containerized freight shipments, distribution centers, and freight consolidation programs.
3. Become increasingly diligent concerning favorable mode and carrier cost-service trade-offs.
4. Balance customer service and inventory costs; resist the tendency to overservice.
5. Institute automatic or mechanized inventory systems, including optical scanners; UPCs; and automatic labeling devices.
6. Expect continued product modification; integrate product planning and inventory planning.
7. Phase out older, inefficient facilities.
8. Consider sale-leaseback arrangements to finance some locations.
9. Establish distribution centers to meet required customer service levels.
10. Coordinate packaging changes with both packaging material suppliers and equipment manufacturers.
11. Evaluate new package designs and materials to secure improved handling and lower transportation costs (for example, increase density, improve cube, and reduce package weight).
12. Consider packaging implications of product modifications initiated by the production and marketing functions.
13. Eliminate unnecessary materials handling by proper planning, scheduling, and dispatching.
14. Redesign the materials handling system to accommodate changes in the product and/or packaging and the company's marketing strategy.
15. Install EDP equipment to speed up ordering and produce operating information.
16. Insist on adherence to ordering policies by middlemen.
17. Monitor product development; overcome resistance to improve products.

In the fifth and final stage, decline, Kaminski and Rink recommend the following physical distribution strategies and tactics:

1. Simplify the transportation system to maintain flexibility.
2. Commit unneeded private transportation capacity to other burgeoning products.
3. Establish a contingency plan for transporting future product recalls.
4. Employ rigid cost control; stockouts unimportant; fill orders as necessary.
5. Consider form postponement; that is, hold final manufacturing, assembly and packaging until customer specifications are known.
6. Coordinate reduced inventory requirements with the production and purchasing functions.
7. Withdraw from marginal markets in an orderly fashion.
8. Adopt a centralized facility policy.
9. Establish a location for spare parts, repairs, and recalls.
10. Maintain inventory of general (multi-) purpose packaging materials.
11. Anticipate that uneven work flow in packaging lines will create high labor costs and delays.
12. Commit unneeded packaging capacity to other burgeoning products.
13. Maintain the lowest possible ratio of equipment investment to units of material handled.
14. Consider a salvage plan for specialized automated equipment.
15. Minimize idle time and reassign unnecessary personnel.
16. Maintain close communication with major customers; advise them of impending abandonment.
17. Adopt flexible ordering policies to take advantage of loyal customer segments and other marketing opportunities.

PLC ADVERTISING STRATEGY MODEL

In this model, the PLC theory is related to the function of advertising. For each stage of a five-stage PLC—precommercialization, introduction, growth, maturity, and decline—Dodge and Rink[42] prescribe advertising strategies that managers should adopt.

In the precommercialization stage, Dodge and Rink suggest the following advertising strategies:

1. Formulate promotional objective(s) and purpose(s).
2. Determine promotional message, focus, and copy.
3. Determine promotional target markets.
4. Establish desired promotional coverage.
5. Establish desired promotional blend (personal selling, publicity, sales promotion, and advertising).
6. Determine desired promotional media mix.
7. Allocate promotional dollars to products, target markets, territories, media, and so forth.
8. Establish promotional budget.
9. Forecast the size of the potential market.
10. Develop and register the product's trademark.
11. Pretest promotional campaign to insure compliance with federal, state, and local laws, regulations, and ordinances.
12. Same as 11 but in terms of acceptability to environmentalists, consumerism groups, and so forth (for example, socially responsible, safety, and nonviolent).
13. After selecting desired channel system and middlemen, hold a training session to: (a) explain initiatory promotional plans and service policies; (b) assist in forecasting demand requirements, proper ordering, and maintaining adequate inventory; (c) distribute sales promotion materials; and (d) develop timely and effective promotional programs.
14. Develop unique selling proposition (USP) for promoting the product.
15. Same as 13a and 13c but for firm's salespersons or sales representatives.
16. Hold a press conference regarding the product to gain widespread, free publicity.
17. Prepare news release regarding the product and forward it to appropriate print and broadcast media.
18. Develop an unusual promotional scheme or gimmick to pick up some free publicity for the product (for example, publicity stunt).
19. Same as 11 but in terms of consumers to determine whether promotional campaign is effective, acceptable, and credible (realistic).
20. Same as 19 but in terms of middlemen.
21. Develop point-of-purchase displays and other sales promotion items.
22. Develop and offer advertising allowances to middlemen to achieve the eventual desired promotional effort.
23. Develop and enter into cooperative advertising efforts with middlemen to entice them to carry the firm's product.
24. Detail and properly sequence all phases of the advertising mix in the promotional campaign.
25. Locate and acquire most credible media type(s) for customers.
26. Advertise in all mass media types to build consumer interest and to make the public aware of the product (for example, inform).
27. Develop institutional ads to foster goodwill toward firm and/or industry.
28. Design advertisements that will draw inquiries.
29. Develop pioneering advertisements.

Within the introduction stage, Dodge and Rink recommend these advertising strategies:

1. Distribute free samples of product.
2. Offer special price discounts to customers and middlemen.
3. Institute sales contests among middlemen and salespersons.
4. Affix games, recipes, baseball cards, contests, and so forth to packaging or add small gifts to contents of package.
5. Set up point-of-purchase displays in stores and other outlets.
6. Offer free home trial period for product.
7. Develop and use sweepstakes program.
8. Have demonstrations of product at fairs, trade shows, exhibits, and so forth.
9. Set up booths in shopping centers, grocery stores, and so forth to answer questions about the product as well as to distribute booklets and pamphlets describing the product.
10. Provide middlemen with attractive window displays, store decorations, and point-of-purchase displays.
11. Check consumer complaints and suggestions to find ways to improve the product or its use (as a public relations tool).
12. To stimulate primary demand, rely heavily on mass selling as opposed to personal selling.
13. Implement pioneering advertisements.
14. Use push money or price money allowances.
15. Post-test promotional campaign to see if desired consumer and middlemen reactions and responses were attained.
16. Employ mass media and wide dispersion to educate and inform consumers of product.
17. Inform consumers that product exists, what it does (for example, benefits), how it can be used, and where to purchase it.

18. Use advertising to create awareness and interest in the product and its unique properties (for example, gain attention of consumers).
19. In an initiatory campaign, spend either about twice the desired market share on promotion or as much as the firm can possibly afford in order to reach as many people as possible.
20. Issue cents-off or trial coupons via magazine and newspaper media to encourage purchase of product.
21. Develop promotional campaign to create and foster goodwill and publicity for the product.

In the growth stage, Dodge and Rink emphasize the adoption of these advertising strategies:

1. Alter advertising message to be slightly more persuasive.
2. Use advertisements to develop brand preference by stressing firm's brand and qualities of firm's product relative to competitors'.
3. Spend slightly less than twice the desired market share on advertising.
4. In advertisements, stress benefits and utilities of product to consumers.
5. Employ a larger variety of media types (for example, TV, radio and billboards) to reach more potential consumers.
6. Advertisements should stress availability of different product models, features, options, extras, colors, sizes, and so forth.
7. Buy more spots on various media so that advertisements are repeated continually to build: (a) conviction by creating a favorable product attitude; and (b) retention of the product in consumers' minds.
8. Enlist the services of leading personalities, professionals, and experts to endorse the product and its use.
9. Establish or join a trade association.
10. Increase warranty coverage of product in terms of components covered and/or time span.
11. To stimulate selective demand, rely on *both* mass selling and personal selling.
12. Develop and implement competitive advertisements.
13. Aim for the mass market by buying media that produce the widest coverage or dispersion.

With the advent of the maturity stage, Dodge and Rink prescribe these advertising strategies:

1. Promote new product uses, more frequent product use, variations of basic product, appeal to new users, and so forth.
2. Develop and implement reminder advertising campaign via a mass media approach that will appeal only to loyal product purchasers.
3. Reissue cents-off coupons via magazine and newspaper media to encourage purchase of product.
4. Reoffer special price discounts to customers.
5. Reinstitute sales contests among middlemen and salespersons to sell slow-moving items.
6. Reinstitute sweepstakes program.
7. Use approximately the same media mix, promotional campaign, and advertising expenditure level as major competitors, if possible resource-wise.
8. Offer stamps (for example, S & H) to customers in addition to merchandise purchased.
9. Plan for (or expect) inefficiencies in buying media to get maximum dispersion.
10. Place advertisements in every conceivable media type.
11. Stress product diversification and uniqueness as well as service in advertisements.
12. Develop advertisements that will attract new segments of the market (or develop different advertising programs to appeal to different market segments).
13. Advertisements should be both aggressive and persuasive in comparing the product's favorable attributes to the competition (for example, superior quality, product improvements, service).
14. Advertisements should reinforce consumers' impressions and attitudes in order to maintain loyal customers.
15. Brand preference should be stressed in promotional campaign.
16. Advertisements may become more newsworthy in promoting new uses of the product.
17. Ease payment terms on dealers' and middlemen's purchases.
18. Implement an aggressive sales promotion program to attract other brand users through cash rebates, mark-downs, and other cash incentives.
19. Offer trade-in allowances on older models to middlemen and customers.
20. Make use of a horizontal cooperative strategy between manufacturers and noncompeting products that go well together.
21. Allocate more dollars to the firm's promotional program as competitors' promotional efforts increase substantially and become more persuasive.
22. Develop and implement comparative advertisements.

23. Advertisements by middlemen should stress service, delivery, and credit extension, whereas those by manufacturers should stress price.
24. The promotional blend should be dominated by mass selling (advertising, publicity, and sales promotion) relative to personal selling.
25. Reoffer or repromote special or free premiums to consumers that they can obtain by submitting proof of purchase (for example, seal from package or cash register receipt) to the firm. (self-liquidating premiums)
26. Match advertising to a variety of market segments instead of concentrating on only one theme for the market.

In the final stage, decline, Dodge and Rink recommend the following advertising strategies:

1. Use specialized media to reach the remaining consumers in the market.
2. Eliminate mass advertising media (for example, TV and radio).
3. Use a wave advertising strategy (that is, arrange media schedule so that advertising is intermittently placed on and off).
4. Budget adequate promotional dollars to appeal to one or two remaining market segments.
5. If a strong customer franchise has been achieved, use only reminder-type advertising to stimulate repeat purchases.
6. If resources have permitted the firm to remain in the market, gear up an advertising campaign to recapture some of the lost market share by enticing consumers away from competitors' brands, which are no longer on the market. (This assumes such consumers will not switch to other, newer products.)
7. Promote going-out-of-business or liquidation sale.
8. Promote inventory clearance sale to get rid of slow-moving items.
9. Establish give-away program to get the product to the last remaining customers.
10. If product can be revived, use substantial promotion. Otherwise, eliminate all promotion.
11. Eliminate discounts, special deals, cooperative advertisements, advertising allowances, and so forth.
12. Subcontract or sell out service aspect of business and inform middlemen and customers of this event.
13. Sell parts aspect of business to some interested party and inform middlemen and customers of this event.
14. Subcontract out or sell product to another firm.
15. Substantially decrease total amount of promotion in an attempt to cut costs and remain profitable.

DO BUSINESS MANAGERS USE THE PLC CONCEPT?

Most of the published material on the PLC has been written by academicians, not business executives. The business person would like to know if his or her peers have used the PLC concept. If so, the idea of employing the PLC would gain credibility. Unfortunately, it is not easy to determine exactly how much the PLC is used because that question has received little research attention.[43] In brief, the few studies that have been conducted have found evidence that the PLC concept is used in practice.

To the best of our knowledge, only five studies have investigated whether practitioners use PLC-strategy models. Investigating the domestic sales and advertising for a household cleaner, Parsons[44] found empirical support in the case of advertising for Mickwitz's hypothesis that demand elasticities of the marketing mix variables change over the PLC.[45] In measuring the advertising elasticities for several products at various PLC phases, a packaged goods company found that advertising elasticity tends to decrease as the product progresses through the PLC.[46] Interviewing industrial procurement executives, Rink[47] found that most were intuitively using Berenson's PLC-Organizational Procurement Strategy (OPS) model[48] in formulating and implementing their firms' organizational buying strategy across the PLC. After expanding Berenson's PLC-OPS model to include 83 procurement strategies,[49] Rink and Fox[50] found that over 60 percent of their model was confirmed by industrial procurement executives. Finally, after modifying and expanding earlier PLC-physical distribution strategy models to focus exclusively on transportation,[51] Rink and Kaminski[52] found that almost 75 percent of their 78 transportation strategies and tactics were substantiated by corporate industrial distribution executives.

SUMMARY

In this chapter, we present a set of suggestions for business strategy and tactics to use as the life cycle for a product progresses from precommercialization to introduction to growth to maturity to decline. PLC-strategy models for general marketing, personal selling, advertising, physical distribution, transportation, purchasing, and finance have been presented. The business executive can select the model (or models) most applicable to his or her individual situation. Having chosen a specific model, this will provide the manager with a concise yet panoramic perspective of which strategy(ies) to employ in a given circumstance(s). Of course, the prescriptive model must be adjusted by the executive relative to his or her particular case (for example, industry, market, and product). The overall model and corresponding strategies and tactics are *not* universals but a point of departure for custom tailoring to the conditions confronting the company.[53]

Our final word to the business executive is: Why reinvent the wheel? A rich set of suggestions for actions to be taken as the PLC unfolds has been developed. Much of that material is available in this chapter.

NOTES AND REFERENCES

1. R. Buzzell, "Competitive Behavior and Product Life Cycles," in *New Ideas for Successful Marketing,* eds. J. Wright and J. Goldstucker (Chicago, Ill.: American Marketing Association, 1966), 46-68.
2. E. Scheuing, "The Product Life Cycle as an Aid in Strategy Decisions," *Management International Review* **9**(1969):111-124.
3. J. Swan and D. Rink, "Fitting Market Strategy to Varying Product Life Cycles," *Business Horizons* **25**(1982):72-76.
4. C. Berenson, "The Purchasing Executive's Adaptation to the Product Life Cycle," *Journal of Purchasing* **3**(1967):62-68; M. T. Cunningham, "The Application of Product Life Cycles to Corporate Strategy: Some Research Findings," *British Journal of Marketing* (Spring 1969):32-44.
5. Suggested in part by C. F. Rassweiler, "Product Strategy and Future Profits," *Research Review* (April 1961):1-8; A. Patton, "Stretch Your Product's Earning Years: Top Management's Stake in the Product Life Cycle," *Management Review* **48**(1959):9-14.
6. Swan and Rink, "Fitting Market Strategy to Varying Product Life Cycles," 72-76.
7. H. Thorelli and S. Burnett, "The Nature of Product Life Cycles for Industrial Goods Businesses," *Journal of Marketing* **45**(1981):97-108.
8. Buzzell, "Competitive Behavior and Product Life Cycles," 46-68; D. Clifford, Jr., "Managing the Product Life Cycle," *Management Review* **54**(1965):34-38; J. Forrester, "Industrial Dynamics," *Harvard Business Review* **36**(1958):37-66; D. Luck and A. Prell, *Marketing Strategy* (New York: Appleton-Century-Crofts, 1968), 176-183; G. Mickwitz, *Marketing and Competition* (Helsingfors, Finland: Centraltryckeriet, 1959), 87-89; Patton, "Stretch Your Product's Earning Years," 9-14; Scheuing, "Product Life Cycle as an Aid in Strategy Decisions," 111-124; H. Thorelli, "Market Strategy over the Market Life Cycle," *Bulletin of the Bureau of Market Research* **26**(1967):10-22.
9. T. Staudt and D. Taylor, *A Managerial Introduction to Marketing,* 2nd ed. (Englewood Cliffs, N.J.: Prentice-Hall, 1970), 166-182.
10. E. J. McCarthy, *Basic Marketing: A Managerial Approach,* 4th ed. (Homewood, Ill.: Richard D. Irwin, 1971), 345-348, 527-528, 670-672.
11. P. Kotler, *Marketing Management: Analysis, Planning, and Control,* 2nd ed. (Englewood Cliffs, N.J.: Prentice-Hall, 1972), 434-437, 659-660.
12. C. Wasson, *Dynamic Competitive Strategy and Product Life Cycles,* rev. ed. (St. Charles, Ill.: Challenge Books, 1974), 247-248.
13. Staudt and Taylor, *Managerial Introduction to Marketing,* 168-171.
14. McCarthy, *Basic Marketing,* 345-348, 527-528.
15. Wasson, *Dynamic Competitive Strategy,* 247-248.
16. Staudt and Taylor, *Managerial Introduction to Marketing,* 171-174.
17. McCarthy, *Basic Marketing,* 527-528.
18. Wasson, *Dynamic Competitive Strategy,* 247-248.
19. Staudt and Taylor, *Managerial Introduction to Marketing,* 174-178.

20. Wasson, *Dynamic Competitive Strategy,* 247-248.
21. Kotler, *Marketing Management,* 436.
22. McCarthy, *Basic Marketing,* 345-348, 527-528.
23. Until recently most writers divided the maturity stage into two separate stages: maturity and saturation. In the maturity phase, unit sales increased but at a decreasing rate. Profits reached a plateau and declined slightly. In the saturation phase, unit sales reached a plateau and declined slightly. Profits decreased at an increasing rate.
24. Staudt and Taylor, *Managerial Introduction to Marketing,* 178-180.
25. Kotler, *Marketing Management,* 434-437, 659-660.
26. Wasson, *Dynamic Competitive Strategy,* 247-248.
27. Staudt and Taylor, *Managerial Introduction to Marketing,* 180-182.
28. Kotler, *Marketing Management,* 437.
29. Wasson, *Dynamic Competitive Strategy,* 247-248.
30. H. Fox, "Financial Strategy across the Product Life Cycle," a working paper, 1978.
31. P. Kaminski and D. Rink, "Industrial Transportation in a Systems Perspective," *Transportation Journal* **21**(1981):67-76.
32. H. R. Dodge and D. Rink, "Phasing Sales Strategies and Tactics in Accordance with the Product Life Cycle Dimension rather Than Calendar Periods," in *Research Frontiers in Marketing: Dialogs and Directions,* ed. S. Jain (Chicago, Ill.: American Marketing Association, 1978), 249-252.
33. Berenson, "Purchasing Executive's Adaptation to Varying Product Life Cycles," 62-68.
34. Ibid., 62.
35. H. Fox, "PLC: Pattern for Purchasing," *Purchasing* (19 June 1973):**57,** 59-60; D. Farmer, "Corporate Planning and Procurement: A Checklist Guide," in *Corporate Planning and Procurement,* eds. D. Farmer and B. Taylor (London: Heineman, 1975), 1-27.
36. H. Fox and D. Rink, "Coordination of Purchasing with Sales Trends," *Journal of Purchasing and Materials Management* **13** (1977):10-18.
37. Ibid., 18.
38. D. Bowersox, *Logistical Management* (New York: Macmillan, 1974), 63-68; D. Bowersox, E. Smykay, and B. LaLonde, *Physical Distribution Management,* 2nd ed. (New York: Macmillan, 1968), 5; G. Davis and S. Brown, *Logistics Management* (Lexington, Mass.: D. C. Heath, 1974), 222-225.
39. P. Lynagh and R. Poist, "Physical Distribution and the Product Life Cycle: An Examination of Theoretical Relationships," in *Proceedings of the Southern Marketing Association* (New Orleans: Southern Marketing Association, 1977), 34-37.
40. Kaminski and Rink, "Industrial Transportation in a Systems Perspective," 67-76.
41. P. Kaminski and D. Rink, "PLC: The Missing Link between Physical Distribution and Marketing Planning," *International Journal of Physical Distribution and Materials Management* **14**(1984):77-92.
42. H. R. Dodge and D. Rink, "Advertising Strategy Across the Product Life Cycle," a working paper, 1979.
43. D. Rink and J. Swan, "Product Life Cycle Research: A Literature Review," *Journal of Business Research* **7**(1979):219-242.
44. L. Parsons, "The Product Life Cycle and Time-Varying Advertising Elasticities," *Journal of Marketing Research* **12**(1975):476-480.
45. Mickwitz, *Marketing and Competition,* 87-89.
46. P. Kotler, *Marketing Decision Making: A Model Building Approach* (New York: Holt, Rinehart, and Winston, 1971), 181-182.
47. D. Rink, "The Product Life Cycle in Formulating Purchasing Strategy," *Industrial Marketing Management* **5**(1976):231-242.
48. Berenson, "Purchasing Executive's Adaptation to the Product Life Cycle," 62-68.
49. Fox and Rink, "Coordination of Purchasing with Sales Trends," 10-18.
50. D. Rink and H. Fox, "The Role of Product Life Cycle Theory in Formulating Industrial Procurement Strategy: An Empirical Analysis," in *1984 AMA Educator's Proceedings,* Series No. 50, eds. R. Belk et al. (Chicago, Ill.: American Marketing Association, 1984), 275-278.
51. Kaminski and Rink, "Industrial Transportation," 67-76.
52. D. Rink and P. Kaminski, "Using the Product Life Cycle Theory to Formulate Industrial Transportation Strategies and Tactics: A Survey of Corporate Industrial Distribution Executives," a working paper, 1985.
53. Fox and Rink, "Coordination of Purchasing with Sales Trends," 10-18.

process, including the strategic choices concerning strengths and weaknesses.

SWA Is an Integral Part of Planning

The second premise is that *the strength-weakness assessment is, in fact, meant to be an integral and useful element of strategic planning.* Although this may seem obvious, there is evidence to suggest that it is not always the case. Many strength-weakness studies are prepared and thereafter never referred to in the planning process. Of course, this lack of reference to the study does not, in itself, demonstrate its lack of impact. It may well be that the assessment process or its product has created an enhanced level of awareness of strengths and weaknesses that more subtly influences the planning process. However, when the analysis is recorded in voluminous reports of studies performed by staff rather than the planner-managers, explicit reference to the study undoubtedly would normally be required by the managers who are doing planning. If such reference is not made, the analysis is probably not really being used.

More important perhaps is the fact that the process of evaluating alternative strategies is so complex that in order to operationalize the impact of SWA in planning, it is necessary for alternative strategies to be *explicitly* tested against the important strengths and weaknesses of the firm. Thus, for an SWA to be *useful,* it probably will be explicitly *used.*

SWAs Produce Information

The third premise of SWA is the recognition that the results of a *strength-weakness assessment represent strategic information and not merely data.* This distinction may seem to be one of semantics to some, but it is of crucial importance.

Information is data that have been evaluated for some purpose.[9] The telephone directory and many reports of traditional strength-weakness studies contain data in the form of charts, graphs, and numbers. These are not information until they have been evaluated in terms of their strategic significance. Only then can they become an integral element of the development of business strategy.

Perceptions of Strengths and Weaknesses Are Most Important

The fourth premise of SWA is *that, although many organizational strengths and weaknesses may be factual in nature and therefore provable, it is more important that they be perceived than that they be proved.* Some firms have strengths and weaknesses that can be empirically demonstrated, but their behavior belies them as being perceived by the firm's managers.[10] It is perceptions, not necessarily truth, that guide the behavior of people. Since people must implement plans and control them over time, it is more important that they perceive the strengths and weaknesses on which they are based than it is that those strengths and weaknesses be conclusively demonstrated.

Indeed, it may even be more important that they be perceived than that they be valid. Since real strategic planning is an approach to influencing the future, even invalid strengths and weaknesses have the potential to become self-fulfilling prophecies *if they are perceived.* If they are not perceived, they have no likelihood of influencing anything, no matter how valid they may be.

SWA Is Future Oriented

The fifth SWA premise is that *strengths and weaknesses must be assessed in terms that go beyond mere static and historical descriptions of the firm itself; they must, to some degree, be predictions or assessments of potential.* If a particular strength is not truly significant, but if the firm has the ability to significantly enhance the strength in the future, it can become a strategy-relevant *opportunity*. Just as external opportunities are routinely dealt with in planning, so too should internal opportunities that are the result of SWAs be an integral element of strategy development.

Strengths and Weaknesses Are Relative

The sixth premise of SWA is that to be useful, strengths and weaknesses must also be comparative in nature; they must be assessed *relative to those of competition*. Some firms suffer from the tunnel vision that comes from not knowing enough about competition and their comparative advantages or lack thereof. They may assess a strength to be much more important than will be reflected by a comparative assessment. Good engineering talent, for instance, may be thought of as a strength, but it may recede in significance when assessed against the technical personnel of another firm.[11]

On the other hand, important strengths and weaknesses, particularly those that emphasize opportunities for the future, may be rather subtle in nature and therefore difficult to assess in absolute terms. However subtle, when compared with similar attributes of competitors, their importance may become more apparent.

IMPLICATIONS OF THE PREMISES

The above-noted premises may be used to derive a set of implications that are useful for the design of an improved SWA procedure. These implications are reflected in the SWA process presented below.

Planner-Managers Should Perform SWAs

Just as it has become increasingly accepted that planning should be performed by managers rather than by analysts, so too should SWAs be performed by the managers who are the organization's real planners. This implication is a rather drastic departure from established practice in some firms as well as from the generally accepted division of responsibilities, even in those firms where many of the other elements of strategic planning are indeed performed by managers.

However, the premises dictate that if managers perform the SWA they will understand it, perceive its implications, and be more likely to see it as an integral part of the planning process and use it as a guide for their strategic decisions. Moreover, managerial perceptions and judgment are necessary for the identification and prioritization of the most significant strengths and weaknesses, especially those that are subtle in nature and those that emphasize comparative assessments of the firm's characteristics relative to competitors.

Groups of Managers Should Negotiate SWAs

If managers are to perform SWA, a natural question is Which managers should do this?, since strengths and weaknesses, to be useful, must reflect the firm's *overall* situation. Since no single manager can reasonably be expected to address such overall breadth, the implication is that groups of managers, chosen to reflect the various functional areas and areas of interest in the firm, must work to develop negotiated SWAs.

This is the only way to ensure that SWAs prepared by managers will not be parochial and reflect only their own narrow set of values and biases. If various managers argue their own viewpoints in a consensus-seeking environment of negotiation, the result may not be truth, but it is likely to be realistic and credible. Moreover, it is likely to be reflective of the fragile combination of truth and political reality that probably provides the best practical basis for effective strategy in any organization.

This view of how SWAs should be per-

Table 22-1. Illustrative SWA Summary

Major Strengths	*Major Weaknesses*
Technical expertise in centrifugal area	Low market share
International sales force	Lack of product standardization
Heavy machining capability	Poor manufacturing nonfragmented product line
Business systems	Weak domestic distribution network
Puerto Rico facilities	High price image (?)
Technically superior image among customers	

weakness planning guidebook that was prepared as an integral part of the planning process. The remainder of the book contained exhibits that explained and elaborated on these conclusions.

Table 22-1 is the result of an evaluative process that began with the compilation of a long list of potential strengths and weaknesses and proceeded, through the efforts of a task force composed of upper-middle managers from all functional units of the business, with evaluation of those strengths and weaknesses that were, or should have been, most crucial in the determination of the firm's strategy.

This list was then referred to in the latter phases of the planning process to suggest strategies, to aid in evaluating strategies, and to guide the development of plans for implementing the chosen strategies.

Some of the items in Table 22-1 might be thought of as simply describing the status of the firm or as problems that need to be corrected, rather than strategic weaknesses. However, in the firm's detailed explanation of these areas in their planning guidebook, they were treated as *weaknesses that must be used to guide strategic decisions,* rather than merely as problems to be overcome. For instance, the detailed explanation of the low-market-share weakness emphasized the strategies that were precluded as a result of low market share rather than emphasizing low market share merely as a deficiency to be corrected. In this way attention was directed toward the strategic implications of each strength and weakness rather than to the familiar sort of exhortations for the firm to do better by increasing market share.

Two of the entries in Table 22-1 refer to the business's image. These indicate that the unit believed that its customers perceived it positively in terms of its technical superiority and (perhaps), negatively in terms of the relatively high price of its products. Such image assessments are commonly incorporated into statements of business objectives as well as into SWAs. Yet often the firm is, to some degree, uncertain of just how it is perceived. In this instance, the company was not certain if its high technical quality was perceived as justifying its relatively higher price.

Such a resolution of the image situation in a state of partial uncertainty is a common and valuable outcome of SWA. It points up the need for more detailed and specific information than that which the task force may be readily able to collect. This result can lead to a special study or to the development of a more detailed information system that will provide the firm with the answers that it wants in order to do effective strategic planning.

Another benefit invariably realized from the SWA approach should not be overlooked. The participative SWA process will usually involve a variety of middle-level managers from each of the organization's functional and product units. The SWA task force process is thereby a way of involving middle managers in the strategic thinking of the *total* organization long before they might normally become so involved by virtue of their operational job responsibilities. Thus, the SWA approach serves as a training ground for the development of those elusive strategic thinking abilities so necessary to successful general managers.

FRAMEWORK FOR CONDUCTING THE SWA PROCESS

If the SWA task force process is undertaken in an unstructured fashion, it probably will not be as effective as it might be. An unstructured task force process inevitably will begin with and be constrained by the existing wisdom of its members. This existing wisdom may reflect primarily the past, personal biases, and individual parochial viewpoints more than it reflects the future, its opportunities, and fresh viewpoints. Therefore, the SWA task force should have a conceptual framework that will facilitate the thinking through of a wide range of potential strengths and weaknesses.

One such conceptual framework used by some firms is that of Porter.[13] Porter identifies five competitive forces that drive industry competition and that should therefore influence strategy:

Threat of new entrants.
Bargaining power of buyers (customers).
Bargaining power of suppliers.
Threat of substitute products or services.
Rivalry among existing firms.

Although these five categories are too broad to be operationally useful, they can form the basis for a task force agenda. Each one can be broken down into its constituent elements, which can, in turn, be used as a guide for the preliminary identification of strengths and weaknesses.

For instance, the threat of entry category has been analyzed as shown in tree-diagram form in Figure 22-1.[14] The primary elements that operationally define the threat of entry are capital requirements, accessibility of distribution channels, economies of scale, product differentiation, cost advantage, and government policy. In Figure 22-1, each of these major elements is, in turn, broken down into successively more specific categories about which data may be collected and assessed.

This is the key to the use of such a framework to guide the SWA process. It can be used as a checklist by the task force in identifying potential strengths and weaknesses. Moreover, the level of detail in the example in Figure 22-1 is such that any elements at the bottom of the tree that are believed to be related to potential strengths or weaknesses can be investigated through the collection of data. For instance, product and service features that are believed to be particularly good relative to competition can be investigated and documented, as can presumed location, and cost advantage or disadvantages.

Thus, the use of a framework can ensure that relevant issues are comprehensively, clearly, and specifically identified and considered by the SWA task force. Although determination of those things that will finally be identified as a major strength or weakness is judgmental, the use of a framework such as this ensures that the judgmental determination has a factual base and that important issues have not been omitted from consideration.

SUMMARY

SWAs can be transformed from academic exercises into important integral elements of strategic decision making. The SWA procedure described here makes use of a participative process, involving task forces that are made up of managers representing the diverse interests of the organization.

The SWAs that are thus developed thereby become important informational elements of

VI

Information Systems for Strategic Management

The strategic importance of information has become apparent in recent years. Strategic planning processes have become information intensive as have the strategies of some businesses.

In Chapter 23, Lawrence C. Rhyne describes information systems support for various kinds of planning systems using the results of a survey of *Fortune 1000* companies. He focuses on the nature and sources of information as they relate to the several levels of the planning system.

In Chapter 24, Malcolm C. Munro discusses a process for identifying the critical information for strategic management. The process makes use of the critical success factor idea discussed in Chapter 20. He describes a field study in which the approach was used and cautions against overuse of the powerful CSF construct, particularly if management does not have a good mental model of the dynamics of their business.

In Chapter 25, William R. King describes decision support systems (DSSs) as they apply to the strategic management context by developing the notion of a DSS and developing criteria for judging whether a DSS is truly being employed in a strategic mode. Despite the many claims that have been made concerning the application of DSSs in strategic management, he finds that not many systems fully meet the criteria that are implicit in the basic notion of a strategic management DSS.

OVERVIEW

Classification of Planning Systems

Planning systems can be expected to differ depending on the objectives of management, the level of organization at which the planning takes place, and the time horizon of the plans produced. Lorange and Vancil described an evolutionary development of planning systems within organizations.[3] They suggested that as the planning effort matured the characteristics of that effort would change, reflecting management's increased familiarity with the planning process. Higgins found that mature planning systems addressed different problems in contrast to recently introduced systems.[4] Miles, et al. and Ansoff have suggested that organizations may conduct their planning at various levels of openness to their environment with success dependent on whether the level of openness was appropriate given the external turbulence facing the organization.[5]

Strategic management theory has held that the fundamental goal of planning is to produce a strategy that achieves a "match," "fit," or "alignment" of internal resources and capabilities with external opportunities and threats.[6] However, Eliasson identified two distinct aspects of planning: analysis and control.[7] Lorange made a similar distinction between adaptive and integrative aspects of planning.[8] Leontiades characterized these aspects as evolutionary planning and steady-state planning.[9] The analysis or adaptive dimension addressed external factors and responded to environmental-change, whereas the control or integrative dimension focused on the coordination of internal resources. Strategic management theorists have tended to emphasize the analysis-adaptive dimension, whereas practitioners have tended to emphasize the control-integrative dimension.[10]

Considerable debate has taken place on whether it is desirable or even possible to conduct strategic decision making in a formal planning system. A number of authors have held that planning in large, complex, and diverse organizations can be managed effectively only by the use of formal systems.[11] However, others have argued that strategy formulation primarily takes place on an informal basis and is not necessarily communicated to the organization as a whole.[12] At this time, the issue remains unresolved.

Regardless of the process by which it takes place, the analysis-adaptive (openness) dimension must be addressed for true strategic planning to take place. Within this context *strategic* planning is qualitatively different from *long-range* planning. In strategic planning the firm's domain or area of operations remains open to examination and redefinition. Long-range planning assumes the firm's domain to be essentially given.

Executives also address strategy formulation at a number of levels within a firm.[13] Primary or corporate-level strategies define the domain of an organization, and secondary or business-level strategies address the manner in which the organization will compete in that domain.[14] A one-year operating plan or budget is probably the most prevalent form of planning. However, in order for planning to be most effective, a longer-term perspective, corresponding to the length of time necessary to executive a strategy, is required.

In this study, corporate-level planning systems were assumed to exist on a continuum of increasing sophistication, with planning sophistication defined as a combination of the level of planning openness and the planning horizon. This definition conformed to the basic elements of strategic planning models presented by a number of authors.[15] Five key points on that continuum were defined:

Short-term forecasting. Planning concentrates on operating results over the next three to six months and is often conducted on an informal basis. The information required normally exists within the firm; however, some means of reporting sharp external changes affecting the enterprise is required.

Budgeting. Planning consists of detailed cost and revenue projections of the past and current situations. The time horizon is based on the enterprise's operating cycle and is probably no more than one year. The information system must supply detailed cost and revenue figures generally available within the firm. A reporting system capable of identifying variances from budget is usually present.

Annual planning. Planning includes some attempt to review the environment and to identify problems, opportunities, and turning points within a one-year time horizon. The external information may be obtained on an informal basis. This type of planning normally overlays budgeting type information and reporting systems.

Long-range planning. Planning attempts to identify external opportunities and threats facing the firm and to decide in which direction the firm should move to maximize results. The options considered are closely related to current operations. The time horizon may cover five, ten, or fifteen years, depending on the lead time of the firm's operations and investments. The information system becomes more sophisticated in an attempt to identify future trends. External as well as some environmental information is collected and evaluated.

Strategic planning. This classification of planning also reflects a long-term perspective. However, rather than concentrating only on areas closely related to current operations, the planning effort involves a wide-range scanning of external and environmental factors. This review attempts to identify new areas in which the organization's skills and comparative advantages may be applied as well as threats to the organization's current operations.

Although the organization may choose to remain in areas closely related to current operations, it is nevertheless aware of a broad range of trends and options that might affect operations in the future. This classification of planning requires information of a strategic nature on internal, external, and environmental factors. Such information is often qualitative, and the organization must be willing to react to relatively weak signals of possible future events.

These points were not considered to be discrete categories with individual planning systems moving from one category to another in a step-wise fashion. Rather, a company was expected to gradually adopt an increasing number of the characteristics associated with another point on the continuum until that point more accurately described its planning effort. As a result, the planning systems identified are considered to be key points on the planning continuum rather than mutually exclusive categories. Of course, a number of planning types may exist at the same time at various levels of an organization. Characteristics of each type of planning and the related information requirements as defined in this study are summarized in Table 23-1.

Senior planning executives were asked to indicate on a scale of one to five the level of emphasis placed on a number of items that had been recommended by various authors as possible characteristics of planning systems. Analysis of the responses indicated that eight items formed a single dimension that closely resembled the *a priori* definition of strategic planning. A firm's average score on these items was used to evaluate the level of openness of the planning effort. The planning openness rating and the firm's planning horizon were used to categorize the firm's planning system. See Table 23-2 and Figure 23-1 for the individual items and the classification scheme.

Types and Sources of Information

The external search or environmental scan required by strategic planning leads the organization to secure new and different types of information not readily available from the organization's transactional information system.[16] This type of information deals with customer attitudes, competitors' actions, regulatory patterns, technological trends, and so forth.[17] Ansoff held that the degree to which an organization is successful in integrating such information into its planning will determine its level of strategic thrust.[18]

A number of previous studies have pro-

established in very large multinational companies.[22] However, Jain found that only 29 percent of the *Fortune 500* companies responding to his survey had developed formal environmental scanning systems.[23] The quality of these systems evolved over time and the existence of a "formalized system of strategic planning" was found to be a prerequisite to the development of a structured environmental scanning system.

Palia et al. found that the importance of internal functional areas varied depending on a firm's overall strategy.[24] Hitt et al. concluded that a proper match between the importance of certain internal functions and a firm's grand strategy was related to performance.[25] The importance of external and environmental factors also can be expected to vary with the nature of a firm's planning system.

Based on the work of a number of authors and the results from a preliminary field study, three categories of information and information systems were defined in this study:

1. Internal. Operations within the enterprise.
2. External. Information on factors outside the enterprise with which the enterprise interacted directly on a regular basis.
3. Environmental. External factors with which the enterprise did not directly interact but that might effect operations.

Senior planning executives were asked to rate the importance of 60 types of information to the outcome of their firm's planning process. Analysis of the responses indicated that 41 of the individual information types could be grouped into 10 information factors. See Table 23-3 for the information factors and types.

Table 23-3. Information Factors and Types

Information Factors	*Information Types*	*Information Category*
Political	Legislative trends	Environmental
	Pressure groups	Environmental
	Regulatory trends	External
Market size	Total market dollars	External
	Total market units	External
	Market share	External
Environmental match	Market segments	External
	Pricing trends	External
	Basic strategy or mission	Internal
	Distinctive competence	Internal
	Key vulnerabilities	Internal
Social	Trend setters	External
	Demographic trends	Environmental
	Life-style changes	Environmental
Operating	Detailed manufacturing costs	Internal
	Service policies	Internal
	Credit policies	Internal
	Detailed sales forecast	Internal
	Detailed administrative expense	Internal
	Customer location	External
	Service requirements	External
Technical	Technological trends	Environmental
	Obsolescence	Environmental

Table 23-3 (*Continued*)

Information Factors	*Information Types*	*Information Category*
Industry	Ease of entry and exit	External
	Financial traits	External
	Capacity and utilization	External
Financial	Financial controls	Internal
	Long-term financing	External
	Cash management	Internal
Organizational	Distribution	Internal
	Sales force type	Internal
	Sales force size	Internal
	Management availability	External
	Information systems	Internal
	Organization structure	Internal
Customer-competitor	Customer type	External
	Product use	External
	Customer strategies	External
	Competitors, type	External
	Competitors, number	External
	Competitors, abilities	External
Information Types Not Included in Information Factors		
	Plant capacity	Internal
	Workforce availability	External
	Unionization	External
	Raw material sources	External
	Market research policies	Internal
	Product quality	Internal
	Advertising and promotion policies	Internal
	Basic research emphasis	Internal
	Applied research emphasis	Internal
	Engineering capability	Internal
	Cyclicality-seasonality	External
	Industry growth rate	External
	Firm's public image	External
	Inflation-deflation	External
	Revaluation-devaluation	External
	Short-term economic forecast	External
	Mid- long-term economic forecast	Environmental

Source: L. C. Rhyne, "Strategic Information: The Key to Effective Planning," *Managerial Planning* **32**(1984):6. Reprinted by permission of Planning Executives Institute.

Table 23-5. Information Factors and Types Very Strongly Related to Planning Classification

Information Factor and Type	*Significance*	
Technical	***	
Technological trends	***	Environmental
Obsolescence	***	Environmental
Environmental match	***	
Market segments	**	External
Pricing trends	**	External
Basic strategy or mission	***	Internal
Distinctive competence	***	Internal
Key vulnerabilities	***	Internal
Industry	***	
Ease of entry and exit	***	External
Financial traits	***	External
Capacity and utilization	***	External
Political	***	
Legislative trends	**	Environmental
Pressure groups	*	Environmental
Regulatory trends	***	External
Workforce availability	***	External
Product quality	***	Internal
Engineering capability	***	Internal
Management availability	***	External
Information systems	***	Internal
Cyclicality-seasonality	***	External
Industry growth rate	***	Internal
Potential competitors	***	Environmental

Source: L. C. Rhyne, "The Relationship of Information Usage Characteristics to Planning System Sophistication: An Empirical Examination," *Strategic Management Journal* **6**(1985):331. Reprinted by permission of John Wiley and Sons, Ltd.
Note: Significance: * = $p < .10$; ** = $p < .05$; *** = $p < .01$.

The annual planning category focused on external opportunities. Four of the top ten items dealt with customer traits (product use, service required, types of customers, and customer location), and three more items were externally directed (competitor types, detailed sales forecast, and firm's public image). Although representing an outward-looking approach to planning, these items did not appear to reflect a long-term orientation.

The budgeting category, as expected, emphasized control of operating costs. Inflation-deflation was the most important item, and detailed manufacturing costs was ranked second. Distribution, detailed sales forecast, plant capacity, and raw material sources also were ranked in the top ten items and completed the pattern of emphasis on operating detail.

Other patterns appeared to relate to the planning horizon. Pricing trends, market share, product quality, and regulatory trends were ranked in the ten most important items in both long-range and strategic planning but were ranked lower in the shorter time horizon classifications. Plant capacity was ranked among the top ten items for budgeting, annual planning, and long-range planning but was ranked only twenty-first in strategic plan-

Table 23-6. Ten Most Important Information Types by Planning Type

Rank	*All Firms*	*Budgeting*	*Annual Planning*	*Long-range Planning*	*Strategic Planning*
1	Inflation-deflation	Inflation-deflation	Cash management	Long-term financing	Basic strategy or mission
2	Long-term financing	Detailed manufacturing controls	Product use	Plant capacity	Inflation-deflation
3	Cash management	Long-term financing	Service required	Pricing trends	Pricing trends
4	Pricing trends	Cash management	Competitors, types	Cash management	Cash management
5	Plant capacity	Distribution	Plant capacity	Inflation-deflation	Competitors' abilities
6	Market share	Detailed sales forecast	Detailed sales forecast	Market share	Market share
7	Competitors' abilities	Types of customers	Financial controls	Product quality	Regulatory trends
8	Product quality	Plant capacity	Types of customers	Financial controls	Product quality
9	Regulatory trends	Raw material sources	Customer location	Regulatory trends	Technological trends
10	Financial controls	Competitors' abilities	Firm's public image	Competitors' abilities	Distinctive competence

Source: L. C. Rhyne, "Strategic Information: The Key to Effective Planning," *Managerial Planning*, **32**(January-February 1984):8. Reprinted by permission of Planning Executives Institute.

Table 23-8. Importance of Source of Information by Planning Type

Rank	*All Firms*	*Budgeting*	*Annual Planning*	*Long-range Planning*	*Strategic planning*
1	Superiors	Superiors	Superiors (tied)	Superiors	Superiors
2	Accounting system	Accounting system	Accounting system (tied)	Accounting system	Subordinates
3	Subordinates	Subordinates	Subordinates	Subordinates	Inside reports
4	Inside reports	Inside reports	Specific MIS	Inside reports	Accounting system
5	Specific MIS	Outside studies	Inside reports	Specific MIS	Specific MIS
6	Outside publications	Specific MIS	Outsiders	Outside publications	Outsiders
7	Outsiders	Outsiders	Outside publications	Outsiders	Outside studies
8	Outside studies	Outside publication	Outside studies	Outside studies	Outside publications

Source: L. C. Rhyne, "Strategic Information: The Key to Effective Planning," *Managerial Planning* **32**(1984):4. Reprinted by permission of Planning Executives Institute.

acts to emphasize the significance of the planning process to the organization.

Furthermore, the predisposition to explore unfamiliar courses of action implicit in strategic planning will exist only with senior management's vigorous support. Although information gathering may take place on many levels, an effective planning and information system will include procedures to insure that significant information and business alternatives reach the highest decision makers in the corporation. Senior management's active participation in the planning process appeared to be critical to the successful implementation of strategic planning.

Financial Performance

Firms with planning systems more closely resembling strategic management theory were found to exhibit superior long-term financial performance both relative to their industry and in absolute terms. Not only did ten-year total returns to investors increase with the planning continuum, but, in addition, strategic planning was found to have higher returns than other planning categories.

Strategic planning appeared to be a discipline adopted by companies that had been successful prior to the time of this study as evidenced by the 1979 ten-year performance figures. Whether strategic planning resulted in superior performance or superior performance permitted strategic planning remains difficult to specify. However, those firms practicing strategic planning in 1980 did continue to exhibit higher ten-year returns to investors for the period 1980 to 1983.

In addition, the four-year average of absolute annual returns to investors also was higher for strategic planning companies, indicating that these companies were able to maintain their superior performance. Although the specific details of a firm's plan are likely to vary from year to year and certainly would not be expected to remain constant over a five-year planning horizon, a firm's overall strategy could remain constant for a number of years if successful. A combination of openness to the environment coupled with a long-term perspective appears to be a characteristic of successful management. The association of relative results with strategic planning was only slightly more significant than the association of absolute results. The firms in this study exhibiting superior performance apparently were able to identify industries with higher potential profits and then exploit those situations successfully.

In retrospect, this appears to be a perfectly logical result given that the purpose of strategic planning is to position the firm in industries most favored by environmental conditions. However, given the importance of the industry on an individual firm's performance and the difficulty in shifting industries quickly, *a priori,* this result was not predicted.

A wide variance in performance was evident in firms practicing long-range planning, whereas firms conducting strategic planning produced more consistent results. Even though the mean long-term results of strategic planning firms were higher than firms in the other categories, in most years the best and the worst results for the long-range planners were beyond those of the strategic planners. Firms performing at an extremely high level may not have the time, resources, or inclination to investigate new areas. Alternatively, firms with very poor results may find that they do not have sufficient resources to execute a major redirection in strategic effort.

An important limitation should be noted in that this study focused on formal planning systems. The characteristics of informal planning processes were not necessarily captured. The study also found that firms in the strategic planning category tended to utilize the formal planning system for strategic decision making. It is possible that firms in the other planning categories carry out that process on an informal basis.

A SIMPLE, EFFECTIVE ENVIRONMENTAL SCANNING SYSTEM

Given the wide range of factors necessary to support strategic planning and the volatile nature of these factors, many managers find it difficult to decide how to initiate an environmental scanning effort. In addition, they are usually concerned that such an effort may be costly, time consuming, and still not produce any measurable or useful results. In other words, how can a firm obtain the information it needs for strategic decision making without expending excessive quantities of time and money?

The solution is to focus on those factors likely to have a major impact on the firm and to use current personnel to collect most of the data. A simple but effective environmental scanning system can be implemented in four phases:

1. *Identification* of factors to be scanned.
2. *Collection* of data related to those factors identified.
3. *Evaluation* of the data and conversion to useful information.
4. *Communication* of the information to the appropriate members of the organization.

Identification

Factors Critical to Success for All Firms. Within a given industry, there are usually a few key factors on which all firms depend for success. These may be a widespread level of manufacturing technology, certain characteristics of customers, the availability of low-cost raw materials, or the presence of protective regulation. Changes in these factors may dramatically affect the industry as a whole or may provide a major competitive advantage if an individual firm achieves a breakthrough. For example, if all firms are using the same basic manufacturing technology, a new method of production can give one competitor or a totally new firm a competitive advantage while rendering other firms' production obsolete.

Factors Essential for Success Given the Firms' Individual Strategy. As the firm develops a specific strategy additional factors will become important to success. The firm may decide to concentrate on high quality or per-

2. K. Radford, "Some Initial Specifications for a Strategic Information System," *OMEGA* **4**(1978):139-144.
3. P. Lorange and R. F. Vancil, "How to Design a Strategic Planning System," *Harvard Business Review* **54**(1976):75-81.
4. R. B. Higgins, "Long-range Planning in the Mature Corporation, *Strategic Management Journal* **2**(1981):235-250.
5. R. E. Miles, C. C. Snow, and J. Pfeffer, "Organization-Environment Concepts and Issues," *Industrial Relations;* **3**(1974):244-264. H. I. Ansoff, *Strategic Management* (New York: Wiley, 1979).
6. K. R. Andrews, *The Concept of Corporate Strategy* (Homewood, Ill.: Richard D. Irwin, 1971); C. W. Hofer and D. Schendel, *Strategy Formulation: Analytical Concepts* (St. Paul, Minn.: West Publishing, 1978); Ansoff, *Strategic Management;* C. E. Summer, *Strategic Behavior in Business and Government* (Boston: Little, Brown, 1980).
7. G. Eliasson, *Business Economic Planning* (New York: Wiley, 1976).
8. P. Lorange, *Corporate Planning* (Englewood Cliffs, N.J.: Prentice-Hall, 1980).
9. M. Leontiades, *Strategies for Diversification and Change* (Boston: Little, Brown, 1980).
10. Eliasson, *Business Economic Planning.*
11. F. J. Aguilar, R. A. Howell, and R. F. Vancil, *Formal Planning Systems—1970* (Cambridge, Mass.: Harvard Business School, 1970); H. E. R. Uyterhoeven, R. W. Ackerman, and J. W. Rosenblum, *Strategy and Organization* (Homewood, Ill.: Richard D. Irwin, 1973; Hofer and Schendel, *Strategy Formulation.*
12. H. E. Wrapp, "Good Managers Don't Make Policy Decisions," *Harvard Business Review* **45**(1967):91-99; H. Mintzberg, *The Nature of Managerial Work* (New York: Harper and Row, 1973): J. B. Quinn, "Managing Strategic Change," *Sloan Management Review* **21**(1980):3-20.
13. T. J. McNichols, *Executive Policy and Strategic Planning* (New York: McGraw-Hill, 1977); Hofer and Schendel, *Strategy Formulation.*
14. L. J. Bourgeois, "Strategy and Environment: A Conceptual Integration," *Academy of Management Review* **5**(1980):25-40.
15. H. I. Ansoff, *Corporate Strategy* (New York: McGraw-Hill, 1965); Ansoff, *Strategic Management;* Andrews, *Concept of Corporate Strategy;* McNichols, *Executive Policy and Strategic Planning;* Summer, *Strategic Behavior in Business and Government.*
16. G. B. Davis, *Management Information Systems: Conceptual Foundations, Structures, and Development* (New York: McGraw-Hill, 1974); Radford, "Some Initial Specifications for a Strategic Information System.
17. Hayes and Radosevich, "Designing Information Systems for Strategic Decisions"; H. Mintzberg, D. Raisinghani, and A. Theoret, "The Structure of "Unstructured" Decision Processes," *Administrative Science Quarterly* **21**(1976):246-275; L. A. Gordon, D. F. Larcker, and F. D. Tuggle, "Strategic Decision Processes and the Design of Accounting Information Systems: Conceptual Linkages," *Accounting Organizations and Society* **10**(1978):203-313.
18. Ansoff, *Strategic Management.*
19. Eliasson, *Business Economic Planning.*
20. A. Kefalas and P. D. Schoderbek, "Scanning the Business Environment—Some Empirical Results," *Decision Sciences* **4**(1973):63-74; W. J. Keegan, "Multinational Scanning: A Study of the Information Sources Utilized by Headquarters Executives in Multinational Companies," *Administrative Science Quarterly* **19**(1974):411-421.
21. F. J. Aguilar, *Scanning the Business Environment* (New York: Macmillan, 1967); Keegan, "Multinational Scanning"; L. Fahey and W. R. King, "Environmental Scanning for Corporate Planning," *Business Horizons* **20**(1977):61-71.
22. P. S. Thomas, "Environmental Scanning—The State of the Art," *Long Range Planning Journal* **13**(1980):20-28.
23. S. C. Jain, "Environmental Scanning in U.S. Corporations," *Long Range Planning Journal* **17**(1984):117-128.
24. K. A. Palia, M. A. Hitt, and R. D. Ireland, "The Relationship of Grand Corporate Strategy to the Importance of Major Organizational Functions: Moderating Effects of Production System and Perceived Environmental Uncertainty." Paper presented at the annual meeting of the Academy of Management, Detroit, (August) 1980.
25. M. A. Hitt, R. D. Ireland, and G. Stadler, "Functional Importance and Company Performance: Moderating Effects of Grand Strategy and Industry Type," *Strategic Management Journal* **3**(1982):315-330.
26. F. E. Emery and E. L. Trist, "The Causal Texture of Organizational Environments," *Human Relations* **18**(1965):21-32; J. D. Thompson, *Organizations in Action* (New York: McGraw-Hill, 1967); R. Duncan, "Characteristics of Organizational Environments and Perceived Environmental Uncertainty," *Administrative Science Quarterly* **17**(1972):313-327; W. M. Lindsay and L. W. Rue, "Impact of the Business Environment on

the Long-range Planning Process: A Contingency View," 38th meeting, *Academy of Management Proceedings* (1978):116-120; Ansoff, *Strategic Management.*

27. J. Child, "Organizational Structure, Environment and Performance: The Role of Strategic Choice," *Sociology* **6**(1972):1-22.
28. Andrews, *Concept of Corporate Strategy;* P. F. Drucker, *Management* (New York: Harper & Row, 1974); Eliasson, *Business Economic Planning;* McNichols, *Executive Policy and Strategic Planning;* Hofer and Schendel, *Strategy Formulation.*

CHAPTER **24**

Identifying Critical Information for Strategic Management

Malcolm C. Munro

Malcolm C. Munro is an associate professor with the Faculty of Management at The University of Calgary, in Calgary, Alberta, Canada. He is a consulting editor of *Canadian Journal of Administrative Sciences,* co-editor of *MIS Interrupt,* and former chairman of the Management Science and Information Systems Area. He holds a B.Comm. with Honors from the University of Saskatchewan and the M.S. and Ph.D. in management information systems from the University of Minnesota. Professor Munro's research interests are in the areas of critical success factors, information requirements analysis, and the organizational processes for the assessment and adoption of information technology.

A 1982 book was titled *The Logic of Strategic Planning.*[1] The title was well chosen in that planning is a singularly logical and rational activity. Yet planning is frequently neglected, especially by senior managers. This chapter begins with a brief discussion of some subtle reasons why planning is so challenging and continues with description and discussion of an effective (and enjoyable) method for involving senior managers in planning. Referred to as the critical success factor (CSF) approach, the method is gaining increasing acceptance by senior managers as a planning aid and for identifying critical information for strategic management.

Drawing on a full-scale corporate application, the steps in the CSF method are described, with particular attention paid to how the method is integrated into the strategic planning process. Next, an analysis of the advantages and disadvantages of the method is presented, followed by a description of the use of the CSF method with a *group* of senior executives, as opposed to the usual one-on-one technique. Finally, two examples of the use of corporate system modeling are given that demonstrate that caution in the use of the CSF approach should prevail when the corporation is already experiencing significant difficulties.

WHY IS PLANNING SOMETIMES NEGLECTED?

The reason why managers neglect planning is related to the nature of a manager's work, the nature of the planning activity, and senior

A portion of this chapter is drawn from M. C. Munro and B. R. Wheeler, "Planning, Critical Success Factors, and Management's Information Requirements," *MIS Quarterly* **4**(1980):27-38.

managers' mental processes. An immediate operational difficulty is that planning requires the discipline to extricate oneself from the press of daily events, to temporarily ignore the pressures of the moment. The senior executive, accustomed to the fast track, must gear down and attend to matters from which results may not be seen for some considerable time. This behavior would seem to run against the grain, to be inconsistent with the prevailing character of managerial work. Mintzberg has pointed out that it is folklore that managers are reflective, systematic planners.[2] "Study after study has shown that managers work at an unrelenting pace, that their activities are characterized by brevity, variety, and discontinuity, and that they are strongly oriented to action and dislike reflective activities."[3]

Thus, the way managers work does not seem to mesh well with the nature of the planning activity. The following characteristics of planning have been identified as reasons for its neglect.[4]

1. Planning is a very difficult cognitive activity. It is hard mental work. Because of the cognitive strain involved in doing planning work, people avoid planning.
2. Planning makes evident the uncertainty of future events. By making explicit the various uncertainties, the future may appear more uncertain after planning than before. There is a human tendency to avoid uncertainty, and this may be reflected in planning avoidance.
3. Planning reduces perceived freedom of action. When plans are made, individuals are committed to a narrower range of actions than when no formal plans are made.
4. Planning is computationally tedious. Each change in planning assumptions affects other figures in the plans. Analysis of past data and current expectations requires significant computational work.
5. Plans often are made and then ignored. One reason they may be ignored is that they do not represent real agreement. However, if they are ignored, people become reluctant to be involved in planning.

In general, planning seems to have too little relevance to present reality and does not seem to fit naturally into the organizational picture.

Although it is apparent that the nature of managerial work and the nature of planning both play an important role in understanding why planning is often neglected, the phenomenon also has much to do with how senior managers think. One of the peculiar limitations of the human mind is that we are apparently "not very good at assessing the degree of relationship among variables—even though the skill is critical for successful management".[5] Instead, we apparently have a tendency to rely on preconceptions and to see correlations that do not exist in reality unless relationships are obvious. In fact, the conclusion from hundreds of laboratory and field studies is that the human mind is imperfectly rational and that "we should curtail our impractical and overly ambitious expectations of managerial rationality."[6]

Still, as Isenberg points out, companies need to strive toward rational action in the attainment of corporate goals, and strategic planning is one of the rational processes that businesses may employ. "Programming rationality into the organizational functioning is important [because] rational systems free senior executives to tackle the ambiguous, ill-defined tasks that the human mind is uniquely capable of addressing."

PROGRAMMING RATIONALITY

Notwithstanding the many difficulties of planning, it is entirely essential that various means be found by which this rationality may be programmed into an organization's functioning. The means chosen must somehow recognize the difficulties of planning as a human activity while maintaining compatibility with the nature of managerial work and the thinking processes of senior managers. To some considerable degree, the CSF method accomplishes these objectives by focusing on and conceptualizing issues critical to the success of an organization, a discussion topic that is highly stimulating for most senior executives. The CSF method is certainly not a substitute for planning; rather, it may be used as a mechanism for stimulating thoughts about plans to be formulated, or it may be a device to aid in the implementation of plans. Consequently, the CSF method is ideal in overcoming many of the problems managers experience with regard to planning and is an effective way to identify critical information needs.

CSF

The CSF method was first formulated by John F. Rockart and researchers at the Center for Information Systems Research at M.I.T.[8] In its simplest form, the CSF approach involves two or three highly focused discussions between an analyst or consultant and a senior executive, each discussion lasting about two hours. The product of the exercise is a list of CSFs for this specific manager in this specific position, along with certain generic CSFs. (The actual technique is described in more detail later).

CSFs are those that determine success for a company or business unit, those tasks that must be done well to ensure success. As such it is these "areas of activity that should receive constant and careful attention from management. The current status of performance in each area should be continually measured, and that information should be made available.[9] In essence, Rockart indicates that ensuring the attainment of an organization's goals necessitates good performance in these areas, therefore, it is imperative that management receive constant feedback regarding them.

The distinction was made above between generic CSFs and specific or personal CSFs. Generic CSFs are those that are industry dependent and would be common to any manager in the industry; personal CSFs depend on the individual manager's situation and decision-making style. King and Cleland have described a "Business and Industry Criteria Strategic Data Base," which is a forerunner of the generic CSF concept.[10] Business and industry criteria are the key elements of what it takes to be successful. "While successful top executives may have a good feel for these elements, a rational planning process will make the elements explicit and available to all planning participants. Such planning inputs help to guide the choices which are made.[11]

In the following section a field study is described in which managers involved in planning participated in the development of CSFs, performance measures, and, in doing so, defined their information needs for strategic management. The field study facilitated development of the various tasks required to carry out a CSF study and to develop critical information needs. The description of the field study is followed by an analysis of the CSF approach from the viewpoints of the manager and analyst.

THE CSF FIELD STUDY

Prior to and during the field study, I was engaged as a seminar leader in a corporate training seminar for senior middle-level managers of a large natural resources company.[12] Since the seminar was concerned with improving the participants' skills and executive performance, several on-the-job assignments were contrived in cooperation with one of the corporation's vice-presidents. Briefly, the information systems assignment required each manager to:

Derive the key objectives for the business unit by analysis of the corporate strategy statement.

Develop appropriate performance measures for the business unit, including standards for comparison.

Several weeks after the seminar was completed, I undertook to interview individually some of the participants. The purpose of the interviews was to review and assist with the assignment, assess progress, and, in particular, attempt to discover the strategies used to execute the assignment. From observation and analysis of the strategies employed, I sought to develop a general strategy that might be useful to other managers faced with the problem of defining information that supports strategic planning and control functions.

Interviews lasted up to two and one half hours depending on the nature of each manager's responsibilities. Those with relatively unstructured decision responsibilities necessitated a longer interview than did those whose decision responsibilities were less complex and more highly structured. Time committed to direct completion of the assignment by the participants, exclusive of the interviews, ranged from as much as two full days with extensive subordinate participation in the case of a senior executive to as little as three or four hours when the executive acted independently. The latter situation was most common.

A significant environmental factor in this field study was the corporation's high level of commitment to short- and long-range planning. Planning was a serious activity for all involved, and individual performance was assessed in relation to the successful completion of developed plans. Consequently, the participants approached the assignment aggressively, which greatly assisted me in deriving the approach described in the following section.

The Approach

The process of determining critical information needs consists of five major activities:

- Understand business unit objectives.
- Identify CSFs.
- Identify specific performance measures and standards.
- Identify data required to measure performance.
- Identify decisions and information required to implement the plan.

Understanding this process is enhanced by an examination of conventional corporate planning processes and the manager's role in these processes.[13]

In a typical case, the corporation will distinguish between a business plan and a corporate plan. The business plan is usually concerned with the strategies for particular products, groups of products, functional activities, and so forth. For example, an oil company might choose to develop distinct business plans for exploration, production, and marketing activities, respectively. By contrast, the corporate plan seeks to coordinate all of the activities of a company so that superordinate goals are achieved. The business plan is normally concerned with a time frame as short as one to five years, whereas the corporate plan may look ahead as far as twenty years, though ten years is typical.

The process described below is an example of a general approach to top-down corporate planning in a decentralized firm.[14] In this example, the corporate plan or strategy is produced by a top management council consisting of the CEO and vice-presidents. The vice-presidents, in turn, work with their respective subordinates to identify those aspects of the corporate plan that are relevant to the business units, divisions, or departments for which they are responsible. The products of this activity are business unit strategy statements that specify key objectives, strategies, specific objectives, and so forth for each business unit. Each of the business unit strategy statements developed are, in turn, presented for approval to the planning council or top management prior to implementation.

Once the various business unit strategy statements have been approved, senior managers in charge of each business unit work with their subordinates to develop specific business unit plans that are as quantified as possible. The figures generated for each business unit plan, are, in turn, provided to a staff group that combines all business unit plans to generate a business plan for the corporation as a whole. Once again the business plan is presented to the planning council for approval. From this point on, it becomes the responsibility of the executive in charge of each business unit to achieve the objectives specified and implied in each business unit plan. This is accomplished by developing detailed programs that implement the strategies indicated in their business plan. The corporate planning process as described is summarized in Table 24-1.

Understand Business Unit Objectives

The first activity in defining required information is begun following approval of the corporate business plan and the development of programs to implement the plan. It is at this stage that an information analyst becomes involved. This activity requires ensuring that the manager and the analyst both understand the overall corporate plan, its goals, objectives, and general thrust. Even more important to the task at hand, both must have a full understanding of the objectives of the manager's business unit in relation to the corporate plan. If the manager was an active participant in the formulation of the business unit plan, an understanding of objectives already will be present. Otherwise careful discussion of the plan is required.

Developing this understanding will be relatively straightforward if an explicit up-to-date corporate plan is available; this is increasingly the case as long-range planning becomes more in vogue. On the other hand, where such a plan is not in existence, research, discussion, and analysis of the business may be required. In general, the manager's perception and judgment of corporate directions and business unit objectives should prevail.

Identify CSFs

For each of the objectives identified for the manager's business unit, the manager and analyst now identify the CSFs that the manager must address if each business unit objective is to be achieved. These CSFs can be identified during one or two manager-analyst discussions. CSFs are best described by a short label or expression (as opposed to a narrative statement) that effectively communicates the area of the activity. For example, a CSF identified by all managers in the field study was *human resources.* Ensuring that sufficiently qualified staff is available appeared critical to the success of each business unit. In Rockart's article, some CSFs mentioned were technological reputation, company morale, and liquidity position.[15]

Table 24-1. Summary of Corporate Planning Process Activities

Agent	*Activity*
Planning council (president, vice-presidents)	Produces a corporate plan (corporate strategy statement)
Vice-president, business unit managers	Develop business unit strategy statements
Planning council	Approves business unit strategy statements
Business unit managers	Develop business unit plans by detailing and quantifying business unit strategy statements
Staff group	Develops corporate business plan by combining business unit plans
Planning council	Approves corporate business plan
Business unit managers	Develop detailed programs to implement strategies

Identify Specific Performance Measures and Standards

Having identified CSFs, the manager and analyst attempt to identify specific performance measures for each one. To be useful, the measures should be quantitative as opposed to qualitative in nature. In the field study, the dimensions of quantity, quality, cost, and time were used as general performance measures by three of the managers interviewed. There was no requirement to choose these dimensions, but they served well in many areas and proved useful as a starting point for specific performance measure generation. For example, performance measures identified for the human resources area were percentage of staff turnover (quantity), number of staff qualified for promotion (quality), and amount of money to be spent on personnel training (cost).

However, in some cases quantification may not be easy. At such times the use of indirect surrogate measures that infer progress toward an objective may be considered. For example, part of one manager's responsibilities involved developing and maintaining a favorable image for the corporation in certain sensitive areas. His activities in this regard were directed toward influencing the opinions of individuals such as legislators, government administrators, and research personnel—persons normally sought out by the media when incidents or accidents associated with oil operation occurred. Keeping opinion leaders informed of company efforts to avoid accidents, maintain standards, and so forth leads to a more balanced public view of the company. The obvious measure of performance in this delicate area would be the changing condition of the corporate image; yet such a measurement is neither regularly available nor useful for assessing the performance of a single manager. Instead the manager chose the surrogate measure *number of visits,* reasoning that his talents at persuasion and interpersonal communication were the reasons why he held his position. Therefore, if his information exchange plans were being executed, it would result in an improved corporate image or at least a better protected one. The number of visits was used as an indirect surrogate measure to infer progress in improving the corporate image.

Once performance measures have been identified, the actual performance standards to be met can be set. Frequently, performance standards will be derived directly from the business plan. In other cases, a manager may develop performance standards in consultation with his superior or subordinates or may wish to set the performance standards on his own. Obviously, where past performance figures are available, such figures may provide some assistance in setting realistic standards.

Identify Data Required to Measure Performance

Generally, once the performance measures and standards are determined, the data required to evaluate progress in relation to these standards is easily identified. For example, if percentage of staff turnover is chosen as a performance measure, it is obvious that the information needed is the total number of staff and the number of staff who have departed during a given period. Similarly, assessments of quantity, quality, cost, or time required with respect to performance in other areas are not difficult to generate, collect, and record and seldom require additional computer systems development.

Taken together, business unit objectives, CSFs, performance measures, and performance standards constitute a work plan for the managers in question. To ensure that both the managers and their superiors agree on the work plan developed, a document should be prepared that specifies the critical factors identified, the performance measures developed, and their desired levels. This document should be regarded as a contract between the managers and their superiors that will guide the managers in the performance of their responsibilities over the designated period and facilitate more objective evaluation of their performance. This does not preclude making adjustments to the work plan but contributes to a better understanding of what is expected of the managers by both themselves and their superiors.

Once performance measures and standards have been decided on, it is important for the corporation to establish a performance review process. If such a process is not established, the performance measurement exercise will be of little value. When the concept of performance measurement is first implemented, all concerned are likely to be conscientious about evaluating performance. However, as time passes there is a natural tendency for such systems to be forgotten and no longer used. Preventing this is not an easy task. Active rather than passive procedures are required. That is, procedures must be designed to require superiors to take some action or make some decision with respect to individual managerial performance. For example, managers might be required to submit copies of their subordinates' work plans to their superiors, along with periodically written evidence that progress in relation to these work plans has been reviewed.

ADVANTAGES AND DISADVANTAGES OF THE CSF APPROACH

Manager's Viewpoint

A strong dislike for the reflective nature of planning was cited as one of many reasons for the disinclination of some managers to plan. Thus, an unavoidable requirement for any approach to be successful is that it overcomes this disinclination. The experience of a number of practitioners suggests that the "CSF method generates user acceptance at the senior managerial level. Senior managers seem to intuitively understand the thrust of the CSF method, and consequently, they strongly endorse its application as a means of identifying important areas that need attention."[16] In fact, the CSF method beams directly at the heart of a manager's existence, that is success and the factors to achieve it; hence managers draw naturally to the subject.

Defining CSFs also provides a high-level design strategy for a manager's information support system. Zani argues that a truly effective information system can be designed only from the top down.[17] A clear connection between the goals and objectives of an organization and managers' information is deemed essential. Using the approach described in this chapter, information provided by the system, whether manual or computerized, is used directly in the management control function, that is, to compare performance against standards for the purpose of identifying the need for corrective action. Performance standards, in turn, are tailored to ensure that the manager meets key responsibilities to achieve the objectives and carry out the strategies of the corporate business plan.

By utilizing the corporate planning process, the information analyst designs the information system from the top down and is therefore able to determine the information most critical to senior management. "Since an organization's CSFs are those factors that must go well for the organization to succeed, a link is provided between a corporation's tactical and strategic planning objectives. CSFs then, provide a means of explicitly relating information resources to an organization's strategic planning efforts.[18] The concept of linking organizational and MIS strategic planning was also articulated by King and Zmud.[19]

In a financial services firm, another application of the CSF approach proved to provide senior managers with important information for strategic planning efforts. "The explicit identification of CSFs sharpened management's understanding of those factors central to the firm's success. Corporate reorganization efforts were accelerated by the CSF exercise, and they are likely to be more successful because of it."[20].

Thus, this approach provides information

tailored to the manager's needs. As a result, the manager is not only more effective but also more efficient. Since only information directly associated with the manager's planning and control processes is provided, the volume of information the manager must absorb and analyze is significantly reduced, saving time. In the field study, the information required by one manager monitoring the progress of several different projects was reduced from several pages to a single page. Furthermore, the time spent analyzing the information was reduced from about three hours to less than half an hour. Brevity, fragmentation, and verbal communication characterize a manager's work according to Mintzberg.[21] An information reporting system that ignores this will not be used. Information about a manager's CSFs, permitting rapid monitoring of performance in the key areas and hence more effective strategic management, would seem to better accommodate the time pressures of the senior executive.

The process of engaging in a detailed analysis of business unit objectives, CSFs, and performance standards with the information analyst will aid the manager in understanding the role that both the manager and the business unit play in achieving corporate objectives. This, in turn, should lead to better performance by the manager and better direction for the manager's subordinates. If the manager in question sees the wisdom of having subordinates participate in the process, improved performance, communication, and productivity of the entire business unit should result. In addition, the more explicit the measures of performance, the more able the manager will be in appraising subordinates' performance.

This approach also may provide structure to some managerial jobs previously thought to be relatively *free-form*. For example, in the field study, one of the managers involved was a director of environmental affairs. His responsibilities involved assessing the environmental impact of the corporation's exploration and production activities and maintaining constant liaison with environmental groups. This manager had been of the opinion that the qualitative nature of his work would not lend itself to CSF analysis. However, he found the approach rewarding as it forced him to analyze his responsibilities in detail and to identify the key tasks on which to focus his efforts. He came to view his activities as being relatively structured through involvement in this activity.

Although more structure may be welcomed by some managers, it is likely that others may view such job structuring negatively. Those executives whose managerial style is highly entrepreneurial and whose successes are based on heuristic talents may dislike the constraints implied by having to specify CSFs and work plans. Such individuals may well possess executive talent of the greatest value to the corporation. Their ability to perform successfully in highly unstructured environments may mean that they are more suitable than others for top management positions where major decisions must often be based on fragments of seemingly unrelated data. For this kind of manager, imposing a structure that runs counter to his or her cognitive style may be dysfunctional. If the situation becomes too uncomfortable or the manager's productivity declines, the company and the manager may part ways. There is also the risk that the maverick is passed over at promotion time because of these real but obscure difficulties engendered by the planning process. Care must be taken to preclude use of the above argument as a means of opting out, however, a solution to the dilemma might be to develop CSFs and performance measures for the executive's use only. In other words, such information might be used by the executive to monitor performance of the business unit but not by the company to measure the performance of the executive.

Analyst's Viewpoint

The normally difficult challenge of determining a manager's information needs becomes much more manageable from the viewpoint of the information analyst. Natural guidelines as to relevance, accuracy, timeliness, and other information characteristics are inherent in the process by operating within the planning context. Moreover, utilizing CSFs as an anchor point, the resulting information is likely to have greater utility for the manager. "If the system is designed on the basis of those strategic issues which are identified by use managers to be those that are most urgently needed in supporting their strategic choices, the system is likely to be used.[22]

Utilizing the planning processes of the organization also provides much more structure to the analyst's interviews, a feature that often is inadequate. An unstructured interview usually leads to educated guesswork as to what information to provide the manager, and the product is frequently unsatisfactory. In the absence of structure the analyst becomes dependent on the manager to articulate the information required, an ability that the manager often may not have. For example, most managers have a real need for both hard (quantitative) and soft (qualitative) information. Hard information, most often derived from the organization's financial reporting systems, is the easiest to articulate and obtain. However, the ready availability of such information often leads to overlooking the need for identifying critical soft information. This problem can be overcome by an approach that is driven and structured by the requirement to identify information for CSFs, whether associated with hard or soft areas of the manager's responsibilities.

Identifying CSFs in the context of an organization's planning processes seems to overcome a potential difficulty noted by Davis.[23] Davis suggests that human information-processing limitations might cause managers to reduce the number of CSFs they actively consider to a range of five to nine, whereas the number of CSFs in the actual situation may be greater. In other words, managers might inadvertently overlook or fail to recall some CSFs. This is a real possibility when the manger-analyst interviews are unstructured. However, using the approach in this chapter, the manager-analyst discussion is structured by the presence of goals and objectives identified earlier by the planning process. As a consequence, CSFs are generated in response to stimuli, that is, goals and objectives, as opposed to the analyst relying solely on the individual manager's limited information-processing capacities. Since the number of CSFs for a particular objective seldom exceeds three or four, the structure provided by the goals and objectives identified in planning should ensure that no CSFs are overlooked.

A second difficulty exists with respect to the problem of generating performance measures in soft areas. The use of surrogate measures from which performance might be inferred in these areas was referred to earlier. Yet surrogate measures obviously are not completely satisfactory. If performance measures are to assist the manager in achieving some measure of control over assigned responsibility areas, the more direct must be the connection between the performance being measured and the organizational component being controlled.

Third, an approach to determining information requirements that depends on a commitment to planning may be difficult to apply in a corporation in which organized planning does not occur. The greatest problem in this regard will be the absence of corporate strategy from which to derive the necessary goals and objectives that apply to the individual business units. Still, the information requirements may be defined reasonably well if the manager in question holds a clear understanding of the unit's organizational mandate.

Finally, as the organization and its environment changes, so too will corporate goals, objectives, and management's CSFs. Adoption of the approach described in this chapter will ensure that through the mechanism of the planning cycle CSFs, performance measures, performance standards, and related information are regularly updated.

CSFS USED TO INVOLVE TOP MANAGEMENT GROUP

In the foregoing field study, the CSF approach was used in *one-on-one* fashion with a series of managers. However, the CSF approach is also effective when used with a group of managers. An excellent example of this application occurred at Southwestern Ohio Steel (SOS).[24]

In this situation, SOS had first turned to a consulting firm to undertake a major review of its information systems capability. However, the executives of SOS were shocked to learn that the consultants recommended a $2.4 million conventional system design and implementation process with major results and benefits not apparent until after four years. As a result, SOS turned to a different consulting group for a more creative approach toward assisting top management to understand its systems needs. The result was a three-phase process based on three major concepts: CSF *group* discussions, decision scenarios, and prototyping. Each phase, in turn, had two or three parts.

The first step in phase one began with an introductory workshop in which the mission or strategy of the corporation was discussed and primary objectives for the organization were generated. The second step involved an individual CSF interview with each senior manager in which the specific factors most responsible for the achievement of the objectives were identified. Finally, a third step involved a focusing workshop in which the results of the interviews and their implications were evaluated and thoroughly discussed simultaneously by the entire group of senior managers.

The reactions of the participants to this first phase were at least as important as the objectives and CSFs generated. In the words of one of the vice-presidents, "We all knew what was critical for our company, but the discussion—sharing and agreeing—was really important. What came out of it was a minor revelation. Seeing it on the blackboard in black and white is much more significant than carrying around a set of ideas which are merely intuitively felt."[25] Another executive commented that focusing on 'what makes the company a success' intrigued almost all of top management. It appealed to a group of top managers, allowing them to engage in a discussion of what they knew best and what seemed important to them."[26]

In the second phase, another workshop was held to define the performance measures that would be used to evaluate the CSFs identified earlier. In particular, the project team studied the business in more depth and identified the distinct systems priorities that would be required to support the fundamental managerial processes identified. To ensure that management fully understood the connection between the CSFs and the systems to be developed, the project team worked with the managers to establish the recurring decisions management faced in coping with these situations. Decision scenarios or situations were developed, each one of which was concerned with a particular managerial event. By working through these scenarios, managers were able to gain much better insight into the workings of the proposed systems. They were able to determine what the systems would do and would not do.

The third and final phase involved creating initial prototype designs and actual systems development. The prototypes developed were in a sense incomplete systems that would facilitate the demonstration of what a fully developed system could do.

Rockart and Crescenzi explained that not all of the key executives became committed until the prototyping activity was well underway. One executive in particular had remained skeptical throughout the process because of his having heard a number of computer horror stories. He had considerable apprehension that control of the inventory might be lost in the conversion process and furthermore that the system in general would not support his particular inventory needs. It was only after an early prototype design was well underway that the executive became fully comfortable with the prototype approach. The prototype was finally seen by the executive as a means of lowering the company's and his risks to a acceptable level. "I would have slept better at night if they [the consultants] had fully communicated the prototype concept from the beginning. Once the idea finally struck me, it really turned me on. I went from negative to highly enthusiastic."[27]

In general, coming the CSFs approach with group workshops and prototyping proved to be a particularly effective method of solving a major planning problem while educating senior executives about their own corporation. The CSF approach has also been successfully applied in the areas of small business[28] and information systems management.[29]

A MAJOR CAUTION

As intriguing and conceptually powerful as the CSF approach appears, it may not work at all times in all companies.[30] The existence of certain environmental factors can make a great difference. For example, the outcome of the process may depend on whether or not senior management is ready to be involved. In the case of Southwestern Ohio Steel, the prospect of spending several millions of dollars for systems development created that readiness. Other organizational and environmental factors can have a similar or opposite effect. In any case, a climate for change must exist for such organizational changes to take place.[31]

Whereas the advantage of an enthusiastic and receptive management group is obvious, a subtle but highly important factor may well result in the CSF approach causing management to work more effectively in the wrong directions. The factor in question is the adequacy of management's mental model of their organization, that is, the quality of their understanding of the dynamics of their business.

The importance and value of a good analytical model to the CSF method has been identified by Davis. "What is needed is an analytical model of the business unit that the analyst can use in eliciting executive responses and with which to evaluate the critical success factors obtained from executives for relevance, correctness, and completeness."[32] This is a vital matter because the factors expressed as "critical to success" by a decision maker are of necessity a reflection of that manager's understanding of the system in which he or she operates. The quality or accuracy of the mental map of the decision maker will affect what is perceived to be important. If the decision maker's mental map is inaccurate or incomplete, perhaps because his or her cognitive complexity causes him or her to simplify his or her mental map and resolve away any unwanted uncertainty, the CSFs specified may be inaccurate or incomplete.

Unfortunately, evidence exists that managers commonly hold an inadequate mental model of their environment. As Axelrod has stated: "The picture of a decision maker that emerges from the analysis of cognitive maps is that of one who has more beliefs than he can handle, who employs a simplified image of the policy environment that is structurally easy to operate with, and who then acts rationally within the context of his simplified image."[33] In other words, it is becoming increasingly apparent that managers simplify the mental models that they use to structure their decisions, they ignore uncertainties, and in general they ignore feedback that runs contrary to their existing beliefs.

For the capable, self-confident senior manager sometimes with decades of experience to draw on, the suggestion that he or she may not fully comprehend the dynamics of his or her corporation is difficult to accept. The notion that somehow his or her understanding of the effects of his or her decisions on the organization may be faulty is hard to swallow. The idea that he or she may not *really* know what is critical to success for the firm is probably seen as highly unlikely. In the following section, two examples (among several) are described that illustrate the need for an analytical model and raise a major caution regarding use of the CSF method.

Failing to Understand

Recent published research in the area of corporate system modeling provides several examples of the failure by management to understand organizational dynamics. Consider the demise of certain mass circulation magazines such as *Life*, *Look*, and the *Saturday Evening Post*.

> At the time of its initial crisis, each of these magazines reported its highest circulation and largest revenue. There must be an explanation for such a paradoxical situation wherein a record circulation and revenue is associated with poor profit performance. It is hardly credible that a large number of the leading magazines were being mismanaged simultaneously. These magazines continued to grow in spite of keen competition with other large circulation magazines and other mass communications media until they reached a critical point in their history. This suggests that the pathology of magazine publishing is, perhaps a complex phenomenon.[34]

By using system dynamics, a model was built to represent a working hypothesis of how a magazine publishing system functions. An extensive empirical study, using data covering a twenty-year period of the operations of the *Saturday Evening Post*, lent credence to the model as a reasonable representation of reality. The model was used to understand the rise and fall of the *Saturday Evening Post* in terms of the dynamics of the entire system, including the management decision processes.

The model revealed a number of variables associated with magazine publishing. These variables, which seem intuitively appropriate even to a layperson, are as follows:

Subscription rate, that is, cost of a subscription
Trial subscriptions sold
Yield of regular readers converting from trial readers
Regular subscribers who resubscribe
Number and growth of total readers
Annual volume of magazine pages
Annual volume of advertising pages
Advertising rate
Production costs
Promotional expenditures
Annual revenues
Profit margin

In addition to the above, the model revealed that management failed to understand the relationship between some of the variables and consequently made costly mistakes. For example, at one point, management's action to counteract a drop in profit margin was to substantially increase the subscription rate. The effect of increasing the subscription rate, however, was shown by the model to have a detrimental effect on the fraction of regular subscribers who resubscribe. Consequently, the growth of readers of the magazine leveled off. The company found itself in the peculiar position of having unrestrained growth leading to a depressed profit margin and eventual stagnation in the growth of readers due to the drop in renewal rate of regular subscribers.

Another example drawn from the magazine publishing study indicates yet other evidence of management's inability to see cause-and-effect relationships. Perhaps as a reaction to the situation in the earlier example, management decided that readership was the key to unlock future growth and consequently reduced the magazine's subscription rate and massively increased promotional expenditures. To pay for the ever-increasing promotional expenditure, the only available alternative was to increase the advertising rate. However, the rate of increase in readers did not match the rate of increase in advertising rate, and the advertisers purchased fewer pages in the magazine. In addition, the yield of regular subscribers from trial subscribers also dropped. By this time, almost one-half of the total readers were trial readers, and only a small proportion of these readers was taking out regular subscriptions as the annual volume (in pages) of the magazine declined in response to the ever-increasing advertising rate. During this period, readership grew, annual revenue grew, but profit margin fell from 8 percent of revenues to a loss position. "The company was on the brink of bankruptcy and never really recovered from this policy cul-desac of too high a subscription rate, too high an advertising rate, a declining annual volume, and too high a promotional expenditure to solicit trial readers to replace the defecting readership."[35] Following a strike, editorial revolt, resignation of the president, and successful libel action against the magazine, this senior managerial nightmare ended with discontinuance of the *Saturday Evening Post.*[36]

This example demonstrates that management did not have a full understanding of the dynamics of their business. The absence of a proper analytical model likely would have resulted in the articulation of an incorrect set of CSFs, attention to which would probably have accelerated the demise of the firm.

Averting a Crisis

In a totally different context, a corporate system model of a sports club was used to avert a crisis caused by inadequate management understanding.[37] The club in question had been experiencing a steady decline in membership over several years. Combined with loose control over financial matters in the past, this trend nearly resulted in the collapse of the club.

By working with the executive of the club, a model was built to describe club operations. The model described the essential features of the club in terms of club membership, various revenues flowing from membership, rental income, bar and food revenues, and quality of management of the club facilities. Together, these affect the profit of the club, the fee structure and quality of service that, in turn, influence the level of membership. The model explained the role of each of these features in affecting membership levels.

The traditional approach to the problem of declining club revenues had been for management to respond by increasing membership fees. Unfortunately, this produced a further debilitating effect on membership. The model demonstrated that the process of increasing fees to cover deficits led at best to an exclusive club or at worst to the collapse of the club. Reducing membership fees indicated a beneficial effect on clubhouse memberships but did nothing other than slow the inevitable decline of active members (as opposed to social members). The model showed that the manipulation of membership fees alone would not reverse the downward spiral of events.

The CSF in this situation proved to be the member's perceptions of the quality of the club's services vis-à-vis the fees being charged. The newly elected executive of the club attacked the cancellation rate problem by increased promotion, expansion and improvement of services, and use of other devices aimed at altering members' image of their club and perceptions of the services available relative to fees charged. The result was an improvement in total membership and a dramatic improvement in the financial situation.

It should be readily apparent that the operation of a sports club is significantly less complex than the operation of a major corporation. However, those who had been intimately involved in the management of the club for many years seemed unable to grasp its organizational dynamics, the manipulation of which could lead to either success or failure. Again, a CSF analysis probably would have led to an entirely erroneous list of the matters to be attended to. The set of CSFs generated would have reflected the inadequate understanding of management of the dynamics of the organization. Similar examples may be examined in Roberts and Roos.[38]

SUMMARY

Defining CSFs can be a highly effective method for identifying the information critical to successful strategic management. By integrating the CSF approach into the organization's planning process, the entire planning activity benefits. An additional product of the exercise is a high-level strategy for information systems redesign and possibly redesign of the organization. Overall, the CSF approach can assist in identifying and realizing an organizational strategy. These relationships are illustrated in Figure 24-1.

The cautionary examples should lead both senior management and analysts or consultants employing a CSF approach to proceed with care. Serious corporate difficulties may be a reflection of management's failure to sort out the powerful cause-and-effect forces within the organization. Applying the CSF approach in such a situation may in fact exacerbate the difficulties.

It should not be construed, however, that one cannot successfully use the CSF approach in the absence of the use of systems dynamics or some relatively involved form of corporate system modeling. A history of success for a particular organization over time should constitute *prima facie* evidence that senior management does in fact adequately grasp the intricacies of the organization. One should ensure, however, that the managers involved in the CSF activity participated in the success of the organization. In other words, the CSFs generated must reflect the collective wisdom of some relatively homegeneous group of experienced managers rather than the CSFs as perceived by any individual or by any group of inexperienced managers.[39]

Ideally, of course, the analyst would employ system dynamics or other similar techniques to secure an adequate model with which to assist the process of generating CSFs. Given the importance of the activity, taking the extra time to develop the model would seem well justified. Fortunately, development of a model using system dynamics is entirely feasible, though several weeks may be required for very large organizations. However, in the absence of such an effort, analysts and strategic managers should at least be sensitive to the fact that complex interrelations and counterintuitive feedback effects may be present without their awareness.

In general, the CSF method is highly effective in engaging senior management in the strategic planning activity and in generating information for strategic management purposes. Senior managers find the process stim-

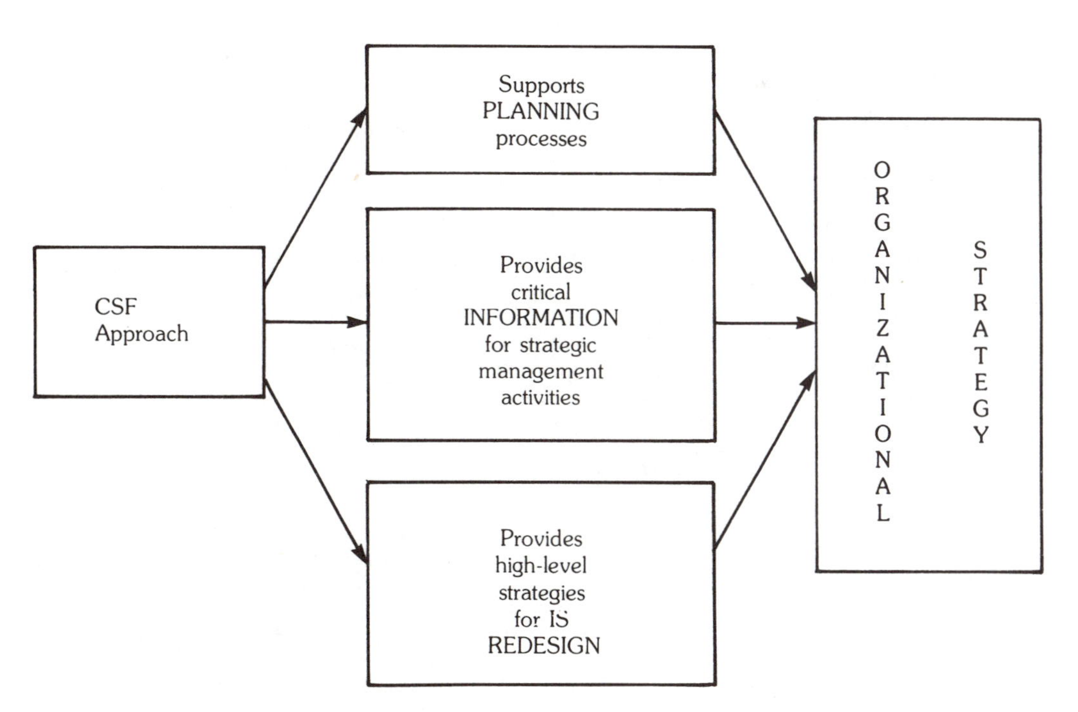

Figure 24-1. CSF approach as it relates to organizational strategy.

ulating and invigorating and perceive the entire thrust of it to have high face validity. In some situations, corporate system modeling may be required to pin down the more obscure organizational causes and effects; in other cases, direct application of the CSF approach may occur.

NOTES AND REFERENCES

1. J. H. Grant and W. R. King, *The Logic of Strategic Planning* (Boston: Little, Brown, 1982).
2. H. Mintzberg, "The Manager's Job: Folklore and Fact," *Harvard Business Review* **53**(1975):49-61.
3. Ibid., 50.
4. G. B. Davis and M. H. Olson, *Management Information Systems: Conceptual Foundations, Structure, and Development,* 2nd ed. (New York: McGraw-Hill, 1985).
5. D. J. Isenberg, "How Senior Managers Think," *Harvard Business Review* **62**(1984):83.
6. Ibid., 88.
7. Ibid., 89.
8. J. F. Rockart, "Chief Executives Define Their Own Data Needs," *Harvard Business Review* **57**(1979):81-92. Conceptual predecessors include W. R. King and D. I. Cleland, "Information for More Effective Strategic Planning," *Long Range Planning Journal* **10**(1977):59-64; W. M. Zani, "Blueprint for MIS," *Harvard Business Review* **48**(1970):85-90.
9. Rockart, "Chief Executives Define Their Own Data Needs," 85.
10. King and Cleland, "Information for More Effective Strategic Planning."
11. Ibid., 59.
12. Although the field study described involved senior middle-level managers, the conclusions are equally applicable to top executives, especially given the size of the business units involved and the style of organization.
13. G. A. Steiner, "Comprehensive Managerial Planning," in *Strategic Planning for MIS,* eds. E. R. McLean and J. V. Soden (New York: Wiley, 1977), 247-265.
14. Ibid.
15. Rockart, "Chief Executives Define Their Own Data Needs."
16. A. C. Boynton and R. W. Zmud, "An Assessment of Critical Success Factors," *Sloan Management Review* **25**(1984):18.
17. Zani, "Blueprint for MIS."
18. Boynton and Zmud, "Assessment of Critical Success Factors," 20.
19. W. R. King and R. W. Zmud, "Management Information Systems: Policy Planning, Strategic Planning and Operational Planning," in *Proceedings of the Second International Conference on Information Systems,* (Cambridge, Mass.: 1981), 299-308.
20. Boynton and Zmud, "Assessment of Critical Success Factors," 23.
21. Mintzberg, "Manager's Job."
22. J. I. Rodriguez and W. R. King, "Competitive Information Systems," *Long Range Planning Journal* **10**(1977):46.
23. G. B. Davis, "Comments on the Critical Success Factors Method for Obtaining Management Information Requirements in Article by John F. Rockart, 'Chief Executives Define Their Own Data Needs' *Harvard Business Review,* March-April 1979," *MIS Quarterly* **3**(1979):57-58.
24. J. F. Rockart and A. D. Crescenzi, "Engaging Top Management in Information Technology," *Sloan Management Review* **25**(1984):3-16.
25. Ibid., 8.
26. Ibid., 8.
27. Ibid, 8.
28. R. A. Dickinson, C. R. Ferguson, and S. Sircar, "Critical Success Factors and Small Business," *American Journal of Small Business,* **3**(1984):49-57.
29. E. W. Martin, "Critical Success Factors of Chief MIS/DP Executives," *MIS Quarterly* **6**(1982):1-9; E. W. Martin, "Information Needs of Top MIS Managers," *MIS Quarterly* **7**(1983):1-11; J. F. Rockart, "The Changing Role of the Information Systems Executive: A Critical Success Factors Perspective," *Sloan Management Review* (23 1982):3-13.
30. Boynton and Zmud, "Assessment of Critical Success Factors."
31. K. Lewin, "Frontiers in Dynamics," *Human Relations* **1**(1947)5-41.
32. Davis, "Comments on the Critical Success Factors Method"; *MIS Quarterly* **3**(1980)8.
33. R. Axelrod, *The Structure of Decision: The Cognitive Maps of Political Elites* (Princeton, N.J.: Princeton University Press, 1976), 238.
34. R. I. Hall, "A System Pathology of an Organization: The Rise and Fall of the Old *Saturday Evening Post,*" *Administrative Science Quarterly* **21**(1976):185-211.
35. Ibid., 203.
36. M. J. Culligan, *The Curtis-Culligan Story* (New York: Crown, 1970); O. Friedrich, *Decline and Fall* (New York: Harper & Row, 1970).

37. R. I. Hall and W. B. Menzies, "A Corporate System Model of a Sports Club: Using Simulation as an Aid to Policy Making in a Crisis," *Management Science* **29**(1983):52-64.
38. F. S. Roberts, "Strategy for the Energy Crisis: The Case for Commuter Transport Policy," in *The Structure of Decision: The Cognitive Maps of Political Elites*, ed. R. Axelrod (Princeton, N.J.: Princeton University Press, 1976), 142-179; L. L. Roos, Jr. and R. I. Hall, "Influence Diagrams and Organizational Power," *Administrative Science Quarterly* **25**(1980):57-72.
39. K. I. Ghymn and W. R. King, "Design of a Strategic Planning Management Information System," *OMEGA* **4**(1976):595-607.

CHAPTER 25

Decision Support Systems for Strategic Management

William R. King

William R. King is University Professor of business administration in the Graduate School of Business at the University of Pittsburgh. He is the author, co-author, or co-editor of more than a dozen books and 150 articles that have appeared in the leading journals in the fields of strategic management, information systems, and management science. He has served as senior editor of *Management Information Systems Quarterly* and serves as associate editor of *Management Science, Strategic Management Journal OMEGA: The International Journal of Management Science*, and various other journals. He frequently consults in these areas with corporations and government agencies in the United States and Europe.

The notion of a decision support system (DSS) became popular in the late seventies.[1] In the early eighties, the concept of a strategic decision support system (SDSS) or strategic management decision support system began to emerge.[2]

This chapter reviews the DSS concept and evaluates its applicability and utility to strategic management. The approach is one of critical evaluation that focuses on whether SDSS is an area of thought and research that is distinguishable and that is likely to be important to strategic management. A variety of criteria are developed for assessing the SDSS area in this regard.

These criteria for evaluating the salience of SDSS to strategic management draw on basic notions of a DSS, of the role of information in strategic decision making, the concept of information-based strategic comparative advantage, and of organizational processes for achieving such advantage. The early sections of the chapter deal with these building blocks on which the SDSS criteria are based. Then, the criteria are developed and supported with illustrations of systems that meet, or nearly meet, the rigorous standards for SDSSs.

NOTION OF A DSS

There are a variety of definitions and usages for the term *decision support system* (DSS). After more than a decade in which the notion has been popular,[3] it might be surprising to some to learn that the term is used in different ways and is only imprecisely defined. As recently as 1982, the title of an article in *Computing Newsletter* clearly indicates this to be the case: "Question Yet Unresolved: What is a Decision Support System?"[4]

There are at least three different ways in which the term *DSS* is commonly used. In the most general sense, it can be used generically to describe any system having the objective of supporting decision making.[5] This usage views the DSS concept as one that encompasses other more traditional notions such as *management systems* and *planning systems*.[6] Thus, this generic use of the term cannot be argued to be incorrect; but it is so broad that it is not very useful.

At the next level of generality, the term *DSS* is used to describe a computer-related system that has the general purpose of aiding in decision making. In this context, it may be as comprehensive as a management information system (MIS), since the decision-support notion was certainly the basis for distinguishing that variety of system from electronic data processing (EDP) systems in the 1960s.[7] Alternately, it may be used as specifically as by a software vendor describing a particular user friendly language such as IFPS or EMPIRE.[8]

A still more specific use of *DSS* describes it in terms of a number of characteristics or attributes. For instance Sprague and Carlson identify four such characteristics:[9]

> They tend to be aimed at the less well-structured, underspecified problems that upper-level managers typically face.
>
> They attempt to combine the use of models or analytic techniques with traditional data access and retrieval functions.
>
> They specifically focus on features that make them easy to use by noncomputer people in an interactive mode.
>
> They emphasize flexibility and adaptability to accommodate changes in the environment and decision-making approach of the user.

Similarly, King has identified the typical DSS to be an integrated system that is made up of various subsystems:[10]

1. Decision models
2. Interactive computer hardware and software
3. A data base
4. A data management system
5. Graphical and other sophisticated displays
6. A user friendly modeling language

In this view, there is no minimum set of such subsystems that must be incorporated into a system to make it a DSS. However, most such systems involve many of these subsystems.

These characteristics and attributes definitions are accepted and used by most DSS specialists since they are more precise, specific, and operational than the more general ways of using the term *DSS*.

Thus, from this viewpoint, a DSS is an interactive computer-based system that is used by upper-level managers to deal with relatively unstructured and underspecified problems, using decision models, easy and efficient access to significant data bases, and various display possibilities. It also incorporates a modeling language and other user friendly attributes to give the user the opportunity to go beyond preprogrammed models to construct and use his or her own decision-aiding constructs to accommodate changing situations or user needs.

INFORMATION, STRATEGIC MANAGEMENT, AND DSS

The more specific and operational varieties of definitions of a DSS clearly serve to suggest their applicability to strategic decision making, where problems are unstructured and underspecified and planners need to experiment with various formulations and approaches rather than merely with alternative strategies and solutions.

However, it is important to make this connection between strategic management and computer-based systems on some more substantive basis. The notion of *information* provides such a basis.

All computer systems, whether DSS or EDP, involve information. Indeed the term *decision and information system* was an early attempt to describe systems for which the term *DSS* later became popular.[11]

Alter distinguishes between the traditional computer system (EDP) and the DSS in a fashion that emphasizes the nature of the information, the criteria used, and the use to which information is put.[12] Table 25-1 shows an adaptation of this description that clearly relates DSS to the strategic management context through information. The table shows a

Table 25-1. Comparison of EDP and MIS

EDP	*DSS*
Passive use	Active use
Clerical activities	Line, staff, and management activities
Orientation toward mechanical efficiency	Orientation toward overall effectiveness
Focus on the past	Focus on the present and future
Emphasis on consistency	Emphasis on flexibility and *ad hoc* utilization

Source: Adapted from S. L. Alter, *Decision Support Systems.* (Reading, Mass.: Addison Wesley, 1980), 2.

DSS to be oriented toward information related to effectiveness rather than efficiency, information with a future outlook, and information that can be used actively and flexibly by both line and staff people. Although he did not develop this description in the context of strategic management, the fit is nearly perfect.

An EDP system may provide data in rather raw form, with only things such as aggregates and averages to make the data useful to decision makers. Moreover, the data are generally operational-level. However, a DSS generally provides information that is of two different varieties: (1) evaluated data; and (2) model outputs. Both represent *transformed data.* These transformations are performed by the system to benefit the decision maker, who is the user of the system.

Data that have been evaluated to be *relevant* and *useful* for making of particular decisions or classes of decisions are called evaluated data. Unlike a management information system, which may facilitate access to much of the basic operational data of the organization, the data bases that the DSS user can call on are generally more focused and directed toward the specific uses for which the DSS is designed. For instance a strategic DSS would not normally access *operating* data. Thus, one of the varieties of transformed data that the SDSS entails is sets of data that have been collected, assessed for relevance to a particular set of decisions, for quality, and for validity and made accessible to the SDSS user. In effect, this process transforms *data* into information,[13] and, despite the fact that the user of the system may be given direct access to a data base, the data therein have previously been substantially evaluated and transformed.

For instance, the data bases that most traditional corporate models access are of this variety in that they entail financial, market, and other data at a fairly high level of aggregation.[14] Their use does not normally require, and the DSS usually does not readily permit, access to detailed operational-level data such as those that concern business transactions.

The second variety of information that a DSS provides is the *output* of decision models. Models, be they of the descriptive, predictive, or optimization variety, may be viewed as *devices for transforming data into more useful information.* For instance, a descriptive model, such as one of the business position matrices that are often used in planning,[15] summarizes a great deal of diverse data concerning a product or a business's competitive position in a pictorial form that serves to enhance greatly the value of the information.

Similarly a predictive model, such as an econometric model of the economy,[16] transforms data on the current and past states of the economy and on leading indicators into predictions of future economic situations. An optimization model carries this information transformation process a step further through the evaluation of the alternative actions that are available and the prescription of an alternative that is best according to some criterion such as profit maximization.

Any or all of these models may be incorporated into a DSS, although the descriptive and predictive varieties are most commonly so used. In all cases, they provide information to the user that has had value added in the transformation accomplished by the model.

There is much discussion in the literature of DSS concerning the role of decision support systems in aiding the user to identify and structure decision problems, as opposed to solving them. Although this is a valid objective and many such systems are extensively oriented to this purpose, problem identification and formulation are simply other *uses* to which the two previously noted varieties of DSS output information may be applied. They do not represent generically different ways of processing information.

Varieties of problem formulation systems that go under the title of *artificial intelligence*[17] represent a special case of information that is the output of models. In these cases, the models are not themselves so highly structured as are many predictive or optimization models, but they are nonetheless models in that they represent systematic and prescribed ways of manipulating data. The basis for these models may sometimes be hueristic, but they are only different in degree, not in substance, from other more familiar varieties of models.

Thus, although some may argue that the problem identification and formulation *use* of a DSS is important in distinguishing DSSs from other varieties of systems, from the point of view of information transformations, these uses are not significantly different from other uses for the DSS.

EVOLVING ROLE OF INFORMATION IN STRATEGIC MANAGEMENT

The importance of information to effective strategic management hardly needs to be argued. In the sense just described, most of the developments in the methodology of strategic planning made in the recent past are represented by models for transforming data into more highly valued information. For instance, the bubble charts and PIMS analyses pervasively used in modern strategic analysis are models and techniques for evaluating data for a purpose and transforming it into useful information.[18]

However, there is a new role for information that is emerging in strategic management—that of *information as a source of strategic comparative advantage.*

To understand why this notion is, in fact, new and different from the roles that have been played by information in the organization previously, one must recognize that computers have indeed been the primary means of processing data for many years. However, although computers have already created substantial societal changes through the creation of new computer- and information-related businesses[19] and through their application in myriad everyday contexts such as credit billing and traffic signal control,[20] until very recently they have not had the comprehensive revolutionary impact that had so often been predicted. Indeed, various studies of the actual impact of the computer have shown that impact to be substantially less in many respects than many had anticipated.[21]

Even in business, which depends so much on computers, one might characterize the overall role and impact of the computer prior to the era of microcomputers to be a rather modest one.[22] The mainframe computer had been used to automate many clerical functions to the degree that the business is dependent on computers in much the same way as we all depend on them in our daily lives. The nuts and bolts of the business—billing, payments, paychecks, and so forth—would quickly halt if computers were to stop functioning, just as would the telephone system and many of the other commonplace services of daily life.

However, prior to the microprocessor era many businesses were dependent on computers *only* in these mechanistic operational ways. The mainframe computer and its attendant information systems provide support to operations and to management that is faster, more accurate, and more efficient than was previously provided or that now could be provided by other means. However, in most instances, the computer system and the information it processes had not yet come to serve as an integral and critical resource to business success.

The introduction and use of computers generally has been a low-risk, and consequently low-payoff, activity for most business firms over the past several decades. Companies had commonly invested an amount equivalent to 1 percent of sales into a service function that generally had more than paid for itself but that had not radically changed the business or the way in which it was managed.[23] They came to depend on it mechanically, just as they depend on heat, light, and water, but computer systems had little more to do with the core of the business than those other systems, that is, they have little to do with the critical strategic resources that made the business what it was and that determined the strategies it followed.

Moreover, the literature of MIS makes it clear that computer systems really had not intruded into the daily lives of most *managers* to any great extent. In many firms, only a relatively few computer technicians directly interacted with the computer on a day-to-day basis. Only a few professionals and managers who were so inclined tended to be directly involved with computer systems in such firms. As an advertisement of Wang Laboratories eloquently states, "For 25 years, the most powerful tool of the twentieth century was kept in the back room."

Thus, the computer had not been used effectively as an instrument of strategy by most business firms. Rather, it primarily had been used as a tactical tool for achieving cost savings and greater efficiency.

The emerging role of computer technology in the strategic management of business is easy to envision. One need only view a major business firm to see that new computer-based technology is being introduced in quantity at many locations—word processing systems, electronic mail and filing, electronic communications networks, desk top computers, and so forth. Moreover, these technologies are rapidly being linked together into DSSs of the variety just discussed.

A major implication of this explosion of technology is that far more people at higher levels in the organization are directly involved with the computer system than ever before. Computers and appurtenant technology are no longer relegated to the back room as a specialized service function that has little to do with the day-to-day activities of most people in the firm. They are out in front in virtually every executive office, workstation, and production line.

As mentioned earlier, in the past, only a small number of computer specialists were in direct contact with, and were direct users of, the computer system. In many organizations, one member of each department or unit was specifically identified to perform this role.[24]

With the technological revolution that is occurring, virtually everyone will be a user of the computer system. This widespread intrusion of computers into the lives of so many will have a profound impact, and, although it will not be without problems, it will tend to increase understanding, reduce apprehensions, and enable many more people to better envision the widespread potential for computer applications. Heretofore, many such applications were envisioned only by computer specialists, who often had difficulty selling their ideas to management, or remained unseen because computer specialists lacked the requisite business knowledge and experience to relate computer capabilities to business needs.[25]

This pervasiveness of computers and the increasing familiarity with them of people at all levels of the organization will inevitably lead to a wide variety of new computer applications. More importantly, however, it will lead to the amplification and acceleration of a phenomenon that is already beginning to be experienced—*the creation of a comparative business advantage through information.*

INFORMATION-BASED COMPARATIVE BUSINESS ADVANTAGE

The concept of a comparative business advantage is certainly not new. However, the notion of an information-based comparative business advantage is neither well developed nor understood. For instance, Porter describes three generic business strategies—overall cost leadership, differentiation, and focus—that may be used to achieve a comparative business advantage.[26] Two of these generic strategies represent ways of achieving a competitive edge using traditional business factors that are *not* generally highly information intensive. For instance, *overall cost leadership* reflects a comparative advantage in pricing or profitability that is based on a product cost advantage, which may, in turn, be derived from experience-curve effects or lower-than-average wage rates. Similarly, *differentiation* is often achieved based on a quality advantage that may be derived from superior design or technology. Neither are information-intensive comparative advantages (although if such advantages are systematically conceived, they are certainly based, in part, on information and knowledge).

However, firms have achieved information-based comparative advantages—that is, the development of a relative advantage in the competitive marketplace on the basis of superior information or knowledge rather than by the more traditional means for directly creating comparative advantage (such as price advantage based on a cost advantage or a quality advantage based on a technology advantage).[27]

For instance, Porter's third generic strategy, *focus,* may readily be achieved or implemented primarily on the basis of superior information. This strategy is one in which a firm concentrates on some particular product subline, group of customers, geographic area, or set of distribution channels to obtain a *local* comparative advantage rather than a global one. Clearly, such a strategy *can* be the result of careful analytical market segmentation.[28] If so, it would be knowledge or information based.

Two generic ways in which an information-based strategic comparative advantage may be achieved are by using information and/or systems as an integral element of the product or its delivery or by using information as a basis for decision making in a context in which the information and systems are invisible to anyone outside the organization.

Illustrations of the former way to gain an information-based strategic comparative advantage are becoming rather common in this information intensive era. For instance, Merrill Lynch's Cash Management Account was created by the development of integrated information systems that made it possible to connect a number of existing financial products—brokerage accounts, checking accounts, lines of credit, credit cards, and so forth—into a single integrated product. This development of information capability created a new product that gained many hundreds of thousands of new accounts and many billions in newly managed assets. The innovation has been patented and, even if the patent protections do not apply, gained Merrill Lynch a four-year lead on its competitors.

Various airlines have developed their own computerized reservation systems that have been placed in travel agencies and that operate to show favoritism to the airline in question, thus enhancing the likelihood of the system generating additional business for the airline offering the information service. If this design bias works as it is clearly intended to do, it should produce significant strategic gains for the airlines involved.[29]

Foremost McKesson, the nation's largest wholesale distributor, virtually has redefined the function of middleman by implementing an information-based strategy.[30] Foremost has placed computer terminals in the hands of stock clerks in the aisles and microcomputers in the offices of its drug store customers to permit them to do analyses and to order. It also handles third-party billing (to insurance companies) for drugs sold over the counter in its customer's stores. In effect, it has transformed its business into a broader one that is based on information rather than solely on the distribution of products.

Information-based strategic comparative advantages may also be developed in contexts of purely management information.

Since such information is invisible to the outside world, these cases are both more difficult to identify and more confidential in nature. Among the examples known to me is a firm that has developed a sophisticated cost information system that enables it to determine the actual cost of its diverse products and to accurately estimate the cost of proposed new products and of custom-ordered ones. It has integrated this into an overall system that permits it to design consistent pricing strategies and sales incentives. This system has given it a significant advantage over its competitors, most of whom do not even have a good system for calculating the true cost of the various items in their complex product lines. It has achieved this advantage despite the fact that, in the opinion of the key managers, *it does not have a cost advantage over competition.* Rather, it has an information advantage.

Another firm has a similar system allowing it to choose orders based on "incremental profit through the bottleneck" in a production system experiencing capacity constraints. Yet another firm has used competitive information directly through the development of a sophisticated competitor information system that allows it to infer competitor reactions and to anticipate them. Still another firm has made use of its sophisticated information on technology to develop product enhancements on a planned timely basis, giving it a continuing competitive edge.

Each of these examples illustrates a firm that has achieved strategic comparative advantage on the basis of information. In many of these cases, this achievement was somewhat fortuitous in the sense that the firm may have set out to achieve a more modest goal, for example, the development of an improved cost information system.

If information-based comparative advantages are to be an important element of strategic management, however, their genesis must go beyond fortune and the organizational effects of the pervasiveness of computer technology. The next step that must be taken is that of developing systematic processes and techniques for creating information-based strategic comparative advantage.

THE STRATEGIC DSS

Having developed the basic notions of DSSs, the nature of strategic information, and information-based strategic comparative advantage, we can now turn to the question of the SDSS. Does the notion of an SDSS really have meaning or is it merely one of an almost endless set of contextual applications of a DSS? The answer to that question as developed here is that it *can be* a significant and unique variety of system that deserves to be thought of as an element of the strategic management field rather than merely a computer tool for strategic management to use.

The case for strategic decisions as an area of application for DSSs can readily be made at the superficial level. DSSs would appear to have obvious potential for direct application in the strategy arena, where problems are not initially well defined and perhaps not even well recognized or understood. The flexible capabilities of a DSS give the user (the manager or planner) the opportunity to ask for information, to test alternative ways of viewing the problem, subsequently to ask for different information, to use preprogrammed models, to construct his or her own decision-aiding models, and so forth. Such a flexible iterative process is much like the way in which real-world strategic decision making is conducted. Many of the computerized models that were proposed for use in planning before the DSS era were not realistic in that they presumed the problem to be well understood and well formulated—a characteristic seldom present in real strategic decision situations.

However, such a generic linkage between characteristics of strategic problems and attributes of DSSs is an inadequate basis to argue for the potential significance of SDSSs. To do so requires a more detailed assessment as well as illustrations of how specific systems can meet various strategic decision support needs.

Criteria for SDSS

A number of criteria appear to be necessary if the notion of an SDSS is to have an appeal and meaningfulness that goes beyond that of a particular application area for DSS technology.

First, to qualify as an SDSS, a system must truly be a DSS, that is, it must be a system that qualifies under the characteristics or attributes definitions of King[31] or Sprague and Carlson[32] as described earlier.

Second, it must be directed toward an objective or class of problems that normally are judged to be strategic. There are various definitions of *strategic,* but the one to be used must involve notions of the scope, importance, timeframe of the impact of the decision, and degree of its reversibility.[33]

Third, an SDSS must be used to seek strategic comparative advantage through the manipulation of information. Without this objective, a computer-based system that operates at the strategic level is little more than a data processing system for high-level information.

The fourth criterion to be applied to SDSSs that have the objective of supporting specific strategic decisions is that they must directly address the salient and unique features of such decisions, such as their lack of structure.

Certainly, all strategic decision problems are not totally unstructured. What is generally meant by the association between lack of structure and strategic decisions is that strategic decisions are first identified in unstructured form. During the process of formulating and solving the strategic decision problem, a structure is identified or imposed. As similar problems are repeatedly dealt with, the structure is refined and revised until the initially unstructured problem becomes a rather structured one.

Thus, the degree of structure of a problem depends on the degree to which a structure has been developed and codified. For instance, acquisition decisions were considered to be rather unstructured at one time, but, because they appear frequently in some firms, the various stages of the acquisition decision-making process, from problem identification to solution, are highly structured and codified. Other decisions, such as those involving the changing of a corporate mission,[34] appear only infrequently and have therefore had less structure identified or imposed on them. In most instances, they are dealt with in a rather uncodified fashion.

Moreover, a particular decision is less structured when it is in the early intelligence or design phase of the problem-solving process than it is in the choice phase after alternatives have been identified, effectiveness measures have been validated, and comparative assessments have been made.[35]

Therefore, the problem-solving type of SDSS should have the capability of dealing with this variety of lack of structure. In other words, the SDSS should provide support for operations such as objective identification, problem diagnosis, identification and quantification of relevant issues and variables, generation of alternatives, and assessment of risk and benefits as opposed to the mere comparison and evaluation of alternatives.[36]

This may be accomplished in either of two ways—through a system that is a general problem solver or through one that has the capability of supporting the early stages of the problem-solving process in which the specific problem is lacking in structure.

The notion of a general problem-solving DSS—one that can deal with a wide variety of different strategic decision problems—is well beyond the state of the art, although such systems have been discussed.[37]

However, the criterion that requires a true SDSS to be able to provide support for the early formulation steps in problem solving rather than the later evaluation and comparison ones is a rigorous but practical test. If a system has *only* the capability to perform comparisons of alternative strategies, it is not significantly different from many systems developed for various managerial control applications.[38] For an SDSS to have some uniqueness that warrants the use of a special term to describe it, it should provide the capability for supporting the intelligence and design phases of problem solving.

Applying the Criteria

Most of the DSSs developed for strategic applications meet some but not all of these criteria. For instance, Equitable Life Assurance Society has developed CAUSE, a DSS that provides computer assistance in the making of insurance underwriting decisions.[39] IBM's research division has developed GADS (Geodata Analysis and Display System) that permits great flexibility to the user in viewing spatial arrangements of data in a wide variety of situations. This system has proved useful in analyzing various configurations of census tracts for political redistricting, police beat design, and a variety of other strategic choices that involve spatial configurations as strategic alternatives. Others, such as RCA, Citibank, Louisiana National Bank, American Airlines, and the First National Bank of Chicago have reported the successful implementation of SDSSs.[40] However, these systems are generally focused toward the later—comparison and evaluation—stages of the problem-solving process, and they do not have the strategic comparative advantage notion as an integral system objective.

If the SDSS notion is to be more than an objective to be devoutly sought, it is important to give examples of systems that do meet the various criteria.

SICIS

Rodriguez and King developed a Strategic Issue Competitive Information System (SICIS) that is used interactively by a manager who is trying to learn more about a competitor-related strategic issue.[41] It does not provide answers, but merely chauffered access to competitor information that has been collected and stored. The essence of the system is an issue tree such as is described in Figure 25-1. The user enters the tree by selecting the strategic issue from a predefined set and then is led through the tree by the system. This is done by identifying the nature of the next level of nodes in the tree and allowing the user to refine his or her definition of the issue by selecting from them.

For instance a competitor-related strategic issue might be: What is the capability of our competitor to introduce a new product before 1989? The resolution of such an issue could be a basis for achieving a strategic competitive advantage.

The SICIS user interested in this issue begins using the system by identifying this particular issue. He or she then is presented with a description of each of the nodes on the next level of the issue tree for that issue. For instance, these might be the competitor's financial capability, production capability, marketing capability, and technological capability.

The user then can identify specific areas of interest or indicate that he or she desires the total picture related to his or her initial question. A user indication of interest in marketing capability might produce a system response indicating the availability of data on competitive distribution channel capacity, field sales capability, service capability, technical sales expertise, and a variety of other marketing-related areas. The system also would indicate its ability to provide projections of future market growth.

Indications of interest in other areas would produce similar system responses that would, after each new user response, indicate successively more detailed subissues. Ultimately, when the user reaches the end of any of the tree limbs, he or she obtains the information deemed relevant to that branch.

Obviously, such a system is limited by the fact that the set of issues, the issue trees, and the relevant information are predetermined. It is system *use* that is flexible rather than the system itself. The technology of artificial intelligence offers promise of the development of more sophisticated systems that have even

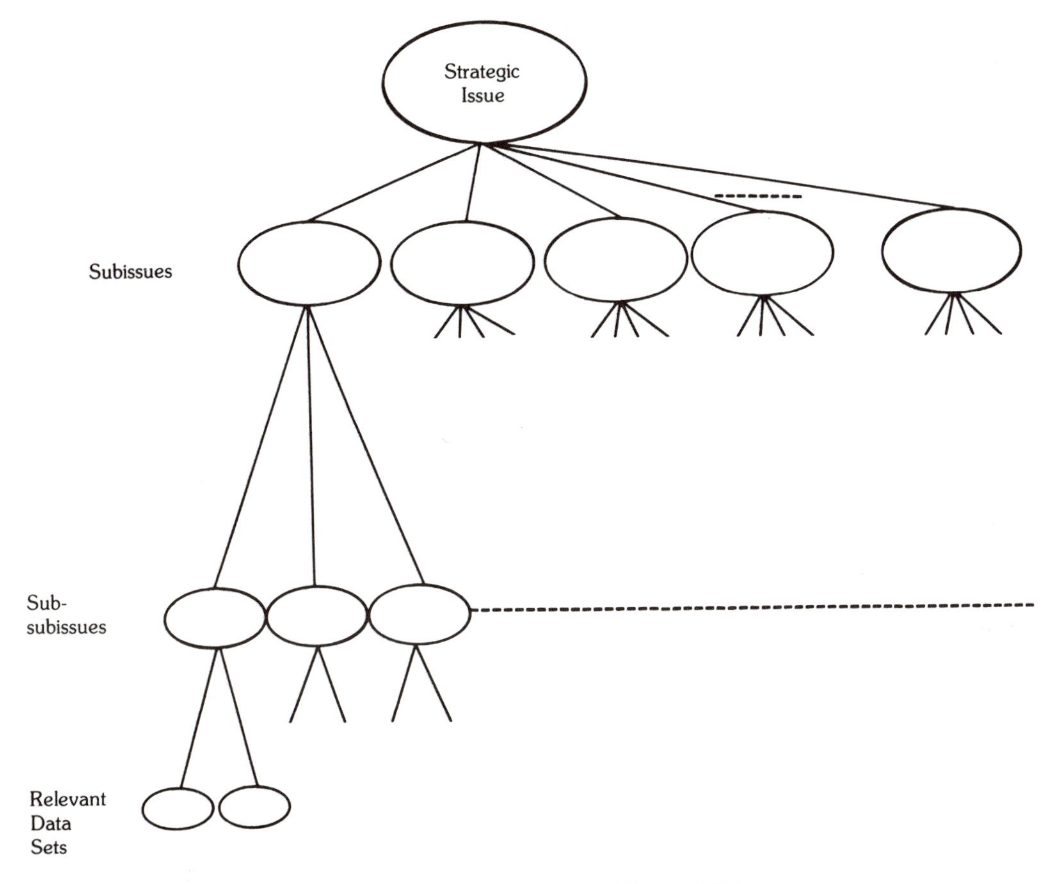

Figure 25-1. Generic SICIS issue tree (after Rodriguez and King, 1977).

greater flexibility.[42] However, the condition that human judgment will need to be used to structure the model and the data in critical ways is inevitable.

However, a system such as SICIS does have the crucial property of enabling the user to deal with issues that represent the *strategic use* that he or she wishes to make of data. Many other systems do not have this property. It also has its primary use in the earlier more creative stages of the problem-solving process.

To illustrate the difference, consider another much more technically sophisticated system that deals with the same general area—the domain of strategic competition. This system, called COSMOS for Competitive Scenario Modeling System, developed by King and Dutta, permits its user to attempt to out-think his or her competitor by testing alternative strategies against those that the competitor might use as well as those strategic revisions that the competitor might make to respond to new strategies.[43] The COSMOS system is conceptually and technically sophisticated and it addresses an area of competitive strategy that is an innovative application of computer technology.

However, it is not an SDSS because it does not meet the criteria just set forth. In particular, since it addresses a particular strategic decision-making context, it must deal with the early stages of the problem-solving process if it is to qualify. Since COSMOS is a very sophisticated "What if?" system, it requires that each contemplated strategy be originated by the user. It serves to compare and evaluate each strategy in a very complex and sophisticated fashion, but it does not deal with the earlier intelligence and design phases of the process.

SPIRE

Another system that is suggestive of one that meets the SDSS criteria is SPIRE (Systematic Procedure for Identifying Relevant Environments). As described by Klein and Newman, SPIRE is not a DSS, but rather a batch-mode computer-based procedure.[44] However, it is conducive to being implemented as a DSS, and its other attributes are such that it would qualify as an SDSS if this were done.

SPIRE provides a link between environmental assessment and the later stages of strategic decision making. It uses computer technology to relate strategically impacting environmental factors to strategically impacted components of the organization.[45] The SPIRE computer analysis identifies groups of interdependent factors of these two varieties and describes but does not produce diagrams for presenting them to the decision maker. The decision maker then reviews the diagrams and feeds back into the computer program any changes that are desired or indicated.

It is easy to see how this process lends itself well to implementation in the DSS mode in which the user would have diagrams presented to him or her on a screen and be able to request new configurations, to provide new inputs, and so forth.

Klein and Newman describe the use of SPIRE in the context of developing and understanding the relationship between groups of environmental factors and the business areas that would require strategic-level thinking.[46] Thus, the system clearly addresses the early stages of the problem-solving process rather than the later ones. Therefore, with the exception of it being implemented in a mode that does not utilize DSS technology, it meets the SDSS criteria. Because it could readily be implemented in a DSS mode, it is used here to illustrate the SDSS notion.

HSJ

The HSJ (Hop, Skip, and Jump) Method is designed to produce alternative solutions that are distinctly different from previously generated solutions.[47] Thus, it directly addresses the formulation phase of problem solving. The developers of the technique describe it in a multiobjective linear programming problem context and as a "man-machine interactive process," although they do not claim that it is a DSS. However, like SPIRE, the feedback characteristics that are an essential element of the model-based process suggest that it is a good candidate for implementation in the DSS mode.

The alternatives generated by the HSJ approach represent a set of strategies that may be reasonably considered to be representative of the domain of strategies. In complex decision situations, such a representative set may be difficult or impossible to otherwise identify.

Artificial Intelligence-based Systems

The rapidly developing field of artificial intelligence (AI) has led to the development of systems that show great potential for dealing with unstructuredness.[48] Perhaps the best way to describe the potential of AI is to use the medical diagnosis context in which a variety of systems exist. The objective of these systems is to provide a valid medical diagnosis based on available symptoms. The analogy to the diagnosis of strategic business problems and to the understanding of strategic opportunities is direct.

Most such systems qualify as DSSs in many respects. Often, they are expert knowledge based in the sense that the knowledge of experts has been codified and programmed into the system. Artificial intelligence technology provides the basis for identifying the relevant knowledge and of aggregating it to provide diagnoses.

An SDSS Prognosis

The application of DSS technology to other aspects of the formulation phase of strategic decision making is rapidly expanding. For instance, Sanders et al. have developed a system to identify critical success factors—those factors of critical importance to the success of a business.[49] This notion previously has been used in strategic management[50] but is now being adapted to DSS technology.

It appears that this will be the path that will be followed in the near future. Problem formulation techniques such as those described, as well as others first will be computerized in a batch mode.[51] Then as the potential advantages of the instant feedback to the individual or group is realized, DSS technology will be applied. The result will be a wide range of systems that meet the stringent requirements for an SDSS as developed here.

Thus, the overall conclusion of the critical assessment of the SDSS notion that has been conducted is positive. It would appear that the idea of an SDSS is unique and sufficiently substantive to warrant attention as a significant subarea of strategic management. As a result of the ongoing proliferation of computer technology in society, it is probably not going too far to suggest that it may be a very important subarea in the not-too-distant future.

SUMMARY

This chapter deals with the topic of strategic decision support systems in a critical fashion. Rather than accepting the notion of an SDSS as being inherently meaningful, it develops criteria that are intended to provide the basis for assessing whether the term *SDSS* meaningfully describes a substantial body of thought that is important to strategic management.

The criteria developed are:

1. That an SDSS must have the attributes and characteristics of a DSS and not be merely a computer-based system.
2. That an SDSS must address a class of problems that are strategic in nature.
3. That the system objective must be that of achieving an information-based strategic comparative advantage.
4. That an SDSS must directly address the unstructured nature of such decisions either in being capable of dealing with unique problems that have not been structured and codified in the fashion of a general problem solver or being capable of supporting the intelligence and design phases of the problem-solving process rather than the later choice phase.

Some systems have been developed that meet, or very nearly meet, these stringent criteria, which leads to the conclusion that the SDSS notion is more than a fad and more than merely another in the large set of areas for the potential application of DSS technology.

REFERENCES

1. P. G. Keen and M. S. Scott Morton, *Decision Support Systems: An Organizational Perspective* (Reading, Mass.: Addison Wesley, 1978).
2. W. R. King, "Achieving the Potential of Decision Support Systems," *Journal of Business Strategy* **3**(1983):84-91.
3. G. A. Gorry and M. S. Scott Morton, "A Framework for Management Information Systems," *Sloan Management Review* **13**(1971):55-70.
4. "Question Yet Unresolved: What Is a Decision Support System?" *Computing Newsletter* **16**(1982):1.
5. R. H. Sprague and E. D. Carlson, *Building Effective Decision Support Systems* (Englewood Cliffs, N.J.: Prentice-Hall, 1982), 4.
6. W. R. King and D. I. Cleland, *Strategic Planning and Policy* (New York: Van Nostrand Reinhold, 1978).
7. G. B. Davis, *Management Information Systems* (New York: Wiley, 1974).
8. T. H. Naylor and M. H. Mann, *Computer Based Planning Systems* (New York: Planning Executives Institute, 1982).
9. Sprague and Carlson, *Building Effective Decision Support Systems.*

10. King, "Achieving the Potential of Decision Support Systems."
11. W. R. King and D. I. Cleland, "A New Method for Strategic Systems Planning," Business Horizons **3**(1973):55-64.
12. S. L. Alter, *Decision Support Systems* (Reading, Mass.: Addison Wesley, 1980).
13. W. R. King, *Marketing Management Information Systems* (New York: Van Nostrand Reinhold, 1977).
14. T. H. Naylor and H. Schauland, "A Survey of Users of Corporate Planning Models," *Management Science* **9**(1976):927-937.
15. J. H. Grant and W. R. King, *The Logic of Strategic Planning* (Boston: Little, Brown, 1982).
16. O. Eckstein, "Decision Support Systems for Corporate Planning," U.S. Review, Data Resources, Inc., February 1981.
17. R. Davis and D. B. Lenat, *Knowledge-based Systems in Artificial Intelligence* (New York: McGraw-Hill, 1982).
18. Grant and King, *Logic of Strategic Planning;* W. R. King, "Information as a Strategic Resource," working paper, 1981.
19. W. R. King and G. Zaltman, *Marketing Scientific and Technical Information* (Boulder, Colo.: Westview Press, 1979).
20. Davis, Management Information Systems.
21. K. Kaiser and W. R. King, "The Manager-Analyst Interface in Systems Development" *MIS Quarterly,* **6**(1982):49-60; T. L. Whisler, *The Impact of Computers on the Organization* (New York: Praeger, 1970).
22. Note that the microcomputer revolution now occurring will dramatically alter this situation.
23. Kaiser and King, "Manager-Analyst Interface in Systems Development.
24. Ibid.
25. M. H. Olson and B. Ives, "Measuring User Involvement in Information Systems Development," *Proceedings of the First International Conference on Information Systems* (1980), 130-143.
26. M. E. Porter, *Competitive Strategy* (New York: Free Press, 1980).
27. King, "Information as a Strategic Resource."
28. D. F. Abell and J. S. Hammond, *Strategic Market Planning* (Englewood Cliffs, N.J.: Prentice-Hall, 1979).
29. "How Airlines Duel with Their Computers," *Business Week* (23 August 1982):68-74.
30. "Foremost-McKesson: The Computer Moves Distribution to Center Stage," *Business Week* (7 December 1981):115-118.
31. King, "Achieving the Potential of Decision Support Systems."
32. Sprague and Carlson, *Building Effective Decision Support Systems.*
33. King and Cleland, *Strategic Planning and Policy.*
34. Ibid.
35. H. A. Simon, *The New Science of Management Decisions* (New York: Harper & Row, 1960).
36. E. D. Carlson, "An Approach for Designing Decision Support Systems," in *Building Decision Support Systems,* ed. J. L. Bennett (Reading, Mass.: Addison Wesley, 1983), 15-40.
37. R. H. Bonczek, C. W. Holsapple, and A. B. Whinston, "Future Directions for Developing Decision Support Systems," *Decision Sciences* **11**(1980):616-631.
38. Keen and Scott Morton, *Decision Support Systems.*
39. Alter, *Decision Support Systems.*
40. Ibid.; R. H. Sprague and R. L. Olson, "The Financial Planning System at Louisiana National Bank," *MIS Quarterly* (1979); G. M. Welsch, "Successful Implementation of Decision Support Systems: Pre-Installation Factors, Service Characteristics and the Role of the Information Transfer Specialist," Ph.D. diss., Northwestern University, 1980.
41. J. I. Rodriguez and W. R. King, "Competitive Information Systems" *Long Range Planning* **10**(1977):59-64.
42. Davis and Lenat, *Knowledge-based Systems.*
43. W. R. King and B. P. Dutta, "A Competitive Scenario Modeling System," *Management Science* **26**(1980):261-273.
44. H. Klein and W. Newman, "How to Use SPIRE: A Systematic Procedure for Identifying Relevant Environments for Strategic Planning," *Journal of Business Strategy* (Summer 1980):32-45.
45. Ibid.
46. Ibid.
47. E. D. Brill, Jr., S. Chang, and L. D. Hopkins, "Modeling to Generate Alternatives: The HSJ Approach and an Illustration," *Management Science* **28**(1982):221-235.
48. Davis and Lenat, *Knowledge-based Systems.*
49. G. L. Sanders, J. F. Courtney, and J. R. Burns, "A Decision Support System for Identifying Critical Success Factors," in *Proceedings of the National AIDS Conference* (1982), 209-211.
50. W. R. King and D. I. Cleland, "Information for More Effective Strategic Planning," *Long Range Planning Journal* **10**(1977):59-64.
51. R. L. Ackoff and E. Vergara, "Creativity in Problem Solving and Planning: A Review," *European Journal of Operations Research* **7**(1981):1-13; A. L. Delbecq and A. H. Van de Ven, "A Group Process Model for Problem Identification and Program Planning," *Journal*

of Applied Behavioral Science **7**(1971); W. J. Gordon, *Synectics* (New York: Harper & Row, 1961); H. E. Klein and R. E. Linneman, "The Use of Scenarios in Corporate Planning: Eight Case Histories," *Long Range Planning Journal* (October 1981); I. I. Mitroff, Jr., R. Emshoff, and R. H. Kilmann, "Assumption Analysis: A Methodology for Strategic Problem Solving," *Management Science* (June 1979); J. Summers and D. E. White, "Creativity Techniques: Toward Improvement of the Decision Process," *Academy of Management Review* (April 1976):99-107; M. Turoff, "The Design of a Policy Delphi," *Technological Forecasting and Social Change* **2**(1970):149-171.

VII

Implementing Strategy

After strategies have been developed, assessed, and selected, they must be put into action. It is in this implementation phase of strategic management that many planning failures occur. Part VII deals with the process of implementation and explores a variety of ways in which successful strategy implementation may be achieved.

In Chapter 26 Roy Wernham discusses implementation issues in terms of some of the underlying concepts and ideas that have been applied to implementation situations in a variety of contexts.

In Chapter 27 Ian C. MacMillan and Patricia E. Jones focus on the design of the organization as a means of implementing strategy. They emphasize that such designs must be chosen in the context of the competitive environment.

William H. Davidson, in Chapter 28, extends these ideas to the international domain in terms of the evolving roles of businesses and the new environmental pressures in international markets. He concludes by describing the effective international organization.

In Chapter 29 Russell D. Archibald views strategic projects and programs as vehicles for implementing strategy. He describes project management notions as they apply in this strategic implementation context.

Milton Leontiades treats managerial selection as a key element of successful implementation in Chapter 30. He deals with the judgments that must be made in selection at all levels, from lower organizational levels to the CEO.

In Chapter 31 A. Graham Sterling describes a bonus plan for a growth company that has the objective of sharing the value added and motivating improved performance through tying in bonuses to key organizational results.

CHAPTER 26

Implementation: The Things That Matter

Roy Wernham

Roy Wernham is deputy corporate planner for British Telecommunications plc, London, England. He taught in East Africa before joining Post Office Telecommunications in 1969. A graduate in natural science and law from Cambridge University, he took a diploma in management studies in 1971 and completed a doctorate in management science in 1982 at City University Business School, London. He also works as a tutor at the Open University, Milton Keynes, England.

Implementation is a problem. The key to understanding the problem and doing something about it lies in recognizing that some present approaches to understanding organizations and how they work are very much simplified. A simplified view can be useful as a teaching aid to further understanding but also can mislead if carried beyond its original purpose. In this chapter the classical pyramidal view of organizational structure is contrasted with the center-periphery or network model. The comparison generates useful insights that are used to supplement the conventional view of how the implementation process works. Factors are identified that can either help or hinder implementation.

A GLANCE AT THE THEORY

This chapter sets out to be a help to the practitioner as well as, and perhaps more than, the academic. If we are prepared to allow that there is nothing more practical than a good theory, then a logical point of departure is to consider what basis modern theory has to offer for practical action. How often is it said "This is a good idea in theory, but it won't work in practice"? The implications of that statement are twofold:

1. The action-centered, practical person has little time for theorizing. This is a road that quickly leads to action being treated as a substitute for

This chapter draws on parts of the author's Ph.D. dissertation at City University Business School. The views expressed are those of the author and do not necessarily represent those of British Telecommunications, British Telecommunications plc, or the City University. The author gratefully acknowledges the assistance of British Telecommunications and British Telecommunications plc and A. J. B. Scholefield for helpful suggestions.

thought. This usually is not very helpful for, as Winston Churchill once remarked, "There is nothing more frightful than ignorance in action."
2. Theories are by their nature naive and oversimple attempts to model only the key aspects of the real world. By definition they fall far short of capturing the complexity of the real world by failing to take cognizance of things that people of affairs must take into account. That is a more serious criticism of theories as well as of those who construct them. Certainly, in the management field it cannot be denied that many of the early attempts at theory—and not a few of the more recent ones—have seemed woefully inadequate in the light of later developments.

However, let us ask the doubters to suspend judgment and accept for the moment that there is indeed nothing more practical than a good theory and see what insights research into implementation has to offer. The bad news is that the academic literature itself is rather fragmented. Several clusters of academics have studied implementation problems from within the fields of public administration, management science and operational research, strategic management, and organizational behavior. It is as if the same building were being floodlit from several different directions at once. The good news is that concepts from one can be imported to the others, illuminating darker areas, allowing cross-fertilization of ideas as well as challenging entrenched ideas and unrecognized assumptions.

Traditionally, management theory draws a clear distinction between policy formulation and implementation.[1] Figure 26-1 indicates the traditional view of the process. Although useful conceptually and as a teaching device, in practice this apparent dichotomy may be misleading as policy planning is then conceived to be a linear, sequential process. An alternative view is that it is an interactive process whereby the perceived outcomes of implementation react back on policy formulation and become a part of the formulation process.[2] This discussion focuses on implementation, but an underlying premise is that implementation is an integral part of a highly complex and interactive planning and policy-formulating process. Much of the literature assumes implementation to be a top-down process in which those at the top of the organization ensure that their wishes are carried into effect lower down. However, it seems that implementation at the bottom of the organization is, at least in part, a proactive process, not merely a reactive working out of the plans of others. There has developed a certain amount of confusion over terminology. The term *implementation* can refer to one or more of the following three things:

1. The *output,* that is, the desired end result, the outcome originators intend should happen at the outset.
2. The *outcome,* what actually happens warts and all, allowing that in translation from the top of the organization to the place of implementation there is a possibility that the intended plan may be transformed, adapted, or subverted.
3. The *process* by which an attempt is made to put the plan or policy into effect by its initiators through other people in the same or other organizations, that is, the transmission and transformation mechanism.

For our purposes, it is enough to assume that it comprises both an intended outcome (or outcomes) and an action designed to achieve that outcome.

MODELS OF ORGANIZATION

Let us now look at the ways in which organizations are commonly assumed to be structured or modeled. The prevalent, pyramid model or view takes a look at the organization as set out in a conventional organization chart (Figure 26-2), the sort of chart that every large organization usually draws up. It is a familiar, convenient, and efficient device for showing the reporting arrangements and the reward systems in use. However, as will become apparent, it is less efficient as a device for demonstrating the distribution of power to control resources within the organization. To show all things flowing up and down the organizational pyramid is a device that is convenient and at the same time congenial to

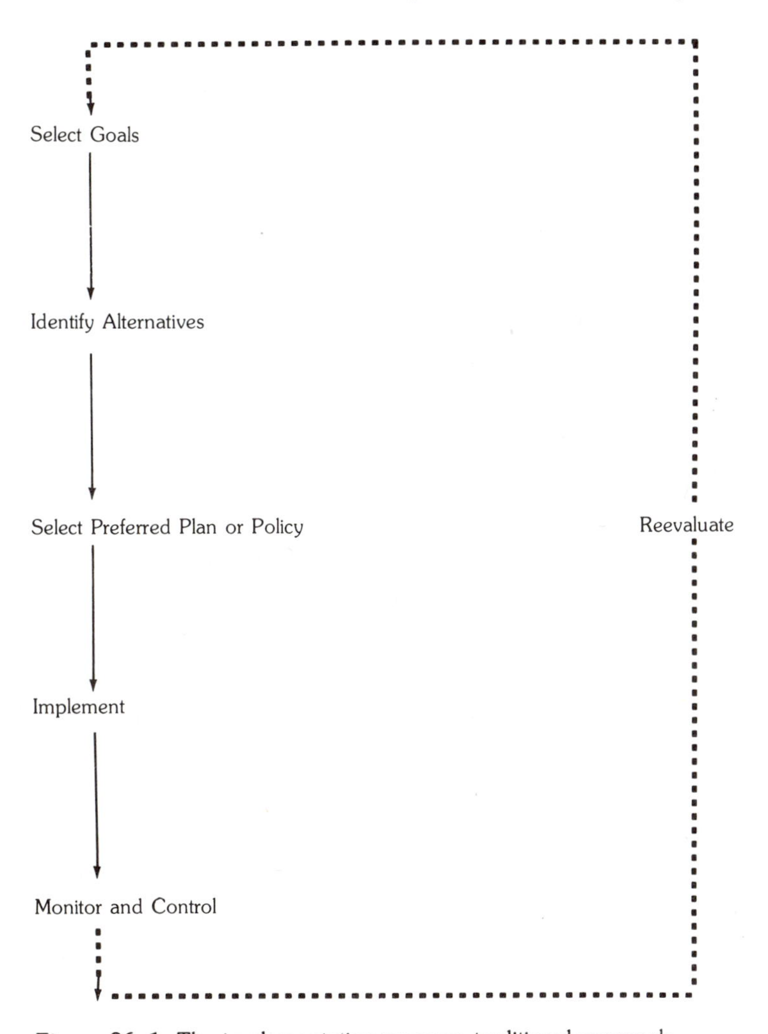

Figure 26-1. The implementation process: traditional approach.

those who have worked their way nearer to the pinnacle. It shows the distribution of power in terms of patronage—for making and withholding appointments and direct personal rewards—but does not so readily show how power is distributed in the sense of the ability to control resources. The practice of withholding or even withdrawing rewards is at its most effective in its *potential* for use rather than in its being applied. An imperial commander, as the empire's man on the spot, has considerable freedom to act as he chooses. The famous incident in which Lord Nelson put his telescope to his blind eye in order to avoid reading a signal he preferred to ignore is an outstanding but by no means isolated example, nor is it the sole prerogative of the military! Although the man on the spot can always be recalled and replaced, carrying out the draconian step is not always easy or convenient. Moreover, the successor then necessarily enjoys a honeymoon period in which there is considerable *de facto* freedom from outside interference.

It is instructive, in looking for a contrast to this paradigm, to look to the center-periphery model used in the public administration literature (Figure 26-3).[3] In this model the units at the periphery depend on the central organization for the provision of certain resources,

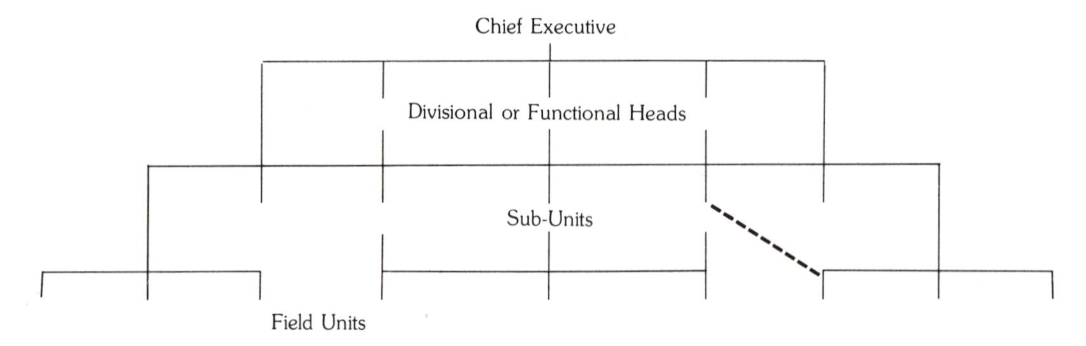

Figure 26-2. Typical organizational chart: the pyramid view.

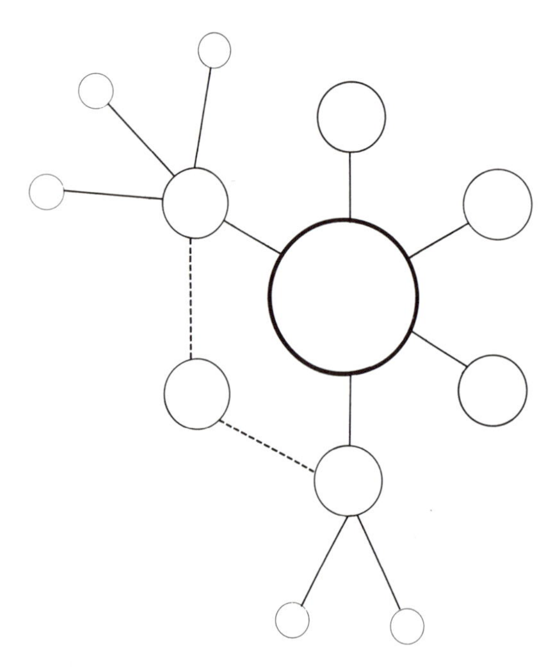

Figure 26-3. Center-periphery model: the network view.

for example, funding, and perhaps in other ways but are constitutionally separate from it. They have their own, separately elected or appointed officers (that is, people who are not appointed by the central organization) who therefore have a measure of official freedom of action and a separate power base from those at the center—they are part of a separate reward system—but depend on the center for certain resources. Examples are common in the field of local government, in churches, and in sports and leisure associations. A typical instance is the relationship between the UK central government and local authorities who have their own independently elected councils but are given central funds and have certain other demands made on them by those at the center. The relationships between a local authority and a school with its own governing body having the power to make appointments or the relationship between a football league and its constituent clubs provide further examples.

There is about the relationship a delicate balance. The center has considerable power and might reasonably be described as *primus inter pares,* but its power is clearly circumscribed, not least because the usual range of rewards and sanctions available within an organizational context are absent. The periphery, too, has its own sources of independent power, often jealousy guarded, and yet must conform to certain requirements from the center if it is to receive the resources it needs to do what it wants. The result is not anarchy, but neither is it monarchy. It is a complex balance in which power is shared, with the balance often shifting over time as circumstances change. Independence is guarded to a considerable extent and perceived threats from the center may call forth a strenuous defensive reaction from the periphery. It is significant that this model is often pictured "flat" as a network and not vertical as a pyramid. Relationships also may develop between the peripheral units themselves, quite independently from their relationships with the center. This may simply be to share resources, to compare notes on common problems, or for the purpose of attempting to subvert or influence the goals or actions of the central agency. It is the stuff of local politics in the widest sense of the term and may embrace an *ad hoc* coalition of school heads or of football club managers who want to see a change in the rules, as well as the more usual cabals formed by those more commonly perceived to be politicians. It is not really such a big step to suggest that the reality of organizational life is partly but not wholly captured in the pyramid view of organization but owes not a little to the center-periphery model as well. The key to the paradox lies in what we mean by "the organization."

The convenient, simplifying assumption that is often made that the organization can be seen as a thing in its own right, distinct from the people within it and with a coherent set of aims able to be attributed to it, is a half truth that beguiles and merely serves to gloss over the pitfalls that await the implementation process. Subordinate managers may have differing values and priorities or may favor different aims and channel resources earmarked for one use to another.[4] Even the top or central management of an organization may not speak with one voice, leaving interpretation to others whose actions will decisively influence implementation and who are then, by default, policy makers. Where there are many voices, it becomes easier to select some and ignore others while appearing to heed all—at least some of the time—as any politician knows. It becomes easier to please most of the people some of the time.

THE PROBLEM OF IMPLEMENTATION

An important contributor to the implementation problem in organizations is that plans or policies (we are speaking of both throughout) are carried out, at least in part, by people not themselves involved in the original formulation. The goals of the top or central planners are interpreted by others at the bottom or periphery, or somewhere—perhaps several places—in between. This represents at one and the same time both a loss of control over, and adaptive flexibility by, the periphery; they are opposite sides of the same coin for one person's flexibility is another's loss of control.

Implementation theorists have used either a pyramid or network view of organization. The pyramid protagonists belong to the same school of thought as the early classical management writers.This view, as Dunsire notes, reflects the view of a hierarchy of authority;[5] the implementation process flows down the organization with orders being passed and obeyed from one level to the next, from superior to subordinate, in a cascade or chain reaction. The center-periphery model is typified by the work of Pressman and Wildavsky.[6] The pyramidal model with its cascade of vertical layers is replaced by a network analogue with analysis being concerned with the nature and operation of the linkages. The implementation process must either forge new chains or links in an appropriate form to carry out the planner's will, or it must be carefully designed so that the existing networks of linkages can cope with it, so that *it* will fit into *them*.

A not uncommon failing is to make a simplifying assumption that implementation is not a problem. Once this is brought out into the open and its implications recognized, the problems often associated with implementation can be put into context and understood much more clearly. The classical management textbook view of implementation, as we have seen in Figure 26-1, is of a simple straightforward linear process. Plans are made, tasks and resources allocated, progress monitored; it is as simple as that. There is what Van Meter and Van Horn call a "naive assumption" that all will be well.[7] Wilensky reduces the difficulty to "the problem of getting work done and securing compliance with organizational rules."[8] The possibility tends to be overlooked that these twin strands may be in conflict or may be seen as such by some of the actors.

In the classical management tradition, lower-order participants hold delegated power as agents of higher-level managers.[9] The degree of power they wield is held at the whim of higher-level managers and can be varied or revoked at will. There is reason to view some of these ideas with scepticism and to consider that lower-order participants have resources at their disposal that give them considerable power as Crozier demonstrated in his study of French governmental bureaucracy.[10] We can even go so far as to suggest that in larger organizations it becomes progressively more difficult, and may well even not be appropriate, for top management to try and rule by fiat because they are too far removed in time and space from the action. Power becomes diffused between groups whose consent is needed and who accordingly have a power base built on factors including their control of resources and their ability to form powerful coalitions with other groups.[11] Although higher management can indeed wield considerable power through rewards and sanctions, the management information and decision processes, and the manipulation of certain resources, that power is very far from absolute. Indeed the reward system may not even encourage the "right" patterns of behavior!

The corollary is that top management may well lack the power to guarantee successful implementation even when keen to see it through. The process of implementation involves other groups whose recognition and involvement contribute critically to success or failure.[12]

The key insight that we have drawn out here is the recognition that those down the organization or at its periphery are often not the passive recipients of someone else's ideas, waiting machinelike to put them into effect. They are active formulators of policy in their own right. Thus, they do not simply take in the washing of their colleagues at the top of the organization and process it according to the rule book. Very often they will choose their own soap powder and develop their own programs and style of ironing as well. The planner or policy maker who does not recognize that possibility is heading for disappointment.

Of course, the classical view of implementation, as outlined in Figure 26-1, is helpful and contains a good deal of truth, which is why it has survived so long. However, it contains some important simplifications and half truths that mean that the reality is often rather different from the simple pattern described there.

To spell out the implications of the classical management approach to implementation, rooted as it is in the pyramid view of organization structure, is to call them severely into question as a guide for action. There are other realities, which so many who work in organizations recognize instinctively a lot of the time, but that the academic community would seem to find so elusive and hard to build into their models.

The implication of the pyramid view is that the organization is homogeneous—a unitary creation imbued from top to bottom with a single, coherent set of goals. All power and ideas flow from the top. Role holders at the lower levels in the hierarchy are supine and always do exactly what they are told by head office. Having been chosen for their powers of judgment and initiative, they will never use either in dealings with higher levels of the organization! Naturally, they are assumed and expected to adopt a very different stance with the outside world and stand up for the organization against all comers! Simply to make explicit such assumptions—convenient though they are—is to demonstrate their naïvety and to indicate how it is that many of the problems of implementation arise. Given then, that we have challenged the traditional view of organizations and their structure and presented an alternative, let us see what this has to tell us about implementation. These contrasting views imply that certain factors exist, which research has suggested and underpinned and which can both help and hinder the implementation process. These we will explore shortly but first a word about measuring the results of implementation.

MEASURING SUCCESS AND FAILURE

It is worth pausing, as we leave this part of the discussion, to ponder just what it is that we *mean* by success or failure in implementation. In the classical pyramid approach to organization, implementation is regarded as successful if the policies and plans conceived at the top of the organization have been faithfully reproduced in the activity taking place at the grassroots without significant distortions. However, for the organization as a whole, this can only be counted a success if the behavior at the grassroots is appropriate in the circumstances ruling there. If not, some distortion of the original policy may prove to be desirable and even vital for the continuing health of the organization. The moral here is to consider carefully what is to be understood by *success* or *failure* in implementation and what will be used to measure it. A policy decision to cut costs by abandoning the firm's delivery fleet may be implemented successfully but can still be a failure in terms of the original objective of cutting costs if vehicle hiring and other transportation costs more than outweigh the savings made. The success may in fact be a failure on the bottom line if its repercussions are traced through fully.

In measuring the outcome of the implementation process, we can use objective measures, subjective measures, or some combination of both. With some justice, there tends to be a preference for objective measures, based on hard, factual data such as costs, numbers of units produced, and time taken. There is a solidity in such data that is reassuring. However, are the right things being concentrated on, or is it a matter of measuring what is measurable rather than what is important?

Executives often will set great store by their ability and that of others to make judgments, to size up a situation and make a decision when objective data is inadequate, unreliable, or not available. This places great reliance on the more subjective elements in a decision. The subjective element of judging implementation should certainly not be ignored, especially if, as is often the case, the objective measures do not really concentrate on the essentials. What is the general perception of the plan or policy? Do those concerned generally rate it a success? Have they taken to it? Is someone's heart really in it? If not, be very cautious about rating it a success whatever the objective measures may say! As the, perhaps apocryphal, planner remarked, "It's been implemented all right, it's just that no one's using it"! It is much easier, in cases of doubt, to choose objective measures that give an illusion of success than to convince those at the sharp end that things really are better when their gut feel is that they are not.

Just what constitutes success can vary according to the standpoint of the judge. For the engineer or programmer, technical operation may constitute success. The user may have a wider perspective and see success as the improved implementation of a policy to which the project or machine has contributed. Management may adopt another, wider perspective and see success depending on the contribution made to the unit's effectiveness or profitability. Top management may look for the impact on overall organizational effectiveness. In a multilevel organization further combinations are possible, giving shades of meaning to each of these.

WHAT HELPS IMPLEMENTATION

Let us now turn to those factors that have emerged from our discussion and from the research that has been inspired by it. If you are at the implementation stage, there are things to avoid as we shall see, but let us start with being positive, looking at the other side of the coin. What is there that will help the process along and make it more successful?

The unkindest observation to make is that if you are ready to proceed to implementation but have only just started thinking about potential problems you are almost certain to have difficulty! Implementation needs to be an integral part of the management process, not a discrete module somehow glued on at the end after all the *real* work has been gone through and finished—we are back to the Figure 26-1 linear model! So when you are thinking about implementation, and when you are going live and doing your homework, what factors do you need to bear in mind that will help the process along?

The Plan Itself

The nature and form of the plan itself can be important as they usually serve a number of purposes. Plans and the information they contain are important for they often are the means by which information about the policy is transmitted from one part of the organization to another.

Thus the form of presentation to those at the lower levels can be important in helping the process forward. Should details be on paper or would a personal presentation be preferable? An opportunity to ask questions, to criticize, to air doubts and to see that the originators of the plan are warm and lovable—well, human anyway—can help to generate acceptance. What goes down the line is information and perhaps instruction but it is also a sales pitch. As such there is an opportunity to convince and persuade the waiverers and maybe even the opponents, to generate excitement, goodwill, and positive expectations rather than dampen them through demanding conformity from the security of a head office 50 miles from criticism.

Existence of a Problem

Often a policy change or planned development occurs in response to an existing or incipient problem being perceived. There is a large normative literature mostly based on experiences of problems with past plans. The resulting fragmented and anecdotal evidence is valuable but not coherent.

A powerful engine of change is the perception of a need for change, an acceptance that existing stable behaviors are no longer adequate and that change is required. Thus, the planner who wishes his or her plan to find acceptance needs to ensure that there are willing users who believe that, even if not the answer to the maiden's prayer, it will make a contribution to solving their problem.[13] Thus, the existence of a felt need, *a recognized problem,* and acceptance that what is becoming available may offer at least a partial solution is a help to success. The potential user is likely to adopt a more favorable attitude and be willing to try out the solution offered as a means of easing the problem. It is about motivation of the implementors and assumes some freedom of choice on their part.

The major company that reported its first ever loss, after years of seemingly being able to do no wrong in terms of its financial results, found its response to the crisis achieved some startling changes. At all levels of the company there was a high level of commitment to attacking the problem. Solutions were sought at all levels and even drastic measures to cut labor and other costs were accepted philosophically. The spur of a universally recognized and accepted goal, to which staff at all levels of the organization could subscribe, was of tremendous assistance in implementing the many remedial measures needed, even those likely to have been unpalatable in happier circumstances.

Other companies have found that the emergence of competition into previously sheltered markets has had an electrifying effect on even the most somnolent.

Organizational Mission

A closely related point is that plans or policies that are readily recognized as contributing directly to a major aim or mission of the organization will be more likely to succeed than things that appear more as a fringe activity, that can be taken or left without any real loss to the overall mission of the organization. If, say, our mission is to serve our customers' requirements for a service, plans that are perceived to contribute directly toward their achievement will be more likely to succeed.

Resources

The availability of adequate resources appears to be a necessary but not a sufficient condition of success. You may still not succeed even if the resources are there, but if they are missing failure is sure. Resources might include materials, technical, organizational, and marketing back-up and advice, or appropriate skills, knowledge, and staff training arrangements. As one manager put it rather ruefully, "I may not like the priorities or what I'm asked to do—but if I've got the bits it's a lot harder to duck."

Money generally will be required to fund a new development or continue an existing one. It may be wise to ensure that the funds cannot be diverted to other more pressing uses once they arrive at their destination. Simply granting a budget increase without conditions can still lead to disappointment. Personnel, both numbers of people to carry out tasks and people with the requisite skills, knowledge, and motivation, are needed. Material resources may be needed in terms of consumable stores or machinery or space in buildings. Unless adequate material and personnel resources are available, implementation will not succeed. There is evidence though that a surfeit of resources does not guarantee success. Hence our view that resource availability is a necessary but not a sufficient condition for success.[14]

Intellectual resources of skill and knowledge are just as important as provision of adequate numbers of people. If the members of the organization lack the capability to do what is asked of them, the plan is very likely to fail without some radical redefinition. Intellectual resources are hard to transfer and may find expression in recorded form through standard instructions or procedures governing the way in which operations are carried on or possibly through the medium of a program of training or graded experience.

Allocation of resources initially occurs through the process of project approval, which has been characterized as a bottom-up process beginning with a felt need occurring in an operating division. After definition, it gains divisional backing and acquires impetus. Eventual approval by the upper echelons is little more than a formality but one that confers on it the resources sought.[15]

There is some sense in regarding *time* as a resource. The implementation of a major policy shift in large organizations can be an extremely complex process taking a great deal of time. Time frames of six or seven years or even longer have been cited by Steiner and Miner and Eveland et al. as not untoward to implement major changes in very large organizations.[16] The demands on financial and human resources of such a time frame are considerable but may well be a necessary price to pay where a major discontinuity is occurring. Given time, people can adapt and adjust themselves and their expectations to a considerable extent, but time is itself an expensive resource, especially in a race with competitors.

Turning from resources, there is the problem of matching: Matching hardware performance to expectation; market performance to need; organizational procedure to current practices.

Technical Validity

A new plan or policy frequently involves the use of new equipment, new hardware, perhaps new working methods or software developments as well. The expert or architect of a new piece of equipment or software is primarily concerned with providing the organization with a product or policy that performs certain tasks. The expert's concern is that, in a purely mechanistic sense, the technology works; it is a technically valid solution to the problem as the expert perceives it. Implementation, as perceived by practicing management scientists and operational researchers, understandably has been greatly concerned with achieving technical validity.[17]

Of course, even a solution that is technically valid needs to be workable by its users. Inscrutability is not a helpful characteristic in new technology! The relevant expertise has to be transferred to the users. Thought needs to be given to any training that may be necessary and experts need to be on hand to provide an after-sales service to assist users and help to iron out any teething troubles. Ergonomic factors are important as is understanding the existing user skills and know how. Simplicity in operation is a factor that has also been associated with success in that it provides ease of use and understanding and is therefore a component of technical validity. Moreover, to have an impact, there must be willing users who are prepared to accept that the plan or policy is a valid response to problems they perceive.[18] Technical validity is vital to successful implementation, lack of it—the plan or policy is inadequate to the task or is applied in such a way that it does not produce a feasible solution for the user—is the high road to failure in implementation.

Organizational Validity

Users—tiresome though it may be—start from where they are. It is like the advice given by the countryman to the traveler to Birmingham, "I wouldn't start from here if I was you." Implementors start out from where they are, not where the planners are, or even where the planners would like them to be.

There has grown up the realization that what the technical expert may see as a valid solution to the problem—technical validity—may not address the problem as perceived by the user. From this has developed the concept of mutual understanding between user and technical expert or planner in an attempt to bridge the gap in understanding and aims between them. The expert tends to be concerned to install a system that works on day one, after which his or her interest flags. For the user that phase is only the end of the beginning; his or her interest is in a trouble-free system that is robust, easy to handle, and goes on working.

The need for good relations to be cultivated between planners and practitioners cannot be overstated; the planners have a sales task and a duty to listen as well as to talk.

Building onto that perspective, Schultz and Slevin developed the concept of *organizational validity,* that the solution the specialist regards as technically satisfactory is also culturally satisfactory from the viewpoint of the user.[19] Few plans or policies can be written on a blank sheet. The implementing units already have their own organizational subculture, their existing methods and procedures, and a general sense of "the way we do things here." This has important implications for the planner who needs to consider carefully how what he or she wants to achieve will mesh in with what is already in place. What changes in behavioral patterns are required of users? How easy will it be for them to adapt? Can the new be adapted to fit in with the existing "way we do things"? If so, then the chances of its being accepted are improved. If not, then the difficulties of implementation will multiply, more time will be needed in addition to some form of educational program—formal or informal—with no guarantee of acceptance at the end of it.

Top Management Support

One of the few points on which virtually all observers of the implementation process regularly agree is the importance of top management support. The functions of top management can be outlined as recognizing a need for action, giving it direction by providing resources, acquiring the necessary effort and expertise, and legitimating the efforts of the implementors. All of this may be done by top management without commitment to the results; indeed they may well distance themselves initially, keeping their powder dry, until some results indicating likely success are emerging from trials.

At a later stage they will accept and fund the development, reward success or penalize failure, and resolve any conflicts that may arise. Support by top management may involve a measure of fiat, invoking, or threatening to invoke, sanctions to try and push through measures. As we have seen, although fiat may be successful at times, where societal pressure and values may be in opposition to the expressed wishes of higher management whose control can never be total, it may not always flow through. Moreover, there may be limits to which top management will be prepared to back the planners; they certainly will not write a blank check and may withdraw support if things begin to go wrong. Although the *support* of top management is needed, heavy reliance on their patronage is best avoided if it is seen as a substitute for establishing good relationships with implementing units. At least one research study suggests that, although successful implementation is indeed associated with top management support, this factor may well be less critical than many of the textbooks suggest—we need only to look back at the earlier discussion on the center-periphery model to see why.[20]

Track Record

A key but elusive element is the confidence the implementors have in the planners and in what is being implemented. Their confidence is, in turn, a function of information, training, and experience of the planners' track record. Bismarck's famous remark that "nothing succeeds like success" finds its echo in implementation. Those who are able to establish a reputation for competence and being successful have an inside track. This may be from previous experience on earlier schemes or from building up a steady record of success—perhaps in meeting time schedules—in doing what we say when we say. A record of success breeds confidence and expectations of further success that generate a momentum that, in turn, helps to carry implementation through—proven success breeds expectations of further success.

Size

At least two studies indicate that, perhaps surprisingly, one of the better predictors of success in both technical and organizational change is the size of the implementing unit;[21] there is a positive association, that is, the larger the unit, the better the chances of success. This may seem a surprising conclusion in some ways, and the reasons for it are not altogether clear. Perhaps it is the importance to the organization of putting things right in its larger units. Perhaps a larger unit is better able to find the resources to see through implementation, for example, by allocating people full time to the venture. Perhaps the more dynamic management is located there; certainly managers of large units seem to be younger than managers of small ones.

We have considered some of the plus factors for implementation. Let us now look at the things that can hinder implementation.

WHAT IMPEDES IMPLEMENTATION?

When a plan is ready to be put into action, are there features you can look to as possible tripping points, things that have been stumbling blocks for others and are worth keeping an eye open for? Indeed there are; some are essentially the obverse of the positive factors already discussed, others are distinct. Here are some of the more common impediments:

Values

The values and general approach of people to past plans and projects—the converse of their confidence in the implementators—can hinder implementation. A person who is enthusiastic about one is likely to be enthusiastic about others and vice versa. So check their track record, as well as examining your conscience and the record on your own, what has been their approach in the past? How have others fared here before you?

Resources

If the necessary resources of money, materials, and people in terms of their numbers and the knowledge and skill base they possess are missing, then implementation will, at best, be delayed or will not succeed at all. Of course, resources intended for implementing plan *A* can, on occasion, be diverted to plan *B* unless measures are taken to prevent it, especially if the person on the spot perceives *B* to be the more important. Measures that can be taken to reduce the risk of such diversion include committing resources outside the control of the implementing unit; for example, provision of materials in kit form.

Track Record

Historical performance of current and past plans and the expectations of success or failure they have created are important as we have already seen. We are back to the confidence of those involved in the implementors and in their abilities. Just as "nothing succeeds like success," so it appears that "nothing fails like failure." A record of missed targets and failures can so easily generate mistrust and expectations of further failure, an atmosphere that is hard to dispel once established; give a dog a bad name and it tends to stick. People's expectations are a vital but frequently neglected feature of implementation. Neglecting them, and passing by opportunities to fashion and reinforce their development, is an important contributor to failure in implementation.

Technical and Organizational Validity

We have already discussed the general features of technical and organizational validity. A lack of fit between the actual hardware or software performance and the general level of expectations of it is a good way to fail. Beware, too, of the organizational demands imposed by both the new plan or policy and strains in matching existing practices. Organizational validity characterizes how well the new policy fits with the organization, that is, is compatible with existing practices and does not require substantial changes or changes not accompanied by a real and direct benefit to the users. The more drastic the changes demanded of people's behavior patterns and approach, the more chance there is of failure.

In any organization, people tend to be well attuned to a limited repertoire of responses to stimuli; none has infinite flexibility even if some have more than others. Accordingly, a plan or policy that requires users to depart substantially from an accustomed repertoire runs a higher risk of failure to be implemented than one that does not. It is essentially "not possible to lead an organization in any arbitrary direction that may be desired."[22] An elephant may learn to jump, but it will never become like a gazelle, and it may well not want to make the attempt.

Information Transfer

Another good way to fail is to ensure that there are deficiencies in the transfer of information between units at different levels. Closely related to the timely provision of adequate resources is provision of information—it could even be regarded as a form of resource—both information needed to carry through implementation and, more generally, information about what is going on. Your implementors will begin to *feel* like mushrooms if they are always kept in the dark. Moreover, in the absence of reliable and timely information, the grapevine will fill the gap with information that is not always accurate and is rarely flattering to the official view!

Goal Conflict

It is quite rare in practice for the totality of an organization's goals and overall mission to be in harmony with one another. If conflict is perceived between the goals of implementation and other organizational goals, then the chances of successful implementation are much reduced. Again the pitfall to be avoided is in appearing to force those at the periphery of the organization to choose between sets of conflicting priorities. It puts them in an invidious position and invites them to use the existence of the one as a heaven-sent excuse for ignoring the other.

Table 26-1. Review of Factors That Either Help or Hinder Implementation

Helpful Factors	*Impeding Factors*
The plan itself	Values
Existence of a problem	Resources, lack of
	Track record
Organizational mission	
	Technical validity
Resources, availability	
Track record	Organizational validity
Technical validity	Information transfer
Organizational validity	Goal conflict
Top management support	
Size	

ATTITUDES TO CHANGE

These positive and negative factors are listed for ease of reference in Table 26-1. Just one more new point before we draw the threads together. All of the foregoing implies someone somewhere having to make changes in their patterns of behavior, in the way in which they do their job. So who wants to see change? No one, according to conventional wisdom. On this view, all change is resisted tooth and nail as people want to stay set in their cosy rut. However, that is a gross oversimplification. The whole history of the development of civilization, of the development of Western industrial and postindustrial society, has been one of wholesale and continuing change. Can all these changes have been resisted stoically by all concerned? No, they have not. There is plenty of change going on all around, by no means all of it is being opposed. To be sure many changes *are* resisted, the Luddites and their fellow travelers have long been with us; they are alive and well today.

However, many other changes have been made and accepted readily. Why the difference? The key to it lies in the way in which the consequences of change are perceived. If change is perceived to be for the worse, it will be resisted. Communities that depend very heavily on a single entity for their existence have been seen to resist strongly attempts at industry rationalization that would reduce job opportunities and hence be perceived as a threat to the continued existence of such communities.

Change that is perceived as being for the better is a different story! There change is welcomed and indeed even pressed for by those who see net benefits coming to them from it. As one manager put it:

> I run a retreat every year. I take the Chairman, Secretary and branch representatives of (union) away for a day and we go through the 5 year manpower review. . . . When I started that it was greeted, with utmost suspicion . . . [but] now the pressure is "When are we going to have our seminar?"

The moral here is that the agent of change, quite apart from the intellectual, planning side of the task, has a marketing role to play as well in pondering the benefits to those affected and ensuring that they are both positive and presented in a positive light.

GUIDELINES

Let us try to pull the discussion together, to summarize and suggest what the busy planner or manager who anticipates an implementation problem might *do* about it? Some of the lessons of recent research have been discussed and can now be summarized in a few straightforward guidelines. Many of them stem from recognition of the very obvious, very simple, but often overlooked fact that implementation is more than just an intellectual process. It involves getting other people doing and cooperating in things that make it into a social and political process as well (that is, about power and people relationships). Managers in large organizations like to present themselves as acting in a wholly rational way. Much academic writing is founded on the premise that they do just that. However, there is a good deal of evidence that, important though the intellectual element is, it is not the sole contributor to the decision-making process, nor is it necessarily decisive. Personal preferences, personal relationships, and emotions all have their part to play. Use of such factors tends to come under the general heading of judgment, of "gut feel" or "flying by the seat of the pants," implying a commendable willingness to be decisive and to take risks. However, they also suggest a major element in the decision process that cannot readily be supported in a way that would be accepted widely as rational. Of course, much of the information needed to support a rational judgment will often be missing where new ground is being explored, but to say that is to emphasize the importance of the irrational element, not remove it altogether.

We have discussed the standard way of looking at a company's organization in terms of the traditional view as set out in its organization chart, that is, as a layered wedding cake or pyramid. We have suggested that this is a good representation of the system of rewards and sanctions within the organization but have found it wanting as a description of how the power to obtain and apply resources is dispersed. We have looked outside the mainstream of academic writings on management theory for inspiration and looked at the center-periphery model, which seeks to describe the relationships between constitutionally distinct organizational entities, on the face of it an unpromising opening, yet it is an analogy that nonetheless has dependency relationships—often those of bilateral monopolies—in respect of the flow of resources and perhaps in other ways, too. Although this model, if narrowly applied, would imply that the subunits of a large organization had more freedom than their constitutions suggest, it still contained a number of useful insights about the dispersal of power over resources in the organization and the reasons for that dispersal. It served thereafter to point to a number of positive and negative features in implementation that the pyramid model does not illustrate.

Having established the social as well as the intellectual or planning content of implementation, what factors appear that need to be taken into account? Perhaps the most obvious point, but one that sometimes goes unrecognized, is that in anything other than a wholly new organization—if there even—strategy cannot be formulated on a blank sheet of paper as if no constraints existed. The real world is complex and the planner or policy maker who ignores its complexity will suffer from problems of implementation. Important components of this complex reality are:

The plan. The plan is itself a device for communicating goals and the manner of their achievement. When translated into written form it becomes an important scene setter for those at the receiving end. It is not only a planning device, it is a communicating and marketing tool as well.

A problem. It is commonplace that those being

communicated with start from where they are, not from where the planner is or would like them to be. If the plan or policy can be presented as contributing to easing or solving a real and immediate problem for the implementors, it stands a much better chance of being accepted and acted upon. A cautionary note is needed here too, as, faced with conflicting goals, managers select whichever best suits their circumstances at the expense of the others and seize upon the existence of the conflict to justify achieving neither!

Resources must be adequate. By this we mean resources in the broadest sense to include materials and money as well as numbers of people with the requisite skills and knowledge. There is some merit in seeing time as a resource, albeit in part a subset of the

Table 26-2. Checklist for Implementation

The Plan Itself
- Is the plan clearly formulated and expressed in communicable form?
- What is the best way to communicate its contents?

Existence of a Problem
- Does the target population perceive a problem that the plan addresses?
- Do they realize it? If not, how do we tell them?

Organizational Mission
- How does the plan integrate with the organization's major mission so that people are motivated to put it into action?

Resources
- Do we all have access to the personnel, money, and materials needed?
- Are the skills and knowledge needed available and in the right places?
- Can the resources be diverted to something else?

Track Record
- Are the implementors confident in you and your plan?
- How will they perceive your past record and that of those with whom they identify you?

Technical Validity
- Does the thing work?
- Does it work properly and do what is claimed for it? (Avoid overselling and generating expectations that cannot be satisfied.)

Organizational Validity
- How do the procedures tie into what people are doing now?
- How good is the fit?

Top Management Support
- Does top management support the venture?
- How far are they prepared to go in giving visible support?
- Are they prepared to risk their reputations, too?

Size
- How large are the implementing units within the organization? (The bigger the better.)

Values
- How confident are *we* on the basis of *their* track record?

Information Transfer
- What do people need to know and want to know?
- How do we tell them and keep the information up to date?

Goal Conflict
- Is there conflict in the goals we are asking people to achieve?
- Is there conflict with what others are pressing them to achieve? Will one be played off against the other?

money resource, the need to fund development for the necessary time period.

Technical validity. The technology and related arrangements must fit the purpose to which they are applied. This will include, besides the hardware, the software, and perhaps marketing arrangements if there is a market-related aspect. It is a question of ensuring competence at the task level, of assurance that the thing works.

Organizational validity. Organizations find it easier to go on doing the kind of things they have done before; the better a new plan or policy fits into the old and familiar way of doing things, the better its chances of success. Even when radical change in behavioral patterns or culture is required, much of the time what emerges is likely to be recognizable as older forms with a different gloss on them. This is not to underrate the need for, or importance of, change but a recognition that neither organizations nor people are infinitely adaptable. New patterns of behavior generally owe a lot to their antecedents.

Track record. This feeds through into people's confidence, an elusive but important ingredient in implementation. Past experience conditions attitudes to future developments and generates expectations in people of likely success or failure. Such expectations once formed are not easy to reshape but are very important. We are all prisoners of our past.

Information transfer. Those implementing need to know what is happening and when. Even when they do not *need* to know, there is an understandable wish to be kept in the picture, to be told where the situation rests—the bad news as well as the good.

Top management support. The attitudes of senior staff are important though not perhaps quite as important as their coverage in the academic literature would imply.

Size. The size of the organization or subunit appears to have an influence on implementation with implementation generally proceeding earlier and being more successful in larger units for reasons that are not well understood.

Values. The values or prevailing culture of those doing the implementing need to be taken into account with history providing the guide. Both stick-in-the muds and restless innovators are likely to remain true to type rather than change their spots for an isolated occasion.

Attitudes toward change depend on how people perceive the change affecting them in their jobs. There is evidence that change for the better is welcomed and not resisted. The challenge to the planner then becomes ensuring that change will indeed be seen as for the better from the standpoint of its victims. That involves a two-stage task for the planner or policy maker, of analyzing the change from the viewpoint of its recipients and then marketing its positive features. Resistance to change is not inevitable. Table 26-2 is a checklist that summarizes many of the points made in this chapter and points the practitioner toward ways of absorbing their impact.

REFERENCES

1. See, for example, K. J. Cohen and R. M. Cyert, "Strategy: Formulation, Implementation and Monitoring," *Journal of Business* **46**(1973):349-367.
2. H. Mintzberg, "Strategy Formulation as a Historical Process," *International Studies of Management and Organization* **7**(1977):28-40; J. B. Quinn, "Managing Strategic Change," *Sloan Management Review* **21**(1980):3-20; R. P. Nielsen, "Toward a Method for Building Consensus During Strategic Planning," *Sloan Management Review* **22**(1981):29-40.
3. Social Science Research Council, *Central-Local Government Relationships* (London: Social Science Research Council, 1979).
4. E. Bardach, *The Implementation Game* (Cambridge, Mass.: MIT Press, 1977); M. J. Hill, "Implementation and the Central-Local Relationship," in *Central-Local Government Relationships* (London: Social Science Research Council, 1979), 2-26.
5. A. Dunsire, *Implementation in a Bureaucracy* (London: St. Martin's Press, 1978).
6. J. L. Pressman and A. B. Wildavsky, *Implementation* (Berkeley and Los Angeles: University of California Press, 1973).
7. D. S. Van Meter and C. E. Van Horn, "The Policy Implementation Process: A Conceptual Framework," *Administration and Society* **6**(1975):445-488.
8. H. L. Wilensky, *Organizational Intelligence* (New York: Basic Books, 1967).
9. See, for example, H. Fayol, *General and Industrial Management* (London: Pitman, 1949).
10. M. Crozier, *The Bureaucratic Phenomenon* (London: Tavistock, 1963).
11. C. I. Barnard, *The Functions of the Executive* (Cambridge, Mass.: Harvard University Press, 1938); J. G. March, "Footnotes to Organi-

zational Change," *Administrative Science Quarterly* **26**(1981):563-577.

12. J. R. Galbraith, "A Change Process for the Introduction of Management Information Systems: A Successful Case," *TIMS Studies in Management Science* **13**(1979):219-233.
13. W. M. Evan and G. Black, "Innovation in Business Organizations," *Journal of Business* **40**(1967):519-530.
14. Wilensky, *Organizational Intelligence.*
15. J. L. Bower, *Managing the Resource Allocation Process* (Cambridge, Mass.: Harvard University Press, 1970).
16. G. A. Steiner and J. Miner, *Management Policy and Strategy* (New York: Macmillan, 1977); J. D. Eveland, E. M. Rogers, and C. Klepper, "The Innovation Process in Public Organizations." Paper prepared for the National Science Foundation, Washington, D.C., 1977.
17. C. W. Churchman and A. H. Schainblatt, "The Researcher and the Manager: A Dialectic of Implementation," *Management Science* **11**(1965):B69-B87; R. L. Schultz and D. P. Slevin, "Introduction: The Implementation Problem," in *The Implementation of Management,* ed. R. Doktor, R. L. Schultz, and D. P. Slevin (Amsterdam: North Holland, 1979), 1-15.
18. Evan and Black, "Innovation in Business Organizations."
19. R. L. Schultz and D. P. Slevin, *Implementing Operations Research/Management Science* (New York: Elsevier, 1975).
20. R. Wernham, "A Study of Strategy Implementation in a Major U.K. Nationalised Industry." Ph.D. diss., City University Business School, London, 1982.
21. Ibid.; J. R. Kimberley and M. J. Evanisko, "Organizational Innovation: The Influence of Individual, Organizational and Contextual Factors on Hospital Adoption of Technological and Administrative Innovations," *Academy of Management Journal* **24**(1981):689-712.
22. R. L. March, "Footnotes to Organizational Change," *Administrative Science Quarterly* **26**(1981):575.

CHAPTER 27

Successful Strategy Execution via Competitive Organization Design

Ian C. MacMillan and Patricia E. Jones

Ian C. MacMillan, director of the Center for Entrepreneurial Studies and professor of management, Graduate School of Business, New York University, received the B.Sc. from the University of Witwatersrand and the M.B.A. and D.B.A. from the University of South Africa. He has been a chemical engineer, a director of companies in the travel and import/export business and a consultant with a number of companies including IBM, Citibank, and General Electric. He has published numerous articles and books on organizational politics, new ventures, and strategy formulation.

Patricia E. Jones holds a degree in psychology from Goucher College and the M.S. in business policy from Columbia University. She gained an extensive background in management consulting, product management, and training development working at AT&T. Currently she is an independent consultant.

In 1979 Kiechel claimed that 90 percent of U.S. corporations were unable to develop and execute successful strategies.[1] Based on our own experience, the key word here is *execute*. Most managers are capable of developing technically correct business strategies; strategic failures increasingly are being traced to poor implementation.[2]

Changes in strategy may well affect the organization's work, its formal structure, its informal working arrangements, and the type of individuals needed to do the work. Failure to adapt successfully all four of these organizational components to the new direction can easily contribute to failure in implementation. Without forgetting the importance of the other three components, we concentrate here on the importance of competitive organizational design for successful strategic execution.

By design we mean not only the selection of the organizational structure but also the design of the support, planning, and control systems that deliver the strategy via this formal structure. The challenge of developing such a design can be likened to the challenge facing the general who has prepared a superb campaign strategy and must now design the army that will execute that strategy. Without the correct assembly of battle units and the appropriate logistical support services, the best planned campaign cannot succeed.

This chapter elaborates on material developed in I. C. MacMillan and P. E. Jones, "Designing Organizations to Compete," *Journal of Business Strategy* **4**(1984):11-26.

Among others, Galbraith and Nathanson and Nadler and colleagues have done much to develop guidelines for organizational design.[3]
Their three basic steps are:

1. Determine organization design *imperatives*, that is, the demands and constraints placed on the organization by its environment and its strategy.
2. Design an organizational structure to meet these imperatives.
3. Carefully and systematically manage the implementation of the design.

It is critical that these steps be carried out in a competitive context. Nearly all organizations are engaged in some form of competition, if not for customers then for scarce resources such as funds or staff. An organization design carried out without the competitive overlay can easily be one with a homogenized structure—similar structural units with common policies, procedures, and measurement and reward systems—developed with an eye toward efficiency, which is of little avail if the different parts of the organization are under attack by very different competitors in very different environments.

Competitive design is an extension of the above three basic steps, expanded to take into account the competitive nature of the environment.

THE COMPETITIVE DESIGN PROCESS

Building on the three design steps above, we present here a methodology for designing competitive organizations. This process is presented schematically in Figure 27-1. First, four analytical steps are needed to determine the design imperatives. These four steps help in the analysis of the organization's strategy, environment, competitors, current design strengths and weaknesses, and the work requirements dictated by the strategy. The de-

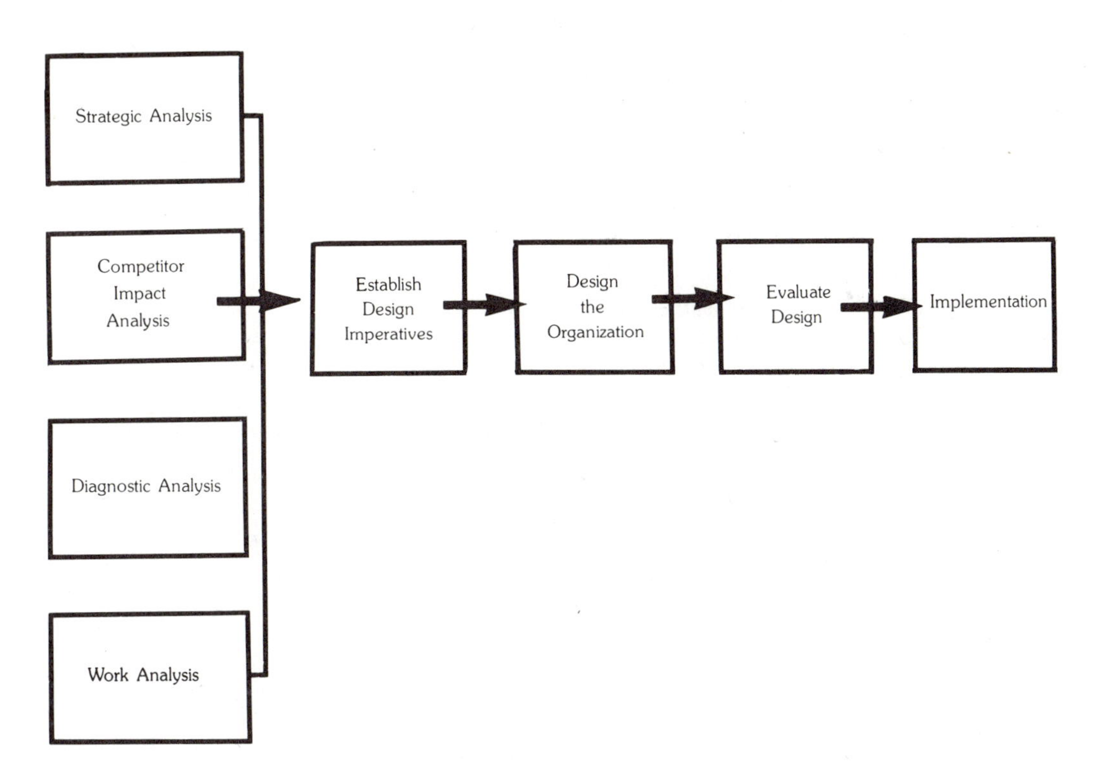

Figure 27-1. Competitive design process.

sign imperatives culled from the analytical steps serve as critical input to the organizational design decisions: the grouping and linking of organizational units, including the design of the necessary support systems. Finally, the design is evaluated and implementation carefully planned and managed to ensure successful strategy execution.

The remainder of the chapter is devoted to a detailed discussion of each step of the design process. The questions used to guide the process are summarized in Tables 27-1 and 27-2, which also present a format for documenting design imperatives and work functions during analysis and design decisions later on.

Before we begin, however, a note on design imperatives and work functions is in order. What do we mean, for instance, by design imperatives? Generally, these are

Table 27-1. Questions for the Identification of Design Imperatives

Analysis	*Questions to Ask*	*Design Imperatives*	*Work Functions*
Strategic	1. What is the organization's strategy? What is its contribution to society? What is its strategic role? By what strategy will the above be accomplished? What is its strategic role? What are the key success factors? 2. How will we know that the strategy has been accomplished? What strategic accomplishments spell success? In pursuing this strategy, what ideology should shape our decisions? What changes are necessary to achieve the strategic accomplishments?		
Competitor	3. How will competitors be impacted? Who are target competitors? What are their strategic strengths and weaknesses? What are their design weaknesses?		
Diagnostic	4. What are our own strategic and design strengths and weaknesses? 5. What else must change?		
Work	6. What is the nature of the organization's work?		

Table 27-2. Questions for the Design, Evaluation, and Implementation Decisions

Questions to Ask	*Decisions and Implications*
7. What major work groupings are feasible design alternatives? What groupings address the needs of the marketplace? What groupings address the competitive advantages? What are the vulnerabilities of each of these feasible groupings? Can we accomplish the strategy without major reorganization?	
8. What linkages are necessary between groupings? What mechanisms link key groupings to critical functions? What linkages address key vulnerabilities?	
9. What support systems are needed? Have necessary support systems been identified? What new management and functional skills are needed?	
10. Will it work? Have key events been simulated?	
11. What execution problems can be anticipated? Has stakeholder impact been identified and managed? Have problem owners been assigned?	

statements that specify the design requirements resulting from each analysis. They can best be formulated as the completion to the sentence that begins, "Design an organization that..." For example, a midwestern bank analyzed its strategy and found that its customers' needs varied greatly, most often based on geographical area. The resulting design imperative read as follows: design an organization that allows and fosters branch autonomy in responding to customer needs. As a result, the bank gave greater freedom to its branch managers with regard to product offerings, while responding to another design imperative of profitability by also holding them accountable for their branch's bottom line. Authority, control, and reward systems—all formal design mechanisms—were used to ensure that both design imperatives were met. A well-conceived design imperative will not simply reiterate the condition that fostered it (the bank did not say: design an organization that recognizes market diversity); it will translate that condition to a design potential (branch autonomy).

Furthermore, the analyses and design imperatives will reveal specific work functions (existing or needed) that the organization must carry out if its strategy is to be properly implemented. Although this seems obvious, too often the *additional* work required as a result of a strategic shift is not formally recognized and therefore not done, let alone considered in the work analysis step. It is important, then, throughout the first three analytical steps to document the work functions, which are then addressed in the fourth step.

STRATEGIC ANALYSIS

The strategic thrust provides the first important input to the design decisions that will be made after all analyses are complete.

The organization's *competitive strategy* includes statements regarding its functional and strategic roles, how it plans to compete, key success factors, action plans, goals, and objectives. It delineates how financial, physical, and human resources must be allocated in response to the threats, opportunities, and constraints facing the organization.

Output from this and the next two analytical steps should include not only design imperatives but also a growing list of work functions implicit in the analyses for input to the fourth and last analytical step: the analysis of the work itself.

What Is the Organization's Strategy?

Rothschild has developed a detailed and helpful list of questions to aid in the strategic analysis process.[4] Any analysis, however, should answer at least the following questions:

What will the organization's contribution be to society? In other words, what gives it the right to exist in society? It is important to identify what need is being satisfied by the organization from the point of view of society. For instance, in its early days, IBM recognized that its societal contribution was not to manufacture and sell computers but to solve data storage, retrieval, and processing problems of large organizations. This question not only puts the business into a societal perspective, it also helps the business to recognize that there are many ways of satisfying the societal need. It is also a useful question for support departments to ask with respect to their corporation.

What is the strategic role? In other words, what strategic purpose must the organization accomplish for itself? MacMillan has identified eight role choices, depending on the organization's position in a larger corporate portfolio: build aggressively, build gradually, build selectively, maintain aggressively, maintain selectively, prove viability, divest or liquidate, and competitively harass.[5] (The missions associated with each role are summarized in Table 27-3.)

By what strategy will the above be accomplished? A business strategy must be selected to accomplish the above challenges. This involves selecting an array of products or services, targeting them at selected markets, and making use of selected competitive advantages. Thus, in the 1950s IBM elected to build aggressively in the medium-to-large business market, the government market, and the medium-to-large educational market by supplying an array of mainframe computers with modest modularity and with a full range of peripheral devices. They elected advanced (but not cutting-edge) products, cutting-edge technology, and high operational reliability of product as driving competitive advantages.

Such strategic decisions are made in the context of the organization's own skills and the opportunities and constraints present in the environment. Just as the strategic planner must consider environmental conditions and trends in making product and market mix selections for the organization, the organization designed must also be cognizant of these environmental factors in making design decisions.

A thorough analysis of the environment should consider the nature of at least the following factors: buyers, suppliers, distributors, competitors, labor pool, government regulation, technology, and the economy. If it had not recognized the geographical difference in customers needs, the midwestern bank men-

Table 27-3. Possible Strategic Accomplishments for Strategic Roles

Role	*Action Required*	*Relative Accomplishments*	*Absolute Accomplishments*
Build aggressively	Build share on all fronts as rapidly as possible.	Rapid growth in share—all markets. Leadership technology, service.	Limits on losses and negative cash flow.
Build gradually	Steady sustained increase in share of entire market.	Sustained growth in share—all markets. Leadership in quality, service.	Limited losses. Sustained cost reductions.
Build selectively	Increased share in carefully selected markets.	Share growth in selected markets. Leadership in customer satisfaction. Superiority in market research.	Growth in profits and profitability. Growth in cash flow.
Maintain aggressively	Hold position in all markets and generate profits.	Hold market share in all markets. Relative cost leadership—fixed and variable. Technology leadership—in product and process.	Improve asset utilization. Growth in profitability and cash flow. Improve expense-to-revenue ratio. Reduce force levels.
Maintain selectively	Select high-profit markets and secure position.	Overall share reduction. Hold market share in selected markets. Improve relative profitability. Distribution, service leadership.	Minimum investment. Improve asset utilization and cash flow. Reduce fixed cost/sales.
Prove viability	If there are any viable segments, maintain selectively, divest rest.	In this exhibit, see the sections to the left called "Maintain Selectively" and "Divest/Liquidate."	Minimize drag/risk to organization. Growth in profits and cash flow.
Divest or liquidate	Seek exit and sell off at best price.	Reduce share except for highly selective segments. Enhance value added via technical leadership.	Minimize investment. Reduce fixed costs. Improve profitability. Maximize selling price. Reduce work force levels.
Competitively harass	Use as vehicle to deny revenues to competitors.	Attack competitor's high-share business but do not gain share. Relative price never above that of target competitors.	Minimize fixed costs. Sustained reduction of variable costs. Limits on losses and negative cash flows.

Source: From I. C. MacMillan and P. E. Jones, "Designing Organizations to Compete," *Journal of Business Strategy* **4**(1984):13. Reprinted with permission.

tioned earlier could easily have designed homogeneous control and measurement systems for its branches, thereby crippling the branch managers ability to compete locally.

Furthermore, to the extent that we can anticipate likely changes, both at the micro and macro level, we can, to some extent, design for the future. Rapid technological change in the electronics industry, for instance, demands more fluid and dynamic organizational structures that will allow companies competing in that environment to respond quickly to change. The organizational designer should consider industry and society trends as well as current conditions and their possible impact on the organization.

What critical functions will drive the strategy? If the strategy is to be achieved, a limited number of functions must be performed and performed well, as Rothschild indicates.[6] In fact, these functions must be performed well for the company to attain the competitive advantage it seeks, and it is these functions that dictate where major resource allocations and trade-offs are to be made. For instance, IBM's commitment to advanced technology and the operational reliability of the product on site caused development and service functions to take a critical role in the early days. Marketing and finance, although aggressive and powerful, generally had to take a back seat in trade-offs involving performance reliability of their computers versus introduction of new technology, new product releases, or inventory allocations to customers. As a result, IBM continues to extract a premium price based on its image of reliability and superb service.

What are the key success factors? These will vary from company to company, but generally they include exceptional management of several of the following: product design, market segmentation, distribution and promotion, pricing, financing, securing of key personnel, research and development, production, servicing, maintenance of quality and/or value, or securing key suppliers. (For example, companies in the grain-trading business simply will not survive if they do not have exceptional sourcing and delivery logistics as well as superior options and contract skills.)

Once the key strategic decisions are analyzed, answers to the next important set of questions define the major design parameters of the strategic control system.

How Will We Know When the Strategy Is Accomplished?

The critical challenge in designing a strategic control system is to create one that is self-controlling; one that can respond autonomously to environmental challenges but one in which the various subunits still pursue their assigned strategic roles according to their designated strategy. It has been our experience that this is accomplished by detailed attention to strategic accomplishments and ideology.

What strategic accomplishments spell success? In competitive design, those charged with executing strategy should be able to gauge when they have been successful. Such success is best monitored by specifying the set of accomplishments that will demonstrate that the organization has achieved what it set out to do—namely, pursue the strategic role via the selected strategy. Therefore, we need to specify clearly what will indicate the means by which it was accomplished. For instance, IBM's accomplishment of an aggressive build role via its selected strategy could be demonstrated in the late 1960s by the following accomplishments:

It had a major share of all markets it had elected to serve.
Growth had exceeded that of all major competitors (in other words, IBM was gaining share).
Profits were increasing and cash flows were positive.
It was acknowledged worldwide as having the most up-to-date equipment—both mainframe and peripherals.
It was acknowledged worldwide as having superior product reliability and service.

Thus, there was no question that IBM has been successful in its aggressive build role (a role that calls for substantial gains in share in all markets) and that this had been done via IBM's selected strategy of providing advanced products with unequaled operational reliability and service.

An imperative in the specification of strategic accomplishments is the selection of the minimum number of key criteria that management will pursue in the actual execution of strategy. Having too many criteria is a drawback since each additional criterion inhibits the flexibility and adaptability of those charged with strategy execution. Yet there also should be a sufficient number of criteria to ensure that the single-minded pursuit of one or two of them is not carried out to the detriment of the long-term health of the organization.

Of particular importance is the selection of at least one relative criterion and at least one leadership criterion. Absolute criteria are inclined to encourage an inward orientation, with attention to performance compared to past history rather than current competition. By relative criteria we mean measuring performance relative to competitors', whereas leadership criteria are those in which our performance exceeds that of all competitors. Criteria should be selected in such a way as to steer the organization toward fulfilling its selected role. In our experience, the relative criteria (such as: gain share versus largest competitors) set the strategic direction by forcing the organization to focus on its competitive performance, whereas the absolute criteria (such as achieve 15 percent ROI) set limits on the extent to which relative performance can be single-mindedly pursued to the long-run detriment of the organization. Finally, the leadership criteria (such as be the leader in quality) force the organization to focus on the desired competitive advantage. Table 27-3 lists some typical criteria selected by firms in past competitive designs.

Having selected the criteria, the final challenge is the selection of appropriate measures of accomplishment. Absolute criteria and relative criteria generally pose less of a problem than leadership criteria, but for all criteria the important thing is to find measures that are objective or external rather than subjective and internal. It is all too easy for the organization to convince itself that it is a technology, quality, or service leader if it listens only to itself or its loyal distributors and customers. However, leadership is in the eyes of the *market,* and it may be necessary to survey the market to establish whether the claim of leadership is actually wishful thinking or a fabrication of the organization's imagination.

Problems can arise with the need for appropriate measures. Often a direct measure cannot be obtained. For example, cost leadership and, particularly, technology leadership are difficult to assess because hard data about competitors are rarely available. It may be necessary to come up with several imaginative surrogate measures that are indicators of the actual performance. One high-technology equipment manufacturer used the following as indicators of technology leadership:

Ratings of its R & D department compared to those of its two leading competitors. Ratings were done by customers, suppliers, investment bank analysts, and three leading universities.

Relative number of patents (compared to leading competitors) applied for in the past five years and past two years.

Relative number of new models introduced by the company compared to its leading competitors, and ratings of these models by key customers (not only its own) in the served market.

Number of recent scientific postgraduate R & D job applicants that were lost to competitors, compared to number of R & D job applicants captured from competitors.

Note that it may be possible for each of these measures to be subverted by a devious development department, or for each of these measures to give false signals as to the true status of the firm's technology leadership. However, senior management stressed that the spirit of the measures was more important than each individual measure itself. The firm's development staff was aware that in spirit it was expected to achieve technology leadership and would be judged more on the gestalt of the above measures than on any individual measure.

For strategic control it is direction and commitment that are important, not precision of measurement. By limiting the number of criteria to guide the organization in the desired strategic direction, management is free to pursue this direction with maximum flexibility, responding autonomously to competitive conditions as they occur.

In order to create a *self*-controlling organization in which relatively autonomous decisions can be made, it also is necessary to develop and disseminate suitable ideology.

What should the ideology be? Development of a suitable ideology has been discussed by many authors. By ideology we do not mean the corporate culture, which we regard as a passive manifestation of corporate beliefs. Ideology is an explicitly generated, consciously managed, and clearly disseminated set of values by which senior management will judge the quality of internal behavior and attitudes in the organization and the quality of external interactions of the organization and its members.

Peters and Waterman in their study of excellent companies identified a number of factors that need to be included in ideology.[7] These components of ideology were made more explicit in the analysis of several companies by MacMillan.[8] The fundamental ideological principles held by the organization should broadly specify the following:

Scope. What constitutes desirable types of products and markets?
Drivers. Which shall be the critical functions?
Style. What style of management shall prevail?
Ethics. What are the determinants of ethics?
Attitude to risk-taking. How much risk is encouraged?
Attitude to competition. How aggressively should competition be pursued?
Attitude to customers and channels. How will customers and distributors be treated?
Attitude to employees. What are the fundamental principles by which employees will be treated?
Attitude to external groups. What is the attitude toward various special interest groups, government, and so forth?
Self-image. How does the company feel about its control over its density?

The underlying philosophy is that once managers in the organization have internalized these key principles, they will be able to make autonomous decisions in response to competitive or environmental challenges that will stay within the boundaries of behavior desired by the organization as a whole while none the less addressing the specific challenge.

This set of explicit principles, plus the strategic accomplishments, creates a framework for autonomous but self-controlling responses by the managers to the competitive environment.

With the foundation of the strategic control system in place, it is necessary to consider what changes are required to effect these accomplishments.

What changes are necessary to achieve the strategic accomplishments? If major strategic shifts must be made, this will certainly necessitate major changes in staffing, in support systems, and particularly in skills required. It is important to identify what changes will be necessary for two reasons: (1) change must be formally managed, or the organization will revert to past practices; and (2) such changes have to be supported by the key external and internal stakeholders, and the process of generating their support also will have to be managed. Therefore the organization designer must identify what critical changes must occur if the strategy is to be accomplished; these changes then become the focus of the change management process.

COMPETITOR IMPACT ANALYSIS

A key requirement for designing a military campaign is a thorough assessment of the impact of the battle plans on the main units of the enemy. Although an analysis of the competitors will most certainly have been done in the strategy formulation process, the organization designer must now consider the impact of strategy execution on the competitors' strengths and weaknesses.

What Will the Impact Be on Individual Competitors?

In reality, no strategy has the same impact on all competitors. In fact, a good indicator that a strategy has not been thoroughly thought through is when the strategists cannot pinpoint which of the competitors will bear the main brunt of the impact of their strategy.

Which are the target competitors? Even if the strategy formulation process did not start with target competitors in mind, the analysis of the impact of strategy on the competitors will highlight those companies that will be target competitors—namely, those that bear the brunt of the strategic attack. Their capacity to respond needs to be evaluated and their responses predicted. Hence the next three questions.

What are the target competitors' strategic weaknesses? This question ensures that we have in fact identified the real target competitors. These are competitors whose strategic weaknesses render them most vulnerable to our strategy. If our strategy is based on a competitive advantage in distribution and service, then the brunt of our attack will be borne initially by those firms that are weak in these areas, not those that are strong. (If *none* of the competitors are weak, then we are not undertaking an attack, but defense; essentially, we are playing catch-up.)

What are the design vulnerabilities of the target competitors? It also is important to analyze the organization design of the target competitors, particularly their formal structure, since this could tease out some interesting opportunities for an aggressive competitive design that directly attacks a competitor where it is most vulnerable. At each organizational level, work can be grouped in one of three basic ways: by activity (typically, a functional design), by output (typically, a product division), or by client (typically, a market segment division). The first few levels of design are critical; each combination of groupings at these levels brings with it substantial strengths and weaknesses. For example, a computer firm for years had been successfully organized by activity and output. Within the marketing department (activity), sales of its two product (output) groups—large and small computers—were organized separately. However, in the 1970s, the distinction between large and small computers started to blur, resulting in salespeople from each product group calling on the same customers, creating customer confusion and irritation. Turf wars started between competing sales forces, service difficulties increased, and eventually the company became vulnerable to a small-computer competitor that had designed its structure around key accounts under the marketing function (thus using an activity-client structure).

Competitors are in business to defend or take market share from the organization, and their ability to do so varies with their individual strengths and weaknesses, many of which spring from their own structure. One tool for assessing potential design advantages is presented in Appendix 27A. First, consider how target competitors have grouped work at the top two layers and then identify the potential strengths and weaknesses inherent in these particular groupings. Then, when considering alternative groupings for our own design, we can select combinations of groupings that recognize the competition's strengths, and, if we are fortunate, we can also take advantage of their weaknesses.

For example, if the target competitor is organized by activity-activity, then by grouping by output client we will create a much higher sensitivity to market needs. Because of its organization, the competitor inherently will have a much slower response to these needs. This is the situation in which AT & T found itself in the late 1970s when MCI attacked specific high-density routes and focused on specific client segments on these routes (business and consumer).

Note that the more usual case where the target competitor does happen to be grouped effectively does not preclude us from grouping our own organization in the same way. In fact, we are even more obliged to design effectively. However, there may be weaknesses in other parts of the competitor's design, such as their linking mechanisms (those specific arrangements the competitor has designed to facilitate and control the flow of critical information between groupings). For instance, direct responsibility to respond to our strategy may not fall within the jurisdiction of one particular department and there may be no formal vehicle for communications *between* departments that should respond to our strategies. This happened when Merrill Lynch attacked the commercial banks' big retail customers by offering one-stop cash management accounts. Although several bank departments were affected, no single one of which had the authority to respond and no vehicle existed for coordinating effectively between departments. As a result, Merrill Lynch was able to build up a $40 billion CMA base before effective responses could be generated.

Other competitor support systems, such as planning, control, and reward systems, standard operating procedures, or even ideology, also shape the rate at which they become aware of and respond to strategic moves we make.[9] The organization can build superior linking mechanisms or major support systems into its own structure to capitalize on the vulnerabilities created by a competitor that has poor linking or support systems.

With the competitors' strengths and vulnerabilities identified, it is time to turn to a diagnosis of our own organization. Many considerations go into designing the formal organizational structure: work grouping, linking, job design, methods and practices, standards and measurements, physical work environment, human resource management systems, reward systems, and support systems. All of these must reflect both the organization's strategy and its competitive environment and it is here that it becomes all too easy for the *competitive* imperatives to be lost in the details of design.

DIAGNOSTIC ANALYSIS

Diagnosing your own strengths and weaknesses is as important as knowing those of your competitors. Care must be taken not only to eliminate or strengthen current weaknesses but also to preserve design strengths congruent with the new strategic direction.

What Are Our Own Strategic and Design Strengths and Weaknesses?

Even the most well-conceived strategy is based on a certain amount of guesswork involving management's best judgment regarding the future, subject to all the uncertainty associated with predictions. To the extent that a structure can be put in place that will allow adaptation without subjecting the organization to upheaval, the designer will improve the odds of successful implementation. This calls for a review of the assumptions on which the strategy was based and an assessment of what problems may materialize if the assumptions prove unfounded.

The current organizational structure, like that of the competitor's, must also be analyzed. Structural strengths that are in line with the new strategy should be preserved. Potential vulnerabilities on the other hand may indicate a need for design changes to reduce these vulnerabilities. Again, Appendix 27A should prove helpful in such assessments.

What Else Must Change?

Not all strengths and weaknesses will reside in the organization's formal structure. As mentioned earlier, the organization's informal structure and the individuals who people it must also be examined if we are to ensure that all components important to implementation are considered. Issues such as internal power groups, the current management style, the inventory of skills, the organization's current ideology, and its own history of change may also surface and should be considered for implementation.

Here, too, should surface the reality of financial consequences or restraints. How much can we actually spend on the new design? To avoid designing a Cadillac when, in fact, we can only afford a Pinto, we must add such constraints to the design imperatives now rather than later.

WORK ANALYSIS

A review of the above questions and the documented work functions provides the basis for a thorough work analysis—the actual work that must be done if the strategy is to be accomplished.

What Is the Nature of the Organization's Work?

For our purposes we define *work* as those major functions and subfunctions that must be done if the organization is to carry out its mission. *Interfaces* are the critical links that tie together the work of the organization.

The work analysis identifies (1) the major functions to be performed as the strategies are executed; and (2) the key tasks and responsibilities associated with each of these functions.

The analysis raises two types of issues. The first arises from the nature of the work itself, for instance, the need for a new market planning function may give rise to pricing responsibilities that may not have existed before. It is important then to spell this out in the design imperatives. The second type of issue arises due to interfaces, both between internal functions and with functions outside the organization. For example, the creation of an interdepartmental product task force implies an interface between it and the various departmental product planning functions within the organization.

Characteristics of these work flow interdependencies and the critical interfaces have important design implications. A good way to begin identifying these is to trace how the work moves through the organization. By examining the work flow, one can identify critical interdependencies between work and work units. How much uncertainty is associated with the work at different points? Where do (or might) problems such as breakdowns in communications, missed targets, or delays occur? These may all be indicators of unmet information-processing needs.

ORGANIZATION DESIGN

Once analyses are complete and design imperatives documented, it finally is possible to address the organization design itself. Design decisions fall into three major categories.

Grouping decisions address which major work functions are to be grouped together or apart.
Linking decisions address how those work units, once grouped, will again be tied together to ensure the integrity of the work and information flow.
System decisions, a subset of linking, ensure that necessary support systems are in place.

Which Major Work Groupings Are Feasible Design Alternatives?

The first key decision addresses major work groupings.[10] At this point in the process, the key inputs for the grouping decision are at hand: the design imperatives and the work to be performed. The grouping decision is based on both. The work functions should be grouped in a way that best addresses the marketplace and the organization's competitive advantages. It is here that the organization designer may be tempted to design with the focus on the internal activities rather than on the external challenges facing the organization. The questions that follow force an external focus.

Which groupings address the needs of the marketplace? The organization is in business to deliver products or services to its target markets, therefore the first step in grouping should address how the organization plans to serve and compete in the selected markets. The strategic and competitive analyses recommended above suggest what factors should be considered.

For example, an industrial foods firm decided that to compete with its larger competitors it had to keep prices competitive. At the top level, this called for a functional design (activity). However, the firm's business and institutional markets were drastically different, and each required very different and specialized marketing skills. Management therefore elected to segment the marketing efforts by client (see Figure 27-2) at the next level. At the third level, the marketing for each segment was grouped geographically to reflect discerned regional differences on the East and West coasts. Thus, the marketing activity was grouped according to a client-client grouping (see Appendix 27A). Had the regional sensitivity been identified in the strategy as the key difference, their second- and third-level groupings in marketing might have been reversed.

Which groupings address the competitive advantages? The next grouping decision must address the organization's desired competitive advantages. Continuing our example in Figure 27-2, the industrial foods firm selected two areas in which it felt it had to develop a competitive advantage: product development and cost leadership. In order to maintain cost competitiveness, the firm decided on an activity-client design for R & D and another form of activity-client design for manufactur-

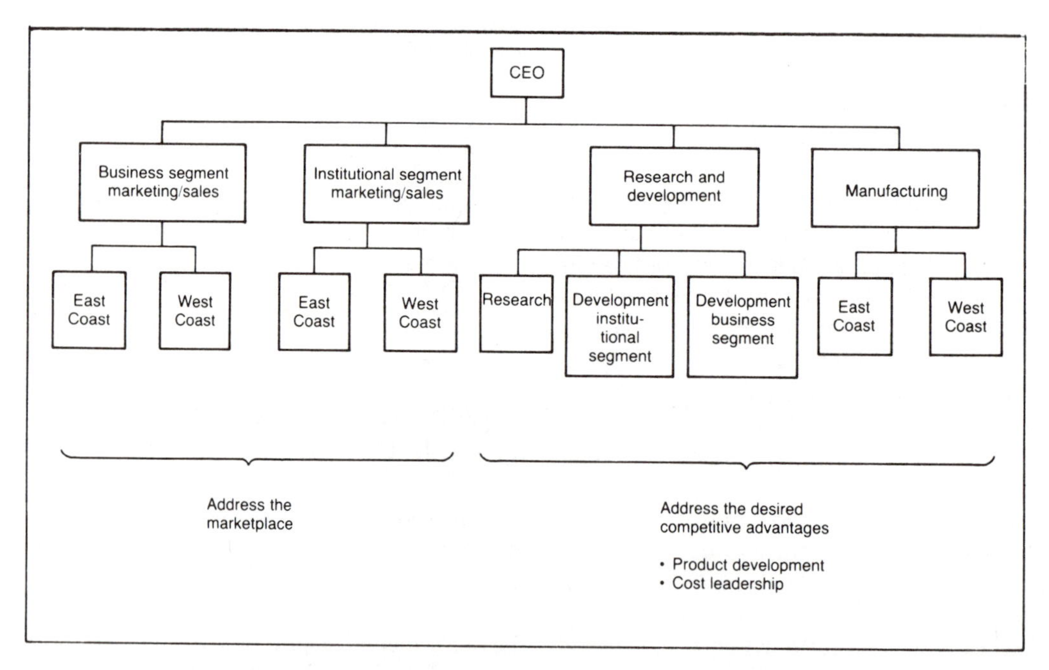

Figure 27-2. Key groupings for industrial foods manufacturer (from I. C. MacMillan and P. E. Jones, "Designing Organizations to Compete," *Journal of Business Strategy,* **4,** 1984:17. Reprinted with permission).

ing. Some interesting points emerge as we look at these subgroupings. Under R & D, two development functions were established: one in support of the business segment, the other in support of the institutional segment. Although it may be argued that such a grouping reduces efficiencies available from economies of scale, it does produce an organization capable of responding quickly to needs and changes in each market segment, thereby designing in effectiveness. If serving particular market segments well with rapid and market-responsive product introductions is key to the success of the organization, then effectiveness must take priority over efficiency. Also, note that the research function was separated from the development function The "blue sky" nature of basic research would be ill-served by grouping it with the shorter-term development function. This decision was based on work analysis design imperatives.

However, there remained the problem of keeping the research function in line with the rest of the organization and with the demands of this environment. This called for effective linkages[11] and the company created a formal linking position: product managers for each market who interacted constantly with the development function on the subject of new products.

The next area of interest was in manufacturing. The geographical groupings at the second level were in response to the need for a cost-leadership position: economies of scale were essential, so each regional plant manufactured products for both markets. However, the market focus of these plants was maintained once more by the product managers, whose linking role here was to manage conflicts between the needs of marketing and those of manufacturing.

Note that if the firm had been a much smaller one, such a structure may have highlighted the impossibility of cost leadership and product leadership as competitive advantages in a small firm. Such a discovery in the design phase is not at all inconceivable and may lead the organization to reexamine its selected strategy.

What are the vulnerabilities of each of these feasible groupings? The best grouping mix is one that enables the organization to maximize most effectively its strengths and to minimize its vulnerabilities vis-à-vis the competition. At this point in the design process, several structures might seem feasible. However, some alternatives can be eliminated by weighing the advantages and disadvantages of the various groupings, as reflected in Appendix 27A. Some of these vulnerabilities could be defused with appropriate linking mechanisms (which will be addressed in the next section), but none should be tolerated that interfere with the organization's ability to effectively address the needs of marketplace.

Can we accomplish our strategy without major reorganization? If one of the feasible groupings closely resembles the current method of grouping, it should become a prime candidate for final selection, particularly if this grouping structure seems to have relatively minor vulnerabilities vis-à-vis those of the key competitors. The reason for this is that a major reorganization is a tremendous source of trauma for the company. Enormous amounts of time, energy, talent, and competitive vigor are consumed as the new structure is developed under the new set of relationships. It is estimated that it takes at least two years to recover from a major reorganization. Therefore, if at all possible, it is preferable to try to use the existing groupings, supported if necessary by well-conceived new linking mechanisms in order to avoid the disruption created by reorganizing. Once again, the pragmatic challenge is not to design systems that are perfect but systems that are merely better than those of the competitors and that are appropriate to the strategy.

What Linkages Are Necessary between Groupings?

The very act of grouping certain functions together also will mean that certain natural work flows are separated. For instance, it may be appropriate to divide production into two geographically separate plants, even if some of the production of one plant is needed at the other. However, if there are no coordinating or linking mechanisms, disruptions or exceptions at the first plant can create problems for the second. Linking mechanisms must be designed to coordinate and control critical interfaces. The focus at this stage should be on those interfaces that are critical to the organization's strategic and competitive success. Table 27-4 lists the key mechanisms used to link interfaces. The details of how to use such linking mechanisms are not discussed here since they are available in the literature.[12] We have, however, tried to indicate the major *circumstances* under which each should be employed by focusing on the nature of the disruptions-exceptions conditions that create the need for intergroup coordination. For instance, if the expected disruptions are not likely to be serious but will occur frequently and their nature is fairly predictable, the most appropriate linking mechanism is the appointment of formal liaison people in each department who will coordinate actions when these disruptions occur. A typical example would be to assign a plant supervisor the responsibility of coordinating the acquisition of supplies from another of the company's plants. To do this, he or she would coordinate activities with those of the supervisors in the other plant.

In Table 27-4 sequential dependence means that one group depends on the input of another, such as in a production line. With reciprocal dependence, the output of one group becomes the input of another, which in time provides input back to the first. Typical reciprocal situations are relations between units that use equipment and units that maintain equipment (airlines, railroads). Another reciprocal situation is the relationship between the design, production, and marketing departments as new products and their supporting production systems are developed.

For the purposes of Table 27-4, the following definitions apply:

Task force. A group staffed with members from the various departments involved is formed to tackle a specific intergroup problem. It is automatically disbanded after the problem is solved.

Team. A group is selected from various departments in the organization to respond to recurring problems that cross over group boundaries. It is a permanent coordinative arrangement that is disbanded only by a higher level in the hierarchy.

Integrating role. An individual is charged with formal responsibilities for coordinating the flow of information between two or more groups. A common example is the product

Table 27-4. Key Linking Mechanisms

Exception Conditions				
Impact	*Predictability*	*Frequency*	*Grouping Interdependency*	*Linking Mechanism*
Low	High			Rules
Low	Low	Low		Contact
Low	Low	High		Liaison
High	High	Low		Contingency plan
High	Low	Low		Task force
High	High	High		Teams
High	Low	High	Sequential	Integrating role
High	Low	High	Reciprocal	Matrix

Source: From I. C. MacMillan and P. E. Jones, "Designing Organizations to Compete," *Journal of Business Strategy* **4**(1984):18. Reprinted with permission.
Note: Critical to success of linking mechanisms is to ensure that they take place at the correct level with correct delegation.

manager whose task is to see that specific products get adequate attention from the marketing, production, and service functions.

Integrating department. A department with independent resources and staff whose task is to ensure coordination between different functions. A typical example is an expediting department in a manufacturing firm that coordinates marketing and production. In a defense contracting firm, an expediting department might coordinate specific projects with the various technical departments supplying skills to execute the projects.

Matrix structure. A person simultaneously reports to and has responsibility for a number of managers, each in charge of different activities or resources that must be coordinated. In addition, below this matrix manager is a structure in which competing tasks must be executed. The classic example is the project manager in a construction company who is required to report both to operations (for project programs) and to the technical manager (for staffing requirements) while managing a specific project under his or her control. In effect, the matrix manager absorbs conflicts between groupings and thus shields the subordinates from the intergroup conflicts, thereby allowing the subordinates to pursue their tasks without such disruptions. The matrix structure is extremely difficult to operate and should generally be avoided in favor of a simpler linking mechanism if possible.

By looking at the nature of anticipated exception conditions in the firm's environment according to the dimensions of Table 27-4, it is feasible to use the table to decide what the lowest-cost, yet appropriate, linking mechanism is. For example, suppose the exception condition is the anticipation that competitors will introduce new products but relatively infrequently. This condition will have low frequency, low predictability but high impact, and thus calls for a task force to be pulled together whenever it occurs, with representatives drawn from all those functions that will be involved in responding to the new product introduction.

What mechanisms will link key groupings to critical functions? As mentioned above, the critical functions are those that must deliver the competitive advantage and those that deliver the strategic role. Here, the designer must build linking mechanisms to ensure that these critical functions perform their tasks and at the same time remain responsible to the markets being served. In our example of the industrial foods company, it is futile for the development department to create new products that the market does not want. It is also fatal for the manufacturing department to seek cost reductions by reducing inventory and scheduling costs if this means the company would have to deny an appropriate variety of products to the market or reduce the availability or reliability of supply to the customers.

The R & D function must deliver new products appropriate to the market, and the manufacturing function must accomplish cost reductions appropriate to the marketing effort. To accomplish this, the company in our example created permanent product development and product cost teams in which a senior marketing manager from each market segment worked with a senior development manager and a senior manufacturing manager. It was the charter of these teams, together with any *ad hoc* members co-opted as deemed appropriate, to drive the coordinated development of new products and processes and the coordinated development of cost-reduction programs in directions that were appropriate to the markets being served.

What linkages address key vulnerabilities? Every structure has its vulnerabilities. Vulnerabilities arising from the organization's structure can be reduced, if not eliminated, with the design and installation of additional linkages. For instance, the industrial foods company was aware that since its fundamental design was an activity-client one, the key vulnerability was from low-cost competitors who might undercut prices to secure price-sensitive clients. This is why the firm had to select continuous cost reduction as a key strategic imperative, and why the teams of marketing, development, and production managers were created to pursue cost reductions on a *continuous* basis.

What Support Systems Are Needed?

The grouping and linking decisions that shape the organization's formal structure still do not adequately address the requirements for a complete design. Nadler et al. and Peters and Waterman among others point out the need for additional, highly interrelated support systems and the need for congruence between these components if the organization is to perform effectively.[13]

Have necessary support systems been identified? If the above questions have been answered adequately, it now is possible to use the more traditional approaches to organization design to plan the necessary support systems. The idea is to design detailed planning, control, and reward systems focused on the pursuit of the strategic accomplishments and the ideology established in earlier stages.

What new skills will be needed to produce the accomplishments? It is critically important to identify the changes in skill mix that may be required if the strategy is to be accomplished since there is usually a substantial lag between recognition of a required change in skill mix and its accomplishment. Furthermore, the required skill mix may extend well beyond the traditional boundaries of the business: During its aggressive build phase in the 1950s and 1960s, IBM was astute enough to recognize that it was vital to develop the programming skills of potential clients. It therefore spent millions of dollars educating and training its clients' employees worldwide. To accomplish this, IBM's own capabilities to devise and deliver such training required development. The result was a massive internal program to train programming instructors and software applications instructors.

The management attitudes and skills needed to carry out a particular strategic role may also call for a change in the human resource management systems. Selection, promotion, and compensation guidelines must be designed and installed to encourage the development of the needed attitudes and skills. Without systems that support the strategic role, traditional American business values such as the tendency to maximize short-term profits and minimize risk, will quite naturally take precedence, whether or not they suit the role.[14]

If the strategy is to be effective, it also may be important to recognize the skill requirements of suppliers and distributors. For decades, the large U.S. auto manufacturers had a formidable advantage over foreign competition because of their support systems of dealers who offered service and repair capabilities. It was only by designing highly reliable vehicles that Japanese auto producers considerably reduced the need for such service skills, thus defusing the support system advantages of domestic producers.

To get a better idea of the importance to implementation of required change in the skill mix, it is useful to ask what the impact on strategic accomplishment will be if there is a shortfall in the skill categories needed to support a particular strategy. One electrical equipment manufacturer estimated that a 10-percent shortfall in trained repair technicians would be enough both to damage severely its service reputation and virtually to destroy a proposed aggressive maintenance strategy based on distribution and service leadership. This discovery precipitated a major reallocation of senior management attention to internal technical training and a substantial redirection of corporate resources and support for technical training in regional public school systems.

EVALUATION

Although the design will have been tested throughout its inception against the design imperatives, it is critical that it be subjected to one more test before it is implemented. This last test will ensure that all functions have been defined, all critical interfaces identified and resolved, and all possible design imperatives met.

Will It Work?

One way to ensure that the design will be effective is to simulate the flow of major inputs or triggering situations through the new organization. This is the best way to ensure that all contingencies will be dealt with in the new design.

Have key events been simulated? By simulation we mean *walking through* the organization major events that it must be able to handle if its strategy is to be executed. This is in essence a reality check, an attempt to answer the question Will it work? This analysis must be done in detail if the design is to implement the strategy.

The first step is to identify and list the key challenges the organization is likely to encounter. A good place to look for the identification of these key challenges is in the organization's strategy. If, for example, the organization seeks to achieve market leadership via new product introductions, then a new product is a major event and should be walked through the proposed design to make sure that the design is set up to handle the identification, development, and introduction of new product ideas as well as the responses of distributors, customers, and competitors to its introduction.

Another important check is to identify competitive or technological triggers. For example, if a target competitor counterattacks, how will the proposed design identify and respond to this move? Once the key inputs are identified, the progress of each can be traced through each design alternative.

The simulations will uncover conflicts, redundancies, or misunderstandings about groupings, questions about how coordination and control mechanisms will function, and any missing or incomplete functions. At this point, if more than one design alternative remains, the final choice can be made by selecting the one that best meets the resulting strategic and competitive imperatives. Specifically ask: Have all the strategic and competitive imperatives been met?

IMPLEMENTATION

The last issue to address is the implementation of the design itself.

What Execution Problems Can be Anticipated?

Answers to this last set of questions reduces the organization's exposure to major implementation pitfalls to which so many of the most well-conceived strategies and designs fall prey.

Has stakeholder impact been identified and managed? The impact of the design on stakeholders cannot be ignored. Strategic accomplishment may be delayed considerably by resistance from key stakeholders or considerably facilitated by their active support. The extent to which stakeholders will be affected as the strategy is executed must be analyzed, and major threats or opportunities emanating from stakeholder reaction must be identified and managed. Specific plans for managing both positive and negative stakeholder reactions should be formulated and appropriate action taken before the first move, as well as throughout the implementation phase.

Have problem owners been assigned? A very effective mechanism, invented by IBM, has been the identification and appointment of problem owners for implementation. All major implementation problem areas identified in the prior simulation—for example, technical difficulties, motivation problems, human resource management challenges, and stakeholder resistance—are assigned to a problem owner who acknowledges and accepts the responsibility for managing this problem area. Problem owners are not necessarily given resources and authority, but they are held accountable for managing all problems that may arise in their problem area during implementation. They do have the right to veto (if they *have* to) any move that has a negative impact on their problem area. Two benefits are obtained with this mechanism: The problem owners themselves receive important developmental experience, and the organization receives the smoothest possible change implementation. It is interesting to note that in IBM the problem ownership role is eagerly sought by those executives who seek to demonstrate that they can manage their problems despite lack of formal authority and official resources, since senior management regards such skills as important indicators of promotion potential.

SUMMARY

This chapter poses the key questions that shape competitive organization design. More detailed design considerations are handled by the conventional design approaches.

Adequate answers to the above questions lead to a design that addresses the need to compete as well as to seek efficiency. There is no question that such designs are somewhat more complex and less efficient than designs that focus primarily on efficiency. However, in actuality, competitive design is concerned with effectiveness—after all, it is not essential to deliver a perfect design, but rather a design that is better than that of the best competitor.

Another way of looking at competitive design is to return to the military analogy with which we started. In designing the army to conduct the campaign, the general can ill afford to seek perfection on the parade ground. Rather, he must assemble his forces and materials to create an army superior to that of the enemy in the actual terrain. Our questions have consistently assisted managers in achieving this more modest, but pragmatic, purpose of creating competitive, if imperfect, designs. They go a long way toward delivering designs that are competitively superior—a key challenge of the 1980s.

APPENDIX 27A

DESIGN STRENGTHS AND WEAKNESSES

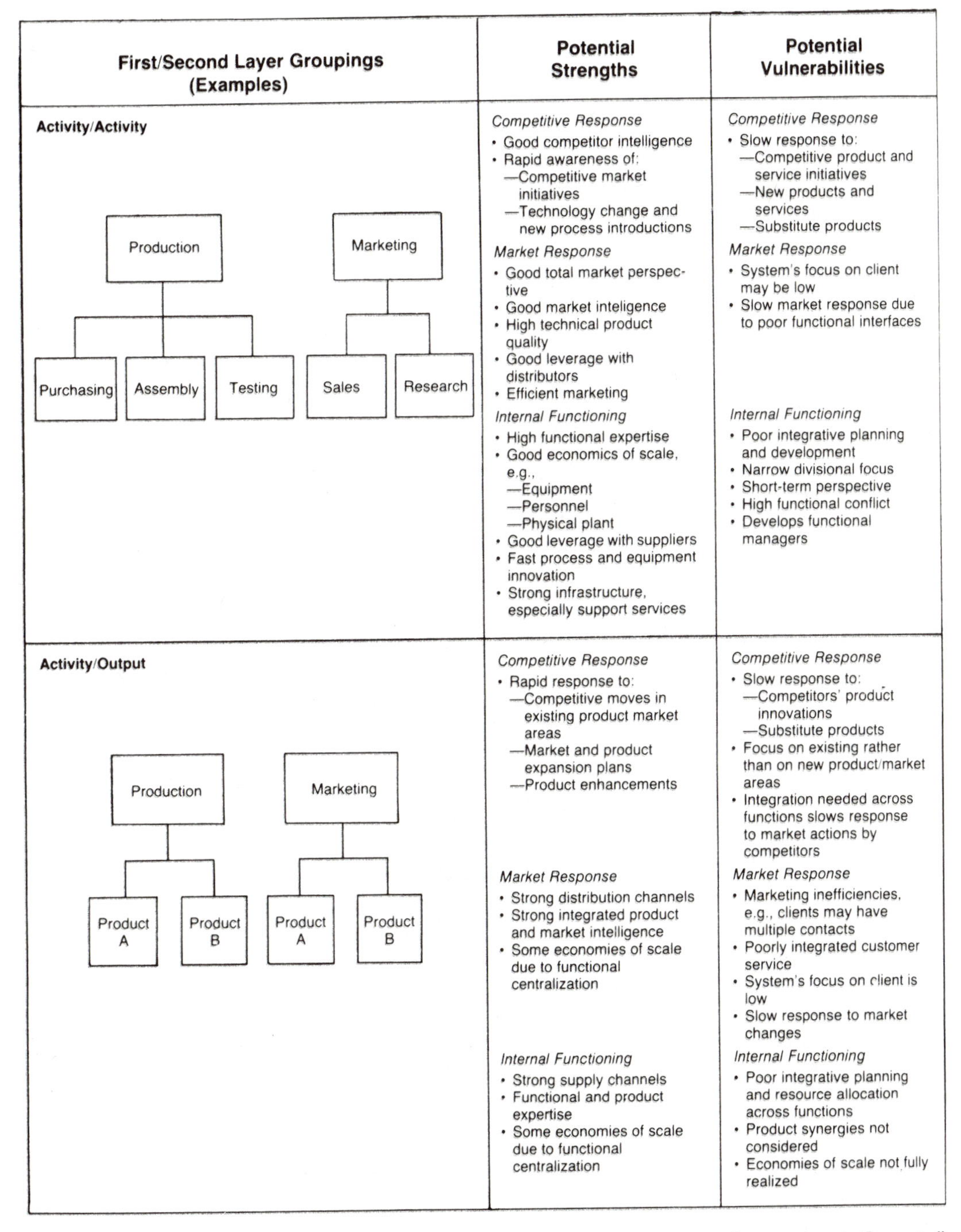

First/Second Layer Groupings (Examples)	Potential Strengths	Potential Vulnerabilities
Activity/Activity Production: Purchasing, Assembly, Testing Marketing: Sales, Research	*Competitive Response* • Good competitor intelligence • Rapid awareness of: —Competitive market initiatives —Technology change and new process introductions *Market Response* • Good total market perspective • Good market inteligence • High technical product quality • Good leverage with distributors • Efficient marketing *Internal Functioning* • High functional expertise • Good economics of scale, e.g., —Equipment —Personnel —Physical plant • Good leverage with suppliers • Fast process and equipment innovation • Strong infrastructure, especially support services	*Competitive Response* • Slow response to: —Competitive product and service initiatives —New products and services —Substitute products *Market Response* • System's focus on client may be low • Slow market response due to poor functional interfaces *Internal Functioning* • Poor integrative planning and development • Narrow divisional focus • Short-term perspective • High functional conflict • Develops functional managers
Activity/Output Production: Product A, Product B Marketing: Product A, Product B	*Competitive Response* • Rapid response to: —Competitive moves in existing product market areas —Market and product expansion plans —Product enhancements *Market Response* • Strong distribution channels • Strong integrated product and market intelligence • Some economies of scale due to functional centralization *Internal Functioning* • Strong supply channels • Functional and product expertise • Some economies of scale due to functional centralization	*Competitive Response* • Slow response to: —Competitors' product innovations —Substitute products • Focus on existing rather than on new product/market areas • Integration needed across functions slows response to market actions by competitors *Market Response* • Marketing inefficiencies, e.g., clients may have multiple contacts • Poorly integrated customer service • System's focus on client is low • Slow response to market changes *Internal Functioning* • Poor integrative planning and resource allocation across functions • Product synergies not considered • Economies of scale not fully realized

From I. C. MacMillan and P. E. Jones, "Designing Organizations to Compete," *Journal of Business Strategy* **4**(1984):22-26. Reprinted with permission.

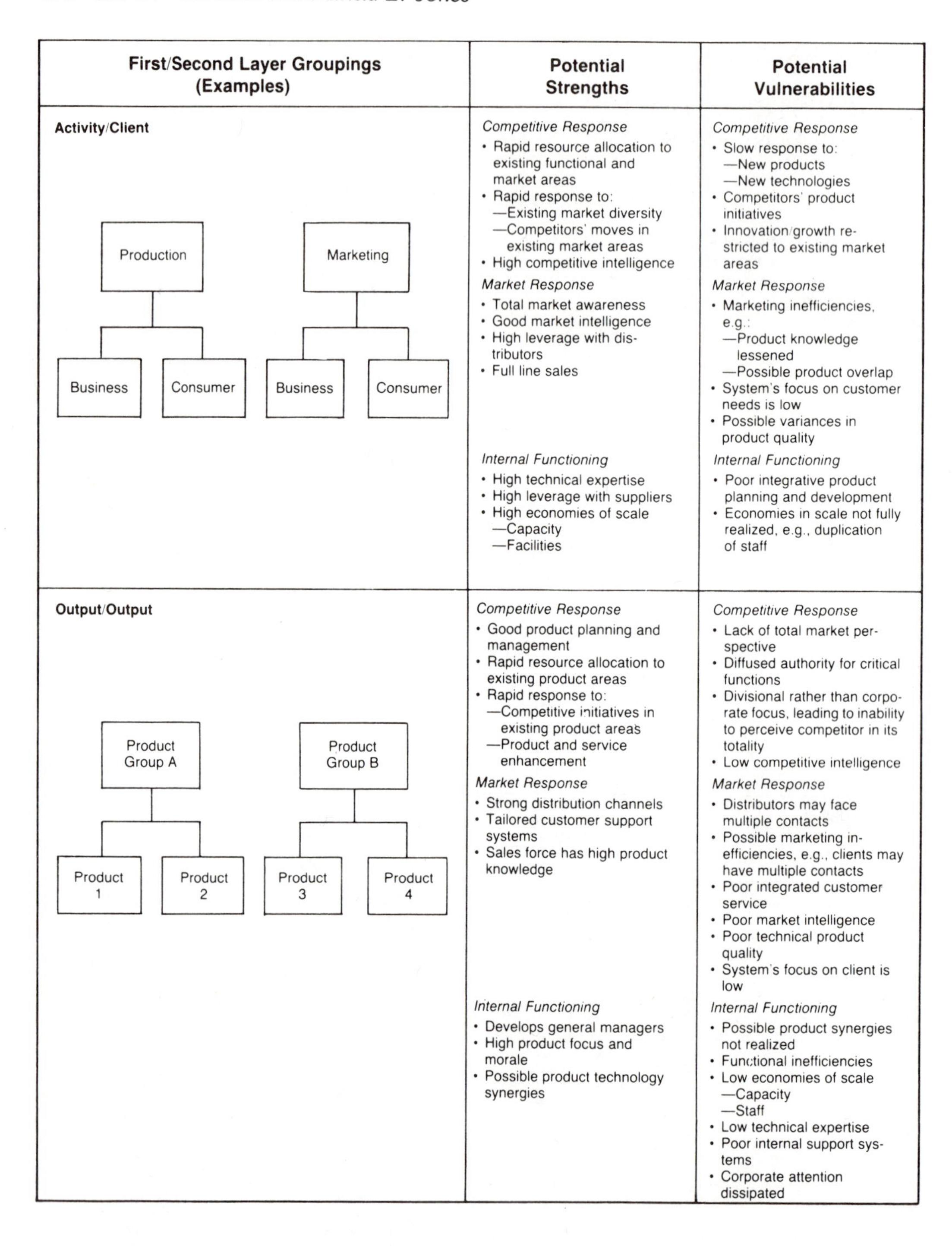

First/Second Layer Groupings (Examples)	Potential Strengths	Potential Vulnerabilities
Activity/Client Production: Business, Consumer Marketing: Business, Consumer	*Competitive Response* • Rapid resource allocation to existing functional and market areas • Rapid response to: —Existing market diversity —Competitors' moves in existing market areas • High competitive intelligence *Market Response* • Total market awareness • Good market intelligence • High leverage with distributors • Full line sales *Internal Functioning* • High technical expertise • High leverage with suppliers • High economies of scale —Capacity —Facilities	*Competitive Response* • Slow response to: —New products —New technologies • Competitors' product initiatives • Innovation/growth restricted to existing market areas *Market Response* • Marketing inefficiencies, e.g.: —Product knowledge lessened —Possible product overlap • System's focus on customer needs is low • Possible variances in product quality *Internal Functioning* • Poor integrative product planning and development • Economies in scale not fully realized, e.g., duplication of staff
Output/Output Product Group A: Product 1, Product 2 Product Group B: Product 3, Product 4	*Competitive Response* • Good product planning and management • Rapid resource allocation to existing product areas • Rapid response to: —Competitive initiatives in existing product areas —Product and service enhancement *Market Response* • Strong distribution channels • Tailored customer support systems • Sales force has high product knowledge *Internal Functioning* • Develops general managers • High product focus and morale • Possible product technology synergies	*Competitive Response* • Lack of total market perspective • Diffused authority for critical functions • Divisional rather than corporate focus, leading to inability to perceive competitor in its totality • Low competitive intelligence *Market Response* • Distributors may face multiple contacts • Possible marketing inefficiencies, e.g., clients may have multiple contacts • Poor integrated customer service • Poor market intelligence • Poor technical product quality • System's focus on client is low *Internal Functioning* • Possible product synergies not realized • Functional inefficiencies • Low economies of scale —Capacity —Staff • Low technical expertise • Poor internal support systems • Corporate attention dissipated

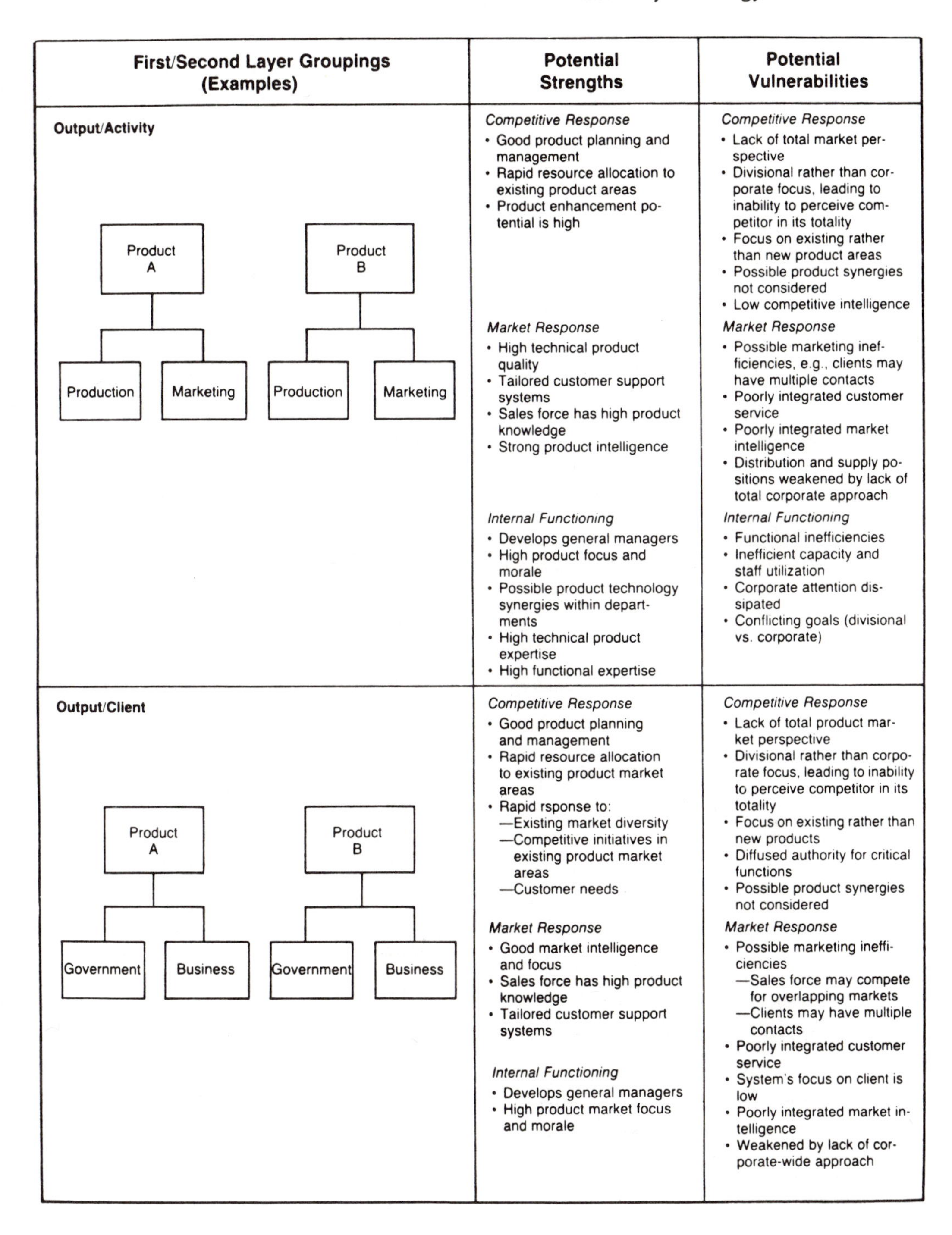

First/Second Layer Groupings (Examples)	Potential Strengths	Potential Vulnerabilities
Output/Activity Product A: Production, Marketing Product B: Production, Marketing	*Competitive Response* • Good product planning and management • Rapid resource allocation to existing product areas • Product enhancement potential is high	*Competitive Response* • Lack of total market perspective • Divisional rather than corporate focus, leading to inability to perceive competitor in its totality • Focus on existing rather than new product areas • Possible product synergies not considered • Low competitive intelligence
	Market Response • High technical product quality • Tailored customer support systems • Sales force has high product knowledge • Strong product intelligence	*Market Response* • Possible marketing inefficiencies, e.g., clients may have multiple contacts • Poorly integrated customer service • Poorly integrated market intelligence • Distribution and supply positions weakened by lack of total corporate approach
	Internal Functioning • Develops general managers • High product focus and morale • Possible product technology synergies within departments • High technical product expertise • High functional expertise	*Internal Functioning* • Functional inefficiencies • Inefficient capacity and staff utilization • Corporate attention dissipated • Conflicting goals (divisional vs. corporate)
Output/Client Product A: Government, Business Product B: Government, Business	*Competitive Response* • Good product planning and management • Rapid resource allocation to existing product market areas • Rapid rsponse to: —Existing market diversity —Competitive initiatives in existing product market areas —Customer needs	*Competitive Response* • Lack of total product market perspective • Divisional rather than corporate focus, leading to inability to perceive competitor in its totality • Focus on existing rather than new products • Diffused authority for critical functions • Possible product synergies not considered
	Market Response • Good market intelligence and focus • Sales force has high product knowledge • Tailored customer support systems	*Market Response* • Possible marketing inefficiencies —Sales force may compete for overlapping markets —Clients may have multiple contacts • Poorly integrated customer service • System's focus on client is low • Poorly integrated market intelligence • Weakened by lack of corporate-wide approach
	Internal Functioning • Develops general managers • High product market focus and morale	

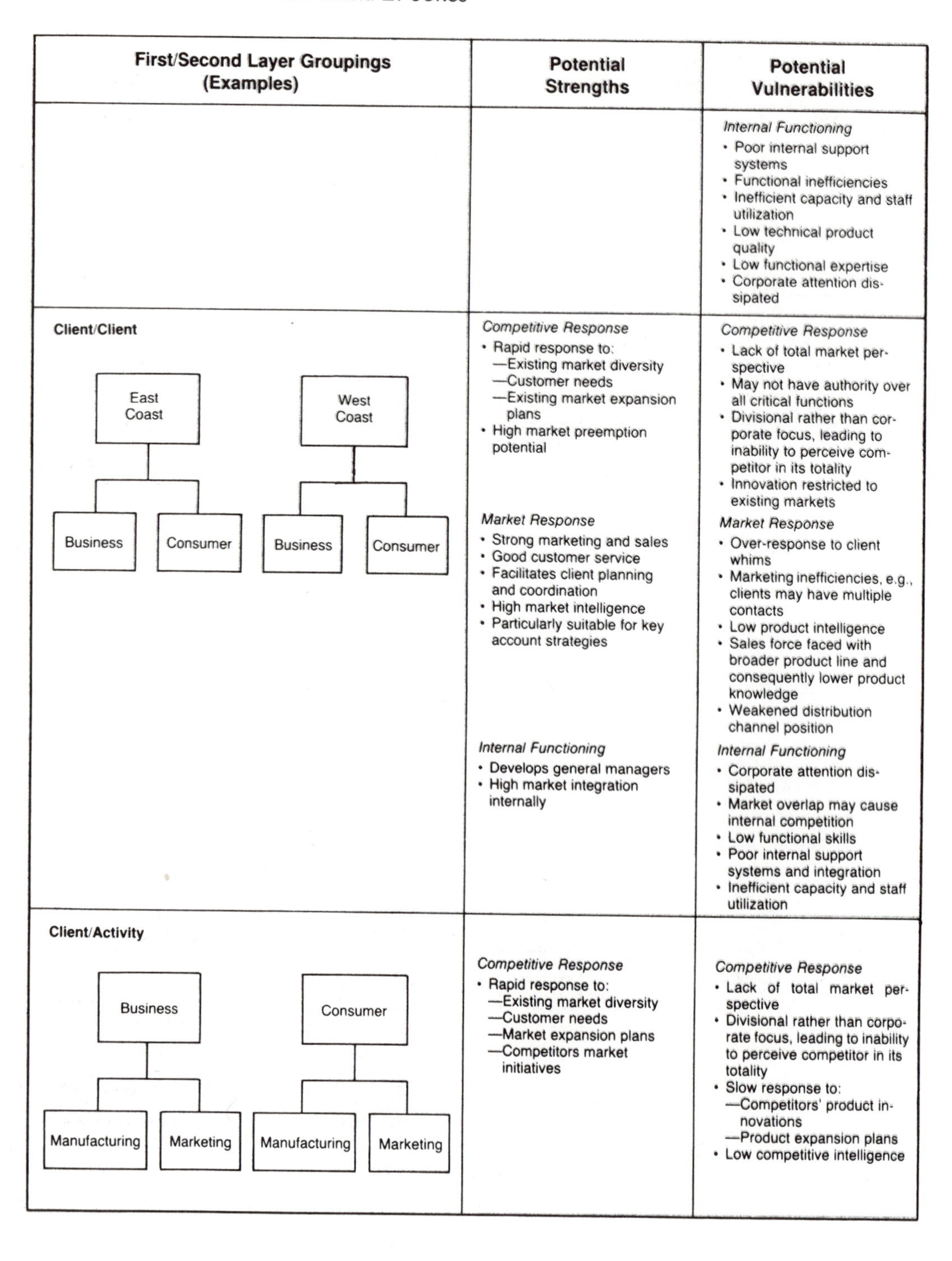

First/Second Layer Groupings (Examples)	Potential Strengths	Potential Vulnerabilities
		Internal Functioning • Poor internal support systems • Functional inefficiencies • Inefficient capacity and staff utilization • Low technical product quality • Low functional expertise • Corporate attention dissipated
Client/Client East Coast: Business, Consumer West Coast: Business, Consumer	*Competitive Response* • Rapid response to: —Existing market diversity —Customer needs —Existing market expansion plans • High market preemption potential	*Competitive Response* • Lack of total market perspective • May not have authority over all critical functions • Divisional rather than corporate focus, leading to inability to perceive competitor in its totality • Innovation restricted to existing markets
	Market Response • Strong marketing and sales • Good customer service • Facilitates client planning and coordination • High market intelligence • Particularly suitable for key account strategies	*Market Response* • Over-response to client whims • Marketing inefficiencies, e.g., clients may have multiple contacts • Low product intelligence • Sales force faced with broader product line and consequently lower product knowledge • Weakened distribution channel position
	Internal Functioning • Develops general managers • High market integration internally	*Internal Functioning* • Corporate attention dissipated • Market overlap may cause internal competition • Low functional skills • Poor internal support systems and integration • Inefficient capacity and staff utilization
Client/Activity Business: Manufacturing, Marketing Consumer: Manufacturing, Marketing	*Competitive Response* • Rapid response to: —Existing market diversity —Customer needs —Market expansion plans —Competitors market initiatives	*Competitive Response* • Lack of total market perspective • Divisional rather than corporate focus, leading to inability to perceive competitor in its totality • Slow response to: —Competitors' product innovations —Product expansion plans • Low competitive intelligence

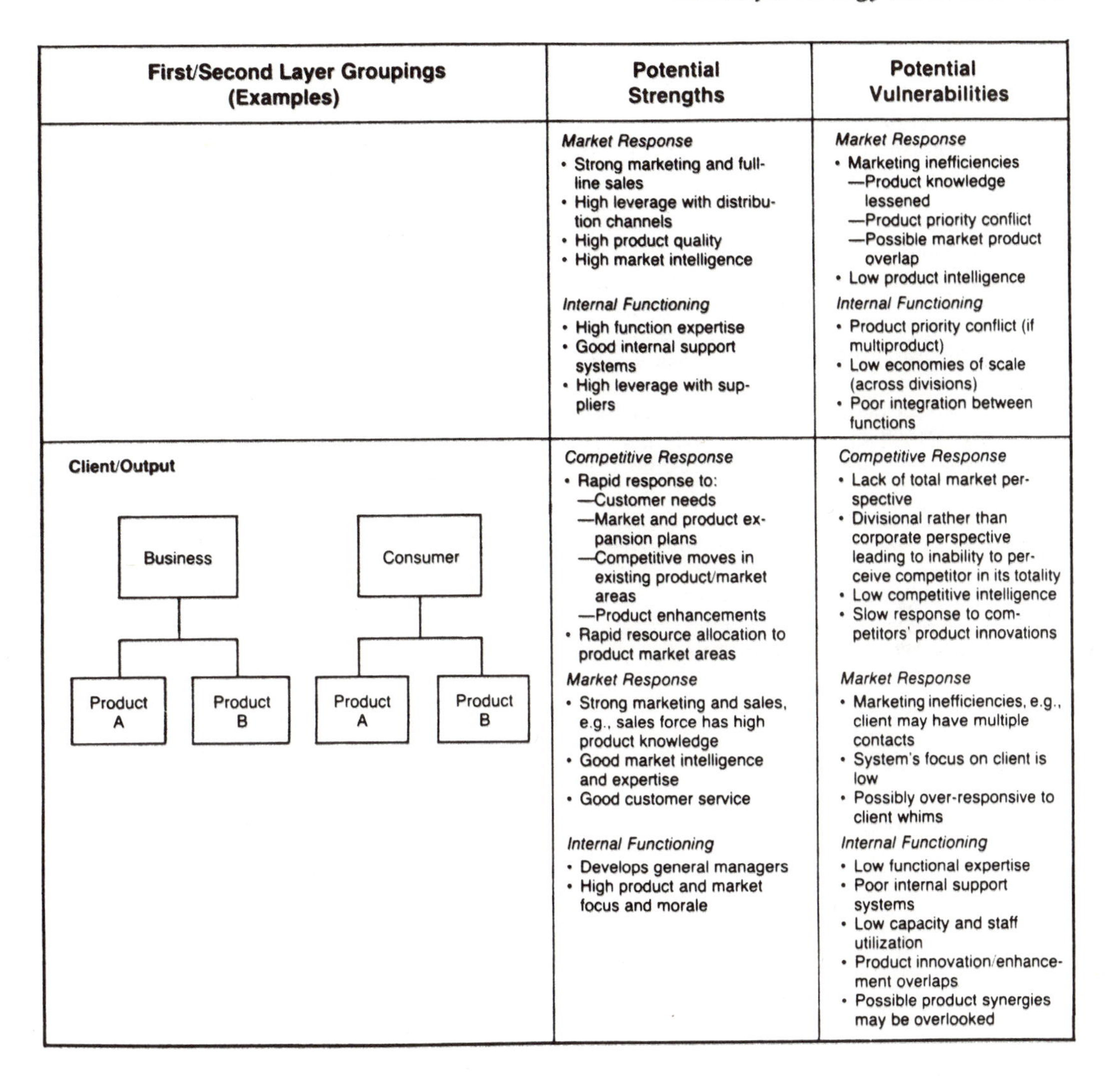

First/Second Layer Groupings (Examples)	Potential Strengths	Potential Vulnerabilities
	Market Response • Strong marketing and full-line sales • High leverage with distribution channels • High product quality • High market intelligence *Internal Functioning* • High function expertise • Good internal support systems • High leverage with suppliers	*Market Response* • Marketing inefficiencies —Product knowledge lessened —Product priority conflict —Possible market product overlap • Low product intelligence *Internal Functioning* • Product priority conflict (if multiproduct) • Low economies of scale (across divisions) • Poor integration between functions
Client/Output Business: Product A, Product B Consumer: Product A, Product B	*Competitive Response* • Rapid response to: —Customer needs —Market and product expansion plans —Competitive moves in existing product/market areas —Product enhancements • Rapid resource allocation to product market areas *Market Response* • Strong marketing and sales, e.g., sales force has high product knowledge • Good market intelligence and expertise • Good customer service *Internal Functioning* • Develops general managers • High product and market focus and morale	*Competitive Response* • Lack of total market perspective • Divisional rather than corporate perspective leading to inability to perceive competitor in its totality • Low competitive intelligence • Slow response to competitors' product innovations *Market Response* • Marketing inefficiencies, e.g., client may have multiple contacts • System's focus on client is low • Possibly over-responsive to client whims *Internal Functioning* • Low functional expertise • Poor internal support systems • Low capacity and staff utilization • Product innovation/enhancement overlaps • Possible product synergies may be overlooked

REFERENCES

1. W. Kiechel III, "Playing by the Rules of the Corporate Strategy Game," *Fortune* (24 September 1979):110-115.
2. "The Future Catches Up with a Strategic Planner," *Business Week* (27 June 1983):62.
3. J. R. Galbraith and D. A. Nathanson, *Strategy Implementation: The Role of Structure and Process* (St. Paul, Minn.: West Publishing, 1978); D. A. Nadler, J. R. Hackman, and E. E. Lawler III, *Managing Organizational Behavior* (Boston: Little, Brown, 1979).
4. W. E. Rothschild, *Putting It All Together* (New York: AMACOM, 1976); W. E. Rothschild, *Strategic Alternatives* (New York: AMACOM, 1979).
5. I. C. MacMillan, "Seizing the Competitive Initiative," *Journal of Business Strategy* **2**(1982): 43-51.
6. Rothschild, *Strategic Alternatives.*
7. T. J. Peters and R. H. Waterman, *In Search of Excellence: Lessons from America's Best-Run Companies* (New York: Harper & Row, 1982).
8. I. C. MacMillan, "Corporate Ideology and Strategic Delegation," *Journal of Business Strategy* **3**(1983):71-76.
9. MacMillan, "Seizing the Competitive Initiative."
10. H. Mintzberg, *The Structuring of Organizations* (Englewood Cliffs, N.J.: Prentice-Hall, 1979).
11. J. R. Galbraith, *Designing Complex Organizations* (Reading, Mass.: Addison-Wesley, 1973).
12. MacMillan, "Corporate Ideology and Strategic Delegation"; Galbraith, *Designing Complex Organizations.*
13. Nadler et al., *Managing Organizational Behavior;* Peters and Waterman, *In Search of Excellence.*
14. R. H. Hayes and W. J. Abernathy, "Managing Our Way to Economic Decline," *Harvard Business Review* **58**(1980):67-77.

CHAPTER 28

International Business Development and Administration: An Organizational Perspective

William H. Davidson

William H. Davidson is an associate professor of management and organization at the University of Southern California. Formerly on the faculty of the Amos Tuck School (Dartmouth College), and the University of Virginia, he also has been a visiting professor at INSEAD in Fontainbleau, France. He holds an A.B. in economics from Harvard College, where he was a full-scholarship student, and an M.B.A. and D.B.A. from Harvard Business School. Dr. Davidson is presently serving as a consultant to the Canadian government and to several firms in the information technology sector.

The management of international business encompasses a variety of skills and activities, but all the challenges of international management are encompassed in two broad responsibilities: business development and business administration. Although they seem inseparable, these two functions are distinct in terms of objectives and activities. It is possible to distinguish periods of time in any company's history when one of these two functions dominates the strategy, organization, and management of the firm, often at the expense of poor performance in the other dimension.

Issues of international business administration, not business development, dominate the current concerns of most U.S. corporations. The era of initial international expansion for U.S. corporations has largely run its course. The leading firms in the move abroad possess far-flung networks of established foreign affiliates. For these firms, international expansion is no longer the primary management objective. A new set of issues and challenges has come to the forefront. The key issue for most firms today is not in building but in running a successful global enterprise. Efforts to meet this challenge typically focus on two key undertakings.

The first broad thrust focuses on the integration of international operations. The goal of such efforts is to increase operating efficiency through the consolidation and coordination of manufacturing, marketing, engineering, purchasing, and other functions on a global scale. Such efforts typically involve a rationalization of physical facilities and operations. They also frequently involve a less tangible shift from what has been called a multidomestic corporate orientation to an integrated global posture. Although the transition is more readily perceived in operating areas, both of these shifts hold important organizational implications.

The second broad managerial thrust focuses on the elimination of marginal product lines and affiliates to improve the overall profitability of international operations. This effort is manifested in the sale or closure of lines of business, facilities, and affiliates. Part of this retrenchment stems from the consolidation of manufacturing activities, but the primary force behind divestment activity has been the elimination of nonperforming units. International business portfolios have been pruned extensively in many cases as part of a general reassessment of international activities. The magnitude of this consolidation movement can be seen in the aggregate statistics for U.S. foreign direct investment.

INTERNATIONAL CONSOLIDATION

Foreign investment by U.S. corporations grew at an annual rate of 11 percent during the 1960s and 1970s. In 1982, for the first time, U.S. companies actually reduced their level of investment in foreign affiliates (Table 28-1). These aggregate statistics reflect a fundamental shift in U.S. corporations' strategies for world markets. The decline in investment masks the fact that many new U.S. companies continue to enter world markets and build networks of affiliates. Increases in foreign investment by such firms were more than offset by divestments in established firms. This dramatic reversal is due to a number of developments in the external environment of world business and in management philosophies and practices of U.S. multinational corporations.

NEW ENVIRONMENTAL PRESSURES

Externally, the world business environment has become more difficult. Among the many factors cited by managers of multinational companies as reasons for reduced international activity, several seem particularly important. Heightened investment controls restrict access to many markets. Extensive restrictions on equity investment have been imposed by a range of host countries. Japan, India, and the Andean Pact countries were among the first to erect such barriers to foreign direct investment, but such barriers now exist in a wide range of host countries, including Mexico, Brazil, Indonesia, and many others. Even Canada, the largest recipient of U.S. investment, created a barrier to investment in the form of FIRA, the Foreign Investment Review Agency, which has powers to review all proposed and existing foreign investments in Canada.

Investment controls reflect but one dimension of increased host country pressures on multinational companies. Host governments have become increasingly sensitive to the costs and benefits of foreign direct investment

Table 28-1. Trends in U.S. Foreign Investment ($ billion)

	U.S. Foreign Direct Investment								
	1960	*1965*	*1970*	*1975*	*1980*	*1981*	*1982*	*1983*	*1984(e)*
Total	32.7	49.2	75.5	124.2	215.4	226.4-	221.3	226.1	229.0
Manufacturing	11.1	19.2	31.1	55.9	89.2	92.4	90.6	90.1	—
Petroleum	10.9	15.3	19.7	26.2	51.2	51.2	55.7	56.6	—
Other	10.7	14.7	24.7	42.1	78.5	82.7	75.0	74.2	—

Source: Survey of Current Business, U.S. Department of Commerce

and increasingly sophisticated in their control of new investment activity. Many also have initiated efforts to restructure the activities of existing affiliates. In either case, such efforts can significantly reduce the profitability and attractiveness of new or existing foreign investments. In the extreme, pressures on existing affiliates can lead to expropriation or forced divestment. Such political pressures contribute to divestment activity and discourage foreign investment.[1]

Increased competition from indigenous firms—and from new global rivals—also reduces the attractiveness of foreign operations and investment. In virtually every sector of the world economy, the number of firms competing for a share of foreign markets has increased substantially in recent years. This fact reinforces host governments' abilities to extract a higher price for market access because of the larger number of qualified suppliers contending for the market. Even without host government intervention, however, rising competition reduces the attractiveness of ongoing foreign operations and new foreign investment opportunities.

PORTFOLIO BUSINESS PLANNING PRESSURES

These external developments coincided with the widespread adoption of new business planning techniques in U.S industry. These techniques focused on the review of individual business units within a company, primarily for investment and divestment decision purposes. The best-known of these tools, the Boston Consulting Group (BCG) portfolio planning framework, appraises business units in terms of their market growth and market share characteristics within this framework: high-growth, high-share businesses received priority in resource allocation and low-growth, low-share businesses were prime candidates for divestiture.

The application of this framework holds several important implications for international activities. The framework is generally applied to business units that are defined in terms of product lines or divisions. A product line that is growing slowly in the United States may be fast-growing or have great growth potential abroad. Strategies for such business units derived from portfolio planning frameworks may not place adequate emphasis on international growth rates or opportunities.[2]

This problem can be addressed, of course, by examining geographic subunits for each product line or division. Each subunit can then receive the appropriate strategic emphasis rather than applying a uniform approach to the business unit for all world markets. However, it is difficult to establish and maintain a balance between a single, global mission for a business and a set of varied missions for geographic subunits. Even if this refinement can be achieved, the application of these planning techniques influences domestic investment decisions regarding R & D and staffing that may reduce support for international operations in otherwise attractive markets.

The addition of geographic considerations to the planning framework may influence the choice of host country investment sites. Host country growth rates have always been linked to U.S. foreign investment activity, but the use of formal planning processes greatly reinforces that relationship. U.S. direct investment has in fact been rising rapidly in the faster-growing countries of Latin America and Asia.

Stopford and Dunning report that

> The share of U.S. foreign direct investment capital steered toward developing countries steadily declined until the mid-1970s. Since the mid-1970s, the share of newer investment going to the developing countries has increased. Between 1974 and 1980, for example, 36.4% of new U.S. foreign direct investment occurred in developing countries compared with 18.7% in the preceding seven years.[3]

It is interesting to note that the share of foreign direct investment committed to developing countries by British and Japanese corporations declined substantially over the same time period.[4] Portfolio planning appears to have had an impact on foreign investment activity.

DIVESTMENT

The most immediate result of planning activities appears to be not so much a redirection of investment flows, however, but an increase in divestment activity. Portfolio planning is used widely to help weed out losers, and many firms engage in widespread divestitures following adoption of this approach. If adoption of portfolio planning resulted in a similar effect for international activities, high divestment rates can be expected in low-growth host markets. Divestment can be expected to be particularly concentrated in low-growth markets with many competitors, where market share is widely dispersed.

The primary candidates for disinvestment might in fact appear in older, mature markets such as Canada and the United Kingdom. These markets have low growth rates and a very large base of U.S. affiliates. U.S. firms invariably have started their foreign expansion thrusts in Canada and the United Kingdom. These countries were attractive because of common language, culture, economics, tastes, and demand. Products that were successful in the United States could be marketed there without much adaptation, and consumer response was relatively predictable. In fact a study of the international expansion sequences of over three thousand U.S. product lines revealed that over three-fourths were introduced initially in either Canada or the United Kingdom.[5] However, as the first market in the international introduction sequence, these markets were also the first to mature, and they attracted far more foreign direct investment than other countries, which meant more competition for market share. Many affiliates in these countries would be identified as divestiture candidates in portfolio planning exercises.

In fact, a recent study shows that divestment has been particularly intense in Canada.[6] The study examined 396 U.S manufacturing affiliates of *Fortune 500* parent corporations in Canada. These affiliates represented about half of total U.S. direct investment in Canada, and their experience can be assumed to reflect the mainstream of U.S. corporations' activities. Each of these affiliates was active and profitable in 1974. By 1984, 46 of these 396 affiliates had been liquidated, 91 had been sold and 34 had experienced reorganizations involving the termination of major product lines. One-third of this affiliate group had been divested by their U.S. parent corporations.

Among those who remained in existence, extensive consolidations were common. The recent experience of General Electric in Canada provides insights into the current concerns and initiatives of U.S. multinational corporations. Canadian General Electric, a billion dollar affiliate, sold nine business units between 1978 and 1983. *Business International* described GE's recent steps in Canada as an example of how to cope with the planning challenges of "mature firms in crowded markets . . . with significantly lower projected rates of growth."[7] According to *Business International,*

> Every CGE product has been examined to determine its potential for market leadership. Businesses with . . . low potential . . . have been sold. Other businesses have been rationalized on a global basis. Since 1980 management has rationalized the company's housewares line, eliminating 22 items and retaining 7 with global product mandates from the parent company. For example, in 1978, CGE manufactured 30,000 skillets for the Canadian market; last year 250,000 were produced, mostly for U.S. consumers. Stronger production and automation has also helped raise CGE's share of the Canadian market from 10% to a position where it outsells all its competitors combined.[8]

It is interesting to note that in GE's case, funds raised from divestments represented only one-quarter of capital outlays in Canada during this period. The consolidation of GE's operations did not result in net divestment as measured in financial terms. A large proportion of the other parent companies in this sample, of course, experienced a significant reduction or termination of their investment position in Canada.

The experience of these U.S. corporations in Canada reflects the impact of new external pressures and new internal planning frameworks. In the most fundamental sense, however, their experience reflects a broad shift in the international strategies of these parent corporations. Reassessment, pruning, and consolidation of international operations has moved into the mainstream of corporate strategies. Rising divestment activity represents a primary result of this movement. A second primary result appears in the rationalization of existing operations.

RATIONALIZING INTERNATIONAL OPERATIONS

At the same time that many firms were divesting businesses highlighted by planning exercises, efforts to rationalize international operations became increasingly important. The primary focus of these efforts has fallen on manufacturing activities. Efforts to achieve cost reduction through integrated, global-scale production has driven many firms to restructure their manufacturing activities. Rather than serving market requirements through local manufacturing sites in each country, world-scale facilities are developed to serve a set of markets.

General Electric's recent activities in Canada provide a useful example. As mentioned above, GE Canada eliminated local production of 22 houseware products, but retained production responsibility for 7 items, for which the affiliate assumed *global product mandates*. A global product mandate refers to expanded manufacturing responsibility for a given line. For example, GE Canada increased its production of electric skillets ninefold in order to serve not only the Canadian market but the U.S. and other markets as well. The 22 items no longer produced in Canada may still be sold there, but will be supplied by large-scale facilities in other countries.

One of the most advanced examples of a rationalized production system can be seen at IBM Corporation. IBM's international manufacturing facilities are highly specialized; each typically produces a small number of products in high volumes for shipment to IBM's national marketing organizations around the world. Ford's world car project is another well-known example of such efforts. Engines for the car were produced in a single world-scale facility in Brazil. Other parts were produced in large-scale facilities in various countries and shipped to assembly points around the world. Under this approach, existing low-volume, multiproduct facilities would be phased out or specialized to manufacture a limited number of components or products in large volumes.

Rationalization strategies have become an important theme in many multinational corporations. The goal of these strategies is cost reduction, an essential objective in an increasingly competitive world. In order to achieve the goal, however, significant organizational changes are required. Efforts to reduce costs through rationalization programs carry important organizational implications.

In order to achieve real cost reduction, functions such as capacity management, purchasing, and logistics need to be overhauled and managed centrally. Extensive transshipment of products and components between affiliates characterize this approach to operations, necessitating extensive coordination within the network of affiliates. Requirements for management of these and other activities on a global, systemic basis create pressures for reorganization of responsibilities within the parent corporation. Increased central responsibility and authority can be viewed as critical if firms are to achieve the desired benefits of cost reduction. The new strategies of consolidation and rationalization have stimulated a parallel shift in the organization of U. S. multinational corporations.

REORGANIZATION: THE GLOBAL IMPERATIVE

In the initial phase of international expansion, managerial autonomy and entrepreneurship often leads to an organization based on highly independent international units. In some cases, extensive strategic independence can exist at the affiliate level. Affiliate managers enjoy great discretion in key business decision areas. Minimal controls and policy constraints permit a fast, flexible, and responsive style of management. This orientation supports sales growth and business development, but it also can lead in many cases to problems of product proliferation, poor capacity management, inconsistent and overlapping marketing efforts, and ultimately to poor financial performance.

As autonomous country managers introduce new products and adapt existing lines to local conditions, the company's overall product line may expand dramatically, resulting in higher inventory carrying costs, shorter production runs, and higher unit costs. As independent local managers add or expand local manufacturing capacity for these products, further inefficiencies can be built into the firm's operating base. A fragmented manufacturing network, characterized by short runs of many products in many facilities, contributes to a noncompetitive cost position. Independent marketing efforts can result in problems of internal competition between sister affiliates and in inconsistent offerings to global customers. All these problems eventually affect the firm's financial performance.

International expansion need not lead to problems of this sort, but it frequently if not universally does. The linkages between the strategy of international expansion and negative performance are rooted in organizational choices. Organization here refers not only to the structure, but also the systems, and culture established by the firm. In creating an international organization, choices regarding the extent and use of management information and control systems are important. The prevailing decision environment established by the firm also is critical. Who makes key strategic decisions in the firm? Decentralized decision environments can enhance the speed and market responsiveness of operating units but can pose problems in other respects.

In many firms, the expansionary phase is driven by an organization with limited management systems, a decentralized structure, and an entrepreneurial culture. With a firm commitment to international expansion, a permissive organization, and entrepreneurial management, many of the problems cited above can emerge. Figure 28-1 presents the vicious cycle that can attend international expansion.

In the scenario presented in the figure, the initial sales growth generated from international activities may be followed by a deterioration of costs, margins, and returns on investment. These problems, combined with concerns about the ability of foreign affiliates to cope adequately with the unique requirements of diverse product lines and the need to compete on a coordinated basis against global rivals, has led many firms to attempt to reorganize their international operations. Although a variety of approaches can be observed, the primary thrust of the reorganization phenomenon has been the displacement of independent international divisions by new global product divisions. The motives for adoption of global product divisions include desires for increased control and efficiency in capacity management, sourcing, manufacturing, product development, and marketing.

In addition, global product divisions are thought to offer benefits in improved technology transfer from product divisions to foreign affiliates, more consistent marketing efforts, and better responsiveness to global competitive conditions. Adoption of global product divisions is expected to initiate the virtuous cycle presented in Figure 28-2.

Many firms that have chosen to emphasize more efficient operations and business administration, as opposed to business development, have moved to a global product division format. Among the largest 180 U.S. multinational firms, at least 107 had moved

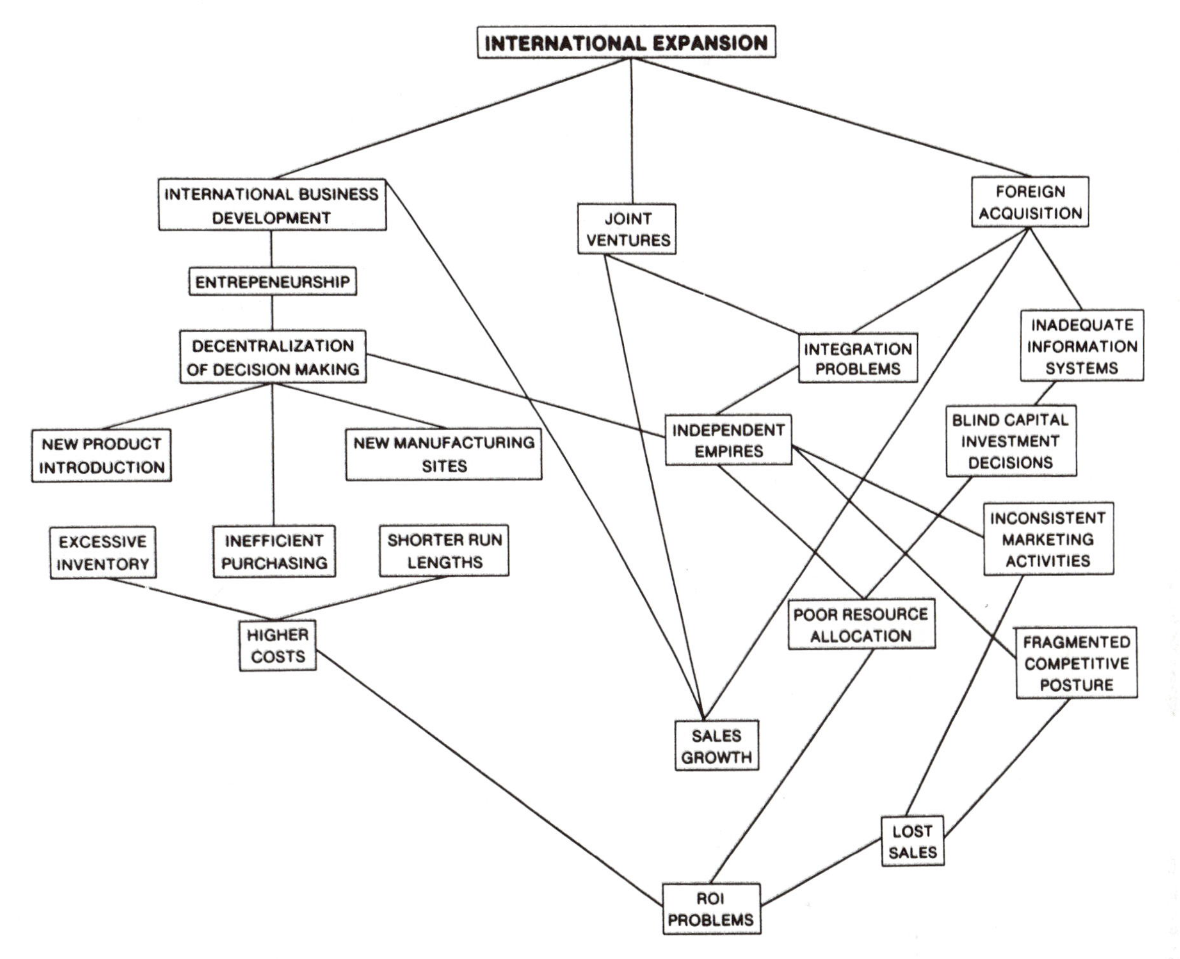

Figure 28-1. Decentralized organizational structure.

to the format by 1985. A significant shift to this form of global structure continues to dominate management efforts to improve international performance. However, the expected results of the shift generally have not been forthcoming.

COST EFFICIENCY

Companies universally cite improved cost efficiency as the global product structure's single most important advantage. Centralization of capacity and sourcing management creates a significant impetus for cost reduction. Yet this improved cost efficiency has a price. To the extent that it results in single-plant sourcing, it makes a company more vulnerable to such risks as transportation problems and labor disputes. In moving from an international division organization to a global product structure, for example, International Harvester created a new division that centralized worldwide manufacturing for components such as axles, engines, and transmissions in single plants. When the UAW struck the company in 1979 over the issue of voluntary overtime, the company was pushed to the brink of bankruptcy.

The strategy also runs afoul of important restrictions imposed by host governments to solve balance-of-payments problems. Competitors who are willing to produce within foreign markets can often displace companies

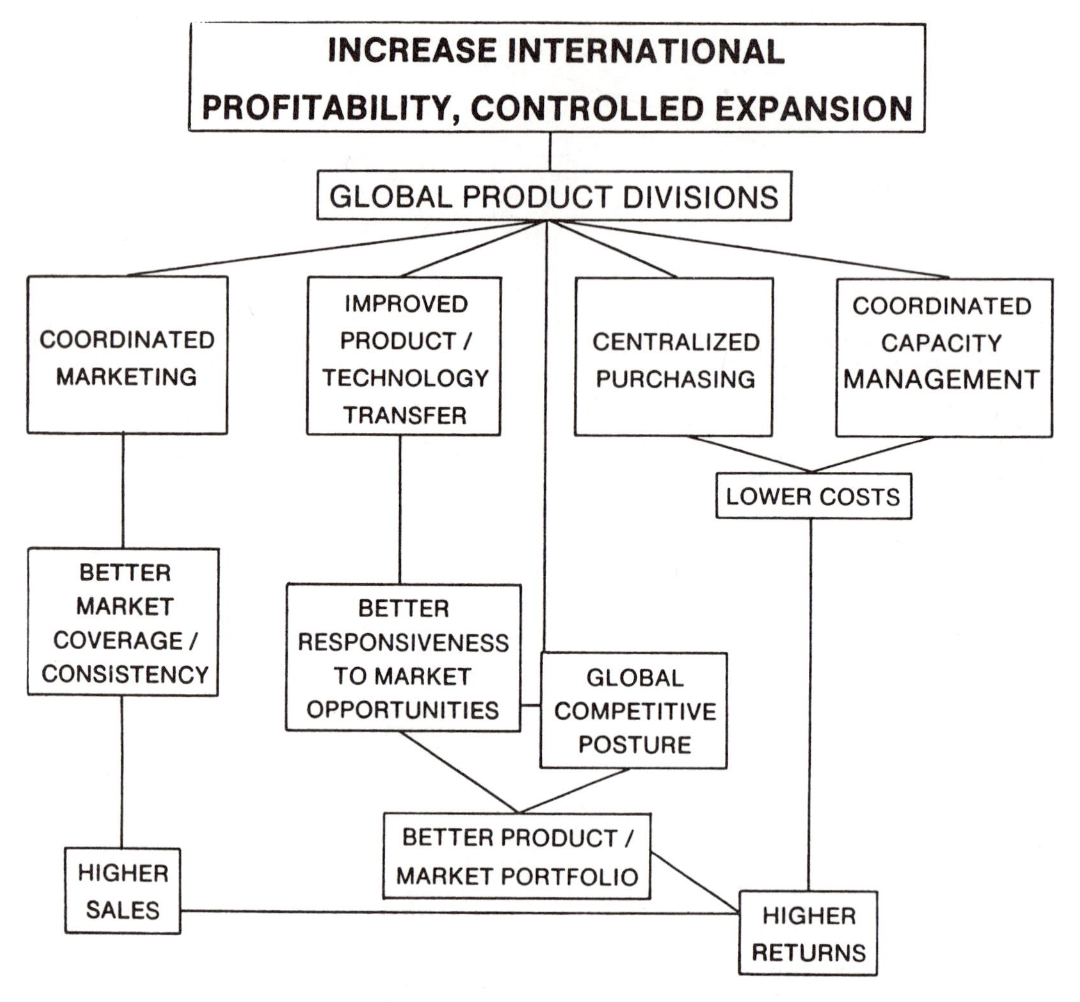

Figure 28-2. Organization by global product divisions.

that serve markets by export from centralized plants. Centralized manufacturing may be an inferior approach in a world of rising protectionism.

Volatile foreign exchange rates also disrupt centralized sourcing systems. Sales and profits can be affected dramatically by changes in exchange rates, causing pricing, planning, and financial problems for local managers. S. C. Johnson & Son, Inc., which used a central sourcing strategy in Europe, lost ground to competitors who sourced locally when exchange rates began to float in the 1970s.

Most important, the benefits, when they do appear, may not last long. Rationalization provides an optimal production configuration only in a static market. Market shifts can make existing production facilities outmoded. Philip Caldwell, CEO of Ford Motor Co., admitted that the company's shift to single-model assembly lines for world markets seriously reduced its ability to adapt to fluctuations in demand.[9] Changes in the geographic composition of the demand can also reduce the benefits of centralized sourcing.

TRANSFER OF RESOURCES

Most companies assume that the shift to a global product structure removes the international division as a barrier to the transfer of technology to foreign markets. They contend that global business units will facilitate direct communication between the domestic source and foreign users of the new technology, resulting in better identification of opportunities and faster execution of projects.

Improvement of technology transfer skills is an important objective for many multinational companies since their international performance is closely linked to the exploitation of proprietary technologies. Many observers believe that adoption of the global product division structure will enhance a firm's ability to identify and execute technology transfer opportunities. Adoption of this structure should result in faster and greater transfer activity levels. Parent companies in this structure might be expected to exhibit increased levels of foreign investment activity as well since investment is directly linked to technology transfer activity.

My study of 57 large U.S. multinationals found that technologies (as embodied in new products) in fact spread abroad more slowly in companies organized along worldwide product lines.[10] This issue was examined by first identifying all commercially and technically significant new products introduced by these 57 companies between 1945 and 1978. Each case of foreign technology transfer was recorded for this sample of products. With this data, it was possible to measure the speed and scope of technology transfer for individual products and companies. One of the most important findings of this study revealed that the speed and scope of international technology transfer varied dramatically depending on the organizational structure of the parent corporation. Table 28-2 classifies the sample of 954 new products by their speed and level of foreign introduction for companies in different structures.

Companies organized along global product lines transferred fewer new products abroad than companies in other structures. That result is unexpected, and the magnitude of the differential in speed and level of activity is striking. Products introduced by companies organized along global product lines spread abroad even more slowly and infrequently than products in companies without any formal international management structure. More importantly, when viewed individually, companies moving to the global product structure show a significant decline in the number of product transfers made after the transition (Table 28-3). These firms were more active in transferring technologies abroad *before* the transition to the global product division structure. This reduction occurred despite a pervasive tendency within this sample of firms for transfer activity to accelerate over time.

Could the reduction in transfer activity for

Table 28-2. International Transfer Patterns for New Products by Structure of Parent at Time of U.S. Introduction for the Product

Parents Organizational Structure	*Number of New Products*	*Percent First Transferred Abroad*	
		Within One Year	*Within Five Years*
Domestic only (no formal international structure	212	16.0	40.8
International division	496	19.9	45.3
Global product divisions	140	8.1	30.2
Global matrix	106	23.6	57.5
Total	954	17.7	43.5

Source: W. H. Davidson, *Experience Effects in International Investment and Technology Transfer* (Ann Arbor, Mich.: UMI Research Press, 1980), p. 57.

Table 28-3. International Transfer Patterns for New Products for 12 Companies before and after Transition to a Global Product Structure

Parent Structure	*Number of Products*	*Percent First Transferred Abroad*	
		Within One Year	*Within Five Years*
Global product	140	8.1	30.2
Previous structure	101	13.9	39.7
Total	241	10.5	34.1

Source: W. H. Davidson, *Experience Effects in International Investment and Technology Transfer* (Ann Arbor, Mich.: UMI Research Press, 1980, p. 59.

companies in the global product structure be a reflection of success in rationalizing overseas production facilities? On the surface, that interpretation might be consistent with the data. With centralized manufacturing, fewer transfers of technology would be required. However, a more detailed examination of transfer patterns suggests that other forces drive the reduction in international technology transfer activities.

It is widely believed that companies pursuing a strategy of global rationalization will be extremely averse to outside ownership in their manufacturing affiliates.[11] The benefits of outside ownership are negligible in such cases because markets for the affiliates' output are largely internal to the parent. In addition, the need for outside funding or politically motivated partnerships is low because host governments welcome export-oriented facilities with generous financial incentives. In addition, the risks of outside ownership are relatively high. Conflicts over operating policies with outside owners in a single affiliate could disrupt the entire system. Such concerns encourage the company to maintain complete ownership and control of manufacturing affiliates.

If companies organized by global product lines are in fact pursuing rationalization of production activities, outside ownership of manufacturing affiliates would not be expected. In fact, the group of companies organized along global product lines exhibits the highest level of outside ownership in the sample. Particularly striking is the rate of licensing by such companies. Table 28-4 shows that these companies use licensing in slightly more than 30 percent of all transfers, compared to only 5 percent for companies in the matrix

Table 28-4. Licensing Rates for Companies with Different Organizational Structures

Parents' Organizational Structure at Time of Transfer	*Percent of Transfers to Independent Licensees*
Domestic only	30.0
International division	24.8
Global product divisions	30.3
Global matrix	5.1
Total	25.9

Source: W. H. Davidson, *Experience Effects in International Investment and Technology Transfer* (Ann Arbor, Mich.: UMI Research Press, 1980), p. 61.

structure. Why such a dramatic difference? The answer focuses on several inherent shortcomings of the global product structure.

ORGANIZATION, ECONOMICS, AND BEHAVIOR

In most companies, the change to a global product organization has reduced significantly the scope of international commitment. Attention shifts from reliance on investment to emphasis on licensing. Licensing is viewed more favorably than other forms of international involvement by terms with global product divisions. The reason is that this organizational structure affects the valuation of foreign direct investment projects negatively in two ways: first, through its impact on the economics of these projects and, second, through a change in the decision-making environment of the managers who make international commitments.

The profitability of any foreign investment project is directly related to the costs and risks associated with the project. Different organizational structures view the costs and risks of foreign investment very differently. For example, resource sharing, in the form of joint manufacturing, materials management, marketing, or administrative overhead with other units, can significantly reduce project costs. Managerial experience, in the form of information and expertise, also reduces costs and risks. A firm's ability to realize the benefits of resource sharing and to utilize existing managerial experience depends greatly on its organizational structure. Firms with an international division possess a central repository for managerial expertise and information. The international division, which typically administers a network of regional and country managers, is particularly effective at realizing the benefits of resource sharing, at least at the host country level.

The transition to global product divisions systematically erodes the company's ability to realize these benefits. The global product structure fragments the international resources of the company previously embodied in the international division and in the foreign affiliates. Where a single country manager oversees activities at the host country level in the international division format, several independent product division managers will be active under the global product division structure. Coordination and resource sharing become increasingly difficult. Duplication of overhead begins to occur at the affiliate level, raising costs and reducing the attractiveness of foreign operations. A similar phenomenon occurs at corporate headquarters. The overhead costs previously born by the international division will now be duplicated by the global product divisions, raising the costs of international operations. These higher cost levels influence the economics of new investment projects under evaluation by divisional management, and they affect the attractiveness of ongoing operations.

The result can be lower levels of investment activity and a greater preference for the use of licensing to transfer technology abroad. Licensing becomes more attractive as the estimated return from direct investment declines. Licensing requires little or no investment, and it provides immediate financial returns. However, the short-term positive impact on divisional ROI is just that—short term. Licensing agreements typically terminate after three to six years, and the parent corporation will have foregone an opportunity for a longer-term presence in an attractive foreign market. The licensing of technology to foreign companies may also stimulate the emergence of formidable rivals in foreign markets or even in the parent corporation's domestic market.

The impact of fragmented administrative structures on the economics of foreign project evaluation is compounded by a change in the managerial setting in which foreign projects are evaluated. Foreign investment projects, like all others, are shaped largely by the managers who define, propose, and support them. Elimination of the international division can lead to a decline in managerial support for foreign projects. Experienced international personnel are often lost once the company decides to adopt a global product structure. Top international managers may leave rather than accept diminished status in the new structure.

In practice, the choice of managers for new global product divisions tends to aggravate this problem. Worldwide business unit managers are usually taken from the ranks of former domestic division managers. These new managers often are not well versed in foreign matters, but they now hold the decision-making reins. These managers, who have far less international experience than their predecessors, face far higher levels of personal uncertainty in evaluating foreign opportunities. They are less likely to be effective decision makers in international matters. These new decision makers may postpone decisions or take the path of least risk in pursuing foreign opportunities. These tendencies could result in lower levels of foreign investment and a higher reliance on licensing. There may be an important linkage between the declining rates of foreign direct investment observed in the aggregate statistics and the widespread adoption of global product division structures.

A GLOBAL COMPETITIVE ORIENTATION?

One of the arguments in favor of the global product structure that holds great attraction for most companies is its promise of a worldwide perspective on competition. In a decentralized organization driven largely by autonomous country managers, it can be difficult to observe and counterattack the advances of global rivals on a coordinated basis. The global product division structure will improve competitive monitoring and response, ultimately making the parent corporation a more vigorous coordinated competitor.

Country managers are less able to fashion the integrated response that is more natural to a worldwide business manager, but they are quick to perceive changes in the local competitive environment. Even in industries where the economics of competition are completely global, there is always a need for local adaptability. Responsiveness to local conditions and developments will be diminished under the new structure.

In addition to problems of local responsiveness, the risk is great that global business unit managers will become increasingly averse to risk in their outlook on foreign operations. Where the strategic intent is to grow and be more aggressive internationally, the worldwide product structure tends to be counterproductive for economic and behavioral reasons. Many companies lose foreign market share and see their rate of international growth slowed after the transition to global product divisions.

The foreign sales growth histories of a large sample of U.S. companies also condemns the global product structure (see Table 28-5). Of the 85 companies in the Harvard Multinational Enterprise Project sample that remained in the same structure from 1970 to 1980,[12] companies with an international division increased foreign sales by an average of 435 percent; those with a global matrix structure saw international revenues rise by 371 percent. However, the foreign sales of companies organized by global product divisions grew only 259 percent—even though they started from a lower base of sales.

The global product division format generally has not delivered its promised benefits. Not only have firms adopting this approach generally experienced a slower rate of growth in international sales, but they have not experienced higher rates of profitability as compensation.

A recent study of the international performance of 85 U.S. multinationals compared growth rates in sales and profits for firms organized along global product lines with those who had retained an independent international division. The comparison involved calculation of the percentage increases in international sales and profits between 1977 and 1983, and the percentage increases in domestic sales and profits over the same time frame. These percentages were then com-

Table 28-5. International Sales Growth for Firms in Different Organizational Structure (1970-1980)

Parent's Sales Structure	*Number of Companies*	*Average Percentage Increase in Foreign Sales 1970-1979*	*Average Foreign Sales 1970 ($ million)*
International division	34	435	$424.3
Global product division	27	259	354.7
Global matrix	24	371	600.1

Source: Company *Annual Reports*

pared to examine how fast each company's international sales and profits had grown in contrast to domestic performance. The focus here was identification of those firms who grew most rapidly in foreign markets relative to their domestic growth rates. The results showed that firms with international divisions increased international sales and profits substantially faster than their domestic results. Firms in the global product division format, by contrast, increased *domestic* sales and profits more rapidly than international results (see Table 28-6). These data reinforce earlier findings and suggest that the international division format can be effective in providing its widely recognized business development benefits without sacrificing profitability.[13]

The international division format may in fact be far from obsolete as a means of organizing international operations. This approach does have significant shortcomings, however. One of the most important issues facing American management today involves a transition to thinking about the world as a global competitive environment. In the past, many firms viewed the world in terms of the domestic market first and a separate set of international markets of varying attractiveness second. Today, it is imperative to view the world from a geocentric perspective: there is only one world in business competition. The rising tide of foreign competition in U.S. markets has greatly reinforced needs to incorporate this perspective into business planning and practice.

The international division format generally does not support the shift to a global concept of the firm's activities. The global product division format is designed to support this shift in orientation, but it may do more to inhibit than stimulate effective global activities. These concerns have led to an alternative approach to organization; one that appears to offer superior performance to its users.

AN ALTERNATIVE APPROACH

Not all firms experience the vicious cycles of international expansion and consolidation described above. These experiences are not in-

Table 28-6. International Performance and Organizational Structure

Parents' Structure	*No. of Firms*	*Foreign Sales Growth Minus Domestic Sales Growth, 1977-1982*	*Foreign Profit Growth Minus Domestic Profit Growth, 1977-82*
Global product division	31	−9.7%	−24.7%
International division	34	16.81%	24.1%

Source: W. H. Davidson, "Administrative Orientation and International Performance," *Journal of International Business Studies* **15**(1984):11-23.

evitable, although the forces underlying them affect all companies venturing abroad. These forces are particularly intense for firms with diverse business lines. In diverse companies, the basic hazards of international management are compounded by the fact that different businesses with a range of managerial requirements are being run in essentially the same manner. Yet it is clear that different types of businesses require different approaches to organization and management. In coming to grips with this issue, insights into means of coping with the underlying problems of international management can also be gained.

Insights into this issue can be gained by examining three businesses that exhibit significantly distinct characteristics; semiconductors, consumer tableware, and watches. The semiconductor industry offers highly standardized products to its customers around the world. These products are purchased and used in essentially the same way in all world markets. Consumer tableware by contrast exhibits strikingly different design, purchase, and use characteristics across markets. Preferred materials differ sharply. In some countries, tableware is expected to move easily from freezer to oven to table; in some it is ornamental in nature, in others, functional. It must be designed for dishwashers in many markets and cleaning without water in others. Purchase patterns can also differ greatly; street merchants are common in many markets, gift shops in others. The third industry, watches, exhibits a common usage but significant differences in design and purchase patterns. Preferred materials, colors, sizes, and design vary sharply from market to market, and distribution channels for watches can also be quite different across markets.

Customer and competitor bases also differ sharply in at least one key respect in these industries. In the semiconductor industry, both customers and competitors tend to be large, multinational companies. In tableware, the customers and competitors tend to be smaller, national concerns. For watches, the competitors are largely global, but the industry sells to local, national customers. These distinctions, combined with several other operational characteristics, can be used to help define the appropriate strategy and organization for these businesses, whether they exist in a single-business company or within a diversified firm.

The primary characteristics of each business can be used to gauge the appropriate organization. Table 28-7 classifies the three businesses in terms of the benefits of centralization. Semiconductors clearly require a centralized approach in order to provide consistent products, prices, quality, and delivery to global customers, to best utilize technological resources and capital-intensive production facilities, and to adopt a global view of customer and competitors. The tableware business requires decentralized, locally autonomous management to respond to unique local market conditions and customer requirements. The tableware business experiences minimal technological turnover and exhibits low capital intensity and economics of scale, permitting decentralized manufacturing capacity management.

The watch business, because of high technology content and significant economies of scale, requires centralized research and manufacturing. However, the watch business also requires some responsiveness to unique local marketing requirements. In this case the appropriate organization can be defined to include centralized engineering and manufacturing but decentralized marketing. In other words, engineering and manufacturing would be managed centrally, at division headquarters, reporting to divisional general management, and marketing would report to a local manager in each host country. In terms of an organization chart, this framework would lead to the structure shown in Figure 28-3, assuming for the moment that these three businesses were all within one company.

This approach emphasizes the unique characteristics and needs of each business to create the optimal organization. Evidence from a small sample of firms indicates this customized approach appears to offer superior performance. Diverse firms with customized ap-

Table 28-7. Industry Characteristics and Organizational Requirements

	Benefits of Centralization		
Business Characteristics	*Semiconductors*	*Tableware*	*Watches*
Marketing			
Degree of global standardization:			
Product design	High	Low	Low
Product use	High	Low	High
Buying behavior	High	Low	Low
Distribution	High	Low	Low
Nature of customer			
Global/national ratio	High	Low	Low
Nature of competition			
Global/national ratio	High	Low	High
Production and Technology			
Rate of technical change	High	Low	High
Capital intensity	High	Low	High
Minimum efficient scale	High	Low	High
Slope of cost curve	High	Low	High
Value to weight	High	Low	High

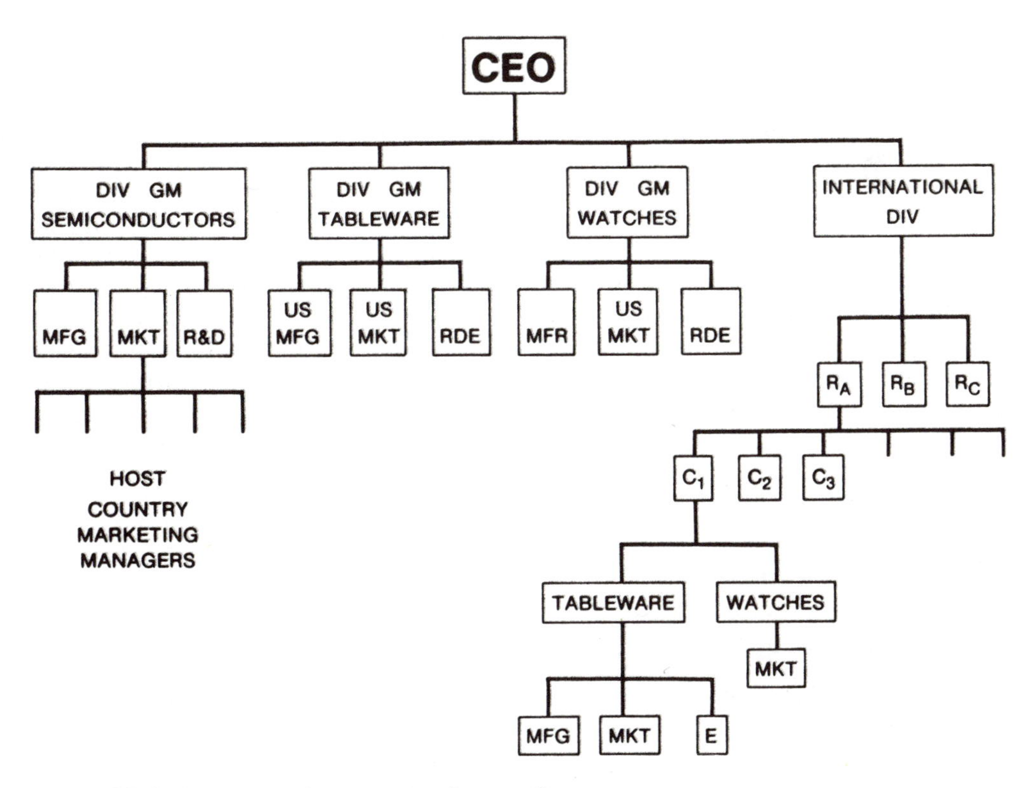

Figure 28-3. A customized organizational approach.

proaches to organization—those who adapt unique structures for each business unit as opposed to one structure for all units—exhibited significantly higher international sales and profit growth than firms in either the pure international division or global product division formats. In contrast to the firms cited in Table 28-6, firms organized along customized lines experienced relative foreign sales growth similar to firms organized along international division lines (see Table 28-8). For firms in the international division format, foreign sales grew 16.8 percent faster than domestic sales growth. The figure for firms with customized organization was virtually identical. However, the relative growth in foreign profits found among the customized firms was almost three times greater than that of the international division firms.

This approach appears to offer both international business development capability and the increased control and profitability that are the primary goals of international business administration. How are these benefits realized, and what are the mechanisms that support superior performance?

THE EFFECTIVE INTERNATIONAL ORGANIZATION

Effective international companies appear to possess an ability to adapt their organizations to the unique requirements of individual businesses, functions, and decisions. These companies can operate in a variety of modes, and they are able to identify and implement the mode that will be most effective in any given situation. This suggests, for example, that certain businesses within such companies will be managed in a highly centralized manner, whereas others are run in a decentralized fashion.

Adaptability can appear at the functional level as well. Within a given business, manufacturing management might be centralized whereas marketing is decentralized, for example. IBM manages research and engineering on a global basis, with very strong central direction. Manufacturing is managed largely on a regional basis. The marketing function, however, is dominated by very strong local, national management.

Effective international organizations differentiate administrative structures for individual businesses and functions. Many organizations even differentiate the way in which individual decisions are addressed. Within the marketing function for a given business, pricing and credit terms may be determined by headquarters, whereas packaging and advertising are directed locally, for example.

Many of the most effective international companies exhibit highly customized organizations. In addition, however, they also generally exhibit high levels of flexibility, permitting management approaches to vary not only across areas of responsibility but also over time in any given area.

The matrix structure employed at 3M provides an example of organizational flexibility. Within 3M, independent host country mar-

Table 28-8. International Performance for Firms Organized Along Custom Lines

Number of Firms	*Foreign Sales Growth Minus Domestic Sales Growth 1977-82*	*Foreign Profit Growth Minus Domestic 1977-82*
27	16.6%	71.7%

Source: Company Annual Reports

keting managers interact with global product managers in the United States to review business opportunities and establish priorities for international activities. The review process is formal, but the decision process is very informal. Either side, market or product managers, can dominate decisions regarding particular product and/or market opportunities.

This approach, at its best, can offer substantial benefits. Country managers, who may miss market opportunities because of a lack of product and technology-related expertise, will be exposed to product champions intent on pushing their products into new markets. Product managers, who may face significant uncertainties about international markets, will find ready sources of expertise in their country manager counterparts. The review process can strengthen the weak spots of all the participants, allow the most qualified participants to drive decisions, and lead to superior decisions about the direction and level of international activities.

Foreign investment decisions made in such organizations can differ substantially from those made in companies organized along global product division lines. Risk-averse decision patterns, characterized by low levels of investment activity and a reliance on licensing, can be displaced by a more aggressive pattern of decisions. The matrix structure encourages the sharing of resources and application of information to improve project economics. It also tends to reduce the role of risk-averse managers, where risk aversion is due to inexperience, by heightening the role of the experienced managers in the decision process. In a true matrix, decisions should be driven by the participants with the greatest expertise in the area in question. A flexible organization can encourage expertise to rise to the surface in all decision areas. Matrix structures formally recognize that product, country, and functional managers all play roles in shaping the company's international activities. By permitting each dimension of the organization to participate in key decision areas, firms employing such structures enhance their chances of making enlightened decisions.

Flexibility, as permitted in matrix organizations, can support superior management of international operations, but it may lead to problems of control and consistency. Such organizations are frequently criticized because of their ambiguity of authority. Decision processes are lengthened substantially because of communications requirements and the democratic nature of the decision process. Deadlocks between equal participants can last indefinitely. Many managers consider matrix management an unworkable solution to the challenges of international business.

Success with the matrix structure does not come easily. The effectiveness of this structure, as with any other, depends on other elements of the firm's strategy, organization, and management. The matrix decision process works best when all the participants share clear, common objectives. A clearly defined corporate strategy for international activities provides direction and focus for management.

The matrix structure also requires sophisticated management control systems. Mechanisms for review and guiding the decision process itself are particularly important. Corporate planning staff can be invaluable in the area of business reviews and project evaluations. Functional staff members can help guide decisions affecting ongoing operations. Performance evaluation systems must be sensitive to the roles of all the managers in the matrix. Financial results should be double-posted to the books of product and country line units. A management-by-objectives system can allow other contributions and results to be defined and measured for individual managers.

Finally, the corporate culture must support the new philosophies and practices embodied in the matrix structure. Many corporations find the matrix unworkable because it embodies principles that conflict with basic beliefs in the existing culture. Companies with strong, vertical superior-subordinate relationships find it difficult to adjust to the collegial, horizontal relationships fostered in the matrix structure.

Accountability may be lost in the transition, and the ambiguity of the matrix format may be unacceptable to managers accustomed to clearly defined reporting and responsibility relationships.

Corporate cultures are not easily changed, but they can be molded to better support international activities. Management training plays a critical role in supporting an effective international organization. Training can help develop the personal skills and sensitivities necessary to make the matrix structure work. Training efforts can also help clarify corporate strategies and objectives and present a broader view of organizational deficiencies and problems. The hiring and training of effective managers is critical to the success of any organization but particularly in matrix organizations.

In order to be effective, international organizations must mold their staff, systems, structure, and culture to support efficient, responsive international activities. These companies must constantly assess new business opportunities and aggressively pursue high-priority markets. Existing operations must be scrutinized continuously for possible improvement. Local responsiveness should be maintained by country managers while a global perspective on customers and competitors is represented by product managers.

The cost of these activities is borne every working day in the extensive communications requirements of the matrix organization and in the conflicts between different managerial perspectives inherent in this approach. These costs seem justifiable, however, in terms of performance benefits. Organizations that follow this path may realize the benefits of increased efficiency and profitability without sacrificing new business development. Such organizations avoid the self-destructive effects that plague so many companies who move from the international division format to the global product division structure. Matrix structures are expensive, complex, and unstable, but they hold great promise in the field of international management.

REFERENCES

1. R. Vernon, *Storm Over the Multinationals* (Cambridge, Mass.: Harvard University Press, 1977).
2. P. Haspeslegh, "Portfolio Planning: Uses and Limits," *Harvard Business Review* **60**(1982): 59-73.
3. J. M. Stopford and J. H. Dunning, *Multinationals: Company Performance and Global Trends* (London, Macmillan, 1983), 12.
4. Ibid.
5. W. H. Davidson, *Experience Effects in International Investment and Technology Transfer* (Ann Arbor, Mich.: UMI Research Press, 1980).
6. W. H. Davidson and D. G. McFetridge, "Recent Directions in Global Strategy," *Columbia Journal of World Business* **19** (1984):95-101.
7. "General Electric in Canada," *Business International* (10 April 1984):121-124.
8. Ibid., 123.
9. Philip Caldwell, speech at the Harvard Business School, May 1980.
10. Davidson, *Experience Effects.*
11. See J. M. Stopford and L. T. Wells, Jr., *Managing the Multinational Enterprise* (New York: Basic Books, 1972), especially 113-117.
12. For a description of this sample, see J. C. Curan et al., *Tracing the Multinationals* (Cambridge, Mass.: Ballinger, 1977).
13. W. Egelhoff, "Strategy and Structure in Multinational Corporations." Ph.D. diss., New York University, 1980.

CHAPTER 29

Implementing Business Strategies through Projects

Russell D. Archibald

Russell D. Archibald consults with clients in the United States, Europe, Asia, and Latin America in strategic management, project management, and international business development. He was formerly vice-president of international planning with The Bendix Corp. and has held management and engineering positions with ITT Corp., Booz, Allen & Hamilton, Inc., Hughes Aircraft Co., Aerojet General Corp., and the U.S. Air Force (captain, senior pilot). He is an internationally recognized authority on project management and is listed in *Who's Who in the World*.

Strategically managing a company, agency, institution, or other human enterprise requires

A *vision* of the future of the organization at the top level.
Consensus and commitment within the power structure of the organization on the mission and future direction of the organization.
Documentation of the key objectives and strategies to be employed in fulfilling the mission and moving toward the future direction.
Implementation or execution of specific programs and projects to carry out the stated strategies and reach the desired objectives.

Objectives are specific descriptions of where we want to go. *Strategies* are statements of how we will get there. Strategies are carried out and objectives are reached through the execution of programs and projects. Projects translate plans into actions and objectives into realities.

The purpose of this chapter is to convey an understanding of how objectives, strategies, programs, and projects are linked and to describe an effective way to assure the implementation of strategies through defining, authorizing, and executing projects. (Also see chapter 5.)

Many organizations have good or even excellent strategic plans produced by an otherwise effective planning process, but the plans may not actually be implemented: they sit on the shelf. The approach described here links the objectives, strategies, and projects to assure that the growth plans of the organization are actually achieved.[1]

HIERARCHY OF OBJECTIVES, STRATEGIES, AND PROGRAMS AND PROJECTS

It is important to recognize that objectives and strategies exist in a hierarchy, not just at one level, in most organizations. A useful way to describe this hierarchy is to define three levels:

Level 1: Policy.
Level 2: Strategy.
Level 3: Operational Programs and Projects.

The specific terms used here (objectives vs. goals, and so forth) are not important and can be changed to reflect the current terminology used within a given organization. Rather, the important point is to recognize the several-level hierarchy of objectives, strategies, and operational programs and projects. The precise number of levels also is not too significant: some situations produce only two levels, others may require four.

POLICY OBJECTIVES

Policy objectives define where the company (or division) wants to be in the longer-term future. These derive from and are consistant with the organization's *mission and purpose statement*. Figure 29-1 gives some typical examples.

STRATEGIC OBJECTIVES

The strategy statements at the policy level usually are so broad that they cannot be implemented without breaking them down further. In fact, these policy-level strategies usually become the *strategic objectives* at the next level down. More specific *programs and projects* are then defined to achieve each of these objectives.

Using a part of the examples in Figure 29-1 to illustrate this concept produces the result shown in Figure 29-2.

Policy Objectives	Strategies for Their Achievement
1. Assure the long-term survival of the organization.	1.1 Continue to reinvest 4 percent of sales to develop profitable new products for our traditional markets and customers.
	1.2 Acquire and retain well-qualified people at all levels of the organization.
	1.3 . . .
	1.4 . . .
2. Become a leader in the global market for our products by achieving at least a 20-percent share each in the North American, European and Asia/Pacific markets.	2.1 Increase our penetration of the European market to 20 percent through the proper combination of U.S. exports and local manufacture and/or assembly of product lines A and B.
	2.2 Enter the Japanese market through acquisition of, or a joint venture with, a Japanese company in our field.
	2.3 . . .
	2.4 . . .
	3.1 . . .
3. . . .	3.2 . . .

Figure 29-1. Examples of policy objectives and strategies.

Strategic Objectives	Strategies for Their Achievement
1.1 Continue to reinvest 4 percent of sales to develop profitable new products for our traditional markets and customers.	1.1.1 Expand product line A to fill existing gaps as identified in Marketing Report No. 12, through in-licensing the circuit technology from Europe and Japan.
	1.1.2 Complete the development of required new process technology for expansion of product line B and invest the capital required to produce an additional $10 million in U.S. sales in year three.
	1.1.3 . . .

Figure 29-2. Examples of strategic objectives and strategies.

OPERATIONAL OBJECTIVES OF PROGRAMS AND PROJECTS

The strategy statements for achieving the Level 2 strategic objectives now become the more specific *operational objectives* for well-defined programs and projects. Programs and projects are very similar in nature, with the difference relating to scope and duration. Generally accepted usage of these terms is as follows:

A *program* is a fairly broad, medium- or long-term effort made up of two or more *projects*. A program may be devoted completely to one strategic objective, or it may consist of a logical collection of projects that are related to several strategic objectives. An example of the latter case would be a Capital Investment Program, which brings together a number of specific projects for new, expanded, or improved facilities, each related to a different strategic objective.

A *project* is a short- or medium-term effort (generally up to three years in duration) directed toward achieving one specific strategic objective (or portion thereof). A project is made up of interrelated tasks performed by various organizations. It has well-defined objectives, a target schedule, and a target cost (or budget).

A *task* is a short-term effort (generally not more than a few months) performed by, or under the responsibility of, one organization. It has a clear statement of work, an established schedule, and an approved budget. Task schedules and budgets are integrated to meet the project schedule and budget, and achievement of task objectives will in turn achieve the project objective.

An organization's growth strategies are thus realized through execution of programs and projects. The hierarchical linkage leading from policy-level objectives to growth strategies that become project objectives is shown schematically in Figure 29-3.

CONCLUSIONS AND ASSUMPTIONS THAT UNDERLIE OBJECTIVES

For each objective at any level, good strategic management practice requires writing down the underlying conclusions and assumptions that have led to the establishment of the objectives.

These conclusions and assumptions usually will fall into one of three categories:

The industry and its environment, which includes the markets in which the company is, or wishes to be, active.

The company's competitive situation.

The company itself (its products, services, resources, strengths, and weaknesses).

Conclusions are statements, based on available information, about the current or fu-

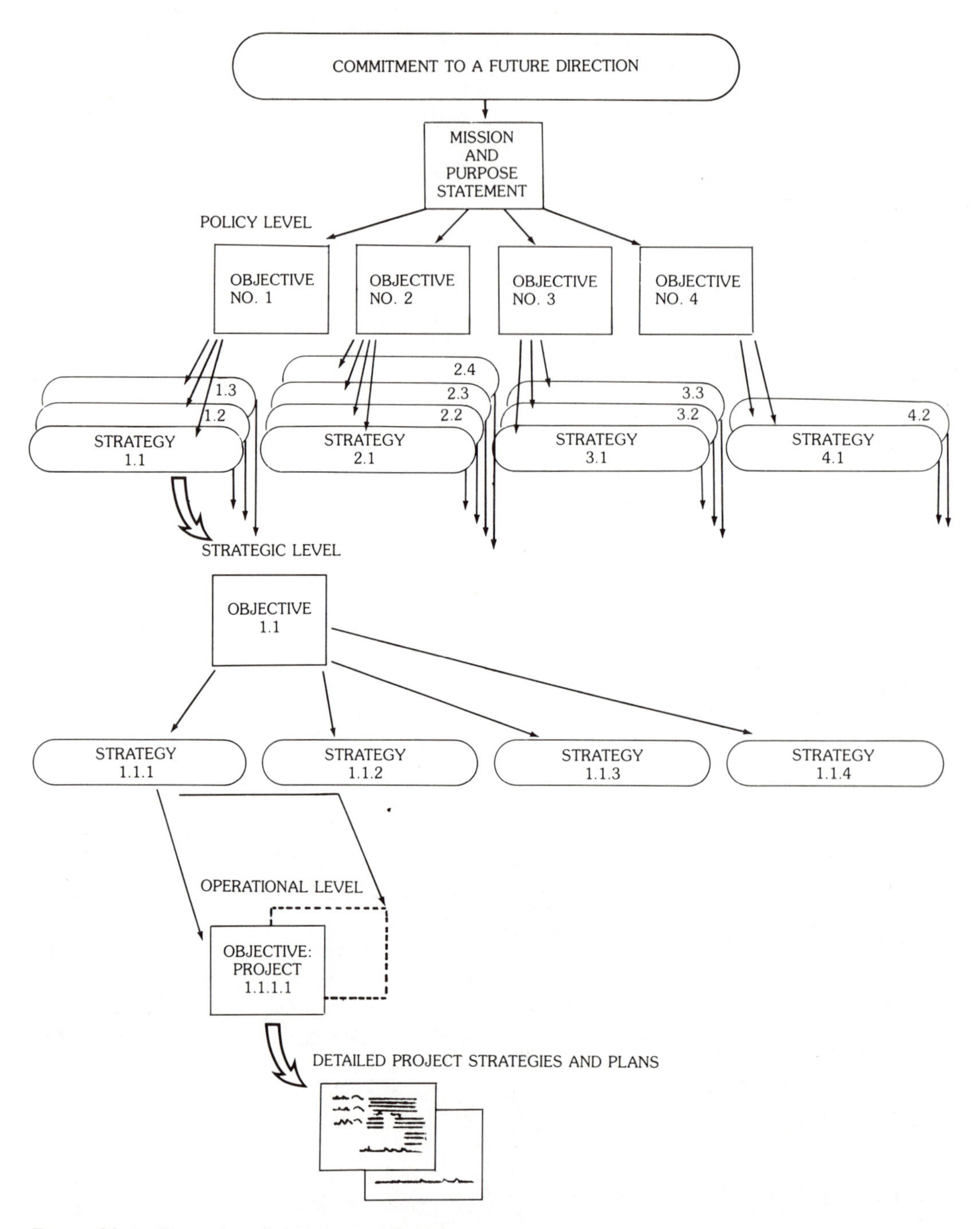

Figure 29-3. Hierarchy of objectives and strategies.

ture situation in the above categories regarding issues, threats, opportunities, problems, or strengths.

Assumptions are similar statements but are made in the absence of adequate information. Assumptions often relate to the future: competitive actions or reactions, market conditions, economic conditions, and so on.

Figure 29-4 illustrates a listing of typical examples of conclusions and assumptions un-

Strategic Objective	Conclusions	Assumptions
Increase U.S. market share to 20 percent from the current 12 percent for product line A.	1. Competitor X, with current 45 percent market share, is vulnerable due to quality and delivery problems.	1. Increased volume can be produced without capital investment for plant expansion by operating third shift.
	2. Our market share can be increased through small price reductions with emphasis on quality, short response time, and prompt, dependable delivery (based on Chicago market test.)	2. Unions will agree to third shift operation.
		3. Pricing levels will not fall more than 5 percent.

Figure 29-4. Examples of conclusions and assumptions.

derlying a strategic objective. Obviously, continuing effort should be expended to gather sufficient information or to monitor pertinent events or data to confirm or deny the assumptions: ideally they should be converted into conclusions as soon as possible.

Very little is permanent in this world, and changes could occur that make either the assumptions or the conclusions invalid. Hence, in addition to the effort to substantiate the assumptions, there should be an ongoing process to ensure that the conclusions and the objectives they support are modified as required when new information indicates that changes are needed.

Measurable Indicators of Success

For any objective to be worthy of the name, it must be possible to measure if, when, and how well it has been achieved. It also is very desirable to be able to measure along the way how we are doing and to predict what the likelihood is of reaching the objective successfully.

Conversely, if it is not possible to tell *whether or not* we have achieved a given objective or *when* we achieved it (if we did), then further effort is required to define a real, bona fide objective. Quantifiable, measurable indicators of success must be identified for each objective at each level.

Figure 29-5 lists some typical measurable indicators of success for strategic objectives.

OBJECTIVES-STRATEGIES MATRIX

Figure 29-6 presents a useful format for documenting a set of objectives, strategies, the underlying conclusions and assumptions, and the identified indicators of success for each. The final column, Potential Problems, provides for a correlated listing of possible adverse conditions, potential competitive or market reactions or changes, or other potential problems that bear continual scrutiny during implementation of the strategies. A few representative examples of conclusions and assumptions are given in Figure 29-7 to illustrate the use of this matrix.

MANAGING PROGRAMS AND PROJECTS

Having defined and linked the objectives and strategies as described above, the principles and practices of project management must now be brought to bear on the specific projects identified. In the absence of good project management practices, experience shows that excessive time and cost will be incurred in ex-

Financial	Market	Operations	Products
Rates of Return	Market Share	Productivity	"New"
on income on equity on assets	Market Size, Location, Newness	Head Count	"Improved"
Ratios	Sales and Distribution Methods Sales Volume Prices Other	People Turnover	Quality
balance sheet inventory turns investment		Costs	Performance
R & D capital		Capacity	Returns, complaints, claims, law suits
Cash Flow		Other	
Other			

CHANGES IN ANY OF THESE

Figure 29-5. Examples of measurable indicators of success.

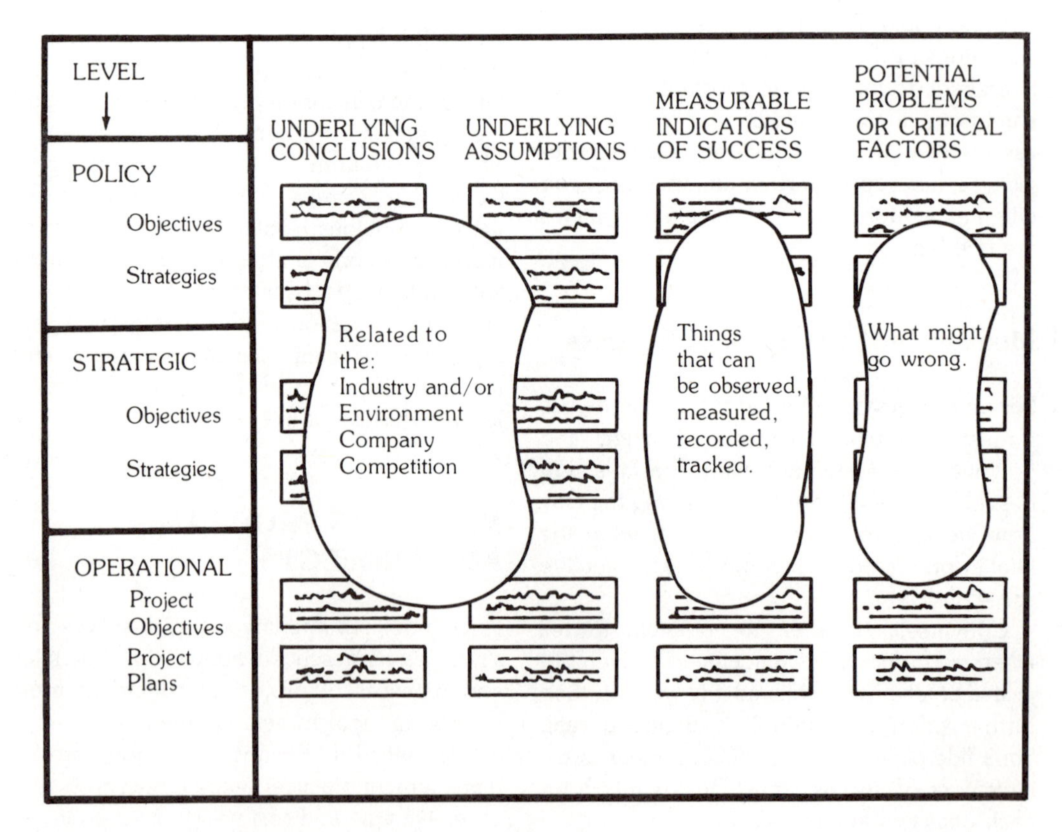

Figure 29-6. Objectives-strategies-projects matrix.

	CONCLUSIONS	ASSUMPTIONS	MEASURABLE INDICATORS OF SUCCESS
POLICY LEVEL OBJECTIVE NO. 3: OBJECTIVE NO. 2: OBJECTIVE NO. 1: Assure the long-term survival of the organization. STRATEGY 1.3: STRATEGY 1.2: STRATEGY 1.1: Continue to reinvest 4% of sales to develop profitable new products for our traditional markets and customers.	More benefit to shareholders to continue rather than merge or liquidate.	Long-term market will continue to grow 3% per year for next 5 years.	Organization continues independent existence. PE multiple not below 8. PBIT not less than 8%.
STRATEGIC LEVEL OBJECTIVE 1.1: Continue to reinvest 4% of sales to develop profitable new products for our traditional markets and customers. STRATEGY 1.1.3 STRATEGY 1.1.2 STRATEGY 1.1.1: Expand Product Line A to fill existing gaps as identified in Marketing Report No. 12, through in-licensing the circuit technology from Europe or Japan.	This rate of investment in new products will maintain or increase our competitive position. Technology can be licensed from one of three identified sources. Existing gaps offer easy entry to foreign competitors	No quantum-leap technological change will occur in our industry in next 5 years. Complete coverage in product line will increase sales & market share. We can move on this before competitor brings in foreign technology.	Number of new products launched successfully each year. Percent of sales from new products launched in past three years. Increase in sales volume for Product Line A. U.S. market share increase.

Figure 29-7. Examples of conclusions, assumptions, and measurable indicators of success for objectives and strategies.

ecuting the projects, or worse, they will not achieve the intended objectives in time to support the planned strategies.

The concepts described below apply equally to programs and projects. However, the discussion is related at the project level, since most organizations find that the application of these concepts is most effective at this level. In some cases, it is useful to appoint a program manager, but frequently the program level is more of an administrative convenience, as in the case of a Capital Investment Program, which includes a dozen plant and equipment projects in several geographic areas supporting a number of product lines.

Key Concepts of Project Management

The basic concepts that underlie the project management approach are:

Project Manager: A single point of integrative responsibility for a given project.
Project Team: All contributors to the project must work together as a team and understand each others' roles and responsibilities.
Integrated, Predictive Planning and Control Systems: All essential elements of information must be correlated and made available on a timely basis for project planning and control purposes.

Project Manager Role. When a project is created, the *role* of project manager is automatically created as well. This role is to the project what the general manager role is to a division or corporation: to integrate, coordinate, and orchestrate the efforts of all persons contributing to the organization, in the case of the general manager, and to the project, in the case of the project manager. How best to assign one person to this role of project manager is a complex question because most projects cut across many organizational lines. There is extensive project management literature dealing with this question, which is beyond the scope of this chapter. Suffice it to say that this assignment usually results in creation of some form of a matrix organization, with the project lines of direction intersecting with the traditional functional lines of direction at various levels in the organization. In brief, the project manager's job is to make sure that the project objectives are achieved on schedule and within the approved budget.

Project Team. All persons who contribute directly to the project, whether from within or outside the parent organization, are considered to be members of the project team. If all members can be made aware of their roles, if the project manager is able to build them into a team, and if good planning and control information is made available in a timely fashion using well-integrated systems, then remarkable success in project execution can be produced.

Integrated Project Planning and Predictive Control. Effective methods, procedures, and systems have been developed over the past several decades to plan, monitor, and control projects of any size and complexity. These systems are based on systematic definition of the project to produce one framework, which is used to correlate the time, cost, and technical information related to the project. Network planning and critical path analysis (PERT: *P*rogram *E*valuation and *R*eview *T*echnique; CPM: *C*ritical *P*ath *M*ethod; PDM: *P*recedence *D*iagram *M*ethod) is a technique that enables large or small projects to be planned, scheduled, monitored, and controlled in a very effective manner. Many computer software packages are available for mainframe computers, minicomputers, and microcomputers to assist with the planning and control of projects using these methods and integrating the time, cost, resources, and technical information.

Application of Concepts

The basic steps in applying these concepts are as follows. These steps are listed in general order of initiation, but most of them overlap each other and carry on to the end of the project:

1. *Define the Project*
   Develop the matrix of objectives and strategies for the project.
   Develop the project and work breakdown structure (the basic framework to be used for project control).
2. *Organize and Build the Project Team*
   Assign or plan the assignment of responsibility for each of the project elements.
   Initiate team building, leadership, and motivation.
3. *Plan, Schedule, and Budget the Work*
   Prepare the project summary plan.
   Develop the project master schedule and budget.
   Develop detailed task plans, schedules, and budgets, including contracts, subcontracts, and purchase orders.
   Select the information system to be used for project planning and control and indoctrinate all concerned in its use.
   Develop a list of potential problems and prepare contingency plans for their resolution.
4. *Authorize the Work and Start Up the Execution Activities*
   Authorize the work (work orders, directives, contracts, and so forth).
   Launch the initial tasks.
5. *Control the Work, Schedule, and Costs*
   Put into operation the project planning and control information system.
   Initiate monthly (or more frequent) project reporting procedures for physical progress.
   Initiate monthly cost control procedures.
   Initiate monthly (or more frequent) technical progress measurement procedures.
   Prepare and disseminate monthly schedule, cost, and technical progress reports to appropriate managers.
6. *Evaluate Progress and Direct the Project*
   Identify current or future variances in schedule, cost, or technical performance.
   Initiate corrective actions to prevent or resolve the variances or problems.
   Follow up on previous corrective actions.
   Report to higher management and others as required.
7. *Close Out the Project*
   Verify completion of each task, delivery of each deliverable end item, and achievement of each project objective.
   Plan and manage the phase-over transition period jointly with the person who will manage the ongoing operation of the project results.
   Complete the project close-out check lists.
   Prepare and submit the project final report and properly dispose of all project documentation and records.

CONCLUSION

During execution of any project, unexpected conditions will be encountered, unforeseen events will occur, and things may not always go according to plan. The project manager and his or her superiors are frequently called upon to make trade-off decisions between the technical, cost, and schedule objectives of the project. If the hierarchy of objectives and strategies is well understood, these trade-off decisions can be made so that the higher-level objectives are best achieved, even though one or more of the lower-level objectives must be sacrificed.

Unless the organization defines specific programs and projects to execute its growth strategies at the operational level and then executes these projects with good management practices, the strategies will remain on the shelf, collecting dust with other dreams of the future.

REFERENCE

1. W. R. King. "Implementing Strategic Plans through Strategic Program Evaluation," *Omega: The International Journal of Money Science* **8**(1980):173-181.

CHAPTER 30

The Strategic Importance of Managerial Selection

Milton Leontiades

Milton Leontiades is professor of management at Rutgers University, the State University of New Jersey. Dr. Leontiades has published numerous articles as well as several books on business planning and strategy. His latest book is *Managing the Unmanageable: Strategies for Success in the Conglomerate* (Addison-Wesley, 1986). His business experience includes strategic planning and development positions with two *Fortune 500* industrial companies. He also was a management consultant for the international CPA firm of Touche Ross & Co. and has held senior research positions with the U.S. Treasury and the New York Stock Exchange. He is senior editor of *Planning Review* and a member of the editorial board of *The Journal of Management Case Studies*.

In a 1979 survey of chief executives, 85 percent listed manpower planning as one of the most critical management challenges for the 1980s.[1]

The paramount importance of selecting the right managers has drawn the attention of managers and business experts. A tremendous practical and intellectual effort has focused on getting the right person to fit the job. Many articles and studies have emphasized the matching of certain managerial types to correspond with a particular company's needs.

A match-up between managers and duties is a logical association to make. The implication that certain traits are better suited for certain positions, or that good managerial selection will improve company performance, or that strategy should influence the choice of managers to implement it is instinctively logical.

The categorization of managers according to primary characteristics is one by-product of the continuing research on management selection. For example, some researchers trace the distinction between entrepreneurship and professionalism back to the sociologist Max Weber, whereas others see the distinction reflected in James MacGregor Burns's identification of transformative leaders who make things happen and transactional managers who attack problems methodically. Leaders and managers also are the two types used by Abraham Zaleznik of Harvard to contrast basic psychological differences between creative leaders and analytical managers.

Carl Jung, on the other hand, claimed that everybody is a primary type—either a feeler, a sensor, or an intuitor. The thinker is objective, the feeler is guided by emotions, and the sensor is the most instinctive and least calculating. More recent studies have included a colorful lexicon of managerial types best suited to lead specific types of companies: that is, financial controllers, statesmen, gamesmen, superstars, rescue artists, harvesters, visionaries, and so forth.

In addition, there have been a number of

classification systems developed to define ideal relationships between managerial types and managerial strategies. I suggested one myself.[2] It relied primarily on basic strategy alternatives chosen by the company at both the corporate and business levels to determine the choice among several types of managers. The idea was to combine common concepts in the literature of strategic management with familiar, if generalized, managerial prototypes.

A number of other variations on associating tasks and talents are feasible. Two such possibilities are given as examples in Figures 30-1 and 30-2. They take a slightly different orientation to a fundamentally similar objective: putting the best person in the right job.

Each of the above approaches, and other possibilities too numerous to mention, act as useful guides. Given a company's specific circumstances, a model of management selection can be constructed to best fit its needs. In conjunction with other evaluative techniques—such as psychological tests, interview screening, use of search firms, employee data files—a company can reduce the guesswork in making personnel decisions.

Despite every precaution, however, mistakes are made. No matter how scientific the tools, the wrong manager sometimes gets promoted. Nor can a perfect analysis of the job guarantee that the right type of person is available to fill it. Also, with the best intent, people make errors of judgment. Even with an ideal technique for making personnel choices, the motives of the selector may bias the outcome. In brief, manpower planning depends on the integrity of the selection process, yet this aspect has drawn scant reviews. Although ever better analytical tools for identifying good managers reduce the scope of the problem, they do not necessarily address the right issues. In some companies, internal

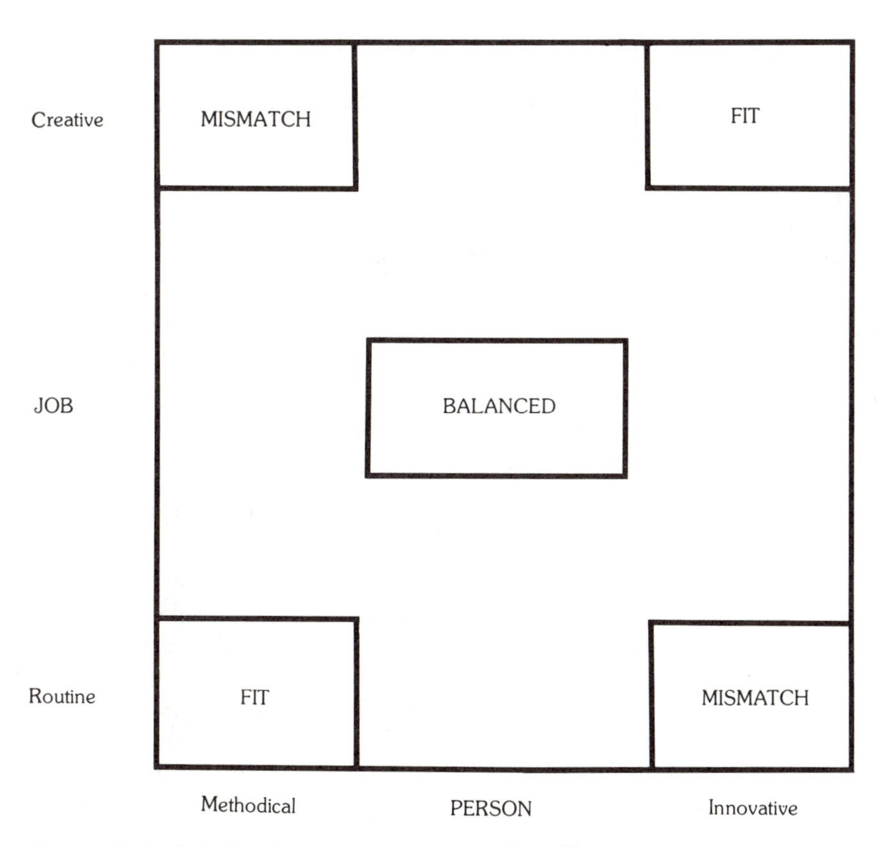

Figure 30-1. Job characteristics versus personal qualifications.

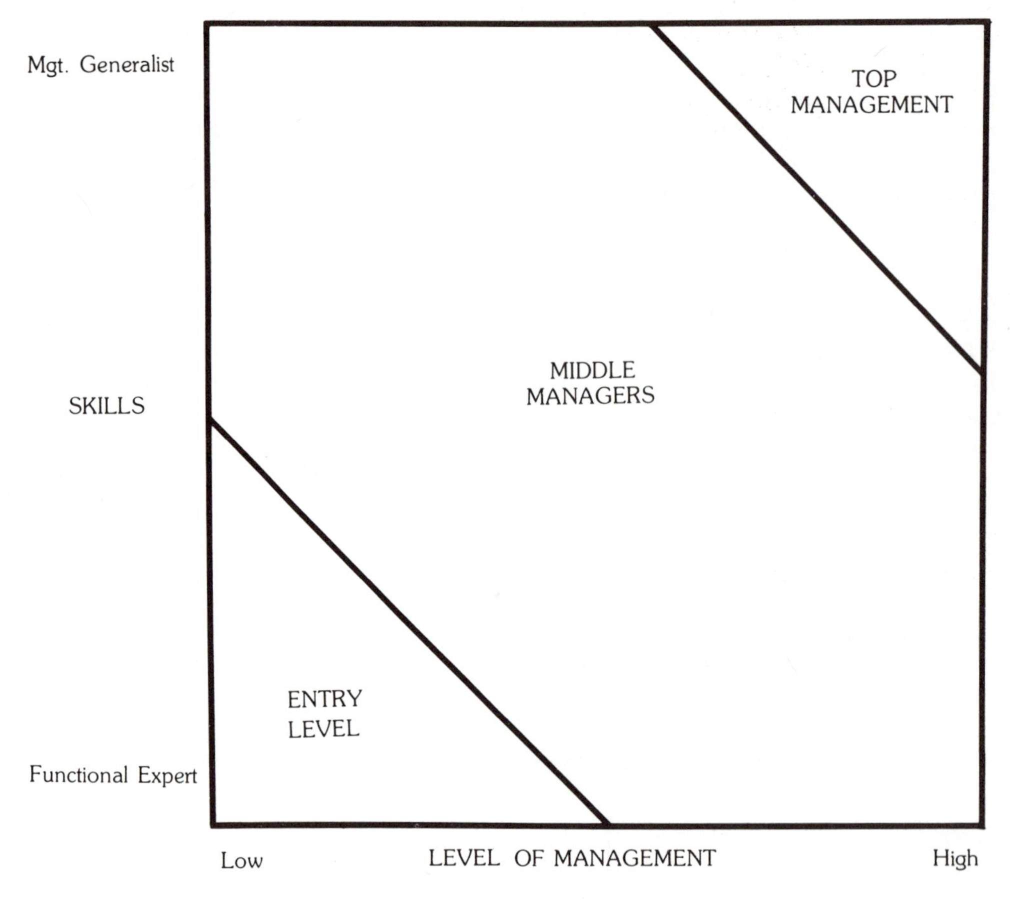

Figure 30-2. Relating management skills and management position.

rigidities impede an objective approach to staffing decisions.

In this chapter, attention is given to common barriers that militate against effective and impartial manpower planning decisions. Rather than proposing yet another conceptual design for managerial selection that might add marginally to those already available, the focus is on practical obstacles that can stand in the way of any design succeeding. There are, in other words, imperfections in the organizational *process* of making choices that can frustrate the ability of an ideal selection *model* to work as intended.

The first two problem areas discussed are oriented to the chief executive's office, because the leader sets the style for the rest of the organization. Without the right person in the top job, the quality of the people he or she attracts will be affected. Similarly, without the proper selection of the top strategist, the effectiveness of strategic management of the firm as a whole will suffer.

THE ONE-PERSON SHOW

The risk of having a less-than-effective CEO is a chance every company takes. However, in firms where the top executive is an autocratic leader and intends to stay, the effect on personnel planning throughout the company can be suffocating. Worthwhile successors can be driven away, the ranks of qualified middle managers can thin, and the remaining managers can become conditioned "yes" people.

Perhaps most critical, the strategy of the company translates into one person's ideas or whims. Is there any question that at Occidental Petroleum strategy is what Dr. Armand Hammer says it is, or that Teledyne's major strategy is influenced by anyone other than Dr. Henry Singleton? In cases of one-person rule, the impact of formal strategy, sophisticated tools of analysis, and consensual decision making are all secondary. The primary force for major decisions is raw power.

Contemporary business history is replete with cases of one-person rule where the company stumbled after that ruler departed. To name just a few: Charles Revson of Revlon, Jerry O'Neil of General Tire, Gerald Trautman of Greyhound, and Dr. Edwin Land of Polaroid. These strong-willed leaders never groomed a successor. A formal procedure for managerial succession was not in their nature. As one associate of Dr. Edwin Land of Polaroid said: "To talk about a successor is to acknowledge that Land himself will die. That is not spoken of at Polaroid."[3] Dr. Land was 70 at the time.

A number of CEOs currently are hanging on who are past normal retirement age. The list includes William Norris (Control Data), Peter Grace (W. R. Grace), Harry Gray (United Technologies), John Connelly (Crown Cork & Seal), Leon Hess (Amerada Hess) and Milton Petrie (Petrie Stores). Nor is this just a transitory concern. When this crop of autocrats dies or retires, the problem will not disappear. Replacements always seem to emerge and make this a continuing phenomenon. At present, many additional CEOs are in unquestioned control of corporations, not yet near retirement, and will continue to dominate their companies for many years to come.

The intended message is not that autocratic leaders necessarily are poor managers. A passionate devotion to business can compensate for an unstructured management style. The extraordinary drive and single-mindedness of the entrepreneurial chief executive can embarrass presumably better run companies enjoying all the modern techniques of modern management.

Under William LaPorte's long reign at American Home Products, for example, the company performed brilliantly on all the traditional financial measures. The unbroken string of sales and earnings increases stretched far back. Yet no formal strategic design was driving the company. Strategic management, as practiced at American Home Products, had not progressed even to the point of long-term planning. "We have no long-range strategy," William LaPorte said in 1980. "Our vision, our plan is for whatever period is necessary. We have nothing beyond a one-year budget but hope."[4] The strategy was simple. Every division was expected to increase sales and earnings by 10 percent compared to the same period in the previous year.

The point is not, therefore, that strong-willed CEOs are necessarily better or worse managers than other CEOs. It is that theirs is a unique management style centered on the vision of a single individual. Whereas other CEOs arrive at decisions through communications with other top executives and try to build consensus before taking actions, the lonely CEO relies on his or her instincts and accumulated experience. The process of decision making becomes personal—and sometimes arbitrary—rather than carefully structured.

This has two important implications for the subject at hand. First, in those companies guided by entrepreneurs rather than professional managers, standard management principles seldom are applicable. No matter how elegant its logic, the science of management cannot breach that part of industry that is not receptive to its message. Secondly, the importance of management succession planning will make scant impression in companies whose top executives do not practice it.

Thus, the notion of a systematic approach to selecting managers to fit the needs of the company and complement its strategy, although an eminently sound idea, is not necessarily viable in every case. Without sponsorship from the top, neither strategy nor succession planning can be optimized. And where companies are in the grip of a domineering CEO, the chances for strategy or succession planning going according to formula is slim.

UNQUALIFIED SELECTORS

If CEOs are part of the problem, who is responsible for selecting them in the first place?

Boards of directors or the current CEO generally pick a successor CEO. Either way, the decision is not always impartial or informed.

In theory, boards of directors are empowered with the legal authority to hire CEOs and to fire those that are not performing. That is their ultimate exercise of power and means of expressing dissatisfaction with the top manager and his or her strategy. It presumes to strike a balance of power between the CEO and the board. The chief executive manages the company and makes operational decisions. He or she confers with the board on long-term strategy and formulation of basic new initiatives. Members of the board review the facts and give their opinions. Theirs will be an unbiased counsel, based on objective evaluation of the evidence.

Properly selected, board members will represent a diversity of talents and specializations. A chief executive should be able to call on a financially capable board member when the numbers and consistency of a major investment or strategy need to be verified. Similarly, if the topic was marketing or production or human resources, the CEO should be able to turn to a member of the board with expertise in each of these areas.

In order to give objective counsel, however, board members must be independent. They must feel free to disagree with management. If they think a policy is flawed or unsound, they must protest in order to represent the shareholders and provide management with guidance.

In sum, an ideal board would be comprised of independent members chosen for a particular specialty that makes a specific contribution to the company. How well does this accord with the facts? Sad to say, the evidence is not too heartening on any of these scores.

Instead of giving the chief executive independent counsel, too many directors owe him or her their allegiance. Although nominally elected by shareholders, shareholders seldom bother to vote and, if they vote, overwhelmingly support the nominations placed on the ballot. Those persons nominated for the board often have been handpicked by the CEOs, with the criteria for who stands for election depending not merely on qualifications but on an agreeable manner.

Management experts and executives alike point to many faults in the present system. Peter Drucker bluntly states that lawyers, bankers, and consultants do not belong on the board: "No one who receives payments for services except as a board member belongs on the board. Retired people don't belong there; if you want to retain them as consultants, fine."[5]

President Reagan's former White House personnel director, E. Pendleton James, a specialist in recruiting directors, lays a lot of the blame for succession problems on cronyism among directors. "I'm negative on boards as they're now constituted," he says. They're ossified and incestuous." There is a tendency, he feels, for many directors to "stay too long at the fair."[6]

A tendency to vote in management's favor also is influenced by the compensation for directors that managements control, such as directors' fees and various perks, not to mention the status of the job. These tangible benefits are not readily forsaken for the satisfaction of being right. If a particular director disagrees too persistently or too frequently, he or she may be tagged as a troublemaker and lose the support of other members of the board. Over time, the pressures to conform and go along with management accumulate, especially if there is a strong desire by the CEO for a harmonious rather than deliberative atmosphere in the boardroom.

An anecdotal account on this score was provided by Lee Iaccoca. After assuming command at Chrysler, during the depth of its troubles, a first priority was to unravel what had gone on before. He was astounded by the oversights he uncovered. Although Chrysler had been led by former accountants, this expertise apparently did not filter down to Chrysler's financial system. According to Iaccoca, financial controls were almost non-existent. "I couldn't find out *anything.* This was probably the greatest jolt I've ever had in my business career."[7] Similar disasters were uncovered in such areas as inventory controls, dealer relations, and product quality. As for the board of directors, Iaccoca could only wonder where they were when all of this was going on. He was not so impolitic as to accuse those who had just hired him, but once or twice he did ask, as politely as he could: "How did management ever get their plans past such a distinguished group of businessmen? Didn't you guys get any information?"[8]

If the beginning point for improving managerial selection throughout a firm is to assure a better selection of the CEO, then this in turn suggests reforms in the system for picking members of the board of directors, who have the principal responsibility for choosing a good CEO and dismissing a bad one.

The second likeliest influence on who gets picked to be the next chief executive is the current chief executive. In well organized, professionally managed companies, the participation of the CEO can be of real assistance in picking his or her successor.

A classic illustration of a CEO unselfishly structuring the process for choosing his successor is the example of Reginald Jones of General Electric picking Jack Welch to succeed him. Two guiding principles for Jones reportedly were to make the decision with the future strategy of the company a foremost consideration and to surround the new CEO with a team of top executives who could work with him. The process entailed a painstaking review of each candidate for the top job, including a series of background interviews to determine the chemistry of each candidate.

Each prospect was asked who he would pick to lead GE in the case of a plane crash in which he and Jones were killed; asking each candidate in effect to rate his competition. Sometime later, a second interview plumbed even deeper to resolve questions of who was best qualified to assume command, and how each candidate would work with other top managers. In the end, the person selected was unlike Jones in character and style, but uniquely fitted to the new challenges that GE was to face. In short, a textbook pattern of a carefully structured process of management selection.

Returning to situations where the process is unstructured and personalized, the results can be largely fortuitous. Objectivity is far from assured when it runs counter to the CEO's personality, or conflicts with his or her unshakable personal vision of the firm's strategy.

A domineering CEO, as previously noted, does not share power and therefore does not tend to cultivate a strong successor. He or she may clone himself or herself unthinkingly or promote an assistant who has worked under his or her shadow. Neither is likely to impose the management discipline most needed by a company after the tenure of a freewheeling entrepreneur. Alternatively, strong CEOs who pick their own successor may not look favorably on any drastic changes in strategy, even if new strategic initiatives are indicated. A too-independent new voice is not a quality single-minded CEOs are likely to seek or tolerate if, after retiring, they maintain an influential role as a board member or voting stockholder—a not infrequent happening.

Some companies thus find themselves committed to the strategies of the past by an incumbent CEO who is blithely unaware or indifferent to changes in the company's environment. Even if shifts in strategy are obvious, he or she may be unable to choose the best candidate to carry them out, or, upon choosing a successor, he or she may be unable to divorce himself or herself from involvement in corporate affairs. Unsatisfactory transitions of this nature are all too common. Harold Geneen's high energy personality, for instance, clashed with Lyman Hamilton's deliberate style, and William Paley of CBS proved unable to surrender the reins to Arthur Taylor, his successor. When the selection process turns on the whim of a single person, personality invariably plays an important part in the final decision. Thus, companies can be burdened with a present CEO's eccentricities as well as an ineffective board of directors as obstacles to making rational and proper judgments. In either case, the selection of a qualified CEO will be a gamble.

Even with good procedures and qualified selectors, the choice of a new CEO is not as unemotional as the models of management selection would suggest. Neither CEOs nor board members view this decision with the same detachment as, say, the purchase of a new plant or the opening of a new store.

INCORRECT MATCHING OF NEEDS

Not all important decisions are made by the CEO. Nor are all the crucial managerial selections made at the corporate level. Many more jobs are filled at the business level and, correspondingly, care must be given in evaluating employees throughout the organization. Some precautions are noted below.

Structural Defects in Rating Procedures

A difficult problem to resolve is the issue of objectivity. Managers often find it difficult to give anyone low marks. Knowing full well a candidate's marginal qualifications, a favorable recommendation is still forthcoming. In some cases, the supervisor does not wish to be stuck with the employee and therefore does not prejudice the chances of moving him or her on to some other department.

Who has not puffed a letter of referral? Psychologically, it is hard to turn down requests for recommendations and relatively easy to gloss over deficiencies and stress favorable traits. As for personnel reviews, these generally must be presented and justified in person to the individual being evaluated. This makes it even more difficult to write a negative review and even more painful for managers to convey the bad news.

Managers do not like to be put on the spot. They often work and interact closely with the people they rate. Tensions can arise if a manager is too bluntly honest with too many subordinates. To escape such sensitive decisions, an unwritten code can develop where the most senior person is automatically the next to be advanced. This often results in passing over better qualified candidates. It invariably will frustrate and discourage younger, aggressive employees. However, once the practice of selecting according to seniority becomes routine, it is a difficult habit to break. Promoting a less senior person means starting reverberations in the work force. In the end, managers may tend to bend to the pressure of expectations by going along. It does not make for good personnel decisions, but it becomes justified, over time, as the least disruptive recourse.

The higher up in the organization, the more damage such structural defects in promotional practices can cause. To avoid this trap, a number of checks and balances can be built in. One company limits the percentage of persons reviewed each year who can fall in each ratings category with, say, 10 percent at the top, 20 percent in the next bracket, and so forth. This resembles a grade distribution, and it has the drawbacks of any mechanical approach to judgmental decisions.

Another, more effective rule is to avoid unilateral ratings of key employees. Instead, multiple assessments are provided by several managers. This type of check is important to assure the quality of the inventory on up-and-coming managers. The success of an organization will depend on its future managers, so it is the concern of more than the immediate boss that the ratings are fair and objective. As the employee moves up the organization, more of the responsibility for career tracking and advancement may be shifted to a company committee, rather than a single superior.

Qualifications Are Incorrectly Perceived

A major electronics firm that prided itself on management assessments nevertheless failed to capitalize on an early lead in integrated circuits because it allowed the wrong type of manager to head the new product line. A brilliant scientist responsible for the technology was put in charge. Favoring his strengths, he spent heavily for R & D. Unfortunately, he had almost no regard for cost controls. Alarmed at the lack of financial discipline, senior management eased the scientist out and replaced him with a new manager who introduced a strong accountability system.

It took some time before management discovered that the scientific manager's stress on R & D was right all along. Strong controls were helpful but a high level of developmental spending was absolutely essential. Fortunately, the error was caught in time and the company, under a new manager, was able to regain its momentum in this fast-evolving field of technology.

Appraising a manager's style and then finding the specific job tailored to it are imprecise tasks at best. Making an improper judgment risks losing valuable creative leaders straddled with routine jobs or forcing competent managers into positions demanding aggressive personalities.

When selecting managers, many companies go after the best and most ambitious candidates. Yet the company itself may be operating a routine business with no dynamic new strategies in view. In these cases, the company's style does not accord with the types of managers it tries to recruit.

Going back to Figure 30-1, a company must first be honest in evaluating what it has to offer an employee before asking the employee what he or she can do for the company. Not every situation requires a managerial genius in command. There are only so many challenges to overcome in conservatively run businesses. Some companies or their divisions date back to the last century and have remained basically committed to the same business and a consistent strategy. This does not mean they must settle for less than the best, *given the demands of the job itself.* It simply means choosing the right criteria to define their needs before setting personal qualifications.

A misjudgment of strategic needs is a conspicuous mistake often made by emerging high-tech enterprises. Skilled in product technology, they do not adequately anticipate the financial or marketing expertise needed to assure success and, in some cases, survival. The background of the CEO often determines the overriding emphasis of the firm, to the exclusion of other needs. This can mean reluctance to spend money on sales and marketing, or the top executives may be preoccupied with raising needed capital, while letting quality and design considerations slip. Whatever the particular cause, a number of promising start-ups have proved deficient in one or more parts of their strategy formulation, which, in turn, led to inadequate or improper staffing decisions.

Unanticipated Shifts in Industry Structures and Competitive Assumptions

Emerging growth companies are not the only ones tripped up by an inability to perceive strategic shifts. Companies experiencing turning points in their history, such as International Harvester, have been punished for their reluctance to change and for an inappropriate response to staffing for change once the need became obvious. Cyclical reversals in long-term trends have always been treacherous for companies. In addition, the post-World War II era has introduced a new and equally unpredictable phenomenon: industrial restructuring. As companies leave old industries and enter unrelated ones, familiar competitive assumptions become obsolete. New competitors force new strategies, and new styles of managers are needed to deal with them. These managerial realignments are especially disruptive in mature industries where management styles have become ingrained and resistant to innovation.

One example of restructuring is in the newly forming sector of financial services. A decade or so ago, financial services had a fairly well-defined meaning according to which segment of the market a firm served: commercial banking, stock brokerage, insurance, consumer financing, or savings and loan. Now these segments are being intermingled. Industry boundaries are being blurred as companies combine one or more of these services under one corporate management. Also, companies outside the conventional boundaries, such as Sears, American Express, and General Electric are acquiring entry into several of these areas in order to become significant players in a giant financial services market of the future that they foresee.

In the process of restructuring, strategies must change, and accompanying the new strategies must be managers prepared to implement them. This poses a painful and time-consuming problem. It can mean turning bankers into strategists, managers into leaders, and transforming specialists into visionaries.

Addressing a conference of executives recently, I asked a banker how his company was coping with this new dimension of managing. "Not too well," he replied. "Previously, our bank managers were primarily concerned if the books balanced. Now they have a whole new competitive marketplace to comprehend. Our reorganization still has a long way to go."

What the bank executive did not go on to explain but what is probably inevitable is that not all the present managers will make the transition. New managers will have to be hired from the outside and some will be nonbankers. Current employees too entrenched by habit, age, or orientation will have to give way in companies hoping to succeed in this quickly restructuring market. If history is a reliable guide, there will be fewer but larger survivors as financial services begin to take shape as a new service sector in the economy. The urgency for companies to formulate new strategic initiatives is obvious and critical. Just as important will be the ability of companies to develop or hire managers to undertake these new directions.

Two other significant forces in changing the way U.S. companies will do business in the future are foreign competition and deregulation. In the former case, the Japanese have been the major influence. In particular, their concern for the people side of management is permeating U.S. industry. At first, a few U.S. companies were primarily exposed to the notion of quality circles, where groups of employees work together on project teams. First experiments were tentative and some half-hearted attempts failed. However, the tendency toward a more participatory management style seems more than a fad. Increasingly, it is being seen as a way to boost productivity of the work force and, indirectly, profits.

The old hierarchical style of communications, with commands handed down from the top, no longer meets the needs of modern industry. It is too structured for reacting to rapid changes in today's markets. Moreover, it stifles the individual initiative that companies are looking for in managers of their fastest growing divisions. These managers, in turn, are seeking to motivate employees as a means of increasing output. Thus, although the focus on the bottom line has not changed, the best path to get there is undergoing serious reevaluation.

Deregulation is another relatively recent variable that has disrupted traditional routines in certain businesses. Long experience in regulated markets, as defined by the federal government, dulled the competitive instincts of firms in industries such as airlines, trucking, and railroads. What was the sense of devising a least-cost strategy if costs could simply be passed on to consumers? Likewise, prices could not be unilaterally lowered or raised, so few strategic initiatives in this direction were encouraged. Entry into the markets also was restricted, so that the competitive balances, once established, were seldom disturbed. Overall, these industry conditions fostered a complacent group of managers.

Deregulation in the early 1980s abruptly altered the rules of the game. Companies previously shielded from competition now found themselves vulnerable. New firms sprang up to grab shares of market. Old companies vied for each others' customers or merged to forge larger and stronger partnerships. Lower cost structures in one company pressured competitors to follow suit. In the airline industry, for example, a two-tier wage pattern set by American Airlines, whereby new hires were paid less than current employees holding comparable jobs, caused United Airlines to press its unions for similar wage settlements. Fat that had been tolerable under the old rules was being exposed quickly under the fierce new terms of competition.

As a consequence, the emphasis in managerial styles also shifted—from courtliness to competitiveness. Management compensation based on mere competence could in the future be a recipe for disaster. Performance is the new key to success. And the managers who can produce are not likely to have the same management traits as those who operated under the protected shield of regulation.

As in any major reorganization, whether forced or planned, the personnel consequences tend to reflect the dramatics of the new strategy. In each of the above circumstances, strategy and management style ideally go hand in hand. Using strategic assumptions as a guide, the required types of managers can be inferred. The critical point is *first* to define strategic needs, *then* to take the proper management actions. Models of management selection do not make explicit the dynamics that can change one or the other of these criteria. Without this appreciation, well-laid plans can fail for lack of effective managers, or companies can get stuck in a particular groove even as changing conditions suggest a fundamentally different competitive orientation.

CULTURAL RESISTANCE TO STRATEGIC SHIFTS

Observing a relationship between style and strategy is one thing; trying to keep them in alignment is another. An organizational style, or corporate culture as it is often referred to, is by far the most difficult of the two to revise. Whereas new strategy can be formulated almost overnight, organizations are not so changeable or adaptable. Because strategy depends on managers to make it work, this aspect is given further attention below.

Culture can be a major strength where it reinforces a company's strategy. Firms pursuing a single grand strategy, for instance, develop a set method of operation. Things are done in a standard way and in a certain pattern. These mannerisms impart a particular style to the company. Refined and fine-tuned over time, they become instinctive and known as a company's culture.

The longer a company stays with a particular culture, the more proficient its operations become. This reflects the familiar learning curve effect. J. C. Penney, for example, has the reputation of a conservative management style. Building long-term loyalty is prized over quick victories. The same aggressiveness that may be valued at an ITT would be out of place at J. C. Penney. Following its cultivated inward-focused style serves J. C. Penney well as long as it sticks to its basic retailing line of business.

Should it venture afield, however, like its major rival Sears has done, the same attributes now prized would fetter the company from making the personnel shifts needed to correspond with the new strategy. Sears has, in fact, discovered that the culture of its acquired securities business—Dean Witter Reynolds—demands a fundamentally different perspective from the retailing end. Sears is taking a cautious approach toward melding the managers of the various companies it has acquired. Yet, until a unified management team can be knit together, the individual strategic acquisitions will lack the full synergy of working as a single coherent organization.

With the massive movement to acquire unrelated businesses since the mid-1950s, more and more firms have run into the problem of culture shock. As a company diversifies afield, the ability to conform hard-to-change values to the existing organization's culture can spell success or failure of overall strategy. It is, after all, demanding quite a bit to plan the multiplicity of personnel moves when a company is participating in a familiar competitive arena. When a number of new businesses are entered, and a host of new managers and their different values are acquired, the personnel job becomes immense.

Moreover, the rewards of successful adaptation are long term, whereas the personnel actions can be drastic and immediate. A major services company, for example, decided to bite the bullet and make the management changes that its new diversification strategy dictated. One-third of the top management team was gone in less than a year. The compensation and incentive system was overhauled, with greater emphasis on rewarding overachievers. Younger managers were encouraged to take advanced educational courses. Older employees were given bonuses to take early retirement. In all, the culture of the company was reshaped to recognize superior performers, rather than competent managers, and to change from a familiar and comfortable management culture by cultivating an aggressive and hungry attitude.

Such a sweeping reorganization suggests a complete rethinking of prior management assessment procedures and techniques. When companies move to reposition themselves in new markets or depart old ones, they must consider what effects such moves have for managerial selection. It may be that top managements will have to decide whether their strategies are best kept intact and in tune with well-established cultures, as in the case of J. C. Penney, or that the risks of cultural adjustments are justified by the potential rewards perceived in a completely new strategy, as in the case of Sears.

One of the major difficulties in effective personnel planning is its oversight at the time new strategies are being designed. Too often the strategy is engaged with little or no thought given to the effect it will have on old traditions. In company after company and industry after industry, resistance of old managements is one of the major roadblocks to successfully executing new ideas or integrating new businesses. Such resistance cannot be overcome by appealing to reason alone. The logic of a major change may be irrefutable. The financial aspects of the strategy can show fantastic financial results. However, unless the people side of management is addressed and dealt with as meticulously as the strategy itself, the personnel difficulties of making it work can render the other considerations moot.

TRAINING TO BECOME A MORE EFFECTIVE CEO[9]

The strategic importance of managerial selection ultimately comes down to doing the best with what you have. It does not mean every company can achieve equal success in managerial selection, no more than companies can be said to be equally matched in many other competitive aspects. Those staffed with superior managers are likely to attract a better quality of future managers. There is, however, one job that is key for making a quantum jump in raising the company in all other respects: the job of the CEO. Only in this position is there leverage to make a significant difference in every other part of the firm. The right choice at this level can compensate for less-than-perfect staffing decisions at less strategic points.

Given that the right CEO can make a difference, are there ways to improve the way he or she is selected? Maybe. However, each situation is likely to differ. A more tractable question to try and answer is whether a CEO can be better prepared to take over the top job. Is there special training that has not been tried or widely applied in preparing the newly designated CEO to be a more effective strategist?

The rotation of executives among various top posts within a firm is common enough, of course, but this primarily gives a feel for running existing operations rather than managing them from the top. It also is inadequate preparation for guiding a company in need of new thinking. A new CEO should have time free from daily operating pressures in order to systematically plan for the future. A suggested innovation in this direction is to choose the new CEO well in advance of his or her taking office, giving him or her adequate opportunity to develop a strategy for the firm he or she will soon lead. Once in office, everyday pressures tend to consume all available time. For this most important personnel decision, any small improvement can be multiplied many times over in other parts of the firm.

A phase-in of new CEOs was tried in the two companies described below, both of which then performed exceptionally. As in all prescriptive advice, no future guarantees apply, but the stories of Harry Cunningham of K-Mart and Donald Kelley of Esmark provide two examples of success worth considering.

Cunningham's entry into the top position at K-Mart (called the S. S. Kresge Co. then) was not a masterpiece of planned execution. The story has it that the prior chief executive did not want Cunningham around the office. Cunningham himself suggested travel as an alternative to hanging around corporate headquarters. "I had no mandate," said Cunningham, "I was simply turned loose." In an impressionistic encounter, he met Eugene Ferkauf who had just opened an E. J. Korvette store on Long Island, a new model for discount stores at that time. Spending an entire day there, Cunningham talked to buyers, salespeople, customers—anyone who would listen to him. He came away impressed with the basic discount concept: a free-standing store in the suburbs with plenty of parking space. The margins would be kept low, helped by customer self-service, open-rack displays, and strong advertising support. The key ingredient for Cunningham was "that every item must move—turnover, turnover, turnover."

Determined from this and other experiences that discounting was the way to go, Cunningham kept his own counsel at first. At the time, discounting was not respectable. Even after becoming CEO, Cunningham made no immediate sweeping changes. Even though he had the authority, Cunningham realized that "if you haven't sold the people in your organizations, you'll fall flat on your face." He had to convince them they had an important part to play in this new venture. Also, Cunningham's enthusiasm and charisma were no small help in selling the idea.

In the end, Cunningham succeeded in innovating a new and formerly suspect discounting strategy. He cut prices, insisted on good value for the money, stressed customer service and a clean and orderly store. Each store was standardized in size and central merchandising policies but, within that context, store managers were given latitude in running their operations.

Cunningham's leadership style may have made him a successful chief executive even without the K-Mart concept. However, without the time off to reflect and develop an idea of what he wanted to do as CEO, he would not have been as prepared or as innovative. One reason this story played out so well is because the chief actor learned his part before being asked to perform. Having strong ideas of what to do, he was able to concentrate on the equally important aspect of execution, persuading others to help implement his vision.

Don Kelley spent a similar period preparing for his succession, albeit for a very different kind of company. Kelley was and is foremost a numbers man. He is the classic financial controller type of entrepreneur. His promotion, like Cunningham's, was probably assured by his innate management genius. Also like Cunningham he was undoubtedly helped by the opportunity to study intensively the organization he was later to head.

In the mid-sixties, conglomerates were busily acquiring businesses of all kinds. One chief executive who was determined not to be taken over was the chairman of Swift, the old-line meatpacker. He sent Kelley, then the controller, on an exhaustive review of the company's situation and alternatives. It was, according to Kelley, a "painstaking, nine-month, seven-day-a-week, thirty-six-hour-a-day" assignment.

He evidently learned what he needed to know. From his analysis, Swift transformed itself into Esmark, a holding company with Swift as a meatpacking division. In 1977, Kelley became CEO of Esmark. The rest is history. Donald Kelley succeeded in restructuring Esmark away from oil and chemicals and much of the meatpacking business into consumer products such as International Playtex, eventually acquiring Norton-Simon, itself a diversified consumer goods business, and finally selling Esmark to Beatrice, a conglomerate operating in many businesses, at a fancy price. Not at all similar to what Cunningham did for Kresge, but in one respect both CEOs were alike. They both hit the floor running, aided by a clear idea of where they wanted to go and given the opportunity and exposure to know best how to get the job done. This may not be the perfect solution in all cases, but it is clearly better than muddling through. To quote Lyman Hamilton, a former CEO of ITT, "Every CEO says he plans for a long term. But every CEO lies."

SUMMARY

Management selection attempts to put the right person in the right job at the right time. One essential ingredient in making these decisions is to be able to evaluate which employees are best suited for specific jobs. In this regard, better tools of analysis are welcome. And a considerable amount of thought and study have gone into refining this aspect of management selection.

This chapter looks at another critical component of management selection: the judgmental aspect. In choosing new employees, who decides obviously influences who gets picked. In a perfect world, these decisions would follow a model set of guidelines. In the imperfect world in which we live, however, there are a number of missteps possible between the ideal and reality. Pragmatic decisions are not always made for the right strategic reasons, may be swayed by personal feelings, or may simply involve mistaken judgment.

There are a number of instances in which the *procedures* of management selection can work contrary to an ideal model's intent. Shortcomings in specific companies may be extremely difficult to resolve if, for example, the boards of directors or chief executives are part of the problem and therefore outside the reach of normal remedies or threats. In these cases, time and external pressures from stakeholders or competitors may be the only real hope for reform.

Some precautions are suggested that would improve the chances of an efficient internal selection process at lower organizational levels. In addition in selecting of the CEO, a proposal is made that could raise the quality of his or her performance before he or she takes office. In the end, although the combination of good tools and their intelligent application are key, the final outcomes still involve subjective judgments and luck.

REFERENCES

1. *Wall Street Journal* (4 September 1979):1.
2. M. Leontiades, "Choosing the Right Manager to Fit the Strategy," *Journal of Business Strategy* (Fall 1982):55-69.
3. *Forbes* (2 October 1978):44.
4. *Business Week* (20 October 1980):81.
5. "Conversations . . . with Peter F. Drucker," *Organizational Dynamics* (Spring 1974):50.
6. *Fortune* (2 May 1983):58.
7. *Fortune* (26 November 1984):224.
8. Ibid.
9. This section is based on excerpts from M. Leontiades, *Managing the Unmanageable: Strategies for Success in the Conglomerate* (Reading, Mass.: Addison-Wesley, 1986).

CHAPTER 31

Sharing the Value Added: A Bonus Plan for a Growth Company

A. Graham Sterling

A. Graham Sterling is vice-president of Analog Devices, Inc., a leading firm in the design, manufacture, and distribution of electronic components and systems for real-time acquisition and processing of real-world signals. The firm also is widely recognized as an outstanding company within which to work and to develop one's career. Mr. Sterling earned B.S. and M.S. degrees at The Massachusetts Institute of Technology, where he also was elected to membership in the Eta Kappa Nu and Tau Beta Pi Engineering Honor Societies.

The quintessential function of the firm is to provide secure and rewarding employment to people who together create valuable goods and services for the whole society.

To perform this function, the firm must be self-sustaining and self-renewing, which is to say it must persuade its customers to pay, in return for the goods and services supplied, a stream of cash equal to the raw cost plus a fair assessment of the value added by the firm and must persuade its employees to remain loyal and diligent and to assimilate additions and replacements as necessary. In order to sustain and to renew itself continuously and to expand its capacity as necessary to satisfy the demand for its products, the firm must offer returns to its investors in proportion to the risks they bear. Therefore, a goal of the firm is to maximize the market value of its equity, which value represents the invested wealth of its shareholders and the ultimate security of its creditors.

The quintessential function of the firm and all its prerequisites and subordinate goals are supported by a well conceived employee bonus plan.

A MODEL OF THE FIRM

An operating model of the firm is an instrument for self-control, enabling the employees to visualize the problem of self-sustaining, self-renewing growth and to manage the short- and long-cycle processes required to solve that problem. An operating model consists of an internally consistent set of scale-invariant ratios defining timeless performance standards for the firm in dynamic equilibrium with its markets. The minimum set of conditions to be described by an operating model are the annual rates of growth of sales and productivity and of return on assets (ROA).

Refer to Table 31-1 for an outline of the operating model of a representative growth company. This minimum set of ratios is usu-

Table 31-1. Operating Model of a Hypothetical Growth Firm

Operating Statement		Balance Sheet	
Sales	100.0%		
Purchases	48.0%		
Base pay and benefits	31.0%	Current assets	165.0%
Bonus pay	3.5%	Fixed assets	120.0%
Operating Profit	17.5%	Total assets	285.0%
Interest	3.2%	Accounts payable	61.0%
Taxes	5.7%	Debt	96.0%
Net Income	8.6%	Equity	128.0%

Other Operating Ratios[a]	
Annual of growth of sales and profit	32.0%
Annual growth of productivity (Sales per Employee)	15.0%
Operating pretax return on average assets	25.4%
Annual net return on starting equity	32.0%
Leverage = ratio of debt to equity	75.0%

Note: Ratios are percentages of quarterly sales.
[a]All these ratios are average or trend-line goals over the business cycle, not fixed ratios to be maintained at every stage of the cycle.

ally expanded to define norms for key functions within the firm. For example, purchases may be analyzed into variable, fixed, and depreciation components, pay and benefits may be analyzed into direct, indirect, and expense components, and everything between sales and operating profit may be analyzed functionally, for example, manufacturing cost, engineering expense, marketing expense. On the Balance Sheet side, current assets may be analyzed into cash, inventories and receivables and fixed assets into plant and equipment components. Note that this model is self-funding in the steady state in that the ratio of net income to beginning equity is equal to the rate of growth of sales, while interest expense is not too high in comparison to operating pretax profit and debt is not too high in comparison to the equity or to the net working capital. Thus, as the firm grows 32 percent per year, so does its equity, through retained earnings, and its debt/equity ratio and all its other model ratios are trendless.

Also note that when this firm is operating right on model, its value added (the sum of pay and benefits plus operating pretax profit) is 52 percent of sales. Two-thirds of the value added is paid out to the employees who created it while the other third is used for the benefit of the government, the creditors, and the shareholders.

If the firm enjoys steady growth of profits on a self-funding basis, returns to shareholders are closely related to the annual rate of growth of sales (32 percent per year) while returns to employees are 34.5 percent of sales, 90 percent of this being base pay and 10 percent being bonus.

PURPOSES OF THE BONUS PLAN

The purposes of the bonus plan are communications and feedback: communicating the standards of organization performance, sharing the fruits of the firm's successes with the employees and requiring them to absorb some of the pain of the low value added during downturns of the business cycle. As in all well-designed feedback systems, this arrangement works for stability near the model ratios.

The communications function of the bonus plan is fulfilled by the functional relationship between the bonus payoffs and specified key operating results of the company, and by the opportunity for the management of the firm to explain the formula and to discuss each bonus payoff in the context of the recent results of and the outlook for the firm.

The bonus plan also works to moderate the swings in the ratio of pay and benefits to total value added over the business cycle. In other words, the ratio of pay and benefits to value added should not rise too much in bad times and should not fall too much in good times.

Fairly wide swings of the ratio of pay and benefits to value added are inescapable over the business cycle, and if there were no systematic way of clipping the peaks of this ratio that occur at the troughs of the business cycle, the firm would be hard pressed to assure stability of employment. Similarly, if there were no systematic way to increase total compensation during boom periods, the motivation of the employees might suffer for lack of tangible feedback.

Some mechanism for creating variable or contingent compensation is inevitable in a free society. One archaic but effective way is to discharge a fraction of the work force during downturns of the business cycle, thus varying the compensation of those particular people from base pay to zero pay. But this method is inconsistent with the duty of the firm to provide secure employment.

Another way to vary compensation over the business cycle is to cut pay during recessions, but this is difficult to administer equitably and inflicts psychic as well as financial pain on the employees without an offsetting psychic benefit to the firm. Each employee's sense of personal worth is related to his or her base pay, so even a temporary cut in base pay is psychically reductive as well as financially painful.

A well-designed bonus plan provides a more systematic thus more equitable and less painful way to vary compensation with the business cycle. For the hypothetical model firm pictured in Table 31-1, the model bonus expense is 3.5 percent of sales, but this ratio can be significantly higher in boom times and can fall to zero during recessions, thus making the operating profit ratio significantly more stable than would otherwise be the case.

Although fluctuations in bonus pay are financially painful to the employees, that pain is moderated if it is known that it is shared systematically and equitably and that it is in the interests of the health of the firm.

BONUS FORMULA TIED TO KEY ORGANIZATION RESULTS

To be fully acceptable to the employees and then most effective in accomplishing its purposes, the determination of bonus payoffs should be systematically objective and should be based on meaningful, timeless, scale-invariant operating measures of the performances of teams, rather than on individual performance. (There are other methods for recognizing individual performance, so team performance can be kept clearly separate.)

A firm in the for-profit sector of the economy is viable only if both its financial capital and its human capital are adequately productive. The productivity of financial capital is measured by the rate of return on capital, which is a composite of the rates of returns to creditors and to shareholders. The return to creditors is virtually always paid in cash (as interest payments) and is more or less standard for all comparable corporations, therefore is not one of the dimensions of excellence. Return to shareholders does, however, vary from firm to firm and therefore is properly one of the dimensions of excellence.

In growth segments of the economy, most of the return to shareholders is in the form of capital gains on the market value of the shares, and more than 80 percent of the difference in returns among such companies is explainable by the differences in their trend line rates of growth of sales. It follows that the rate of growth of sales should be one factor in the bonus formula determining the conditional part of the compensation returns to employees.

A firm is viable over the long term only if its return on capital is competitive, that is, is in adequate proportion to the perceived riskiness of the investment. This competitive rate varies among industries, with most of the differences being explainable by the differences in sales growth opportunities. In any case, a firm whose return on capital is not competitive finds itself unable to attract the influx of debt and/or equity capital needed to grow in pace with its markets and/or becomes vulnerable to acquisition by some other firm believing itself able to make that capital more productive. The internal operating result most closely linked to return on capital is pretax returns on assets, so it follows that the rate of ROA should also be a factor in the formula determining bonus payments to employees.

Finally, a firm can be successful only if it can secure an adequate influx of human capital. In growth industries, the available pool of adequately talented and trained people is finite, and the annual growth of this pool is the fundamental factor limiting the rate of growth of the industry and its ability to compete with foreign producers. It follows that the rate of growth of productivity of human capital is a necessary condition for viability of the firm and should be considered as a third factor in the formula determining bonus compensation.

Given the principle that we wish to determine bonus payoffs on the basis of team performance and given the above three factors that might enter into the formula, the question is how to define "team results." From the point of view of the outside world, the team consists of all the employees of the firm, particularly management. Also the viability of the firm as a whole is important to every employee, and every employee has some part to play in assuring the health of the firm as an institution. For these reasons, the consolidated operating results of the firm best describe the performance of the whole team.

However, many firms have diverse product lines and distribution channels, therefore they are articulated into groups, divisions, and profit centers specializing in coherent subsets of the firm's total domain. Under these conditions, the employees of a division, for example, should visualize themselves as a team within the larger team constituting the firm. This is a persuasive reason for having profit center results influence the bonus payoffs to the employees of the profit center. On balance, it seems desirable, whenever the accounting system is sophisticated enough, to make the bonus payoff to each employee partly dependent on consolidated results and partly on the results of his or her profit center. In effect, each employee's pay may be divided, with one part receiving a bonus based on the consolidated results of the firm and the other part on profit center results.

In summary, the bonus payoffs should be a function of a small number of timeless and scale-invariant measures of the operating results of the consolidated firm and of the profit centers within the firm. The best candidate variables seem to be:

Annual rate of growth of sales.
Annual rate of return on assets.
Annual rate of growth of productivity.

Not accidentally, these three factors are the principal determinants of the market value of the firm, so are essential to our goal of maximizing the market value of the equity.

Rate of ROA is a good proxy for rate of return on capital, and the excess, if any, of the rate of return on capital over the exogenously determined cost of capital has a great deal to do with the excess of market value over book value of the firm.

Rate of growth of sales and profits is the primary determinant of the present value of growth opportunities, which can account for 60 percent or more of the total market value of a growth company.

Rate of growth of productivity is the factor which enables the firm to grow its sales and profits faster than its employment, thus prospering in spite of shortages of adequately educated and experienced people.

RATIONALE FOR THE EXACT BONUS FORMULA

Suppose we want our bonus payoff factor to be some monotonically increasing function of sales growth (S) productivity growth (P) and ROA (R).

$$\text{Payoff Factor} = f\,(S, P, R).$$

What should be the form of the functional relationship? There are infinite possibilities, so let us go about a process of convergence. Do we think the function should be built around some weighted sum of the three factors or around some product or quotient of the factors?

If we feel that a given level of ROA, for example, would be more valuable at higher than at lower rates of growth of sales, then the functional relationship should include the product of sales growth times ROA. However, if we feel that each of the three performance factors is independently valuable, that is, equally valuable at all levels of the other factors, then the functional relationship should be built around the weighted sum of the three factors.

In Analog Devices our instinct is that over the ranges of interest the three factors are independently valuable, therefore we conclude that the functional relationship should be of the form

$$\text{Payoff Factor} = f\,(aS + bP + cR),$$

wherein the coefficients a, b, and c are constant weighting factors representing our judgments of the relative importance of the three factors.

Now we want the payoff factor to increase monotonically with increasing weighted sums of the factors. The simplest such function is the linear relationship:

$$\text{Payoff Factor} = K \times (aS + bP + cR).$$

However, we feel that the rate of increase of the payoff factor should not be constant but should itself increase with increasing sums of the performance factors. In other words, our instinct tells us that the function should not only increase monotonically but should also be concave upward. To make the function concave upward, we can raise the bracketed expression to some power greater than unity, causing the function to become

$$\text{Payoff Factor} = K \times (aS + bP + l\, cR)^d,$$

where d is some constant greater than unity.

Converging another step, let us restrict the exponent d to the set of odd integers greater than unity. This assures us that the payoff factor would be negative if the sum of the weighted factor were negative, whereas an even exponent would cause the computed payoff factor to be positive if the weighted sum of the performance factors dropped below zero. (Of course, we do not expect the sum to be negative very often and we do not intend to pay negative bonuses, so if we wished we could deal with negative performance factors with an arbitrary cutoff.)

The exponent could be any odd integer; the greater the exponent the more sharp the departure of the payoff function from linearity. Let us try the exponent 5, and adjust it if necessary. Then the functional relationship becomes

$$\text{Payoff Factor} = K \times (aS + bP + cR)^5,$$

which is equivalent to

$$\text{Payoff Factor} = \left(\frac{aS + bP + cR}{M}\right)^5,$$

where M is another constant, a transformation of the original constant K, simplifying the payoff function and making its form more intuitively pleasing.

Deciding the values of the coefficients a, b, and c is necessarily intuitive. For Analog Devices, we consider the ROA ratio somewhat more difficult to achieve, so we set coefficient c equal to 1.00 and coefficient a equal to 0.40, and it seems not unreasonable to make coefficient b also 0.40. The payoff function then becomes

$$\text{Payoff Factor} = \left(\frac{R + 0.4\ (S + P)}{M}\right)^5$$

Setting the coefficient M is also judgmental. The question is, What do you want the payoff factor to be when sales and productivity growth rates and ROA are at some specified goal levels.

In Analog Devices, S = 25 percent, P = 15 percent, and R = 23 percent are aggressive expectations, therefore let us set the coefficient M so that when these levels of performance are achieved the payoff factor is 1.0. The value of M that satisfies the condition is 39.0, so the final bonus formula becomes

$$\text{Payoff Factor} = \left(\frac{R + 0.4\ (S + P)}{39}\right)^5$$

If performance rises to 35 percent sales growth, 15 percent productivity growth, and 25 percent ROA, the payoff factor becomes 2.05, which seems to be adequate incentive to try to keep the performance at model level, on average, over the business cycle.

The final question is whether to cap the payoff factor so it cannot exceed a ceiling level no matter how strong the performance, and, if so, at what level to set the cap. We might set the cap at 3.0 payoff factor, but there would still remain an incentive to perform at higher levels during boom periods so as to stabilize payoffs between 2.0 and 3.0 throughout the normal business cycle. (At Analog Devices the payoff is capped at 2.0.)

Since excessively rapid growth of sales can be counterproductive unless accompanied by adequate ROA, and very high ROA would be sterile without growth of sales, we also cap the payoff at a certain rate of growth of sales and at a certain rate of ROA and require that any increases in the payoff factor must be based on better balanced performance. Also, there should be some level of rate of growth of sales below which no bonus is warranted.

Table 31-2 and Figure 31-1 show how the hypothetical bonus formula would pay off for various combinations of performance factors. In particular, the graphic presentation shows that if performance drops below the heavy border on the south or moves to the left of the heavy border on the west, the payoff fac-

Table 31-2. Bonus Formula: Payoff Factor = [(R + .4 × G)/39]⁵

Sum of percent annual rates of growth of sales and productivity (G = S + P)	*Percent Annual Rate of Return on Assets (R)*										
	11	*13*	*15*	*17*	*19*	*21*	*23*	*25*	*27*	*29*	*31*
0	0.0	0.0	.01	.02	.03	.05	.07	.11	.16	.23	.32
5	0.0	.01	.02	.03	.05	.07	.11	.16	.23	.32	.43
10	.01	.02	.03	.05	.07	.11	.16	.23	.32	.43	.58
15	.02	.03	.05	.07	.11	.16	.23	.32	.43	.58	.77
20	.03	.05	.07	.11	.16	.23	.32	.43	.58	.77	1.00
25	.05	.07	.11	.16	.23	.32	.43	.58	.77	1.00	1.28
30	.07	.11	.16	.23	.32	.43	.58	.77	1.00	1.28	1.63
35	.11	.16	.23	.32	.43	.58	.77	1.00	1.28	1.63	2.05
40	.16	.23	.32	.43	.58	.77	1.00	1.28	1.63	2.05	2.54
45	.23	.32	.43	.58	.77	1.00	1.28	1.63	2.05	2.54	3.00
50	.32	.43	.58	.77	1.00	1.28	1.63	2.05	2.54	3.00	3.00
55	.43	.58	.77	1.00	1.28	1.63	2.05	2.54	3.00	3.00	3.00

Bonus Payoff Factor = $[(R + .4 \times G)/39]^5$ (see Table 31-2 for data points)

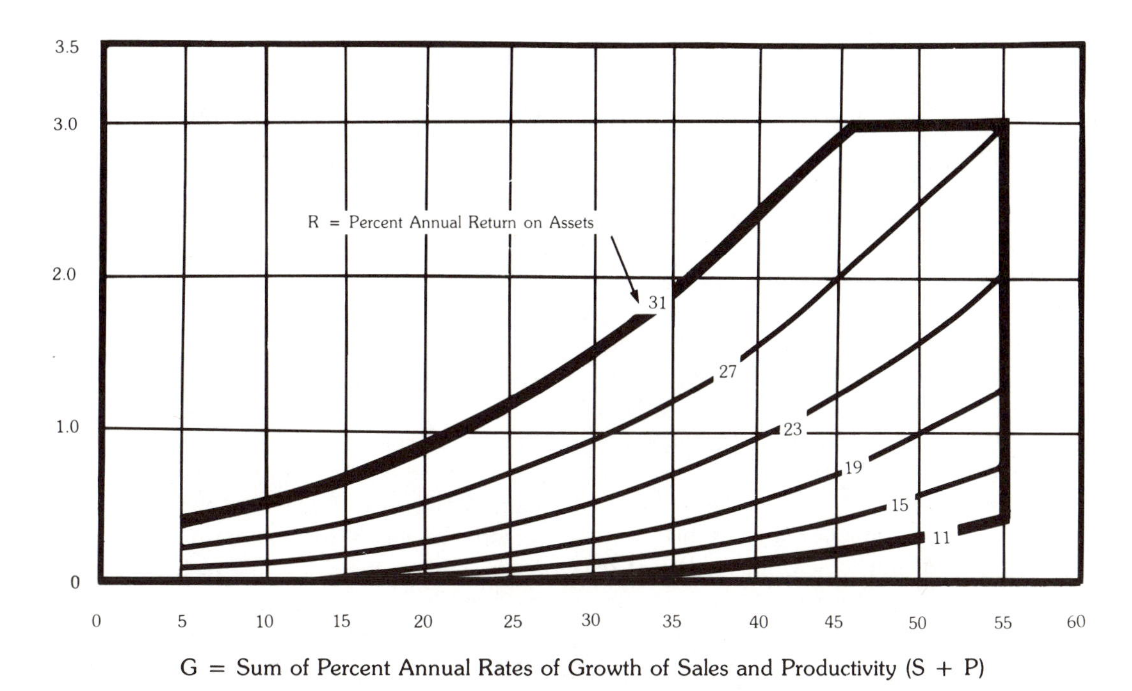

Figure 31-1. Bonus payoff factor as function of sales and productivity growth and return on assets with limits at 55 percent growth and 3.0 payoff factor.

tor becomes zero, whereas the payoff factor may not exceed the limit implied by the heavy borders on the north and east sides of the payoff region.

AVERAGING PERFORMANCE FACTORS TO STABILIZE PAYOFFS OVER THE BUSINESS CYCLE

The three key measures of the firm's performance (rate of growth of sales, rate of growth of productivity, and rate of return on assets) are computed and published quarterly but should be averaged over several quarters for use in computing the quarterly payoff factors. The reason for averaging is to help shield the payout factors from transient events and to ease the pain when the industry slides into recession. Another practical effect of averaging is to reward the employees when they are able to achieve superperformance during boom periods, by allowing some of that superperformance to slide out from under the cap and into subsequent more stringent periods.

For the Analog Devices plan, we use only rate of growth of sales and rate of ROA. Analog Devices' definitions of average performance factors for bonus purposes are:

Average Rate of Growth of Sales is the simple average of the three most recent determinations of quarter-from-same-quarter-last-year growth rates. (If we used productivity growth as one of our bonus factors, we would probably average it similarly.)

Average Rate of ROA is the ratio of the annualized sum of the most recent three quarters' operating profits to the average operating assets employed during those three most recent quarters.

Three-Quarter-Moving-Average-Assets means the average of the three separate quarters' beginning and ending assets.

ELIGIBILITY FOR THE BONUS PLAN

One purpose of a bonus plan should be to share the value added of the firm among the people who created it, providing feedback to help assure that the trend-line rate of growth of that value added proceeds at close to the model rate.

That each employee of the firm creates his or her share of the value added is not simply a figurative statement but is literally true. Normally the pay and benefits of employees constitute about two-thirds of the value added of the average manufacturing firm. Operating profit makes up the remainder of the value added of the firm.

Since every employee creates his or her small share of the total value added of the firm, it is logical that essentially every employee should be covered by the bonus system that makes some fraction of the employees' share of the total value added conditional on the performance of the firm.

Since there is always the possibility that a newly hired employee will not yet be productive and since the rate of turnover during the first year on the job is generally higher than in later years, it is good practice to consider a new employee to be in a probationary or training mode during his or her first year with the firm and to restrict participation in the bonus plan to those employees with one year or more of service.

MODEL RATIO OF BONUS TO BASE PAY

It is generally understood that the higher one's rank (base pay) in an organization, the more influence one has over results and the more risky one's position. Since the higher one's compensation, the farther one lives from the poverty line, many also feel that higher-ranked people can tolerate having higher percentages of their total compensation at risk, that is, contingent on the operating performance of the firm. Table 31-3 shows how the personal payoff factor may be used to put a larger share of the risk at the higher levels of the firm.

There is also the question of whether or not to cap the individual payoffs and, if so, at what level. One option is to say that the payoff factor may never exceed the model level, however high the performance of the firm. The other option is to set some higher cap, say, 3.0 or 4.0, over the payoff factors for the higher-ranking people and to slant the cap down to 2.0 for lower-ranking people. (At Analog Devices, we use the same 2.0 cap for everyone.)

Finally, the model payoff factor need not be

Table 31-3. Risk Increases with Rank.

Base Pay					*Bonus*
Range	*Average*	*Personal Factor*	*Model Payoff Factor*	*Average Bonus Compensation*	*Percent of Base Pay*
100,000-150,000	125,000	.25	2.0	62,500	50
70,000- 99,000	85,000	.20	2.0	34,000	40
50,000- 69,000	65,000	.15	2.0	19,500	30
30,000- 49,000	40,000	.10	2.0	8,000	20
Below 35,000	20,000	.05	2.0	2,000	10
Average of all employees					11.3[a]

[a]Not a ceiling, merely a reconciliation to the model of the firm, assuming base pay is 31 percent of sales.

2.0 but could as well be 1.0 or 3.0. The important thing is to decide what percentage of compensation should be at risk and to adjust the personal factors and the payoff function to yield a meaningful bonus (say 5 percent to the lowest ranking employees) at some conservative expectation of performance and a significantly higher bonus for model performance.

TIMING OF THE BONUS PAYOUTS

A bonus plan should be used to communicate and recommunicate corporate and division performance standards and to give all employees useful feedback on the most recent operating results of the company as well as information about the outlook for the future. The factors determining the bonus payoff factor are few and easily understood—sales and productivity growth and ROA—and since every employee has at least a little influence over all of these factors, prompt and frequent (that is, quarterly) bonus payoffs can be very efficient communications.

For public corporations, at least, operating results are published quarterly; these quarterly events should also be the occasion for managers to address all employees and, among other things, to announce the bonus payoff factors that will result in bonus checks a few days thereafter.

If bonus payments were made more frequently than quarterly, they could become confused with disbursements of base pay. This would not be desirable because it would weaken the feedback message and result in more intense withdrawal pains during the downswings of the business cycle. On the other hand, if the payments were made less frequently than quarterly, the communications and feedback effects would also be weakened, and, in an extreme case, the firm might find itself in a steep downturn at the time the bonus payoff was due and might not find the courage to make the payment, thus depriving the system of one of its key elements—its firm foundation on recent past operating results.

ACCOUNTING FOR THE BONUS EXPENSE

The operating profits used to compute bonus ROAs should be worldwide consolidated profits, thus not affected by transfer prices. The operating profits reported to the outside world and used to compute the ROAs for bonus purposes should be themselves net of accruals for bonus liabilities.

The accruals should not be based on averaged performance factors but on performance factors calculated from the current quarter's results and should not be capped. However, the actual cash bonus payouts should be based on averaged performance factors and should occur within a few days of the publication of each quarter's results. These quarterly disbursements should not be charged to current operating expense but to the accrued reserve for bonus liabilities.

These accounting practices will cause the accrued bonus liability to exceed the cash payouts during boom periods and the payouts to exceed the accruals during downturns of the business, thus smoothing both reported earnings and bonus payoffs. Under these conditions, it is conceivable that a firm could ride through a brief recession with bonus payoff factors remaining at the ceiling level throughout.

The data tabulated in Table 31-4 depict a hypothetical business cycle composed of sets of three strong sales growth quarters followed by two flat quarters and show how the prompt accrual of bonus expense and the variability of the bonus reserve can operate to stabilize the actual bonus payments in the 6-12-percent range throughout. If the profit percentage of sales had been driven higher during the boom periods, the bonus payments could have been virtually constant at about 12 percent of average base pay.

Table 31-4. Bonus Over a Hypothetical Business Cycle

					Operating Profit		This Quarter		3-Qtr Average		Bonus Payoff		
Quarter	Assets	Sales	Cost and Expense	Bonus Accrual	Amount	% of Sls	Sls Growth %	ROA %	Sls Growth %	ROA %	Amount	% of Pay	Bonus Reserve
11	5,934	2,082	1,645	103	334	16.1	44.3	23.9	33.2	25.7	74	11.5	88
12	5,934	2,082	1,700	72	310	14.9	44.3	20.9	38.8	23.7	76	11.8	24
13	5,934	2,082	1,800	18	264	12.7	27.7	17.8	38.8	20.8	52	8.1	54
14	6,705	2,353	1,859	71	423	18.0	27.7	26.8	33.2	21.9	50	6.9	45
15	7,577	2,658	2,100	80	478	18.0	27.7	26.8	27.7	24.0	56	6.8	78
16	8,562	3,004	2,373	148	483	16.1	44.3	23.9	33.2	25.7	107	11.5	85
17	8,562	3,004	2,450	103	451	15.0	44.3	21.1	38.8	23.8	111	11.9	70
18	8,562	3,004	2,550	33	421	14.0	27.7	19.7	38.8	21.5	83	8.9	36
19	9,675	3,395	2,682	103	610	18.0	27.7	26.8	33.2	22.6	80	7.6	93
20	10,932	3,836	3,030	117	689	17.9	27.7	26.7	27.7	24.6	88	7.4	65
21	12,353	4,335	3,424	215	695	16.0	44.3	23.9	33.2	25.7	154	11.5	154

SUMMARY

A bonus plan should be designed to focus organization attention on and to reward sustained high productivity of the firm's human and financial resources, said productivity being measured against the key performance standards of the firm. A bonus plan is a feedback loop, assuring a more equitable sharing of the value added than would be the case if employees' compensation were fixed at base pay levels in good times as well as bad.

A good bonus plan provides positive, invigorating feedback to the employees while simultaneously providing negative, stabilizing feedback to the firm, working to moderate both the peaks and the valleys of the business cycle and to foster job security and steadiness of management of those internal activities that are the driving forces behind the firm's successes.

For a growth company, the key measures of success are rapid growth of sales, proportionate growth of productivity of the employees, and high return on assets. The intent of the bonus plan should be to encourage team effort and to share the rewards of team success in achieving or exceeding the growth rate and return on assets goals of the firm. These measures of success are also the most important determinants of the market value of the equity of the firm, which we wish to continuously maximize.

APPENDIX 31A

STEP-BY-STEP DESIGN OF A BONUS PLAN

1. List the performance measures of the firm that should become the basis for the bonus plan and record precise definitions, including rules for averaging over two or more quarters.
   Suggestions: Rate of growth of sales = S
   Rate of growth of productivity = P
   Rate of return on assets = R
2. What relative weights a, b, c, and so forth should be applied to the factors selected? For example:
   $0.5R$; $0.3S$; $0.2P$
3. Decide on a goal (conservative expectation) combination of levels of these factors that, when achieved, will justify a bonus of 5 percent of base pay for the lowest ranking employees. For example: R = 24 percent, S = 30 percent, P = 15 percent. The weighted sum of these goal measures becomes the denominator of the bonus payoff function

$$\text{Payoff Function} = \frac{0.5R + 0.3S + 0.2P^k}{12 + 9 + 3}$$

4. Choose an exponent k that will impart an appropriate degree of nonlinearity for performances above the goal levels. Suggestions: k = 3 or 5 or 7
5. Decide whether the payoff function should be capped and, if so, at what level, and what other limits should be imposed. For example, payoff factor may not exceed 4.0. Increments of sales growth rates above 50 percent per year do not count for bonus purposes.
6. Decide who should be eligible to participate in the bonus plan. Suggestion: All regular full-time employees not on probation, with one year or more of service at the start of a fiscal quarter and also on the payroll at the end of the quarter should be eligible for bonuses on account of results through that quarter.
7. Articulate the rule for computing the individual employees' bonuses. For example, the bonus payment shall be the product of the employee's quarterly base pay at the start of the quarter, times his or her personal factor times the payoff factor computed from the suitably averaged results of the quarter.
8. Assign personal factors and payoff factor caps by job grade or some other suitable symbol of relative rank. For example,

Grade	Personal Factor	Cap on Payoff Factor
5	.25	4.0
4	.20	3.5
3	.15	3.0
2	.10	2.5
1	.05	2.0

9. Announce and explain the plan.

APPENDIX 31B

ANALOG DEVICES, INC.

ABSTRACTS FROM THE GENERAL RULES FOR THE 1985 BONUS PLAN

1. Description and purpose of the bonus plan

The bonus plan is designed to focus the attention of the organization on and to reward sustained high productivity of the human and financial resources invested in the company. The intent is to encourage team effort and to share the rewards of team success in achieving or exceeding the worldwide consolidated growth rate and return on assets goals of the company.

There is a corporate bonus payoff function centered at 25 percent growth and 23 percent return on assets and yielding a bonus payment payoff factor of 1.00 for that combination and different payoff factors for other combinations of these two performance measures. This same model is used for computing divisional payoff factors.

The payoff function is deliberately nonlinear, causing the payoffs to increase more rapidly than performance, up to specified limits.

The president at his discretion may invite a division to participate in the corporate plan. Normally this invitation would only be made to start-up divisions or early-stage product lines.

2. Definition of Terms

Sales Growth Rate is the average of the three most recent observations of quarter-from-prior-year's-quarter growth rates.

Return on Assets is a three-quarter average computed as operating pretax profit reported in the three most recent quarters, annualized, divided by the average of those three quarter's average net operating assets. Operating PBT is net of the actual bonus payments and accruals.

Operating Pretax Profit for the consolidated corporation is defined, for bonus purposes, as operating income as reported in the company's public statements, adjusted to eliminate the effects of fixed overhead expense deferred to or relieved from inventories. Amortization of deferred startup costs and the amounts paid and accrued under the bonus plan are operating expenses above the operating profit line.

Net Operating Assets of the consolidated corporation include all assets except cash and short-term investments in excess of model requirements and investments and goodwill.

Three-Quarter Average Net Operating Assets means one-third of the sum of the averages of each of the three quarters' beginning and ending net operating assets.

Summary Performance Measure (*Q*) is the weighted sum of return on assets and the growth rate. The weighting factors have been set at 1.00 for ROA and 0.40 for sales growth.

$$Q = \text{ROA} + 0.40 \times \text{sales growth}.$$

Summary Performance Standard (Q^s) is the weighted sum of 23 percent return on assets, 25 percent sales growth rate:

$$Q^s = 23 \text{ percent ROA} + 0.40 \times 25 \text{ percent sales growth} = 33.0 \text{ percent}.$$

Payoff Factor (B) is computed from the summary performance measure (*Q*) divided by the constant 33, with this quotient then raised to the 4.5th power, except that the resulting factor may not exceed 2.00

$$\text{Thus } B = \left(\frac{Q}{33}\right)^{4.5}, \text{ but not to exceed 2.00.}$$

At 23 percent ROA and 25 percent sales growth, a payoff of 1 results from the bonus formula.

$$\left(\frac{23 \text{ percent ROA} + .4 \text{ percent } 25 \text{ percent sales growth}}{33}\right)^{4.5} = \left(\frac{33}{33}\right)^{4.5} = 1.00$$

At model of 24.5 percent ROA and 35 percent sales growth, a payoff of 2 results from the bonus formula.

$$\left(\frac{\begin{array}{c}24.5 \text{ percent} \\ \text{ROA} + .4 \text{ percent} \\ 35 \text{ percent sales growth}\end{array}}{33}\right)^{4.5} = \left(\frac{38.5}{33.0}\right)^{4.5} = 2.00$$

3. Personal Bonus Percent

Personal bonus percent is the percent of base pay payable as bonus to a participant. This is established according to job grade and content and ranges from 5 percent to 25 percent.

4. Bonus Calculations

The payoff factor, computed each quarter based on operating results through that quarter, is multiplied by each participant's personal bonus percent and beginning of quarter base salary, including shift differential, if any, to arrive at the cash bonus payable to that participant for that quarter's performance. For divisional participants, bonus payout factors are based 50 percent on divisional and 50 percent on corporate results.

5. Participation Eligibility

All regular employees, except those on probation, shall participate in the bonus plan in the first full fiscal quarter following their first anniversary date as employees and every quarter thereafter.

For employees in exempt labor grade 8 and up, participation in the plan shall begin in that fiscal quarter immediately following the fiscal quarter in which such participant was hired or promoted into exempt labor grade 8.

6. Definitions of Payment Terms

Every participant is paid a quarterly bonus computed 50 percent on corporate results and 50 percent on divisional results, except for corporate participants and members of the executive committee who are paid bonuses based solely on corporate results.

Changes in personal bonus percent as a result of promotions and so forth that occur in the middle of a quarter will not be effective until the subsequent quarter.

Every participant's total bonus is charged to the same division as his or her base pay, except that bonuses to the executive committee and to corporate staff members are charged to the president's cost center.

No participant shall be entitled to a bonus award for any particular fiscal quarter if he or she is not employed by the company at the beginning and end of such fiscal quarter. This provision may be waived by the board of directors of the company

Employees on an authorized leave of absence or disability of over four weeks' duration will receive bonus payments on a prorated basis. Proration will be based on whole work weeks. If the absence spans two quarters, each quarter's bonus will be prorated according to the amount of time the employee was actively at work during each quarter. Bonus checks for employees on leave are not released until the employee returns to work.

7. Cutoff Levels of the Bonus Plan

A saturation point is built into the bonus formula at factors of 40 percent sales growth and 29 percent ROA. Division or company performance beyond either of these points will not result in a higher payout unless movement along the other axis is also achieved. This is consistent with the objective of Analog's strategy for balanced performance where neither of the performance variables will be pursued or rewarded at the expense of the other.

At the beginning of each fiscal year, the executive committee and/or the board of directors will specify the levels of consolidated quarterly earnings below which no bonuses will be earned by any employee anywhere in the corporation. Similar cut-off levels may be established for each entity's bonus plan, standing independently of the rest of the company. In FY 1985, no bonuses will be paid for any quarter in which consolidated corporate operating pretax profit is below 50 percent of the Benchmark Plan OPBT for that quarter.

8. Communicating The Approved Bonus Plans

Copies of applicable general and special rules with a list of the approved participants, their levels of participation, and their personal bonus percents will be communicated to corporate, division, and group managers in writing. The division manager and controller are responsible for administering the plan. The system is subject to post-audit by the manager of internal auditing.

9. Miscellaneous

A. No employee nor his or her spouse or other designee shall have any right to commute, sell, assign, pledge, transfer or otherwise convey any interest he or she may have under the plan.
B. Benefits payable under the plan shall be independent of and in addition to any employment agreement that may exist from time to time between a participant and the company. The plan shall not be deemed to constitute a contract of employment between a participant and the company, nor shall any provision hereof restrict the right of the company to discharge any participant or restrict the right of any participant to terminate employment.
C. The rights of any participant under the plan shall be solely those of an unsecured creditor of the company.
D. The invalidity or illegality of any provision of the plan shall not impair or affect the validity or enforceability of any other provision of the plan, and such other provisions shall remain in full force and effect in accordance with their terms.

10. Amendment and Termination

The company may amend the bonus plan by appropriate action of the board of directors. There is no obligation on the part of the company to continue the plan after the end of the 1985 fiscal year. However, the company may continue or amend the plan for any subsequent fiscal year or years by appropriate action of the board of directors.

VIII

Organizing and Evaluating Planning

The manner in which strategic planning and management is itself organized and evaluated ultimately will determine the degree to which the promise of the field is achieved. Part VIII deals descriptively with the organization of the planning function and prescriptively with the evaluation of planning.

In Chapter 32, Ravi R. Chinta, Per V. Jenster, and William R. King present the results of a survey of members of the North American Society for Corporate Planning (now The Planning Forum), the major professional society in the strategic planning profession. They describe the different roles that planners actually play at various levels in the firm and in different kinds of businesses.

In Chapter 33, William R. King prescribes an approach to evaluating the effectiveness of a strategic planning system. The approach involves the development of an overall profile made up of a variety of individual assessments of the input, output, and process elements of strategic planning.

CHAPTER 32

Functions of Strategic Planners

Ravi R. Chinta, Per V. Jenster, and William R. King

Ravi R. Chinta, born in India, holds a B.S. in chemical engineering, an M.B.A., and a Ph.D. from the University of Pittsburgh. He has four years of experience as a planner in the chemical industry. Currently he is assistant professor at the College of Business Administration, Louisiana State University.

Per V. Jenster is a faculty member at the McIntire School of Commerce, University of Virginia. He received the Ph.D. in strategic management and management information systems from the University of Pittsburgh. Dr. Jenster's current writing is in the areas of strategic planning and information systems design. His research and consulting activities have involved new ventures as well as large national and international corporations.

William R. King is University Professor of business administration in the Graduate School of Business at the University of Pittsburgh. He is the author, co-author, or co-editor of more than a dozen books and 150 articles that have appeared in the leading journals in the fields of strategic management, information systems, and management science. He has served as senior editor of the *Management Information Systems Quarterly* and serves as associate editor for *Management Science, Strategic Management Journal, OMEGA: The International Journal of Management Science,* and various other journals. He frequently consults in these areas with corporations and government agencies in the United States and Europe.

There is general acknowledgment in the field of strategic management that the nature of strategy is different at the corporate, business unit, and functional levels.[1] At the corporate level, strategic management involves issues concerning the businesses in which to engage and the allocation of resources among them. At the business or divisional level, strategy primarily is concerned with how to compete in a particular industry or product-market segment. At the functional level, strategy focuses on how to maximize effectiveness within each functional area.

Because of these differences in the nature of strategy issues, the tasks of the strategic planner or the planning executive at each level also might reasonably be expected to be different. However, the *processes* of planning at the various levels are, in fact, somewhat generic,[2] so that it is not at all obvious that this expectation is borne out in practice, particularly in terms of the perceptions of the planners who perform the planning tasks.

Despite the general agreement concerning the different issues that comprise this hierarchy of organizational strategies, the hierarchy's implications to the planner's tasks at the various levels has not been well documented.

The objective of this chapter is to present

empirical evidence concerning the task actually performed by planners at the three levels. In addition, the influence of corporate diversity, company size, geographic scope, and business focus on planning roles is addressed. Data for the analyses were collected from members of a large North American professional planning society and have been analyzed by the authors under terms of a contract between that organization and the Strategic Management Institute at the University of Pittsburgh.

HYPOTHESES DEVELOPMENT

The basic working hypothesis is that *the tasks performed by the corporate, group, and divisional planner are different.*

The hypothesis derives from the notion of a strategy hierarchy discussed earlier. In addition, in a case study of three corporations, Lorange demonstrated how adaptation and integration planning needs can vary across levels.[3] If this is generally true, planning tasks should also differ.

The influence of four other potentially important variables also is addressed: (1) corporate diversity; (2) company size; (3) company geographic scope; and (4) primary business focus. The importance of each of these variables has considerable substantiation in the literature of strategic management.

Corporate Diversity

Dundas and Richardson, Calingo et. al., and Leontiades suggest a relationship between the corporation's diversity profile and its strategic planning tasks.[4] Dundas and Richardson proposed that firms of varying diversity (for example, single business vs. multibusiness) exploit different market failures[5] and develop their distinctive competences accordingly. Leontiades also argued that the corporate planning task is determined in part by the state of corporate development (that is, single product, related products, or diversified). The diversity profile is, therefore, a possible determinant of planning needs.

Company Size

Many authors and studies have postulated or demonstrated the importance of size as a determinant of strategy and organizational form. Pugh et al. propose size to be the driving force generating changes in organizational form.[6] Grinyer and Yasai-Ardekani found that the link between strategy and structure in a sample of 40 companies was via size.[7] Lorange and Vancil carried this further to indicate that the size of a company is an important situational factor in determining the type of strategic planning system that an organization needs.[8] Robinson and Pearce found that small banks do not greatly benefit from formalized planning systems.[9] Calingo found size to be related to the comprehensiveness of the planning process and the completeness of the plans produced.[10]

Mintzberg proposed that the larger the organization, the more elaborate is its structure, the more specialized are its tasks, the more differentiated are its units, the more developed is its administrative unit, and the more formalized is its behavior.[11]

Size also is considered an important determinant of "capability to innovate successfully."[12] Therefore, it is often studied in other areas in which the nature of some organizational processes (for example, processes for information system development) is taken to be the dependent variable.[13]

Company Geographic Scope

Geographic scope is commonly used as a moderating variable in studying the relationship between strategy and structure. For instance, Schendel and Patton use geographic scope in their study of strategy in the brewing industry.[14] The international level of geographic scope also has been a major factor in a variety of studies done over the years at Harvard.[15]

Primary Business Focus

Probably the most common taxonomic variable in such studies is one that indicates the nature of the business. This variable is basic to the PIMS data base.[16] PIMS newsletters use a business focus categorization (durables, nondurables, and so forth) as a primary variable.

METHODOLOGY OF THE STUDY

The data for this study were obtained from a national membership survey conducted by the North American Society for Corporate Planning (NASCP). NASCP was a large organization of about four thousand planners and planning executives in business and government. (It has subsequently merged with another organization to form the Planning Forum.) Questionnaire responses were received from 1,373 members; the response rate was 36 percent. Thirty-five percent of the respondents work in firms in the manufacturing industries, 13 percent in financial institutions, and 31 percent in the mining, oil, and chemicals industries.

A total of 1,184 usable responses were reached after adjustment for missing data and invalid answers to the items relevant to this study. Other characteristics of the respondents are shown in Table 32-1

The data on the planning tasks are expressed in terms of frequencies with which each of 12 planning responsibilities were noted by each respondent. The planning tasks are listed by number in Table 32-2.

Three complementary methods of analysis were used to address the issues involved in the hypotheses.

1. Cross-tabulations were prepared and analyzed.
2. The tasks were ranked in terms of the importance attributed to each at each organizational level, and correlation analyses were conducted.
3. Regression analyses were used to confirm the prior analyses as well as to confirm the directionality of prior expectations.

PLANNING TASKS AT VARIOUS ORGANIZATIONAL LEVELS

With respect to the basic hypothesized relationship between planning tasks and organizational level, the data confirm that there is a substantial difference between the relative emphasis given to different planning tasks at the various organizational levels.

Table 32-3 shows the frequency, ranks, and the statistical significance levels. Thus, 8 of the 12 planning tasks can be regarded as

Table 32-1 Number of Respondents by Organizational Level and Diversity Profile of Respondent's Company

	Diversity Profile		
Organizational Level	*Single Business*	*Multibusiness*	*Total*
Corporate	408	409	817
Group	54	92	146
Division	111	110	221
Total	673	611	1,184

Table 32-2. List of Planning Tasks

1. Design and administer strategic planning process (DSPP)
2. Develop corporate plans (Corp Plg)
3. Develop SBU strategic plans (SBU Plg)
4. Provide training and education of management (Trg)
5. Provide internal consulting on strategic planning (Consult)
6. Support acquisitions and divestitures (Acq)
7. Identify business development opportunities and implement (Bus Dev)
8. Review and evaluate strategies (Review)
9. Monitor implementation actions (Impl)
10. Conduct environmental scans and forecasting (Scan)
11. Conduct financial analysis and construct financial plans (Analysis)
12. Participate in functional planning (Func Plg)

Note: An "other" category also was used in the survey instrument and was completed by 73 respondents. However, no significant patterns or results were associated with this category, and it is ignored in this report of the study.

Table 32-3. Frequency, Rank, and Statistical Significance of Planning Tasks at Various Organizational Levels

	Organizational Level					
Planning Task	*Corp*	*Group*	*Div*	*Other*	*Row Total/ Sample Total*	*Significance Level*
1. DSPP	466(1)	77(1)	97(2)	25(1)	665(1)/913	0.0087*
2. Corp Plg	413(2)	28(12)	35(12)	14(5)	490(4)/914	0.0006*
3. SBU Plg	111(12)	58(6)	81(5)	15(3)	265(12)/914	0.0001*
4. Trg	238(8)	34(10)	44(11)	14(5)	330(9)/916	0.0578*
5. Consult	405(4)	75(2)	84(4)	14(5)	528(3)/917	0.0026*
6. Acq	314(6)	49(7)	69(8)	13(7.5)	445(5)/916	0.2066
7. Bus Dev	259(7)	60(4.5)	72(6.5)	13(7.5)	404(7)/917	0.1060
8. Review	406(3)	73(3)	105(1)	17(2)	601(2)/917	0.2471
9. Impl	206(10)	38(9)	72(615)	11(9.5)	327(9)/916	0.0196*
10. Scan	380(5)	60(4.5)	63(9)	8(11.5)	439(6)/917	0.0036*
11. Analysis	219(9)	30(11)	60(10)	11(9.5)	320(11)/915	0.1947
12. Func Plg	191(11)	44(8)	89(3)	8(11.5)	332(8)/917	0.0001*

Notes: Entries are frequencies. Number in parentheses indicates the rank in each column. Statistical significance (*) is based on a chi-square test at the 0.01 level for each task as it relates to the organizational levels.

Table 32-4. Summary of Planning Tasks and Organizational Levels

Planning Tasks Performed at Higher Organizational Levels	*Planning Tasks Performed at Lower Organizational Levels*
Design and administer strategic planning process	Develop SBU strategic plans
Develop corporate plans	Monitor implementation actions
Provide training and education of management	Participate in functional planning

Note: Task 10's level was not confirmed.

being significantly different in terms of the time spent on them at the various organizational levels, as indicated by the asterisks in the right-most column.

Design and administer strategic planning process
Develop corporate plans
Develop SBU strategic plans
Provide training and education of management
Provide internal consulting on strategic planning
Monitor implementation actions
Conduct environmental scans and forecasting
Participate in functional planning

Further analyses were conducted to confirm both these results and the hypothesized directionality. Details of these analyses are contained in Chinta et al.[17] These analyses confirm that tasks 1, 2, and 4, tend to be performed significantly more at higher organizational levels (for example, corporate) and tasks 3, 9, and 12 at lower levels. This is shown in Table 32-4. Task 10's (Scan) level was not confirmed.

INFLUENCE OF CORPORATE DIVERSITY ON PLANNING TASKS

Corporate diversity was assessed with a dichotomous variable—single business versus multibusiness—in the survey instrument. Table 32-5 sums up the results of the several analyses that were conducted in this regard.

Table 32-5. Influence of Corporate Diversity on Planning Tasks

Most Important for Multibusiness Firm	*Most Important for Single Business Firm*
Develop SBU strategic plans	Develop corporate plans
Support acquisitions and divestitures	Participate in functional planning
Review and evaluate strategies	

INFLUENCE OF COMPANY SIZE ON PLANNING TASKS

When company size is assessed in terms of its relationship to planning tasks, the results are those shown in Table 32-6. Company size was measured using a total-sales surrogate with the following categories:

Under $50 million
$50 million to $300 million
$300 million to $1.5 billion
$1.5 billion to $7 billion
Over $7 billion

The array of tasks in Table 32-6 portrays larger firms as being more involved in the more sophisticated aspects of planning (for example, review and scanning), in acquisition analysis, and in SBU planning, whereas smaller firms are more involved in basic design of planning systems and in lower-level

Table 32-6. Influence of Company Size on Planning Tasks

Most Important for Larger Firms	*Most Important for Smaller Firms*
Develop SBU strategic plans	Design and administer strategic planning process
Support acquisitions and divestitures	Develop corporate plans
Review and evaluate strategies	Identify business development opportunities and implement
	Monitor implementation actions
	Participate in functional planning

(functional and implementation) planning concerns.

INFLUENCE OF GEOGRAPHIC SCOPE ON PLANNING TASKS

Firms were categorized as being either Regional/Local, National, or International in the survey instrument. When the influence of this variable was assessed, 4 of the 12 planning tasks were significantly different.

Table 32-7 shows these tasks and the direction of their relationship in terms of the scope of the firm. Thus, firms that were greater in scope placed more emphasis on acquisition and review than those that were narrower in scope.

Table 32-7. Influence of Geographic Scope on Planning Tasks

Most Important for Firms with Broader Scope	*Most Important for Firms with Narrower Scope*
Support acquisitions and divestitures	Develop corporate plans
Review and evaluate strategies	Participate in functional planning

These results strongly suggest that for firms that are international in scope, the doing of the planning requires less time than does the review of plans and acquisition activities.

INFLUENCE OF BUSINESS FOCUS ON PLANNING TASKS

All planning tasks were found to differ significantly across the industry categories that are shown at the left of Table 32-8.

The tasks listed in the body of Table 32-8 are those that the confirmatory regression analysis showed to be the most positively and most negatively associated with each industry. These may be thought of as those tasks in which the most time is spent by planners and managers in that industry and those in which the least is spent, respectively.

The blank rows for four industry categories indicate that no such strongly positive or negative associations were found for those industries.

Thus, in durables, the most time is spent on SBU planning and acquisition analysis and the least on financial analysis and functional planning. In banking the most is spent on designing and administering the planning system, implementation, and scanning and the least on SBU planning, acquisitions, and planning review.

Utilities spend the most time developing the corporate plan and scanning and the least on acquisitions, training, and designing and administering the system.

CONCLUSIONS

These results confirm that the functions of the strategic planner are indeed different at the various levels of the business hierarchy. At the corporate level, the tasks carried on, such as designing and administering the planning system, developing corporate plans, training and internal consulting, differ significantly from the development of SBU plans, implementation, and functional planning that is

Table 32-8. Influence of Business Focus on Planning Tasks

Business Focus	*Positively Associated Planning Tasks*	*Negatively Associated Planning Tasks*
1. Durables	#3 (SBU Plg) #6 (Acq)	#11 (Analysis) #12 (Func Plg)
2. Nondurables	—	—
3. Banking	#1 (DSPP) #9 (Impl) #10 (Scan)	#3 (SBU Plg) #6 (Acq) #8 (Review)
4. Utilities	#2 (Corp Plg) #10 (Scan)	#1 (DSPP) #4 (Trg) #6 (Acq)
5. Wholesale	—	—
6. Mining	—	—
7. Services, Government, and Education	—	—

carried out at lower levels. This confirms the previously untested mythology of the field.

The analyses also suggest that there are not significant unanticipated differences between single-business and multibusiness firms in terms of the planning tasks performed.

When company size is analyzed, larger firms are found to perform the more sophisticated planning tasks (review, scanning, acquisition analysis) with significantly greater frequency than smaller firms. International firms appear to spend less time in the doing of planning and more in review and acquisition activities than do firms of lesser scope.

The analysis of business focus suggests substantial differences across industries with respect to the planning tasks performed. For these industry categories—durables, banking, and utilities—it was possible to develop profiles of the tasks most positively and most negatively associated with the specific industry. These profiles suggest differences that are potentially important in the prescriptions that may be made concerning planning methodologies in these categories of businesses.

Although none of these results are dramatically different than anticipated, they do represent large-sample empirical verification of many widely held beliefs in planning. In their confirmation of the basic notion that planning activities are different at the various levels of the business hierarchy, they validate a fundamental theoretical construct of strategic planning.

REFERENCES

1. J. H. Grant and W. R. King, *The Logic of Strategic Planning* (Boston: Little, Brown, 1982); D. E. Schendel and C. W. Hofer, eds., *Strategic Management: A New View of Business Policy and Planning* (Boston: Little, Brown, 1979).
2. Grant and King, *Logic of Strategic Planning.*
3. P. Lorange, "Formal Planning Systems: Their Role in Strategy Formulation and Implementation," in *Strategic Management: A New View of Business Policy and Planning,* ed. D. E. Schendel and C. W. Hofer (Boston: Little, Brown, 1979), 226-240.
4. K. N. M. Dundas and P. R. Richardson, "Corporate Strategy and the Concept of Market Failure," *Strategic Management Journal* **1**(1980):177-188; L. M. R. Calingo, J. C. Camillus, P. V. Jenster, and T. S. Raghunathan, "Strategic Management and Organizational Action: A Conceptual Synthesis," Working Paper #527, Graduate School of

Business, University of Pittsburgh, 1982; M. Leontiades, "A Diagnostic Framework for Planning," *Strategic Management Journal* **4**(1983):11-26.

5. O. E. Williamson, *Markets and Hierarchies: Analysis and Antitrust Implications* (New York: Free Press, 1975).
6. D. S. Pugh, D. J. Hickson, C. R. Hinings, and C. Turner, "An Empirical Taxonomy of Structures of Work Organization," *Administrative Science Quarterly* **14**(1969):115-126.
7. P. H. Grinyer and M. Yasai-Ardekani, "Strategy, Structure, Size and Bureaucracy," *Academy of Management Journal* **24**(1982):471-486.
8. P. Lorange and R. P. Vancil, "How to Design a Strategic Planning System," *Harvard Business Review* **54**(1976):75-81; Lorange, "Formal Planning Systems."
9. R. B. Robinson and J. A. Pearce II, "The Impact of Formalized Strategic Planning on Financial Performance in Small Organizations," *Strategic Management Journal* **4**(1983):197-207.
10. L. M. R. Calingo, "Tailoring Strategic Planning Systems Design to Corporate Strategy: Effect of Situational Fit on System Performance," Ph.D. diss., Graduate School of Business, University of Pittsburgh, 1984.
11. H. Mintzberg, *The Structuring of Organizations* (Englewood Cliffs, N.J.: Prentice-Hall, 1979).
12. R. W. Zmud, "Process Innovations: An Examination of 'Push Pull' Theory and the Role of External Information Channels," Working Paper, School of Business Administration, University of North Carolina, June 1982.
13. J. Hage, *Theories of Organizations: Form, Process, and Transformation* (New York: Wiley, 1980); M. K. Moch, "Structure and Organizational Resource Allocation," *Administrative Science Quarterly* **21**(1976):661-674.
14. D. E. Schendel and G. R. Patton, "A Simultaneous Equation Model of Corporate Strategy," *Management Science* **24**(1978):1611-1621.
15. L. E. Fouraker and J. M. Stopford, "Organizational Structure and Multinational Strategy," *Administrative Science Quarterly* **13**(1968):47-64; J. Stopford, "Growth and Organizational Change in the Multinational Field," Ph.D. diss., Harvard University, 1968; J. Stopford and L. Wells, *Managing the Multinational Enterprise* (New York: Basic Books, 1972); L. Franko, "The Move toward a Multidivisional Structure in European Organizations," *Administrative Science Quarterly* **19**(1974):493-506; L. Franko, *The European Multinationals* (Stamford, Conn.: Greylock Publishers, 1976).
16. S. Schoeffler, R. D. Buzzell, and D. F. Heany, "Impact of Strategic Planning on Profit Performance," *Harvard Business Review* **52**(1974):137-145; R. D. Buzzell, B. T. Gale, and R. G. M. Sultan, "Market Share—A Key to Profitability," *Harvard Business Review* **53**(1975):97-106; S. Schoeffler, "Cross-sectional Study of Strategy, Structure and Performance: Aspects of the PIMS Program," in *Strategy and Structure = Performance*, ed. H. B. Thorelli (Bloomington, Ind.: Indiana University Press, 1977), 108-121.
17. R. Chinta, P. Jenster, and W. R. King, "The Functions of the Strategic Planner: An Empirical Analysis," Working Paper #590, Graduate School of Business, University of Pittsburgh, 1985.

CHAPTER 33

Evaluating the Effectiveness of Planning Systems

William R. King

William R. King is University Professor of business administration in the Graduate School of Business at the University of Pittsburgh. He is the author, co-author, or co-editor of more than a dozen books and 150 articles that have appeared in the leading journals in the fields of strategic management, information systems, and management science. He has served as senior editor of the *Management Information Systems Quarterly* and serves as associate editor for *Management Science, Strategic Management Journal, OMEGA: The International Journal of Management Science,* and various other journals. He frequently consults in these areas with corporations and government agencies in the United States and Europe.

Most of the evaluations of planning effectiveness that are conducted are either so impressionistic as to be of questionable validity or so global and academic as to be of no practical utility. The former variety is exemplified by a corporate planning director who recently said that the only evaluation of planning effectiveness that was relevant was, "Does it keep the CEO happy?" The latter approach is represented by a host of academic research studies that purport to relate planning to overall business performance (profitability, growth, and so forth).[1]

If the field of strategic planning is to continue to develop, it must have some evaluative base that is more objective and reliable than the boss's or a collection of bosses' approval. If these evaluations are to be useful, they must be more specific than general conclusions that planning does or does not lead to improved business performance.

EVALUATING PLANNING

Planning is a process that represents only one element of a strategic planning system (SPS).[2] Therefore, a comprehensive evaluation must include the other elements of the system as well—the plans, their content, and so forth.

Figure 33-1 shows a simple mode of planning in an organization. This planning function is accomplished through an SPS, as shown in the double-boxed rectangle. The SPS is the complete set of processes, techniques, staff organizations, computer support systems, and so forth through which the organization performs planning.

The figure shows two inputs to the SPS. The Resource Inputs are the people, funds, computer time, and so forth that are consumed in the planning function. It also includes the line managers' time that is used in planning. The Planning Goals represent the

Adapted, with permission, from W. R. King, "Evaluating the Effectiveness of Your Planning," *Managerial Planning* **33**(1984):4-9.

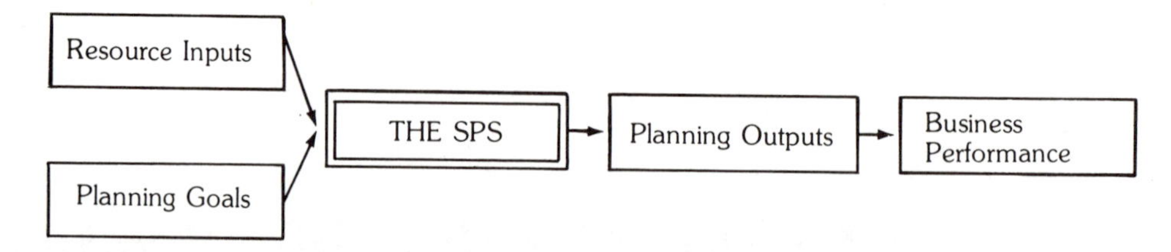

Figure 33-1. Schematic diagram of planning in an organization.

specific purposes for which the SPS was developed (and against which its efficacy must therefore, in part, be assessed).

An SPS produces outputs called *plans*. Figure 33-1 depicts Planning Outputs that represent the content of those plans—the strategies, objectives, and programs selected, via the SPS, to guide the organization toward its future.

At the right, Figure 33-1 shows the outputs as having an impact on business performance as measured by profitability, growth, and so forth.

Figure 33-2 shows an expanded version of Figure 33-1 that can be used as a framework for evaluating planning. Only one basic element has been added to those in Figure 33-1—the large block at the top labeled External Standards. The other arrows and letters represent elements of the planning evaluation methodology that will be explained subsequently.

The External Standards block in Figure 33-2 represents those bodies of external standards that may be applied to various phases of planning and elements of the output of planning. For instance, an external standard might be that which is considered *good planning practices* in the field, or it might be an industry norm. Although planning must be evaluated in terms of internal standards such as the goals established for the SPS, such external standards are also important.

The various elements of the planning evaluation framework are represented with circled letter designations in Figure 33-2. The various assessments are of the:

A. Effectiveness of Planning
B. Relative Worth of the SPS
C. Role and Impact of the SPS
D. Performance of Plans
E. Relative Worth of Strategy
F. Adaptive Value of the SPS
G. Relative Efficiency
H. Adequacy of Resources
I. Allocation of Planning Resources
J. Appropriateness of Planning Goals

Each of these elements of the direct assessment framework are hereafter discussed and illustrated. After each has been separately discussed, the *overall* evaluation of planning will be treated.

Effectiveness of Planning (A)

This represents a set of measurements that can be made to address the issue of how well the firm's SPS has met *its* goals. Every firm begins to do strategic planning, then changes to a new SPS or to a new approach to planning with some *planning goals* in mind. Often, these are communicated in the form of a proposal for a new system or for the hiring of a consultant.

Although the logic of evaluating a system in terms of the goals that were established for it is irrefutable, the fact is that it is seldom formally done. In fact, this is a common problem in complex systems of many varieties.[3]

In one company, this assessment was performed by first having various stakeholder groups identified:

Corporate top management
Corporate planning staff
Other corporate staff groups
Business unit top management
Business unit planning staff
Other business unit staff
Other business unit line management

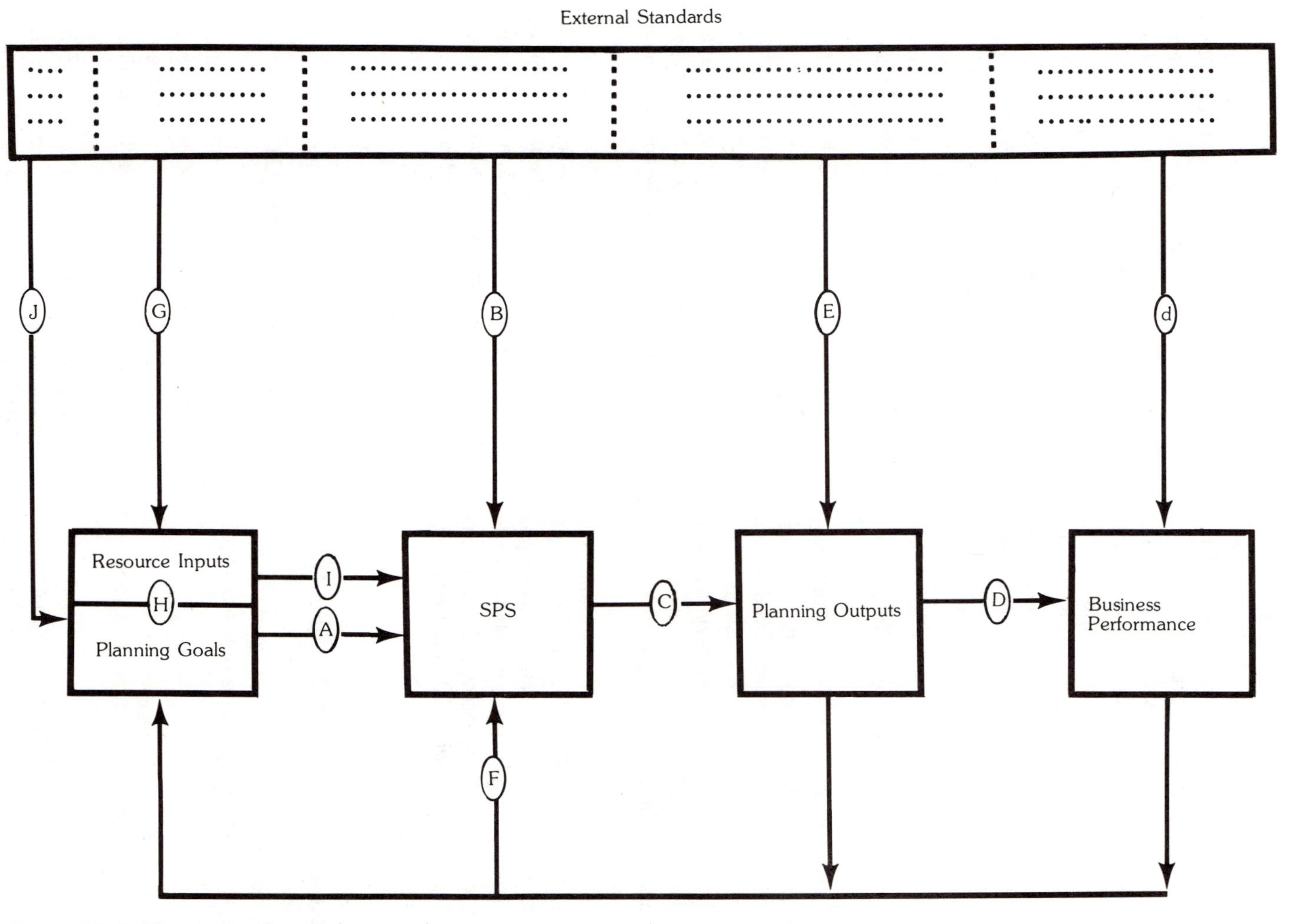

Figure 33-2. Planning evaluation framework.

Representatives of each of these groups were identified and asked to complete an instrument in which they evaluated the SPS in terms of the degree to which they perceived it to have attained the goals that had been (some years previously) prescribed for it. Personal interviews were conducted to ensure that revisions in the goals were not ignored and to give each individual the opportunity to elaborate on the responses given in the instrument. The resulting evaluations were tabulated and reported in the form of an effectiveness profile for the SPS.

Relative Worth of the SPS (B)

This element of the assessment relates the features and characteristics of the SPS to *external standards for modern planning*. Such standards can be based on those developed by Steiner, for instance,[4] or they can be related to the benefits that have been prescribed as attributable to planning—for example, see Warren.[5] Making such a comparison serves to relate the planning system to the state of the art in the field, whereas the previous measures relate it only to those goals that were desired by the firm. Holloway and King present a methodology that has been applied, in a somewhat different form, to the making of such assessments.[6]

Role and Impact of the SPS (C)

This assessment addresses a set of issues that are best thought of in terms of two questions:

1. Are the strategic plans produced by the system *implemented* in the organization?
2. Is the plan really used to guide the strategic direction of the firm?

The first question deals with a characteristic of the SPS's operation, which may have already been captured in terms of the external standards of B. However, it is so important that it is specifically addressed as an element of C. The issue of planning implementation, as well as implementation in other contexts, has received much attention in recent years and various instruments have been developed to assess the degree of implementation and to diagnose implementation difficulties.[7]

The second question has to do with the degree to which the strategic choices, which are the outputs of the SPS, truly guide the firm. In some firms, a sophisticated planning process is carried on, but many of the firm's strategic moves—such as acquisitions—are, in fact, opportunistically arrived at through the actions of a chief executive or other high official. In such cases, planners often believe that the SPS, however good it may intrinsically be, is devalued because it is not truly guiding the most important choices of the firm. This clearly must be judged in perceptual terms as in A. In one case in which this was done, the perceptions of top managers and those of middle managers and planners were found to be quite different. Top managers believed that the SPS was, in fact, guiding the firm's strategy and its strategic actions, whereas middle managers (to some degree) and staff planners (to a very large degree) felt that it was not. This resulted in the identification of a need for better feedback from top management to those who were intimately involved in the process of formulating alternative strategies and making recommendations based on them.

Performance of Plans (D)

The ultimate test of the strategic choices made by a firm is business performance. However, unlike the black box relationship postulated in the indirect evaluation approach, the direct approach involves a more detailed and specific view of the strategy-performance relationship. In particular, it (1) focuses on the specific impact of *each* strategic choice element, not merely on aggregate business performance; and (2) assesses performance relative to specific business goals and objectives, rather than merely in terms of aggregate organizational measures.

In making such assessments, such things as the specific goals and targets for strategic programs need to be established and used as a standard for comparison. A company that has a system such as that of Texas Instruments, in which various levels of strategy are delineated and managed has a clear basis for doing this. It is more difficult to do at the higher levels of strategy, but it is certainly feasible at the strategic program level.[8]

However, such aggregate measures as earnings per share and the value of the business, are indeed relevant. For this reason, the lower-case letter "d" is shown in Figure 33-2 to indicate that the direct measurement of the performance of each strategic choice element relative to its goals (D) must be complemented by the assessment of business performance relative to external industry and market standards (d). The two assessments, taken together constitute a complete assessment of the performance of plans.

Relative Worth of Strategy (E)

In addition to the performance basis for assessing strategy, there are some external standards that may be applied to strategy. For instance, internal consistency is an important standard that is widely applied.[9]

In addition, one important external standard that is not directly captured by business performance is that which has to do with *strategic opportunities*. To illustrate, in one company an assessment was made on the basis of the question: What opportunities have been available or considered that have been foregone and not enacted into strategy? Two varieties of such opportunities are relevant: (1) those good opportunities that were either not identified or not recognized to be good; and (2) those bad opportunities that were foregone after having been identified and assessed. The former are clearly missed opportunities that serve to devalue the performance of the SPS. The latter represent an important output of formal planning that is seldom discussed or identified—those bad ideas that the SPS led the firm *not* to pursue.

In the company in question, I did the assessment in this area as a consultant because it was believed that outside objectivity was necessary. The assessment was necessarily incomplete since complete records were unavailable and because the former category involves the identification of opportunities not previously identified. However, through a process of interviews, of searching through the files of some parties who had retained documents regarding project and strategy proposals, and by assessing the innovative strategies of competitors during the relevant time period, an assessment was made possible.

Adaptive Value of the SPS (F)

One important dimension of an SPS is its ability to change. Although this may be considered one of the external standards for B, it is so important that it is specifically identified on the feedback loop portion of Figure 33-2.

That feedback loop represents the SPS's ability to adjust its own characteristics or inputs to its outputs and the business performance they produce. For instance, if an SPS produces high-risk strategies as outputs and if those strategies do not produce the desired business performance, can the system be adapted so that it does not continue to do so?

In one situation in which this was the case, it was determined that a group decision-making process was producing risky recommendations that, when enacted by the firm, did not produce the anticipated results. When this was discovered and the process was audited with respect to the poor strategies, it was determined that the group process should be changed.

This variety of assessment can be made directly by tracing through the steps that produced a bad strategy. If the system leaves such an audit trail, the system itself or its inputs or goals can be adjusted to attempt to correct the problem so that it will not be repeated.

Of course, every bad strategy is not the result of a system deficiency. Yet, it is necessary to audit specific strategies and programs. To do otherwise will lead only to a morass in which no responsibility is assessable!

Relative Efficiency (G)

Little attention is paid in planning circles to the resources that are consumed by planning. Questions of the efficiency with which resources are employed are often shunted aside. One planner was recently heard to remark that to apply such a criterion to an activity that deals with such important issues is "nit picking." Yet, at the same time, planning departments' resources often suffer inordinately during periods of business downturn.

There are readily developed external standards for the relative efficiency of planning. For example, how many man-years are put into preparing the plan for each million dollars of sales? It is rather easy to estimate industry standards as well as standards for well-managed firms to which a firm might wish to be compared. Indeed, in one such case in which I participated, direct requests to other firms resulted in the provision of a wealth of data concerning their planning resources!

Adequacy of Resources (H)

A related measure is that of the adequacy of resources relative to the goals for planning. Sometimes unrealistic expectations are implicitly established for planning in terms of both the quality and quantity of SPS outputs. For instance, if staff planners are frequently assigned to perform special projects, as they are in many firms, the planning resource base may be so diluted that planning goals cannot be met.

In one case, the mere quantity of pro forma statements that were required in the planning document were so great as to tax the capabilities of the planning department's clerical and secretarial staff. When this was formally demonstrated through such an assessment, a commercial computerized financial planning language was purchased to serve (initially, at least) merely as a high-speed printer for those documents that were to be included in the plans.

Allocation of Planning Resources (I)

A third variety of planning resource assessment has to do with the *allocation* of the resources to the various functions and activities of planning. One way of assessing this is to develop a description of the various planning activities and to survey both planners and managers in terms of the time and resources spent on each activity.[10] In one case, this was done in terms of *actual* versus *desirable* proportions of total planning effort with some revealing results that implied severe misallocations of planning effort.

Appropriateness of Planning Goals (J)

The evaluation procedure of Figure 33-2 ends where it began—with planning goals. The appropriateness of these goals, in the context of the firm's environment and position, should be subjected to scrutiny using external standards.

Although specific and precise standards of this kind may not be readily available, a good deal of research has been conducted in terms of the level of uncertainty in the firm's environment as well as in delineating realistic expectations for planning.[11] This can provide the basis for a judgmental assessment, probably best conducted by someone external to the firm, of the reality and appropriateness of its planning goals.

Overall Assessment of Planning

There is great appeal to the notion of assessing planning in multidimensional terms, such as in A-J, and then of combining the various assessments into one overall measure. However, there are significant problems involved in doing so.[12] Thus the approach taken here is to synthesize the various measurements on a judgmental basis only.

For instance, as one planning officer who reviewed this approach noted, "How can one tell whether using more resources than is deemed adequate by external standards is good or bad?" The answer, of course, is that it is *neither* per se. *It is only relatively good or bad when viewed in the light of the other assessments.*

If one were to make highly favorable assessments on each of the other dimensions, it is clear that an unfavorable assessment of Relative Efficiency (G) would not be damning. One would expect that such overall effectiveness would be attained only at a price in terms of the resources consumed by planning.

Thus, an overall assessment of the effectiveness of planning requires that a *profile* of the 12 planning assessments be developed. It is only on the basis of this profile that an overall judgment may be made.

SUMMARY

A new approach to the evaluation of planning is presented in terms of a methodological framework. Each of the dozen separate assessments that make up the methodology have been applied in business firms, although the overall methodology has not been applied in a single firm.

The methodology employs both the collection of objective data and the making of subjective judgments. Its operating precept is that both can be made and used as long as the judgments are obtained in a structured fashion (for example, using structured interviews or a predetermined questionnaire) that facilitates aggregation and comparison.

The approach ultimately requires that judgments be made of *overall* planning effectiveness. However, that judgment is not merely impressionistic. Rather, it is guided by a series of prescribed measurement points and prescribed assessments, which, although the details may vary from firm to firm, are generic in nature.

REFERENCES

1. For instance, H. I. Ansoff et al., "Does Planning Pay? The Effect of Planning on Success of Acquisitions in American Firms," *Long Range Planning Journal* **3**(1970):2-8; P. M. Grinyer and D. Norburn, "Strategic Planning in 21 UK Companies," *Long Range Planning Journal* **7** (1974): D. M. Herold, "Long Range Planning and Organizational Performance: A Cross Validation Study," *Academy of Management Journal* **15**(1972):91-102; R. J. Kudla, "The Effects of Strategic Planning on Common Stock Returns," *Academy of Management Journal* **23**(1980):5-20; Z. A. Malik and D. W. Karger, "Long Range Planning and Organizational Performance," *Long Range Planning Journal* **8**(1975):60-64; L. W. Rue, "Theoretical and Operational Implications of Long Range Planning on Selected Measures of Financial Performance in U.S. Industry," Ph.D. diss., Georgia State University, 1973; L. W. Rue and R. Fulmer, "Is Long Range Planning Profitable?" *Proceedings of Academy of Management* (1972); G. A. Sheehan, "Long Range Planning and Its Relationship to Firm Size, Firm Growth and Firm Growth Variability," Ph.D. diss., University of Western Ontario, 1975; S. Thune and R. House, "Where Long Range Planning Pays Off," *Business Horizons* **13**(1970):81-87; D. R. Wood, Jr., and R. L. LaForge, "The Impact of Comprehensive Planning on Financial Performance," *Academy of Management Journal* **22**(1979).
2. W. R. King and D. I. Cleland, *Strategic Planning and Policy* (New York: Van Nostrand Reinhold, 1978).
3. W. R. King and J. I. Rodriguez, "Evaluating Management Information Systems," *MIS Quarterly* **3**(1978):43-51.
4. G. A. Steiner, *Top Management Planning* (New York: Macmillan, 1969).

5. E. K. Warren, *Long Range Planning: The Executive Viewpoint* (Englewood Cliffs, N.J.: Prentice-Hall, 1965).
6. C. Holloway and W. R. King, "Evaluating Alternative Approaches to Strategic Planning," *Long Range Planning Journal* **12**(1979):74-78.
7. R. L. Schultz and D. P. Slevin, *Implementation of OR/MS* (Amsterdam: North Holland, 1975).
8. W. R. King, "Implementing Strategic Plans through Strategic Program Evaluation," *OMEGA: The International Journal of Management Science* **8**(1980):173-181.
9. J. H. Grant and W. R. King, *The Logic of Strategic Planning* (Boston: Little, Brown, 1982).
10. R. K. Zutshi, "Strategic Planning Systems Evaluation: A Systems Approach," Ph.D. diss. Graduate School of Business, University of Pittsburgh, 1981.
11. Warren, *Long Range Planning.*
12. Zutshi, "Strategic Planning Systems Evaluation."

IX

Strategic Management in Small Business and the Public Arena

Part IX deals with the issues surrounding strategic management in two contexts that are related to, but different from, those that have been discussed in prior chapters.

In Chapter 34, Harold W. Fox discusses strategic planning in small firms and in Chapter 35, Marianne Moody Jennings and Frank Shipper discuss the development of strategies and tactics for the public arena.

CHAPTER 34

Strategic Planning in Small Firms

Harold W. Fox

Harold W. Fox, a former entrepreneur, is professor of marketing at Pan American University in Edinburg, Texas. He also is active as a consultant to small and large businesses, a member of quasi-boards of directors, and a leader of executive seminars. An energetic researcher, Dr. Fox has authored some three hundred publications on business strategies and functional policies.

The U.S. Small Business Act defines a small firm as "independently owned and operated and not dominant in its field of operations." In the United States some 99 percent of all businesses qualify as small on such dimensions as total sales, number of employees, or value of assets. Specific criteria include annual revenues under $12 million or employment below 250 persons,[1] but official dimensions differ by industry and purpose of the classification. Qualitatively, a firm is small if its activities are not complex. Strategies and operations are intertwined.

This chapter analyzes various key factors and practices that, taken together, delineate the future course of a firm with annual sales under, say, $12 million. In what activities will the enterprise engage and toward what end? This question epitomizes the subject of strategic planning. It requires the chief executive to specify objectives and chart their attainment within a stated period of time. Systematically, a plan seeks to assure effective use of available and potential resources toward full exploitation of existing and created opportunities. Thus strategic planning concerns deliberate decision making on a company's critical options.

Although its importance is manifest, the writing of strategic plans is not universal. Studies pinpoint lack of effective planning as a major cause of the very high rate of small business failures.[2] Firms managed by personal overview (instead of formal delegation) are particularly apt to be remiss in strategic planning.[3]

Managers' reactions to strategic planning are mixed. Why? Negatives and positives will illuminate this issue. Next, a description of the planning process leads to consideration of strategic decisions. Finally, a comparison of small companies' planning with large companies' provides a transition to the rest of this handbook.

DISADVANTAGES OF STRATEGIC PLANNING

Strategic planning is hard work. It is highly demanding of a person's imagination and knowledge. Planning must be learned. This effort can seem unnecessary to an owner-manager who, day after day, solves operating problems intuitively. Personal involvement in operations renders business realities immediate and direct—and deceptively easy.

No wonder many entrepreneurs dismiss strategic planning as theoretical. They are risk takers who revel in combat with sudden challenges. As long as business is satisfactory, planning seems unnecessary. If the situation is perilous, tactics for current survival demand precedence.

Moreover, the future is unknown, so how can planning improve long-range performance? In fact, modifications in a plan will usually be necessary in the course of execution. A small firm must preserve spontaneity.

However, many plans are misapplied. Falsely, some managers use this tool as the predeterminant of daily actions. In those cases, plans are, indeed, restrictive and impractical.

In addition, even though a plan is not a blueprint of future operations, planning can engender internal conflict. Tentative decisions on the future course of business automatically affect the roles and careers of major executives and the importance of different markets, products, plant operations, and technical research. These are emotion-laden issues. Executives fear adverse repercussions on their career if actual results fall short of the goals in the plan. For these and other reasons, resistance to planning is not unusual. Hence sensitive planners tread gingerly. Unless top management enforces full cooperation and demonstrates its own continuous commitment, the project collapses.

A case in point is the experience of an equipment distributor. Recent growth had outstripped the managers' administrative capacities. They failed to avert preventable setbacks, and some customers were not properly served. The owner recognized that formal planning was overdue. Because everybody in the firm was busy and lacked planning skills, the owner assigned strategic planning to a bright M.B.A. graduate. The latter developed new strategies, which were carefully explained in a briefing book. What happened? Nothing. Frequent emergencies continued to preoccupy the managers as before; the M.B.A.'s report was generally ignored. Eventually the owner despaired of planning and discharged the M.B.A.

Top-level performance of the planning function consumes much management time. Interruptions beckon insistently. Operational pressures are more urgent. Results of daily problem solving (or neglect) appear promptly. Time is money. Strategic planning infringes on executive and secretarial agendas. Further, explicit outlays are often necessary for supplies, telephone and travel, a meeting site, research studies, typing of proposals and plans, and consultants. In a minor firm, such expenditures can loom as a financial burden.

Indeed, many entrepreneurs are secretive about their business. They abhor the very idea of baring their vision—the key to their independence and livelihood. Finally, if planning documents fall into the hands of competitors, regulatory authorities, or adversaries in a lawsuit, they could actually harm a company. Safeguards are necessary to preserve the confidentiality of written plans.

Altogether, the arduous, time-consuming, and somewhat costly efforts, uncertain results, potentials for conflict, possible leaks, and the antipathy of some entrepreneurs put strategic planning on its mettle. To capture its benefits, astute owner-managers specify purposes of strategic planning and make sure it delivers the desired advantages.

ADVANTAGES OF STRATEGIC PLANNING

Strategic planning is a vehicle for providing direction instead of drift. Why do small firms need it? Certainly, many survive for a while without this tool, but its absence deprives a firm of opportunities and stability. A relatively minor misstep can have severe consequences—at worst, destroy the precarious enterprise.

In that sense, strategic planning is somewhat akin to insurance. Sooner or later, an adversity that strategic planning could have averted threatens the continuity of the firm or its management. In short, *effective planning is just as vital in a small firm as in a giant corporation*. (Procedures, of course, differ. They must fit a firm's unique characteristics.)

Organizations on a limited budget typically do not maintain such programs as environmental scanning, management by objectives, or zero-base budgeting. Close corporations often have a nominal, not working, board of directors. In partnerships and proprietorships this institution does not even exist. Absence of such management aids makes strategic planning all the more essential.

Alternatives are simply unworkable. Extrapolation, reaction to crises, accommodation to unforeseen pressures, subjective improvisation, and optimism may do for a brief interval but are inadequate for coping with problems and opportunities over a longer period of time. Also, illegal or impractical proposals are nipped in the bud. A planner needs to know enough about business law and about trends in regulatory and judicial decisions to recognize when to enlist the assistance of a lawyer.

In a fast-changing environment, the commercial viability of the founder's vision requires periodic review. Strategic planning guards against complacency that sometimes follows a period of success. A fresh diagnosis of unfavorable and favorable trends—as well as formal consideration of options, identification of needed resources and lead time, and provisions for timely action toward new goals—prepares a firm to face new conditions. Bankruptcies of such meteors as Osborne Computer Co. and Pizza Time Theatres dramatize the danger of inadequate planning. Less publicized are the thousands of unspectacular ventures that fall victim to lack of preparedness every week.

Who gains from orderly planning? Some firms and nonprofit organizations direct their plans toward satisfying their customers or clients, community, and other stakeholders, but usually the chief beneficiary is the person doing the planning: the owner-manager. Depending on the owner's temperament and the caliber of the subordinates, these benefits may extend to the executives via some style of participative management. In any event, strategic planning changes a manager's perspective from a firm's current activities to its long-range potential. Important trends supersede daily pressures. The planner appraises likely effects of external developments. Also, he or she discovers interdependencies of processes within the firm. As a by-product, the planning project clarifies and reflects the owner's philosophy. Even if the tangible result—the written plan—is ignored, all this learning makes strategic planning worthwhile. However, the plan should not be ignored. This tool can guide employees, equipment, and procedures to productive pursuits and prevent waste.

Recently, for example, the proprietor of a metal fabrication plant began a program of strategic planning. During the assessment stage he wondered why a waste-collection service was receiving exorbitant payments. His inquiry brought to light that an old, temporary agreement had never been renegotiated. Financial reports had split this expense departmentally, obscuring the magnitude of the total. Thus, even in a small concern with hands-on management, a substantial squander of money had escaped the owner's notice until he instituted formal planning.

Strategic planning discloses informational gaps, leading to modernization of data sources, processing, and reports. It helps identify the critical variables—what really matters in the firm's success. Accordingly, a new set of priorities is adopted and unproductive activities are dropped. The incidental net effect may be to conserve time. Planning and systematization free an owner to concentrate on nonprogrammable challenges that must be handled extemporaneously.

Another managerial aid is improvement in internal coordination. Strategic planning paves the way for a streamlined set of policies to help achieve the new goal smoothly. It establishes procedures for quickly reaching agreement on unforeseen future challenges. Explanations to all concerned forestall rumors. If conflict is inevitable, calm unhurried resolution is preferable to crisis-forced expediency. A plan should unify the firm. Also, it establishes accountability.

Reviews built into the planning process expose glib executives who improvise plausible explanations for shortfalls in their work. On the other hand, planning fosters personal growth and facilitates recognition of merit. Participants sharpen their vision and logic. Solid proposals bring an employee's latent abilities to the owner's attention, planting a seed for the former's advancement.

The advisability of realigning key personnel may become evident. With the plan as a guide, the chief executive can delegate without losing control and without running high risks of unsound operating decisions.

Realistically, management cannot avoid future effects of current actions, whether by decision or default. Which will it be? Systematic decision making in a planning framework enables management to anticipate and influence the events. "The key to success . . . is to integrate futuristic thinking with careful analysis."[4] Conducted persistently, strategic planning should result in higher profit or better realization of other goals.

Some firms use completed plans for orientation and training of new personnel. Plans also document a company's efforts to comply with various regulations. An ounce of precaution is worth a pound of court costs. In some transactions, such as applications for loans or sales of equity, plans are mandatory supports. Thus strategic planning offers many benefits to a modest-sized firm, justifying the self-discipline that conduct of a planning process imposes on the owner.

A listing such as Table 34-1 can help the

Table 34-1. Disadvantages and Advantages of Strategic Planning

Possible Disadvantages	*Possible Advantages*
Usurps much time	Eliminates unproductive activities; focuses on the key factors for long-range success
Incurs expenses	Repays in higher profits and substantial savings
Seems abstract	Substitutes realism for hope about the future
Encroaches on improvisation	Provides direction instead of drift
Entails hard work	Yields worthwhile insight into business structure
Has to overcome doubt in the efficacy of planning	Is essential under dynamic conditions and for loans or formal proposals
Requires top management's personal involvement	Prepares participants for greater responsibility
Triggers postponable conflicts	Stresses objectivity over emotions
Causes upheavals	Improves internal coordination; allows smooth adjustments
Tempts misuse as a blueprint	Produces an instrument for flexibility
Creates danger of leaks	Aborts courses of action that are illegal or unethical, or whose disclosure would be embarrassing
Calls for specialized knowledge and imagination	Constitutes a possible training tool

designer of a planning agenda avoid pitfalls and capture benefits. The key is: simple, yet adequate. Management should obtain the services wanted without the previously mentioned disadvantages by developing a wieldy process of strategic planning.

PROCESS OF STRATEGIC PLANNING

The process of strategic planning should result in a written work guide covering the following topics:

Assessment of the company's situation
Problems and opportunities
Explicit assumptions and objectives
Brief itemization of feasible options
Tentative strategic decisions
Provision for implementation
Strategic controls

These informational requirements are quite common, but the emphasis, depth of coverage, nomenclature, format, and sequence of topic development differ. Most concerns start with a standard format—such as the above list or the items in Table 34-2—and modify it until it suits their wants exactly.

Table 34-2. Manager's Check List for Strategic Planning

1. Where are you?
   a. Business
   b. Environment
   c. Capabilities and opportunities
2. Where do you want to go?
   a. Assumptions and forecasts
   b. Possibilities
   c. Objectives
3. How do you get there?
   a. Policies
   b. Strategies
   c. Resources
4. Who is responsible?
   a. Organization
   b. Staffing
   c. Delegation
5. How do you monitor results?
   a. Priorities and schedules
   b. Budgets
   c. Controls

If planning is decentralized, some official acts as coordinator. He or she distributes an agenda and makes sure that all activities are performed on time. Tailor-made forms reduce the writing burden and promote uniformity. The coordinator checks all contributions for conformance with objectives, for consistency across functions and with the avowed purpose of the business, and for reasonableness.

A Typical Process

Although this discussion will follow the sequence in Table 34-2, the process of strategic planning is not linear. A graphic model (Figure 34-1) shows that each step relates to all others. To dovetail all elements, much backtracking is usually necessary. With experience, planning efficiency improves.

WHERE ARE YOU?

Assessment of the firm's present situation serves as the point of departure for strategic planning. The nature of the business helps determine what external factors require analysis. In turn, this analysis reveals what types of capabilities, threats, and opportunities are relevant.

Nature of the Business

A seemingly simplistic question actually deserves much thought: What should be our future business? Obvious answers in terms of present or future products should be rejected, especially in a tiny enterprise that has to be flexible. Ideally, the definition of the business describes *why* (for what uses or wants) targeted prospects buy the firm's output.

For purposes of strategic planning, a processor of vegetables is not in the business of packaging string beans. Examples of more useful conceptualizations include: To appease hunger, provide nutrition, improve health, supply dietary food, balance meals, offer a change of pace, serve as outlet for farmers,

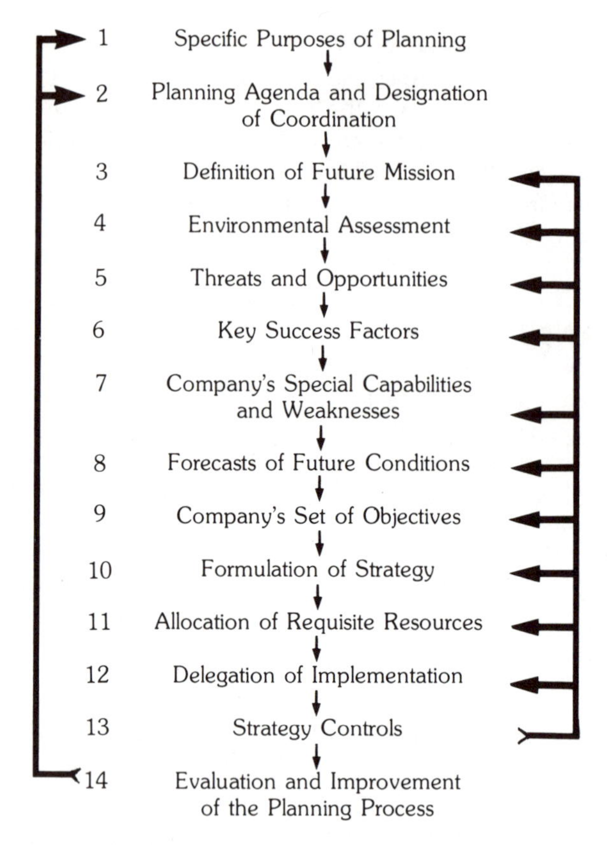

Figure 34-1. Model of the strategic planning process.

transport fresh (or frozen) produce. Each definition requires a different strategy. Each broadens the owner's vision to substitute or create additional ways of supplying the targeted buyers' needs or utilizing the firm's techology.

A tractable creed must be chosen. Statements such as to process and sell food or to supply supermarkets and groceries are so general as to be unusable. Presumably, the small firm could enhance its beans with vitamins or switch to peas but has no facilities for meat or candy (other types of food) or for stationery or beauty aids (sold to supermarkets).

External Factors

The envisioned mission of the business delineates its pertinent environment. Of greatest importance are buyers whose generic needs the firm supplies. What are the company's sales opportunities? Competitors that currently or potentially vie for this business also receive scrutiny.

Further, the planner itemizes key suppliers and prospective relationships. In the political-regulatory arena, the Congress, U.S. Department of Agriculture, Food and Drug Administration, and analogous state bodies require watching. New technologies—from harvesting to discard of residues—may revolutionize the industry.

Moreover, what is the impact of recession or prosperity on the company's sales, employment, cash flow, and so forth? Also, actions by consumerists, conservationists, and other societal interest groups could help or hinder the firm's future accomplishments.

Some large corporations and consultants maintain elaborate environmental scanning systems, but most small firms get by with negligible expenditures. At a small manufacturer of vending machines the personnel manager monitors external developments in her spare time. Most of her sources are internal. The credit and collection department provides financial data on present and potential buyers and on competitors. Review of specially designed salesforce reports supplements this information. Purchasing relays comments from vendors' representatives about competitors' purchases and new products. The quality control supervisor dissects rival products. Inquiries from executives provide guidance on what to look for. An additional source of competitive intelligence is revelations volunteered by job applicants.

This personnel manager owns one share of each publicly held competitor, so that all announcements and reports to stockholders come to her automatically. She maintains a good relationship with the local librarian. Additionally, subscriptions to trade journals and general business magazines, membership in the industry association, visits to trade shows, and general sensitivity to usable news suffice for her data base, formerly a file of handwritten notes but now on computer disks. At present, these diverse activities are evolving into systematic information management. Instead of compiling and distributing whatever comes along, this manager has ascertained strategic needs and now tries to fill them in an orderly manner.

Analysis of the environment helps identify threats and opportunities. Are consumption patterns changing? Perhaps major retailers are integrating backwards or farmers' cooperatives will take over the processing. Bills before a legislature; changes in the weather, domestic or abroad; and new strategies by competitors can affect the small firm in coming periods. These examples are illustrative.

Critical threats and opportunities also depend on the particularity of the business. The nature and environment of the business determine the special capabilities essential for success. A processor of prestige-brand beans must serve up a succession of arcane dishes, advertise heavily, and gain shelf space in supermarkets. On the other hand, if the beans are unbranded, the critical factor is efficiency. Market targets are distributors and institutions. Which path is the company better qualified to pursue?

On a simple form such as Figure 34-2 the planner cross-classifies the company's strengths and weaknesses in essential competences and resources with environmental threats and opportunities. The firm will try to concentrate its special strengths on the most suitable opportunities. It will plan to circumvent or meet threats, and it will either shore up its weaknesses or adopt strategies in which these weaknesses do not matter.

Informational needs are always in flux. New sources must be developed and obsolete data discarded. Wary of excessive data, planners compile sufficient information to help determine the firm's direction.

WHERE DO YOU WANT TO GO?

A sound information base narrows uncertainties about future conditions and allows the planner to infer reasonable assumptions. (If the data are so ambiguous or contradictory that no single inference is plausible, two plans reflecting different premises should be prepared.) Assumptions must be explicit, in written form. They are the axioms on which forecasts—indeed, the strategic plans—are built.

A forecast of next year's sales of beans, for example, entails specific assumptions about the aforementioned external factors as well as about the firm's marketing program. Ideally, an analyst prepares a set of forecasts using at least two independent methods for each variable to be predicted. One might be quantitative, the other qualitative. If predictions from separate forecasts are similar, they seem acceptable.

However, if results differ substantially, the reasons for the deviation should be sought. Interpretations of data are reviewed, or assumptions must be clarified. Perhaps additional facts would be of help. The reconcilia-

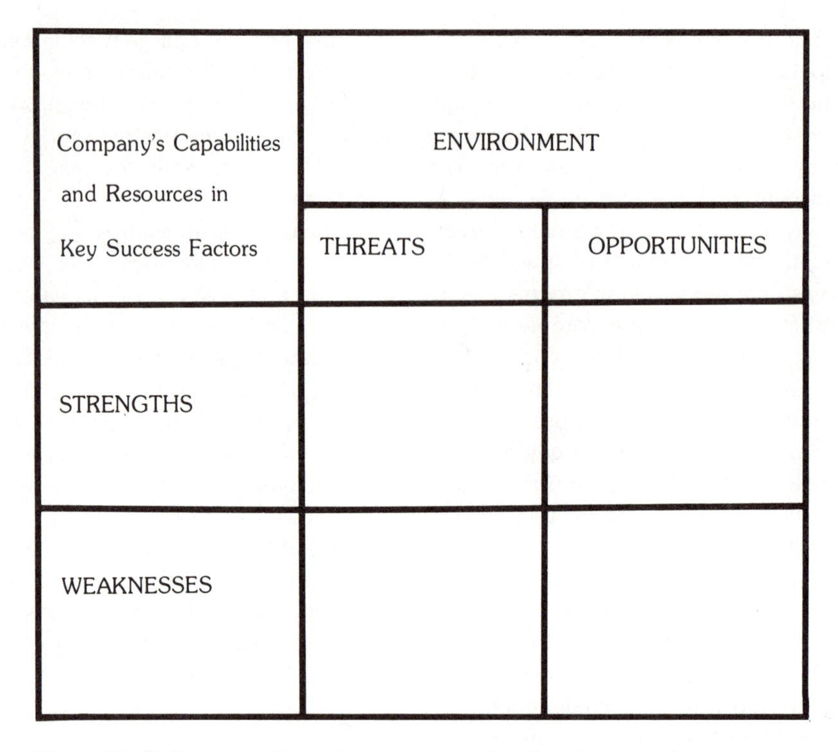

Figure 34-2. Key strengths and weaknesses related to threats and opportunities.

tion continues until a confidence-inspiring set ensues. As an additional precaution, forecasts of related variables are checked for consistency.

Accepted forecasts of future conditions along with above-cited inputs circumscribe what the enterprise seems able to attain. A limited number of feasible options is considered in depth. Top management selects its preference, and then objectives can be formalized. Objectives crystallize an entrepreneur's dream. They demarcate the firm's character and dimensions to which the owner-manager aspires.

The approved set of objectives is the keystone in a strategic plan. The main objective specifies the company's overall intended result in a testable manner. An example is, to earn $100,000 after taxes in the next fiscal year. Here is a numerical goal with a time limit. Hence one can verify whether or not it has been reached.

Attainment depends in part on whether the personnel believe the objective to be reasonable. If they deem the required outcome unreachable, trying will appear useless. If they regard the objective as too lax, they will feel that no effort is necessary. Top management has to make sure that objectives have general acceptance and support.

All subobjectives are integrated into the main objective to contribute towards its accomplishment. For example, if the net profit margin is 10 percent, the sales department's objective should be $1,000,000 revenues during the same fiscal year as above. In turn, this sales target is broken down into quotas by salesperson and product. After adjustment for inventory changes, the production schedule and the cash flows must also tie in to the main objective and the other subobjectives.

To recapitulate, a proper structure of objectives is workable and testable, and aligned horizontally and vertically, but, instead of stopping to admire it as an elegant piece of art, the analyst should heed Drucker's dictum that objectives are commitments to action.[5]

HOW DO YOU GET THERE?

Policy decisions, strategy formulation, and resource allocation implement approved objectives consistent with management's prior determination and findings. Policies in this context are guides that indicate the arena of the firm's coming actions. Will the processor of beans orient the business toward gourmet dishes or nondescript hospital fare? Does it forgo certain spices, perhaps because of the owner's religious restrictions?

Subject to policy directives, an appropriate strategy moves the firm's distinctive capabilities to the chosen arena's key success factors. A small company might specialize, for example, in some product or market. Concentrating on a field with little competition, a modest-sized firm could become that tiny industry's leader. The bean processor might be the major supplier to a small ethnic group.

Alternatively, it could skim several areas of operations. The processor of beans could have a factory outlet (retailing), sell its output along with others' produce to food stores (wholesaling), and establish contact between large buyers and sellers (brokerage). This strategy exploits the firm's flexibility.

Another strategy, especially appropriate for some small enterprises, is to develop a stream of new products, for sale or licensing to larger manufacturers and/or distributors. Likewise, a firm could engage in applications engineering. It supplies the unique needs of some manufacturers who finance these efforts. Less imaginatively, many firms offer products similar to their competitors. Even in the last case, differentiation is often feasible through a previously unavailable level of quality, price, or services.

Some fringe firms move from one activity to another in response to changing opportunities. Others cultivate an insignificant portion of the market, hoping to escape competitive retaliation. The limitation of small size is turned into a basis for survival. Like all planning, strategy formulation is more art than science. Small firms prize alacrity over system.

Whatever strategy is adopted must comply with legal and ethical norms. An explicit statement of deportment, prepared with the help of a lawyer, summarizes the imperative of complying with all applicable laws and with the owner's philosophy. Actually, an informally administered enterprise may have an even greater need for a code of conduct than a larger corporation in which committee deliberations precede decisions and audited documentation follows.

Next, management allocates or procures physical and financial resources necessary to reach the firm's objectives. This step leads to one of the many reviews throughout the process. Can the firm afford the devised strategy? A different strategy will have to be developed unless the firm can provide the essential needs of research and development, production, marketing, and support functions. Even more important are the human resources.

WHO IS RESPONSIBLE?

People make things happen—not techniques, money, equipment, or strategic planning, but people who are properly organized and motivated. The work necessary "to get there" must be divided into jurisdictions that are reasonably homogeneous, manageable, and controllable. Coordination can be informal. Employees know each other, work together, share a common objective, and understand the total operation. Hence formal meetings are usually unnecessary.

In most small firms a simple organization of several functional departments suffices. Each department is headed by a specialized manager who is responsible for achieving specified subobjectives.

As noted earlier, each subobjective contributes toward the accomplishment of the company's overall goal. An organization chart showing names and titles, with corresponding position descriptions, can help everybody visualize formal reporting relationships and work responsibilities, but an owner should not mistake boxes on paper for the administrative tasks of identifying and delegating the work to be done without duplications and discontinuities. Leadership is more powerful than management.

Intermediate supervisors (between management and operatives) should be used sparingly. However, when designing an organization, prudent owners take the capacities of the executives into account. The ideal division of work yields to the practical consideration of personnel availabilities, skills, and interests. Interference with customary personal (informal) contacts is avoided.

Staffing top slots often is a severe problem, especially for minor firms in industries with ease of entry. Capable employees are apt to leave and set up rival enterprises. One nonfinancial incentive to foster their loyalty is participation in strategic planning. Personal involvement is particularly important in a small business. If employees care about their work and associates, they can be more resourceful than bureaucratic functionaries in a giant corporation. Quite likely, evident respect for employees' suggestions and adoption of at least some of their recommendations will strengthen their commitment to successful implementation.

Such sharing of strategic planning is not just a concession to placate powerful executives. Lower-level employees often have superior knowledge of a firm's weaknesses and are adept at sizing up external threats and opportunities. Under astute leadership, employees generate ideas and make decisions that improve the effectiveness of the strategy. Then they will work hard to prove the merit of their contributions.

Nevertheless, dysfunctional enthusiasm for parochial courses of action harbors dangers. The sales manager of a small electronics parts manufacturer advocated a price reduction of 10 percent. His schedule showed that a lucrative doubling of volume would ensue. However, the chief executive tactfully pointed out that the company lacked the cash or credit to purchase equipment, hire additional employees, carry receivables, and so on. More important, large competitors should be expected to match or undercut the price reduction, particularly if it causes some of their customers to transfer their patronage to the attacking firm. Strategic planning favors a long-term view, including thorough examination of foreseeable repercussions. Applied to more familiar activities, this lesson helped the sales manager administer his department.

Rigorous scrutiny encompasses not only imaginative assessments of likely internal and external consequences, but also considers the personal reactions of all executives concerned. Whole-hearted acceptance by each manager charged with executing a portion of an adopted strategy is a prerequisite to successful delegation of the tasks.

Concurrence embraces both goal and wherewithal. Most managers believe that their department's resource allocation is inadequate. In small organizations this suspicion is usually correct. After some sparring, the owner and each manager agree on the latter's mission, resources, and authority.

The delegation makes an executive responsible for achieving specified subobjectives, permits decision making within the executive's jurisdiction, places assistants, equipment, and budgetary allowances at the executive's disposal, and may provide other facilities. Again, reviews ensure that all elements dovetail. On the finished plan all delegations, domains, and deadlines will be consistent.

HOW DO YOU MONITOR RESULTS?

The next step is to divide the long-term objectives and strategies into years, quarters, and months. Management establishes priorities for hiring of key executives, purchasing of equipment, and manufacturing (or buying) of semifinished or finished goods.

Most businesses are seasonal, so that inventories must be built in anticipation of sales. Meantime, employees may be paid weekly and vendors, monthly. Sales generate receivables that will be liquidated eventually. Clerks prepare schedules that reconcile these temporal disparities.

Accountants translate the first year's portion of the strategic plan into dollars. After approval, it is divided into quarterly or monthly budgets. The budget includes end-of-year pro forma financial statements. Key figures and ratios are compared with norms provided by the industry association or by Dun & Bradstreet.

Only selected data—such as revenues and profits before taxes—are estimated for more distant periods. Progressively more detail is desirable as the budgetary period nears. About two-thirds through the first year, the following year will become the new budgetary period.

Meantime, feedback informs management of the company's progress toward realizing the objectives. Particularly in small operations, informal observations and contacts can be more pertinent than accounting reports. The plan lists the sources and timing of interim information, along with the evaluative criteria.

Schedules must be kept current. Priorities change. Flexible work weeks, executive sicknesses or turnover, uneven cash flows, sudden transitory opportunities, and unpredicted environmental developments require constant attention. Minor changes, overlooked by competitors, may connote worthwhile opportunities for a marginal operation. Revisions and updating will be necessary. Management may have to alter the strategies.

Strategy feedback serves at least four purposes: (1) It informs management whether results are on track or seriously at variance from the plan. (2) If the latter, it triggers an operational investigation to pinpoint the causes. Is performance deficient or is the plan no longer valid? (3) Feedback makes the planning process credible. Serious attention to strategic planning is apt to depend on the foreknowledge that outcomes will be traced to the tentative decisions—and to the decision makers. (4) Finally, it is the input for strategy control.

Quarterly or as necessary, the chief compares actual occurrences with the assumptions in the plan. Errors in assumptions are an early warning signal that the plan is wrong. Mistakes are admitted and rectified.

Typically some results exceed interim objectives, and some are deficient. Management can focus its attention on full exploitation of unexpected opportunities. Next, shortcomings should be handled constructively. If the plan is still realistic, what tactics will henceforth restore the intended direction and results? The last step is to prevent a recurrence of the inadequate performance. This includes identification of the specific operational details and institution of generally accepted guides and requirements that will improve henceforward accomplishments. Future use of operational and planning lessons will strengthen the enterprise.

All involved can help to make strategic planning less cumbersome and more constructive. Ultimate responsibility devolves upon the owner-manager. Annually, he or she reviews the purposes of the planning activity and their realization. Sound planning draws on both logic and intuition. It retains the firm's simplicity and flexibility. Above all, it provides direction and enhances outcomes. Unless evaluated and improved, planning deteriorates into a perfunctory exercise with many of the disadvantages noted at the outset of this chapter. However, with proper administration, the role of strategic planning can evolve from (1) instrument of learning to (2) guide of operations to (3) way of business life.

WAY OF BUSINESS LIFE

Because a fringe business is precarious, many decisions have strategic consequences. Functional and administrative choices—in the realms of marketing, production, personnel, office management, and finance, which a large corporation can routinize and delegate—may be of strategic import in a small firm. The typical owner-manager confronts dozens of such critical options every week or possibly every day. Responses are urgent. Time (and inclination) for a protracted analysis of each problem are lacking.

Management styles differ. Some owner-managers shoot from the hip. A few of these are served well for a while by their ingenuity. Often, the high profitability of their invention or service shields the firm's inefficiency. Examples include ventures in electronics, toys, apparel, and medical supplies. Many prospered for a few years, followed by failure. Functional interdependencies are not always obvious. In the wake of solving an urgent problem, expedient management is apt to create, unwittingly, another one. As a result, employees work at cross-purposes. Customers are confused. Eventually a radical change in conduct or structure becomes inescapable.

It is exactly because of the plenitude of pressures converging on a single decision maker that an overall strategic plan is so useful. An up-to-date valid plan serves the proprietor as the reference to which all major decisions and revisions conform. The aim is a consistent set of actions that will help attain the company's objective. A sound plan obviates the need for a fresh study of most daily challenges and paves the way for an enlightened resolution of a great diversity of business problems. Typical of such strategic decisions are the following functional examples.

MARKETING

The key decisions in marketing pertain to the target(s) of buyers and the program of cultivating them. The operator of a small firm usually knows many of its customers personally. If the company sells to other organizations, the number of customers is apt to be small. The principal might deal with each directly. A retailer, on the other hand, usually has a larger number of customers, and they are in constant flux. Even here, however, a shop owner usually works in the store and meets the consumers. Whether marketing to organizations or to consumers, many proprietors act as the sales manager or as a salesperson.

A related decision concerns the company's geographic scope of operations. It ranges from worldwide operations of some import-export firms or mail-order houses to neighborhood trading areas of most retail stores. The point is to determine in advance whom to target—who and where the qualified prospects are.

Demarcation of the prospects depends in large part on the product line and, of course, vice versa. Eventually a company's goods or services and the rest of the marketing program must be tailored to the buying criteria of the customers. That is why planlessly selling to just anybody can be inefficient and unprofitable. All functions are involved in the effort to satisfy customers.

Many small companies specialize. Specialization is efficient but risky and restraining. Customers depart. Products become obsolete. Perhaps business is seasonal. To counteract such disadvantages, some firms diversify their markets or products. Diversification adds complexity to daily operations. Even huge corporations often stumble when their activities are dissimilar. However, in many lines of business a variety of product-market strategies on a modest scale is a viable option.

One form of product diversification that is essential in many fields—and at which some small firms excel—is the development of new products. Strategic choices abound. Many fringe firms conduct technical research and development in-house, especially if the founder is an inventor.

One alternative is to acquire product concepts, technology, or completed projects from others. For example, many large companies are eager to dispose of projects that have too little potential for them, but the potential is in the proper range for a Lilliputian firm. Arrangements include outright purchase, exclusive license, nonexclusive license, and purchase of finished goods. Marketing strategies for new products include: (1) first to market; (2) emulation of the pioneering company; (3) imitation of existing products; and (4) abstention from new products. Another key decision is whether to innovate minor improvements or radical changes.

One major aspect of product-market decisions is a strategy for reaching buyers. Actually, there are two parts. One is an effective program of communicating with prospects. The other is arrangements for transfer of purchased products.

Communications with buyers can be direct through a company's own salesforce, advertisements, and supporting promotions. Alternatively, a tiny firm may prefer to leave the marketing to a sales agent or to specialized middlemen such as distributors and wholesalers. (The firm would still have to sell to the intermediaries.) Some companies do both—direct and indirect.

If a small enterprise performs the tasks of promotion, it must decide the size of its salesforce, compensation, and employment and training policies. Advertising decisions include the thrust of the messages, the particular vehicles, and the frequency. In many small firms advertising decisions are made by the president who has never been initiated into this arcane art. Larger companies use an advertising agency. It may also help a company get favorable publicity. Other promotional decisions involve participation in trade shows, distribution of free samples, use of incentives and coupons, and preparation of catalogs and brochures.

The other part of reaching buyers, physical distribution, entails key decisions on size, location, and management of inventories, use of warehouses, arrangements with transportation companies, and perhaps even the terms of sale (where title passes). Manufacturers also decide whether to strive for intensive or selective availability of their goods. A regional firm, for example, focuses on a limited geographic area and sells to as many outlets or end users in that area as possible. Under this strategy, both communications and shipments are less costly than if a small number of buyers were serviced all over the country.

Some satellite firms locate and organize to serve just one large customer such as General Motors Corp. They arrange their operations for just-in-time delivery. Of course, they are highly vulnerable to changes in that single customer's demand or policies. Location is critical for retail stores, too.

Coupons, mentioned above as a part of promotion, and terms of sale in connection with physical distribution are really elements of pricing strategy. If buyers are price-sensitive, a key strategic choice is the level relative to competition. Should the company's prices exceed those of near competitors, or be the same, lower, or span all levels, or perhaps bracket a rival's price? Further, credit has strategic power, especially in international business. Another type of strategic decision involves disposal of old merchandise. Options include systematic markdowns or close-outs to a special segment of buyers.

The foregoing decisions relating to the target(s) of buyers, the line of products, the promotional strategy, distribution, and pricing are interrelated. In addition, marketing strategies also affect production, personnel, office management, and finance—and vice versa. All functions must be considered simultaneously.

PRODUCTION

Just as companies determine the extent of their involvement in marketing, they can elect to produce a large or small portion of their merchandise in-house or to buy finished goods. Most resellers (which are mainly small firms) opt for the latter. It follows that the choice of vendors is of strategic import to them. Of course, procurement strategies are very important for manufacturers, too.

Typical manufacturing decisions include plant location, kind of facilities and modernization policies, manufacturing processes, layout, staffing, safety programs, scheduling the types and quantities of complete goods and spare parts, quality control, and operational controls.

Preproduction entails strategies concerning vendors and inventories of raw materials and components. Postproduction involves inventories of finished goods, traffic, warehousing, and similar operations that overlap with marketing. In fact, postsale technical service may be part of production or marketing.

During recent decades many manufacturers have begun to lease instead of buy the equipment they use. (Plants and stores have always been available on either a rental or ownership basis.) A large variety of leasing arrangements can be negotiated. The amount of periodic payments to the lessor depends partly on the size of the contract and how the various benefits and burdens—with respect to taxes, maintenance and service, risks of obsolescence, destruction, variations in interest rates and so forth, cancellation privilege, equipment disposal—are allocated.

Basically, a user can choose an operating or a financial lease. Operating leases are rentals for a user's temporary needs. Financial leases are commitments that provide for spreading the outlays over a longer period, similar to an installment sale.

Lease arrangements are in wide use on many types of durables. Companies can modernize their facilities without depleting their cash. Perhaps they lack the requisite amount of money. Undercapitalized firms favor lease arrangements because other sources of financing may not be available to them. Ownership of the venture is kept intact. Return on investment seems higher than with an outright purchase. Sometimes, state property taxes are avoided. Thus, tax benefits, interest costs, and lessor's services will often make it more advantageous for a small firm to lease instead of buy. However, leases usually are much more expensive than a cash purchase.

A noncancellable lease can deprive a user of flexibility. The financial lessee has to depend on the lessor for maintenance and repairs. Altogether, leasing decisions have many strategic ramifications with respect to production and finance.

A small manufacturer's strategic problems often include cut-off from vendors during periods of shortages, insufficient trade credit, obsolete equipment, inefficient production, and high cost of shipments. However, the biggest strategic challenge is recruitment and retention of competent employees.

PERSONNEL

Employees are a strategic resource. With adroit leadership and well-planned administration, employees can be productive and cause a firm to prosper. Treated haphazardly, however, personnel can be a financial drain and cause discord. Every company has to determine what work is essential, the requisite number and proficiencies of full-time and part-time employees together with an effective organizational design, and policies on recruitment and hiring, compensation including benefits, and training, employee relations, advancement, and termination.

Distinctive Requirements

Carefully thought-out personnel strategies reflect the characteristics of the business. Ideally, the chief is a leader who inspires an almost missionary zeal. A small firm's success depends more on the initiative, ingenuity, and loyalty of the personnel than on job descriptions and standard procedures. It needs people who are flexible. They contribute toward attainment of the firm's objectives in the spirit of enjoying their work. They are eager to learn and to perform new skills and to deal with unprecedented situations.

People with this type of disposition are apt to harbor aspirations and ambitions. Hence in many small firms, personnel turnover is high. Yesterday's employee may be today's competitor—working for another employer or as an independent businessperson. Many marginal firms are vulnerable to competition from an ex-employee because they are in industries with ease of entry and seeming ease of operation. Hence prudent protection of trade secrets is a strategic necessity for small firms.

Excessive personnel turnover is expensive for companies of all sizes, but it is particularly distressful to small firms because in the latter, employees have no backups. Each incumbent is indispensable.

Distinctive factors such as these illustrate the small firms' strategic dependence on their personnel, especially on their skilled employees. Granted that small employers need a particular type of individual, what do they offer to candidates? Again, the mix of pecuniary and psychic inducements differs from the blandishments of large corporations.

Distinctive Inducements

With few exceptions, small organizations do not match the large companies' prestige, benefit packages, and amenities. Opportunities for advancement to general management and part ownership through stock options are limited. In some closely held firms, nepotism prevails. Members of the owner's family hold the influential decision making and supervisory spots. If they are ineffective, the business suffers. So does morale. Employees also feel insecure: what will happen to them after the owner withdraws from the firm? An outsider's part-ownership, even if granted, is illusionary. A minority interest in a corporation with non-publicly traded stock has virtually no rights and no market value. However, these negatives can be outweighed by sound planning and administration. For example, some companies provide for gradual transfer of the business to the personnel through an employee stock ownership trust.

Needing people with distinctive personalities and distinctive work styles, a small enterprise can appeal to their distinctive motivations. While satisfying their basic physiological and safety needs, a diminutive firm, by virtue of its smallness, has an almost unique opportunity to gratify their higher-order yearnings for belongingness, esteem, and self-fulfillment. Just a few persons, respecting each other, working close together toward a common goal in a harmonious, achievement-oriented atmosphere, joining in top-level decisions, striving to master new challenges—this could be an ideal setting for certain individuals. Admittedly, many enterprises fall short of providing such idyllic conditions, but strategic planning can map conscious efforts toward institutionalizing this kind of work environment. For example, strategy formulation forces an owner-manager to determine respective priorities for proprietary secrecy and participative style.

Especially if much of the personnel strategy is informal, frequent open discussions between boss and employees plus other feedback about their achievements and feelings are essential. Excessive turnover, absenteeism, drug abuse, defective work, and unproductive conflict are symptomatic of problems that must be identified, ventilated, and remedied.

In short, a small firm's personnel strategy is to employ qualified people who welcome wide variety in their work and who respond at least as much to psychic motivators as to pecuniary incentives. Owner-managers can individualize employment arrangements to foster commitment and loyalty.

Shrewd principals balance the imperative of flexibility with pragmatic accommodation to legal pressures for formality. Helped by a lawyer, small firms publish and practice equal employment opportunity and other systems of fairness such as employee performance evaluations, grievance procedures, and disciplinary penalties. They comply with legal requirements pertaining to minimum wages, withholding and social security taxes, occupational safety and health, workmen's compensation insurance, labor's right to organize, and so forth. Hiring candidates who support its personnel strategy, the enterprise reduces the risks of discord, lawsuits, and unionization.

Strategies toward unions range from support to hostility. Some firms welcome them. A few make illegal "sweetheart" deals. Others go through a development process from legalistic sparring to cooperative agreements on wages and working conditions. Some concerns fight a union until one or the other gives up. Their owners suspect that formally negotiated labor contracts will deprive their company of its familylike atmosphere, its flexible work style, and its competitiveness. Also, they may view formation of a collective bargaining unit as a personal repudiation. Whatever the philosophy, a plan is needed to protect the principal's interest and the workers' right. Planning is especially vital during a time of change, such as has been occurring in offices since the mid-1980s.

OFFICE MANAGEMENT

In small firms the office is a hub of support activities such as materials control, production scheduling, inventory records on finished goods, order processing, customer service, correspondence and memos, and payroll. Although all of these administrative and clerical functions are necessary, output is far below the scale justifying specialized departments. Versatile employees and a mix of equipment tend to these diverse tasks.

Efficient performance of routine and emergency white-collar work poses a challenge in planning, coordination, and control. In recent years the cost of office services has soared. As in other areas, excessive expense threatens the profitability and survival of a small firm. Risks, too, require management in a strategic perspective. Another strategic decision in office management is whether to work in-house or to use outside service companies for all or part of the clerical tasks. Some contractors even furnish personnel who perform the tasks on the firm's premises. Off-premises, numerous manual and computerized processing arrangements are available.

Information: A Strategic Resource

Much information has strategic significance. If a product-market strategy is not feasible—as in the case of some commodities—a firm can attain a superior position over competitors through helpful inside sales personnel, efficient interactive computers, prescreening of customer credit, uniform speed of shipments, courteous clerical personnel and friendly telephone operators, and prompt postsale service. When customers perceive competitive offerings as equal in quality and price and on other major criteria, vendor selection is apt to turn on the effectiveness of the seller's office activities.

Also in its relations with other parties, a firm's white-collar functions can have a critical role. The companies on which the marginal firm depends for raw materials or finished goods prefer organizations that send out clear orders and pay promptly. Employees care about the condition of their records and the firm's discharge of its obligations. Timely transmittal of news to all concerned is important. Reports to various regulatory agencies are a costly burden; mistakes require rework and may result in fines. In general, poor documentation puts a firm at a disadvantage in cases of claims, lawsuits, and other disputes.

Since the mid-1980s information management has become a strategic priority. Technology has become available that can make a company more effective toward customers and other parties. Knowledge is power! Information is a key resource for strategic decision making and control.

Automation of Office Work

Many large corporations are far along in automating their white-collar functions. This is a continuing project. Knowledge requirements change and new equipment becomes operational. Small firms are planning conversion to electronic offices as a competitive necessity.

Strategic planners take a systems approach, reflecting the strategic value of information. Top executives identify the types and timing of facts needed for strategic and operating purposes. An office systems specialist who understands small firms in that particular industry can be very helpful in this assessment phase and at later stages.

The specialist proposes alternative systems: configurations of technology and people to conduct the input, processing, output, and distribution of the desired entity of information and allow for future modifications. Various analyses to reduce costs and complexity follow. A sound installation meets strategic needs and, at least long-term, decreases expenses.

Next is a compromise between the technically possible and the administratively practical. Seasoned executives who will not operate the proposed equipment must be accommodated. The capacity of present clerks to learn new skills and procedures is limited. So is the availability of proficient replacements. Present, functionally obsolete machines are a sunk cost; management is loath to abandon them while they are still operable. But patchwork changes merely postpone outlays and staffing decisions, condemning the firm to strategic inferiority until a reasonable degree of office automation is working.

Planners of strategic information management aim for compatible interactive computer systems with standard software. Prepared for users who design their own applications, this information equipment also is simple to operate. In systems with multiple access to data bases, the integrity of the data and the security of proprietary information need protection. Additional objectives include resolution of equipment limitations, control over user errors, and coordination of projects.

Altogether, integration of strategy and information management will force substantial changes in staffing, organization, and operations. But in many diminutive firms, office automation will be less difficult to institute than in huge corporations because small-business executives are accustomed to hands-on management. Operation of their own personal computers or terminals will not be so drastic a change in work style for them. And the analytical needs in a small organization are less complex. As training catches up with current technological capabilities, strategic use of information will become reality. Progressive firms are planning a systematic conversion now.

FINANCIAL MANAGEMENT

All of these strategic decisions—in marketing, production, personnel, office management, and elsewhere—have monetary implications. Financial planning parallels the entire panoply of strategic planning from Table 34-2. Balance sheets depict where a business is; statements of income and of funds, how it got there. Profit plans show where the business expects to go. Cash budgets summarize how it will get there. Control systems follow assignments of responsibility and control reports monitor financial results. Increasingly, this process is computerized.

Financial Monitoring and Planning

Financial assessment includes comparisons of latest results with corresponding data from the past. Further, selected items from a balance sheet, income statement, and funds statement are put into ratio form for clues to the concern's profitability, liquidity, and solvency. An astute owner-manager examines reports and ratios comprehensively. Analysis of cash is a case in point. It is necessary; firms can fail at a time when profit is high because they are illiquid. Another analytical technique is common-size statements, which facilitate examination of statements from various companies.

Although differences in accounting methods and administrative policies make intercompany comparisons tenuous, it can be instructive to note major similarities and differences in financial results between one's firm and its competitors. Financial statements of small competitors are often published in credit reports. Composites are available from Robert Morris Associates, Dun & Bradstreet, Inc., Internal Revenue Service Statistics of Income, industry associations, trade journals, and other sources.

From this assessment the owner-manager detects the venture's financial strengths and weaknesses. Large borrowing capacity represents an opportunity, whereas insufficient cash is a limitation. The latter might be rectified through an additional investment, or solving the cash stringency could be one of the objectives for the planning period.

Usually the financial objective of an independent businessperson is a specific amount of profit. This intended bottom line is the point of departure for the profit planning process. It ends with a budget that details the revenues and expenses corresponding to the firm's strategic plan. A complete set of pro forma statements will forecast the firm's financial position and history at the end of the planning period. Much juggling of the strategic plan and the profit plan is necessary to hit the profit amount that is both satisfactory to the owner and feasible strategically.

Profit plans are budgets on an accrual basis, meaning they time revenues and expenses as they occur. However, because of prepayments, credit arrangements, and periodic expensing of fixed assets, an accrual basis is not useful for cash management. Many small ventures are undercapitalized. They must monitor cash receipts and payments daily. Hence the profit plan is translated into a cash budget, which further tests the feasibility of the strategic plan. For example, seasonal shortfalls of cash revealed by the cash budget can be resolved by negotiating a line of credit in advance.

During the period of the strategic plan, all revenues and expenses are classified according to executive responsibility as a basis for subsequent evaluation and improvement of financial performance. Interim reports monitor results by comparing actual (accounting figures) with profit-plan budgets. Major variances alert the chief to areas requiring personal attention, detailed investigation, and corrective action.

Financial Strategies

Like financial monitoring, financial strategies are interdependent with overall business strategies. The adequacy and manner of capitalization are an example. How much money is invested by the owner(s)? The larger the proportion of debt, the higher is the financial leverage. Positive and negative returns on investment are magnified. Low investments usually equate with low credit capacity, low growth potential, and low staying power.

Further, does the ownership estate put its personal holdings at risk? It does, by operation of law, in proprietorships and partnerships (except for limited partners). In closely held corporations, conventional or Subchapter S, stockholders risk only the amount of their stock subscriptions, but bankers and other financiers often require the owners to subordinate their personal holdings to creditors' claims. Subordination is less common in trade credit. When money is raised by sale or pledge of receivables and equipment, the respective assets are the creditors' security. Also, as noted under Production, a firm can avoid initial large outlays by leasing.

A common strategy is to start a business on a "shoestring." Money at risk is miniscule. Operations are conducted in the entrepreneur's garage, perhaps only part-time. Relatives supply cash and labor without recompense. Eager vendors deliver essential equipment and merchandise on deferred payment terms. All sales are for cash or for cash in advance. If the venture proves viable, the owner expands the investment and operations to a regular commercial scale. Funds are easier to raise after high earnings potential has been demonstrated.

The tables are turned if a small firm deals with other undercapitalized firms. It will be under pressure to give liberal credit terms, transfer merchandise on consignment, and advance money to vendors. Large customers and vendors may afford to be more generous. The degree of their cash support of small businesses differs. In any event, a firm has to determine its credit strategies vis-à-vis suppliers and customers.

Something as picayune as cash discounts can be of strategic import to a small enterprise. Many wholesalers, for example, pursue a strategy of voluntary arrangements whereby they sell at the same price as the cost invoiced to them, retaining only the cash discount.

In a few other fields, cash discounts are substantial. Rates range up to 8 percent; discount periods up to 120 days.

Even for tiny firms, tax strategy can be very complicated. Tax implications often override business strategies. Typically, small firms depend on net profit for growth (or survival) capital. Income taxes constrict this lifeline. Hence strategic planning of taxation is always critical.

At inception, tax considerations are a major determinant of the form of business organization. During a firm's active operation, taxes govern in large part the incidence and accounting treatment of owner's compensation, acquisition and expensing of fixed assets, inventory policy, and outlays for research and experiments. Looking ahead to eventual exit from business, the principal plans to maximize capital gain and to reduce estate and gift taxes.

Assistance from a tax adviser is essential. As the owner's situation and the tax laws change, new arrangements may be in order.

Strategic planning of expenses entails decisions about deployment and cost behavior. How much should the concern spend for wages, other production costs, advertising, and so on? Theoretically, the company should spend the minimum necessary to get the job done. Pragmatically, however, cash limitations will almost always force a small firm to fall short of this ideal. Thus financial stringency constrains strategies and operations.

Many firms neglect strategic planning of cost behavior. It entails structuring of expenses into two categories: responsive to changes in volume and unresponsive. If responsive (like commission), the expense in total varies in proportion to changes in production or sales, units or dollars. If unresponsive or fixed (like salary), the expense in total remains the same within a particular range of volumes. This classification should not be delegated to an accountant. It is a strategic decision in which the owner-manager makes policies that, in turn, determine the firm's operating leverage.

The higher the fixed costs as a proportion of all costs, the wider are the swings of profit as sales vary. High operating or financial leverage, although risky, is advisable if the firm can increase its sales profitably by reducing prices or increasing promotional efforts. The opposite obtains for low leverage. To summarize strategic planning of cost behavior: Fixed costs are preferable if a firm's prospects for sales growth are high. Conversely, costs should be responsive (variable) if a firm expects to remain small or to retrench.

Finally, each firm needs financial strategies for growth and disposition. Some proprietors withdraw as much money as their firm's cash flow permits. Many owner-managers aim for the largest size that preserves their full control through personal overview. Others seek growth. Fast growth requires much money; retention of earnings may be just a foundation. Usually, many ownership prerogatives must be surrendered when a company raises capital through sale of stock, long-term loans from banks, arrangements with venture capitalists, or other external financing.

If the plan is sale of the business to outsiders, the usual financial strategy is to manage for consistent growth in profits. Outlays and write downs are deferred wherever the results of this policy are not obvious. An alternative strategy is to amass large sales, even at a loss, and persuade the buyer that superior management would make the firm profitable. Some prospective owners, especially inexperienced ones, fancy themselves superior managers.

Timing is important. The most lucrative opportunity for selling a business occurs just as a period of fast, profitable growth is about to end. A savvy owner-manager may foresee satiation of demand, product obsolescence, entry of competitors, or expiration of patents. No wonder outsiders are suspicious. Sale of a small firm to strangers is difficult.

More often, a retiring owner intends to transfer the business to associates, to the next generation, or to the employees. Both parties to the transaction are familiar with the firm and know each other personally. Prudent owners protect their family by prearranging an orderly disposition of the firm in case of their sudden death or disability. However distasteful personally, plans for these contingencies are essential lest the small firm end its existence when the owner departs.

SIMPLICITY: HALLMARK OF SMALLNESS

Indeed, many small firms have a much shorter life span than huge corporations. Discontinuance does not necessarily reflect failure. Some owners wind up affairs to pursue different or greater opportunities elsewhere. Entrepreneurs are restless. Often the strategic plan at inception provides for a stopgap between other ventures or between jobs. Many times nobody is known or is qualified to continue the business.

Not only is the typical small firm more ephemeral than giant corporations, but its ability to influence the environment is more limited. Constraints such as these discourage some proprietors from strategic planning. Entrepreneurs insist on keeping their options open. They resist sophisticated methodology that might straitjacket them. Many of the other differences from large companies itemized in Table 34-3 likewise inhibit strategic planning in minor firms.

Table 34-3. Strategic Planning in Large and Small Companies

Strategic Factor	*Typical Large Company*	*Typical Small Company*
Company Administration and Strategic Decisions		
Strategy, operations, and management	Complex	Relatively simple
Operations	Tactics (distinct from strategy)	Intertwined with strategy
Dealings with customers and vendors	Via lower-level employees	Personally by owner
Administration of company	Formal planning and delegation	Owner's personal overview
Organizational design	Based on tasks to be accomplished	Based on capabilities of incumbents
Desired qualification of employees	Specialized expertise	Mainly versatility
Attractiveness to employees	Maintenance factors exceed motivational incentives	Motivational incentives exceed maintenance factors
Regulations	Heavy	Though burdensome, enjoys many exemptions
Typical life span	Perpetual	Brief
Strategic Planning Procedures		
Pervasiveness of strategic planning	In general use	Most firms abstain
Where used, plan's format	Written (formal)	Oral
Organization	Decentralized planning. Coordinator needed	Owner conducts centralized planning
Strategic subunits	Divisions or SBUs	Functions

Table 34-3. (*Continued*)

Strategic Factor	*Typical Large Company*	*Typical Small Company*
Perceived financial burden	Almost negligible	Substantial
Specialized assistance	Board of directors and planning experts	None
Tools, techniques	Management information systems, sophisticated models, portfolio analyses, scenarios, and other planning routines	Much improvisation
Alignment of functions and projects	Numerous special meetings	Frequent routine contact
Financial analysis	In-depth cost accounting, adjustments for time differences, inflation, etc., many other techniques	Superficial
Beneficiary	All stockholders	Owner-manager
Strategic Planning Substance		
Company mission	Formal statement	Not defined
Environment	Proactive	Passive
Environmental scanning	Formal efforts	Haphazard, mainly owner's chance contacts
Assessment of strengths and weaknesses	Realism (documented)	Optimism (opinion)
Forecasts	Causal methods	Extrapolation
Options	Analyses in depth	Few considered; little analysis
Feedback	Reports	Boss's inspection
Financial objective	Return on investment	Amount of profit
Financial planning	Main quest is quarterly growth in sales and profit	Main problem is insufficient cash
Type of strategy	Multiple approaches (differentiation strategies)	Concentration (focus strategy)
New products	Strong in marketing	Strong in development

Some small firms have devised ingenious ways to overcome their deficiencies in strategic planning. An example is the use of a quasi-board of directors to help formulate strategies and to simulate accountability of the owner to a superior body. A quasi-board exerts psychological pressure on the chief executive to set aside sufficient time for strategic planning and to do an impressive, useful job.

However, the key distinction between large and small companies, as has been demonstrated throughout this chapter, is the latter's relative simplicity. It often makes strategic planning in fringe firms seem unnecessary. In the minority of small firms that do practice strategic planning, the lack of complexity and resources allows or forces many abridgements, as Table 34-3 exemplifies. Other differences between small and large enterprises with respect to strategic planning will become apparent from a perusal of other chapters.

REFERENCES

1. U.S. Small Business Administration, mid-1980s.
2. R. Wyant, *The Business Failure Record* (New York: Dun and Bradstreet, 1977).
3. D. L. Sexton and P. M. Van Auken, "Prevalence of Strategic Planning in the Small Business," *Journal of Small Business Management* **20**(1982):20-26.
4. W. R. King and D. I. Cleland, *Strategic Planning and Policy* (New York: Van Nostrand Reinhold, 1978), 87.
5. P. F. Drucker, *Management* (New York: Harper & Row, 1974), 99, 102, 119.

CHAPTER 35

Developing Strategies and Tactics for the Public Arena

Marianne Moody Jennings and Frank Shipper

Marianne Moody Jennings was graduated from Brigham Young University with a B.S. in business management and finance in 1974 and a J.D. in 1977. She has been a professor of business law at Arizona State University since 1977, and in 1984 she served as Governor Bruce Babbitt's appointee to the Arizona Corporation Commission. She has worked as a consultant to a number of firms in the fields of corporate and real estate law. Professor Jennings has been guest speaker at over fifty seminars and has had over eighty professional publications, including eight textbooks. She has received numerous awards, including Professor of the Year—Arizona State University College of Business and Outstanding Young Woman of the Year in Arizona.

Frank Shipper is a member of the Department of Management at Arizona State University. He received the B.S. in engineering from West Virginia University, the M.B.A. and Ph.D. in business from the University of Utah. In 1980, he was awarded an American Assembly of Collegiate Schools of Business Federal Faculty Fellow. He continues to pursue his teaching, research, consulting, and writing on the relationship between business strategy and the political arena.

Developing business strategies and tactics for the political-legal arena has been neglected by both academics and people in business. In 1980 one scholar argued that there exists an intrinsic importance to the study of business political activity.[1] However, more recently information has come to the forefront that indicates that the economic viability of a corporation may depend on how well it interacts with political-legal forces. For example, one recent study found that the degree of political intervention in a corporation's affairs had a greater influence on that company's financial risk than either the rapidity of technological change, the variability of the economy, or the competitiveness of the markets in which the company participated.[2] In an internal report one company estimated that $49,237,000 in 1983 and $29,897,000 in 1984 were saved due to its efforts to enact, amend, or defeat legislation.[3] Although such knowledge has not been widely disseminated in the business world, many managers through experience have become more cognizant of the impact of political forces and have begun to respond.

Old line companies in the United States did

Portions of this chapter are excerpted and edited from F. Shipper and M. M. Jennings, *Business Strategy for the Political* Arena (Westport, Conn.: Quorum Books, 1984); and M. M. Jennings and F. Shipper, "Strategic Planning for Managerial-Political Interaction," *Business Horizons* July-August(1981):44-51. Both used with permission.

not have an ongoing forum for executive-level discussions on political issues from which actions could be taken until 1972 when the Business Roundtable was founded. The American Electronics Association did not have a Washington office until 1980. However, the new corporate elite, as identified by *Business Week,* formed the Amerian Business Conference in 1979 to lobby for issues of concern for high-growth companies. Executives of both old and new line corporations are now spending inordinate amounts of time, energy, and corporate resources to both learn about and influence the political-legal arena.

To be effective in the political arena requires knowledge of the structure of that arena, its pressure points, the players who occupy it, the power players who control it, the strategies that can be used, and the strategies that are most effective in a particular situation. The political arena contains the huge government bureaucracy, but the power structure is amorphous, and ever changing. One must be familiar with the entire structure because any part can be the critical element that can either support or thwart the strategies of the corporation. The power players change, especially from one presidential administration to the next. However, one unspoken rule of the arena is that the early supporters have greater influence than the Johnny-come-lately tag-alongs. Finally, the effectiveness of the political-legal strategies varies based on the particular situation and the cost of the various strategies. Thus, an effective strategy considers the structure of the political-legal arena, the key players necessary for implementation, the various strategies that can be employed, the available tactics for each strategy, the tangible and intangible costs in pursuing each strategy, and the probability of a successful outcome.

STRUCTURE OF THE POLITICAL-LEGAL ARENA

The structure of the political-legal arena has frequently been oversimplified to either the nefarious Iron Triangle and/or the courts. The Iron Triangle is a term used frequently by outsiders to refer to the bureaucratic structure made up of the legislative, and administrative branches of government plus the special interest groups and their representatives (also known as lobbyists, lawyers, and public relation specialists) who seek to influence the political and legal processes. The same structure is also referred to as the Golden Triangle since it controls the legislative, budgetary, and administrative processes that determine how much and who receives what. The courts are also an integral part of the political-legal arena for two reasons. First, the judicial branch has the right to declare any legislation or act of the other branches of government unconstitutional. Second, the courts are a proper venue for settling differences between organizations. The final element of the political-legal arena is the press. The power of the press to influence the public who can then elect or bring pressure on public officials, who can then influence both legislation and administrative and judicial appointments cannot be overlooked. Only when all five parts of the political-legal arena are considered can effective action be taken by those hoping to influence the political-legal processes.

Congressional Apex

The congressional apex of the Iron Triangle consists of both the House of Representatives and the Senate. The former is probably more amenable to lobbying efforts than the latter for several reasons. First, representatives have to run for reelection every two years. Thus, they are interested in short-term results, such as a new plant or a public construction project, in their home districts. They then can tell the voters they provided *x* number of jobs. Second, a representative is elected from a smaller identifiable population than a Senator. Therefore, fewer voters are needed to either change their vote, show-up, or not show-up in order to affect an election.

An example of the power of representatives on governmental contracting can be seen in the battle over which company will get the contract for a jet engine for a fighter plane. A spokesperson for Pratt and Whitney, the jet engine builder, acknowledges that General Electric has an edge because GE has plants in over two hundred congressional districts, whereas P&W has influence in only about twenty.[4] With a total of only 435 representatives, GE would appear to have the advantage.

Congress does not award a contract directly, even those of this magnitude. Contracts are supposedly awarded in an impartial competitive bidding process conducted by an administrative agency. However, on major contracts a political appointee in the administrative agency will have the final say on the awarding of the contract. Congressional members and their staffers affect the process in both the formation of the bid proposal and the evaluation of the bid. Documents in Washington are frequently written by consensus. A draft copy of the proposal is first written and circulated to all interested parties. This means to people both inside and outside the agency. The outsiders are frequently congressional members on influential committees. The staffers of the congressional member may, in turn, pass a copy on to a company representative in the home district. Sometimes the agency will directly circulate the draft to potential contractors. The representative can then suggest specifications that may be in the interest of the public but are certainly easier for the representative's company to meet than for the competition. (In fact, many proposals for contracts are even initiated by company representatives.) The suggestions may or may not be forwarded to the administrative employee in charge of the proposal for suggested incorporation in the document. If the congressional member is a frequent supporter of the initiating agency or if the agency wants to increase its influence with the congressional representative, the suggestions are likely to be incorporated.

Just how effective a company can be in obtaining prior knowledge of the specifications on a new contract may be illustrated by the following example. In 1977, General Motors began building a bus called the Rapid Transit Series II. In 1978, the federal government released certain engineering specifications required for all buses to be purchased with federal funds. The GM bus, which had been in production for a year, met these requirements.[5] Some individuals may say that this was pure coincidence. More likely it was the product of good Washington representation and lobbying by GM.

Effective representation, as in the GM example, is a matter of a series of activities. The words frequently used to describe Washington power ploys are soft in tone such as *suggested* and *access*. Washington is a town of subtleties, especially between the apexes of the Iron Triangle. A special interest representative suggests to opposing congressional members that they reconsider the issue. Likewise, a member of Congress does not blatantly oppose a bill unless gathering political hay-votes. Everyone in the Washington arena seems to be for something that begins with congressional support. Members of Congress also want congressional support for their bills, plus they want their own reelection. Special interest supporters can frequently supply a congressional member either with the funds or the campaign workers who will enhance the probability of reelection. Thus, the diverse power structure outlined earlier creates a need for mutual cooperation if congressional support is to be obtained and maintained by both the members and special interest representatives.

The Insiders and the Outsiders. Both insiders and outsiders are at every apex of the Iron Triangle. The outsiders of the congressional apex are the elected members of Congress. They can grant access to the insiders, but they do not control access. The insiders of the congressional apex are committee and personal staffers.

Those new to the Washington arena often are surprised to hear that the elected representatives are referred to as outsiders, but a few simple explanations exist. First, the "old-boy" network and party loyalty ties of yesteryear are less effective. The seniority system was lessened by congressional reform in the 1970s that allowed effective first-termers to rise quickly in power. Senator Orrin Hatch (R, Utah), a right-wing conservative, has demonstrated how far a first-termer can go. Party loyalty is no longer the strong club it used to be. Single-issue (for example, balanced-budgeters, right-to-lifers) members of Congress understand exactly what got them there. In addition, the old coalitions, especially of the Democratic party prior to the McGovern era, no longer have the clout they once did.

A second reason congressional members are considered outsiders is that their tasks are beyond belief. During a session the number of bills introduced is in the ten thousands, and the number of pages in the *Congressional Record* in the hundred thousands. No one, not even a speed reader, can even scan all of this material. During a typical busy day a congressional office receives hundreds of phone calls and hundreds of visitors. Because of the huge workload, members of Congress reported spending fewer than 11 minutes per day in 1981 in research and reading versus approximately one day a week in 1977.[6]

The new-breed congressional members are often depicted as public relations figureheads with good media personalities. Although this may be partially true, they are also smart enough to know when to rely on their staff. The new breed evolved both because of the media and because of a more vocal constituency. The people from back home now ask more of their congressional representatives and more frequently want political favors. Some segment of the public now expects congressional help with everything from Small Business Administration (SBA) loans to veterans' benefits. In addition, members of Congress are asked constantly to deal with conflicting issues. The same people who ask for a balanced budget are likely to turn around and ask for higher social security benefits or educational assistance. Thus, a primary attribute a member must have is the ability to be a verbal tap dancer.

Congressional members emerge as outsiders for a third reason: the effective staffers are the permanent residents of the District of Columbia. When a member of Congress decides to leave Capitol Hill whether through personal choice or voter disapproval, the highly regarded staffers will be in great demand both on and off the hill. A person who knows the pathway through the legislative process, the bureaucratic process, and the corridors of Washington (for example, the Capitol, the White House, the Executive Office Building, the Pentagon, the Social Security Administration) is invaluable. Many long-time staffers have served in all three areas of the Iron Triangle and thus know the "games" better than any congressional member.

Finally, the perennial staffer gains even more of an insider track than an elected official because the rate of turnover in Congress appears to be on the upswing. Thus, fewer congressional members gain the experience required to be effective. Some leave for personal reasons, such as one ex-Congressman who left for religious reasons. Others, once they gain the experience, leave because of monetary reasons. If the economic pressures of trying to maintain two residences—one in the home state and the other in the high-priced Washington area—do not persuade a congressional member to leave, then the lucrative outside job offers may. After a member of Congress learns the ins and outs of the committee process and demonstrates an ability to successfully usher legislation through Congress, various outside job opportunities become available. For example, the private lobbyist groups will be more than willing to hire an ex-congressional member at pay higher than the congressional salaries. Clark MacGregor, an ex-representative and currently a United Technologist lobbyist, is just one example of an ex-congressional member who has made the transition.[7]

Another reason for the apparent increase in turnover is frustration and disillusionment. Some members of Congress find that they do not have the temperament to deal with the intricate, entangled, and entrenched Washington bureaucracy, and so they leave. Regardless of the reason for leaving, including losing an election, without considerable longevity a congressional member may never become an insider in the Iron Triangle.

The Committee Staffers. The special interest representative must realize that committees are the emerging power structure of Congress, and committees and their staffs are the growth business of Congress. The personal staffs of all congressional members total approximately 3,600. On the other hand, the number of staffers serving on and supporting 200-plus committees and subcommittees exceeds 15,000.[8]

The committee system has become the brains and guts of the legislative process. The majority of the research on and the drafting of bills is now overseen by committee staffers and not by the congressional member's personal staff as in the past. The approximately 1,000 individuals who serve directly on committees are key supporting players. The days of the great floor debates are also over. What happens in the committee by and large determines the fate of a piece of legislation.

Each committee has a staff director, a minority counsel, and a chief clerk. They make the detailed work assignments for the other staffers. The direction of these assignments can determine the tenor of ensuing legislation. The staff director and minority counsel are responsible for overseeing budget considerations, drafting legislative plans, and holding hearings. Committee staffers determine who gets to present a point of view in the committee hearing—a national forum covered by the national press—and in what order.

Committee staffers should not, however, be regarded as independent fact finders and unbiased researchers. The staffers are hired by either the committee chair or the ranking minority member and serve at their pleasure. The hiring is not always done directly by a congressional member. Sometimes the legislative or administrative aide of the congressional member or a previously hired staff director or minority counsel will do the actual hiring. Nonetheless, no one should interpret this to mean that the staffer will not be loyal to the congressional member in whose name the hiring was done.

The staffers' duty—self-preservation—is to take care of their employers. Thus, how receptive a staffer may be to various ideas depends on who hired that individual and on the constituents of the employer. Staffers who lose sight of the fact that their employment is contingent on remaining in the good graces of whoever hired them, or their surrogates, will not be around long.

In the past, both personal and committee staffers may have been a form of patronage positions. Today, this is unlikely. Staffers today are most likely to be young (under 35) with a college degree plus some graduate school. Obviously many of them are lawyers, but political science, public relations, and, increasingly, technical degrees are held by the various staffers. They work long hours, especially when Congress is in session, and prior to an election. Their advancement is tied primarily to the success of the congressional member who hired them.

Typically, the staffer views the present position as a stepping stone or an apprenticeship. Some staffers have political ambitions of their own and use this as a way to gain experience and credibility. Others see their position as a way to move into the upper reaches of the administrative branch as a political appointee. Still others want to move into law, public relations, and other firms representing the special interests. As a whole, the congressional staff should be regarded as an ambitious group, and the effective staffers should be regarded as bright and politically astute.

Special Interest Apex

The special interest apex has evolved over a number of years as an additional branch of government to those prescribed in the Constitution. This special interest apex includes all lobbyists—registered and unregistered—public relations experts, lawyers, and other nomenclature given to individuals trying to influence legislation. Also, it includes people of all political persuasions. In total, over 15,000 representatives of various special interest groups are estimated to inhabit the Washington area.[9]

Whether an individual representing an organization needs to be a federally registered lobbyist is a matter of the legal interpretation of vague wording. In a not-too-atypical fashion, in 1946 when Congress wrote the law requiring lobbyists to be registered, the final wording was selected to please the majority of the supporters of the bill while displeasing the fewest opponents. As a result, the wording leaves a loophole through which most of the people attempting to influence legislation can walk. The basic requirement is that, if influencing legislation is the primary function of a job, then the individual should register. One organization and its lawyers have interpreted this to mean primarily the number of hours per week actually spent on Capitol Hill. However, no known definition exists listing the requirements for registering as a lobbyist.

Barry Commoner and the organization he founded (Common Cause) used to issue lengthy reports that were chronicled on the six o'clock news listing the number of registered and unregistered lobbyists, but apparently this issue no longer receives public scrutiny or cries of outrage.

One of the reasons given by many who choose not to register is that a stigma is attached to being a registered lobbyist. This may be a false worry. To insiders the treatment accorded an individual representing an organization will hardly depend on whether the individual is registered or not. If the individual has political clout (with a large number of active constituents) and/or credibility (the ability to produce political results) the title of the individual will have little effect. At the same time, if the individual is a political gadfly (and there are many of them), he or she will be shuffled to the newest congressional staffer. By contrast, when dealing with the public, the argument that a stigma is attached may be realistic. At the same time, the distinction between a registered and an unregistered lobbyist probably escapes the general public. Thus, whether to register or not should be based more on an interpretation of the law than on perceptions of the term.

The number one rule for all individuals trying to influence legislation is to be flexible. Because of the number of issues and the breakdown in the party system and its associated loyalty, the Washington arena is more pluralistic than ever before. Power has become fragmented. To succeed, coalitions are needed to assemble the required clout and votes to pass a bill through Congress. Coalition formation is an ongoing, temporary phenomenon. The coalition that serves a purpose today may be ineffectual tomorrow. The current bedfellow may become an arch enemy on the next issue.

The Insiders and the Outsiders. As with the congressional apex, special interest lobbyists can be categorized into two groups: insiders and outsiders. The insiders are the professional lobbyists and long-time residents of the Washington area; the outsiders are the corporate lobbyists to whom this is another three- or four-year rotational assignment to broaden their experience. The number of corporate offices in the area exceeds 500, with staffs numbering from 2 to over 100.[10] Corporate lobbyists are not on the whole held in high respect by the Washington power brokers. A veteran Capitol Hill staffer characterized them: "Generally, Washington offices are a mess. Much of what they do is simply gross. They tend to be excited, ignorant, hysterical. They live by rumor."[11]

Despite the reputation of corporate lobbyists, an organization can have effective corporate-political relations. The first requirement is to hire a strong insider to head the Washington office. United Technologies Corporation has done just that. It hired Clark MacGregor, an ex-representative with 10 years of congressional experience. In addition, MacGregor served in the White House as a Capitol Hill lobbyist. When United Technologies hired him, MacGregor already had the contacts and understood the do's and don't of Washington that are essential to do an effective job. The United Technologies Washington office is regarded as lean and mean and also as one of the better corporate offices.

A second way to represent effectively an organization's interest is to hire a lobbying or public relations firm that already has a number of experienced insiders. For example, Occidental Petroleum, Lazard Frere, and Warner Communications all have hired Grey and Company to partially represent their interests.[12] Grey and Company is a public relations company headed by Robert K. Grey, past secretary of President Eisenhower's cabinet. Another example is the firm of Patton, Boggs & Blow, one of Washington's fastest-growing law firms, also known as a "full-service lobbying firm." It has represented Chrysler, MCI Communications, and Marathon Oil, among others. Thomas H. Boggs, Jr., a senior partner in the firm, is considered to be one of the most powerful lobbyists or "rainmakers" in Washington.

A third way to be successful in Washington is to find an effective public advocate with interests coaligned with those of the firm. Then the corporation provides the public advocate with the research, statistics, and other technical information needed to present its case. Funds are sometimes provided to the public advocate to assist in gathering independent data or to do other work that is required. In turn, the public advocate does the public lobbying; the company, or an association representing it, plus the public advocate will then engage in private lobbying. The automobile insurance industry for years has been using this approach with Ralph Nader as the public advocate. As a result, the public got an added cost to their new cars—for the seat-belt interlock system and the 5-mph bumpers. The former was short-lived when the public rebelled against it. However, the latter saved the insurance industry millions of dollars.

Hiring an individual or firm that specializes in a particular legislative area to gain access to the insiders is a fourth approach. Such representatives will have better connections with the appropriate legislative communities and administrative offices than firms that try to cover all areas. For example, General Electric Company has hired J. D. Williams to oppose the use of tax "leasing" although GE already has a Washington lobbying office.[13] Williams is considered by many to be one of the top-ranked tax lobbyists in Washington. The supporters of tax leasing in a countermove, hired as representatives Charles E. Walker and Ernest S. Christian, Jr., two other Washington tax lobbyists considered to be among the most effective.

One could ask, if the Washington offices of the different organizations are hiring outsiders, is that office earning its pay? The answer is a definite yes. The current effective insiders for this year, for this two-year term between congressional elections, and for this four-year presidential term, can change quite quickly with the waxing and waning of the various political stars. Also, whenever the party or individual in the White House or the majority in either congressional House changes, the insiders with the ear of the new movers and shakers also change. The meteoric rise, fall, and reemergence of David Stockman may illustrate the need to be tuned in to a number of political ears and the ability to move quickly to the next issue or political advocate. A good corporate office always keeps abreast of a number of issues, picking up clues, and finding out who the key players are before the issues become page one news.

Administrative Apex

Of the three areas of the Iron Triangle, the administrative apex is probably the hardest to penetrate as an outsider. First, the vastness of the administration arena can easily overwhelm any one person. By one count in 1976 there were 11 cabinet departments; 59 independent agencies; 1,240 advisory boards, committees, councils, and commissions; plus 1,026 aid programs; and almost 400 programs affecting higher education, 228 health programs, 158 income security programs, and

83 housing programs.[14] Obviously, this count is already out of date because under President Jimmy Carter two additional cabinet posts—energy and education—were created. But even the most up-to-date count can only begin to convey the immense size of the administrative bureaucracy. Because of its size, the administrative bureaucracy has overwhelmed a number of modern attempts by both Congress and the president to reform it. The administrative bureaucracy continues to grow.

In addition to its vastness, the inertia associated with an organization of such size has caused it to take on a life of its own. The direction of this life is relatively unaffected by a new political appointee at the cabinet level, by one Congress or another, or by one presidential administration or another. One indication of its own direction and life is that 75 percent of the yearly budget that the bureaucracy administers is considered by Mace Broide, executive director of the House Budget Committee, to be uncontrollable.[15]

At the same time, at the working level of the bureaucracy the number one complaint among the high-ranking civil servants is a lack of direction.[16] Another major complaint is micromanagement or overmanagement. At first these two complaints appear incongruent. The first complaint, however, is aimed at the appalling lack of leadership and proactive management provided by both the elected and appointed political officials in Congress and the administration and the high-ranking civil servants. The second complaint is aimed at the tendency of both Congress through the General Accounting Office (GAO) and the presidential administration through the Office of Management and Budget (OMB) to overengage in reactive managerial practices.

The overuse of reactive forms of management begins during the budget phase and lasts until the audit phase. During the budget phase everyone wants the programs they are associated with to grow. Each program, from subsidies for new home owners to grants for research, is advanced as the savior of the country and as a program that would produce results if only funded at a higher level. Once a program is funded, there is little review to discover if it accomplished its stated purpose. For example, after all the funds spent on the so-called War on Poverty, certain indicators are that more poverty exists than ever. After all the increases in educational funding on a per-student basis, the average high school student is less prepared for college now than a student was 20 years ago. The cry of the administrators of such programs and the congressional supporters is: if only more money were allocated. This effect is known as the wedge effect. Once a program is established, it is extremely hard to abolish. Nixon with all his power, and ruthless use of it, could not abolish even the National Tea Tasting Board.[17]

In contrast to the established programs, new programs are much harder to fund during the budget process than old programs. New programs stand out like sore thumbs during the budget process because no line existed for them the previous year. The question is asked of new programs, "Why is this needed?" instead of asking the old programs, "Why are these funds needed again?" During the Carter administration, some flirtation occurred with zero-base budgeting and sunset laws to prevent the recurrent funding of ineffective or outdated programs. Unfortunately, the implementation of these ideas went awry.

During the audit phase new programs tend to be singled out and examined with closer scrutiny than older programs. If the program receives any publicity, unfavorable or not, the auditors descend in droves. For example, during the first year of the Quality Circles program at Norfolk, Virginia, visiting teams of management analysts and auditors from the U.S. Naval Material Command, the Office of Management and Budget (OMB), the Chief of Naval Operations Office, the General Accounting Office (GAO), and the Office of Personnel Management (OPM) all descended on the base. Fortunately, the program withstood the scrutiny and subsequently was named as an outstanding practice by OPM. Nonetheless, the question remains, does any one program especially a successful one, need this much auditing from so many ranking agencies?

The budgeting and auditing cycles are the two processes all members of the administrative apex must be aware of and capable of handling adroitly if they are to succeed: first, in obtaining funds for their programs and second, in avoiding the many bureaucratic equivalents of the Golden Fleece Award of Senator William Proxmire (D, Wisconsin).

The Insiders and the Outsiders. As in the other apexes the administrative apex has insiders and outsiders due to the complexity of the essential process. The insiders are usually high-ranking, career civil service and military employees, and the outsiders are the political appointees. The major advantage the career employees have over the political appointees is experience within the bureaucracy. The typical SES (Senior Executive Service) manager has at least 20 years of experience, whereas the political appointee has less than 18 months. In fact, 18 months is the average tenure in office for assistant and under secretaries.[18] Thus, communication through and from the political appointees tends to lose continuity every 18 months, or even more frequently. As a result, the career civil servant never wants to become overcommitted to the programs of one appointee, especially near an election. Within civil service, a slogan exists that sums up the attitude toward the revolving door parade of appointees, "We were here when you came and we will be here when you go."

The resistance of the career civil service and military managers to programs advocated by political appointees can be fierce. A story attributed to Retired Admiral Elmo Zumwalt after he left the office of Chief of Naval Operations, a political appointment, can illustrate this point. He is supposed to have said after leaving the office that he felt like he was riding a bicycle pedaling like hell and steering like hell, but no one had told him that the chain had been disconnected and the handlebars had been loosened.[19] Thus, the many reforms Admiral Zumwalt wanted to make, especially in living conditions and civil rights for the navy personnel, were never realized. The upper- and middle-level managers of the administration apex hold the key to effective implementation of programs, and their support is surely needed if a new program is to be carried forth.

The key to success in the Washington arena is to build relationships among the insiders—the congressional staffers, the professional representatives, and the career officials—of the Iron Triangle. A network of relationships with the insiders will provide continuous access to the outsiders—the members of Congress, the corporate representatives, and the political appointees—even when there is frequent turnover. With an effective insiders' network, coalitions can be formed to provide the clout needed to influence legislation, contract proposals, regulations, and other governmental actions.

The Press

At least since 1833 when Thomas Carlyle credited the press with having "wide world-embracing influences," the press has been recognized as a force with which to be reckoned. In response, members of Congress, administrative officials, and representatives of special interest groups hold press conferences and have press releases prepared. Congress has even furnished itself a complete studio to film television releases.

To better understand the role and effect of the press on the political-legal arena one must differentiate between the national and the local press because each emphasizes different kinds of issues. The national press tends to focus on foreign and domestic issues that have a large emotional component such as disarmament, pollution control, and food safety; whereas, the local press tends to focus on more parochial issues with tangible impacts such as the building of a water project or the funding of a defense contract in which a local plant will participate.

Corporations have for years had public relations departments to interact with the press.

These departments usually have done an effective job working with local issues and press, but corporate America has had a less than desirable relationship with the national press. Many reasons could be given for this situation, such as the one found by one study—the members of the press do not interpret information as neutrally as people in business.[20] In fact, the same study indicated that members of the press almost invariably interpret information from a liberal point of view. However, such finger pointing does not resolve the problem.

Another study of the national press corps has indicated that its members tend to be youthful with frequent turnover.[21] Thus, the individual member of the press corps has little historical memory. In addition, the members of the press corps were found to have little knowledge of economics. Therefore, one should not be surprised that information simply handed out in the form of a press release or provided in an infrequent press conference is scrambled once it reaches the morning paper or the evening news.

Fortune proposes that there are four basic strategies that may be used in interacting with the press: (1) stubbornly closemouthed; (2) willing to respond, but not right now; (3) willing to cooperate when prospects of immediate gain are apparent; and (4) candid, cooperative, and accessible.[22] The first three strategies are considered to be ineffective. The first two strategies tend to raise the curiosity level of the reporters. The second and third strategies tend to make the reporters feel like they are being used and over the long run might prejudice them. Journalists become wary of any release, feeling it may be just another puff piece. The fourth strategy according to *Fortune* is the most appropriate for today's environment. Since the press devotes more time and resources than ever to the use of outside sources such as court records, Securities and Exchange Commission filings, competitors, security analysts, and disaffected employees, any story is likely to get out. Kenneth Bilby, recently retired executive vice president for corporate affairs at RCA states, "You might as well try to get your point of view across, no matter what the outcome is."

No one is suggesting that press relations are a one-way street. When a story does appear that is in factual error or has questionable conclusions, a number of tactics can be used such as a telephone call to the reporter, a letter of protest to the editor, and a personal visit to the editor-in-chief. Other tactics that have been used in more extreme cases are not responding to requests from certain reporters and cutting off all contact or cooperation with the specific media organization. Mobil went so far as withdrawing its advertisements from the *Wall Street Journal* and General Motors stopped responding to *60 Minutes*. The latter two tactics should be used only with great caution as both guarantee that the organizations' points of view will not be heard.

One final tactic that can be used is the filing of a lawsuit against the offending media. Former Mobil Oil Corporation President William Tavoulatreas had his suit against *The Washington Post* for reckless disregard for the truth upheld in federal court.[23] This occurred approximately six years after the article appeared. Whether the press will become more responsible due to such actions is not known. However, following an open, candid, and cooperative strategy with the press may prevent the need for lawsuits.

Thus, the press is just one more area where business must establish long-term working relations. Since the turnover rate among reporters tends to be high and the knowledge of economics low, business may need to spend considerable time and effort educating new reporters in an informal manner. Of course the cost may be high, but when the next issue arises business may have a chance to have its side of the story heard.

The Judicial System

The judicial system can have tremendous and immediate impact on the fortunes of a company. It can declare a merger invalid or award millions in damages to a consumer or a competitor. Even more far reaching it can declare a widespread practice such as disparate impact unconstitutional.

Judicial Players. The judicial side of political planning is unique. The players are the parties themselves and their lawyers along with the judge. In a true judicial setting there is little that insiders or outsiders can do to influence the judicial mind. When the judicial role is part of an administrative agency, the influences that work for agencies work for administrative decisions.

Outside Influences on the Judiciary. For the long term, the best that a business can do is influence the structure of the judiciary. Since the times of Franklin Roosevelt, it has been clear that presidential appointees to the Supreme Court will influence the decisions and philosophy of Court decisions. For the first time since 1963, the famous Miranda decision was modified and restrained by the 1985 Supreme Court, and the change was attributed to the conservative majority created with the Reagan appointment of O'Connor. Likewise, each president leaves his mark on the federal judiciary with a typical four-year president appointing nearly 60 judges.

STRATEGIC PLANNING FOR THE POLITICAL ARENA

Whereas political intervention into commercial enterprise has become pervasive, the corporation is disenfranchised from direct involvement in the political process. Obviously, a corporation cannot vote. In addition, it cannot make political contributions as individuals can. Thus, the corporation must find other strategies if it is to cope with the increase in political intervention.

By utilizing an effective and timely political strategy, management can often gain a competitive edge. This advantage can be just as great as one gained from an innovative marketing campaign or a technologically advanced production process.

Postures as Effective Strategies

Since the degrees of government intervention vary, businesses must adopt strategies contingent on the extent of intervention present or anticipated. These contingent responses can be categorized into three main postures: sensing, progressive, and aggressive. The relationships between these postures and varying levels of intervention are depicted in Figure 35-1.

The Sensing Posture. When government intervention is general, a sensing posture is most appropriate. This posture entails having a well-developed early warning system that recognizes attempts to change or initiate legislation that would have had an adverse effect. One method of developing this awareness is to participate actively in organizations such as the Better Business Bureau, the Chamber of Commerce of the United States, and industrywide associations. Informed businesses can work together against efforts they do not like. For example, in 1978 a bill before Congress not only proposed that social security taxes be raised, but also that the employer pay a higher proportional share. Only through intensive lobbying by business interests did the equal-proportional payment basis remain the same.

One industry that needs a sensing posture is the retail clothing industry. In general, this industry is relatively unfettered by specific regulations, but it is subject to the vagaries of short-term issues: For instance, retailers had to resolve—both economically and politically—how best to dispose of their stock of children's pajamas that did not meet new inflammability regulations. Within a short time, the issue became how to dispose of the stock of pajamas treated with a fire retardant that was suspected to be carcinogenic. Without appropriate channels of political communication such as trade associations, these types of issues could not be resolved effectively or in a timely manner.

A Progressive Posture. When political intervention begins to affect directly either the product or the production process and, in

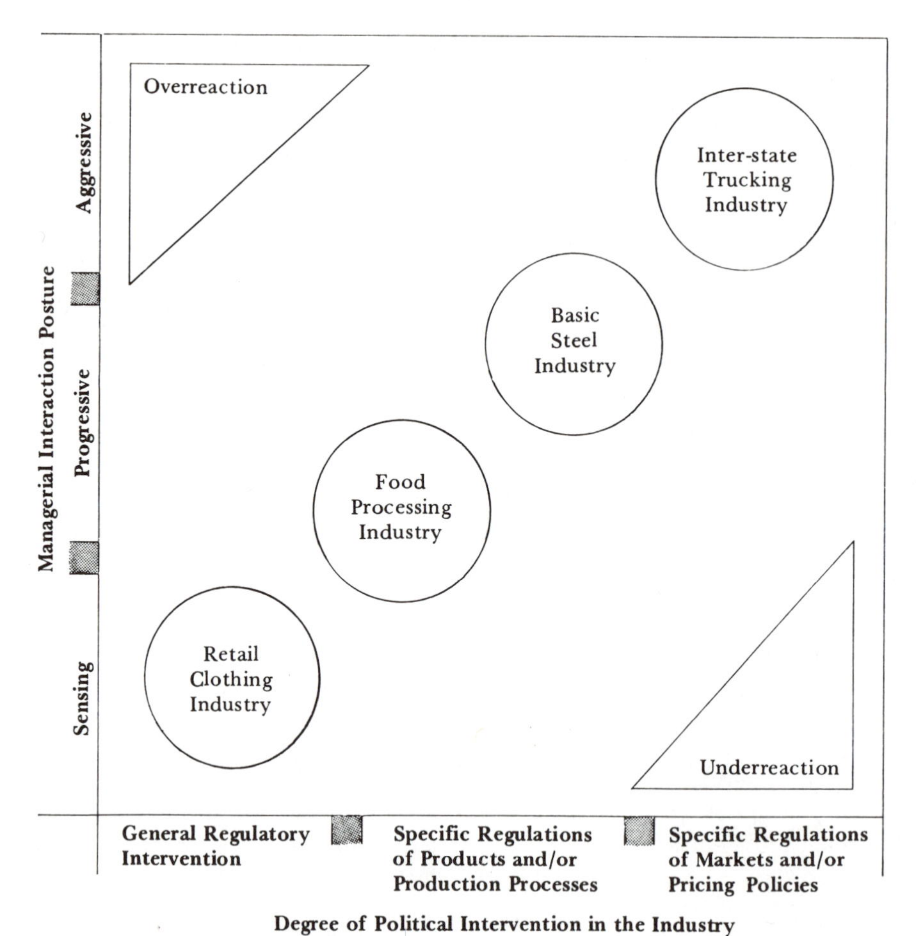

Figure 35-1. A profile of management-political intervention (from M. M. Jennings and F. Shipper, "Strategic Planning for Managerial-Political Interaction," *Business Horizons,* July-August 1981:47. Used with permission).

turn, the competitive ability, a firm must take a progressive posture. A progressive posture includes all the elements of the sensing posture plus such tactics as direct corporate lobbying and proactive involvement in administrative proceedings for the promulgation of regulations. For example, Ray Kroc, the builder of McDonalds, attempted to have an amendment placed on the Minimum Wage Act that would exclude the part-time teenagers employed by the hamburger chain. Had this amendment not been defeated, it would have greatly reduced the labor costs at the chain throughout the country.

The steel companies have formed an alliance with the steel workers' union to lobby for lower trigger price levels on steel. This would reduce the dumping of imports, which can benefit both groups by allowing the industry to sell more of its products at higher prices and thus provide more jobs for domestic workers. In addition, the sheer number of steel workers constitutes a powerful voting block, especially in the key presidential voting states of Pennsylvania and Ohio, thereby increasing the political effectiveness of the alliance. Thus, the company gains the synergistic effect of more political clout through the assistance of the union.

Education is another progressive tactic

commonly used in basic industries. Although the governmental administrators who write regulations may know and understand the harmful effects of the various pollutants, their knowledge of the production process can be limited. Many companies have offered tours of their facilities to the regulators in order to inform them on the intricate chemical and physical processes that produce pollutants. The steel industry countered the initial regulations, which applied to each specific producer and outlet on a given plant site, with a proposal of a bubble concept that would regulate all the pollutants for the entire site. The industry then had to convince the regulators that by its counterproposal more pollutants could be captured per dollar of expenditure, and the resulting overall level of pollution would be lower. The industry was successful in this endeavor and capital expenditures for environmental protection equipment were considerably reduced.

A fourth progressive tactic is to influence the promulgation of regulations by placing the right person in the right position. Since many industry regulations are promulgated and enforced, not by Congress, but by administrative agencies, the structure and composition of these agencies have become increasingly important, particularly since action by an administrative agency can have devastating effects on an industry. The structure of an agency can be such that political tactics effective for Congress may be ineffective in the agency. For example, the history of the Federal Communications Commission (FCC) illustrates that lobbying for the right person in the right position can save business expense later. Appointments in administrative agencies can affect industry profits, for appointed heads influence the direction of the regulatory agency and its attitude toward industry and profits. Throughout the FCC history, those commissioners appointed from the broadcast industry have been less stringent in their policies and enforcement, whereas those heads appointed from outside the industry have been more harsh. When Charles Ferris was appointed head of the FCC, he stated that it was valid for the FCC to consider the problems with children's TV ads, whereas prior to his appointment the FCC had taken the position that this regulation was best accomplished privately and voluntarily through the broadcasters' code. This same issue of children's ads was avoided by the Federal Trade Commission (FTC) until Michael Pertschuk was appointed chairman. Pertschuk initiated and supported a proposal that would substantially restrict ads during children's hours on prime television. The controversial subject matter had been brought to the attention of the FTC by parents' groups over a 10-year period, but it was not until Pertschuk's 1977 arrival that the proposal came to life.

The effects of ad regulation might include an increase in retail price or a decrease in sales and profits. Exercising influence before the appointment of an administrative head may alleviate problems of attempting to avert regulations proposed once those heads are appointed. Thus, influencing the revolving-door process between private enterprise and political appointments can be a preventive tactic.

The Aggressive Posture. When political intervention has grown to the point that both market availability and pricing policy are affected, then the industry and the companies within that industry must take an aggressive posture to protect their interests. One avenue for self-protection is the judicial system. For instance, in *California Motor Transport Company v. Trucking Unlimited,* an association of truckers instituted federal, state, administrative, and judicial proceedings to oppose applications for operating rights by those outside the association. The result was an increase in association business and an antitrust suit by the applicants who had been denied. However, the Court held that such access to the courts was protected by the First Amendment and not a violation of the Sherman Anti-Trust Act.

The use of lobbying action also is a strategic tool in the aggressive posture. When one industry has neglected or been ineffective in lobbying, lobbying by other industries or firms for the same markets can have a devastating effect. For example, in the case *Eastern Railroad President's Conference v. Noerr Motor Freight, Inc.*, a three-decade economic struggle between railroads and truckers came to a head when the railroads in Pennsylvania, in an attempt to improve their economic and competitive edge, began an organized publicity campaign to turn public opinion against the truckers. Lobbying was organized and eventually a bill that would have increased permissible highway weight limits was vetoed by Pennsylvania's governor. The truckers brought suit against the railroad association, alleging that the activities of the railroads were fraudulent, malicious, and violative of the federal antitrust laws. The Supreme Court held that prohibiting such joint lobbying efforts would have a chilling effect on the First Amendment right of the rail industry to make its opinions and wishes known to public officials, and that the antitrust laws were designed to regulate business activity, not political activity. Thus, the railroad industry was able, through legislative influence, to eliminate competition from the trucking industry in one segment of the transportation market. Furthermore, this influence has been afforded the substantial protection of the First Amendment in spite of the anticompetitive effects on other businesses.

On the national level, the trucking industry lobby in combination with the Teamsters Union lobby has been very effective in obtaining and maintaining a protected market environment. Again, the combined lobbying of the industry and union for this mutual self-interest has given the industry more clout than either would have alone. The Teamsters have the largest independent union membership in the country and thus represent a potentially powerful voting block.

At the national level of government, the industry has built a protected pricing and market system in government regulations which has allowed firms to grow and prosper without close regard to internal cost controls. Both Bruce Allen of the University of Pennsylvania and Denis Brien of the University of Washington, in independent studies, have found that intrastate, unregulated truckers can provide the same service for less than regulated, interstate truckers.[24] In addition, Allen has found evidence to suggest that the intrastate trucker is more profitable than the interstate trucker. A prime contributor to the increased costs of the interstate truckers is the wage rate. The Teamsters Union has established itself as having obtained and maintained one of the highest average hourly wages for its members. Since the rate to be charged shippers by all interstate truckers is established in a collective manner, the need for hard bargaining for labor rates at the negotiating table is minimized. The additional costs of higher wages are simply passed on to the consumers through the collective rate-making process. Thus, a profitable and protected environment is maintained by the regulators for the truckers and for their employees. The protected environment has been partially dismantled by deregulation.

INFLUENCING THE LEGISLATIVE PROCESS

In the Washington arena, attempts are made to influence all three areas of the Iron Triangle—the legislative, the executive, and the special interest. However, efforts are frequently concentrated on the legislative area because the power to fund ongoing programs, to initiate enabling legislation for new programs, and to provide continuity of government rests within its purview. As a result, over the years four major forms of recognized strategies have been developed to influence the legislative area. As rated by the legislative assistants in descending order of effectiveness, the four strategies are political, informational, public exposure and/or appeal, and direct pressure.[25] As displayed in Table 35-1, business and industry groups use the most effective political strategies the least, 8.8 percent. Thus, tremendous room for improvement exists for business and industry to increase their ability to influence the political arena.

Table 35-1. Impact and Frequency of Four Major Strategies Used to Influence Legislation

Perceived Impact by Legislative Assistants		*Frequency of Use by Business Groups (in percentages)*
7.05	Political strategies	8.8
4.39	Informational strategies	50.2
1.91	Public exposure and/or appeal strategies	30.0
−1.28	Direct pressure strategies	10.9

Source: From data in J. C. Aplin and W. H. Hegarty, "Political Influence: Strategies Employed by Organizations to Impact Legislation in Business and Economic Matters," *Academy of Management Journal* **23**(1980):446-447.
Note: Impact was measured on a +10 (favorable) to −10 (unfavorable) scale.

The four major strategies can be factored into specific tactics as shown in Table 35-2. As can be seen in the table, the legislative assistants perceive the various tactics to have different magnitudes of impact. The ratings are not mathematically precise enough to say, with any degree of certainty, that the use of specific arguments with members of Congress is twice as effective as public exposure of the voting record. Nonetheless, they do indicate that an organization will be more effective if it has expert witnesses visit the member's office than if it sends a set of technical reports. Therefore, the astute executive examines the range of tactics available, considers other factors such as resources and time available to implement the tactics, and then chooses the one judged to be most appropriate.

Table 35-2. Perceived Impact by Legislative Assistants of Specific Influencing Tactics

Political tactics	
Constituent contact	7.05
Colleague contact	6.10
Informational tactics	
Specific arguments	5.15
Expert witnesses	4.65
Personal visits	4.13
Technical reports	3.62
Public exposure and/or appeal tactics	
Voting record	2.57
Third-party influence	2.12
Letter campaign	1.56
Media campaign	1.38
Direct pressure tactics	
Threat of harm	.09
Financial support	−2.65

Source: From data in J. C. Aplin and W. H. Hegarty, "Political Influence: Strategies Employed by Organizations to Impact Legislation in Business and Economic Matters," *Academy of Management Journal* **23**(1980):447.

Political Strategies

Political strategies rely on direct contact, via the popular media or in person, with either the constituents or colleagues of the members of Congress. Effective contact with constituents is often too difficult, costly, and time consuming to implement; however, it is the most influential of all 12 tactics as viewed by the legislative assistants. When direct contact is cultivated by top management through active participation in civic, charitable, and other public organizations, it can be invaluable.

Constituent contact is definitely one tactic that should be used by business firms for long-range goals, such as educating the public to the special needs of the organization. For example, the oil and gas industry is a major economic part of each congressional district. However, only one of the "Seven Sisters" (the seven fully integrated—exploring, producing, distributing, and marketing—oil companies),

Mobil Oil Company, has tried to educate the public with a massive campaign describing its special problems. However, this effort began only after the oil crisis of the 1970s. The other major oil companies seem content to use image campaigns. If success is to be achieved in the political arena through direct contact with constituents, a proactive, not reactive, strategy needs to be taken.

The second tactic with the highest degree of impact is colleague contact, which occurs when one member tries to directly influence another member of Congress. This tactic also will take careful cultivation to be effective. The probability that a corporate representative, out of the blue, can walk into a congressional member's office and ask that person to try to influence another member's vote is low to nonexistent. Congressional action is almost always based on coalitions of individuals who know each other rather intimately. One member of Congress is only going to ask another member of Congress for assistance if some degree of assurance exists that the response will be favorable. Members of Congress highly value collegial respect. To ask and not receive the assistance would be a loss of face. Thus, a member of Congress will only ask another member for their support if the coalition grapevine assures that the response will be positive.

Information Strategies

Informational strategies are the second most effective group of tactics. They occur when influential attempts rely on the use of either verbal or written materials to affect congressional actions. In comparison to the political strategies, informational strategies can be used in a shorter time frame. Less time is needed to gather the available information, develop the arguments, and choose the appropriate spokesperson for informational strategies than to form the coalitions required for political strategies.

Contrary to what many people involved in the political arena believe, logic appears to be the most effective informational tactic. The legislative assistants rated the use of specific arguments, relative to particular pieces of legislation, as being more effective than the use of expert witnesses, personal visits, or technical reports. Specific arguments may be given to either the member of Congress, the member's staff, or a committee's staff. The effectiveness of this form of political persuasion may be a recent phenomenon as the character of Congress has changed. By the early 1980s, only 50 percent of the members of Congress were lawyers. Recently, more members of Congress have come from business and technical backgrounds, For example, Representative James G. Martin (R, North Carolina) holds a Ph.D. in chemistry from Princeton University and was previously a professor at Davidson College.[26] In addition, more members of the congressional staff have technical and business degrees than 10 years ago. Thus, the use of specific technical arguments on the relative merits of a bill have become more effective.

The second most effective informational tactic, the use of expert witnesses, also appeals to logical analysis. Expert witnesses are frequently used during a formal congressional hearing open to the public and press. At this point, the pro and con sides of the issue may have been already decided by the political considerations. Furthermore, only a very adept and composed expert witness will be persuasive enough and be able to withstand the intimidation of the congressional hearing room setting, the attention of the national press, and the possible grandstanding of a congressional member adamantly opposed to the witness's point of view. Thus, the use of expert witnesses in a congressional hearing may be effective, but the use of specific arguments in informal meetings will have more impact.

Personal visits by key individuals with the members of Congress is seen as the third most effective informational tactic. The Business Roundtable with its 200 CEOs has used this tactic in the past with a high degree of success. However, the Roundtable is frequently acknowledged as the most effective business group in the political arena.[27] Whether the impact of this tactic will carry over to other groups and persons appears questionable. A frequent complaint against business representation in Washington is, "The only time we (the members of Congress) see them is when they want something." Assuredly, the CEO of a *Fortune 500* company who requests a meeting will receive a polite reception from a member of Congress, but such courtesy calls may have only minimal impact on the legislative process. Greater political impact can sometimes be obtained by working through the staff and committees to affect specific pieces of legislation. The CEO visit may be used as a door opener for the corporate staff to gain access to the congressional staff; such combining of tactics can ensure that the strongest possible arguments for the organization's position will be advanced.

The use of technical reports is voted as the least effective informational tactic. Several combined factors explain this position. First, members of Congress and their staffs do not have the time to read long technical reports. Most members of Congress have not personally read and analyzed the merits of the majority of the legislation they vote on, much less any back-up material. Second, approximately 50 percent of the members are lawyers and might not understand the depth of technical reports. The same holds true, probably to a lesser degree, for most of the key staff members. Third, the actions of Congress depend by and large on oral communication. A report can hardly assist in building an interpersonal communication network or a coalition. Unless the report represents the position and has received the endorsement of a powerful coalition, it may be disregarded. Fourth, a report cannot be questioned or assist in its own interpretation. A report sent prior to a visit by a company representative has a higher probability of being assigned to staff members for reading than one received in the mail with a cover letter. Thus, a technical report, to achieve optimal effectiveness, should be combined with one of the other informational tactics.

Public Exposure and/or Appeal Strategies

The public exposure and/or appeal strategies received relatively low ratings as effective methods of influence by the legislative assistants. Yet, they are widely used. The rating and publication of voting records are completed by such diverse groups as the Chamber of Commerce of the United States, the AFL-CIOs, Committee on Political Education, Ralph Nader's Public Citizen organization, Americans for Democratic Action, and Americans for Constitutional Action.[28] As can easily be imagined, the ratings of any one member of Congress by such diverse groups will vary widely. To establish the rating of the congressional member, the executives of the organizations select votes on issues of major concern to them from hundreds of votes cast each year. The congressional members are then evaluated on whether they voted on the right or wrong sides of the issues, and a rating is derived from this tabulation. The ratings are then sent to the supporting members of the organization, to the congressional members, and to the press.

However, the effectiveness of this form of political influence is doubtful. First, voting records may be misleading. Members of Congress vote many different ways on different issues for different measures. As Representative Les Aspin (D, Wisconsin) has said, "If you give Congress a chance to vote on both sides of an issue, it always will do it."[29] For example, to gain or pay back a political IOU, a vote may be cast for a piece of legislation even if the congressional member is personally against it.

Sometimes, the piece of legislation selected to pay back or gain an IOU will have a slim chance of passage. The justification by the members of Congress for such ploys is no harm, no foul. Second, since an array of ratings is published each year, the chances are low that anyone will receive the public exposure required to make it a significant factor in the collective voter consciousness during an election. The elected officials tend to ignore ratings because they are aware of the limited effectiveness they have in swaying voters. Furthermore, if the rating is low on one issue, the rating on another will certainly be high.

Nonetheless, the exposure of voting records should not be discarded as a tactic. On single issues, voting-record disclosure seems to be effective. Although the public may not comprehend or respond to abstract numbers generated by a rating scale, it does seem to respond to single major issues such as abortion, prayer in school, and busing. The strength of single-issue exposure is a phenomenon that emerged as an effective tactic in the late 1970s on the political Right. By 1982, some observers in the political arena were predicting that the efficacy of this tactic was on the wane due to the failure of any of the three issues mentioned to become legislation.

The second most effective public exposure and/or appeal tactic involves the use of a third party in trying to obtain political influence through a public appeal. The third party may be a public relations firm, a hired lobbyist firm, or an industry, a business, or some other special interest association. Labor's committees on political education have used this tactic for years.[30] By contrast, business groups have largely ignored this technique, partly because the legality of funding such campaigns was questionable until a change in the federal election campaign laws was made in the 1970s. Also, the use of third parties may have been dismissed because many business leaders did not recognize the need for long-term involvement in the political arena until the 1970s. Reginald Jones, past president of General Electric and the Business Roundtable, was one of the first business leaders to advocate stepped-up involvement. With the explosive growth of business PACs from 89 in 1974 to 1,557 in 1982, the use of third-party public appeals by business would appear to be on the rise.

The letter campaign is one of the best-known political influence tactics.[31] The nickname of Ma Bell for American Telephone and telegraph was derived from a letter campaign conducted by the stockholders in the late 1940s to suppress an antitrust investigation. Although well known and frequently used, the letter campaign tactic is rated as the second least effective of the public exposure and/or appeal tactics. One of the reasons for this low rating may be that a number of members of Congress and their aides take great pride in relating how they can distinguish between an organized letter-writing campaign and a spontaneous letter from a constituent and that they discount letters generated by organized campaigns. They purport to use such clues as company letterhead vs. plain stationery, typewritten vs. handwritten, postal metermarks vs. stamps, similar or exact duplicate passages vs. unique paragraphs, proper vs. plain English, and correct and lengthy congressional addresses vs. incorrect and short addresses. The former in each instance were deemed to be signs of an orchestrated campaign, whereas the latter were thought to be spontaneous letters. Members of Congress also tend to discount letters that originate outside their districts or states. On the other hand, the ability of the National Rifle Association and other special interest groups representing minority opinions to impose their will on the majority may indicate that letter-writing campaigns cannot be dismissed as ineffective. Rather, letter campaigns may be a good support tactic when used in conjunction with some of the preceding tactics. A legislator may just wonder what all this mail means and devote more attention and time to the issue. In addition, letter campaigns may increase access to key aides and committees.

The final public exposure and/or appeal tactic, the media campaign, is rated as the least effective by the legislative assistants. Media campaigns tend to be least effective because the audience simply does not take any action. Only 35.2 percent of the eligible voters took the minimal political action of participating in the 1978 representative elections, and only 52.5 percent voted in the 1980 presidential elections.[32] The estimates of other political actions taken by the public at large, such as writing letters or working for a candidate, fall in the less-than-5-percent range during any election period when political awareness is highest. Thus, media campaigns tend to utilize resources ineffectively.

In the 1980 election a different form of media tactic emerged—the negative campaign. These campaigns were more against someone than for someone. They were sponsored by groups such as The National Conservative Political Action Committee (NCPAC), and the Moral Majority's National Right-to-Life groups. A postelection analysis indicated that the defeats of Senator George McGovern (D, South Dakota), Senator Frank Church (D, Idaho), and Representative John Culver (D, Iowa) were due in part to negative campaigns.[33] The analysis also suggested that negative campaigns can be effective if they are started early in the campaign, are rational and information oriented, are well documented, and are used against an incumbent. Ironically, in 1982, the antinuclear groups adopted this media tactic in an attempt to defeat a number of Pentagon supporters in Congress. In the future the use of negative media campaigns can be expected to grow, and their ability to influence the outcomes of elections will also grow as better negative campaign techniques are developed.

Direct Pressure Strategies

Direct pressure strategies are rated least effective by the legislative assistants, and we hope that that is a true reflection of their effectiveness. The threat of harm is rated as having almost no effect one way or the other, and the promise of financial support is perceived as having a negative effect. As can be seen from the ABSCAM investigation, representatives and senators who will accept a bribe do exist. However, none of those convicted were power brokers in Congress. In addition, it is doubtful whether the bribe accepters planned to take any action.

Both of these tactics are not only illegal and unethical, but also the backlash from public exposure of the use of such techniques, plus the ensuing multitude of investigations, additional regulations, and reporting requirements, could offset any momentary advantage gained by such techniques. The reason for the tight federal campaign reporting standards and enforcement is due in part to the illegal contributions made to President Richard Nixon's reelection committee. Unethical and illegal activities in the long-run tend not only to come back to haunt the immediate parties involved but also the contaminate the entire segment of society from which they came by guilt through association.

STRATEGIES AND STRATEGIC PLANNING FOR THE LEGAL ARENA

The Inevitable Lawsuit

Ninety-five percent of all lawsuits never make it to trial,[34] but that comforting fact does not make the legal costs any less. You may not get socked for a full-fledged trial, but you will pay for the pregame show—discovery. Most of the money in litigation is spent on discovery.[35] Oddly enough, this part of litigation is the most controllable. Indeed, whether there will or will not be litigation is, to some degree, a controllable element for a business. If the element is not controllable then the disposition and expedition of the case become prime goals for the victim business. The achievement of those goals requires foresight, proper analysis, and strategic planning. Most businesses begin their planning when the lawsuit begins. Rising litigation costs indicate that that type of planning is too little too late. Planning should follow an organized format. This section presents that format.

Step One—Assessment: Areas of Vulnerability. Every business has an area(s) in which they are likely to be sued or have to sue. Areas of vulnerability break down into two categories—the regulatory environment and the civil environment. Vulnerability in the specifics in each category depends on the nature of the business. For example, a manufacturing firm enjoys higher vulnerability in OSHA (Occupational Safety and Health Administration) than does a firm operating retail clothes outlets. A firm with a delivery or sales force has a high degree of vulnerability in the Agency-Employee Negligence category, whereas a catalogue sales firm would not have the problem as a high priority. Real estate syndicators walk a fine line with the SEC (Securities Exchange Commission) but a privately owned firm has no SEC worries. Table 35-3 covers the types of firms significantly af-

Table 35-3. Areas of Legal Vulnerability

Legal Area	*Type of Firm*	*Potential Problems*
Environmental	Manufacturing; utility; mining; industrial developers; contractors	Excess pollutants; factory location; expansion; cost of equipment; business operation restrictions; limited areas of development; historical or archaeological sites
Employment OSHA	All firms covered but high vulnerability in manufacturing, industrial construction, and mining companies	Changing guidelines; excessive and costly projections; OSHA inspections and citations; failure to follow guidelines
Workmen's Compensation	Most employers covered in most states; covered accidents mostly in manufacturing, industrial construction, and mining firms	Inadequate coverage; product liability causation; disputes over compensation amounts
EEOC	Employers with 15 or more employees	Discriminatory hiring, firing, promotion, salary increase practices; sexual harassment; invalid testing instruments
Fair Labor Standards	Nearly all businesses with exceptions for farm, seasonal, family businesses	Minimum wage; accurate overtime compensation; professional and executive exemptions
Union	All businesses with union workers; businesses that become unionized	Union elections; unfair labor practices; boycotts; slow-downs; pickets; strikes; collective bargaining; union rules (seniority and so forth)
SEC	Sales: Any corporation selling partnership or business interests that qualify as securities; *security* = investment in a common enterprise with profits to come solely from the efforts of others. Reporting; Any business with $1,500,000 in assets and 500 shareholders *or* listed on national exchange; Fraud: All businesses selling securities	Exemption filing; registration statements; inadequate disclosure; fraud; false information; faulty audit
State Blue Laws Sky Laws	State exemptions vary; businesses selling securities are governed by some part of the states' securities laws	Failure to file; most merit review; exemptions; fraud

Table 35-3 (*Continued*)

Legal Area	*Type of Firm*	*Potential Problems*
Business Organization	All types—sole proprietorships; partnerships; corporations	Fictitious name registration; filing of partnership agreement; limited partnership requirement; incorporating requirements; annual reports; IRS elections; statutory agent
Patent	Firms with unique process or product	Patent registration; infringement
Tradename, Trademark	Firms with product name or symbol requiring protection	Registration; or avoiding generic use
Copyright	Literary or music publishing firms	Application; protection; enforcement
Product Liability	Firms selling any product or part	Consumer product safety division; Federal Regulation Compliance (Magnuson-Moss)
Warranty	Firms selling products (regardless of whether written warranty is given)	Uniform Commercial Code—Article II—express and implied warranties; efficacy of disclaimers; samples; models
Corporate Crime	All corporations	SEC, antitrust, tax crimes; fraud; check-kiting; failure to correct regulatory violations
Crime, Officer and/ or Director Liability	All corporations	Guilt through responsibility; nondisclosed activities; knowledge of violations
Contracts	All firms	Failure to perform; excuses for nonperformance; breach by other party; improper formation; fraud defenses; statute of frauds (written)
Torts	All firms	Defamatory remarks or publication; advertising defamation; interference with business relationships; collection torts (privacy and/or emotional distress); accidental harms (autos, slip and/or fall)
Product Liability	Manufacturing firms; suppliers to manufacturers	Defective design; inadequate instructions or warnings; quality control; poor distribution system
Agency	All firms	Negligence by employees within scope of employment (auto accidents); employees exceeding their authority; apparent and implied authority; employees leaving to start work for competing firms
Property	All firms	Lease disputes; land ownership disputes; liability to customers on premises; nuisance (operation of plant)

fected by or highly vulnerable to the legal and civil environment areas along with the types of problems.

Step Two—Assessment: Problems and Prevention. It would be too lengthy a task to do a worksheet for every area, but several universal examples covering a broad cross section will illustrate the approach that should be taken as part of Step Two.

The Montgomery Ward Warranty Notebooks

Background. Any firm selling products to consumers (those who buy products for personal or home use) has to worry about and comply with a number of federal laws and regulations. Sometimes the regulations are not clear, and compliance is a case of trial and error. Legal strategy comes into play when a firm has to handle the error part. Montgomery Ward's handling of their errors in attempting to comply with the federal warranty law (Magnuson-Moss Act)[36] is a fascinating study of legal pursuit without legal strategy.

The key to the Magnuson-Moss Act is disclosure. Consumers are entitled to know the full scope and extent of their warranty protections. A seller can use several methods to assure disclosure. The warranty terms can be displayed when the terms are written on the package. Another alternative gives the seller the right to maintain:

> A binder or series of binders which contain(s) copies of the warranties for the products sold in each department in which any consumer product with a written warranty is offered for sale. Such a binder(s) shall be maintained in each such department, or in a location which provides the prospective buyer with ready access to such binder(s) and shall be prominently entitled "Warranties" or other similar title.[37]

Wards, a 650-store chain of retail and catalog stores decided to use this binder alternative to comply with the regulations under the act. Three sets of binders were delivered to each store along with informational signs. The binders were to be kept at an information desk in the automotive department. Generally, Ward Automotive Departments are located in buildings separate from the retail store. The informational signs were placed near the binders and in other prominent areas in the stores.

In 1977, the FTC investigated Montgomery Ward for compliance with the regulations and in 1978 told Ward they were not complying with the regulations. In response, Ward distributed more signs to its stores along with a memo emphasizing the importance of the FTC regulations on notice. The FTC found these actions were not enough and issued a complaint alleging that Ward had failed to make warranty materials available to consumers and was in violation of the Magnuson-Moss Act.

The case went forward with lengthy hearings and the administrative law judge (ALJ) held Ward in violation of both the regulation and the act. A cease and desist order also was issued. The ALJ ruled that the "ready access" mandated in the regulations required at least one set of binders to be available on each selling floor and a notice of the availability of binders in each department. Both parties appealed the ALJ's decision to the full five-member FTC. The FTC staff wanted a binder in every department and Ward wanted the binders just the way they were. The full commission followed the ALJ's decision, and Ward appealed the case to the Ninth Circuit Court of Appeals.[38]

Ward's basis for appeal was that the requirement of one binder per sales floor was an amendment to the regulations that should have been accomplished through proper rule making rather than through a complaint-judicial process. The Ninth Circuit noted a lack of documentation supporting the reasons for the "each floor" requirement but affirmed the decision for "large retailers." However, the Ninth Circuit refused to go along with the imposed requirement of signs "in prominent locations in each department." The language of the regulation requires signs in "prominent locations in the store or department"[39] and the

Court found no basis for the commission's ruling requiring a sign in each department. The Court indicated strongly that the "or" could not be construed by the FTC in such a strained way: "If a violation of a regulation subjects private parties to criminal or civil sanctions, a regulation cannot be construed to mean what an agency intended but did not adequately express."[40]

In a note of chastisement the Court mentioned that the ALJ and the commission could not agree where signs should be placed but went ahead with the department rule, which was indicative that their rulings were "far afield from the plain meaning of the rule."[41] One final victory for Ward came with the Court's conclusion that the ruling on binders on each floor would be applied prospectively only and the cease and desist order against Ward was vacated.

Analysis. After four years of litigation and thousands of dollars in attorneys' and expert witnesses' fees, Ward had its outcome. To an extent, Ward was vindicated but they still had to put the binders on every floor and won only the right to post a few less signs throughout the store. Was it worth it? Ward had the honor of having the first major decision under the Magnuson-Moss Act in its eight years of existence. That eight-year span is significant for two reasons. First, no other large retailer (and others had been found in violation) deemed the issues costly or detrimental enough to pursue. Second, even the most diligent and aggressive shoppers had not complained with very few of them seeking out the binders regardless of their convenient or accessible locations.

Certainly, Ward had a cost basis for its pursuit of the matter. There is nothing wrong with the attempt to minimize the costs of regulatory compliance. Certainly, the FTC was stretching its authority in its interpretation of "ready access." However, the result was just a little less than what the FTC wanted and certainly the difference in cost of five signs as opposed to ten signs was minimal. In addition, Ward looked like the bad guy. It pursued an arguable costly ideal and in the process gave the impression of resisting the spirit of a consumer full-disclosure law. As the Court noted:

> We are sympathetic to Ward's observation that its substantial evidence argument is hampered by the fact that it was presenting evidence of its good faith compliance with the plain language of the pre-sale rule, while the FTC was demonstrating that Wards was in violation of a new version of the rule that never before had been stated.[42]

The bottom line was that Ward was being pressed farther than the statute allowed the FTC to go. Ward could have simply acquiesced and saved itself the cost of a hearing, a commission appeal, ultimately a circuit court of appeals appearance, and a lot of attorney's fees. Clearly the cost of more signs and two or three more binders for each store could not have exceeded the cost of this entire legal affair from start to finish. The arrogance of the FTC cannot be discounted for its level of irritability motivation, but was the victory worth all the costs? One critical note is that from all appearances, Ward faced a particularly hostile FTC that had taken a non-negotiable position. Compromise was not the goal of the staff at the time. The goal was even more stringent standards than the commission itself was willing to accept. An analysis of the Ward predicament (with 20/20 hindsight) is given in Table 35-4.

The Plastic That Crossed Texas[43]

Background. On April 5, 1975, Mobil Chemical Corporation signed a contract to buy plastic at a price of $1.94/pound. Mobil was using the film to make trash can liners. Askco made a warranty in the contract that the film would be free of defects and would conform to Mobil's description and specifications. Mobil was given the right to inspect and approve the shipped product. When the product arrived in Temple, Texas, Mobil tested 45,000 pounds but found that the plastic would not break down properly. Mobil agreed

Table 35-4. Regulatory Action—Settlement vs. Suit

Factor	*Comment*
Cost	Worst possible result, cost of signs in each department vs. legal fees, full hearing; commission appeal; appellate court appeal
Settlement possibilities	Hostile staff; no major case under act after eight years (looking for a test case); had already made changes after notice of violation and complaint still issued; slim likelihood of settlement
Public perceptions	Ward is portrayed as reluctant to disclose warranty rights; FTC is portrayed as white knight; cost of compliance seems minimal
Business perceptions and aid	Nitpicking by FTC; possible aid from retailers' organization to clarify statutes; excessive regulatory action

to pay for the 45,000 pounds ($21,510.28) and shipped back the remainder to Askco. Askco refused to take the plastic and shipped it back to Mobil in Temple. Mobil retested the plastic with the same results and stored the plastic for Askco for about one year. After a year, Mobil carted off and burned the plastic and Askco filed suit for breach of contract. Mobil counterclaimed for the cost of warehousing, freight, loading, and unloading the plastic ($14,079.65). The trial court found for Mobil and after Askco's appeal, the appellate court affirmed.

Analysis. This case is one in which Mobil did everything right (with one exception) and Askco did everything wrong. The contract was clear in its intent that proper plastic be delivered, that Mobil be permitted to test, and that Mobil be allowed to reject the plastic for nonconformity. Mobil tested and rejected because the plastic did not meet its specifications. Under the Uniform Commercial Code, they were entitled to do so. Mobil was not even required to pay for the goods tested, only the costs of testing. Mobil was clearly within its rights and was also permitted to reship the goods. Askco's reshipment was a waste of resources. Their only point was that Mobil could not reject in units of pounds. However, that was the unit for selling and the basis for the price.

In this case, the bottom line was an incensed seller who had no right to be. Table 35-5 is an analysis of the factors in the suit and their application to Mobil and Askco.

The Case of Competing Employees—Alternative Resolution to a Civil Issue in Agency

Background. Intel, an integrated circuits producer, had some of its marketing and technical people leave to form a new company (a competitor) called Seeq Technology Inc. Intel sued the former employees for breach of the anticompetition clause in their contracts and possible trade secret infringement. Seeq settled but Intel was not convinced Seeq would abide by the settlement terms. The solution was to have a college professor monitor Seeq's activities for one year.

Analysis. Although it took a suit to get to the settlement, both sides seem to have acted more rationally in coming up with an alternative resolution to their dispute. Their enforcement mechanism of monitoring saved both sides money, anxiety, and litigation. The key was infringement prevention. Intel took its prevention steps by filing suit and then reaching a settlement that allowed for a watchdog over activities. Table 35-6 is a summary of the issues.

Walt Disney—The Master of Non-Liability in Tort

Background. A Los Angeles attorney has lost only six cases in his 25-year practice and two of them have been to Disneyland

Table 35-5. Breach of Contract Civil Action—Settlement vs. Suit

Factor	*Mobil (buyer)*	*Askco (seller)*
Cost	Worst possible result, required to pay for all plastic; legal fees; cost of reshipping, storing, and buryings; difference between $21,510.28 and full contract price of approx. $56,041.12	Loss of sale; loss of 45,000 pounds of plastic used in testing; cost of shipping and reshipping; legal fees
Settlement possibilities	Hostile seller; not likely	Mobil willing to pay for that amount used; willing to return remaining plastic
Legal precedent	Had right to test and reject; Mobil had right to store; had full warranty and inspection rights on contract	Mobil had significant number of rights; only question of per pound rejection (minor point); deck stacked in Mobil's favor
Jury appeal	Look like the good faith buyer	Look like a seller reneging on a warranty; action of reshipping seems childish

Table 35-6. Infringement Civil Action—Settlement vs. Suit

Factor	*Intel*	*Seeq*
Cost	Loss of competitive edge; attorneys' fees involved in filing suit; cost of settlement supervision; filing suit;	Damages for infringement; probable demise of the company; attorneys fees; limitations on product line
Settlement	High as long as supervision continues	High; desire to avoid injunction and get on with business
Legal precedent	In their favor, former-employee competitive firms looked upon with disfavor	Same
Jury appeal	Same (gives in their favor)	Disloyalty of employees; greedy

Amusement Park.[44] Apart from the inherent charm it enjoys, Disneyland is a shining example of how to work at winning lawsuits.

In 1983, when James Higgins was 18, he rode on Disneyland's Space Mountain. When the roller coaster took a turn, Higgins was thrown from the car. Ordinarily this is the type of sympathy-evoking case that puts gleams in lawyers' eyes. However, Higgins lost his case. Disney presented evidence that the youth had been drinking.

In 1975, a woman and her son were riding in an Autopia car in Disneyland when a 16-foot long branch from a eucalyptus tree fell in front of them. Their stop to avoid hitting the tree caused several cars to hit them from behind. The woman and her son sued on the basis that the tree was rotting, that the roadway was poorly designed, and that the cars had no head rests. After a seven-week trial and Disney's German tree expert, the jury found for Disney in only one and a half hours.

Disney has fewer than 100 lawsuits filed per year and wins those frequently. The odds are not bad considering Disneyland and Disneyworld attract 30 million visitors per year. Disney is a model of excellence in how business planning can minimize legal costs.

Analysis. Disney officials take a quick soothing approach. When someone is injured, hosts and employees descend to comfort and accommodate and get them back out into the park to enjoy themselves.

Also, employees immediately question witnesses to an accident. They also take the opportunity to interview the person injured. With park onlookers, the injured are known, out of pure embarrassment, to make admission about their involvement.

Disney has a hard-line attitude. For those who cannot be soothed and want to sue, Disney takes a full-battle approach. It is clear from the onset that Disney will go to trial.

Once in trial, Disney uses its image to full advantage. Their wholesome employees with the trappings of Disneydom (watches, hats, buttons) appear candid and honest to jurors.

Finally, Disney runs a good business. The parks are well maintained. From the attorney who has lost mostly to Disney comes the following quote: "I have never been as impressed with an operation as I have been with Disneyland Amusement Park. I think Disney wins because it deserves to most of the time."[45]

Step Three—The 5,000 Mile Check: Minimizing Legal Overhauls. The cases to this point discuss resolution after the fact. The problem already exists, the damages have occurred, and it is a matter of minimizing legal fees. Step Three is preventive maintenance. Correcting weak spots or areas of legal vulnerability is a way to prevent the head-to-head combat just covered. For example, putting an arbitration clause in a contract minimizes legal expenses should a problem arise. So long as both sides are required to bargain in good faith; arbitration is a shortcut.

Determining remedies up front can save money. Liquidated damage clauses (damages determined up front and put in contract) can be used to control the effects of litigation.

As a general rule, the more the parties are able to agree upon and put in their contract, the less likely a dispute is to arise. If the problem is covered in clear language, disputes can still arise but the likelihood of settlement increases.

For regulatory areas, the minimization is accomplished simply by reviewing regulations; checking compliance and updating when new regulations are passed. Nothing can prevent a securities fraud suit but there is no excuse for SEC action for noncompliance or incomplete forms.

For employee conduct and liability, nothing beats management review and a system of checks and balances. A hostile, overaggressive salesman was once pointed out to Nabisco but no action was taken. The result was Nabisco's liability to a grocery store owner who was punched out by the salesman when he was refused more shelf space for his Nabisco cookies.[46] A dangerous piece of machinery is pointed out by an employee, but supervisors ignore the warning and send the employee back to the equipment. The employee is severely injured by the hazard pointed out and management is surprised.[47] A system of employee input going past immediate supervisors would help in this type of situation.

For product liability, quality control is legal control. A product is defective if it is mismanufactured. Quality control is not simply to keep production and the number of returns down; it is to keep legal costs down.

Good community relations minimize nuisance, environmental, and other property issues. Good citizenship seems to control when issues of nuisance arise, and settlements are common for businesses with a stake in community affairs.

Step Four—When All Else Fails: Minimizing Litigation Costs. It is a fact of life that even the most cautious and well-maintained business will suffer lawsuits. In some cases, nothing could have prevented the suit. The question then becomes: "How is a lawsuit effectively and efficiently handled?" Once you are in, the best you can do (short of a good settlement) is minimize your damage.

Prelawsuit: the paper blame. When a

lawsuit seems inevitable, call in an attorney. The purpose of this early encounter is to prevent any costly statements or actions. Inevitably, corporate types have a tendency to want to circulate an "I'm incensed" memo to "get to the bottom of things" and "find out who is to blame." The problems with such an approach (although it may be a good management technique) is that it leaves a paper trail. Those in-house memoranda are discoverable and can be used as dangerous admissions at trial. The worst type of memo is one that has the tone that "we're going to lose big on this one." Even if the case never goes to trial, the discovery of such a document can give the other side a known edge in bargaining for a settlement—that you expect to lose. Even highly confidential internal memoranda are discoverable; there is no hiding.

Basically, the prelawsuit strategy is constant but *oral* communication with counsel. The only memo that should circulate is one that halts reports and memos on the lawsuit topic unless cleared first with counsel. Phone conversations and meetings with the other side also should be prohibited.

Prelawsuit: getting the facts together. With counsel's cooperation, the facts and events leading to the lawsuit must be determined. All internal memoranda and letters must be gathered. The relevant documents should be turned over to the attorney. The attorney should have the opportunity to meet with and interview those in the company with knowledge about contract negotiations, performance, trade secrets, or whatever other area is at issue.

Pre- or post-lawsuit: a settlement meeting. It can not hurt to try. Once you have your facts together, arrange for a settlement meeting. Decide who will be there and who will negotiate (lawyer usually). A settlement offer is not admissible in court if the lawsuit proceeds so take comfort in your bargaining position.

A settlement meeting can also serve to alert you to issues and evidence the other side may raise. There can be significant disclosures about the existence of documents and witnesses.

Discovery: pretrial by ordeal. Probably 80 percent of the costs of litigation are attributable to discovery. In the IBM antitrust suit, the government sifted through 30 million documents and then demanded 5 billion more. Depositions can tie up a business if key executives are all deposed for two or three days each. Interrogatories can require 2,000 man-hours for a business to answer in even a simple case. A business involved in supervising discovery can cut litigation costs. Xerox has discovery costs 10 percent of what they were ten years ago because of corporate involvement in litigation. More cases are settled and management demands a battle plan from counsel before discovery begins. Last year Xerox fees for outside lawyers were $3 million—down from the $15-million level of the mid-seventies.[48] The battle plan is based on a premise of settlement with good cooperation on document requests. The "we have nothing to hide" approach foils the other side's intense discovery desires. Resisting discovery only means court time and possible sanctions against use of evidence. Smooth-flowing discovery leads to settlement.

Trial: maximize chances of success. If a trial results, use all that is available. Currently, jury selection experts are being used to help attorneys choose jurors with demographics generally favorable to their client's position. Jury experts are not cheap but their results do have some substantiation.

Make sure your attorney has trial experience and savvy. Many commercial litigators are good at discovery but not at the art of trying a case. If necessary, bring in a trial expert to aid or to have aided by the attorney doing the discovery.

CONCLUSION

Lawsuits happen but some are preventable and most could have their costs reduced. Prior planning, a skill valued by management in every other area of business operations, is ignored in legal areas—the areas that ultimately prove most costly.

NOTES AND REFERENCES

1. E. M. Epstein, "Business Political Activity: Research Approaches and Analytical Approaches," *Research on Corporate Social Performance and Policy* **2**(1980):1-55.
2. J. N. Pearson and F. Shipper, "The Effects of Corporate Strategies and Environmental Factors on Financial Risk and Return," Research report, College of Business, Arizona State University, 1985.
3. Source withheld at request of company.
4. J. M. Perry, "The Power Brokers: Affable Lobbyist Aids United Technologies in Jet Engine Battle," *Wall Street Journal* (1 April 1982):1, 20.
5. K. Tottis, "Malfunctions Plague Fancy New GM Buses, Forcing Cities to Make Expensive Repairs," *Wall Street Journal* (2 June 1982):29.
6. M. J. Malbin, *Unelected Representatives: Congressional Staff and the Future of Representative Government* (New York: Basic Books, 1980).
7. J. M. Perry, "The Power Brokers: A PR Man Uses Access to Influential People, to Lure, Help Clients," *Wall Street Journal* (25 March 1982):1, 21.
8. Malbin, *Unelected Representatives.*
9. H. W. Fox and M. Schnitzer, *Doing Business in Washington: How to Win Friends and Influence Government* (New York: Free Press, 1981).
10. Perry, "Power Brokers: Affable Lobbyist Aids United Technologies."
11. Ibid, 1.
12. Perry, "Power Brokers: PR Man Uses Access to Influential People."
13. T. B. Clark, "Tax Lobbyists Scrambling in the Dark to Fight Taxes That Hit Their Clients," *National Journal* (22 May 1982):896-901.
14. L. Hebron, "Why Bureaucracy Keeps Growing," *Business Week* (15 November 1976):23-24.
15. M. Broide, interview with authors, 22 July 1980.
16. Interviews with 300 civil service managers at the Federal Executive Seminar Center and Federal Executive Institute.
17. R. Steinberg, *Men and the Organization* (New York: Time-Life Books, 1975), 119.
18. H. A. Heclo, *A Government of Strangers: Executive Politics in Washington* (Washington, D.C.: Brookings Institution, 1977).
19. Admiral Daniel J. Murphy, U.S. Navy (Retired), interview with authors, 24 July 1980.
20. C. A. Bridgwater, "Mediamen and Businessmen," *Psychology Today* 73(1984):72.
21. S. Hess, "The Golden Triangle: The Press at the White House, State and Defense," *Brookings Review* (Summer 1983):14-19.
22. W. Guzzard, Jr., "How Much Should Companies Talk?" *Fortune* (4 March 1984):64-68.
23. "Ex-Mobil Chief Wins Libel Appeal against 'Post'," *The Arizona Republic* (10 April 1985):A10.
24. C. G. Burck, "The Pros and Cons of Deregulating the Truckers," *Fortune* (18 June 1979):142.
25. J. C. Aplin and W. H. Hegarty, "Political Influence: Strategies Employed by Organizations to Impact Legislation in Business and Economic Matters," *Academy of Management Journal* **23**(1980):438-450.
26. U.S. Representative James G. Martin, interview with authors, 22 July 1980.
27. "Executives Take a Dim View of their Image-Makers," *Business Week* (7 March 1983):14.
28. "They Grade the Congress," *Congressional Action* (28 March 1980):1-6.
29. "People," *Time* (20 December 1982):67.
30. "Grassroots Lobbying Activities Will Be Expanded by the AFL-CIO," *Wall Street Journal* (15 June 1982):1.
31. "Browbeating Employees into Lobbyists," *Business Week* (10 March 1980):132, 136.
32. J. Citvin, "The Changing American Electorate," in *Politics and the Oval Office,* ed. A. J. Meltsner (San Francisco: Institute for Contemporary Studies, 1981), 31-62.
33. V. L. Tarrance, Jr., *Negative Campaigns and Negative Votes: The 1980 Elections* (Washington, D.C.: The Free Congress Research and Education Foundation, 1982).
34. L. Saunders, "Pretrial by Ordeal," *Forbes* (8 October 1984):105.
35. Ibid. Xerox General Counsel Robert Banks estimates that probably 80 percent of the costs of commercial litigation can be attributed to discovery.
36. 15 U.S. Sections 2301-2312 (1982). The actual name of the statute is the Magnuson-Moss Warranty-Federal Trade Commission Improvement Act.
37. 16 C.F.R. Section 702.3 (a) (1) (ii) (1983).
38. 691 F.2d 1322 (9th cir. 1982).
39. 16 C.F.R. Section 702.3 (d) (i) (ii) (B) (1983).
40. 691 F.2d 1322 quoting *Diamond Roofing Co. Inc. v. Occupational Safety and Health Re-*

view Comm., 528 F.2d 645, 649 (5th cir. 1976).
41. 691 F.2d at 1334.
42. 691 F.2d at 1327.
43. *Askco Engineering Corp. v. Mobil Chemical Corp.*, 535 S.W.2d, 893 (Texas 1976).
44. "No Mickey Mousing Around," *Time* (11 March 1985):54.
45. Ibid.
46. *Lange v. National Biscuit Co.*, 211 N.W.2d 783 (Minnesota 1973).
47. *Scott v. John H. Hampshire, Inc.*, 227 A.2d 751 (Maryland 1967).
48. L. Saunders, "Pretrial by Ordeal" *Forbes* (8 October 1984):105.

X

The Future of Strategic Planning and Management

Many of the chapters in Parts I through IX have attempted to describe the evolution of particular areas of strategic management, but few have been so bold as to forecast the future. In Part X, we deal with some empirical data concerning planning trends as well as some speculation concerning the future of the field.

In Chapter 36, Vasudevan Ramanujam, John C. Camillus, and N. Venkatraman report on the results of a large survey of planning executives concerning key trends in the field of planning. Despite the many criticisms of planning that have appeared in the popular business press, these authors' empirical evidence concerning the future of the field is quite upbeat.

In Chapter 37, Michael A. McGinnis describes the integration of analysis and intuition as the key to effective strategic management. He develops strategic management implications for managers and educators to use in the future development of strategic planning and management in organizations.

CHAPTER 36

Trends in Strategic Planning

Vasudevan Ramanujam, John C. Camillus, and N. Venkatraman

Vasudevan Ramanujam is assistant professor of management policy at The Weatherhead School of Management, Case Western Reserve University. He received the Ph.D. degree in strategic planning and policy from the University of Pittsburgh in 1984, the post-graduate diploma in business administration from the Indian Institute of Management, Ahmedabad, India in 1973, and the bachelor of technology in metallurgy from the Indian Institute of Technology, Madras, India in 1970. He held various management positions in a chemical company in India for over six years before moving to the United States in 1979. His research interests span the fields of strategic planning, technological innovation, and management information systems. Dr. Ramanujam's papers have appeared in journals such as *Academy of Management Journal, Academy of Management Review, Planning Review, Omega, Technological Forecasting and Social Change, Journal of Product Innovation Management, and Human Resource Management.*

John C. Camillus is currently associate dean and associate professor of business administration at the Graduate School of Business of the University of Pittsburgh. He received the Ph.D. in business administration from Harvard Business School in 1972. He has published several books and dozens of articles in professional journals. His latest book, *Budgeting for Profit,* has appeared in multiple editions and has been translated into Spanish and Swedish. Dr. Camillus has thrice received awards from the Foundation for Administrative Research for contributions to corporate and organizational planning. He has consulted extensively, both in the United States and abroad, for a variety of manufacturing, service, and not-for-profit organizations. His research and consulting has focused on strategic planning systems, management control systems, management information systems, and organizational structure.

N. Venkatraman is assistant professor of management at the Alfred P. Sloan School of Management, Massachusetts Institute of Technology. He holds a Ph.D. in business administration (with major area of concentration in strategic management) from the Graduate School of Business, University of Pittsburgh, an M.B.A. from the Indian Institute of Management, and a Bachelor of Technology degree in mechanical engineering from the Indian Institute of Technology. His research interests relate to strategy formulation and implementation issues in business organizations, the relationships between organizational strategy and the use of management systems such as strategic planning systems and information systems, as well as methodological issues in strategy research. His papers have been published (or are forthcoming) in professional journals such as *Academy of Management Journal, Academy of Management Review, Strategic Management Journal, OMEGA, Planning Review, Journal of Management Information Systems,* and *Proceedings of the Academy of Management.* He has served as a consultant to large corporations in India, and is a member of The Institute of Management Science (TIMS), The Academy of Management, Strategic Management Society, and the Institute for Decision Sciences.

This chapter presents a summary of the results of a study on the trends and developments in corporate strategic planning practices in U.S. business organizations. The survey questionnaire was administered to a random sample of about six hundred fifty executives drawn from the *Fortune 500, Fortune 500 Service,* and *Inc. 500* companies. Two hundred and seven of the executives contacted completed and returned the questionnaire.

The participating companies and responding executives are profiled in Figure 36-1. As can be observed, a variety of company sizes, industries, and executive roles are represented in the sample.

This chapter is organized in sections that respectively deal with general trends in planning, changes in functional emphasis in planning, key planning issues, trends in the use of planning techniques, key roles of planning achievement of planning objectives, attitudes toward planning, and changing roles of planners.

Since the data included a highly heterogeneous mix of firms, additional analyses highlighting the differences between the responses of large firms vs. smaller ones, planning executives vs. line executives, more mature planning systems vs. less mature ones, manufacturing vs. service industry firms, and high-performing vs. low-performing firms also are presented.

GENERAL TRENDS IN PLANNING

Responding executives were asked to indicate their perception of the extent of change that has occurred over the last five years in various areas and aspects of formal strategic planning in their organizations. Specifically, 20 statements were rated by each executive, who indicated his or her assessment of the extent of change by selecting one of five alternative responses, namely, significant decrease, moderate decrease, no change, moderate increase, and significant increase. The responses were given scores from one to five, so that scores of four and five indicated an increasing trend and scores of one and two indicated a decreasing trend, with the midpoint score of three indicating no change.

On the basis of the overall average, the five aspects rated to have increased the most were:

1. Overall emphasis on strategic planning systems.
2. Perceived usefulness of strategic planning.
3. Involvement of line managers in strategic planning activities.
4. Time spent by the chief executive in strategic planning.
5. Acceptance of the outputs of strategic planning exercise by top management.

The following three aspects showed the most decrease:

1. Resistance to planning in general.
2. Threats to the continuation of strategic planning.
3. The distance between the CEO and the chief of planning.

Table 36-1 shows the mean scores and the number of responses indicating a perceived increase, decrease, or no change for each item. Taken together, these data indicate an overall positive assessment and a growing acceptance of planning systems. This finding is in marked contrast to the many attacks on formal planning systems that have been appearing in the business periodicals.[1] It is particularly noteworthy that strategic planning increasingly is becoming accepted by line management and by the CEO, without whose active involvement and support no planning system can hope to succeed in an organization.

CHANGES IN FUNCTIONAL EMPHASIS IN PLANNING

The second focal point of the study was the tracking of changes in functional emphasis (that is, marketing vs. finance vs. operations and so forth) in planning. Seven functions

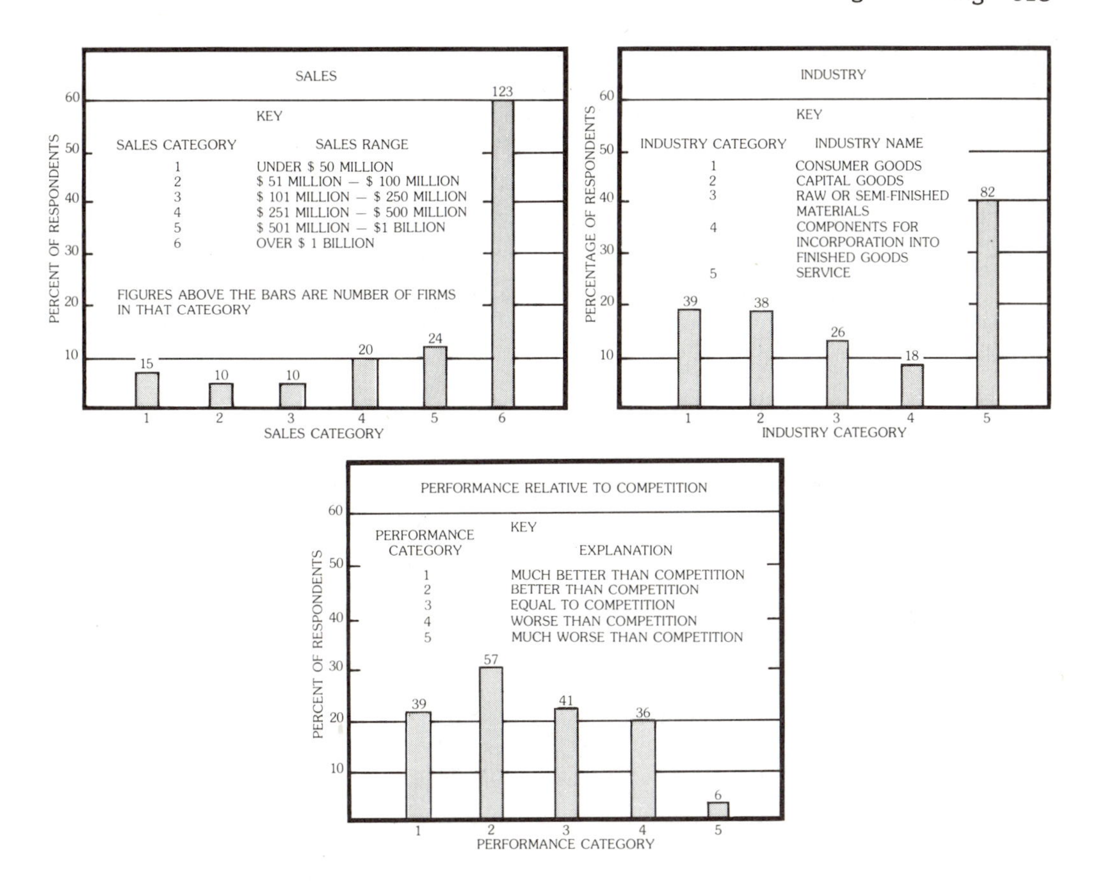

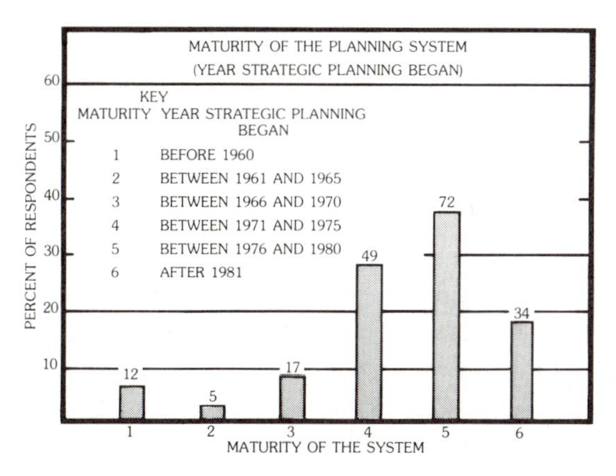

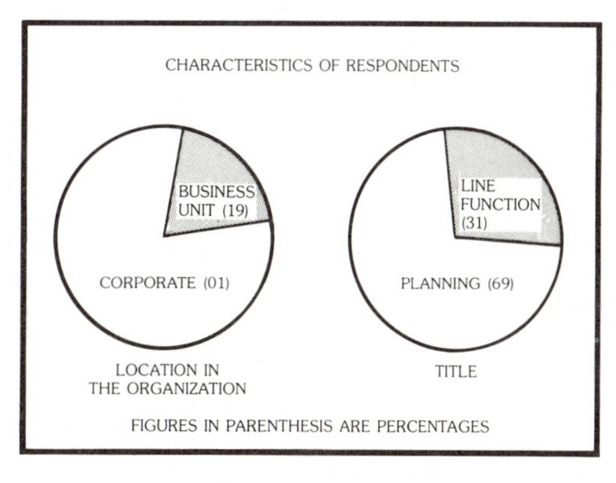

Figure 36-1. Profile of the study sample.

were rated. Their relative ranking with regard to change in emphasis is summarized in Table 36-2.

The increased emphasis on marketing, computers and/or MIS, and R & D is not surprising. For example, an article in *Business Week* spoke of the new emphasis on marketing,[2] whereas the pervasiveness of the concern with computers and MIS is all too evident to see.

However, the low ranking of finance may be interpreted as a statement by responding

Table 36-1. General Trends in Planning

Item	Average Score	Percent of Respondents Indicating: Increase	No Change	Decrease
1. Overall emphasis on strategic planning systems	4.14	81.2	7.7	11.1
2. Perceived usefulness of strategic planning	4.12	82.0	10.2	7.8
3. Involvement of line managers in the strategic planning activities	4.08	75.2	21.4	3.4
4. Time spent by the chief executive in strategic planning	4.06	78.7	17.8	3.5
5. Acceptance of the outputs of the strategic planning exercise by top management	3.94	74.0	20.6	5.4
6. Perceived usefulness of annual planning	3.68	53.9	38.7	7.4
7. Involvement of staff managers in the annual planning exercise	3.64	52.9	39.3	7.8
8. Involvement of the board of directors in strategic planning	3.63	51.4	47.0	1.6
9. Resources provided for strategic planning	3.60	62.9	23.9	13.2
10. Consistency between strategic plans and budgets	3.58	53.4	38.2	8.3
11. Use of annual plans in monthly performance review	3.56	42.3	55.6	2.1
12. Overall satisfaction with the strategic planning system	3.51	57.4	24.5	18.1
13. Number of planners (that is, those management personnel whose primary task is planning)	3.41	52.9	24.8	22.3
14. Attention to stakeholders other than stockholders	3.41	32.8	63.0	4.2
15. Use of planning committees	3.39	40.9	46.1	13.1
16. Attention to societal issues in planning	3.32	33.2	59.8	7.0
17. The planning horizon (that is, the number of years considered in the strategic plan)	3.22	28.8	56.6	14.6
18. The distance between the CEO and the chief of planning	2.62	13.3	45.1	41.5
19. Threats to the continuation of strategic planning	2.61	12.0	47.0	41.0
20. Resistance to planning in general	2.45	10.2	31.7	58.0

executives that corporate planning today is much broader in scope than conventional budgeting and control. The often heard criticism that most planning is merely an accounting or numbers game may be overly harsh and no longer valid.

KEY PLANNING ISSUES

Apart from functional emphasis, we wished to assess the change in relative emphasis on a variety of planning issues. The 14 issues are listed in Table 36-3 in the descending order of their average scores.

The five issues receiving the highest scores for change in emphasis were:

1. Competitive (domestic) trends.
2. Customer and/or end-user preferences.
3. Technological trends.
4. Diversification opportunities.
5. Worldwide or global competition.

In addition, joint venture opportunities appear to be a major current issue in planning.

This pattern supports the trend toward in-

Table 36-2. Changes in Functional Emphasis in Strategic Planning

Function	Average Score	Percent of Respondents Indicating Increase	No Change	Decrease
1. Marketing	4.14	77.8	19.2	3.0
2. Computers and MIS	3.76	62.8	33.2	4.0
3. R & D and technology	3.62	57.0	34.2	8.8
4. Operations and/or manufacturing	3.51	47.7	44.2	8.1
5. Personnel	3.46	45.3	48.8	6.0
6. Finance	3.41	43.3	44.8	11.9
7. Purchasing and/or procurement	3.09	19.8	70.9	9.3

Table 36-3. Key Planning Issues

Issue	Average Score	Percent of Respondents Indicating Increase	No Change	Decrease
1. Competitive (domestic) trends	4.21	83.6	13.5	2.9
2. Customer or end-user preferences	3.96	69.0	29.1	2.0
3. Technological trends	3.89	71.4	25.6	3.0
4. Diversification opportunities	3.74	61.7	30.3	8.0
5. Worldwide or global competition	3.69	59.4	34.4	6.3
6. Internal capabilities	3.65	55.4	40.2	4.4
7. Joint-venture opportunities	3.64	56.6	36.7	6.6
8. Qualitative data	3.60	55.9	38.1	5.9
9. General economic and business conditions	3.52	46.4	47.3	6.3
10. Regulatory issues	3.51	42.8	51.2	6.0
11. Supplier trends	3.25	26.0	69.1	5.0
12. Reasons for past failures	3.22	27.6	62.3	10.1
13. Quantitative data	3.19	36.8	40.7	22.5
14. Past performance	3.11	27.3	51.2	21.5

creasing marketing and technology emphasis noted in the previous section. The least change in emphasis is reported for:

1. Past performance.
2. Quantitative data.
3. Reasons for past failures.

Planning being essentially a prospective exercise, the reluctance of the respondents to dwell on past data and reasons for past failures is understandable. We have argued elsewhere, however, that planning systems can contribute significantly to organizational learning if they would increase their emphasis on historical data analysis.[3] As a matter of fact, as we will show in a later section of this chapter, higher performing firms tended to place much greater emphasis on such retrospection than did the overall sample.

In contrast to the low ranking of quantitative data, which is consistent with the relatively low change in emphasis on the finance function noted above, qualitative data is rated by well over half the respondents as increasing in importance. The low change in emphasis in quantitative data is very likely due to their high initial importance and is not to be interpreted as an indicator of the lessening of the importance of quantitative data. Rather, the increasing importance of qualitative data challenges the stereotypical assertion that planning is merely number crunching.

TRENDS IN THE USE OF PLANNING TECHNIQUES

Changes in the use of nine different and popular planning techniques were another focus of the study. The results are summarized in Table 36-4.

Of particular note is the trend toward decreasing use of BCG-type portfolio planning and the use of PIMS models in planning. This possibly reflects the numerous practical difficulties and disappointments in actually implementing these planning approaches that have been so popular until recently.[4]

The proliferation of computer software packages for forecasting and trend analysis and financial modeling conceivably may have increased their acceptance and use in companies. Similarly, concern for enhancing the value of the firm to stockholders has resulted in an emerging interest in value-based planning, wherein planning is driven by the objective of maximizing stockholder returns and wealth. Whether this proves a viable planning approach or just another fad remains to be seen.

KEY ROLES OF PLANNING

The perceptions of executives concerning the key roles of their planning systems were assessed in terms of 12 roles of planning systems, as shown in Table 36-5. The scale employed varied from significant deterioration (1) to significant improvement (5) for each role.

Although it is encouraging to observe that planning systems generally are not perceived to have led to deterioration with respect to any

Table 36-4. Trends in the Use of Planning Techniques

		Percent of Respondents Indicating		
Planning Technique or Methodology	*Average Score*	*Increase*	*No Change*	*Decrease*
1. Forecasting and trend analysis	3.65	60.4	30.7	8.9
2. Financial modeling	3.61	57.7	32.3	10.0
3. Market-based (value-based) planning	3.43	38.7	57.1	4.2
4. Scenarios and/or Delphi techniques	3.28	33.1	59.5	7.4
5. Stakeholder analysis	3.28	27.2	69.0	3.8
6. Project management techniques (for example, PERT/CPM)	3.24	32.5	57.7	9.8
7. Zero-based budgeting	3.19	28.3	60.4	11.3
8. Portfolio (for example, BCG) planning	2.92	25.1	47.4	27.4
9. PIMS models	2.69	11.5	59.6	28.8

Table 36-5. Roles of the Planning System

Role	Average Score	*Percent of Respondents Indicating* Improvement	No Change	Detrioration
1. Role in identifying key problem areas	3.96	73.5	23.5	2.9
2. As a mechanism for identifying new business opportunities	3.81	66.0	31.5	2.5
3. Ability to communicate top managements' expectations down the line	3.81	68.1	24.5	7.4
4. Flexibility to adapt to unanticipated changes	3.76	67.3	28.3	4.4
5. Generation of new ideas	3.65	58.6	34.5	6.9
6. Ability to anticipate surprises and crises	3.62	60.4	33.2	6.4
7. As a tool for managerial motivation	3.62	54.5	37.5	8.0
8. As a mechanism for integrating diverse functions and/or operations	3.56	50.5	43.3	6.2
9. As a means for fostering organizational learning	3.55	54.8	39.2	6.0
10. As a tool for management control	3.53	50.2	44.3	5.4
11. Ability to communicate line managements' concern to the top level	3.53	54.1	38.8	7.1
12. As a basis for enhancing innovation	3.47	38.0	47.5	4.5

of the 12 roles of planning—all average scores are well in excess of the neutral point 3—some disturbing trends are revealed in the data. Whereas it is often suggested that a key role of planning systems is to serve as a framework for innovation, planning systems do not seem to have facilitated innovation to any greater extent than five years ago. Similarly, despite improvements in the communication of top managements' expectation to the lower levels of the organization, corresponding improvements in the upward communications of line managements' concerns to the top is not reported by participating executives. In other words, planning in most organizations may still be a top-down exercise, despite many respondents' denial of this assertion (see Table 36-6).

The overall emphasis in contemporary strategic planning appears to be on identifying key problem areas and new business opportunities, and on enhancing the ability of the organization to adapt to changes, as opposed to anticipating and precluding them. Given the pervasiveness and discontinuous nature of change in contemporary environments,[5] it is perhaps quite a rational stance to trade off the ability to react to change against the ability to forecast or forestall it. Consistently enough, as we shall see in the next section, for the overall sample, the objective of predicting future trends has not been fulfilled to the same extent as have been some other planning system objectives. However, it is worth noting that high-performing firms tended to report both an increased emphasis on forecasting

Table 36-6. Fulfillment of Planning Objectives

Objective	Average Score	Percent of Respondents Indicating		
		Fulfilled	Neutral	Unfulfilled
1. Evaluating alternatives based on more relevant information	3.78	75.6	17.6	6.8
2. Improvement in long-term performance	3.67	67.5	25.7	6.8
3. Improvement in short-term performance	3.52	54.5	37.1	8.4
4. Predicting future trends	3.51	63.0	21.5	15.5
5. Avoid problem areas	3.39	52.7	32.5	14.8
6. Enhancing management development	3.38	53.8	30.7	15.6

and trend analysis and a greater degree of success in predicting future trends (see Table 36-13).

ACHIEVEMENT OF PLANNING OBJECTIVES

Executives were also asked to rate the extent to which their planning systems have helped them realize six key planning objectives. Higher scores reflect greater levels of fulfillment of the objective, as rated by the respondent. The ranking of objectives in terms of extent of fulfillment was:

1. Evaluating alternatives based on more relevant information.
2. Improving long-term performance.
3. Improving short-term performance.
4. Predicting future trends.
5. Avoiding problem areas.
6. Enhancing management development.

The general perception is that contemporary planning systems are doing a good job with respect to alternative generation and evaluation. Executives are willing to credit improvements in their short-term as well as long-term performance to their planning systems. As far as the objectives of predicting future trends, avoiding problem areas (proactive management), and enhancing management development are concerned, executives are much less enthusiastic about the contribution of their planning systems toward those ends.

ATTITUDES TOWARD PLANNING

Executives' opinions and attitudes toward planning were gauged by having them indicate their degree of agreement (strongly agree = 5) or disagreement (strongly disagree = 1) with 16 opinions or attitudinal statements. The results are summarized in Table 36-7.

The following five statements evoked the strongest expressions of agreement.

1. Reducing emphasis on strategic planning will be detrimental to our long-term performance.
2. Our plans today reflect implementation concerns.
3. We have improved the sophistication of our strategic planning system.
4. Our previous approaches to strategic planning are not appropriate today.
5. Today's system emphasizes creativity among managers more than our previous system.

In contrast, executives tended to disagree most strongly with the following five assertions:

1. The process used for strategic planning has not changed much over the years.
2. Strategic planning in our company or unit is generally viewed as a luxury today.

Table 36-7. Executives' General Opinions and Attitudes

Item	Average Score	*Percent of Respondents Indicating*		
		Agreement	*Neutral*	*Disagreement*
1. Reducing emphasis on strategic planning will be detrimental to our long-term performance.	4.43	88.7	4.9	6.4
2. Our plans today reflect implementation concerns.	3.85	73.6	16.9	9.5
3. We have improved the sophistication of our strategic planning system.	3.82	70.6	18.6	10.8
4. Our previous approaches to strategic planning are not appropriate today.	3.65	64.2	16.2	19.6
5. Today's system emphasizes creativity among managers more than our previous system.	3.57	62.6	20.2	17.2
6. Our strategic planning systems today are more consistent with our organization's culture.	3.49	55.6	30.7	13.7
7. We are more concerned about the evaluation of our strategic planning systems today.	3.45	54.0	29.7	16.3
8. There is more participation from lower-level managers in our strategic planning.	3.41	56.6	18.0	25.4
9. Our tendency to rely on outside consultants for strategic planning has been on the decrease.	3.34	50.8	23.0	26.2
10. Our systems emphasize control more than before.	3.21	41.3	33.0	25.7
11. We now emphasize problem identification more than problem solving.	3.15	33.2	46.7	20.1
12. Our strategic planning approach is best described as a top-down process.	3.04	47.6	11.2	41.3
13. We believe that true strategic planning should be conducted less frequently than annually.	2.93	38.9	19.7	41.4
14. Our plans are becoming more quantitative than qualitative.	2.83	35.5	20.7	43.8
15. Strategic planning in our company or unit is generally viewed as a luxury today.	2.04	15.0	13.0	72.0
16. The process used for strategic planning has not changed much over the years.	1.98	17.2	4.9	77.9

3. Our plans are becoming more quantitative than qualitative.
4. We believe that true strategic planning should be conducted less frequently than annually.
5. Our strategic planning is best described as a top-down process.

Taken together, the general attitude of the executives (both staff planners and line executives) reflects a positive role for strategic planning, which is seen as playing an important role. A general tendency toward qualitative and soft data interwoven with hard or quantitative data can be witnessed, together with a strong emphasis on the implementation as opposed to the creation of plans.

CHANGING ROLE OF PLANNERS

Planners generally play many different roles in organizations.[6] These were assessed by indications of the degree to which each of six possible planner's roles accurately described the actual role of the planner in the respondent's organizations today as well as five years ago.

The rankings in Table 36-8 show that the roles of the planner have not undergone major changes within the last five years—witness the relative stability of the rankings of the roles.

Five years ago, the main roles of the planner are believed to have been those of coordinator and internal consultant. This is true today as well. The least important roles then were those of developer of action plans and active decision maker. Although the rankings of these latter roles are interchanged, they are still the least important (or descriptively least accurate) roles of the planner today. The role of the planner is thus still seen as an advisory and coordinating role.

Table 36-8. Changes in Planners' Roles

	Importance Rank	
Role	*Today*	*Five Years Ago*
1. Coordinator	1	1
2. Internal consultant	2	2
3. Idea generator	3	4
4. Source of external data	4	3
5. Active decision maker	5	6
6. Developer of action plans	6	5

COMPARISONS ACROSS SUBGROUPS OF THE SAMPLE

The sample of 207 responses was found to be heterogeneous in terms of organization size, title of the responding executives, maturity of the planning system being assessed, type of industry represented, and perceived performance of the responding entity relative to its competition.

Since the overall results could be biased by these differences among respondents and may not be generalizable to all organizations, we partitioned our total sample into subgroups and analyzed differences in responses across them. The following subgroupings were created for this purpose.

1. Smaller versus larger ($1 billion or more sales) companies.
2. Planning executives' (planners) responses versus responses of executives with other titles (line executives).
3. Mature (pre-1975) versus relatively recent (post-1976) planning systems.
4. Service versus manufacturing industry.
5. Relatively high-performance versus relatively low-performance organizations.

We will now briefly highlight the major differences in the responses across these subgroupings. Only the statistically meaningful differences are discussed here.

Smaller versus Larger Firms (Table 36-9)

Smaller firms tended to have experienced increased emphasis on the use of plans in monthly performance review and the involvement of staff managers in the annual planning exercise. They reported a lesser degree of change in emphasis on attention to social issues in planning than their larger counterparts.

Smaller firms are also more quantitative in their planning and more retrospective in the

Table 36-9. Significant Differences between the Average Ratings of Smaller and Larger Companies

	Smaller Companies	*Larger Companies*
General Trends		
1. Use of annual plans in monthly performance reviews	3.74	3.41
2. Involvement of staff managers in annual planning exercises	3.85	3.47
3. Attention to societal issues in planning	3.14	3.43
Planning Issues or Themes		
1. Past performance	3.32	2.96
2. Reasons for past failures	3.38	3.08
3. Quantitative data	3.45	3.03
Use of Planning Techniques		
1. Zero-based budgeting	3.36	3.01
2. Financial market-based (value-based) planning	3.19	3.59
3. Forecasting and trend analysis	3.81	3.54
Roles of the Planning System		
1. Ability to anticipate surprises and crises	3.78	3.50
2. Flexibility to adapt to unanticipated changes	3.90	3.66
3. As a tool for managerial motivation	3.78	3.50
4. As a tool for management control	3.71	3.42
Executives' General Opinions and Attitudes		
1. Our plans are becoming more quantitative than qualitative	3.19	2.58
2. Our systems emphasize control more than ever before	3.53	2.98
3. Our tendency to rely on outside consultants is on the decrease	3.66	3.12
4. We believe that true strategic planning should be conducted less frequently than annually	2.68	3.07
5. Our strategic planning approach is best described as a top-down process	3.31	2.89
Profile of the Subgroups		
1. Percent of respondents that are planners	55.3	77.5
2. Percent mature planning systems	31.9	51.8
3. Percent manufacturing companies	57.1	62.8
Performance of the Subgroups		
1. Sales growth	3.81	3.27
2. Net income growth	3.91	3.38
3. Market share changes	3.59	3.24
4. Return on investment	3.90	3.22

sense of focusing on past performance and reasons for past failures.

There is a greater emphasis on the use of value-based planning techniques in the larger firms, which is consistent with their concern with stock market reactions and expectations. On the other hand, zero-based budgeting and forecasting and trend analysis have enjoyed greater popularity among the smaller firms.

Smaller firms reported greater improvement in their ability to anticipate surprises and crises and flexibility to adapt to unanticipated changes. They were also apt to rate their planning systems higher with regard to their use as a motivational and control tool.

Planning in the smaller firm has become relatively more quantitative, more control oriented, and more top down. This tightening of control could well have been a reaction to the greater impact of the recessionary environment on the smaller organization. Smaller firms report less reliance on external consultants in their planning. Small firms also believe that true strategic planning should be an annual exercise.

Planning Executives versus Line Executives (Table 36-10)

Surprisingly, line executives believe that the perceived usefulness of annual planning has increased to a greater degree than do planning executives. Line executives also report a greater degree of staff involvement in planning. The ranking of functions in terms of the change in emphasis is identical for planning executives and line executives, in that marketing and computers and MIS are the top two functions on which planning is now focused.

Line executives are more likely to report a trend toward more use of quantitative data and toward an emphasis on general economic and business conditions in planning. Planning and line executives do not differ much in their assessment of how well planning objectives are being met. Line executives report a higher degree of agreement with the assertion that planning is becoming more control oriented and that it is becoming more top down.

Mature versus Recent Planners (Table 36-11)

Some companies in our sample have been engaging in formal strategic planning since the fifties and sixties, whereas some only recently have begun to plan. The median firm in the sample began its formal strategic planning around 1975. Planners who have been plan-

Table 36-10. Significant Differences between the Average Ratings of Planners and Line Executives

	Planners	*Line Executives*
General Trends		
1. Perceived usefulness of annual planning	3.59	3.89
2. Involvement of staff managers in the annual planning exercise	3.59	3.83
3. The distance between the chief executive officer and the chief of planning	2.55	2.85
Functional Emphasis		
1. Marketing function	4.04	4.35
2. Purchasing and/or procurement function	3.00	3.25
3. Computers and MIS	3.70	3.95
Planning Issues or Themes		
1. General economic and business conditions	3.43	3.68
2. Quantitative data	3.07	3.51
3. Improvement in short-term performance	3.42	3.75
Executives' General Opinions and Attitudes		
1. Our plans are becoming more quantitative than qualitative	2.66	3.22
2. Our systems emphasize control more than before	3.04	3.57
3. Our strategic planning approach is best described as a top-down process	2.91	3.38
Profile of the Subgroups		
1. Percent small companies	31.1	55.7
2. Percent mature planning systems	47.2	38.6
3. Percent manufacturing companies	62.7	50.8
Performance of the Subgroups		
1. Sales growth	3.37	3.73
2. Net income growth	3.45	3.92
3. Return on investment	3.34	3.73

Table 36-11. Significant Differences between the Average Ratings of Mature and Recent Planning Systems

	Mature	Recent
General Trends		
1. Overall emphasis on strategic planning systems	3.84	4.35
2. Number of planners (that is, those management personnel whose primary task is planning)	3.06	3.67
3. Perceived usefulness of strategic planning	3.93	4.26
4. Perceived usefulness of annual planning	3.47	3.83
5. Involvement of line managers in strategic planning activities	3.89	4.26
6. Time spent by the chief executive in strategic planning	3.90	4.19
7. Involvement of staff managers in the annual planning exercise	3.46	3.76
8. The planning horizon (that is, the number of years considered in the strategic plan)	3.05	3.37
9. Acceptance of the outputs of the strategic planning exercise by top management	3.76	4.09
10. Overall satisfaction with the strategic planning system	3.28	3.66
11. Threats to the continuation of strategic planning	2.80	2.47
12. Attention to stakeholders other than stockholders	3.22	3.62
Functional Emphasis		
1. Finance function	3.16	3.55
Planning Issues or Themes		
1. General economic and business conditions	3.35	3.62
2. Regulatory issues	3.29	3.72
3. Internal capabilities	3.53	3.79
4. Quantitative data	2.94	3.34
5. Technological trends	3.77	4.00
Use of Planning Techniques		
1. Portfolio (BCG) approaches	2.76	3.09
2. Financial models	3.39	3.75
3. Zero-based budgeting	3.03	3.36
4. Project management techniques	3.03	3.39
5. Stakeholder analysis	3.16	3.43
6. Forecasting and trend analysis	3.41	3.85
Roles of the Planning System		
1. Role in identifying key problems areas	3.77	4.11
2. As a tool for managerial motivation	3.23	3.92
3. Generation of new ideas	3.35	3.93
4. Ability to communicate top management's expectation down the line	3.60	4.01
5. As a tool for management control	3.28	3.74
6. As a means of fostering organizational learning	3.34	3.74
7. Ability to communicate the line managers' concerns to the top level	3.32	3.71
8. As a mechanism for integrating diverse functions and/or operations	3.35	3.73
9. As a basis for enhancing innovation	3.29	3.67
Fulfillment of Planning Objectives		
1. Enhancing management development	3.18	3.55
2. Avoiding problem areas	3.26	3.50
Executives' General Opinions and Attitudes		
1. Our previous approaches to strategic planning are not appropriate today	3.34	3.89
2. There is more participation from lower-level managers in our strategic planning	3.13	3.64
3. Our strategic planning systems are more consistent with our organization's culture	3.34	3.66

(continued)

Table 36-11. (*Continued*)

	Mature	*Recent*
4. Our plans are becoming more quantitative than qualitative	2.47	3.10
5. We have improved the sophistication of our strategic planning system	3.49	4.06
6. Our systems emphasize control more than before	2.99	3.34
7. Our strategic planning approach is best described as a top-down process	2.74	3.34
Profile of the Subgroups		
1. Percent small companies	28.4	47.6
2. Percent of respondents that are planners	73.2	65.7
3. Percent manufacturing companies	69.1	51.9
Performance of the Subgroups		
1. Sales growth	3.26	3.68
2. Return on investment	3.29	3.67

ning since 1976 are labeled here as *recent planners,* whereas the other planning systems are treated as more *mature* systems.

Not unexpectedly, recent planners have tended to express more positive assessments of the changes in planning. This grouping produced the largest number of statistically significant differences, as shown in Table 36-12.

Of particular note is the tendency of mature planners to back away from a quantitative orientation and from the use of portfolio planning techniques. In addition, mature planners are less apt to describe their systems as top down, implying that their systems are now characterized by more participation from lower levels of management.

Service versus Manufacturing Firms (Table 36-12)

Here again, some interesting differences can be found. Overall, there appears to be a more positive feeling about strategic planning in the service sector than in the manufacturing sector. The service sector is characterized by many recent planners (that is, those that have initiated strategic planning relatively recently) and by relatively smaller firms, operating in the emerging industry of telecommunications and information technology.

However, our sample of service firms was dominated by banks and financial institutions. Hence, it is not surprising that there is less of a trend toward attention to worldwide or global competitive issues and more of an emphasis on regulatory changes, which are radically transforming the financial services industries. The increased emphasis on functional spheres such as marketing, finance, and computers and MIS reflect the prevalent concerns for the attainment of competitive positions through aggressive marketing programs and the expanded use of information technology embodied in innovations such as automated teller machines and electronic banking networks.

High-performance versus Low-performance Firms (Table 36-13)

High-performance firms showed a greater level of commitment to planning.[7] They reported, on average, increases in the number of planners and in perceived usefulness of annual planning. Their CEOs tended to spend more time in planning, and they claimed a higher level of staff involvement in planning. Their planning had become longer term as indicated by modest increases in their planning horizons. They also sought to bring their stra-

Table 36-12. Significant Differences between the Average Ratings of Service and Manufacturing Companies

	Service	*Manufacturing*
General Trends		
1. Overall emphasis on strategic planning systems	4.33	3.99
2. Number of planners (that is, those management personnel whose primary task is planning)	3.73	3.17
3. Perceived usefulness of annual planning	3.91	3.50
4. Use of annual plans in monthly performance review	3.71	3.45
5. Acceptance of the outputs of the strategic planning exercise by top management	4.10	3.80
Functional Emphasis		
1. Marketing function	4.35	4.00
2. Finance function	3.56	3.29
3. Computer and MIS	4.04	3.56
Planning Issues or Themes		
1. Regulatory issues	3.77	3.35
2. Worldwide or global competition	3.52	3.79
3. Internal capabilities	3.83	3.52
4. Quantitative data	3.37	3.05
5. Joint venture opportunities	3.84	3.53
6. Diversification opportunities	3.94	3.64
Use of Planning Techniques		
1. PIMS models	2.95	2.52
2. Financial models	3.78	3.48
3. Forecasting and trend analysis	3.83	3.52
Roles of the Planning System		
1. As a mechanism for identifying new business opportunities	3.94	3.71
2. As a tool for managerial motivation	3.79	3.50
3. Generation of new ideas	3.86	3.52
Fulfillment of Planning Objectives		
1. Improvement in short-term performance	3.74	3.38
Opinions and/or Attitudes Regarding Planning		
1. Our previous approaches to strategic planning are not appropriate today	3.85	3.50
2. Our plans are becoming more quantitative than qualitative	3.06	2.69
3. Our systems emphasize control more than before	3.38	3.08
4. Our strategic planning approach is best described as a top-down process	3.37	2.81
Profile of the Subgroups		
1. Percent small companies	42.3	36.7
2. Percent of respondents that are planners	61.7	72.4
3. Percent mature planning systems	32.9	50.5

tegic plans and annual budgets into closer consistency and showed a willingness to commit greater resources to planning.

Among the high-performance firms, there was a tendency toward increasing emphasis on the manufacturing, R & D and technology, and computers and MIS functions. This can be interpreted as revealing a greater orientation toward productivity on the part of those firms.

Table 36-13. Significant Differences between the Average Ratings of High-Performance and Low-Performance Companies

	High Performance	Low Performance
General Trends		
1. Number of planners (that is, those management personnel whose primary task is planning)	3.59	3.17
2. Perceived usefulness of annual planning	3.89	3.53
3. Time spent by the chief executive officer on strategic planning	4.22	3.80
4. Involvement of staff managers in the annual planning activities	3.88	3.46
5. Planning horizon	3.37	3.04
6. Consistency between strategic plans and budgets	3.68	3.43
7. Resources provided to strategic planning activities	3.72	3.40
Functional Emphasis		
1. Operations and/or manufacturing function	3.69	3.36
2. R & D and technology	3.77	3.45
3. Computer and MIS	3.91	3.62
Planning Issues or Themes		
1. Past performance	3.29	2.95
2. Reasons for past failures	3.34	3.08
3. Diversification opportunities	3.93	3.55
4. Customer or end-user preferences	4.07	3.79
Use of Planning Techniques		
1. Forecasting and trend analysis	3.78	3.50
Roles of the Planning System		
1. Ability to anticipate surprises	3.80	3.45
2. Flexibility to adapt to unanticipated changes	4.00	3.63
3. As a mechanism for identifying new business opportunities	3.94	3.68
4. Generation of new ideas	3.76	3.45
5. As a tool for management control	3.72	3.33
6. As a means for fostering organizational learning	3.66	3.39
7. As a basis for enhancing innovation	3.64	3.26
Fulfillment of Planning Objectives		
1. Enhancing management development	3.54	3.22
2. Predicting future trends	3.70	3.34
3. Improvement in short-term performance	3.73	3.36
4. Improvement in long-term performance	3.81	3.57
Executives' General Opinions and Attitudes		
1. Our plans are becoming more quantitative than qualitative	3.01	2.60
2. We have improved the sophistication of our strategic planning system	3.96	3.62
Profile of the Subgroups		
1. Percent small companies	53.1	26.9
2. Percent respondents that are planners	59.8	72.8
3. Percent mature planning systems	37.2	54.7
4. Percent manufacturing companies	57.8	64.5

Past performance and reasons for past failures apparently preoccupy the high-performance firms to a greater extent. Presumably, they realize the value of their planning in fostering organizational learning. They also report a greater emphasis on diversification opportunities and customer and/or end-user trends.

In terms of use of techniques, contrary to expectations, high-performance firms are not likely to be more favorably inclined toward any planning technique as compared to the low-performance firms. The exception is forecasting and trend analysis, which the high-performance firms tended to emphasize somewhat more. Apparently, the high-performance firms are well aware that ritualistic applications of techniques do not lead to better planning results.

High-performance firms report greater levels of improvement in several major planning roles, namely, the ability to anticipate surprises and crises, the flexibility to adapt to unanticipated changes, the identification of new business opportunities, the generation of new ideas, the accomplishment of management control, the fostering of organizational learning, and the facilitation of innovation. That high-performance firms ascribe a positive value to their planning in enhancing innovation suggests that planning in those firms is achieving a good balance between creativity and control.[8]

High-performance firms tended to report higher levels of objective fulfillment with respect to four of the six planning objectives they were asked to rate. These were, in order of extent of objective fulfillment: improvement in long-term performance, improvement in short-term performance, predicting future trends, and enhancing management development.

High-performance firms did not report a decreased emphasis on quantitative data as did low-performance firms. Instead, they felt they had improved the sophistication of their strategic planning, implying that they were combining both quantitative and qualitative approaches to planning in a more holistic way.

High-performance firms tended to be smaller, with less mature planning systems. They were almost as likely to be service firms as manufacturing. Respondents from the high-performance firms included a near balance between planning and line executives, thus removing the possibility of any systematic bias in the responses.

CONCLUSION

This chapter summarizes the major findings of a large-scale survey on the changes and trends in planning in U.S. organizations. The results indicate a generally positive assessment of planning by participating firms. CEOs are more involved in planning, and it has achieved a greater degree of acceptance in the organization. Of particular note is the increased emphasis on marketing, R & D and technology, and computers and MIS and the growing willingness to incorporate qualitative considerations in planning.

There is considerable variation in the extent of fulfillment of planning objectives, but overall the assessments are again positive rather than negative. The role of the planner shows only modest change.

That high-performance firms differed on numerous aspects of planning from the low-performance firms should give executives and designers of planning systems much food for thought and room for action.

NOTES AND REFERENCES

1. See, for example, W. Kiechel, "Corporate Strategists under Fire," *Fortune* (2 December 1982):34-39; R. B. Lamb, "Is the Attack on Strategy Valid?" *Journal of Business Strategy* **3**(1983):68; R. B. Lamb, "The Future Catches Up with a Strategic Planner," *Business Week* (27 June 1983):62; "The New Breed of Strategic Planner," *Business Week* (17 September 1984):62.
2. "Marketing: The New Priority," *Business Week* (21 November 1983):96; C. D. Burnett, D. P. Yeskey, and D. Richardson, "New Roles for Corporate Planners in the 1980s," *Journal of Business Strategy* **4**(1984):64-68.
3. V. Ramanujam and N. Venkatraman, "Eight Half-truths of Strategic Planning: A Fresh Look," *Planning Review* **13**(1985):27-29.
4. R. Wensley, "PIMS and BCG: New Horizons or False Dawn?" *Strategic Management Journal* **3**(1982):147-158; W. Kiechel, "The Decline of the Experience Curve," *Fortune* (5 October 1981):139-146; W. Kiechel, "Oh, Where, Oh, Where Has My Little Dog Gone? Or My Cash Cow? Or My Star?" *Fortune* (2 November 1981):148-154.

5. H. I. Ansoff, *Implanting Strategic Management* (Englewood Cliffs, N.J.: Prentice-Hall, 1984); J. C. Camillus and N. Venkatraman, "Dimensions of Strategic Choice," *Planning Review* **12**(1984):26-46.
6. J. C. Camillus, "Plotting the Planner's Place," *Planning Review* **7**(1979):13-16.
7. The classification of responses into high-performance versus low-performance categories was made on the basis of a composite index of performance that included sales growth, net income growth, market share changes, and ROI, each of these being relative to the competition and self-assessed by the respondents.
8. J. C. Camillus, "Evaluating the Benefits of Formal Planning Systems," *Long Range Planning Journal* **8**(1975):33-40.

CHAPTER 37

Integrating Analysis and Intuition: The Key to Strategic Management in a Chaotic Environment

Michael A. McGinnis

Michael A. McGinnis is professor of marketing and transportation at Shippensburg University of Pennsylvania. Dr. McGinnis holds B.S. and M.S. degrees from Michigan State University and the D.B.A. from the University of Maryland. His professional interests include strategic management, innovation, and physical distribution management. His articles have appeared in such journals as *Sloan Management Review, Journal of Business Strategy, Business Horizons, Transportation Journal, Journal of Business Logistics,* and *Journal of Marketing Research.*

Since the mid-seventies strategic planning has come under increasing criticism. The criticism focuses on strategic planning's preoccupation with quantitative analysis and the overuse of simplified models to classify a business as a "cash cow," "star," "question mark," or "dog" and then to prescribe standardized strategies based on such classifications. In addition, strategic planning critics have focused on strategic planners' tendencies to simplify the firm's environment and their failure to challenge assumptions underlying corporate strategy.

That many of these criticisms are valid cannot be argued. However, the strategic planning environment of the 1980s has changed greatly from the environment of the 1960s, when much of the current strategic planning thought was developed. The 1960s environment was not inappropriate for traditional planning models. Economic growth had been relatively stable since World War II, interest and inflation rates were low, and foreign competition was a minor threat. In addition, much of the U.S. economy led a sheltered existence. Banking, stock brokerage, telecommunications, and transportation operated under the protection of public utility regulation. The auto, steel, chemical, forest products, rubber, and appliance industries operated much like classical oligopolies. Little wonder that management thought, in general, and strategic planning thought, in particular, as-

Adapted from M. A. McGinnis, "The Key to Strategic Planning: Integrating Analysis and Intuition," *Sloan Management Review* **26**(1984):45-52.

sumed a steady-growth, low-risk, stable environment.

The assumptions of the 1960s began to erode during the 1970s. First the Britton Woods Agreement of 1944 was terminated, allowing foreign currency rates to float. Next, two oil embargoes undermined the assumption of unlimited supplies of cheap raw materials. Third, the erosion of American business competitiveness in foreign and domestic markets undermined long-standing oligopolies. Fourth, a decade of economic stagnation together with high interest rates further eroded the American business-as-usual mentality. Finally, sweeping changes in public policy during the late 1970s and early 1980s greatly reduced public utility regulation of stock brokerage, airlines, banking, trucking, railroading, and telecommunications.

As a result of the changes that occurred in the 1970s, the environment changed from being rather stable and predictable to being unstable and unpredictable. In other words, the strategic planning environment of the 1980s has become chaotic. Questions that come to mind, then, are: What insight does the academic literature provide to enable managers to strategically plan in this environment? How can managers integrate analysis and intuition in order to plan in this environment? What are the implications of an integrated strategic planning process?

INSIGHTS FROM THE LITERATURE

The Environment

The environment that evolved since the 1960s has created a dilemma for the strategic planner. Whereas traditional strategic planning techniques assumed clear goals and a predictable environment, the new environment has become ambiguous and dynamic. The strategic planning implications for this environment were discussed by Thompson.[1] Basically, Thompson argued that both the organizational decision (strategy formulation) and organizational assessment (strategy evaluation) processes change as the environment becomes less certain. This relationship is summarized as Figure 37-1.

Figure 37-1 describes the strategic planning environment in terms of Beliefs about Cause and Effect and Clarity of Goals. Beliefs about Cause and Effect refer to the ability to understand cause-effect relationships in the environment. Clarity of Goals refers to the precision or explicitness of a firm's objectives. For example, econometric forecasters developed fairly accurate forecasting models for predicting new car sales. These models worked reasonably well until the first oil embargo of 1973. Since that time automobile sales have fluctuated widely, and the mix of large and small car demand has been unstable. These changes in the market have been difficult to understand, and industry goals have become increasingly fluid. As a result, strategy formation in the domestic automobile industry has become less computational and more subjective (some combination of judgment, compromise, and inspiration).

In a similar manner, approaches to strategy evaluation have been affected by the environment of the 1980s. The environment of the 1960s enabled firms to evaluate their performance in terms of *Target ROI,* or *X Percentage Growth in Sales,* and *Y Percentage Growth in Earnings.* These quantitative measures of performance of the 1960s have given way to more qualitative measures of performance such as *Protect our core markets, Experience less sales decline than our competitors,* or *Improve our market share over the previous year.* For example, the criteria for strategic evaluation in the domestic automobile industry is no longer return on investment but reducing the break-even point so that the firm can earn a profit when sales are depressed.

In summary, then, organization theory suggests that as management's understanding about environmental cause and effect decreases and/or management's goals become less crystallized then the strategy formation orientation becomes less cut and dried (com-

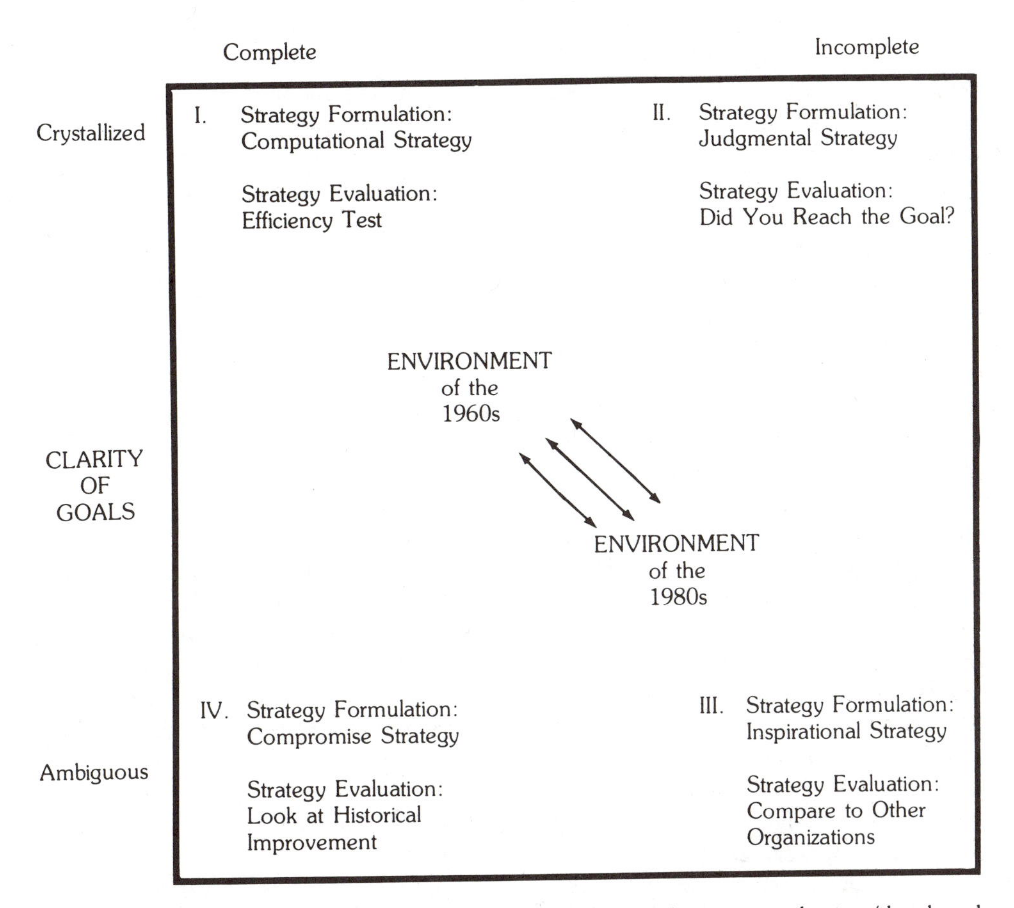

Figure 37-1. Environmental impact on strategy formulation and strategy evaluation (developed from J. D. Thompson, *Organizations in Action,* New York: McGraw-Hill, 1967, especially chapters 7 and 10).

putational or objective) and more intuitive (compromise, judgment, and/or inspirational). In a similar manner, management's approach to strategy evaluation becomes less quantitative (efficiency test) and more subjective (did we reach the goal?; look at historical improvement, compare to other organizations).

Issues in Strategy Analysis and Intuition

A major problem in strategic management lies in both analyzing a chaotic environment and developing a level of understanding that utilizes intuitive skills to create strategic opportunities. At least two teams of researchers have conducted research that provides insights into this problem. Paine and Anderson studied 62 firms over time using the case study method.[2] In particular, they examined the strategic planning process of firms in different environments and compared the strategies of successful and unsuccessful firms. Miller and Friesen studied the strategic behavior of 81 firms using a case methodology.[3] They used three measures to assess the firms' environments: dynamism (amount and predictability of change), heterogeneity (of competitive tactics, customer tastes, product lines, and channels of distribution), and environmental hostility (severity of competition and regulatory restriction, raw material and labor shortages, and unfavorable demographic trends). Examination of the Miller and Friesen findings provides a composite set of perspectives of successful and unsuccessful firms in both moderate and challenging environments. Taken together, these two studies point to six key issues that emerge when strategic planning is done in a chaotic environment. These six issues are intelligence, organizational balance, analysis, innovation, proactivity, and risk taking.

Intelligence. Intelligence is the firm's ability to simultaneously scan and interpret its external environments, monitor itself, and communicate effectively with itself. A weak link in any of these three areas undermines the total corporate intelligence system. For example, Texas Instruments is considered to be very well managed. It usually knows its markets, monitors its own abilities, and communicates well internally. However, in the areas of consumer products, it has constantly been bloodied, first in electronic watches, later in personal computers. Although no single issue has been responsible for these failures, a major contribution seems to have been TI's inability to monitor effectively consumer markets and competitors in those markets. As a result, Texas Instruments has stumbled badly. Although General Motors monitored its markets during the crucial late 1960s and early 1970s, the corporation failed to grasp the magnitude of consumer attitude shifts. These shifts, together with a corporate business-as-usual attitude (which led to a substantial cost disadvantage and poor product quality) resulted in the company losing its market share to foreign competition.

Communications failure is often a key factor in intelligence failures. Communications means the *effective sending* (speaking, writing, and nonverbal cues) *and effective receiving* (listening, reading, and reading nonverbal cues) of information. If both Texas Instruments and General Motors had adequate information concerning their external environments and their own internal strengths and weaknesses, then the intelligence failure was one of inadequate communications (poor sending, poor receiving, or both) rather than one of inadequate information.

Organizational Balance. Organizational balance is the ability to be centralized and decentralized simultaneously. Direction and policy making should be centralized, whereas operating authority should be decentralized. For example, General Motors had proceeded with a strategy of centralizing the design, development, and production of its automobile line during the 1960s when it was the dominant firm in the oligopolistic domestic automobile industry. However, as the environment became more competitive during the 1970s, the company responded with a family of products that was indistinguishable from each other. It has only been very recently that GM began to decentralize decision-making authority for product development, production, and marketing into two divisions. On the other hand, 3M Corporation has a policy of spinning off new divisions rather than allowing its divisions to become too big. This policy keeps 3M's divisions small enough so that they can stay in touch with their markets.

Analysis. Analysis is the ability to evaluate systematically a problem using both quantitative and qualitative information and to develop an array of possible responses. The role of analysis is the clear delineation of choices, not the determination of strategy. For example, a careful analysis of demographic trends reveals that the number of births in the United States is increasing in the 1980s. This quantitative fact suggests that baby food consumption will increase proportionately. However, pediatricians are encouraging mothers to start babies on solid food later in life, and women are more likely to nurse their babies today than they did a generation ago. In addition, today's mothers are more concerned about nutrition than their mothers were, and home appliances make it easier to make homemade baby food.

The impact that the above qualitative issues will have on baby food consumption is not clear. Whether baby food consumption will grow as rapidly as expected will depend on the increase in number of births, the net effect of these qualitative issues, and the strategic response of those in the baby food business. For example, one response by baby food producers has been to lower the salt content of baby food.

Thorough analysis of both quantitative and qualitative issues enables the firm to delineate its strategic choices more clearly. This delineation of choices then enables managers to make strategic decisions in a more enlightened perspective. Even though most (if not all) strategic choices are subjective decisions, thoroughness of analysis enables decision makers to "fly better by the seat of their pants because they are instrument rated."

Innovation. Different individuals use different meanings for the word *innovation*. A broader definition best fits the needs of strategic planning. For purposes of this chapter, innovation occurs when a firm "either learns to do something it did not do before, and then proceeds to do it in a sustained way . . . or learns not to do something that it formerly did, and proceeds not to do it in a sustained way."[4] This means that the innovative firm is willing to learn new ways and is willing to bend. Effective innovation management depends on the firm capitalizing on the external environment, creating an organization and organizational climate favorable to innovation, and effectively managing creative individuals. In particular, top management commitment is crucial in developing an innovative firm, in general, and innovative strategies, in particular.

For example, in developing its personal computer, IBM established an independent task force. This task force was free to do pretty much what it wanted. The PC was developed outside IBM's bureaucratic review process, components supplied by outside suppliers (contradictory to IBM practice) were used rather than IBM-designed components, and the decision was made to encourage outside companies to develop IBM PC compatible peripherals and software. This innovative approach enabled IBM to enter the personal computer market much more successfully than it would have if the normal product development process had been followed.

Proactivity. Proactivity is defined as the firm shaping its environment with new products, technologies, and administrative techniques. In the postregulation era, Piedmont Aviation avoided the fate of most other airlines by pursuing a niche strategy rather than by expanding into highly competitive markets. When Piedmont's Charlotte, North Carolina, hub approached capacity, the management analyzed its options and then decided on a strategy of opening additional hubs rather than expanding the Charlotte hub. As a result of this strategy, additional hubs were opened in Dayton, Ohio, and Baltimore, Maryland. This strategy enabled Piedmont to compete in environments that it could shape, rather than be at the mercy of large competitors in markets that would dictate Piedmont's strategies.

Risk Taking. Risk taking refers to the ability to take bold and venturesome action in the face of uncertainty. A problem with risk taking is that many managers associate risk with change. In some instances, however, a lack of change may be more risky than a bold and venturesome strategy. For example, in the late 1960s, Polaroid was faced with the strategic threat of Eastman Kodak's entrance into the instant photo market. Polaroid's choices were: (1) do little and be swamped by Kodak's technical and marketing power; or (2) develop a superior instant photo system that could withstand the expected threat. Polaroid committed itself to the product, which became the SX-70 system and then proceeded to solve major technical problems. The failure to solve any one of several of these major problems could have doomed the SX-70 system.

Polaroid's decision to proceed with the SX-70 might be considered risky by some; however, failure to proceed with the new product would have doomed the company. The result was that Kodak's instant photo system never achieved a dominant share of the market: Polaroid's instant photo market share eventually stabilized at about 65 percent with Kodak getting the other 35 percent. (In 1986, Kodak lost a patent infringement suit and was ordered to abandon this market.)

Risk can be managed, to some extent, by an effective combination of intelligence, a balance between centralization and decentralization, innovation management, a willingness to be proactive, and a competent use of analysis. However, management must accept the fact that risk exists in any decision. Further, management must accept risk as a part of doing business.

INTEGRATION OF ANALYSIS AND INTUITION

The three issues of intelligence, organizational balance, and analysis are primarily analytical, quantitative, rational, systematically oriented activities. On the other hand, the three issues of innovation, proactivity, and risk taking are primarily qualitative, holistic, intuitive-oriented activities.

Taken together, the three analytical issues impose structure (or understanding) on the unstructured, ill-defined, internal and external environments and identify arrays of alternatives for the firm to consider. For example, analysis techniques from economics, operations research, finance, statistics, and marketing research can impose enough structure on the firm, its markets, and the overall economy to develop arrays of managerially useful insights and to suggest alternate courses of action. In addition, the intelligence system can assess both the firm's internal strengths and weaknesses and the environmental threats and opportunities and then communicate the results of the findings throughout the firm. Finally, the organizational balance of centralized policy making and decentralized operating authority enables each area of the firm to focus on what it does best.

In a similar manner, the three intuitive issues allow the firm to exploit the insights developed by the analytical issues, create opportunities, and develop appropriate strategies that will enable it to exploit the results of its analytical capabilities. This creation of opportunities cannot be exploited without the ability of the firm to be innovative (learn new ways and be willing to bend), proactive (be willing to shape the environment), and willing to accept risk as an integral part of decision making.

Finally, the relationship between the analytical and intuitive issues is an interactive one. This means that the two processes must constantly interact to distill the environment, identify choice alternatives, create opportunities, and develop appropriate strategies. Figure 37-2 summarizes this integration of analysis and intuition in the strategy formulation process.

THE KEY TO EFFECTIVE STRATEGIC MANAGEMENT: INTEGRATION

As suggested earlier, all six issues are important to strategic success. This is especially true in environments that are chaotic (unstable, difficult to predict, complex, and competitive). But the key to effective strategic management is the integration of the analytical and intuitive issues. A potentially fatal problem is the tendency for analytical and intuitive issues to polarize. This polarization leads to strategic planning that will be dominated by either analysis or intuition or to a strategic planning process that lacks coordination among the analytical and intuitive issues.

IMPLICATIONS FOR STRATEGIC MANAGEMENT

What are the implications of an integrated strategic management process?

1. Effective strategic planning is an ongoing process.
2. Organizational openness is crucial to the strategic planning process.
3. Beware of pedestrian (unimaginative) strategies.

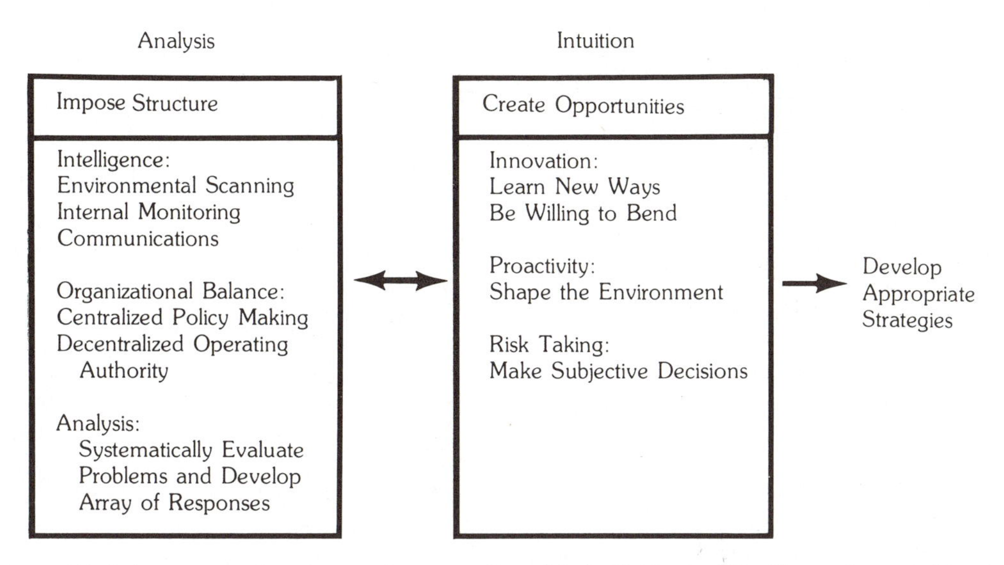

Figure 37-2. Integration of analysis and intuition (from M. A. McGinnis, "The Key to Strategic Planning: Integrating Analysis and Intuition," *Sloan Management Review,* **26,** 1984:48. Copyright © 1984, Sloan Management Review Association, Massachusetts Institute of Technology, Cambridge, MA 02139. Used with permission).

4. The importance of questioning underlying assumptions.

Effective Strategic Planning

A common criticism of strategic planning is that it is conducted periodically by a strategic planning department. In this light, strategic planning is seen as something that is thought about only occasionally. However, the environment does not change at periodic intervals. Technological breakthroughs can occur at any time, or they may even evolve gradually so that they are missed if the firm is not constantly monitoring its environment. Market opportunities may occur at certain unscheduled strategic windows, where strategies must be adjusted if the firm is to capitalize on the opportunity. For example, trucking industry adjustments to deregulation occurred within two years. Prior to deregulation many trucking firms thought that the period of transition would be five years. Trucking firms that have been slow to respond to deregulation have already missed the strategic window that was open during 1981 and 1982.

Organizational Openness

This is the free flow of relevant information, which is both vertical (between organizational levels) and horizontal (between departments). Such communication is essential to detecting strategic opportunities, identifying alternatives, and formulating appropriate strategies. For example, multiple layers of bureaucracy in domestic automobile companies during the 1960s inhibited internal communication. This lack of internal communication contributed to the industry's sluggishness in responding to foreign competition.

Pedestrian Strategies

These strategies create two problems. First, they often are not as effective because they do not capitalize on the firm's strengths and opportunities and/or minimize its weaknesses and threats. Secondly, they are usually predictable. For example, many airlines expanded in competitive markets following deregulation. This, together with the entry of a host of new nonunion low-cost competitors, has resulted in Braniff's bankruptcy together with record losses for many carriers. Piedmont and US Air, on the other hand, exploited market opportunities where conventional wisdom said there were no opportunities. By building effective hub-and-spoke networks in less dense markets, both airlines have avoided price wars and have remained profitable.

The Need to Question Assumptions

Thorough integration of analysis and intuition reduces the chance that previous strategic assumptions will continue to go unchallenged. Bendix's management saw an excellent strategic fit in its proposed acquisition of Martin Marietta; however, Martin Marietta's management did not share the assumption. As a result of Martin Marietta's defense, Bendix was weakened to the point where it was acquired by Allied Corporation, whereas Martin Marietta remained independent.

IMPLICATIONS FOR MANAGERS

American managers are generally well versed in the techniques of operations research, economic analysis, financial analysis, and marketing research analytical techniques. However, these analytical techniques are not well integrated with issues of intuition and communication. Specifically, managers must become competent in the areas of risk taking, management of innovation, and communications.

A substantial amount of research has been published about management of financial risk, however, strategic decision making entails *operational risk*. As far as I can determine, there is no suitable way to assess operational risk before the fact. For example, Braniff's strategy of rapid route expansion after airline deregulation resulted in bankruptcy. In hind sight, Braniff's strategy was very risky. However, if the economy had remained strong and if predictable fare wars had not broken out, Braniff's strategy might have succeeded. On the other hand, Wendy's entered the fast food business in 1969. Conventional wisdom would have given such a late entrant little chance for success. However, by capitalizing on the adult market, Wendy's has established itself as the number three hamburger chain. In retrospect, Wendy's strategy was obvious. At the time, however, it was not. The lesson is clear concerning the management of operational risk. Sooner or later managers have to make subjective decisions and face the consequences—whether they be good, bad, or indifferent.

As defined earlier, innovation means learning new ways and being willing to bend. Innovation management means creating an organizational climate that favors innovation, creating an innovative organization, and harnessing individual innovation.[5] This process of innovation management emphasizes openness to new ideas, meaningful goal setting, employee diversity, a focus on external challenges, flexible personnel policies, and a reward system that encourages innovation. Finally, innovation management requires a genuine commitment to innovation from the top down. Without this last crucial ingredient, the firm will not be innovative.

The final implication for managers is the importance of communication competence. In recruiting graduates and in discussing business school shortcomings, executives emphasize the need for managers to be able to write and speak. Writing and speaking are important as far as they go. However, communication competence also includes reading and listening skills, as well as group interaction skills. When managers learn to receive and send communications free of bias then the analytical issue of intelligence will be able to fulfill its role of keeping the firm in tune with its environment and itself.

IMPLICATIONS FOR EDUCATORS

If managers are to develop their integrative skills, then business schools must train future managers to have an integrative ability. Three suggestions seem relevant to business school faculties:

1. Increase emphasis on strategic integration in the curriculum.
2. Demand more synthesis in academic research and writing.
3. Reassess the organization structure of the business school.

Strategic Integration

This area is independent of functional areas. Even though most business curriculums offer an integrative, case-oriented "capstone" course, often called business policy, the orientation of our business schools is to encourage the development of distinct disciplines. The result, unfortunately, is that functional courses are often taught with one of two biases: either no attempt is made to relate the course to other areas of the firm; or the functional course is taught as being central to the firm's existence. Either bias leaves the student with the mistaken impression that the firm is a group of autonomous functional areas that do not need to relate to one another. These biases limit the ability of students to integrate effectively the functional issues when they reach the capstone course.

One way of reducing functional bias is to stress integration within each functional course. The core marketing course, for example, must recognize that marketing decisions affect the decisions of production, finance, human resources management, and corporate accounting practices. Similarly, decisions made in other functional areas affect marketing. A second way of reducing functional bias is to increase the amount of field work in business school curriculums. Field work, achieved through supervised internships or courses that emphasize field projects, help the student to develop integrative skills. These skills include: developing problem-finding abilities, developing confidence in analyzing real-world problems, and recognizing that real-world problems do not often lend themselves to school solutions.

The last two decades of academic research have been dominated by increasing compartmentalization of functional areas. Sophisticated mathematical and statistical research methodologies have developed without any real attempt to relate these methodologies to real-world problems. Certainly the development of these methodologies has been essential in moving business research from the fly-by-the-seat-of-your-pants era. Unfortunately, there has not been enough emphasis on the synthesis of research findings. This is not to say that additional new research is unnecessary. Rather, in the push for incremental knowledge, applications of research findings to managerial practice have received too little emphasis. Management education is likely to remain weak in the area of integration until greater emphasis is placed by department chairpersons, deans, journal editors, journal reviewers, and doctoral committees on the issues of relevance and integration in academic research and writing.

Reassessment of Business School Structure

The final implication for educators is the organizational structure of the business school. It is unlikely that strategic integration will be given important consideration among business school faculties as long as business schools emphasize a functionally compartmentalized approach to organization. Perhaps business schools should begin to reassess themselves in terms of the 1980s business environment. Are academic departments still relevant in an ear of integration? Should business school faculties continue to teach narrow ranges of courses? Should promotion and tenure decisions emphasize productivity in a narrow specialty, or should the ability to integrate become relatively more important? Many business schools offer a wide range of products including undergraduate and graduate degrees, executive development programs, consulting, and research. Do functional departments still meet this array of needs?

SUMMARY

Strategic success increasingly will depend on the ability of the firm to integrate analysis and intuition. Effective integration depends on the ability to manage effectively the issues of intelligence, organizational balance, analysis, innovation, proactivity, and risk taking. Those firms that learn to manage these fundamental issues will probably enjoy more successes and suffer fewer disappointments than those that do not master these fundamentals.

Managers must become more comfortable with ambiguity. In addition, they must become more accomplished at relating the overview to the specifics and vice versa. Educators too must increasingly conceptualize beyond the limits of their own specialties; academics must begin to relate to an integrative, ambiguous subject matter. Finally, the time may have arrived for academics to rethink what they teach, how they teach, and how they organize to teach.

REFERENCES

1. J. D. Thompson, *Organizations in Action* (New York: McGraw-Hill, 1967), especially Chapters 7 and 10.
2. F. T. Paine and C. R. Anderson, "Contingencies Affecting Strategy Formulation and Effectiveness: An Empirical Study," *Journal of Management Studies* **14**(1977):147-158.
3. D. Miller and P. Friesen, "Archetypes of Strategy Formulation," *Management Science* **24**(1978):921-933.
4. H. A. Shepard, "Innovation-resisting and Innovation-producing Organizations," *Journal of Business* **40**(1967):470.
5. M. A. McGinnis and M. R. Ackelsberg, "Effective Innovation Management: Missing Link in Strategic Planning?" *Journal of Business Strategy* **4**(1983):59-66

Index